DIVERSITY IN UNITY: PERSPECTIVES FROM PSYCHOLOGY AND BEHAVIORAL SCIENCES

PROCEEDINGS OF THE ASIA-PACIFIC RESEARCH IN SOCIAL SCIENCES AND HUMANITIES, DEPOK, INDONESIA, 7–9 NOVEMBER 2016: TOPICS IN PSYCHOLOGY AND BEHAVIORAL SCIENCES

Diversity in Unity: Perspectives from Psychology and Behavioral Sciences

Editors

Amarina A. Ariyanto & Hamdi Muluk

Faculty of Psychology, Universitas Indonesia, Indonesia

Peter Newcombe

School of Psychology, University of Queensland, Australia

Fred P. Piercy

Department of Human Development, Virginia Tech, USA

E. Kristi Poerwandari & Sri Hartati R. Suradijono

Faculty of Psychology, Universitas Indonesia, Indonesia

Routledge
Taylor & Francis Group
LONDON AND NEW YORK

First published 2018 by CRC Press/Balkema

2 Park Square, Milton Park, Abingdon, Oxon, OX14 4RN
605 Third Avenue, New York, NY 10017

Routledge is an imprint of the Taylor & Francis Group, an informa business

First issued in paperback 2020

Typeset by V Publishing Solutions Pvt Ltd., Chennai, India

ISBN: 978-1-138-62665-2 (hbk)
ISBN: 978-0-367-73558-6 (pbk)

Diversity in Unity: Perspectives from Psychology and Behavioral Sciences – Ariyanto et al. (Eds)
© 2018 Taylor & Francis Group, London, ISBN 978-1-138-62665-2

Table of contents

Preface

The 2016 Asia Pacific Research in Social Science and Humanities (APRiSH 2016) was held 7–9 November 2016 at the Margo Hotel, Depok, Indonesia. The theme for this year's conference is "Competition and Cooperation in the Globalized World". This conference is a platform that brings together scholars from Indonesia and other regions in Asia Pacific to share their interdisciplinary perspectives in various areas of social sciences and humanities. The conference tracks covered Psychology and Behavioral sciences, Arts and Humanities, Law and Justice, Economic and Business and Social and Political Science. In this book, we will only cover papers from Psychology and Behavioral sciences.

The keynote speaker in this conference is Professor Philip Zimbardo, a notable figure in the modern psychology, who has lectured more than 50 years on psychology, wrote many journals and books, served as president of the American Psychological Association and was professor emeritus at Stanford until 2008. His new research examined the psychology of heroism, a topic that tries to explain heroic social action, which is important to improve positive behavior among humans. Several prominent figures in social sciences, humanities and psychology served as plenary speakers in this conference. Christina Maslach, a Professor of Psychology (Emerita) at the University of California at Berkeley. She is widely recognized as one of the pioneering researchers on job burnout, has written numerous articles and books, including *The Truth About Burnout*. The other plenary speakers are Prof Adrian Little (University of Melbourne), Prof Mikihiro Moriyama (Nanzan University), Prof. Bambang Shergi Laksmono (Universitas Indonesia), dan Dr. Harkristuti Harkrisnowo (Universitas Indonesia). Dr. Ide Bagus Siaputra (University of Surabaya) has also shared his experience on Meta Analysis and Instrument Adaptation in two workshops.

This conference also aims to create and promote an academic climate of excellence among students, researchers and also practitioners. During the conference which was held from 7–9 November 2016, overall there were 524 papers presented. In the Psychology research area, we received a total of 138 papers. After a rigorous reviewed process, 80 papers were accepted. The papers covered areas of Clinical Psychology (12 papers), Developmental Psychology (10 papers), Educational Psychology (23 papers), Industrial and organizational psychology (12 papers), General Psychology (8 papers), research and methodology (1 paper), and Social Psychology (14 papers).

The editors would like to express their appreciation and gratitude to the scientific committee and the reviewers who have selected and reviewed the papers, and also to the technical editor's team who helped carry out the page layout and check the consistency of the papers with the publisher's template. It is an honor to publish the selected papers in this volume by CRC Press/ Balkema (Taylor & Francis Group). Finally, we would like to thank the steering committee, the chairman of the conference, the members of the organizing committee involved in the preparation and organization of the conference, and for the financial support from Universitas Indonesia.

The Editorial Board of the 1st APRISH Proceedings
for Topics in Psychology and Behavioral Sciences
Amarina A. Ariyanto, Universitas Indonesia
Prof. Dr. Hamdi Muluk, Universitas Indonesia
Peter Newcombe, Ph.D., University of Queensland
Prof. Fred P. Piercy, Ph.D., Virginia Tech
Dr. E. Kristi Poerwandari, Universitas Indonesia
Prof. Dr. Sri Hartati R. Suradijono, Universitas Indonesia

Diversity in Unity: Perspectives from Psychology and Behavioral Sciences – Ariyanto et al. (Eds)
© 2018 Taylor & Francis Group, London, ISBN 978-1-138-62665-2

Organizing committee

STEERING COMMITTEE

Rosari Saleh (*Vice Rector of Research and Innovation, Universitas Indonesia, Indonesia*)
Topo Santoso (*Dean Faculty of Law, Universitas Indonesia, Indonesia*)
Ari Kuncoro (*Dean Faculty of Economics and Business, Universitas Indonesia, Indonesia*)
Adrianus L.G. Waworuntu (*Dean Faculty of Humanities, Universitas Indonesia, Indonesia*)
Arie Setiabudi Soesilo (*Dean Faculty of Social and Political Sciences, Universitas Indonesia, Indonesia*)

INTERNATIONAL ADVISORY BOARD

Peter Newcombe (*University of Queensland, Australia*)
Fred Piercy (*Virginia Tech University, Australia*)
Frieda Mangunsong Siahaan (*Universitas Indonesia, Indonesia*)
James Bartle (*University of New South Wales, Australia*)
Elvia Sunityo Shauki (*University of South Australia, Australia*)

SCIENTIFIC COMMITTEE

Manneke Budiman
Isbandi Rukminto Adi
Beta Yulianita Gitaharie
Surastini Fitriasih
Sri Hartati R Suradijono
Elizabeth Kristi Poerwandari

CONFERENCE DIRECTOR

Tjut Rifameutia Umar Ali

CONFERENCE VICE-DIRECTOR

Turro Wongkaren

ORGANIZING COMMITTEE

Dewi Maulina (*Faculty of Psychology, Universitas Indonesia, Indonesia*)
Intan Wardhani (*Faculty of Psychology, Universitas Indonesia, Indonesia*)
Elok D. Malay (*Faculty of Psychology, Universitas Indonesia, Indonesia*)
Josephine Rosa Marieta (*Faculty of Psychology, Universitas Indonesia, Indonesia*)

Teraya Paramehta (*Faculty of Humanities, Universitas Indonesia, Indonesia*)
Nila Ayu Utami (*Faculty of Humanities, Universitas Indonesia, Indonesia*)
Priskila Pratita Penasthika (*Faculty of Law, Universitas Indonesia, Indonesia*)
Efriyani Djuwita (*Faculty of Psychology, Universitas Indonesia, Indonesia*)
Destri Widaya (*Faculty of Economics and Business, Universitas Indonesia, Indonesia*)

WORKSHOP COMMITTEE

Corina D.S. Riantoputra (*Faculty of Psychology, Universitas Indonesia, Indonesia*)
Fithra Faisal Hastiadi (*Faculty of Economics and Business, Universitas Indonesia, Indonesia*)
Mirra Noormilla (*Faculty of Psychology, Universitas Indonesia, Indonesia*)

TREASURERS

Robby Oka Yuwansa (*Faculty of Psychology, Universitas Indonesia, Indonesia*)
Nurul Husnah (*Faculty of Economics and Business, Universitas Indonesia, Indonesia*)

Keynote speech

Diversity in Unity: Perspectives from Psychology and Behavioral Sciences – Ariyanto et al. (Eds)
© 2018 Taylor & Francis Group, London, ISBN 978-1-138-62665-2

Exploring human nature and inspiring heroic social action

Philip Zimbardo
Department of Psychology, Stanford University, Stanford, USA

ABSTRACT: In this presentation, I describe a variety of research and psychological concepts across many domains in which I have been the pioneering investigator. Beginning with original research on creating evil, in my Stanford Prison Experiment, and then ending with a call for creating new-generation heroes who will help oppose the rise of right-wing totalitarian governments globally, it applies psychological wisdom to real-world problems. This personal journey also takes readers into original research on shyness, time perspectives, the negative impact of video gaming on young men, and finally introduces the Heroic Imagination Project, and its revolutionary educational programme, designed to inspire and train ordinary people to become everyday heroes.

1 INTRODUCTION

The world we are living in is currently undergoing rapid changes in various aspects, whether it is cultural or political, or due to terrorist-created uncertainties. A lot of those changes are moving us towards progress, but there are also setbacks, stemming from hostile stereotypes and prejudice, as seen in numerous countries all over the globe. How can that happen? Why are so many people committing evil acts that are regressing the progress that has been made? When technological progress is going at the fastest rate ever, why are we moving backwards to such an extent? Apparently, the consequence of such negative progress is economic inequality, which then gives rise to poverty. In the USA, one of the global economic superpowers, the poverty rate is appalling at one in five children growing up poor, and that is not only for the USA, but also for other countries globally. It makes children growing up face adversity, and reduces their life expectancy by two years, compared with those who were born more privileged. Thus, poverty is evil; it is a major form of systematic evil.

However, the main themes of my research have been on situational evil instead of poverty. Evil has been the subject of discussions and discourses around the world for centuries, as the core of every religion. Evil is the enemy of goodness. So I ask the question: 'What makes good people go wrong?' Throughout my life, I have been asking that question since I was a little child. Why? My family emigrated from Sicily and I grew up in the South Bronx ghetto in New York City, where poor immigrants from all over the world gathered to started a new life and fulfil their American Dreams. As I got older, I saw some of my friends get involved in numerous shady businesses, running criminal errands for rewards (offered by local evil men) that were petty compared to the risks they would face. However, I knew that my friends were good people, but they gave in to the temptation of money. On the other hand, a few other friends and I were able to resist that temptation and continue with our honest lives. So, I asked another question: 'What is the difference between boys whom I knew were good who gave in to the temptation and those who resisted it?' So, this is a really fundamental question in psychology, a discussion about free will and willpower, and how to exercise them to resist temptation while delaying immediate gratification for further benefits.

However, then comes another question: 'How are poverty and evil interrelated?' Indeed, poverty is systematically evil due to the fact that it makes people not think about far-future consequences of their actions and only in satisfying their immediate needs and wants. It makes people ignore the impact of their decisions and behaviours towards others, and focus too much on themselves in order to achieve their near-sighted goals. However, we should re-examine what evil is from a psychological definition. It is the exercise of power to intentionally harm (psychologically), hurt (physically), or destroy (mortally), and when it is done by nations, they are committing crimes against humanity (genocide by nations).

In psychology, we mainly talk about evil on an individual level, which is attributed to their personal disposition ('bad apple'), but that ignores the situational contexts, which are the influences from social and physical environments ('bad barrels'). We must not solely attribute one's evil deeds to their individual characteristics, but must also take into account the factor of the social environment. One of the examples of evil in situational contexts is my Stanford Prison Experiment.

Meanwhile, evil at the systemic level is seen as the interaction between organisational, political, economic, cultural, and legal consequences ('bad barrel makers'). Although, the most evil deeds are not done by individuals, but by corporations. They often allow corruption, embezzlement, and fraud. It is these systems that have the power to change the situations for better or for worse. However, when we talk about evil, our thorough analysis must be conducted for all three levels, not just focusing on a particular level while ignoring other levels.

Then comes the question: 'Why do good people become evil?' Apparently, there is a series of psychological processes occurring that can make good people commit evil acts. Those processes are dehumanisation, diffusion of responsibility, obedience to authority, group pressure, and anonymity or deindividuation. Dehumanisation occurs when someone is deprived of their human qualities, while diffusion of responsibility happens when someone perceives that a current responsibility is shared with other people, so one does not fully bear the consequences of their decisions. The classical example of diffusion of responsibility is the case of Kitty Genovese, in New York City decades ago, where many people heard her screams as she was being assaulted but did nothing to help—and she died. Moreover, in obedience to authority, as seen from Stanley Milgram's experiments, we can see how someone can blindly follow people in power and commit evil acts. It stems from our childhood when we were trained as a child to be obedient to our parents, teachers and religious leaders. However, we have to differentiate between good authorities who deserve our respect and bad authorities who deserve our defiance because they are not good; they are corrupt. The powerful influence of group pressure to cause someone to act evil comes from our desire to feel accepted by that group. Lastly, anonymity and deindividuation is especially prevalent in the era of the Internet where people can mask their identities and do whatever they want in that dark space, including cyber-bullying others.

An example of systemic evil can be observed in China, where the state kills one million of its people annually by encouraging male citizens to smoke (54.5% men are smokers, whereas only 2% are women). Smoking is portrayed as a symbol of manliness and masculinity. Apparently, the state monopolises the tobacco industry that yields them 605 billion yuan. The Chinese government also controls the media, which prevents any anti-smoking campaigns in the media. The money from the tobacco industry is used to build schools, such as a primary school in Sichuan province, named as The Sichuan Tobacco Primary School. This school has a prominent plaque reading: 'Ingenuity is the fruit of diligence—Tobacco will help you succeed'. Children in this school are raised with the positive acceptance of tobacco, and this is what I call systemic evil—actions by governments that kill its people.

The anticipated results are overwhelming; new data from *The Lancet* (Chen et al., 2015) shows that Chinese men now smoke more than one-third of all the world's cigarettes, while the smoking rate among Chinese women is somewhat less than in previous generations (less than 2%). The number of smoking-related deaths for Chinese men by 2030 will be two million annually, and by 2050, three million Chinese men will die of smoking-related diseases every single year. Many of us are worried about ISIS and terrorism, but the reality of a powerful nation like China passively engaged in killing millions of its population goes unnoticed, and no one is doing anything about it, as far as I know.

Systemic evil worldwide includes wars, genocide, and poverty, but slave labour and sex trafficking are on the rise. Slave labour is where individuals, and sometimes the whole family, work in toxic conditions with minimum pay and are not able to leave. Sex trafficking has become a most profitable business; it is estimated that at least 1 million women and children become victims and the profit from this business is more than 350 billion US dollars. It is usually not legally prosecuted, unlike drug trafficking, so more drug dealers are moving into this better business model!

3 THE STANFORD PRISON EXPERIMENT

An example of situational evil can be seen in the experiment I created in 1961, the Stanford Prison Experiment. This experiment asked: 'What will happen to good people when they are put in an evil place?' College student participants answered a newspaper ad inviting them to be in a study of prison life; 75 applicants then took personality tests, and interviews. The 24 who were most normal, and healthy, were randomly assigned as prisoners or guards. Prisoners wore smocks, and were assigned number IDs as their new names. Guards were given military uniforms, with symbols of power and anonymity, such as reflective sunglasses.

Initial menial tasks escalated daily to be ever more humiliating and degrading. Prisoners were often stripped naked, sexually taunted, and sexually degraded. The first prisoner to be arrested by local police in our simulated arrest at their residences, #8612, was the first to break down from extreme stress reactions in 36 hours, and was then released. After that, the abuse by guards worsened daily—until five prisoners had to be released, one after another from extreme stress reactions. The experiment, intended to be a two-week study, was terminated after only six days.

I wrote several books after completing the Stanford Prison Experiment, such as *Shyness* (1977), *The Lucifer Effect* (2007), *The Time Paradox* (2008) and *Man, Interrupted* (2016). *The Lucifer Effect* is a celebration of the human mind's infinite capacity to make us behave kindly or cruelly, to be caring or indifferent, creative or destructive, and also to make some of us villains; and the good news is that the same human mind directs others to become Everyday Heroes.

I was the first researcher to study shyness in adolescents and adults, starting in 1972. I conceptualised shyness as being a self-imposed psychological prison, in which the shy individual is both guard and reluctant prisoner—limiting freedoms of speech, association, and action. Shyness was then, and is now, a common phenomenon. In the 1970s and 80 s: 40% currently shy, 40% previously shy, 15% situational shy, and 5% never shy. In 2007: 84% shy at some point, 43% currently shy, and 1% never shy. Two-thirds of currently shy people said that shyness was a major personal problem.

Why is shyness increasing? Electronics and the tech revolution are replacing people's direct talk and face-to-face communication. We are not learning or practising basic social skills essential for being able to feel comfortable with others; creating a new kind of shyness—awkward shyness. In the past, shyness was not knowing how to navigate the social landscape and not knowing how to ask for directions (i.e. fear of rejection). I also started a therapeutic Stanford shyness clinic in 1977, which has been very successful in treating all forms of shyness; in fact, it is still in operation at Palo Alto University, 40 years after its inception.

4 MEN DISCONNECTED FROM SOCIETY

New shyness is being unwilling to ask for directions, not wanting to connect. Many young men are preferring to live in virtual reality. This inability to connect has led to a major problem with men in particular.

Why is this such an issue with men and boys? Examining the problem, and using the analysis developed in *The Lucifer Effect*, there are three main contributing elements: (a) Individual factors: shyness, impulsiveness, entitlement; (b) Situational factors: broken families/fatherlessness, video games, pornography; (c) Systemic factors: policies that favour women and alienate men, failing schools, economic inequality, environmental toxins.

In America, a big problem is that 41% of women with children in the US (25% in the UK) are single mothers; the rate is 50% for women under 30 years old. Single mothers are on the rise everywhere; boys need a role model from a father who gives conditional love (to set up a positive standard for their children). There is a negative ripple effect of single, stressed-out mums, which leads to elevated stress hormones and poorer immune systems in children, poorer social development, higher likelihood of Attention-Deficit Hyperactivity Disorder (ADHD) and behavioural problems as an adolescent, and also relationship difficulties later on.

Impulsiveness and entitlement; young men are trapped in a present hedonistic time zone. The instant gratification of gaming and pornography keeps them there, and also creates inflated egos. Young men are 25% more likely than young women to be living at home with parents. Many do not help with chores or bills—they feel entitled to do nothing. Parents enable dependency by not telling them to shape up or ship out. In addition to personal and familial factors, our entire system puts up roadblocks for boys. Boys are failing in schools everywhere (except China and Japan), and dropping out, as well as failing to have friends and girlfriend mates because they have become addicted to playing video games excessively in social isolation, along with a new addiction to freely available online pornography.

Possible solutions for our disconnected males are an increase in male mentorship programmes, to take technology out of our children's bedrooms, limit cell phone use, and more social family time. Teach responsibility and resiliency at home (help boys to have a 'growth-based' mindset, NOT a fixed mindset), and encourage developing a future orientation.

5 TIME PERSPECTIVE

Understanding one's unique time perspective can help with time management and awareness of personal time allocation. I argue that there is a strange 'Time Paradox'. The most important influence on all your decisions and actions is inside your mind, and yet it is something about which you are unaware: your psychology of time perspective. Time perspectives is the study of how individuals in every country divide the flow of human experience into different time frames—or time zones—automatically and non-consciously. These frames vary between cultures, social classes, nations and people. They become biased rather than balanced by learned overuse of some frames, and underuse of others.

I developed a reliable and valid instrument to measure individual differences in different time zones. The Zimbardo Time Perspective Inventory (ZTPI) gives scores for six different time zones for each individual: (1) Past positives TP—Focus on positives; (2) Past negatives TP—Focus on negatives; (3) Present TP—Hedonism; (4) Present TP—Fatalism; (5) Future goal TP—Life—goal-oriented; (6) Future TP—Transcendental—life after death of the mortal body. The most important thing is to have a balanced time perspective—which means high past positive, moderately high future, and moderate present hedonism—used to reward oneself for meeting goals. The key is also in keeping low on past negative and present fatalism factors.

6 INSPIRING EVERYDAY HEROES IN A CHALLENGING NEW WORLD ORDER

Finally, what we have been trying to do for the last ten years is to inspire all youth everywhere to become Everyday Heroes. Most young people in many countries idolise unreal heroes, such as Superman, Batman, Spiderman, Wonder Woman, and many others. To become real-world heroes, we must learn how to use our brain wisely and well to make the world better, and to have the moral courage to do so.

There are many definitions of a hero; of most importance is to act on behalf of others in need, or in defence of a moral cause, while being aware of the risks to life, finances or career. There is a difference between the act of heroism which involves risks, and the act of social altruism which does not involve a risk. Heroes are also usually modest and humble, and usually disown the hero label. We need heroes for many reasons, such as to shift the social norm from passive compliance to pro-social action. True heroes put their best selves forward in service to humanity; they represent ideals we can all aspire to. Heroes are the force of good that opposes evil, which comes in all its many forms—both perpetrators of the Evil Action and the Evils of Inaction (not doing the right thing when opportunity presents a challenge, as revealed in public apathy and indifference).

We need heroes to combat these worldwide threats to democracy in many countries, such as the rise of totalitarian, and neo-fascist political parties, as well as terrorism. But we must not allow governments around the world to use security against terrorism as a reason for restricting our freedoms. Many countries have imposed regulations that trade off our precious freedom with terrorism suppression tactics. The rising threat from major (China/Russia) and minor (Iran/North Korea) dictatorships is worrisome. The slide towards dictatorship in many countries, that until recently seemed to be becoming more open (Turkey/Venezuela), is worrisome. Growing resource and opportunity inequalities, the gap between the rich and the poor, has widened. We need to explore the conditions that enable a potential dictator to spring to power. In the United States, new president Donald Trump will say or do anything to maintain power, whether it is true or not. In Hungary, the right-wing government, headed by Viktor Orban, says that migrants are 'poison' and 'not needed'. He is simply redirecting prejudice and hate to refugees using the media—which is also controlled by the government. The ruling Law and Justice party in Poland wants to amend their already conservative abortion laws to ban all abortions for any reason. They are also changing education in directions that I believe are restricting creativity and innovation, as they focus on traditional forms of training youth.

These regimes are dismantling democracy; true democracy is the counterpoint to authoritarian rule everywhere. We must consider and have dialogue about the following: (a) We need to explore the conditions that enable a potential dictator to spring to power. (b) How malleable is human political behaviour? (c) What are the mechanisms of socialisation for resilience in the family, schools, and communities? (d) What are the psychological characteristics needed to help citizens to be more capable of actively participating in, sustaining, and building effective democracies?

7 HEROIC CHARACTERISTICS

In the olden days, a hero was characterised as being special, unique and gifted. In many countries, this hero image is portrayed as male warriors with their weapons who helped win wars. In my new definition, heroes are usually ordinary, everyday people whose actions in challenging situations are extraordinary. In a video interview our team made with President Obama, he discusses the importance of Standing up, Speaking out, and taking Effective Social Action, always being aware of the 'ripple effect' of our actions on others who observe us. This ripple effect describes how when a person does something (good or bad), they become the model for others to do the same thing. We need more people to be a model of something positive.

8 PROGRAMME FOR POSITIVE PERSONAL CHANGE

I offer Dr Z's 10-step programme for positive personal change towards civic virtue that can promote a positive ripple effect. Compassion (caring for other people), according to the Dalai

Lama, is the highest private virtue, whereas heroism is the highest civic virtue. Compassion has to be converted into heroic action in order to change the world most effectively.

1. Self-Awareness Combined with Situational Sensitivity
   Be rooted in who you are, maintain a vision of your ideal self and be sensitive and aware of situational power.
2. Modesty
   Be modest in self-estimates. It is better to perceive yourself as vulnerable and take necessary precautions than to go where angels fear to tread.
3. Suppress Self in Service to Society
   By maintaining positive self-esteem, you can more easily separate your ego from your actions. Try and laugh at yourself at least once a day.
4. Self-Honesty
   Be accountable to yourself. Be willing to say the following phrases—I was wrong, I made a mistake, I'm sorry, I've changed my mind, I will do better and right the next time around.
5. Mindfulness
   Engage in life as fully as possible, yet be mindful and aware, attuned to the moment, prepare to disengage and think critically when necessary.
6. Question Information
   Separate messenger from message in your mind. Analyse both the source and the content of each message.
7. Combat Gender Stereotypes
   Promote the best performances of all genders, challenge false labels and old-fashioned thinking, so men and women work together as buddies, not always competing or suppressing the other.
8. Confront Calmly
   In all authority confrontations: be polite, individuate yourself and the other, describe the problem objectively. Press the pause button. Allow the other to explain their point of view.
9. Discriminate Ideologies
   Remember all ideologies are just words, abstractions used for particular political, social, or economic purposes. Be wary of following orders and question if their means justifies your end.
10. Don't Follow Blindly
    Many times, rules are abstractions for controlling behaviour and eliciting compliance and conformity. Insist that the rule be made explicit. Challenge rules when necessary. Ask yourself: who made it? What purpose does it serve and who maintains it?

And lastly, practise being a *positive deviant* for a day, A positive deviant in life is very important, and it can be learned, especially when our preferences are mainly based on the opinion of others (conformity). A positive deviant act was shown by Christina Maslach on 18 August 1971 during the Stanford Prison Experiment. She forced me to acknowledge the cruelty and inhumanity that I had allowed when playing my role as the Prison Superintendent. I ended the experiment the next day. Since then I have given up Evil; no more dining in Hell. I will only promote goodness and heroism in my new life! So, she represents what it means to act heroically. I should mention that we were married the next year at the Stanford Chapel and recently had our 44th wedding anniversary, shared by our two daughters.

9 THE HEROIC IMAGINATION PROJECT

The *Heroic Imagination Project* (HIP) is a San Francisco based non-profit organisation teaching people to Stand Up, Speak Out and Change the World. Our approach is based on the premise that ordinary people are capable of taking extraordinary action. Our revolutionary educational programme – *Understanding Human Nature* – consists of six modules (mindset, situational blindness, bystander effect, conformity/peer pressure, adaptive attributions, and stereotype threat, prejudice & discrimination). They are designed for middle and high school students, and also for college students, and for corporate leadership development. Throughout each lesson, students are challenged by provocative videos, as they respond with their partner within organised teams. All lessons share the same eight-step activity framework.

The activity framework:

1. What would you do?
2. Explore the psychology of situations
3. Think of time when you did or did not take action
4. Decide for yourself
5. Develop effective change making strategies
6. Plan for the next challenge
7. Reflect on your Personal Take-Away
8. Spread the Word to others you care about.

What can you do? When one person stops to help in an emergency many others join in; that is the power of ONE. Be that person, you can change the world by being willing to take wise and effective action. Power of two: be an Ally; support those in need. Power of many: create pro-social networks—Hero Squads. Dare to be different; practise being a positive deviant. Heroes are socially focused; act socially to be socio-centric, not egocentric, and who act wisely, well and often.

Start with small steps. Every day focus on others: make someone else feel special every day, give them a genuine compliment, learn and use their name, and hug when appropriate. Ask questions, understand the situation, challenge the rules. Don't blindly obey the authority, practice mindful disobedience. What is your 'Ripple Effect'? Your actions (and inactions) make a difference; they can have an impact on others.

Change your perspective; Me becomes We. I become Us. Be my Hero. The decision to act heroically is a choice that many of us will be called upon to make at some point in time. How can we create a new heroic vision in which silence, apathy and indifference is never an option? Let us work together to create a new generation of heroes in your school, business, community and around the world.

Now HIP educational programmes are in many schools in the western United States, but its global outreach is quite impressive, with training centres in Hungary, Poland, Sicily, Iran, Bali, and soon in the Czech Republic, Ukraine, and Argentina. Hopefully, Indonesia will soon be represented in Jakarta by the University of Indonesia.

REFERENCES

Chen, Zhengming et al., (2005). Contrasting male and female trends in tobacco-attributed mortality in China: Evidence from successive nationwide prospective cohort studies. *The Lancet, 386, 1447–56.*

Zimbardo, P.G. & Boyd, J.N. (2008). *The time paradox: Understanding and using the revolutionary new science of time*. New York, NY: Free Press.

Zimbardo, P.G. & Coulombe, N.D. (2015). *Man disconnected: How technology has sabotaged what it means to be male*. London, UK: Rider.

Zimbardo, P.G. & Haney, C. (2011). Stanford prison experiment. In B.L. Cutler (Ed.), *Encyclopedia of psychology and law*. Thousand Oaks, CA: Sage Publications.

Zimbardo, P.G. & Pilkonis, P. (1978). Shyness. In B.B. Wolman (Ed.), *International encyclopedia of psychiatry, psychology, psychoanalysis and neurology* (Vol. 10, pp. 226–229). New York, NY: Human Sciences Press.

Zimbardo, P.G. (1977). Shyness: The people phobia. *Today's Education, 66*, 47–49.

Zimbardo, P.G. (1977). *Shyness: What it is, what to do about it*. Reading, UK: Addison-Wesley.

Zimbardo, P.G. (2004). A situationist perspective on the psychology of evil: Understanding how good people are transformed into perpetrators. In A.G. Miller (Ed.), *The social psychology of good and evil* (pp. 21–50). New York, NY: Guilford Press.

Zimbardo, P.G. (2007). How the best and the brightest can turn into monsters. *The Chronicle Review*, (The Chronicle of Higher Education). B6–B7.

Zimbardo, P.G. (2008). *The Lucifer effect: Understanding how good people turn evil.* New York, NY: Random House.

Diversity in Unity: Perspectives from Psychology and Behavioral Sciences – Ariyanto et al. (Eds)
© 2018 Taylor & Francis Group, London, ISBN 978-1-138-62665-2

Job burnout in professional and economic contexts

Christina Maslach
Department of Psychology, University of California, Berkeley, USA

ABSTRACT: Job burnout has been receiving more attention in more workplaces around the world. Research in many countries has identified the key characteristics of the burnout experience, and its primary causes and outcomes. Clearly, burnout has high personal costs for individual workers, but it also has high social and economic costs for the organizations in which they do their jobs. A new model of healthy workplaces is proposed to address this problem, by promoting better job-person fit within six areas: workload, control, reward, community, fairness, and values.

1 INTRODUCTION

In this plenary talk, I want to provide a basic overview of the work that has been done on the phenomenon of job burnout, and then discuss some basic trends in how burnout is viewed and dealt with in workplaces around the world. Although burnout has been a topic of debate for many years now, it is only recently that it is beginning to gain more widespread attention about the serious risks that it poses. How might we better understand what is causing this increased concern about job burnout, and what might be constructive responses to this global problem?

Let me begin by describing what the basic problem is. When workers are experiencing burnout, they are overwhelmed, unable to cope, unmotivated, and display negative attitudes and poor performance. They have lost any passion for the work they do, and do not take pride in what they might accomplish. Rather than trying to do their very best, they do the bare minimum (i.e., just enough to get by and still get paid).

2 DEFINING JOB BURNOUT

What has been learned over several decades of research is that burnout is a psychological syndrome emerging as a prolonged response to chronic interpersonal stressors on the job. The three key dimensions of this response are an overwhelming exhaustion, feelings of cynicism and detachment from the job, and a sense of ineffectiveness and lack of accomplishment. Exhaustion refers to feelings of being overextended and depleted of one's emotional and physical resources. Workers feel drained and used up, without any source of replenishment. They lack enough energy to face another day or another person in need. The exhaustion component represents the basic individual stress dimension of burnout. Cynicism refers to a negative, hostile, or excessively detached response to the job, which often includes a loss of idealism. It usually develops in response to the overload of emotional exhaustion, and is self-protective at first—an emotional buffer of "detached concern." But the risk is that the detachment can turn into dehumanization. The cynicism component represents the interpersonal dimension of burnout. Professional inefficacy refers to a decline in feelings of competence and productivity at work. People experience a growing sense of inadequacy about their ability to do the job well, and this may result in a self-imposed verdict of failure. The inefficacy component represents the self-evaluation dimension of burnout. The significance of this three-dimensional model is that it clearly places the individual stress experience within a social context and involves the person's conception of both self and others (Maslach, 1998).

3 OUTCOMES OF BURNOUT

Why should we care about job burnout? It is not uncommon for senior managers in organizations to downplay the importance of employees feeling stressed and burned out. The general view is that if workers are having a bad day, then that is their own personal problem—it is not a big deal for the organization. However, the kinds of issues identified by both researchers and practitioners suggest that burnout should indeed be considered "a big deal" because it can have many costs, both for the organization and for the individual employee. Research has found that job stress is predictive of lowered job performance, problems with family relationships, and poor health, and studies have shown parallel findings with job burnout (see Maslach & Leiter, 2016; Maslach, Schaufeli, & Leiter, 2001; Schaufeli, Leiter, & Maslach, 2008).

Of primary concern to any organization should be the poor quality of work that a burned-out employee can produce. When employees shift to minimum performance, minimum standards of working, and minimum production quality, rather than performing at their best, they make more errors, become less thorough, and have less creativity for solving problems. Burnout has also been associated with various forms of negative responses to the job, including job dissatisfaction, low organizational commitment, absenteeism, intention to leave the job, and turnover. People who are experiencing burnout can have a negative impact on their colleagues, both by causing greater personal conflict and by disrupting job tasks. Thus, burnout can be "contagious" and perpetuate itself through informal interactions on the job. When burnout reaches the high cynicism stage, it can result in higher absenteeism and increased turnover. Employees suffering from burnout don't show up regularly, leave work early, and quit their jobs at higher rates than engaged employees.

The relationship of human stress to health has been at the core of stress research, ever since Selye proposed the original concept. Stress has been shown to have a negative impact on both physical health (especially cardiovascular problems) and psychological well-being. The individual stress dimension of burnout is exhaustion, and, as one would predict, that dimension has been correlated with various physical symptoms of stress: headaches, gastrointestinal disorders, muscle tension, hypertension, cold/flu episodes, and sleep disturbances.

Burnout has also been linked to depression, and there has been much debate about the meaning of that link. A common assumption has been that burnout causes mental dysfunction—that is, it precipitates negative effects in terms of mental health, such as depression, anxiety, and drops in self-esteem. An alternative argument is that burnout is not a precursor to depression but is itself a form of mental illness. Recent research on this issue indicates that burnout is indeed distinguishable from clinical depression, but that it seems to fit the diagnostic criteria for job-related neurasthenia (Maslach, Leiter, & Schaufeli, 2009). The implication of all of this research is that burnout is an important risk factor for mental ill health, and this can have a significant impact on both the family and work life of the affected employee.

Given that most research on burnout has focused on the job environment, there has been relatively less attention devoted to how burnout affects home life. However, the research studies on this topic have found a fairly consistent pattern of a negative "spillover" effect. Employees who are experiencing the exhaustion, cynicism, and inefficacy of burnout are likely to bring home a lot of emotional anger, hostility, and frustration. They are more easily upset by small disruptions, and serious arguments and conflicts erupt over mundane events in the home. But at the same time, they want to get away from people for awhile, and to not hear another voice, or deal with another problem. There is a sense in which the family is "giving at the office," because this person is inaccessible, either emotionally or literally by being on call, or traveling or working much of the time.

Recently, there has been a concern that burnout may be linked to various forms of self-harm, including suicide. Although no research has been done on this specific hypothesis, the discussion about the pressures in tech industries, for example, has suggested that there exist both overwhelming job pressures and expectations, combined with a lack of support for any expressed mental health issues (so that workers are afraid to reveal any perceived weaknesses or to ask for help).

It is interesting to note that tech industries are relative newcomers to discussions about job burnout, as are various customer service organizations. In contrast, professions that traditionally have had a greater concern about problems of burnout include health care and various human service occupations. It is not clear if the greater interest by the latter professions reflect the existence of a greater burnout problem in these fields, or a greater willingness to address issues that can compromise the quality of care provided to patients and clients.

4 THE "BURNOUT SHOP" AS BUSINESS MODEL

All of this evidence seems to point to the conclusion that burnout is bad for business. But, quite to the contrary, many businesses actually operate in a way to promote greater employee burnout. More specifically, they seem to be a modern version of the "burnout shop" that was prevalent in the early years of technology and computer industries. In those times, the burnout shop was usually a start-up, and it advertised for short-term jobs (no more than a few years), which were very intense and demanded a lot of self-sacrifice by the employee for the eventual success of the start-up project. If indeed the projects were successful, the employee who had sacrificed so much for a few years would be rewarded with many financial rewards. The use of the term, "burnout," seemed to have emerged from the field of engineering, and particularly in the space industry, where machines that were pushed to their limits, such as booster rockets, "burned out" and collapsed after they had expended all their force and energy.

In recent times, we are seeing the development of new 'burnout shops" that are not short-term projects, but are long-term models for doing business. Many of the characteristics of the old burnout shops have been applied to this new model, including long hours (often with no extra pay), and sacrifice of one's personal life to the demands of the job. In addition, what has been added is a forced competition between workers (for promotions, or for avoiding getting fired), which has the negative effect of eroding social relationships and teamwork in the workplace.

In these modern burnout shops, a lot of data is collected on metrics for the economic bottom line (e.g., products sold, time to delivery, etc.). But there does not appear to be a parallel commitment to collecting data on the human bottom line. Do these organizations actually track the human costs for their employees, such as physical exhaustion, sleep deprivation, long-term stress and health problems, disruptions of personal life, loss of self-worth and meaningful achievements, or burnout? The underlying assumption seems to be that employees who burn out are not the best ones, so they are expendable and disposable.

Although such assumptions may be the ones in practice, they are not supported by research results. Burnout is not a sign of incompetence or weakness; indeed, even the best and the brightest can experience burnout. The reason is because the key causes of burnout situational and environmental, not personal. However, the common belief that burnout reflects personal failings has led to the proliferation of recommendations for how people can deal with burnout by taking care of themselves. These individual solutions are largely drawn from the stress, coping, and health fields and include: 1) Changing work patterns (work less, take breaks, no overtime, work-family balance); 2) developing coping skills (time management, conflict resolution, cognitive restructuring); 3) obtaining social support (both colleagues and family); 4) utilizing relaxation strategies (meditation, naps); 5) promoting good health and fitness (better diet, exercise, smoking cessation); 6) developing better self-understanding (counseling, therapy). However, although there is value in all of these activities, there is not much evidence that they are actually effective in reducing job burnout (see Maslach & Goldberg, 1998; Leiter & Maslach, 2014).

5 A DIFFERENT APPROACH TO THE BURNOUT PROBLEM

In contrast to this individualistic, "blaming the person for their own problem" approach, the work on job burnout has led to several different conclusions, including the following:

1. Burnout has many economic, as well as social, costs that need to be factored in to the economic bottom line.
2. Building work engagement among employees can prevent, and reduce, the risk of burnout.
3. The causes of burnout are not personal weaknesses, but chronic job stressors in both the physical and social environment of the workplace.
4. The key sources of burnout (or engagement) are six critical areas of job-person mismatch (or match): workload, control, reward, community, fairness, and values (see Leiter & Maslach, 1999; 2004).
   a. Workload. Both qualitative and quantitative work overload contribute to burnout by depleting the capacity of people to meet the demands of the job. When this kind of overload is a chronic job condition, there is little opportunity to rest, recover, and restore balance. A sustainable and manageable workload, in contrast, provides opportunities to use and refine existing skills as well as to become effective in new areas of activity.
   b. Control. Research has identified a clear link between a lack of control and high levels of stress and burnout. However, when employees have the perceived capacity to influence decisions that affect their work, to exercise professional autonomy, and to gain access to the resources necessary to do an effective job, they are more likely to experience job engagement.
   c. Reward. Insufficient recognition and reward (whether financial, institutional, or social) increases people's vulnerability to burnout, because it devalues both the work and the workers, and is closely associated with feelings of inefficacy. In contrast, consistency in the reward dimension between the person and the job means that there are both material rewards and opportunities for intrinsic satisfaction.
   d. Community. Community has to do with the ongoing relationships that employees have with other people on the job. When these relationships are characterized by a lack of support and trust, and by unresolved conflict, then there is a greater risk of burnout. However, when these job-related relationships are working well, there is a great deal of social support, employees have effective means of working out disagreements, and they are more likely to experience job engagement.
   e. Fairness. Fairness is the extent to which decisions at work are perceived as being fair and equitable. People use the quality of the procedures, and their own treatment during the decision-making process, as an index of their place in the community. Cynicism, anger and hostility are likely to arise when people feel they are not being treated with the respect that comes from being treated fairly.
   f. Values. Values are the ideals and motivations that originally attracted people to their job, and thus they are the motivating connection between the worker and the workplace, which goes beyond the utilitarian exchange of time for money or advancement. When there is a values conflict on the job, and thus a gap between individual and organizational values, employees will find themselves making a trade-off between work they want to do and work they have to do, and this can lead to greater burnout.

All of this research argues for the development of a new model of ***healthy workplaces***, in which employees work productively and thrive, and in which burnout rates are low and engagement is high. Based on the six areas of job-person fit, such workplaces would be characterized by a sustainable workload, choice and control over one's work, recognition and reward for a job well done, a supportive work community, fair processes and respect, and clear values and meaningful work. Such healthy workplaces would be more likely to produce the following outcomes: high productivity, high employee satisfaction, good safety records, low frequencies of disability claims and union grievances, low absenteeism, low turnover, and an absence of violence.

6 CONCLUSION

The burnout shop is not a viable, or desirable, future for our workplaces or our societies. The research that psychologists have been doing on job burnout has pointed out clear paths to the better alternative of a healthy workplace. But now psychologists must partner with other experts—in sociology, political science, economics, public health, architecture, etc. – to collaborate in designing the healthy workplaces of the future.

REFERENCES

Leiter, M.P. & Maslach, C. (1999) Six areas of worklife: A model of the organizational context of burnout. *Journal of Health and Human Resources Administration*, 21, 472–489.

Leiter, M.P. & Maslach, C. (2004) Areas of worklife: A structured approach to organizational predictors of job burnout. In: Perrewe, P.L. & Ganster, D.C. (eds.) *Research in occupational stress and well-being volume 3*. Oxford, Elsevier. pp. 91–134.

Leiter, M.P. & Maslach, C. (2014) Interventions to prevent and alleviate burnout. In: Leiter, M.P., Bakker, A.B. & Maslach, C. (eds.) *Burnout at work: A psychological perspective*. London, Psychology Press. pp. 145–167.

Maslach, C. (1998) A multidimensional theory of burnout. In: Cooper, C.L. (ed.) *Theories of organizational stress* (68–85). Oxford, Oxford University Press. pp. 68–85.

Maslach, C. & Goldberg, J. (1998) Prevention of burnout: New perspectives. *Applied and Preventive Psychology*, 7 (1), 63–74.

Maslach, C. & Leiter, M.P. (2016) Understanding the burnout experience: Recent research and its implications for psychiatry. *World Psychiatry*, 15(2), 103–111.

Maslach, C., Leiter, M.P. & Schaufeli, W.B. (2009) Measuring burnout. In: Cooper, C.L. & Cartwright, S. (eds.) *The Oxford handbook of organizational well-being*. Oxford, Oxford University Press. pp. 86–108.

Maslach, C., Schaufeli, W.B. & Leiter, M.P. (2001) Job burnout. *Annual Review of Psychology*, 52, 397–422.

Schaufeli, W.B., Leiter, M.P. & Maslach, C. (2008) Burnout: Thirty-five years of research and practice. *Career Development International*, 14 (3), 204–220.

Contributions

Diversity in Unity: Perspectives from Psychology and Behavioral Sciences – Ariyanto et al. (Eds)
 ISBN 978-1-138-62665-2

The role of family strength on the relationship of parentification and delinquent behaviour in adolescents from poor families

F. Nurwianti, E.K. Poerwandari & A.S. Ginanjar
Faculty of Psychology, Universitas Indonesia, Depok, Indonesia

ABSTRACT: The objective of this study is to investigate the impact of family strength on the relationship of parentification and delinquent behaviour in adolescents from poor families, who are studying at the Centre of Community Learning Activity. Parentification is a concept that represents a process of role reversal experienced by adolescents, where they must carry parents' responsibility without considering their needs and their capabilities. Excessive roles assigned to adolescents could have negative effects on the adolescents' delinquent behaviour. As a part of collective society, parentification could be perceived as being an obligation for their family, and it would be supported by their family. So the impacts of parentification on delinquent behaviour depend on their family. Participants in these studies consist of 442 teenage boys and girls (aged 12–18 years, mean 15.6 years) who come from poor families in Jakarta. The result of this study showed parentification has negative correlation with delinquent behaviour. Moreover, the presence of family strength could lower the negative impact of parentification on delinquent behaviour in adolescents from poor families.

1 INTRODUCTION

Previous research studies have shown that poverty can increase psychological vulnerability by lowering power resources and social support resources. This makes people living in poverty vulnerable to experience physical and mental disruption (Wadsworth et al., 2005). Having limited resources and limited social support makes parents living in poverty push their children to contribute in earning money. This phenomenon of a child carrying parents' responsibility is coined as 'parentified child'. It is also a common thing to express it as adolescents experiencing parentification (Earley & Cushway, 2002). Parentification is a process of maturity in aspects such as contextual, social or individual development, whereby at a young age they must carry roles, gaining knowledge and responsibilities of parents, where these roles are developmentally inappropriate (Jurkovic, 1997; Burton, 2007; Minuchin et al., 1998). Boszormenyi-Nagy and Spark (1973) defined the parentification process as parents' expectation for their children to be able to fulfil the parents' role within the family system. There is a term in the parentification concept called 'role reversal', which depicts children acting as parents for their parents, or acting as a partner for their parents (Hooper, 2007). They are confidants, friends and decision makers for their parents (Earley & Cushway, 2002).

There are two types of parentification: Instrumental parentification and emotional parentification. Instrumental parentification can be defined as the assignment of concrete and practical tasks. Conditions are: where children need to contribute for family income; take care of siblings and family members; and, do chores such as washing, cooking, cleaning, and other practical tasks (Jurkovic, 1997; Hooper, 2007). In addition, children can also become sharing partners, problem resolvers, decision makers, or they can have a role in consoling their parents as well. We refer to those tasks as emotional parentification, which can be defined as children acting as someone who can satisfy their parents' emotional and psychological needs (Jurkovic, 1997; Hooper, 2007).

Research on poor community demonstrated that adolescents from economically deprived families usually experience both instrumental and emotional parentification. Those tasks are not only assigned to older children but are commonly assigned to younger children as well. According to developmental theory, childhood is characterised as playing time (creativity), whereas adolescence is a time to acquire self-autonomy. Moreover, in adolescence children experience fast physical growth which is considered as an indicator of maturity (Papalia et al., 2007). As a result of this perception, parents put higher demands on their adolescent children, give less guidance, and have less tolerance of children's in capabilities in conducting tasks. They lose time for themselves because they do a lot of work to fulfil others' needs while neglecting their own needs. These conditions make the adolescents more likely to experience subjective ageing acceleration compared to adolescents from prosperous families (Benson & Furstenberg, 2007; Foster et al., 2008). This subjective ageing acceleration happens because children from poor families would be more likely to carry household and financial responsibility, and also interact in a mature manner compared with other youngsters around their age (Johnson & Mollborn, 2009).

Some research discovered that parentification can have a negative effect, such as increasing the children's problems at school (Burton, 2007; Fuligini et al., 2005) and increasing external behaviour problems (Stein et al., 2007). Research conducted in the USA and England also suggests that adolescents with the same racial and socio-economic identity, who hold responsibility as caregivers for family members, experience anxiety, antisocial behaviour and lower self-worth compared to those who have none of the caregivers' responsibilities (East, 2010). Seemingly, accumulated task burden makes adolescents vulnerable to be involved in promiscuity in order to release overwhelming stress. It is not surprising that we can easily find a group of adolescents from poor families who possess delinquent behaviour. Cobb (2001) defines adolescents' delinquent behaviour as a behaviour deviating from social norm, such as either a criminal act, breaking family or school rules, or any self-harm or harassment to others, that is committed by children under 18 years old.

Somewhat differently, some other studies suggest that parentification can generate positive influence for adolescents. Parentification increases social competency (Telzer & Fuligni, 2009), problem-solving skills (McMahon & Luthar, 2007), and coping skills (Stein et al., 2007). Jurkovic (1997) argues that parentification can be constructive if adolescents perceive support and appreciation from the family. Existing research also demonstrates that children's high involvement in instrumental task or expressive task can produce positive results only if children conduct parentification to cope with certain distress. A similar result is also produced when their families had high-level cohesion, and when children receive appreciation from their surroundings and receive family support (Telzer & Fuligni, 2009; Jurkovic, 1997; McMahon & Luthar, 2007).

Previous research shows that children can attribute the experience of parentification as a form of trust and appreciation from the family towards their capabilities (Burton, 2007). Parents' compassion, supportive surroundings, and appreciation are characteristics of family strength. Family strength illustrates positive characteristics possessed by families that promote strength for each family member, including the children. Furthermore, factors affecting family strength of parents from poor families are: an environment prone to violence, parents' limited education, and high density because of limited space. Based on these situations, the expected form of family strength that is supposed to be provided by parents from poor families is a close relationship between parents and children, supervision by parents, and support from parents. A study conducted by Valladares and Moore (2009) demonstrates that closeness, supervision, and parents' support towards children are some aspects of family strength that can be provided by poor families. Adolescents that perceive supervision and parents' involvement within their activities regulate negative effects from the environment better than adolescents who do not perceive those family qualities (Moore et al., 2009).

Indonesia is a country that holds collective culture and that emphasises harmonious life between individual's and family's lives more than self-interest (Baron et al., 2008; Matsumoto & Juang, 2008). A study about adolescents from South America, Asia and Europe (Telzer & Fuligni, 2009) found that supportive families do not provoke stress but instead produce happiness.

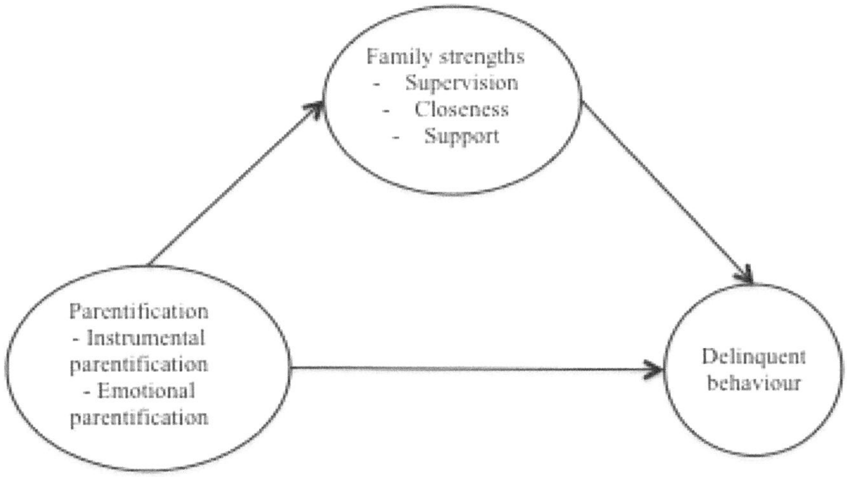

Figure 1. Research framework.

The researchers have speculated that this happiness is related to the meaningfulness they feel after their contributions for the family. Such attribution is more likely to be experienced by adolescents from South America and Asia compared to adolescents from Europe. A cross-cultural study suspected that children from non-Western society are used to cover parents' roles for their family. Furthermore, in some cultures, covering parents' role is recognised as normative and pivotal for a child's development (Orellana, 2001). Other existing studies also support this result, stating that obligation towards family such as pro-sociality, respectful behaviour, and contribution, appear stronger in South American and Asian families compared to European families (Hardway & Fuligni, 2006; Orellana, 2001). Thus, based on these values, parentification in Indonesia and specifically poor families, can be perceived positively as a contribution to fulfil their family needs. Parentification will increase family cohesiveness and closeness, and it will decrease delinquent behaviour.

According to the previous studies outlined above, it can be assumed that parentification can decrease delinquent behaviour, depending on how adolescents perceive their family. There are several questions that arise, such as if family strength would mediate the impact of parentification on adolescents' delinquent behaviour.

1. What is the impact of parentification on delinquent behaviour? Does parentification decrease delinquent behaviour?
2. What is the impact of family strength on the relationship between parentification and delinquent behaviour? Does family strength mediate this relationship?

2 RESEARCH METHOD

2.1 *Participants*

Participants of this study were teenage boys and teenage girls aged between 12 and 18 years old from poor families and members of the Centre of Community Learning Activity in North Jakarta, Central Jakarta, South Jakarta and East Jakarta. The definition of poor family refers to the criteria suggested by the Indonesian Central Statistics Bureau (Badan Pusat Statistik, 2014), namely a family whose maximal expenditure is Rp453,000 per family member per month, and at the same time taking into account the parents' education. The Centre of Community Learning Activity was chosen because it was the most favourable education institution from the parents' point of view, for its low price and time flexibility. Time flexibility was an important aspect because many students have chores to do or side jobs.

2.1.1 *Measures*

This paper used survey as a research method to gather data. Three questionnaires were administered to each participant. The questionnaires used in this study were parentification inventory (Hooper, 2007), self-report about adolescents' delinquent behaviour, and a family strength questionnaire. We started by investigating the validity and reliability of the questionnaires to 179 adolescents aged 12–18 years old. The analysis of all the questionnaires using confirmatory analysis resulted in valid and reliable items, so that data gathering was permitted. The statistical method selected to analyse the mediating role of family strength on the relationship between parentification and delinquent behaviour was SEM (Structural Equation Modelling), which was facilitated by the LISREL 8.8 application.

3 RESULTS

In this study, there are 442 adolescents consisting of 185 teenage boys (41.9%) and 257 teenage girls (58.15). Early adolescents, aged between 12 and 15, contributed to 52.3% of total participants, whereas late adolescents, aged 16–18, contributed to 47.7% of total participants. The education level of participants is junior high school (57.9%) and senior high school (42.1%). The results of the data analysis are presented in the Table below.

This study found that parentification has negative significant correlation with delinquent behaviour ($r = -0.12$, $p < 0.05$*). It means parentification causes more and more decrease of delinquent behaviour. Analysis by SEM showed that family strength affects the correlation between parentification and delinquent behaviour significantly (indirect effect $r = -0.0308$, $p < 0.01$**, direct effect $r = -.09$, $p > 0.05$). So it was proper to claim that family strength was full mediator of the correlation between parentification and delinquent behaviour. Taken together, parentification and family strength gave a contribution of 6.2% to delinquent behaviour ($r = 0.062$). Demographic variables in this study, such as sex, age, education and birth order, do not contribute significantly towards the correlation between parentification

Table 1. Correlation between each research variable.

	Closeness	Supervision	Support	Family strength	Delinquent behaviour
Closeness	1				
Supervision	0.523**	1			
Support	0.429**	0.193**	1		
Family strength	0.998**	0.565**	0.464**	1	
Delinquent behaviour	−0.228**	−0.159**	−0.113*	−0.230**	1
Instrumental parentification	0.116*	0.062	0.048	0.116*	−0.111*
Emotional parentification	0.115*	0.051	0.045	0.114*	−0.081
Parentification	0.119*	0.054	0.046	0.118*	−0.088

P < 0.01**; P < 0.05*.

Table 2. Goodness of fit result of family strength as the mediator of correlation between parentification and delinquent behaviour model.

Model testing	Fit-test	Mean
Model 5: Testing family strength as the mediator between parentification and delinquent behaviour	Chi = 251 df = 8 p = 0.96125 RMSEA = 0.000 SRMR = 0.012 CFI = 1.00 GFI = 1.00	Model fit

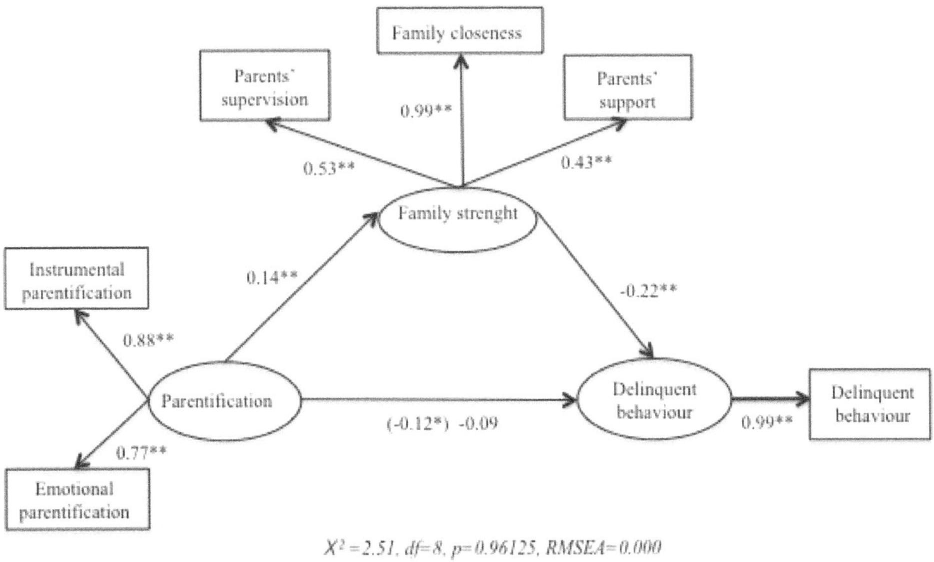

Figure 2. Summary of significance value of family strength as the mediator of correlation between parentification and delinquent behaviour model.

and delinquent behaviour. Besides primary findings, this study also found other important results such as that there are significant differences in delinquent behaviour between age levels ($t = 4.79$, $p < 0.01$**), between education level ($t = -3.226$, $p < 0.01$**), and between sex ($t = 2.921$, $p < 0.01$**). Moreover, tendency of delinquent behaviour is higher in late adolescence than in early adolescence. Comparison between junior high school students and senior high school students showed that senior high school students are more likely to have delinquent behaviour than junior high school students. Additionally, delinquent behaviour is higher in young men than in young women.

4 DISCUSSION

The results of this study showed that family strength is able to reduce the negative effect of parentification on delinquent behaviour. This result supports some research that parentification has a positive impact, such as positive mental health (Shifren, 2001), problem-solving skills (McMahon & Luthar, 2007), and social competency (Tezler & Fuligni, 2009). Furthermore, family strength as a mediator of the relationship between parentification and delinquent behaviour supports the cross-cultural study which suspects that children from non-Western societies are used to cover parents' roles for their family; it is recognised as normative and pivotal for a child's development (Orellana, 2001). Other studies found that in South American and Asian families, parentification is perceived as being an obligation towards the family, such as pro-sociality, respectful behaviour, and contribution (Hardway & Fuligni, 2006; Orellana, 2001). Thus, parentification will increase family cohesiveness and closeness with the family, and it will decrease delinquent behaviour.

In communities of poor families that inhabit big cities such as Jakarta, the neighbourhood is prone to violence and crime. Population density leads adolescents to be overexposed to adult behaviours, both normal and deviant. Adolescents need protections or defences against harmful influence from external environments. Parents, teachers, and peers are important figures who influence adolescents directly. When parents have positive attitude, such as tenacity, patience, honesty and compassion towards adolescents, they will build closeness with their parents and they will be able to empathise with their parents' adversity. A previous study about marginal community in the USA found that adolescents who perceive warmth and empathy from their parents receive positive effects from parentification (Shifren & Kachorek (2003), cited in Hooper et al. (2011)). Adolescents will carry parents' responsibility with enthusiasm so that they are willing to help parents. Jurkovic (1998) also reports that a perception of fairness in the family can increase the positive effects of parentification. Moore et al. (2009) found that monitoring and engagement by parents of their adolescents will control negative effects from the surroundings. Research by Sang et al. (2013) supports the idea that a higher degree of parentification will reduce adolescent intention to commit delinquent behaviour, because they have limited time with their peer group and increased empathy for their parents. Furthermore, Moore et al. (2002) argue that family strength is a group of characteristics in which the family supervise, are close to, and support, in which the characteristics complement each other. The absence of one characteristic can be harmful to children. In conclusion, we can say that adolescents' positive perception towards their parents can prevent them from carrying out delinquent behaviour.

Current findings show correlation with participants' characteristics which are unique characteristics of students from the Centre of Community Learning Activity. For adolescents from poor families, the chance to study at school is an opportunity to take care of themselves because then they have time to play with friends or gain some knowledge that will affect their cognitive development, emotional development and social behaviour. These opportunities, sadly, are not available for adolescents from poor families without school opportunities such as this. This leads us to the remaining question to be addressed in further research, namely the necessity to examine the mediating role of family strength on the correlation of parentification and delinquent behaviour in adolescents from poor families without any school opportunity.

5 CONCLUSION

From the research that has been undertaken, it is possible to conclude that the presence of a family's strength could reduce the likelihood of delinquent behaviour caused by parentification. There are two pivotal findings that support the theory about adolescents from poor families. First, parentification hampers adolescents' delinquent behaviour. Second, as the adolescents' social community, the family has an important role on the adolescents' delinquent behaviour reduction. Particularly for poor Indonesian families, the intensification of the family bond that is facilitated through developing family closeness, family supervision and parental support towards adolescents will protect parentified adolescents from delinquent behaviour.

REFERENCES

Badan Pusat Statistik (BPS). (2014). *Profil Makro Kemiskinan Jakarta*. Retrieved from http://www.bps.go.id.

Baron, R. & Kenny, D. (1986). The moderator-mediator variable distinction in social psychological research. *Journal of Personality and Social Psychology*, *51*(6), 1173–1182. doi:10.1037/0022-3514.51.6.1173.

Baron, R.A., Branscombe, NR. & Byrne, DR. (2008). *Social Psychology* (12th Edition). Allyn & Bacon Publisher.

Benson, J.E. & Furstenberg, F.F. (2007). Entry into adulthood: Are adult role transitions meaningful markers of adult identity? *Advances in Life Course Research*, *11*, 199–224. doi:10.1016/S1040-2608(06)11008-4.

Boszormenyi_Nagi, I. & Spark, G.M. (1973). Invisible loyalties: Reciprocity in intergenerational family therapy, Hagrstown, MD: Harper & Row.

Burton, L. (2007). Childhood adultification in economically disadvantaged families: A conceptual model. *Family Relations*, *56*(4), 329–345. doi:10.1111/j.1741-3729.2007.00463.x.

Cobb, N.J. (2001). *Adolescence.* Mountain View, CA: Mayfield Publishing.
Earley, L. & Cushway, D.J. (2002). The parentified child. *Clinical Child Psychology and Psychiatry*, 7(200204), 163–178. doi:10.1177/1359104502007002917.
East, P.L. (2010). Children's provision of family caregiving: Benefit or burden? *Child Development Perspectives*, *4*(1), 55–61. doi:10.1111/j.1750–8606.2009.00118.x.
Foster, H., Hagan, J. & Brooks-Gunn, J. (2008). Growing up fast: Stress exposure and subjective 'weathering' in emerging adulthood. *Journal of Health and Social Behavior*, *49*(2), 162–177.
Fuligni, A.J., Alvarez, J., Bachman, M. & Diane N. (2005). Family obligation and academic motivation of young children from immigrant families. In C.R. Cooper., C.T.G. Coll, W.T. Bartko, H.M. Davis & C. Chatman (Eds.), *Developmental pathways through middle childhood: Rethinking context and diversity as resources* (pp. 261–282). New York, NY: Routledge.
Gennetian, L.A., Lopoo, L.M. & London, A.S. (2008). Maternal work hours and adolescents' school outcomes among low-income families in four urban counties. *Demography*, *45*(1), 31–53. doi:10.1353/dem.2008.0003.
Hardway, C. & Fuligni, A.J. (2006). Dimensions of family connectedness among adolescents with Mexican, Chinese, and European backgrounds. *Developmental Psychology*, *42*(6), 1246–1258. doi:10.1037/0012-1649.42.6.1246.
Hooper, L.M. & Wallace, S.A. (2010). Evaluating the parentification questionnaire: Psychometric properties and psychopathology correlates. *Contemporary Family Therapy*, *32*(1), 52–68. doi:10.1007/s10591-009-9103-9.
Hooper, L.M. (2007). Expanding the discussion regarding parentification and its varied outcomes: Implications for mental health research and practice. *Journal of Mental Health Counseling*, *29*(4), 322–337.
Hooper, L.M., Decoster, J., White, N. & Voltz, M.L. (2011). Characterizing the magnitude of the relation between self-reported childhood parentification and adult psychopathology: A meta-analysis. *Journal of Clinical Psychology*, *67*, 1–16.
Jankowski, P.J., Hooper, L.M., Sandage, S.J. & Hannah, N.J. (2013). Parentification and mental health symptoms: Mediator effects of perceived unfairness and differentiation of self. *Journal of Family Therapy*, *35*(1), 43–65. doi:10.1111/j.1467-6427.2011.00574.x.
Johnson, M.K. & Mollborn, S. (2009). Growing up faster, feeling older: Hardship in childhood and adolescence. *Social Psychology Quarterly*, *72*(1), 39–60. doi:10.1177/019027250907200105.
Jurkovic, G.J., Morrell, R. & Chase, N.D. (2001). Parentification in the lives of high-profile individuals and their families: A hidden source of strength and distress. In B.E. Robinson & N.D. Chase (Eds.), *High-performing families: Causes, consequences, and clinical solutions* (pp. 92–113). Washington, DC: American Counseling Association.
Jurkovic, G.J. (1997). *Lost childhood: The plight of the parentified child.* New York, NY: Brunner/Mazel.
Jurkovic, G.J. (1998). Destructive parentification in families: Causes and consequences. In L. L'Abate (Ed.), *Family psychopathology* (pp. 237–255). New York, NY: The Guilford Press.
Jurkovic, G.J., Morrel, R. & Thirkield, A. (1999). Assessing childhood parentification: Guidelines for researchers & clinicians. In N.D. Chase (Ed.), *Burdened children.* New York, NY: The Guilford Press.
Kuperminc, G.P., Jurkovic, G.J. & Casey, S. (2009). Relation of filial responsibility to the personal and social adjustment of Latino adolescents from immigrant families. *Journal of Family Psychology: JFP: Journal of the Division of Family Psychology of the American Psychological Association (Division 43)*, *23*(1), 14–22. doi:10.1037/a0014064.
Matsumoto, D.R. & Juang, L.P. (2008). *Culture and psychology*. Belmont, CA: Wadsworth/Thomson.
McMahon, T.J. & Luthar, S.S. (2007). Defining characteristics and potential consequences of caretaking burden among children living in urban poverty. *The American Journal of Orthopsychiatry*, *77*(2), 267–281. doi:10.1037/0002–9432.77.2.267.
Minuchin, S. (1988). Relationships within the family: A systems perspective on development. In R.A. & J. Stevenson-Hinde (Eds.), *Relationshzjx within families: Mutual influences* (pp. 7–26). Oxford, England: Clarendon Press.
Moore, K.A., Papillo, A.R. & Williams, S.W. (2002). Family strengths: Often overlooked, but real. *Child Trends.* Retrieved from https://www.childtrends.org/publications/family-strengths-often-overlooked-but-real/.
Moore, K.A., Whitney, C. & Kinukawa, A. (2009). Exploring the links between family strengths and adolescent outcomes. *Child Trend Research Brief*, *20*(4). Retrieved from https://www.childtrends.org/publications/exploring-the-links-between-family-strengths-and-adolescent-outcomes/.
Orellana, M.F. (2001). The work kids do: Mexican and Central American immigrant children's contributions to households and schools in California. *Harvard Educational Review*, *71*, 366–389.

Papalia, D., Olds, S. & Feldman, R. (2007). *Human Development* (10th ed.). New York, NY: McGraw-Hill.
Sang, J., Cederbaum, JA. & Hurlburt, MS. (2013). Parentification, substance use, and sex among adolescent daughter from ethnic minority families: the moderating role of monitoring. *Family Process*, vol. 53, No. 2, 252–266. doi: 10.1111/famp.12038.
Shifren, K. (2001). Early caregiving and adult depression: Good news for young caregivers, *The Gerontologist*, 41, 188–190.
Stein, J.A., Rotheram-Borus, M.J. & Lester, P. (2007). Impact of parentification on long-term outcomes among children of parents with HIV/AIDS. *Family Process*, *46*(3), 317–333. doi:10.1111/j.1545-5300.2007.00214.x.
Telzer, E.H. & Fuligni, A.J. (2009). A longitudinal daily diary study of family assistance and academic achievement among adolescent from Mexican, Chinese, and European Backgrounds. *Journal Youth Adolescence*, *38*, 560–571. doi:10.1007/s10964-008-9391-7.
Valladares, S. & Moore, K.A. (2009). The strengths of poor families. *Child Trends Research Brief*, 26. Retrieved from http://www.saintlukesfoundation.org/wpcontent/uploads/2013/05/Child_Trends2009_ 5_14_RB_poorfamstrengths1.pdf.
Wadsworth, M.E., Raviv, T., Compas, B.E. & Connor-Smith, J.K. (2005). Parent and adolescent responses to poverty related stress: Tests of mediated and moderated coping models. *Journal of Child and Family Studies*, *14*(2), 283–298. doi:10.1007/s10826-005-5056-2.

Diversity in Unity: Perspectives from Psychology and Behavioral Sciences – Ariyanto et al. (Eds)
© 2018 Taylor & Francis Group, London, ISBN 978-1-138-62665-2

Stress management intervention for women with multiple roles: Case study of working women in post-partum period

T.P. Ningrum, A.S. Ginanjar & Y.R. Sari
Faculty of Psychology, Universitas Indonesia, Depok, Indonesia

ABSTRACT: The rising numbers of working women in Jakarta indicates that women are currently holding many roles in their lives. It increases the risks of these women to experiencing role overload and distress. The distress that are experienced by these women has the potential to bring negative impacts onto themselves, their babies, their partner, family, and even their colleagues. It is also known that the administration of an intervention using stress management techniques has become one of the alternatives to help multiple-role mothers to manage their stress levels. This study was carried out by modifying the Living SMART stress management techniques. The purpose of this intervention is to provide techniques that can assist the process of emotional adjustment and to reduce the distress levels of multiple-role women, particularly for working mothers in their post-partum period. The Living SMART stress management program that has been modified is called the 'Stress Management Program for New Moms'. The program was implemented for five group-intervention sessions, each lasting for 90 minutes. The results indicated that stress management interventions have some positive impacts on the participants, which include improvement in the ability to identify and recognise the stressors experienced during stress reactions, reduced stress levels, and the ability to cope with stressors.

1 INTRODUCTION

Women who live in the modern era decide to work outside their homes for several reasons, some of which are due to their higher education, a need for self-actualization, the rising number of single parents, a need to contribute to their husbands' income, to get socialised (Kassamali & Rattani, 2014), or to get a better quality of life (Kassamali & Rattani, 2014). This rising number of working mothers with infants encouraged a lot of researchers to see the importance of understanding the relationship between a women's job and her roles as a mother, with her psychological well-being. The more roles a person has, the bigger their chances to experience role overload, and this situation could lead to psychological distress. The condition of role overload is a term for an individual who has too many responsibilities or activities for which they are capable of handling (LoCascio, 2009), where an individual perceives that the role pressure exceeds the availability of their time and resources (Pearson, 2008).

Furthermore, the post-partum period or puerperium, is a transitional period that is considered potentially stressful to a woman who has just recently become a mother or experienced childbirth (Blenning & Paladine, 2005; in King, 2009). Women usually experience a post-partum period for up to 12 months after childbirth, and a one-year period is considered as a recovery period for their physical and emotional condition as the result of pregnancy and childbirth (King, 2009). Within a post-partum period, women have to deal with their own expectations for themselves, experiencing dramatic emotional and physical changes, adjusting to expectations given by their surroundings, and at the same time adjusting to their new role as a mother (King, 2009). The researchers emphasised that in addition to the adjustment process, these women are also experiencing dynamic changes in their interpersonal relationships, having greater expectations to receive social support, gaining new responsibilities, experiencing both physical and emotional limitations, as well as having greater economic needs (King, 2009). These huge changes could lead these women to experience post-partum stress, depression, and anxiety (King, 2009).

In addition, a woman who has just experienced childbirth and then gone back to work has to be able to integrate these new roles that have been changed, as an individual, a couple, a parent, and a worker (Lambden, 2001). Hall (Lambden, 2001) identifies the process experienced by women who had just had childbirth and then gone back to work. These women reported that they were experiencing greater stress related to the roles they were doing as a result of guilt, realisation of their inability to fulfil their household responsibilities, rejection of their personal needs, and their perception of other's expectations towards their new roles (Lambden, 2001). In addition, the lack of energy as the result of childbirth, limited time and resources could also increase their stress levels. Role overload may also cause stress, a feeling of being overwhelmed, and even anxiousness, if they cannot handle it effectively. Moreover, the high number of daily chores in a very limited time could also cause anger, depressive mood, and helplessness (Whetstone, 1994).

In addition to job and childcare issues, one other thing that is also important and related to working women stress level is the division of roles and responsibilities with their husbands at home. This transitional period to parenthood could be considered as a crisis period. In this period, the couples are experiencing the impacts of having lots of roles and how it influences their psychological well-being. Furthermore, an unfulfilled expectation of a woman regarding support that they are supposed to receive after childbirth, could cause them to experience difficulties in adjusting to parenthood (Goldberg & Perry-Jenkins, 2004).

Based on statistical data from BPS-Statistics in Indonesia, up until February 2013, there are 114 million working people in Indonesia, and 46.8 million among them are women (38.6%). In the period of 2013–2014 in Jakarta, the capital city of the country, there were 1,978,000 women within productive ages, and 1,742,000 among them were working women. Based on these numbers, we could conclude that there are a very large number of working women in Indonesia, and especially in Jakarta.

Based on the research and treatments given to working women who have children, it shows that the use of intervention using stress management techniques has become one of the alternatives that could help women in that condition to manage their stress level. Stress management refers to the process of identifying and analysing problems that are related to stress and the application of some therapeutic process, through modifying stressors or how the person copes with the stressor they are experiencing (Cotton, 1990). The purpose of stress management is to optimise the individual functioning in a healthy and positive manner (Cotton, 1990).

This study will be conducted in reference to the Living Smart stress management model that was originally conducted by King (2009). This Living Smart programme used a bio psychosocial approach that integrates biological, psychological, and social aspects of the person to deal with the stressors. This could be relevant to be applied in this study because stress is not only limited to biological experiences but also has some impacts on the psychological well-being and behaviour of a person. Moreover, this bio psychosocial approach could also include the needs of social support, which is known to have an important role in reducing and minimising the stress level of working mothers during the post-partum period. A bio psychosocial approach is used to better understand stress symptoms and how to prevent distress that could lead to several physical and emotional impacts, in particular: higher blood pressure, headache, premature ageing of body organs, depression, anxiety, social isolation, unhealthy behaviour, and physical illness.

In this study, the researchers decided to adapt the Living SMART programme and to focus on preventing the development of problems experienced by working women in the post-partum period. The stress management intervention was based on the Living SMART programme that has been modified based on the characteristics and needs of Indonesian women, and it is expected to be able to help those working women in the post-partum period in adjusting to their new roles as a mother, a wife, and a worker. Also, it is expected to successfully reduce the distress levels that are related to the demands and responsibilities of working women. The implementation of this intervention is aimed at having a better knowledge of whether or not this modified Living SMART programme would be beneficial in improving the ability of working women to manage their stress level related to their roles as mother, wife, and worker.

2 METHODS

As researchers in this study, we used a quasi-experimental model as the study design and used before–after study or pretest/post-test design, which means giving manipulation to the participants without using randomisation and strict control (Kerlinger & Lee, 2000; Kumar, 1999). Measurements were performed twice, before the session begins and after the session has ended, to observe the impacts of the treatment given to the participants.

In this study, the criteria for the participants are women who: have a full-time job (working 40 hours/week) outside home and receive income; have a husband who works outside home and receives income; have experienced childbirth within one year (12 months); went back to work after receiving maternity leave; are at least a high school graduate; are willing to attend all five sessions (60–90 minutes per session); and have a husband who agrees to attend one conjoint session.

The sampling technique used in this study is a non-random or non-probability sampling, because not all the individuals in the population have an equal chance to be chosen as a sample (Kerlinger & Lee, 2000).

We gave the participants a pre-intervention assessment to gather some information regarding stress level of the participants before the treatment starts and gave post-intervention assessment after the sessions have ended. The measurements given to the participant include Perceived Stress Questionnaire (PSQ), Subjective Units of Distress Scale (SUDS), and Barkin Index of Maternal Functioning (BIMF), as well as an in-depth interview with each of the participants.

We did some modification to the Living SMART programme and then formed a stress management intervention programme called Stress Management Program for New Moms, with as many as six meetings. The term 'new moms' does not necessarily mean the first birth, but the focus is more on a maximum of 12 months after childbirth. This programme consists of one pre-intervention meeting where prospective participants were interviewed and given the pre-intervention assessment, informed consent form, and a brief explanation of the programme. The selected participants then need to attend a group meeting, every week for the next five weeks (60–90 minutes each session).

Participants were selected after all data of the prospective participants was gathered. In this process, we received 11 registrants who were willing to attend five sessions, including one conjoint session with the husband. The final group consisted of six participants, and these six participants were officially selected to attend the Stress Management Program for New Moms.

To better understand the intervention, Table 1 shows the brief intervention plan of Stress Management Program for New Moms.

Table 1. Intervention plan of Stress Management Program for New Moms.

Session	Planned activities
Pre-intervention (individual)	Briefly informing the programme outline, interview, signing informed consent form, and giving pre-intervention assessment to the prospective participants.
1st session	Introducing group members; Psychoeducation: introducing stress, post-partum stress, coping techniques; Exercise: relaxation techniques, and Relaxation Response.
2nd session	Exercise: guided imagery and creative visualisation, psychoeducation and discussion with gynaecologist regarding stress and common problems experienced by working women in post-partum period.
3rd session	Psychoeducation: stress, Negative Automatic Thoughts (NAT), cognitive distortion; Group discussion: challenging cognitive distortions and NAT; Exercise: meditation (Grounding; I am Grateful).
4th session	Psychoeducation: the importance of husband, family and social support, role division, communication techniques; Couples' exercise: discussion of daily problems and how to handle them effectively.
5th session	Psychoeducation: time management and self-care; Exercise: time management and using daily journal; Debriefing and closing: summarising the programme; Post-intervention assessment and feedback.

3 RESULTS

Based on the results of the Stress Management Program for New Moms, the following are the comparison of pre-intervention assessment and post-intervention assessment scores of all six participants. After their third session, Wita did not continue the remaining two sessions because she had to have surgery; therefore the score is not be added to the final quantitative scoring and is considered in the qualitative results.

Based on the scores of the three measurements above (Table 2), it could be concluded that six of the participants were experiencing lower scores of SUDS and PSQ, as well as higher score of BIMF. This result has shown that the intervention of Stress Management Program for New Moms brings positive impacts in reducing distress level and relatively increases the maternal function of working women in the post-partum period. Based on the BIMF scores, it shows that there are some participants who did not experience a significant change of their BIMF scores, although did experience lower scores of stress.

In addition to using measurements (SUDS, PSQ, and BIMF), we gathered some supporting data in this study, such as interviews, observations, and feedbacks after the sessions ended. Here is some supporting data that has been concluded, based on qualitative analysis:

1. From the scale of 1–10 (1 = minimum benefit, 10 = maximum benefit) on the feedback form, the mean scores from the participants was 8.5 with a range of 7 to 10. The lowest score (7) came from Wita (who was absent on the last two sessions because of her endometriosis surgery), and the highest scores came from Shella and Caca. These results showed that all the participants experienced positive impacts from the Stress Management Program for New Moms.
2. In general, all participants gained benefits from all six sessions, from practising relaxation, identifying stressors, practising coping techniques, the reproductive health session with the gynaecologist, the conjoint session with the partner, and time management. One of the most impactful sessions for the participants was the conjoint session with their partners. At the beginning, Shella thought that the discussion in the group would create a new conflict with her husband after the session ends, but it turned out that her husband's attitude became more positive and they could communicate in a better way after the fourth session. According to Jessica, her husband became very thoughtful and was willing to help her do household chores afterwards. These positive impacts were also felt by Caca, in which case Adi, her husband, became more polite and practised communication techniques at home. At the beginning of the sessions, all of the participants agreed that their husbands are one of the major stressors they experience every day, and therefore it can be concluded that this session has made a great impact for the participants in the relationship with their husbands.
3. Six of the participants worked on their weekly tasks and filled out their journals. Each of the participants has their own favourite techniques for dealing with stress. The Relaxation Response technique from Herbert Benson has proven very effective for Caca and Eka. In addition, creative visualisation was considered useful for Eka and Shella. Caca did not feel

Table 2. Pre-intervention and post-intervention assessments.

	SUDS		PSQ		BIMF	
Name	pre	post	pre	pre	post	pre
Yesi	5	3	0.5	0.26	80	88
Ria	6	3	0.65	0.36	72	89
Caca	8	5	0.9	0.21	31	80
Shella	8	4	0.8	0.36	79	90
Eka	6	4	0.49	0.33	87	90
Mean (group)	6.6	3.8	0.66	0.30	69.80	87.4
Change (group)	42%		54.5%		25.2%	

the impact of grounding technique, while Ria, Shella, and Shella, on the other hand, felt that it has been very useful for them.
4. Of all the techniques that have been taught to the participants, all of them stated that the 'I am Grateful' technique is the most applicable technique for them. Eka practises it every time she prays, and Ria practises it when she is about to sleep every night because it helps her sleep better. All of the participants also stated that this technique is very simple and helps them feel calmer afterwards, so they could practise it every day.
5. Cognitive restructuring is also considered very helpful to the participants. All of them stated that they were often having irrational thoughts and experiencing cognitive distortions. After being given A-B-C techniques, the participants started practising them at home and felt the positive impacts.
6. In terms of participant supports or reactions to another group member and the atmosphere of the group, all of the participants stated that each of the participants acts accordingly, being open, friendly, and able to communicate freely, so that the group sharing sessions could also be beneficial for each of the group members.

4 DISCUSSION

The participants in this study reported that their daily activities cause some physical symptoms such as fatigue and low energy, difficulties in managing their emotions, and also that they experience stress that is impacted to various aspects of their lives, from work tasks, household chores, and irritabilities that contributed to experiencing conflicts with their husbands, even affecting the relationship quality with their babies. The results of this study are consistent with what Bakker et al. (Harlin-Clifton, 2008) state: that, in general, despite the multiple roles of women, they do not have enough time to work, take care of the children, do household chores, take care of their husbands and even take care of themselves. The jobs or careers that these women have, usually require a lot of energy and time and therefore could cause stress. Too many role demands could cause these women with multiple roles to experience problems in adjusting to their current situation, and therefore they will experience stress overload and not have effective coping strategies to manage the situation.

Some of the impacts of stress complained about by the participants in this study are crying, exhaustion, uncontrollable anger, irritability, anxiety, and some physical symptoms, such as muscle tenseness and palpitation. In addition, some of the participants are also receiving complaints from the employer because they were often late for work, trying to go home as soon as possible, and rarely socialising with their colleagues after office hours. Corresponding to the results of this study, role overload has been known to be related to some negative effects, such as stress-related symptoms like anxiety, headache, burnout, even related to lower family and work satisfaction (Bacharach et al., 1991; in Higgins et al., 2010).

The main similarity between the participants in this study is that they are trying to get all of the work done and do their roles as a wife, a mother, a housekeeper, and a successful career woman. It is consistent with a study conducted by Rout et al. (1997) that the stress experienced by working mothers is because they feel that they should master all of their roles and prove to themselves that they are competent in completing tasks and responsibilities, both at home and at their office.

Moreover, based on the results from a study conducted by Savitri (1997) on the sample of first-time working mothers in Jakarta, it shows that problems faced by working women in their transition to becoming a mother is that they do not have a balanced role division between husband and wife. This is because the responsibility and role of taking care of the household and children weighs more on the wives than on the husbands. The results are in line with what was found in this study. From the beginning, the participants stated that their husbands are one of their main stressors; they felt lack of support and attention from their partners after child birth. The participants also complained that their husbands were not taking care of their share of responsibilities in taking care of the household. These results are consistent with Wentling (1998, cited in Pearson, 2008), who states that compared to male workers, female workers have to deal with bigger responsibilities in taking care of their household chores.

Another form of support that is also important to these women is from their surroundings, such as friends and colleagues who could provide them with information and support related to values, lifestyle, and the role as a mother (Terry et al., 1996). Moreover, the results in this study show that all the participants received some support from groups of mothers who also have a baby. Those groups were formed from mailing lists and smart phone applications such as WhatsApp and Black Berry Messenger groups that offer them possibilities for sharing knowledge and receiving new information related to parenting and childcare.

5 CONCLUSION

Based on the results of this study, it can be concluded that the participants experienced reduced levels of distress after attending a series of sessions from the Stress Management Program for New Moms. This was shown in the lower scores of SUDS and PSQ, meaning that the participants are experiencing reduced level of distress. In addition, the higher scores of BIMF in post-intervention assessment show that the reduced level of distress is associated with the increased maternal function experienced by these working mothers. Also, the participants perceived that they were receiving social support, better understanding in implementing self-care and childcare, experiencing improved levels of attachment and bonding with their babies, and understanding better their roles and responsibilities as new mothers.

After attending the Stress Management Program for New Moms, the participants are able to identify their stressors, recognise the symptoms, their stress reactions, and the impacts of stress on their lives. Moreover, they also learn about how to cope with stress by practising relaxation techniques, meditation, identifying and challenging cognitive distortion, learning about time management, and practising effective communication so that they can manage their emotions effectively. Techniques that were given to the participants could be easily applied to their daily lives, although there were different rates of effectiveness on each participant, but in general it could help them reduce stress and anxiety, make them feel relaxed and have a calming effect. Of all the techniques that have been taught, six of the participants stated that the 'I am Grateful' technique is the easiest to do and could be applied in any situation.

REFERENCES

Cotton, D.H.G. (1990). *Stress management: An integrated approach to therapy*. New York, NY: Brunner/Mazel.

Goldberg, A.E. & Perry-Jenkins, M. (2004). Division of labor and working-class women's well-being across the transition to parenthood. *Journal of Family Psychology*, *18*(1), 225–236. doi:10.1037/0893-3200.18.1.225.

Harlin-Clifton, S.M. (2008). *The effects of stress management training on work-stress coping levels for working mothers* (Doctoral dissertation, Walden University, Ann Arbor, MI). Available from ProQuest Dissertations and Theses database (UMI No. 3330545).

Higgins, C.A., Duxbury, L.E. & Lyons, S.T. (2010). Coping with overload and stress: Men and women in dual-earner families. *Journal of Marriage and Family*, *72*(4), 847–859.

Hung, C.H. (2004). Predictors of postpartum women's health status. *Journal of Nursing Scholarship*, *36*(4), 345–351.

Kassamali, N. & Rattani, S.A. (2014). Factors that affect attachment between the employed mother and the child, infancy to two years. *Procedia—Social and Behavioral Sciences*, *159*, 6–15. doi:10.1016/j.sbspro.2014.12.319.

Keadaan Ketenagakerjaan di DKI Jakarta (2014). *Berita Resmi Statistik Provinsi DKI Jakarta No. 26/05/31/Th XVI, 5 Mei 2014*. Downloaded from http://www.jakarta.go.id/jakv1/application/public/download/bankdata/Berita_Resmi_Statistik_Mengenai_Keadaan_Tenagakerja_DKI_Jakartra_Bulan_Februari_2014.pdf.

Kerlinger, F.N. & Lee, H.B. (2000). *Foundations of behavioral research* (4th ed.). Orlando, FL: Harcourt College Publishers.

King, E. (2009). *The effectiveness of an Internet-based stress management program in the prevention of postpartum stress, anxiety and depression for new mothers* (Doctoral dissertation, Walden University, Ann Arbor, MI). Available from ProQuest Dissertations and Theses database (UMI No. 3355047).

Kumar, R. (1999). *Research methodology: A step-by-step guide for beginners*. London, UK: SAGE Publications.

Lambden, M.P. (2001). *The mediational role of working mother perceived self-efficacy*. (Doctoral dissertation, The University of Texas, Austin, TX). Available from ProQuest Dissertations and Theses database (UMI No. 3008375).

LoCascio, S. (2009). *Maternal employment: Factors related to role strain* (Master's thesis, University of North Texas, Denton, TX).

Pearson, Q.M. (2008). Role overload, job satisfaction, leisure satisfaction, and psychological health among employed women. *Journal of Counseling and Development*, *86*(1), 57–63.

Rout, U.R., Cooper, C.L. & Kerslake, H. (1997). Working and nonworking mothers: A comparative study. *Women in Management Review*, *12*(7), 264–275. doi:10.1108/09649429710181234.

Savitri, I. (1997). *Masa transisi sebagai orang tua: Masalah coping dan faktor-faktor yang mempengaruhi penyesuaian diri wanita bekerja saat pertama kali menjadi ibu* (Bachelor's thesis, Universitas Indonesia, Jakarta, Indonesia).

Terry, D.J., Mayocchi, L. & Hynes, G.J. (1996). Depressive symptomatology in new mothers: A stress and coping perspective. *Journal of Abnormal Psychology*, *105*(2), 220–231. doi:http://dx.doi.org/10.1037/0021-843X.105.2.220.

Whetstone, M.L. (1994). The working mother's dilemma. *Ebony*, *49*(7), 26C.

Diversity in Unity: Perspectives from Psychology and Behavioral Sciences – Ariyanto et al. (Eds)
 ISBN 978-1-138-62665-2

Boost outcome expectancies to improve cancer survivors' health behaviours

G.M. Hartono & L.D. Pohan
Faculty of Psychology, Universitas Indonesia, Depok, Indonesia

ABSTRACT: Studies have shown that health behaviours are one of the key determinants of cancer survivors' quality of life, but the real rewards of health—feeling better, living longer, and worrying less—seem to be attached to a distant future. That is why we need to identify and reinforce the intrinsic value of health behaviours that would help cancer survivors maintain those healthy behaviours over time. Therefore, the aim of this study was to investigate the impact of outcome expectancies on health behaviour intention of cancer survivors. It was because, according to the Health Action Process Approach (HAPA), health behaviour intention could be influenced by several factors, including outcome expectancies. This study was conducted on 90 cancer survivors, aged between 15–50 years old. Self-reported instrument and informal interviews were used, such as The Health Behaviour Intention Scale to measure the dependent variable (health behaviours intention), and Life Orientation Test Revised (LOT-R) to measure the independent variable (outcome expectancies). Linear regression analyses indicated that outcome expectancies had positive and significant impact on cancer survivors' health behaviour intention, explained by 54% variance in health behaviours and the β value which reached up to 0.728 with $p > 0.01$. This means that with every 1-point increase of outcome expectancies, health behaviour intention also increased by 0.728. Findings from the current study suggested that, among cancer survivors, outcome expectancies may influence the degree of health behaviour intention and help them adopt health behaviours.

1 INTRODUCTION

Regular healthy behaviours, such as healthy diet, regular exercise, and smoking cessation, are effective interventions to enhance quality of life, reduce fatigue, and potentially reduce recurrence and mortality among cancer survivors (Rogers *et al.*, 2008). The statement was proven by Loh et al. (2013) who found an increasing research evidence demonstrating that being physically active could reduce cancer risks, especially for cholesterol cancer, breast cancer, endometrial cancer, lung cancer, and prostate cancer. Unfortunately, healthy behaviours often declined after cancer diagnosis and might not return to prediagnosis levels. According to Seifert et al. (2012), it is natural that people would engage in new behaviours if they believed there is a reasonable reward (incentive) for it, especially for people diagnosed with cancer; they would need concrete evidence. However, the real reward for healthy behaviour, such as feeling better, living longer, and worrying less, seems to be attached to a distant future.

In order to enhance health behaviours among cancer survivors, a better understanding of the intrinsic value of health behaviours is needed. That is because, according to Seifert et al. (2012), an individual's motivation to change is the most significant stumbling block in health promotion and intrinsically motivated people do not need incentives to motivate themselves to adopt healthy behaviours, as they would do it on their own.

One of the most consistently identified correlates to health behaviour engagement is outcome expectancies, one of the key elements of the Health Action Process Approach (HAPA). Scheier and Carver (1985) referred to the outcome expectancies as being subjective perceptions and impressions of the consequences and effectiveness of particular actions. Indeed, Schwarzer and Luszczynska (2008) argued that outcome expectancies are also extremely crucial to motivating and helping individuals make decisions in the process of performing health behaviours.

Strecher et al. (1986) reported that outcome expectancies are a key component that consists of two predictors of health behaviours in the Health Belief Model: perceived susceptibility and perceived benefits. Perceived susceptibility is a personal assessment on the level of vulnerability to diseases, while perceived benefits is a personal perception on the benefits of health behaviours that they perform in an effort to minimise health risks. That is why a better understanding of the impact of outcome expectancies might serve as a potential contributor to enhance health behaviour intention.

The relatively few studies, all on non-cancer survivor populations, that evaluated correlates of outcome expectancies demonstrated the importance of outcome expectancies on health behaviour intention. Those statements are in line with the results of the study by Snippe et al. (2015), which found that positive outcome expectancies resulted in higher intention among young adults to do sport activities. Other research also suggested that among cancer survivors who realised the potential benefits of physical activity on survival outcomes, they might be more likely to embrace physical activity as an additional means of taking control and preserving hope during the chronic illness experience (Karvinen & Vallance, 2015).

Therefore, the primary objective of this study was to determine the impact of outcome expectancies on health behaviour intention on cancer survivors. Our hypothesis was that outcome expectancies would significantly influence health behaviour intention among cancer survivors. We examined this hypothesis in the context of health behaviours including nutrition, physical exercise, smoking and alcohol resistance, and more.

2 MATERIALS AND METHODS

The participants in this study consisted of: (1) males and females, aged 15–50 years old; (2) cancer survivors who had the required medical treatment; and (3) residents of Greater Jakarta and other major cities in Java. This study used a purposive sampling technique as well as snowball sampling.

The data collection process started in April 2016 and finished in May 2016. This study was conducted at three cancer survivor communities and one hospital: Yayasan Kanker Payudara Indonesia (YKPI), Cancer Information and Support Centre (CISC), Yayasan Kanker Anak Indonesia (YKAI), and Rumah Sakit Cipto Mangunkusumo (RSCM). The data was gathered by asking participants to complete questionnaires, consisting of informed consent, medical record, current health behaviour, and two kinds of instruments to measure the variables.

2.1 *The Life Orientation Test (LOT-R)*

LOT-R is a ten-item scale developed by Scheier et al. (1994), used to measure outcome expectancies. These ten items consisted of three favourable items, three unfavourable items, and four filler items. Participants were asked to rate how they perceived their life since their cancer diagnosis, on a scale from 1 to 4 ('Strongly disagree' to 'Strongly agree'). The scores would automatically be inverted in the three unfavourable items, and the calculation of the filler items were not included in the total score of this instrument. LOT-R had previously been tested for reliability and had demonstrated significant associations with healthy behaviour in cancer survivors. The internal consistency (i.e. Cronbach alpha) in our sample was 0.78.

2.2 *The health behaviour intention scale*

Health behaviour intention was measured using the ten-item scale developed by Renner and Schwarzer (2005). Participants were asked to rate how they intended to change their health risk behaviour to the healthier one, on a scale from 1 to 4 ('Strongly disagree' to 'Strongly agree'). This instrument had previously been tested for reliability, with internal consistency (i.e. Cronbach alpha) of 0.887.

3 RESULTS AND DISCUSSION

Table 1 shows that the total number of participants in this study was 90, with the majority were being female and aged 21–40 years old. The most common types of cancer found among participants were breast cancer and blood cancer (leukaemia), which affected the same number of patients, which was 22 people. Interestingly, there was a unique finding in that the percentage of people who did not know their cancer stage was relatively large. Participants reported that they purposely avoided asking for the information to prevent distress and shock. According to Kubler-Ross (1973), there are five stages of grief, one of them being denial. Denial refers to a defence mechanism that helps individuals cope with grief. Denial could be expressed subconsciously as a refusal to accept reality. This might be the reason why some cancer survivors chose not to know which stage of cancer they had.

Based on the results of linear regression analysis (see Table 2), the F value is 99.501 with $p < 0.01$. It can be concluded from the results that outcome expectancies did in fact have a positive and significant influence on health behaviour intention of cancer survivors, as is also demonstrated by the value of outcome expectancies which reaches up to $\beta = 0.728$. This means that with every 1-point increase of outcome expectancies, health behaviour intention would also increase by 0.728. A positive β value indicates that better outcome expectancies would result in better health behaviour intention. In addition, the R^2 value is 0.531, which means that 53.1% of health behaviour intention variances could be explained by outcome expectancies and the remaining 46.9% by other factors. Therefore, the null hypothesis (H_0) is rejected, and the alternative hypothesis is proved.

These results contradict White et al. (2011), who said that outcome expectancies would not affect health behaviour intention without the role of self-efficacy. That is to say, that outcome expectancies have positive and significant impacts on health behaviour intention. Strecher et al. (1986) also explained that the ability of outcome expectancies to determine health behaviour intention depended completely upon the situation. For instance, outcome expectancies would become crucial in motivating individuals to perform health behaviours

Table 1. Participant demographics.

Participant data		Frequency	Percentage
Gender	Male	29	32.22
	Female	61	67.78
Age	15–20 y.o.	15	16.67
	21–40 y.o.	48	53.33
	41–50 y.o.	27	30.00
Cancer type	Breast	22	24.44
	Uterine	8	8.89
	Blood *(Leukaemia)*	22	24.44
	Brain	3	3.33
	Colorectal	5	5.55
	Nasopharynx	9	10.00
	Lymphoma	8	7.78
	Ovarian	5	5.56
	Bone	6	6.67
	Thyroid	1	1.11
	Hepatic	2	2.22
Stage	1	18	20.00
	2	26	28.89
	3	18	20.00
	4	12	13.33
	No information	16	17.78
Length of life since diagnosis	≤1 year	22	24.44
	1–5 years	26	28.89
	>5 years	42	46.67

Table 2. Linear regression analysis results.

	Health behaviour intention			
Variable	Sig (p)	B	R^2	F
Outcome expectancies	0.000	0.728	0.531	99.501

if the individual is pessimistic towards the positive results of those health behaviours and towards accomplishing even the simplest tasks. In a different case, if the individual is optimistic but the tasks are too complicated, self-efficacy would become crucial to motivate them. That is why, in those cases, outcome expectancies have a bigger role on the health behaviour intention of long-term cancer survivors rather than self-efficacy because the majority characteristics of cancer survivors are being pessimistic and unmotivated.

These results also prove that participant demographics do not significantly affect health behaviour intention. However, mean values between categories of data are still different if compared to each other. For example, the results show that adolescent cancer survivors had the highest intention and outcome expectancies scores of all participants from different age groups, which were young adult stage and middle adulthood. This is rather different from previous studies which argued that young cancer survivors find it the most difficult to perform health behaviours.

Those differences between the results of this study and past studies might lie in the cultural gaps between Indonesia and where other researches took place. Most adolescents in Indonesia still live under the same roof as their parents with maximum supervision, strict family rules, and a myriad of expectations, unlike young adults and middle-aged individuals who can take care of themselves and have full control over their own lives.

Building on prior studies suggesting the usefulness of outcome expectancies to improve health behaviour intention among the population of non-cancer survivors, this is the first study among cancer survivors to assess the outcome of expectancy impact on health behaviour intention. Nevertheless, our results were consistent with the few studies evaluating factors associated with outcome expectancies in populations other than cancer survivors. Positive outcome expectancies represent an individual's confidence in overcoming barriers and physical limitations he or she is currently facing.

REFERENCES

Karvinen, K. & Vallance, J. (2015). Breast and colon cancer survivor's expectations about physical activity for improving survival. *Oncology Nursing Forum*, *42*(5), 527–533. doi:10.1188/15.ONF.527-533.

Kubler-Ross, E. (1973). *Death and dying* (1st ed.). New York, NY: Routledge.

Loh, S.Y., Chew, S.L. & Quek, K.F. (2013). Physical activity engagement after breast cancer: Advancing the health of survivors. *Health*, *5*(5), 838–846. doi:10.4236/health.2013.55111.

Renner, B. & Schwarzer, R. (2005). *Risk and Health Behaviors: Documentation of the scales of the research project "Risk Appraisal Consequences in Korea" (RACK)* (2nd ed.). International University Bremen & Freie Universität Berlin. Retrieved from http://www.gesundheitsrisiko.de/docs/RACK-English.pdf.

Rogers, L.Q., McAuley, E., Courneya, K.S. & Verhulust, S.J. (2008). Correlates of physical activity self-efficacy among breast cancer survivors. *Health Behavior*, *32*(6), 594–603.

Scheier, M.F. & Carver, C.S. (1985). Optimism, coping, and health: Assessment and implication of generalized outcome expectancies. *Health Psychology*, *4*(3), 219–247.

Scheier, M.F., Carver, C. & Bridges, M.W. (1994). Distinguishing optimism from neuroticism (and trait anxiety, self-mastery, and self-esteem): A re-evaluation of the Life Orientation Test. *Journal of Personality and Social Psychology*, *67*(6), 1063–1078. doi:10.1037//0022-3514.67.6.1063.

Schwarzer, R. & Luszczynska, A. (2008). How to overcome health-compromising behaviors: The health action process approach. *European Psychologist*, *13*(2), 141–151. doi:10.1027/1016-9040.13.2.141.

Seifert, C.M., Chapman, L.S., Hart, J.K. & Perez, P. (2012). Enhancing intrinsic motivation in health promotion and wellness. *The Art of Health Promotion*, *26*(3), 1–13. doi:10.4278/ajhp.26.3.tahp.

Snippe, E., Schroever, M.J., Tovote, K.A., Sanderman, R., Emmelkamp, P.M.G. & Fleer, J. (2015). Patient's outcome expectations matter in psychological interventions for patients with diabetes and comorbid depressive symptoms. *Cognitive Therapy Research*, *39*(3), 307–317. doi:10.1007/s10608-014-9667-z.

Stanton, A.L., Rowland, J.H. & Ganz, P.A. (2015). Life after diagnosis and treatment of cancer in adulthood contributions from psychosocial oncology research. *American Psychological Association*, *70*(2), 159–174. doi:10.1037/a0037875.

Strecher, V.J., Devellis, B.M., Becker, M.H. & Rosenstock, I.M. (1986). The role of self-efficacy in achieving health behavior change. *Health Education & Behavior*, *13*(1), 73–91. doi:10.1177/109019818601300108.

Taylor, S.E. (2012). *Health psychology* (9th ed.). New York, NY: McGraw-Hill.

White, K.M., Smith, J.R., Terry, D.J., Greenslade, J.H. & McKimmie, B. (2011). Social influence in the theory of planned behavior: The role of descriptive, injunctive, and in-group norms. *British Journal of Social Psychology*, *48*(1), 135–158. doi:10.1348/014466608X295207.

Diversity in Unity: Perspectives from Psychology and Behavioral Sciences – Ariyanto et al. (Eds)

Social relation of criminals: The analysis of causes and concepts of prevention

W. Kristinawati & E.K. Poerwandari
Faculty of Psychology, Universitas Indonesia, Depok, Indonesia

Z. Abidin
Faculty of Psychology, Padjadjaran University, Bandung, Indonesia

ABSTRACT: Harsh competition and conflicts might perpetuate aggressive behaviour and violence. This research aims to comprehend how social relations and social context play a role in the development of aggressive behaviour and moral disengagement, and how concepts of prevention and intervention can be developed. The study was conducted qualitatively through in-depth interviews of 14 male participants, aged 15–25, who were charged with homicide and were sent to corrective facilities to serve their sentence. Low levels of education and limited insight directed them to pro-aggression solutions as a primary way to resolve conflicts. More problems have evolved in correctional facilities as a result of overcrowding and limited resources for effective programmes. Prevention strategies for youth to not engage in delinquent behaviours should target families as well as peer groups and communities. School-based programmes can prevent delinquency, antisocial behaviour, and early school drop-out in order to prevent the onset of adult criminal careers, thus reducing the cost of crime to society. Most incarcerated individuals are vulnerable participants who are at risk of recidivism. Therefore, analysis for improvement of corrective facilities is also needed to humanise incarcerated people and reduce recidivism.

1 INTRODUCTION

Several causes have been found to be the underlying factors to the presence of aggressive behaviour. Problematic parenting styles (Timomor, 1998) and displays of aggressive behaviour by individuals who have power over the child (Berkowitz, 2003) are factors found to correlate with aggressive behaviour. Peer influences have also been found to be linked to aggressiveness and more so when they are exposed to many aggressive models and fighting becomes a valued attribute (Bartol, 2002). Furthermore, aggressive behaviour will more likely be shown if the perpetrator believes that their behaviour is normal or acceptable. This concept is known as moral disengagement (Bandura, 2002).

The relationship between moral disengagement and aggressive behaviour or violence has been proven in many populations, including in urban African American families (Pelton et al., 2004), Italian school-aged children (Bandura et al., 1996; Bandura, 2002), Chinese adolescents (Yang & Wang, 2012), and Scottish violent juvenile delinquents (Kiriakidis, 2010). However, following up on White-Ajmani and Busik's (2014) study, further research is required to delve into the social context where moral disengagement occurs. Meanwhile, in a group of adolescent convicts, Shahinfar et al. (2001) stated that exposure to heavy aggression relates to the tendency to agree that aggression is a positive solution to reach a goal. If we want to reduce recidivism, this attitude needs to be rectified through development programmes in correctional facilities, amongst other things. According to Kristianingsih (2012), correctional facilities in Indonesia are ineffective and potentially result in high recidivism rates.

Based on the aforementioned background, this study aims to comprehend the social relations and social context of perpetrators of aggression before and after incarceration, as well as how these play a role in the development of aggression and moral disengagement. The understanding of the individual's external situation which contributes to the development of aggression can be used to conceptualise interventions, which can hopefully minimise the amount of violent cases in the community.

1.1 *The role of social context and social relation in the development of aggression*

Nurture is an important contributing factor to a person's development, including the development of aggressive behaviour. Children coming from families with a problematic parenting style tend to become more easily frustrated when facing economic pressures and parental separation or divorce, so much so that their aggressive interests become higher and thus lead them to break the law (Timomor, 1998). Ineffective disciplinary actions by the parents and poor monitoring become the 'basic practice' of the development of aggressive behaviour. According to Mazefsky and Farrell (2005), peer provocation is the mediator between poor nurture and adolescent aggressive behaviour. Adolescents with negative parenting experience will be more likely to act aggressively when provoked; conversely, without provocation—according to Mazefsky and Farrell—aggressive tendencies do not manifest into real actions. Chung and Steinberg (2006) infer that parental monitoring gives a way for adolescents to build relationships with peers who have behaviour problems. Moreover, violent neighbourhoods (Bradshaw et al., 2009; Margolin et al., 2009; Trentracosta et al., 2009), and violence in the media (Krahé et al., 2012), especially video games (Anderson et al., 2010; Huesman, 2010; Lang et al., 2012), are believed to have an impact on the development of aggressive behaviour. Differing from Krahé and colleagues, Kristinawati (2007) found that models of male aggression and peers who are pro-aggression contribute to the development of aggressive behaviour; however, the media was not found to be an influencing factor. Loeber et al. (2005) analysed 63 risk factors that are reputed as being predictors of violent actions and murders by young males, amongst others: parents' effort to seek help, the child's attitude and cognition, psychiatric diagnosis, birth factors, the neighbourhood, and school. It has been found that teenage boys with four or more risk factors are six times more likely to execute violent acts and to commit a murder in the future, compared to teenage boys with fewer risk factors. Furthermore, when an individual makes a moral rationalisation of their violation of the law, aggressive behaviour will tend to persist.

Bandura (2002) states that an action or a behaviour is carried out by a person based on the moral standards they developed through self-regulatory mechanisms as a way to stop doing something bad. On the other hand, self-regulatory systems are also prone to disengagement to avoid internal sanction. By cognitively reconstructing an antisocial act to make it less wrong or even correct, one can disengage the internal emotion checks that usually prevent misconducts (Shulman et al., 2011). There are six mechanisms of moral disengagement: euphemistic labelling, advantage comparison, displacement of responsibility, diffusion of responsibility, distortion of consequences, and dehumanisation (Bandura, 2002). Adolescents with higher levels of moral disengagement tend to be more aggressive and commit more violence when they reach late adolescence (Paciello et al., 2008; Pelton et al., 2004).

1.2 *The impact of imprisonment*

Since 1995, the function of the correctional institution for criminals in Indonesia is no longer for mere imprisonment purposes but also as an effort to establish social rehabilitation and reintegration (Undang-Undang Nomor 12, 1995, Article 1:2). In reality, life inside correctional facilities is a life that is far from ideal. The total number of convicts exceeds capacity in most correctional facilities in Indonesia. This affects the limited use of facilities, and inadequate supervision due to a prisoner-to-officer ratio that is not ideal. Overcrowded correctional facilities creates management conflicts, a lack of activities due to the vast number of users, and increased encounters between convicts who may have behavioural problems. This condition is leading to exposure to aggression inside the prison walls. In addition to that, living in a correctional facility makes institutionalisation possible: the process of assimilation of prison norms to a person's habits, mindset, and behaviour patterns (Vianello, 2013). Although this reaction is relatively normal, the effects can be destructive (Haney, 2002) because the majority of the prison population are participants who have limited personal resource and social support, and are more vulnerable prior to entering the facility (Vianello, 2013).

2 RESEARCH METHODS

2.1 *Participants*

The participants in this study were 14 male convicts (P1, P2, P3... and P14), which included eight premeditated homicides and six non-premeditated (spontaneous) homicides. The participants, whose ages ranged between 14 and 25, were all of Javanese ethnicity, with high school being the highest level of education attained within the group. The selection process was carried out using the purposive sampling strategy.

2.2 *Procedure*

This study was conducted in three locations: Yogyakarta Correctional Facility, Surakarta Detention Centre, and Kutoarjo Correctional Facility for Children. Researchers conducted in-depth personal interviews with each respondent. All interviews took place in rooms accommodated by the respective correctional facility, and audio recordings were made with the respondent's consent. Each interview session lasted between 1–2 hours for 3–6 times. The initial meeting with the research participants began with the stating of the purpose of the study, ethical considerations, and research procedure. Consequently, verbal consents for their participation in the study were obtained. Participants were given care packages which contained consumables, toiletries, and wrapping cloth (sarong) at the end of every interview.

2.3 *Data analysis*

This research used a four-step method for analysing data: (Step 1) interview recordings were turned into text transcript verbatim for content analysis, (Step 2) identifying and listing individual ideas within each interview, (Step 3) organising the individual ideas into meaningful psychological categories, and (Step 4) structuring the ideas and categories into patterns or generalisable themes and making comparisons, displayed in charts, tables, and graphs.

3 RESULTS

Amongst the 14 participants, all of them had some kind of formal education, but only four people finished high school and another four dropped out of elementary school. Some were unemployed prior to serving their sentence, and those who were receiving formal education were often truant and hung out in the streets. All participants were of the middle-lower socio-economic background. Thirteen perpetrators knew their victims personally; some victims were their romantic partners and close friends.

The causes of the development of aggressive behaviour can be categorised into two groups, which are factors that catalyse aggression (risk factors) and factors that hinder it (protective factors). Each factor is divided in two parts: before imprisonment and during imprisonment. See Table 1 for each category.

3.1 *Risk factors of the development of aggression*

3.1.1 *Dysfunctional family*

In general, the participants' family relationships were peripheral and touched more on the physical than emotional needs. For three participants (P10, P11, and P12) a male role model was not obtained from the father and this role was not adequately substituted for by other

Table 1. Risk and protective factors of the development of aggressive behaviour.

	Risk factors	Protective factors
Before imprisonment	Dysfunctional family Aggressive peers History of aggression during childhood Poor problem-solving skills	Positive experience of relating with others
During imprisonment	No remorse, vengeance Relation with other criminals	Fear of imprisonment Concern about the future

family members. With other participants (P1, P4, P6, and P9), a negative father role was manifested in an aggressive type of father who encouraged them to act aggressively.

> 'My father taught me that a man must fight. Back when I was in the third grade, I came home crying because a friend beat me up, and my father was furious. He told me to bring a machete.' (P4, premeditated homicide)

P6 was an only child. He lived with his mother after his parents divorced. In his case, P6 got drunk with six people, including his father, and took turns having sex with a prostitute. Finally, P6 and his father were co-perpetrators in a rape and murder case, and the prostitute was the victim. P4 stabbed a friend at school who had insulted his mother, because according to him, a child should defend their mother by doing what he did to anyone who insulted her.

3.1.2 *Aggressive peers*

According to all participants, the role of peers is more important than parents. Unfortunately, their peers are friends who like to fight and get drunk. When they act aggressively, this act of aggression receives a 'reward' from the aggressive group, in the form of praise and acceptance. Peers who are pro-aggression often encourage the participants to get involved in aggressive acts.

> '...in grade 4 I was invited to join in a gang fight by my senior. He brought a bag filled with blades. My friend also persuaded me to snatch bags and wallets. It became a habit, ever since I realised how nice it felt to have my own money. So I robbed people every night. I brought a sickle to threaten the victims. It feels good, if they feel scared.' (P12, spontaneous homicide)

In cases where aggression was performed together, co-perpetrators blamed each other. Displacement of responsibility, a belief that the self is less guilty than others, is the type of moral disengagement that occurs.

3.1.3 *History of aggression during childhood*

Of the 14 respondents, only P8 and P13 did not have a history of childhood aggression. With the other respondents, violation of rules has been happening since elementary school: getting drunk, stealing, hiring prostitutes, and skipping school. Compared to their childhood days, the types of aggression displayed in their adolescent and young adulthood years are much worse: burglary, looting, beating up their father, and finally murder. The courage to fight and oppose was something to be proud of. When anger appears, they often release their emotion impulsively until they feel satisfied, without worrying about punishment.

3.1.4 *Poor problem-solving skills*

Only two participants said that they have tried to communicate if they have a problem with other people. One participant tended to be quiet and concealed negative emotions, but the other 11 participants resorted to hitting to solve their problems. The blaming-the-victim mechanism tends to be exercised to rationalise their actions. This mechanism appeared in a higher intensity with the premeditated homicides compared to the spontaneous homicides. Participants became so angry at their victim that they planned to murder the victim. With P3, the perpetrator spontaneously grabbed a pair of scissors and ravaged the victim's face not to leave a a trace, but to release his anger.

3.1.5 *No remorse, vengeance*

Imprisonment, for some of the participants, is a way to cover the wrongdoings they have done so that after serving their sentence, they do not have to feel guilty anymore. They built moral disengagement to obscure personal accountability. Displacement and diffusion of responsibility not only represented a cognitive denial mechanism but also the individual's capacity to experience negative social emotions such as shame or remorse. Dehumanisation is a key mechanism that operates by nullifying self-restraints operating through feelings of empathy and compassion.

3.1.6 *Relations with other criminals*

Most of the prison population are weak subjects, with limited personal resources and external support, and are already considerably vulnerable before entering the prison. These personal situations are aggravated when they relate to each other, especially in overcrowded correctional facilities.

3.2 *Factors that hinder the development of aggression*

3.2.1 *Positive experience of relating with others*

Some participants have a positive attitude towards his environment by helping others, such as by serving the elderly. Closeness with mothers is also a factor which hinders participants from behaving aggressively or breaking the law. With P12, his mother always removed the weapons he would use to fight. He also felt sad and remorse when he realised that his actions made his mother sad and her life more difficult.

3.2.2 *Fear of imprisonment*

With the participants who felt remorse and guilt, there is a willingness to serve their sentence. They consider imprisonment to be something they must avoid in the future.

3.2.3 *Concern about the future*

Concern about the future is able to direct a person's actions to become purposeful, in that it increases the chance that he can avoid breaking the law. Unfortunately, according to the researchers, only one participant has a strong desire to become a better person, wherein his child becomes his biggest motivation to seriously think about the future and be more careful with his actions.

4 DISCUSSION

Family factors which contribute to the development of aggression are low levels of warmth and poor parental monitoring. The data supports Bushman et al. (2016) that the tendency to punish, low level of warmth, emotional rejection towards the child, and poor parental monitoring are factors which contribute to the development of aggression. Low socio-economic status seems to worsen the situation, and this corresponds to findings from previous studies (Yoshikawa, 1994; Fabio et al., 2011). Enhancement of parenting skills must be done, with hopes that early identification of aggression can be obtained and resolved in the early stages of development. In addition, mentoring and guidance of aggression cases in young people must also be accompanied by guidance for the family to minimise repeated offence by the same respondent or with other family members. Peer environment which exposes violence around the participants, correlates with the development of aggression, in line with results from Bradshaw et al. (2009), Margolin et al. (2009), and Trentacosta et al. (2009). Therefore, family empowerment still needs to shed a light on improving the social environment as a measure of intervention.

Vengeance is a situational context that increases aggressive tendencies and increases direct perpetrators to dehumanisation with the absence of remorse. This result supports White-Ajmani and Bursik's (2014) finding which states that vengeance is the mediating variable between moral disengagement and aggressive behaviour. Vengeance issues residing in respondents persist before they enter a correctional facility as well as throughout incarceration. This shows that imprisonment is not effective enough in providing guidance and self-development programmes. On top of that, the physical field provided is inadequate when compared to the number of existing convicts. According to Vianello (2013) this is a by-product of overcrowded correctional facilities. Increased encounters between convicts may lead to exposure to aggression inside prison walls, which in turn may engender convicts to transform their aggression into more intense forms of criminal acts. Convicts will be at greater risk of doing more crimes after they are released from prison (Haney, 2002).

It is known that the environmental context of the perpetrators in this study is that they are unemployed, have had inadequate formal education, were often lurking in the streets or were not at home, and that they drank alcohol regularly. All characteristics, similar to findings from a previous study (Kibusi et al., 2013), are factors which significantly links to the tendency to become homicide victims. It can be concluded that these things can potentially cause a person to be involved in homicide, as the perpetrator or victim. Therefore, prevention strategies to avoid involvement with aggression include being engaged with working fields, having a higher level of education and abstaining from alcohol.

Contrary to Mazefsky and Farrel's (2005) findings, data from this study show that provocation is not the main issue in homicide. In the case of several participants, a person can decide to murder and exercise moral disengagement to justify that their action is appropriate, even without being provoked. Poor problem-solving skills paired with limited education are believed to be the cause of opting for aggression as a way to solve problems. Breakthrough preventions for aggression and criminal cases still need to be promoted, possibly in a form of national policy as school programmes. The government needs to provide services to students who tend to be aggressive and have problems controlling their aggression, including training teachers to handle similar cases. The prevention strategy can be applied in conjunction with factors that have been found to hinder the development of aggression, such as cultivating positive experiences with others and the desire to set a goal for the future.

5 CONCLUSION AND RECOMMENDATIONS

People with poor self-control, who dropped out of school and has low socio-economic status, are more susceptible to being involved in aggressive behaviours. Families that do not fully function also increase the risk of aggression in individuals. The risk will increase even more when they have social relations with pro-aggression friends, including when they meet fellow criminal convicts in a correctional facility. On the other side the government can develop community-based peer-mentoring projects: a project to enhance the role of perpetrators and ex-perpetrators as mentors who share a common status and struggle as offenders; they can therefore bridge a gap between staff and service users. Intervention and prevention programmes need to be implanted at all levels: family, neighbourhood, school, and policy. The promotion of non-aggressive behaviours must continue, along with the skill development of the staff. Correctional facilities must concentrate on increasing the quantity and quality of human resources in order to transfer non-aggressive values and improve self-development programmes.

REFERENCES

Anderson, C.A., Shibuya, A., Ihori, N., Swing, E.L., Bushman, B.J., Sakamoto, A., ... Saleem, M. (2010). Violent video game effects on aggression, empathy, and prosocial behaviour in eastern and western countries: A meta-analytic review. *Psychological Bulletin, 136*(2), 151–173. doi:10.1037/a0018251.

Asher, S.R., Rose, A.J. & Gabriel, S.W. (2001). Peer rejection in everyday life. In: M.R. Leary (Ed.), *Interpersonal rejection* (pp. 105–142). New York, NY: Oxford University Press. doi:10.1093/acprof:oso/9780195130157.003.0005.

Bandura, A. (2002). Selective moral disengagement in the exercise of moral agency. *Journal of Moral Education, 31*(2), 101–119. doi:10.1080/0305724022014322.

Bandura, A., Barbaranelli, C., Caprara, G.V. & Pastorelli, C. (1996). Mechanisms of moral disengagement in the exercise of moral agency. *Journal of Personality and Social Psychology, 71*(2), 364–374. doi:10.1037/0022-3514.71.2.364.

Bartol, C.R. (2002). *Criminal behavior: A psychosocial approach* (6th ed.). Upper Saddle River, NJ: Prentice Hall.

Berkowitz, L. (2003). *Emotional behavior: Mengenali perilaku dan tindakan kekerasan di lingkungan sekitar kita dan cara penanggulangannya* [*Aggression: its causes, consequences, and control*]. Jakarta, Indonesia: Penerbit PPM.

Bradshaw, C.P., Rodgers, C.R.R., Ghandour, L.A. & Garbarino, J. (2009). Social-cognitive mediators of the association between community violence exposure and aggressive behaviour. *School Psychology Quarterly, 24*(3), 199–210. doi:10.1037/a0017362.

Bushman, B.J., Newman, K., Calvert, S.L., Downey, G., Dredze, M., Gottfredson, M., ... Webster, D.W. (2016). Youth violence: What we know and what we need to know. *American Psychologist, 71*(1), 17–39. doi:10.1037/a0039687.

Chung, H.L. & Steinberg, L. (2006). Relations between neighbourhood factors, parenting behaviours, peer deviance, and delinquency among serious juvenile offenders. *Developmental Psychology, 42*(2), 319–331. doi:10.1037/0012-1649.42.2.319.

Fabio, A., Tu, L.C., Loeber, R. & Cohen, J. (2011). Neighbourhood socioeconomic disadvantage and the shape of the age-crime curve. *American Journal of Public Health, 101*(1), S325–333. doi:10.2105/AJPH.2010.300034.

Haney, C. (2002) *The psychological impact of incarceration: Implications for post-prison adjustment*. U.S. Department of Health & Human Services, The Urban Institute, pp. 1–18.

Huesmann, L.R. (2010). Nailing the coffin shut on doubts that violent video games stimulate aggression: Comment on Anderson et al. (2010). *Psychological Bulletin, 136(*2), 179–181. http://dx.doi.org/10.1037/a0018567.

Kibusi, S.M., Ohnishi, M., Outwater, A., Seino, K., Kizuki, M. & Takano, T. (2013). Sociocultural factors that reduce risks of homicide in Dar es Salaam: A case control study. *Injury Prevention, 19*, 320–325. doi:10.1136/injuryprev-2012-040492.

Kiriakidis, S.P. (2010). Prediction and explanation of young offenders' intentions to re-offend from behavioural, normative, and control beliefs. *European Psychologist, 15*(3), 211–219. doi:10.1027/1016-9040/a000021.

Krahe, B. (n.d.). *Perilaku agresif.* [Agressive behaviour] Yogyakarta, Indonesia: Pustaka Belajar.

Krahé, B., Busching, R., & Möller, I. (2012). Media violence use and aggression among German adolescents: Associations and trajectories of change in a three-wave longitudinal study. *Psychology of Popular Media Culture, 1*(3), 152–166. doi:10.1037/a0028663.

Kristianingsih, S.A. (2012). Pemenjaraan pada narapidana narkoba di Rumah Tahanan (Rutan) Salatiga. [Imprisonment of drug prisoners at the Salatiga Detention Center]. *Humanitas, 6*(1), 1–15.

Kristinawati, W. (2007). *Keterpaparan agresi dan internalisasi nilai agresi pada remaja pria pelaku pembunuhan.* [Exposure of aggression and internalization of the value of aggression in teenage boys perpetrators of murder]. (Thesis, Universitas Indonesia, Depok, Indonesia).

Lang, A., Bradley, S.D., Schneider, E.F., Kim, S.C. & Mayell, S. (2012). Killing is positive! *Journal of Media Psychology, 24*(4), 154–166. doi:10.1027/1864-1105/a000075.

Loeber, R., Pardini, D., Homish, D.L., Wei, E.H., Crawford, A.M., Farrington, D.P., ... Rosenfeld, R. (2005). The prediction of violence and homicide in young men. *Journal of Consulting and Clinical Psychology, 73*(6), 1074–1088. doi:10.1037/0022-006X.73.6.1074.

Margolin, G., Vickerman, K.A., Ramos, M.C., Serrano, S.D., Gordis, E.B., Iturralde, E.,... Spies, L.A. (2009). Youth exposed to violence: Stability, co-occurrence, and context. *Clinical Child and Family Psychology Review, 12*(1), 39–54. doi:10.1007/s10567-009-0040-9.

Mazefsky, C.A. & Farrell, A.D. (2005). The role of witnessing violence, peer provocation, family support, and parenting practices in the aggressive behavior of rural adolescents. *Journal of Child and Family Studies, 14*(1), 71–85. doi:10.1007/s10826-005-1115-y.

Paciello, M., Fida, R., Tramontano, C., Lupinetti, C. & Caprara, G.V. (2008). Stability and change of moral disengagement and its impact on aggression and violence in late adolescence. *Child Development, 79*(5), 1288–1309. doi:10.1111/j.1467-8624.2008.01189.x.

Pelton, J., Gound, M., Forehand, R. & Brody, G. (2004). The moral disengagement scale: Extension with an American minority sample. *Journal of Psychopathology and Behavioral Assessment*, *26*(1), 31–39. doi:10.1023/B:JOBA.0000007454.34707.a5.

Shahinfar, A., Kupersmidt, J.B. & Matza, L.S. (2001). The relation between exposure to violence and social information processing among incarcerated adolescents. *Journal of Abnormal Psychology*, *110*, 136–141. doi:10.1037/0021-843X.110.1.136.

Shulman, E.P., Cauffman, E., Piquero, A.R., & Fagan, J. (2011). Moral disengagement among serious juvenile offenders: A longitudinal study of the relations between morally disengaged attitudes and offending. Developmental Psychology, 47(6), 1619–1632. doi:http://dx.doi.org/10.1037/a0025404.

Timomor, A. (1998). *Kecenderungan otoriter pola asuh orang tua, konflik keluarga dan kecenderungan agresivitas remaja.* [The authoritarian tendency of parenting, family conflict and adolescent aggressiveness tendencies] (Thesis, Universitas Gajah Mada, Yogyakarta, Indonesia).

Trentacosta, C.J., Hyde, L.W., Shaw, D.S. & Cheong, J.W. (2009). Adolescent dispositions for antisocial behavior in context: The roles of neighborhood dangerousness and parental knowledge. *Journal of Abnormal Psychology*, *118*(3), 564–575. doi:10.1037/a0016394.

Undang-Undang Nomor 12. (1995). *Tentang Pemasyarakatan.* [Law no 12, 1995. About corrections]. Jakarta, Indonesia: Sekretaris Negara Republik Indonesia.

Vianello, F. (2013). Daily life in overcrowded prisons: A convict perspective on Italian detention. *Prison Service Journal*, *207*, 27–33.

White-Ajmani, M.L. & Bursik, K. (2014). Situational context moderates the relationship between moral disengagement and aggression. *Psychology of Violence*, *4*(1), 90–100. doi:10.1037/a0031728.

Yang, J.P. & Wang, X.C. (2012). Effect of moral disengagement on adolescents' aggressive behavior: Moderated mediating effect. *Acta Psychologica Sinica*, *44*(8), 1075–1085. doi:10.3724/SP.J.1041.2012.01075.

Yoshikawa, H. (1994). Prevention as cumulative protection: Effects of early family support and education on chronic delinquency and its risks. *Psychological Bulletin*, *115*(1), 28–54. doi:10.1037/0033-2909.115.1.28.

Diversity in Unity: Perspectives from Psychology and Behavioral Sciences – Ariyanto et al. (Eds)
© 2018 Taylor & Francis Group, London, ISBN 978-1-138-62665-2

Comparison of the marital satisfaction between dual-earner and single-earner couples

C.M. Faisal & Y.R. Sari
Department of Clinical Psychology, Faculty of Psychology, Universitas Indonesia, Depok, Indonesia

ABSTRACT: The number of working women in Indonesia increases every year. Data from the Central Statistics Agency revealed that in 2014, the number of dual-earner couples in Indonesia was 51.2%, while the number of single-earner couples was 39.9%. Several studies have shown that the dual-earner condition has positive and negative impacts on marital satisfaction. This research aims to investigate the comparison of marital satisfaction in dual- and single-earner couples, as well as the comparison of the marital satisfaction between husbands and wives from dual- and single-earner couples. A total of 368 husbands/wives participated in this research. The results show that there is no significant difference in marital satisfaction between dual- and single-earner couples, and neither is there a significant difference in the marital satisfaction between husbands and wives in dual- and single-earner couples. Hence, we can conclude that the wife's working status does not affect marital satisfaction. We suspect that the nature of conjoint agency between husbands and wives in Indonesia might affect this finding. In general, the mean score of marital satisfaction among all participants is high. Some demographic factors, such as similarities of background between the couples, duration of marriage, and number of children, might contribute to this finding.

1 INTRODUCTION

Every year, the number of working women increases drastically; meanwhile, the number of stay-at-home wives keeps decreasing. Based on the National Workforce Survey (Subdirektorat Statistik Ketenagakerjaan, 2015), the number of working women in August 2014 was 43.1 million, while by February 2015 the number had increased to 47.4 million. Meanwhile, the number of stay-at-home wives in August 2014 was 34.2 million, while by February 2015 the number had decreased to 30.8 million. There are several factors that encourage women to work. One of the most dominant factors is equality of opportunity and higher educational background (Barnett, 2008). In addition, the media, especially women's magazines, has had an impact as an agent of change and public opinion-maker for working women (Ardaneshwari, 2013). On the other hand, the increased number of working men is not as high as that of women. In August 2014, the number of working men was 71.4 million. This number had increased to 73.4 million by February 2015.

According to Bird et al. (1990), there are two types of family based on job status of the husband and wife. A dual-earner couple is the couple in which both husband and wife work, whether on a full-time or part-time basis, and both earn income. A single-earner couple is the couple in which only the husband works, while the wife is a stay-at-home housewife. The increased number of working women makes dual-earner couples become more common than the single-earner couples. Based on data processed from the National Social Economy Survey (Badan Pusat Statistik, 2014), it was found that the number of dual-earner couples in Indonesia was 51.2% while the number of single-earner couples was 39.9%.

The conditions of dual-earner and single-earner couples make a great contribution to the marital satisfaction felt by husbands and wives (Ayub & Iqbal, 2012). Marital satisfaction is 'the subjective attitude that individuals have toward their marital relations' (Custer, 2009).

In addition, DeGenova (2008) defined it as 'the extent to which couples are content and fulfilled in their relationship'. Barnett (2008) found that marital satisfaction increases for men and women when financial responsibility is divided equally. However, this finding remains questionable in Indonesia because it is culturally different from Western countries.

The wife's involvement in work is a positive change for gender participation equality, but it also has an impact on family life (Azeez, 2013). Jacobs and Gerson (2001) found that the shift from male-breadwinner to dual-earner couples have created growing concern for balancing work and family. Working wives are the ones most vulnerable to experiencing such problems. In dual-earner couples, besides being in joint charge of finances, the tasks of organising the household and taking care of children should also be shared between the husband and wife. This might not be easy for couples who are not ready to face the transition of traditional roles into more modern ones. Even if the wives work as well, research shows that the division of household activities has still not changed. Even when both husband and wife work full-time, the wife still handles most household tasks (Azeez, 2013). This situation can cause fatigue and stress in the wife. On the other hand, dual-earning is very important to most families in order to maintain better living standards (Walsh, 2003). The opposite condition happens to single-earner couples. Single-earner couples only have one source of finance, but they face smaller challenges in balancing work and family because the task division is clearer.

There are many studies about factors affecting marital satisfaction. The following is a compilation of various research on the subject that is relevant to this study.

1. Financial. According to Williams et al. (2006), bigger income and higher job status owned by the couple implies a higher possibility that they have a good marriage. Dual-earning can give positive impacts on the marital satisfaction, because dual-earning enables the couple to fulfil basic physiological needs, assist them to afford a higher standard and enjoy various leisure and luxuries of life (Mohsin, 2014). Lack of finances and financial problems are likely to become sources of stress, tension, and dissatisfaction in the shared life of any couple and in an individual's life (Mohsin, 2014).
2. Division of household tasks. The husband's contribution to the household can really affect the marital satisfaction of husband and wife (Bagwell, 2006). In the research by Frajerman (2001), wives reported that they did more household tasks than the husbands. The more household tasks handled by wives from dual-earner couples, the lower the marital satisfaction they felt. However, the opposite condition occurred with the husbands; the more household tasks they handled, the higher the marital satisfaction they felt.
3. Gender. There is a significant difference in marital satisfaction between men and women (Jose & Alfons, 2007). Based on a research by Wilkie et al. (1998), when both husband and wife work, there is an advantage for their marital satisfaction, but the husbands feel a higher advantage. This finding is supported by Oshio et al. (2011) who found that the wives have lower marital satisfaction than the husbands because they handle more household tasks.
4. Similarity of background. According to Williams et al. (2006), a successful marriage is marked by having the same backgrounds. Homogamous marriages—marriages between couples who have similar education, ethnicity, race, religion, age, and social class—are more likely to be successful compared to heterogamous marriages (Williams et al., 2006). Duvall and Miller (1985) explained in more detail that those characteristics involve a minimum education of secondary school. Homogamous couples could have a better marriage because they can adapt more easily.
5. Children. The presence of children can improve marital stability, yet can decrease marital satisfaction (Stone & Shackelford, 2007). According to Williams et al. (2006), although the existence of children can bring joy, it also needs huge effort, time, and responsibility. Children need money that could be used for leisure or investment, decrease parent's time for themselves, and demand a lot of needs that may cause stress. Marital satisfaction can also decline because the presence of children forces couples to limit their sexual life (Jose & Alfons, 2007). Wendorf et al. (2011) found that the number of children has a significant negative correlation with marital satisfaction.

6. Duration of marriage. According to the family life cycle, the role and relationship between husband and wife can change based on how they adapt to their children-raising responsibility (Williams et al., 2006). Marital satisfaction is believed to follow a U-shaped trajectory. Couples begin their marriage with satisfaction, then the satisfaction will shrink for a couple of years, and will reach an initial satisfaction in the next years (Stone & Shackelford, 2007). This notion is supported by Jose and Alfons (2007), who found that low marital satisfaction exists in couples with moderate durations of marriage, compared to those couples whose duration of marriage is small or large.
7. Type of family based on job status. Dual-earner and single-earner couples' marriages have different dynamics that can affect their marital satisfaction. Boye (2014) stated that dual-earner couples have better communication and relationship quality because they have similar routines. However, their occupation can decrease their chance to spend quality time together (Voorpostel et al., 2009). Dual-earner couples feel huge pressure to balance work and family which makes them suffer higher stress levels, work–family conflicts, and overload (Elloy & Smith, 2003). On the other hand, single-earner couples have more chance to spend their leisure time together because of the wife's flexible time. They have lower work–family conflicts and overload; hence they tend to feel lower levels of stress. Stress and psychological well-being can predict marital satisfaction (Walker et al., 2013).

This research aims to investigate whether there is a significant difference in the marital satisfaction between dual- and single-earner couples, and between the husbands and wives from both groups.

2 METHODS

This is a non-experimental research in which we did not investigate the causal relationship between variables. The design of this research is between-subject design. We divided participants into two groups based on the work status of husbands and wives. All of the participants of this research are couples with a minimum education of high school, and a minimum one year of marriage. We selected participants with a convenience sampling method.

The instrument used to measure marital satisfaction is the Couple Satisfaction Index (CSI) by Funk and Rogge (2007), adapted to Indonesian. CSI consists of 16 items with a five-point Likert scale, except for item number one which consists of a six-point Likert scale. The CSI has a Cronbach alpha coefficient of 0.898 and an interval validity coefficient between 0.113 and 0.786.

Initially, we surveyed prospective couples who were willing to participate in this study. Then, we gave each couple a serial number and the CSI instrument. The instrument was handed directly to the participants and was also distributed using a Google spreadsheet. The total score of each participant from each group is summed. Next, the independent-sample t-test is used to compare the mean score between the groups.

3 RESULTS

A total of 368 participants joined this research; 352 participants (176 couples) were couples, while the other 16 participants were husbands and wives whose spouses did not fill in the questionnaire. The total number of participants who joined this research from both groups was 95 wives and 89 husbands.

Data processing results show that there is no significant difference in marital satisfaction on all of the questions in this research. First, there is no significant difference in the marital satisfaction between the wives of dual-earner and single-earner couples. Second, there is no significant difference in the marital satisfaction between the husbands of dual-earner and single-earner couples. Third, there is no significant difference between husbands and wives of dual-earner couples. Lastly, there is no significant difference in the marital satisfaction between husbands and wives of single-earner couples. We also conducted an analysis to examine correlation between husbands' and wives'

Table 1. General descriptions of participants.

Personal data	Category	Dual-earner n	Dual-earner %	Single-earner n	Single-earner %
Duration of marriage	1–5 years	97	52.7	67	36.4
	6–10 years	16	8.7	22	12
	11–15 years	26	14.1	32	17.4
	16–20 years	22	12	31	16.8
	21–25 years	14	7.6	30	16.3
	26–31 years	9	4.9	2	1.1
Number of children	None	27	14.7	20	10.9
	1 child	67	36.4	53	28.8
	2 children	61	33.2	63	34.2
	3 children	26	14.1	38	20.7
	4 children	2	1.1	10	5.4
	7 children	1	0.5	0	0
Monthly outcome	< Rp1,000,000	5	2.7	12	6.7
	Rp1.000.000 – Rp3,000,000	61	33.1	54	30.2
	Rp3,000,000 – Rp5,000,000	56	30.4	65	36.3
	Rp5,000,000 – Rp7,000,000	34	18.4	24	13.4
	> Rp7,000,000	28	15.2	24	13.4

Table 2. Results of t-test between dual-earner wives and single-earner wives.

	Wives (Dual-earner)			Wives (Single-earner)				
	M	SD	n	M	SD	n	t	df
Marital satisfaction	66.8	9.2	95	67.4	8.7	95	–0.508	188

Table 3. Results of t-test between dual-earner husbands and single-earner husbands.

	Husbands (Dual-earner)			Husbands (Single-earner)				
	M	SD	n	M	SD	n	t	df
Marital satisfaction	68.07	9.5	89	67.9	9.9	89	0.069	176

Table 4. Results of t-test between husbands and wives from dual-earner couples.

	Husbands			Wives				
	M	SD	n	M	SD	n	t	df
Marital satisfaction	68.07	9.5	89	66.8	9.2	95	0.922	182

Table 5. Results of t-test between husbands and wives from single-earner couples.

	Husbands			Wives				
	M	SD	n	M	SD	n	t	df
Marital satisfaction	67.9	9.9	89	67.4	8.7	95	0.374	182

marital satisfaction. The result shows that there is a significant positive correlation between the husband's marital satisfaction and the wife's marital satisfaction: $r = 0.546$, $n = 176$, $p < 0.01$, two-tailed.

4 DISCUSSION

The main result shows that there is no difference in the marital satisfaction between the dual-earner and single-earner couples. Collectivism in Indonesia could be an alternative explanation for the result. According to de Vries (2011), marriages in a collectivist culture are a conjoint agency. In the conjoint agency, actions are interdependent for others and are responsive to the obligations and expectations of others. When wives want to decide whether to work or not, they highly consider what their spouses want. Therefore, the decision of the wives to work is a mutual agreement. In the end, this agreement is a factor that keeps them satisfied with the marriage. Another explanation might be the similarity of background between the husbands and wives. All participants of this study have the same religion as their spouses, and 63% of them are from a similar ethnicity as their spouses. Similarities of background can make couples adapt to each other more easily (Duvall & Miller, 1985).

The mean scores of the marital satisfaction from all groups range from 66.8 to 68.07 with a maximum score of 81. It shows that the majority of participants have a high level of marital satisfaction. Several demographic factors might contribute to this finding. First, the number of children. According to Twenge et al. (2003), the number of children has significant negative correlation with the marital satisfaction. Couples who do not have children reported to have higher marital satisfaction compared to those who have already got children. About 70% of dual-earner couples and 63% of single-earner couples have one or two children. Therefore, most participants probably have high marital satisfaction because they have relatively small numbers of children. The second contributory demographic factor is the duration of marriage. Around 52.7% participants from the dual-earner and 36.4% participants from the single-earner groups are young couples with a marriage duration of one to five years. This implies that most participants are in the first phase of the family life cycle, and hence have high marital satisfaction (Stone & Shackelford, 2007). The third factor is socioeconomic status. Around 63.5% of participants from the dual-earner group and 66.5% participants from single-earner couples were categorised into the middle-middle socioeconomic class. According to Dakin and Wampler (2008), middle-middle socioeconomic status is a predictor of higher marital satisfaction and lower psychological distress compared to those of lower-middle socioeconomic status.

The majority of the participants come from the middle socioeconomic class, who can more readily fulfil their financial needs. We suggest that future research should be more focused on couples from the lower-middle socioeconomic class whose family dynamics are highly influenced by financial strains. There are also several data that can be beneficial to understanding this finding more profoundly, such as the division of household tasks (Bagwell, 2006) and the wives' reasons and considerations for working (Benin & Nienstedt, 1985; Williams et al., 2006). This information can be collected in the next project or can be investigated more profoundly through qualitative research.

5 CONCLUSION

This study implies that there is no difference in the marital satisfaction between dual-earner and single-earner couples. Both husbands and wives have a high level of marital satisfaction, and their marital satisfactions are positively correlated.

REFERENCES

Ardaneshwari, J. (2013). Portrait of working women's dilemma in Indonesian women's media. *Jurnal Perempuan*, *18*, 37–72.

Ayub, N. & Iqbal, S. (2012). The factors predicting marital satisfaction: A gender difference in Pakistan. *The International Journal of Interdisciplinary Social Sciences*, *6*(7), 65–73.

Azeez, A. (2013). Employed women and marital satisfaction: A study among female nurses. *International Journal of Management and Social Sciences Research*, *2*(11), 17–22.

Badan Pusat Statistik. (2014). *National Social Economy Survey.* Jakarta: Indonesia.

Bagwell, E.K. (2006). Factors influencing marital satisfaction with a specific focus on depression. *Senior Honors Theses, 38.*

Barnett, R.C. (2008). On multiple roles: Past, present, and future. In K. Korabik, D.S. Lero & D.L. Whitehead (Eds.), *Handbook of work-family integration research, theory, and best practices* (pp. 75–93). London, UK: Academic Press.

Benin, M.H. & Nienstedt, B.C. (1985). Happiness in single- and dual-earner families: The effects of marital happiness, job satisfaction, and life cycle. *Journal of Marriage and Family*, *47*(4), 975–984.

Bird, G.A., Day, S. & Cavell, M. (1990). Housing and household characteristics of single- and dual-earner families. *Home Economics Research Journal*, *19*(1), 29–37.

Boye, K. (2014). Dual-earner couples/dual-career couples. In A.C. Michalos (Ed.), *Encyclopedia of quality of life and well-being research* (pp. 1703–1706). Dordrecht, The Netherlands: Springer Science+Business Media.

Custer, L. (2009). Marital satisfaction and quality. In H.T. Reis & S. Sprecher (Eds.), *Encyclopedia of human relationships* (pp. 1030–1034). Thousand Oaks, CA: SAGE Publications.

Dakin, J. & Wampler, R. (2008). Money doesn't buy happiness, but it helps: Marital satisfaction, psychological distress, and demographic differences between low- and middle-income clinic couples. *The American Journal of Family Therapy*, *36*(4), 300–311.

DeGenova, M.K. (2008). *Intimate relationships, marriages, and families*. New York, NY: McGraw-Hill Education.

De Vries, L. (2011). *The similarities and differences between a marriage from collectivistic and individualistic country* (Bachelor's thesis, University of Amsterdam, The Netherlands).

Duvall, E.M. & Miller, B.C. (1985). *Marriage and family development*. New York, NY: Harper & Row.

Elloy, D.F. & Smith, C.R. (2003). Patterns of stress, work-family conflict, role conflict, role ambiguity and overload among dual-career and single-career couples: An Australian study. *Cross Cultural Management: An International Journal*, *10*(1), 55–66.

Frajerman, E.R. (2001). *The relationship between the division of housework, sex roles, and marital satisfaction in dual—career couples* (Dissertation, Pepperdine University, CA).

Funk, J.L. & Rogge, R.D. (2007). Testing the ruler with item response theory: Increasing precision of measurement for relationship satisfaction with the Couples Satisfaction Index. *Journal of Family Psychology, 21*(4), 572–583.

Gravetter, F.J. & Forzano, L.A.B. (2012). *Research methods for the behavioral sciences* (4th ed.). Belmont, CA: Wadsworth.

Gravetter, F.J. & Wallnau, L.B. (2013). *Statistics for the behavioral sciences* (9th ed.). Belmont, CA: Wadsworth.

Jacobs, J.A. & Gerson, K. (2001). Overworked individuals or overworked families? *Work and Occupations, 28*(1), 40–63.

Jose, O. & Alfons, V. (2007). Do demographics affect marital satisfaction? *Journal of Sex and Marital Therapy, 33,* 73–85. doi: 10.1080/00926230600998573.

Mohsin, F.Z. (2014). Marital satisfaction and job satisfaction: A study of dual- and single-earner couples. In *Proceedings of Third Annual Conference on Industrial and Organizational Psychology: Better Organizations Through Collaboration in Education, Research and Practice, December 22, 2012* (pp. 90–107). Karachi, Pakistan: Institute of Business Management. Retrieved from http://www.pbr.iobm.edu.pk/wp-content/uploads/2016/01/PBR-SPECIAL-ISSUE-2014.pdf.

Orathinkal, J. & Vansteenwegen, A. (2007). Do demographics affect marital satisfaction? *Journal of Sex & Marital Therapy*, *33*(1), 73–85.

Oshio, T., Nozaki, K. & Kobayashi, M. (2011). *Division of household labor and marital satisfaction in China, Japan, and Korea*. Report number 502. Hitotsubashi University Project on International Equity Discussion Papers. Retrieved from http://www.ier.hit-u.ac.jp/pie/stage2/Japanese/d_p/dp2010/dp502/text.pdf.

Quek, K.M.T. & Fitzpatrick, J. (2013). Cultural values, self-disclosure, and conflict tactics as predictors of marital satisfaction among Singaporean husbands and wives. *The Family Journal*, *21*(2), 208–216.
Setiawan, B. (2012, June 8). Who is the middle class of Indonesia? *Kompas.com*. Retrieved from http://nasional.kompas. com/read/2012/06/08/13003111/Siapa.Kelas.Menengah.Indonesia.
Stone, E.A. & Shackelford, T.K. (2007). Marital satisfaction. In Baumeister, R.F. & Vohs, K.D. (Eds.), *Encyclopedia of social psychology* (Vol. 2, pp. 541–544). Thousand Oaks, CA: SAGE Publications. Retrieved from http://www. toddkshackelford.com/downloads/Stone-Shackelford-ESP-2007.pdf.
Subdirektorat Statistik Ketenagakerjaan. (2015). *The state of the labor force in Indonesia*. Jakarta, Indonesia: Badan Pusat Statistik.
Twenge, J.M., Campbell, W.K. & Foster, C.A. (2003). Parenthood and marital satisfaction: A meta-analytic review. *Journal of Marriage and Family*, *65*(3), 574–583. doi:10.1111/j.1741-3737.2003.00574.x
Voorpostel, M., van der Lippe, T. & Gershuny, J. (2009). Trends in free time with a partner: A transformation of intimacy? *Social Indicators Research*, *93*(1), 165–169.
Walker, R., Isherwood, L., Burton, C., Kitwe-Magambo, K. & Luszcz, M. (2013). Marital satisfaction among older couples: The role of satisfaction with social networks and psychological well-being. *International Journal of Aging & Human Development*, *76*(2), 123–139.
Walsh, F. (2003). Changing families in a changing world: Reconstructing family normality. In F. Walsh (Ed.), *Normal family processes: Growing diversity and complexity* (3rd ed., pp. 3–26). New York, NY: Guilford Press.
Wendorf, C.A., Lucas, T., Imamoglu, E.O., Weisfeld, C.C. & Weisfeld, G.E. (2011). Marital satisfaction across three cultures: Does the number of children have an impact after accounting for other marital demographics? *Journal of Cross-Cultural Psychology*, *42*(3), 340–354.
Wilkie, J.R., Ferree, M.M. & Ratcliff, K.S. (1998). Gender and fairness: Marital satisfaction in two-earner couples. *Journal of Marriage and Family, 6* (3), 577–594.
Williams, B.K., Sawyer, S.C. & Wahlstrom, C.M. (2006). *Marriages, families, & intimate relationships.* New York, NY: Pearson.

Diversity in Unity: Perspectives from Psychology and Behavioral Sciences – Ariyanto et al. (Eds)
 ISBN 978-1-138-62665-2

The impact of self-efficacy on health behaviour in young adults whose mothers were diagnosed with breast cancer

G. Fatimah & L.D. Pohan
Department of Clinical Psychology, Faculty of Psychology, Universitas Indonesia, Depok, Indonesia

ABSTRACT: The objective of this study is to examine self-efficacy and health behaviour in young adults whose mothers have been diagnosed with breast cancer. The participants in this research were 84 people, aged 18–40 years, who had a mother who was diagnosed with breast cancer. Self-efficacy was measured using the Health Specific Behavior Self-Efficacy Scale (HSB-SES), which was adopted from Penney (2006). Health behaviour was measured using the indicators of health behaviours from Sarafino and Smith (2011). The results show that the null hypothesis was rejected ($F = 14.196$, $p < 0.05$), which means there is a significant impact of self-efficacy on health behaviour.

1 INTRODUCTION

According to the cancer mortality profile released by the World Health Organization (WHO, 2014), the cancer type that causes the most deaths (21.4%) in females is breast cancer. According to Junda (2004), breast cancer is one of the most impactful diseases in comparison to other cancers, not just for the individual, but also for the family as a whole. This is because the family is involved in the treatment and care of the patient: when the family is caring for the sick or keeping others, there will be an impact, either positive or negative. Dellman-Jenkiys and Blankemeyer (cited in Shifren, 2009) explained that when an individual or a child is caring for his or her mother who suffers from a disease, it will build a closer relationship within the family. This is because of the increase in time spent together, for example, when maintaining the house and providing assistance (e.g. in doing household tasks and food preparation).

Meanwhile, research conducted by Potter and Perry (2009) found a negative impact of the situation when a family tries to provide care for cancer patients. This is because families are often the main companion and being a point of contact when discussing the complex health care delivery for patients, and thus the situation indirectly makes the family members feel involved and stressed (Pohan & Basri, 2012). A family that provides care for a family member diagnosed with cancer is required to meet all of his or her daily needs, such as providing support and physical assistance for bathing, eating, and changing clothes. This situation is a stressor, because it puts pressure on the family and can affect the relationship between cancer patients and their families.

A number of studies found that chronic diseases such as breast cancer suffered by the mother will be significantly more influential in family life than other disease, as mothers do most of the housework and parenting (Yong, 1998). According to Forsberg (2003), there is a difference between women and men in young adulthood when their mother is diagnosed with breast cancer. Women and men differ in the general coping style that they tend to use. Men in young adulthood have a tendency to overcome the situation through avoidance strategies, namely the presence of distraction, physical exercise, and removed emotions from the situation. In contrast, women in young adulthood are more likely to be emotionally closer to the situation, and they cope by providing support to the mother, and expressing emotions. Women will do a reversal of roles with the ones previously held by their mothers (e.g. household tasks and child care). They are also more likely to equate their mother's breast cancer as a personal threat against their own bodies and their health when compared to men. When the mother becomes ill, the family will take the role as primary or secondary caregiver that provides direct care. According to Diana et al. (2012), the children of patients will become secondary caregivers, which means that children will indirectly care for their mother diagnosed with breast cancer. This situation can affect the children's self-efficacy to live healthily, especially when one family member is diagnosed with a disease.

Health behaviour is an attempt to maintain one's health by eating nutritious foods, exercising regularly, avoiding harmful behaviours and harmful substances, looking out for symptoms of an illness, and protecting oneself from accidents (Hales, 2013). According to Sarafino and Smith (2011), health behaviour is each individual activity that maintains or improves health condition, which can be done by eating breakfast almost every day, rarely eating snacks, not smoking, not drinking alcohol, doing regular physical activity, and measuring weight regularly. Self-efficacy is an individual's belief about their ability to perform certain behaviours or regulate the behaviour required to produce an outcome (Bandura, 1977).

In the context of health behaviour, self-efficacy refers to an individual's belief that he has the ability to change his risky behaviours by his own actions to implement health behaviours (Schwarzer & Fuchs, 1995). According to Sarafino and Smith (2011), self-efficacy is important in individual health behaviour change because with the absence of self-efficacy, the motivation to change risky behaviours to health behaviours will be low. Besides the self-efficacy factor, many factors could influence individuals to implement health behaviours, such as sex, age, education, history of disease, involvement in health communities, and social support (Erwart, cited in Sarafino & Smith, 2011).

With the exposure of the issues above, we wanted to examine the effect of self-efficacy on health behaviours. Self-efficacy is situational, which allows an individual to have a high self-efficacy in one situation but not in other situations. This is consistent with health behaviours; people can have high scores on physical activity, but they consume alcoholic beverages. Physical activity and consumption of alcohol are indicators of health behaviours. In this study, we wanted to see health behaviours in young adults when they have a mother who has been diagnosed with a chronic illness, because young adults are in a more excellent health condition than those in other age stages, so little attention has been paid to their health condition, such as smoking at the age of 21–25 years and drinking alcohol at the age of 18–25 (Papalia & Feldman, 2012). According to Levinson (1978), there are four stages of young adulthood: age 17–22 years old, which is early adult transition; into adulthood at the age of 22–28 years old; 28–33 years old is age 30 transition period; and 33–40 years, which is the peak of early adult life.

We wanted to examine the behaviour of healthy young adults living with a mother diagnosed with breast cancer, because it specifically describes the behavioural changes in individuals, especially in the health sector when people will see first-hand the behavioural changes experienced by the mother. In accordance with Krauel et al. (2012), many quantitative studies of children with a parent diagnosed with cancer focus on family functioning and psychosocial functioning in children compared to studies that specifically measure the problem or behavioural changes that occur in children. Therefore, this research examined whether self-efficacy can influence health behaviours in young adults with a mother diagnosed with breast cancer.

2 METHODS

According to Kumar (2005), study types can be divided among three perspectives, namely the application of research, research objectives, and methods of data collection. Based on the application, this study represents applied research. Based on the purpose of this research, this study is a correlational study, in which we wanted to see the relationship between self-efficacy and health behaviours. Furthermore, based on the data collection methods, this study uses quantitative research because we obtained scores from each participant.

Samples taken in this study were young adults aged 18–40 years with a mother diagnosed with breast cancer, who had completed at least junior high school, and lived in Jabodetabek (Jakarta, Bogor, Depok, Tangerang, and Bekasi). The number of samples obtained was 84 people. The sampling technique used was a non-probability random sampling with a snowball sampling technique. This was because samples for the study were hard to find, and it was also difficult to obtain access. The snowball sampling technique performed in this study was done by selecting one member of the sample population, who was then asked to provide references of other samples who they knew and who also met the sample characteristics required for the study.

2.1 *Measures*

2.1.1 *Health behaviour*

In this study, we used an instrument to measure health behaviours based on the indicators of health behaviours from Sarafino and Smith (2011). It was created by a team of researchers with thorough provisions in the creation of such measurement tools. The scale used in this study was a Likert scale of 1 to 4. We tested the measuring instrument on 34 young adults who had a parent diagnosed with cancer. From the reliability testing of this measure, the reliability coefficient value obtained was $\alpha = 0.754$.

2.1.2 *Self-efficacy*

In this study, we used an instrument called the Health Specific Behavior Self-Efficacy Scale (HSBSES) to measure self-efficacy. HSBSES was adapted by Penney (2006) to measure the self-efficacy of individuals. It consisted of 22 items and five dimensions, which included the nutrition of diet, physical activity, use of sunscreen, alcohol resistance, and smoking resistance. Only four dimensions were used in this study: the nutrition of the diet, physical activity, alcohol resistance, and smoking resistance. The measuring tool was adapted into Indonesian by a team of researchers. The scale used in this study was a Likert scale of 1–4. We tested the measuring instrument on 34 participants to measure the reliability and validity of the HSBSES. From the overall reliability testing of this measure, the value of reliability coefficient obtained was $\alpha = 0.763$.

Table 1. Example items: Health behaviour.

Number	Example	Subscales answer
1	I had breakfast before starting my activity.	1. Strongly disagree 2. Disagree 3. Agree 4. Strongly agree
6	I'm not smoking.	1. Strongly disagree 2. Disagree 3. Agree 4. Strongly agree

Table 2. Example items: Self-efficacy.

Number	Example	Subscales answer
1	I am able to keep eating healthy foods, although I have tried many times until I managed to eat them regularly.	1. Strongly disagree 2. Disagree 3. Agree 4. Strongly agree
15	I believe that I am able to control myself to not smoke at all.	1. Strongly disagree 2. Disagree 3. Agree 4. Strongly agree

2.2 *Statistical analyses*

2.2.1 *Descriptive statistics*

Descriptive statistical techniques were used to process data from the scores obtained from the participants and the demographic data of the participants. This technique was used to get a general overview of the characteristics of the participants. The results of the demographic data obtained will be included in the discussion of the research.

2.2.2 *Linear regression*

Regression technique was used to obtain the results of the primary analysis, which is the influence of self-efficacy on health behaviours.

3 RESULTS

3.1 *Characteristics of participants*

There were 84 participants: 58 females (N = 58) and 26 males (N = 26) with an age range of 18–40 years. Table 3 explains the overview health behaviour and self-efficacy based on the demographic data.

3.2 *Self-efficacy and health behaviour*

Referring to the issue of research, linear regression analysis obtained an F value of 14.196 with $p < 0.05$. Based on these results, we can conclude that the null hypothesis is rejected, and the alternative hypothesis is accepted. We found that self-efficacy was influential on health behaviours in young adults whose mothers were diagnosed with breast cancer. From these results, we obtained a coefficient determination value of $r^2 = 0.148$, which means that 14.8% of the variance in health behaviour scores could be explained by self-efficacy, and 85.2% could be explained by other factors.

Table 3. Overview of healthy behaviour and self-efficacy based demographic data.

Demographic data		N	Mean PS	Mean SE
Gender	Female	58	56.12	53.71
	Male	26	53.19	50.92
Age	18–22	59	54.77	53.68
	23–28	17	54.41	50.64
	29–33	4	57.00	52.00
	34–40	4	63.25	53.50
Last education	SMP (Junior High School)	1	48.00	54.00
	SMA/SMK (High School/Vocational School)	50	54.66	52.12
	DIPLOMA (Associate Degree)	6	59.16	51.60
	S1 (Bachelor's Degree)	27	55.60	54.40
History of disease	No	42	56.45	53.69
	Yes	42	53.97	52.00
Members of community/ health clubs	No	61	54.81	53.00
	Yes	23	56.26	52.43
Emotional support	Not giving	4	54.00	53.50
	Giving	80	55.27	52.81
Tangible support	Not giving	15	52.66	55.00
	Giving	69	55.76	52.37
Informational support	Not giving	32	53.84	52.68
	Giving	52	56.00	52.94

4 DISCUSSION

The first main result of the study shows that there is an influence of self-efficacy on health behaviours in young adults whose mother was diagnosed with breast cancer. This is consistent with the research of Maddux and Rogers (1983), which explained that in forming a new behaviour, self-efficacy has an important role in predisposing individuals to effect a change in attitude. This change of attitude in the present study was focused on health behaviours. The results in this study are in line with the research compiled by Strecher et al. (1986), which indicated that self-efficacy is a significant predictor of smoking behaviours, weight control, contraceptive use, abuse of alcohol, and exercise. The results are also consistent with the research of Rahmadian (2011), which showed that self-efficacy and health behaviour have a significant connection, based on a study of 195 students from four universities in the area of South Tangerang.

In addition, participants in this study are in a healthy condition, which means they apply behaviours to prevent being diagnosed with a disease in the future. This is consistent with one of the categories of health behaviours, namely preventive health behaviour, or any activity undertaken by individuals that are believed to be healthy, with the aim of preventing or identifying undetected disease (Kasl & Cobb, 1996, cited in Glanz et al., 2008).

Additionally, in this study, it was found that participants in the age group that was switching over to young adulthood (18–22 years old) and in young adulthood (23–28 years old) had a total score on health behaviours that was lower compared to participants who were over 28 years old. This is consistent with Papalia and Fieldman (2012), which stated that in young adulthood the individual is in a health condition that is much better than individuals in other age stages, so little attention is paid to their health condition, and unhealthy behaviours are commonly found, such as smoking at the age of 21–25 years old, and alcohol drinking at the age of 18–25. According to Taylor (2015), health behaviours will increase in older individuals. This is consistent with the results of this study that found that participants in the age range of 34–40 years had an average score of health behaviours that was the highest compared to those in other age groups.

Erwart (1991, cited in Sarafino & Smith, 2011) stated that there are several factors that influence health behaviours: level of education, health condition, interpersonal factors, and community. First, on the level of education, the group of participants who had a diploma as their highest education in this study had an average score that was high. It was less suitable, because in this study the average score for diploma-holders was higher than the average for the group of participants who were recent graduates. The second one is health condition: when people are sick or taking any medication on a regular basis, this condition can affect their mood and energy level, which in turn can affect the cognition and motivation of individuals in adopting health behaviours. This is appropriate because in this study the group of participants who had a history of illness had an average score that is higher compared to that of individuals with no history of disease. Finally, based on interpersonal factors, individuals who provide social support and encouragement to other individuals will affect the lifestyle of those who receive this social support, as well as individuals who provide them. For example, impacts that occur when individuals provide support or care for patients are physical pressure, emotional, disruption of family function, social, and financial pressures (Diana et al., 2012). Many studies have examined the effect of received social support towards health behaviours. According to research by Nollena et al. (2005), social support, defined as the support or the influence felt from family and friends, has been shown to be associated with regulated body weight, fruit and vegetable consumption, physical activity, diet, and cessation of smoking behaviour. In this study, it was proven that the group of participants who provided social support in emotional, tangible, and informational form had the highest average score in health behaviours compared to those who did not provide social support to their mothers.

According to Forsberg (2003), there is a difference in the child when the mother is diagnosed with breast cancer. A man in young adulthood has a tendency to overcome the situation with avoidance strategies, namely the presence of distraction, physical exercise, and emotional abolition of the existing situation. In contrast, women in young adulthood are more likely to be emotionally closer to the situation that occurred and cope by providing support to the mother and expressing their emotions. When linked to the previous research, which stated that individuals who provide more support have a higher average score on health behaviours than those who do not provide support, based on the genders, women participants in this study had an average health behaviour score that was higher than male participants. This might be because the female participants are more likely to equate their mother's breast cancer as a personal threat against their own bodies and their health, compared to male participants. The last factor according to Erwart (1991, cited in Sarafino & Smith, 2011) is the factor of community: individuals will be more likely to implement health behaviours if the behaviours are promoted or encouraged by a community. This is consistent with the results of this study: the group of participants with the highest average score on the health behaviour were those who were members of a health community/club.

5 CONCLUSION

The results show the impact of self-efficacy on health behaviour in young adults whose mothers were diagnosed with breast cancer.

REFERENCES

Badan Penelitian dan Pengembangan Kesehatan. (2013). *Riset Kesehatan Dasar (RISKESDAS) 2013* [*Basic health research (RISKESDAS) 2013*]. Jakarta, Indonesia: Health Research and Development Agency, Ministry of Health.

Bandura, A. (1977). Toward a unifying theory of behavioral change. *Psychological Review*, *84*(2), 191–215. doi:10.1037/0033-295X.84.2.191.

Diana, C.A., Sukarlan, A.D. & Pohan, L.D. (2012). Hubungan antara caregiver strain dan kepuasan pernikahan pada istri sebagai spouse caregiver dari penderita stroke. *Insan Media Psikologi*, *14*(3), 171–178.

Forsberg, J.A. (2003). *Adult children of mothers with breast cancer: A qualitative investigation* (Doctoral dissertation, Colorado State University). Retrieved from http://ovidsp.ovid.com/ovidweb.cgi?T=JS&PAGE=reference&D=psyc4&NEWS=N&AN=2003-95006-140.

Glanz, K., Rimer, B.K., & Viswanath, K. (2008). Health behavior and health education: Theory, research, and practice. San Francisco, CA: Jossey-Bass.

Guilford, J.P. (1956). *Fundamental statistics in psychology and education*. New York, NY: McGraw-Hill.

Hales, D. (2013). *An invitation to health: Build your future* (15th ed.). Belmont, CA: Wadsworth Cengage Learning.

Junda, T. (2004). Our family's experiences: A study of Thai families living with women in the early stages of breast cancer. *Thai Journal of Nursing Research*, *8*(4), 260–275.

Kementrian Kesehatan Republik Indonesia. (2014). *Gaya hidup tidak sehat picu kanker*. Retrieved from http://www.depkes.go.id/article/view/201409240004/gaya-hidup-tidak-sehat-picu-kanker.html.

Krauel, K., Simon, A., Krause-Hebecker, N., Czimbalmos, A., Bottomley, A. & Flechtner, H. (2012). When a parent has cancer: Challenges to patients, their families and health providers. *Expert Review of Pharmacoeconomics & Outcomes Research*, *12*(6), 795–808. doi:http://dx.doi.org/10.1586/erp.12.62.

Kumar, R. (2005). *Research methodology: A step-by-step guide for beginners* (2nd ed.). Thousand Oaks, CA: SAGE Publications.

Levinson, D.J., with Darrow, C. N, Klein, E.B. & Levinson, M. (1978). *Seasons of a Man's Life*. New York: Random House.

Maddux, J.E. & Rogers, R.W. (1983). Protection motivation and self-efficacy: A revised theory of fear appeals and attitude change. *Journal of Experimental Social Psychology*, *19*(5), 469–479. doi:10.1016/0022-1031(83)90023-9.

Nollena, N.L., Catley, D., Davies, G., Hall, M. & Ahluwalia, J.S. (2005). Religiosity, social support, and smoking cessation among urban African American smokers. *Addictive Behaviors*, *30*(6), 1225–1229. doi:http://dx.doi.org/10.1016/j.addbeh.2004.10.004.

Nunnally, J.C. (1978). *Psychometric Theory* (2nd ed.). New York, NY: McGraw-Hill.

Papalia, D.E. & Feldman, R.D. (2012). *Experience human development* (12th ed.). New York, NY: McGraw-Hill.
Penney, A.M. (2006). *The role of self-efficacy in the practice of health behaviors of young adult survivors of childhood cancer* (Master's thesis, Dalhousie University, Canada).
Pohan, L.D. & Basri, A.R. (2012). *Caregiver strain in family with chronical illness patients with undergo medical treatment*. Proceeding of the Fourth Asian Psychological Association Conference. Jakarta, Indonesia.
Potter, P.A. & Perry, A.G. (2009). *Fundamentals of Nursing* (7th ed.). Singapore: Elsevier.
Rahmadian, S (2011). Faktor-faktor Psikologis yang mempengaruhi perilaku sehat mahasiswa beberapa perguruan tinggi di Tangerang Selatan. Skripsi. Fakultas Psikologi UIN Syarif Hidayatullah. Jakarta.
Sarafino, E.P. & Smith, T.W. (2011). *Health psychology: Biopsychosocial interactions* (7th ed.). New York, NY: John Wiley & Sons Inc.
Schwarzer, R. & Fuchs, R. (1995). Changing risk behaviors and adopting health behaviors: The role of self-efficacy beliefs. In A. Bandura (Ed.), *Self-efficacy in changing societies* (pp. 259–258). New York, NY: Cambridge University Press.
Schwarzer, R. & Renner, B. (2007). *Health-specific self-efficacy scales*. Berlin, Germany: Freie Universität Berlin. Retrieved from http://userpage.fu-berlin.de/health/healself.pdf.
Shifren, K. (2009). *How caregiving affects development: Psychological implications for child, adolescent, and adult caregivers*. Washington, DC: American Psychological Association.
Strecher, V.J., Devellis, B.M., Becker, M.H. & Rosenstock, I.M. (1986). The role of self-efficacy in achieving health behavior change. *Health Education & Behavior*, *13*(1), 73–91. doi:10.1177/109019818601300108.
Taylor, S.E. (2015). *Health psychology* (9th ed.). New York, NY: McGraw-Hill.
Turner, J. (2004). Children's and family needs of young women with advanced breast cancer: A review. *Palliative & Supportive Care*, *2*(1), 55–64. doi:10.1017/S1478951504040076.
WHO. (2014). *Cancer mortality profile*. World Health Organization NCD Management Unit. Retrieved from http://www.who.int/cancer/country-profiles/idn_en.pdf?ua=1.
Yong, J.S. (1998). Factors influencing family functioning in families with breast cancer in the mother. *Journal of Korean Academy of Adult Nursing*, *10*(2), 369–384.

Diversity in Unity: Perspectives from Psychology and Behavioral Sciences – Ariyanto et al. (Eds)
 ISBN 978-1-138-62665-2

A comparative study: The effect of self-esteem and anger coping strategies on the level of anger among ordinary teenagers and teenage prisoners

M.A. Putri & P. Hidayah
Faculty of Psychology, University of Pancasila, Jakarta, Indonesia

ABSTRACT: Anger is a variable that connects aggressiveness and violent behaviour. A lot of adolescents cannot manage their anger, and it can cause them to have to deal with lawsuits and also suffer from other negative consequences. The aim of this study was to measure the impact of self-esteem and anger coping strategies on the level of anger. This study also examined the comparison between adolescents who are in prison and those who are not. The subjects of this study were 178 adolescents, 68 outside prison and 110 inside prison. The instruments used were the anger questionnaire developed by Rosellini and Worden (1997), the Rosenberg Self-Esteem Scale (1965) questionnaire, and the Behavioral Anger Response Questionnaire (Linden, 2007). The results show there is a significant relationship between self-esteem and anger coping strategies with anger level. This study found no difference in self-esteem and anger coping strategies between the two groups, and both groups have relatively high self-esteem. Two anger coping strategies significantly related with anger level are direct anger and rumination.

1 INTRODUCTION

It is not uncommon to hear news about violence perpetrated by adolescents. As of April 2016, there have been many cases of violence committed by high school students. Bullying cases have drawn national attention and become a serious concern regarding adolescent development nowadays. Adolescents, according to Papalia et al. (2009), are those individuals in the age range of 11/12 years to 19/20 years. There are many cases of violence that can cause adolescents to have to deal with lawsuits. This is in line with the results of a longitudinal study on adolescents conducted by Elliot (cited in Tremblay, 2000), where he found that acts of violence by boys and girls of the age of 12–17 years increased. It shows that adolescents are prone to behave aggressively, especially if accompanied by other risk factors.

Statistics from the Directorate General of Prison show there are 3,170 juvenile prisoners in 2016 in juvenile prisons, a 2.3% increase on the previous year. It can be said that juvenile delinquency has a lot of forms. Santrock (2002) divides juvenile delinquency into two types. The first is index violation, in the form of criminal acts committed by juveniles, such as robbery, assault, rape, and murder. The second is status violation—a violation frequently committed by juveniles, such as truancy, alcohol consumption under the legal drinking age, and free sex.

Every teenager committing an index violation may have to deal with lawsuits. Punishment or legal charges given to adolescents are regulated in the 2012 Law of the Republic of Indonesia, number 11, regarding the juvenile justice system. Adolescents or juveniles in this context are children aged between 12 and 18 years old who are suspected of having committed a criminal act. The negative consequences these adolescents may face include imprisonment, lack of opportunity to attend school, lack of interaction with parents, and disrupted social life.

Many studies have shown factors related with aggressive behaviour, such as individual, family, environmental, social, economic, and health. According to Scarpa and Raine (1997), aggressiveness or violence has two forms: instrumental or proactive aggression, and hostile or reactive aggression. Instrumental aggressiveness is a form of aggressiveness that is relatively unemotional and is intended to achieve a particular purpose, while hostile aggression is a form of aggressive behaviour associated with a source of anger and has a high emotional intensity.

Based on the above explanation, in this study the variable that connects aggressiveness or violence and violent behaviour is anger. As described by Thampi and Viswanath (2015), anger is a motive behind the occurrence of aggressive behaviour and the subjective experience that accompanies aggressive impulse. Cornell et al. (cited in Siddiqah, 2010) also state that anger is a predisposing factor of aggressive behaviour and anger is parallel with the encouragement of aggression (Berkowitz, cited in Siddiqah, 2010).

According to DeAngelis (2003), there are several factors that can distinguish constructive from destructive anger. The anger can be constructive if someone seeks the settlement of problems or tries to deal with people who become the source of anger, or in other words, appropriate solution-oriented issuing anger. The anger can be destructive if someone issues excessive anger and not in a proper way. Anger directed out may be accompanied by aggressive or violent behaviour, and anger directed at oneself can lead to depression, lack of communication and cause serious health problems.

That is why, when adolescents are not able to manage their anger, at the extreme point it can lead them into having to deal with lawsuits and to suffer poor health, psychological and social consequences. Anger and expressions of anger can cause serious health problems in children and adolescents (Blake & Hamrin, 2007). Anger and the accompanying behaviour can be dangerous and if not managed properly can lead to murder (Nasir & Ghani, 2014). Therefore, research about anger becomes crucial for adolescents in Indonesia, so we can help them recognise anger at different levels and also identify factors necessary to manage the anger.

One of the factors that can cause aggressiveness is self-esteem. Although the relationship between aggressiveness and self-esteem is a subject that is often studied, the relationship between the two is still contradictory (Ostrowsky, 2010; Webster & Kirkpatrick, 2006 cited in Binti-Amad, 2015). Webster (2007) found that one of the causes of aggressiveness in males is low self-esteem or high self-esteem instability, whereas in women, it is a combination of instability in both high and low self-esteem.

Earlier studies also found that anger level is also affected by the coping strategy. As for the impact of anger on the health of children and adolescents, previous studies have shown that children who often complain of physical pain show higher levels of anger than their friends (Jellesma et al., 2006; Rieffe et al., cited in Miers et al., 2007). It is said that the way a person overcomes problems can affect his or her health. Further research by Martin (cited in Miers et al., 2007) explains that increasing levels of uncontrolled anger will affect the physiological system and will eventually bring bad effects on health. Therefore, it becomes another important reason to study how teenagers manage anger, so that they can grow physically and psychologically healthy.

This study also examines the comparative levels of anger, self-esteem and anger coping strategies between adolescents inside and outside prison. The purpose was to obtain a more comprehensive understanding of the dynamics of adolescents and also to contribute to the body of knowledge, both of self-esteem and anger coping strategies, which is still debatable.

2 LITERATURE REVIEW

Anger is defined 'as an emotional response to situations that are perceived as threatening or offensive to oneself or others close to them' (Lazarus, cited in Lochman et al. 2004). Anger is defined 'as an emotional state characterised by feelings of outrage and annoyance' (Pérez-Nieto et al., 2000, cited in Canning, 2011). 'Hostility and aggression represent related constructs to anger, and together they are sometimes considered as a triad: anger–hostility–aggression' (Pérez-Nieto et al., cited in Canning, 2011). Aggression refers to 'overt behaviour' (Harburg et al., 2003, cited in Canning, 2011) 'whereas hostility is defined as a persistent negative attitude towards others' (Pérez-Nieto et al., cited in Canning, 2011).

Self-esteem, according to Santrock (cited in Baron & Byrne, 2004), is a wide description of self-evaluative dimension, which mean a person evaluates his or her definition of self, ranging from positive to negative self-image. According to Coopersmith (cited in Mruk, 2006) there are four components of self-esteem: power, which illustrates the ability of the individual to control or influence others; significance, which refers to the revenues generated by the judgements of others; virtue, which is adherence to the ethical values or moral norms of society; and competence, which means the ability to succeed when encountering the problem.

The term *anger coping style* can be understood as behavioural or cognitive, or both, as a response to the situation that is considered disturbing. According to Spielberger (1985), there are two ways of expressing anger that can damage the body. The first way is to suppress anger inside yourself (also called anger-in), and the second is to outburst anger (also called anger-out). Both ways can be measured using a questionnaire from Spielberger called SAES.

The study of emotion must include the study of cognition, motivation, adaptation and physiological activity. Emotion involves a person's appraisal of the environment, his or her relationships with others, and his or her attempts to cope with these relationships (Lazarus, cited in Plutchik, 2002). According to Lazarus (in Plutchik, 2002), coping 'is [finding] ways to manage and interpret conflicts and emotions'. At the beginning of anger research, it was found there were two ways a person could manage his or her anger, either anger-in or anger-out (Miers et al., 2007).

According to Hogan (1996), there are few studies aimed at developing measurement of anger dimensions with several factor analyses. Linden et al. (cited in Miers et al., 2007) developed a tool to measure anger known as the Behavioral Anger Response Questionnaire (BARQ). Two forms of anger coping strategies in BARQ are direct anger-out and avoidance, similar to anger-out/anger-in from Spielberger (1985).

The BARQ consists of six anger coping styles, that is: aggressive anger-out, defined as aggressively expressing anger to other persons or things; anger diffusion, referring to the way a person overcomes anger that is not aggressive and directs anger at things that are not associated with the source of anger, such as sports or drawing; assertive anger coping, which is about how a person overcomes anger by talking directly to the person who becomes the source of anger without using words or aggressive behaviour; social support-seeking, which is a way in which someone reaches out to other people who are not related to their anger and tries to discuss their feelings with them; avoidant coping, which is a method for forgetting or ignoring the anger; and rumination, which refers to a way to decrease anger by repeating back the memory of the events that became the source of the anger without expressing aggressive behaviour. Research on adolescent mischief usually also considers several factors, including gender, education level, ethnicity and socioeconomic status. Furthermore, it has been found that there are some cultural characteristics of low social class which support adolescent mischief (Thio, cited in Santrock, 2013).

2.1 *Hypotheses*

This research aimed to test the following hypotheses:

1. Hypothesis 1: There is a significant relationship between self-esteem and anger coping strategies and anger level.
2. Hypothesis 2: There is a significant self-esteem difference between adolescents inside and outside prison.
3. Hypothesis 3: There is a significant anger coping strategy difference between adolescents inside and outside prison.
4. Hypothesis 4: There is a significant anger level difference between adolescence inside and outside prison.

3 METHODS

3.1 *Sample description*

The respondents for this study were 178 adolescents, with an age range of 12–20 years old. Because one of the purposes of this research was to compare anger levels between different groups of adolescents, the researchers chose a group of adolescents who had committed juvenile delinquency.

3.2 *Sampling technique*

The sampling method used was multistage sampling. The first stage was using cluster sampling, where the population was divided into two groups: adolescents inside and outside prison. The second stage was choosing population from both locations by random sampling. The respondents of this research were 178 adolescents, 68 outside prison and 110 inside prison.

3.3 *Research variables and measurements*

The dependent variable in this research was anger, which was measured using the questionnaire developed by Rosellini and Worden (1997). The participants had to determine whether the statements in the questionnaire reflected their situation, and they had to answer 'true' or 'false'. Each participant's answers were then calculated, where each 'true' response was counted as one and each 'false' response was counted as zero. The total score of each participant was then grouped into different categories: 0–4 = Low; 5–9 = Normal; 10 or more = High. Further, answering 'true' to any of the statements numbered 21–25 = High. The number of items in this questionnaire was originally 25 but, after the validity test, eight items were removed because they were found to be invalid, so the number of items eventually used in the questionnaire was 17. The reliability coefficient in this research trial was 0.784.

The independent variables in this research were self-esteem and anger coping strategies. The self-esteem variable was measured using the questionnaire for the Rosenberg Self-Esteem Scale. There are four possible answers in this questionnaire, namely, 'Strongly disagree' (= 1), 'Disagree' (= 2), 'Agree' (= 3), and 'Strongly agree' (= 4). This questionnaire consists of ten items, each of which is a statement of an individual's view of their ability, and after the validity test, all the items were declared valid. The reliability coefficient in this research trial was 0.705.

The variable of anger coping strategies was measured using the BARQ questionnaire. This questionnaire consists of 37 items, which are divided into six categories, namely, Assertion (six items), Direct Anger-Out (seven items), Social Support-Seeking (six items), Rumination (six items), Avoidance (six items), and Diffusion (six items). Each category is unidimensional, which means that each stands alone and cannot be combined into one score. After the validity test, some items in some of the categories were removed. The following are the remaining numbers of items for each category: Assertion (six items); Direct Anger-Out (seven items); Social Support-Seeking (six items); Rumination (five items); Avoidance (five items); Diffusion (four items). Each category had a reliability coefficient, as follows: Assertion 0.621, Direct Anger-Out 0.752, Social Support-Seeking 0.656, Rumination 0.602, Avoidance 0.678, and Diffusion 0.629.

The data was analysed using multiple regression analyses, T-test and F-test. An F-test, known as a simultaneous test and the basis for model and Analysis of Variance tests, is a test to see the influence of all independent variables together against the dependent variable. A T-test, known as a partial test, is used to examine the influence of each independent variable individually against the dependent variable.

3.4 *Research procedures*

As the research involved adolescents who were in prison, prior permission from an authorised officer was required. The prison officials helped to identify the adolescents, and the researchers first gave instructions about how to fill out the questionnaire and responded to questions regarding the procedures for answering the questionnaire items. As for the adolescents who were outside prison, the questionnaire was distributed online. An email address of the researchers was provided in case the respondents wanted to ask questions.

4 RESULTS AND DISCUSSION

4.1 *General overview of the respondents*

Of the respondents, 77.7% were male and 22.3% female. The level of education varied: college 27%, senior high school 33.1%, junior high school 31.1%, and elementary school 8.8%. They also came from a variety of ethnic groups: Javanese 37.2%, Sundanese 25.7%, Betawi 25.7%, and Lampung 2.7%, while Buginese, Chinese, Malay and Padang together amounted to 1.4%, and Batak, Palembang and Maluku 0.7%. The respondents' cities of residence were Tangerang 37.2%, Jakarta 25.7%, Depok 25.7%, while Bekasi, Bogor, Banten and Palembang together accounted for 1.4%, Bandung 0.7%, and Lampung 0.7%. The majority of respondents had a father who was self-employed (35.1%), and worked as a labourer (27.7%). Other occupations included artist, military personnel, civil servant, and some were retirees. The majority of respondents had a housewife mother (74.3%).

4.2 *Hypothesis 1*

The multiple regression analysis showed that the influence of the seven Independent Variables (IV) on the Dependent Variable (DV) of anger was 49.5%, while the other 50.5% was influenced by other factors. The influence of all the seven IVs jointly against anger level (DV) was significant, F (7.170) = 15.786, p = 0.000. It means there is a significant relationship between self-esteem and anger coping strategies and anger level.

However, analyses of each independent variable showed that two anger coping strategies had significant influence on the level of anger. These strategies are direct anger coping strategy (p = 0.000), which is defined as aggressively expressing anger to other persons or things, and rumination (p = 0.003) which refers to a way to decrease anger by repeating back the memory of the events that become the source of anger without expressing aggressive behaviour. The other anger coping strategies showed no significant influence on the level of anger.

The analyses showed that self-esteem had influence on anger level and that all the respondents in this study did not have low self-esteem. These results are in line with the findings of Baumeister et al. (cited in Binti-Amad, 2015), which state that 'positively biased self-perceptions may actually be a determinant of aggressive behaviour'. These findings are important for further research; looking at other variables which can balance the influence of self-esteem on anger level, and reduce adolescence agrressive behaviour.

4.3 *Hypothesis 2*

A T-test was used to view the difference in anger level in both groups, namely adolescents who were inside and outside prison. An independent sample T-test showed that scores were significantly higher for adolescents who were in prison (M = 6.96, SD = 3.75) than for adolescents who were outside prison (M = 5.35, SD = 3.62), t(176) = 2.82, p = 0.005.

4.4 *Hypothesis 3*

A T-test was also used to identify the difference between self-esteem and anger coping strategies in both groups, namely adolescents who were inside and outside prison. It was found that there were no significant differences in self-esteem among adolescents who were inside and outside prison. This means that both adolescent groups could have high or low self-esteem. For adolescents inside prison, self-esteem could be high because they were in their peer group (adolescents with delinquency tendencies), and then they might actually get compliments from their peer group, which then gives them aspects of self-esteem (power, significance, virtue, and competence) (Utami, 2014). There are also factors that may influence self-esteem, but which were not included in this study, namely, parenting styles, birth order, and gender. These factors demonstrate that self-esteem is not entirely related to the current state of the adolescents, but has been formed long before they go to jail.

4.5 *Hypothesis 4*

The findings showed that there were no significant differences in every anger coping strategy among adolescents who were inside and outside prison. This condition can be due to the fact that adolescence is a period of transition with rapid changes in cognitive, biological and socio-emotional aspects, which causes adolescents to become unstable, looking for ways to manage a lot of problems, including anger. These findings are in line with those of Phillips-Hershey and Kanagy (cited in Nasir & Ghani, 2014) that said 'anger is difficult for adolescents to manage'.

5 CONCLUSION

There is a significant relationship between self-esteem and anger coping strategies with anger level. This study found no difference in self-esteem and anger coping strategies between the two groups and both groups have relatively high self-esteem. Two anger coping strategies that are significantly related with anger level are direct anger and rumination.

REFERENCES

Baron, R.A. & Byrne, D. (2004). *Psikologi sosial*. Jakarta, Indonesia: Erlangga.

Binti-Amad, S. (2015). *Self-Esteem and aggression: The relationships between explicit-implicit self-esteem, narcissism, and reactive-proactive aggression*. (Doctoral thesis, School of Psychology, Cardiff University, UK). Retrieved from: http://orca.cf.ac.uk/77062/

Blake, C.S. & Hamrin, V. (2007). Current approaches to the assessment and management of anger and aggression in youth: A review. *Journal of Child and Adolescent Psychiatric Nursing*, *20*(4), 209–221.

Canning, A. (2011). *An investigation of the relationship between self-esteem and aggression in care leavers* (Doctoral thesis, Cardiff University, UK). Retrieved from http://orca.cf.ac.uk/8627/.

DeAngelis, T. (2003). When Anger's A Plus. APA *Monitor on Psychology*,34(3), 44–45.

Hogan, B.E. (1996). *Anger coping styles and major personality dimensions: A closer look at the construct validity of the Behavioral Anger Response Questionnaire (BARQ)* (Master's thesis, University of British Columbia, Vancouver, Canada).

Kerlinger, F.N. & Lee, H.B. (1999). *Foundations of behavioral research* (4th ed.). Fort Worth, TX: Wadsworth Publishing.

Lochman, J.E., Palardy, N.R., McElroy, H.K., Phillips, N. & Holmes, K.J. (2004). Anger management interventions. *Journal of Early and Intensive Behavior Intervention, 1*(1), 47–56.

Miers, A.C., Rieffe, C., Terwogt, M.M., Cowan, R. & Linden, W. (2007). The relation between anger coping strategies, anger mood and somatic complaints in children and adolescents. *Journal of Abnormal Child Psychology*, *35*(4), 653–664.

Mruk, C.J.J. (2006). *Self-esteem research, theory, and practice: Toward a positive psychology of self-esteem* (3rd ed.). New York, NY: Springer Publishing.

Nasir, R. & Ghani, N. (2014). Behavioral and emotional effects of anger expression and anger management among adolescents. *Procedia—Social and Behavioral Sciences*, *140*, 565–569.

Papalia, D.E., Olds, S.W., Feldman, R.D. (2009). Human Development (11st ed). Boston: McGraw-Hill.
Plutchik, R. (2002). Emotions and life perspectives from psychology, biology and evolution. Washington, DC: American Psychological Association.

Rosellini, G. & Worden, M. (1997). *Of course you're angry: A guide to dealing with the emotions of substance abuse* (2nd ed.). Center City, MN: Hazelden Publishing.

Santrock, J.W. (2002). *Life span development* (8th ed.). New York: McGraw-Hill.

Santrock, J.W. (2013). *Life span development* (14th ed.). New Delhi, India: McGraw-Hill Education.
Scarpa, A. & Raine, A. (1997). Psychophysiology of anger and violent behavior. *Psychiatric Clinics of North America*, *20*(2), 375–394.
Siddiqah, L. (2010). IPI pencegahan dan penanganan perilaku agresif remaja melalui pengelolaan amarah.*Jurnal Psikologi*, *37*(1), 50–64.
Spielberger (1985), there are two ways of expressing anger that can damage the body Construction and validation of an anger expression scale. In M. A. Chesney & R. H. Rosenman (Eds.), Anger and hostility in cardiovascular and behavioral disorders (pp. 5–30). New York: Hemisphere/ McGraw-Hill.
Thampi, A. & Viswanath, L. (2015). Level of anger and its psychosocial factors among adolescents. *Global Journal of Multidisciplinary Studies, 4*(12).
Tremblay, R.E. (2000). The development of agressive behaviour during childhood: What have we learned in the past century? *International Journal of Behavioral Development 24*(2), 129–141.
Utami, A.R. (2014). *Gambaran self esteem narapidana remaja berdasarkan klasifikasi kenakalan remaja: Studi deskriptif mengenai self esteem pada narapidana remaja di lapas anak Bandung dan lapas wanita Bandung* (Diploma thesis, Faculty of Psychology, Universitas Padjajaran, Bandung, Indonesia).
Webster, G.D. (2007). Is the relationship between self-esteem and physical aggression necessarily u-shaped? *Journal of Research in Personality*, *41*(4), 977–982.

Diversity in Unity: Perspectives from Psychology and Behavioral Sciences – Ariyanto et al. (Eds)
© 2018 Taylor & Francis Group, London, ISBN 978-1-138-62665-2

Solution-focused brief therapy approach intervention for increasing self-esteem of young adult women with cancer who experience chemotherapy-induced alopecia

C. Anakomi, A.D.S. Putri & L.D. Pohan
Faculty of Psychology, University of Indonesia, Depok, Indonesia

ABSTRACT: Having cancer at a young age is always a shocking experience for patients. The stages that must be passed through from diagnosis to treatment burden their psychological conditions. During these stages, the patients experience various changes and losses. One of them is hair loss due to chemotherapy or Chemotherapy-Induced Alopecia (CIA). CIA has been proven to be internalised by women as a profound loss, as the treatment's effect can decrease their self-esteem. This research conducted in-depth analysis of three young adult women with cancer on how they live with hair loss. Next, a Solution-Focused Brief Therapy (SFBT) approach intervention was given individually, consisting of seven sessions (two pre-sessions, four intervention sessions, and a follow-up session). Each session lasted around 90–120 minutes. A Revised Janis and Field Scale questionnaire was used to measure self-esteem at pre-test and post-test. The result of qualitative and quantitative assessment indicated a positive change in their perspectives of themselves and their experiences of hair loss. This research shows that intervention with a SFBT approach could increase the self-esteem of young adult women with cancer who experience CIA.

1 INTRODUCTION

Cancer is a disease that is marked with division and spread of abnormal (uncontrollable) cells, which in time will develop into foreign tissues harmful for the patient (American Cancer Society, 2015). Anyone may suffer this disease; it is not dependent on age, gender, or particular race (Sarafino & Smith, 2011). This is indicated from the prevalence of cancer patient numbers as a whole. Statistical data from the World Health Organization (2012) show that the world's number of people diagnosed with cancer had reached 14 million people as of 2012, increasing from 12.7 million people in 2008. With this global prevalence, Indonesia's cancer patient numbers also show an increase. Data from the Ministry of Health of the Republic of Indonesia (Kementrian Kesehatan Republik Indonesia, 2012) revealed in 2007 acancer prevalence in Indonesia increased up to 4.3 in 1,000 people.

The numbers show that, in recent years, cancer has become one of the most prevalent diseases in Indonesia. As a result, information pertinent to this disease has increasingly spread among the community. The information obtained, however, is inaccurate most of the time, leading to a number of negative perceptions and emotions. For ordinary people, cancer is often perceived as being synonymous with death (Seffrin et al., 1991), due to the belief that this disease is incurable. This perception is indirectly believed by cancer patients and causes a traumatic effect on them (Neilson-Clayton & Brownlee, 2002). Consequently, undesirable anxiety and stress occur and aggravates the patients' psychological condition during critical periods.

A critical period occurs throughout the cancer recovery process with differing stress intensity in each phase (Neilson-Clayton & Brownlee, 2002). The phase in which the medication begins until it ends, according to LaTour (1994), is the culmination of sorrow and crisis experienced by cancer patients. During this phase, cancer patients experience a great deal of loss/

change, and at times they are forced to face all of them simultaneously, starting from change in routine and daily functioning, to changes in physical appearance resulting from the side effects of the medication taken.

Side effects of medication are experienced by nearly all cancer patients. According to Preston (2010), every medical action poses side effects that cause change in the physical appearance of cancer patients, including chemotherapy. As a medical alternative, chemotherapy is deemed to have an advantage in terms of wide drugs absorption. On the other hand, this chemotherapy advantage is also followed by a negative consequence to the patient's body condition. The chemotherapy drug's ability in being absorbed throughout the entire body lets the effect work in an unselective manner (Münstedt et al., 1997).

Both cancer cells and normal cells are affected by the killing effect of the drugs, which causes physical side effects on the patient. These side effects may include pain, nausea and vomiting, extreme fatigue, change in nails and skin, and hair loss (National Cancer Institute, 2009). Among the various physical side effects possibly experienced, Preston (2010) argues that the changes related to physical appearance pose the most harmful effects on the patients' psychological condition, especially on female patients. Female cancer patients consider hair loss as the most burdensome as the change is directly visible to others (McGarvey et al., 2010).

This is supported by the finding of Lemieux et al. (2008). From their review of 38 articles pertinent to hair loss from chemotherapy—which is also commonly known as Chemotherapy-Induced Alopecia (CIA) – they found that CIA is always within the top three among the most troublesome side effects in the majority of studies conducted on female cancer patients. Domestically, a study was conducted by Melia et al. (2012) on 38 patients who were undergoing chemotherapy in RSUP Sanglah, Denpasar. From the results, it was found that alopecia was one of the chemotherapy side effects that caused psychological trauma and changes in social activities, self-image and self-esteem of the patients. Of the three, change in self-esteem was the aspect that left the widest and most fundamental impact on the patients' lives. In cancer patients, according to Curbow et al. (1990), self-esteem is an essential variable in relation to an individual's psychosocial responses to his/her disease. This is evidenced by the finding of Ferrell et al. (1997), which states that the stress from cancer diagnosis, medication side effects, and a lot of uncertainties faced by a patient, give rise to psychological responses in the form of anxiety, depression, and low self-esteem.

This condition is more difficult for women who are diagnosed with cancer at a young age. According to Sebastián et al. (2008), the medication side effects in the initial form of CIA will leave an impact on the body image of female cancer patients, before eventually leading to a decrease in self-esteem. In real-life cases, this is often not appreciated and tends to be neglected by patients as it is not considered to be as important as the physical recovery that they undergo. As a result, this change will persist and develop into a problem that has a significant impact on the way in which an individual goes through the critical period of his/her recovery process (Leary & Baumeister, 2000). Therefore, it is necessary to have professionals help the patients manage their self-esteem decrease due to CIA through an effective psychological intervention.

Thus far, there are a number of psychological interventions that have been done in order to manage the issue of self-esteem decrease in cancer patients. Cognitive Behavioural Therapy (CBT) is one of the approaches frequently applied and the effectiveness of which has been proven. Two pieces of research have provided this proof, namely, studies carried out by Lubis and Othman (2011), and Ando et al. (2011). If further analysed, the CBT approach is basically focused on the problems and their relationship with the patient's past, in which therapists have the key role of directing the process of problem-solving (D'Zurilla, 1986).

One approach that may address this weakness is Solution-Focused Brief Therapy (SFBT). SFBT is an approach that emphasises seeking solutions for the problems rather than merely discussing them (Goldenberg & Goldenberg, 2008). Through this approach, participants are facilitated to set a goal and find solutions for their problems independently. This is in line with what Greenberg et al. (2001) claim, that in SFBT the level of a therapist's role is no different from the participant's, emphasising the knowledge that he/she owns, but treating it as relatively equal to the participant's. Afterwards, Neilson-Clayton and Brownlee (2002) have proven the effectiveness of SFBT application in helping the process of adjustment to the side effects of the disease in cancer patients, along with their families.

Therefore, this study aims to analyse in depth about young adult female cancer patients' subjective perception on how they live with hair loss due to their medical treatment. The SFBT module which has been proven to be effective and used in this research is the module designed by Agustin (2012) in research on quarter-life crisis. This study aims to answer the following research question: 'Will an intervention with a Solution-Focused Brief Therapy approach be able to raise the self-esteem of young adult female cancer patients with chemotherapy-induced alopecia?'

2 THEORETICAL REVIEW

2.1 *Cancer and CIA*

Hair loss is one of the side effects of chemotherapy experienced by the majority of cancer patients. According to Luanpitpong and Rojanasakul (2012), this is due to the fact that the drugs given in chemotherapy cause structural damage on skin where hair grows. The effects vary, starting from change in hair appearance, decreased level of hair growth, to hair loss (chemotherapy-induced alopecia), either partial or total.

Patients with CIA will lose all hair on their body, including the hair that previously grew on the head, eyebrows, and eyelashes, having wide implications, either physically or psychologically (Preston, 2010). Shame, depression, and loss of self-confidence are some of the psychological reactions experienced by cancer patients due to CIA (Moorey, 2007). According to Chassin (2008), this reaction is a reasonable phase which patients go through as part of the process of adjustment to their disease. Specific to young women, this process has the potential to develop into a prolonged sorrow and a harmful threat to their self-esteem balance.

2.2 *Young adult female cancer patients*

The age range of those categorised into this group lies between the ages of 18 or 21 and 45. Based on this, the young adult women defined in this research are those who are between 21 and 45 years old. Furthermore, according to Erikson's stages of psychosocial development (cited in Fleming, 2004), the development issue faced by young adults is intimacy versus isolation. The core of this stage is establishing intimacy and mutual sharing with others. Two relationships belonging to this category are romantic relationship and friendship (Sarafino & Smith, 2011). These relationships encompass the act of establishing long-term commitments with the opposite sex (either in marriage or not) and friends. Within this age range, individuals shift from seeking knowledge and applying it in pursuing a career, to choosing partners to get married (Santrock, 2007).

In female cancer patients who belong to the young adult age group, their development task often faces hurdles. If the diagnosis is received before the individuals get married, the self-esteem decreases due to the fact that physical changes experienced will potentially obstruct the patients from finding partners. On the other hand, if the diagnosis is received after the individuals are married, the changes experienced will potentially disrupt their roles as partners/spouses or parents (Lubkin & Larsen, 2002). Moreover, female patients will also experience changes in terms of their role as workers. According to Beatty and Joffe (2006), cancer patients have to make a number of changes in their career plans pertinent to their physical and psychological conditions.

2.3 *Self-Esteem*

Several definitions of self-esteem are frequently used as a reference in studies that discuss about this topic. Branden (1992, p. 8) defines self-esteem as '...confidence in our ability to think and to cope with the basic challenge of life; confidence in our right to be happy, the feeling of being worthy, deserving, entitled to assert our needs and wants, and to enjoy the fruits of our effort.'

Based on this definition, it can be concluded that self-esteem is an attitude or belief of an individual in themselves (positively or negatively) in relation to the sense of being worthy, capable and appreciative towards their overall being. In the results of the research conducted by Trzesniewski et al. (2003), it is stated that self-esteem is a construct that is relatively stable, resembling characteristics. With the individual's growth, however, a drop or rise in self-esteem is inevitable. According to Mruk (2006), a decrease in self-esteem during adulthood possibly occurs when the individual experiences a meaningful event in his/her life. Furthermore, Wagner et al. (2013) state that self-esteem may decrease when an individual suffers from a serious health problem at a young age, like being diagnosed with chronic disease that causes physical changes.

2.4 *SFBT*

The SFBT model was initially developed in 1986 by Steve De Shazer, Insoo Kim Berg and some of their colleagues (Carr, 2006). According to its name, this therapy is more focused on assisting the clients to find solutions rather than sticking with the problems being experienced. The SFBT approach attempts to facilitate the clients in identifying exceptional situations occurring in the problems they face, which is also known as Infrequent Exceptional (De Shazer et al., 1986). Afterwards, the clients will be invited to increase the frequency of their behaviour in such situations.

Different from any other therapy models, in SFBT the therapy goals are set by the clients. Specific, concrete, realistic goal-setting is a vital part of SFBT (Bavelas et al., 2013). The therapy goals are formulated and improved according to the continuity of conversation carried out by the therapist and the client in relation to what the client seeks to achieve in the future. Furthermore, according to De Shazer and Dolan (2007), some techniques used in SFBT sessions are Miracle Questions, Scaling Questions, Solution-Focused Goals, Exception Questions, Compliments, and Experiments, as well as the assigning of homework.

3 RESEARCH METHODS

This research can be categorised as quasi-experimental research. According to Shaughnessy et al. (2000), in this type of research the researcher administers an intervention to make a comparison between the condition before and after intervention. This is carried out without any strict control on external variables and without selecting participants in a randomised manner. In this research, an intervention in the form of Individual Therapy was given to young adult cancer patients as research participants who were selected based on predetermined characteristics. The participants were selected, based on the criteria, from RSK Dharmais hospital. Afterwards, the participants' levels of self-esteem were measured in both periods. The measurement instrument used was the Revised Janis and Field Scale; the interviews were conducted in two early sessions prior to the intervention session.

4 RESULTS

Table 1 provides an overview of the results of the four sessions each of the participants underwent.

Table 1. Analysis results of all sessions with the three participants.

Session	Participant 1 (SA)	Participant 2 (SK)	Participant 3 (TP)
I	The problem to be solved		
	was the sense of shame and lack of self-confidence due to hair loss experienced that caused anxiety about husband's loyalty, disruption of her role as a mother, and the feeling of reluctance to interact with the people around her.	was the sense of shame and lack of self-confidence due to hair loss experienced that caused the feeling of reluctance to interact with others, tendency to feel offended and anxiety of being disable to find a lifelong partner.	was the feeling of sadness, anxiety, and insecure due to hair loss experienced that caused the feeling of reluctance to see her own reflection in the mirror because of the thought that she would appear ugly.
	The goals to be achieved		
	were to see her husband's reaction to her conditions at that time and to have the confidence to be able to interact with others and able to assume her role as a mother.	was to have the courage to freely interact with others so she would be able to do her daily activities without the help of her family.	were to have the courage to see her own reflection in the mirror and to manage negative thoughts and emotions related to her hair condition at that time.
	The answer to the Miracle Question indicating that the problems will have been solved		
	was when SA is finally able to assume her role as a wife and a mother.	was when SK's confidence backs to normal so she has the courage to meet others without covering her head.	was when TP's hair grows back.
II	The crisis dynamics of self-esteem decrease experienced		
	was that SA's hair loss affected SA in terms of changes in characteristics, habits and belief on overall self-capability. This led to the appreciation that she was no longer perfect as a wife and mother.	was that SK's hair loss affected SK in terms of changes in characteristics, perspective and feeling of being worth. This led to the appreciation that her appearance turned to be strange and not interesting as a woman.	was that TP's hair loss affected TP in terms of changes in perspective, judgment and belief on overall self-capability. This led to the assumption that her appearance turned to be ugly.
	The exceptional condition		
	was when her husband said that what mattered the most was SA's health and he did not demand anything from her either at that time or the following days.	was when SK was interacting with her friends because she would be more focused on the conversation that were taking place and laugh more, and because her appearance with veil covering her head was similar to the appearance of most of her friends.	was when TP was being together with her husband and children.

(*Continued*)

Table 1. (*Continued*)

Session	Participant 1 (SA)	Participant 2 (SK)	Participant 3 (TP)
III	The assessment result on the positive qualities within self		
	was: being nice, affectionate, generous, friendly, fun, helpful, caring, pretty, selfless and eager to face ordeals.	was: being friendly, bright, good-looking, good in listening, smart, adaptive to new environment, optimistic, quick in learning new things, thorough and sociable.	was: being sociable, generous, willing to receive criticisms, accepting thing as it is, admitting mistakes, willing to live humbly, accepting fate, accepting others' opinions, helpful and having beautiful hair.
	The coping skills used in facing undesirable situations		
	were withdrawing from the situation and doing nothing.	were calming herself and reminding herself that her hair would grow back.	were gaining support and suggestions from closest ones and making them a motivation to cope with the problems.
IV	The solutions made		
	were establishing communication more openly with closest ones and attempting to control emotion (especially toward her husband).	were wearing head cover (hat) and veil and asking her mother to accompany her when she wanted to interact with her neighbors.	was having converstions with trusted ones to gain support, care and information that she needed.
Follow-up	The memorable event occuring between sessions		
	experienced by SA was being confident in making interactions without wearing veil when her aunt and her husband's causin visited her. She was sure that, despite being bald, she still looked interesting just like how she was before becoming bald.	experienced was managing to gain insight that despite being in limiting conditions, SK still felt needed or useful for others through the helps she offered/gave.	experienced was successfully achieving the main goal that TP had set, which was seeing her own reflection on a big mirror. This had increased TP's confidence in her own capability which led to subsequent increase in everyday activities
	The results of quantitative evaluation		
	based on the post-test questionnaire showed an increased final score, pointing to each SA, SK and TP's increased confidence on herself, which was related to the feeling of self-worth, self-capability, as well as apprecation on their overall being.		
	Pre-test score: 76 Post-test score: 88 (both are considered into low self-esteem) This is in line with SA's reflection in the end of the session that at that time, she was more confident and passionate in putting the plan for change that she had prepared into practice.	Pre-test score: 118 Post-test score: 132 (both are categorized into moderate self-esteem) This is basically in line with SK's reflection in the end of the session that by taking this counseling, SK felt motivated to bring back the positive sides of hers that she had before.	Pre-test score: 78 Post-test score: 103 (both are categorized into low self-esteem) This is basically in line with TP's reflection in the end of the session that at that time, she had slowly found her self-confidence back and started to think that she was helpful for the people around her.

	SA	SK	TP
The results of qualitative evaluation	showed that based on the entire interview and observation process as well as the administration of scalling questions in every session, SA gave a positive assessment on the development of her own condition as well as the benefits she gained in every session that she had gone through. **The assessment given ranged between 7 to 9 on a scale of 1–10 for every session.**	showed that based on the entire interview and observation process as well as the administration of scalling questions in every session, SK gave a positive assessment on the development of her own condition as well as the benefits she gained in every session that she had gone through. This is marked by the account about her shopping activities with her friends at department store (in session II), her interaction with neighbours at the security post near her house (in session III), and her attendance at her college friend's wedding party (in session III). Besides, SK also expressed her excitement on the results of the My Positive Qualities and My Effective Communication tasks. **The assessment given ranged between 7 to 9 on a scale of 1–10 for every session.**	showed that based on the entire interview and observation process as well as the administration of scalling questions in every session, TP gave a positive assessment on the development of her own condition as well as the benefits she gained in every session that she had gone through. This is marked with her enthusiasm and consistency in achieving small goals in seeing her reflection in the mirror that she had set in session I. **The assessment given ranged between 7 to 9 on a scale of 1–10 for every session.**
	SA mentioned some activities that were useful for her, namely: My Positive Qualities, My Effective Communication and Reconstructing My Solutions Surveys. Meanwhile, according to the results of the observations, a change was seen during the session that took place in her house, in which SA decided not to wear any head cover when interacting with the researcher. Besides, in session IV, SA gained confidence to wear a hat as a head cover (not a veil like she used to wear previously) when she took chemotherapy at RSK Dharmais. **Therefore, it can be concluded that in the early intervension, SA had low self-esteem, but this condition showed an improvement when the session ended.**	Meanwhile, according to the results of the observation, some changes were seen in her expression that turned to be brighter and how she made a positive insight on herself. **Therefore, it can be concluded that in the early intervension, SK had moderate self-esteem, but this condition showed an improvement when the session ended.**	Then, TP mentioned some activities that were useful for her and helped her to gain new insights, namely: My Positive Qualities, My Effective Communication, and Reconstructing My Solutions Surveys. Meanwhile, according to the results of the observations, it could be seen that TP eventually decided to get her hair completely shaved after managing to see herself in the mirror because with only the remaining hair, her appearance looked strange. After doing so, TP did not show any regret for her decision and considered her new look was better than before. Besides, TP's expression and opennes also turned to be more positive to the end of the intervension process. **Therefore, it can be concluded that in the early intervension, TP had moderate self-esteem, but this condition showed an improvement when the session ended.**

5 DISCUSSION

The intervention carried out with the employment of the SFBT approach, by and large, was proven to be able to improve the participants' self-esteem pertinent to the CIA that they were experiencing. The intervention facilitated the gain of different perspectives for the participants in viewing their problems and enabled them to change their negative feeling and appreciation in relation to their CIA. SFBT was selected as an intervention for the problems based on the consideration that this approach would mainly explore the positive qualities owned by the participants and would be able to bring back their sense of worth completely. This is similar to what De Shazer et al. (1986) explained: SFBT makes it easy for individuals to realise their strengths, optimise the use of available resources, the effectiveness of the coping used, and the picture of future goals and possibilities that they have over the problem-solving process.

Moreover, the positive process that was experienced by the researcher and participants was the development of a dynamic relationship, in which openness, sincerity, and thorough understanding of the participants' conditions served as the key modality in the therapeutic relationship. This was intended to create a counselling atmosphere similar to that described by Greenberg et al. (2001), in which the role of the researcher as a counsellor is not at a different level by emphasising all repertoires of knowledge owned, but at the same level as the participants as a mediator to help them develop their goals and solutions on their own. This condition enables the participants to feel comfortable to either share thoughts with the researcher or to design solutions independently while still realising the risks as well as responsibilities in every solution designed.

6 CONCLUSION

According to the results of the research and discussion, it can be concluded that the intervention that employs a SFBT approach was proven to be able to improve participants' self-esteem by facilitating changes of appreciation and perception to be more positive towards themselves and the problems that they were facing. Qualitatively, there was a change in the appreciation towards the participants' selves and behaviours in dealing with CIA, towards a more positive appreciation compared to that before receiving the intervention. Quantitatively, the results can be seen from the improvement of all participants' scores during post-test in comparison to their pre-test score.

7 SUGGESTIONS

The suggestions are as follows:

1. In order to build a deeper picture, it will be better if the characteristics of the participants are focused on one aspect that is more specific. For example, only aiming at participants with a particular type of cancer.
2. For further research, an intervention with a SFBT approach should be continued in the form of couple's therapy or family therapy as a whole by involving the people who act as the participant's significant others.
3. Future researchers are advised to keep visiting the participants periodically until the whole intervention process has come to an end.

REFERENCES

Agustin, I. (2012). *Terapi dengan pendekatan solution-focused pada individu yang mengalami quarterlife crisis* (Master's thesis, Universitas Indonesia, Depok, Indonesia).

American Cancer Society. (2015). *Understanding chemotherapy: A guide for patients and families.* Retrieved from http://www.cancer.org/acs/groups/cid/documents/webcontent/003025-pdf.pdf.
Ando, M., Morita, T. & Oshima, A. (2011). Reminiscence cognitive behavior therapy for spiritual well-being and self-esteem of cancer patients. *Journal of Cancer Therapy*, *2*(2), 105–109.
Bavelas, J., De Jong, P., Franklin, C., Froerer, A., Gingerich, W., Kim, H., ... Trepper, T.S. (2013). *Solution focused therapy treatment manual for working with individuals* (2nd version). New York, NY: Solution-Focused Brief Therapy Association.
Beatty, J.E. & Joffe, R. (2006). An overlooked dimension of diversity: The career effects of chronic illness. *Organizational Dynamics*, *35*(2), 182–195. doi:10.1016/j.orgdyn.2006.03.006.
Branden, N. (1992). *The power of self-esteem.* Deerfield Beach, FL: Health Communications.
Carr, A. (2006). *Family therapy: Concepts, process and practice* (2nd ed.). Chichester, UK: John Wiley & Sons.
Chassin, T. (2008). Treat the whole woman. *Healthy Aging*, *3*(5), 65.
Chon, S.Y., Champion, R.W., Geddes, E.R. & Rashid, R.M. (2012). Chemotherapy-induced alopecia. *Journal of American Academy of Dermatology*, *67*, 37–47. doi:10.1016/j.jaad.2011.02.026.
Curbow, B., Somerfield, M., Legro, M. & Sonnega, J. (1990). Self-concept and cancer in adults: Theoretical and methodological issues. *Social Science & Medicine*, *31*(2), 115–128.
De Shazer, S. & Dolan, Y. (2007). *More than miracles: The state of art of solution-focused brief therapy.* New York, NY: The Haworth Press.
De Shazer, S., Berg, I.K., Lipchik, E., Nunnally, E., Molnar, A., Gingerich, W. & Weiner-Davis, M. (1986). Brief therapy: Focused solution development. *Family Process*, *25*(2), 207–221. doi:10.1111/j.1545-5300.1986.00207.x.
D'Zurilla, T.J. (1986). *Problem-solving therapy: A social competence approach to clinical intervention.* New York, NY: Springer Publishing.
Ferrell, B.R., Grant, M., Funk, B., Green, S.O. & Garcia, N. (1997). Quality of life in breast cancer. Part I: Physical and social well-being. *Cancer Nursing*, *20*(6), 398–408.
Fleming, J.S. (2004). Erikson's Psychosocial Developmental Stages. Retrieved from: http://swppr.org/Textbook/Ch%209%20Erikson.pdf.
Gingerich, W.J. & Peterson, L.T. (2011). Effectiveness of solution-focused brief therapy: A systematic qualitative review of controlled outcome studies. *Research on Social Work Practice*, *23*(3), 266–283. doi:10.1177/1049731512470859.
Goldenberg, H. & Goldenberg, I. (2008). *Family therapy: An overview.* Belmont, CA: Brooks/Cole.
Greenberg, G., Ganshorn, K. & Danilkewich, A. (2001). Solution-focused therapy: Counselling model for busy family physicians. *Canadian Family Physician*, *47*, 2289–2295.
Heatherton, T.F. & Wyland, C.L. (2003). Assessing self-esteem. *Journal of Personality and Social Psychology*, *14*, 219–233.
Holmberg, S., Scott, L., Alexy, W. & Fife, B. (2001). Relationship issues of women with breast cancer. *Journal of Cancer Nursing*, *24*, 53–61.
Kementrian Kesehatan Republik Indonesia. (2012). *Penderita kanker diperkirakan menjadi beban utama ekonomi terus meningkat.* Retrieved from http://www.depkes.go.id/article/view/1937/penderita-kanker-diperkirakan-menjadi-penyebab-utama-beban-ekonomi-terus-meningkat.html.
LaTour, K. (1994). *The breast cancer companion.* New York, NY: Avon Books.
Leary, M.R. & Baumeister, R.F. (2000). The nature and function of self-esteem: Sociometer theory. In M.P. Zanna (Ed.), *Advances in experimental social psychology* (Vol. 32, pp. 1–62). San Diego, CA: Academic Press.
Lemieux, J., Maunsell, E. & Provencher, L. (2008). Chemotherapy-induced alopecia and effects on quality of life among women with breast cancer: A literature review. *Psycho-Oncology*, *17*(4), 317–328.
Luanpitpong, S. & Rojanasakul, Y. (2012). Chemotherapy-induced alopecia. In R. Mohan (Ed.), *Topics in cancer survivorship* (pp. 53–72). Rijeka, Croatia: Intech.
Lubis, N.L. & Othman, M.H. (2011). Dampak intervensi kelompok cognitive behavioral therapy dan kelompok dukungan sosial dan sikap menghargai diri sendiri pada kalangan penderita kanker payudara. *Makara Kesehatan*, *11*(2), 65–72.
Lubkin, I.M. & Larsen, P.D. (2002). *Chronic illness: Impact and interventions.* Boston, MA: Jones & Bartlett.
McGarvey, E.L., MaGuadalupe, L.V., Baum, L., Bloomfield, K., Brenin, D.R., Koopman, C., ... Parker, B.E. (2010). An evaluation of a computer-imaging program to prepare women for chemotherapy-related alopecia. *Psycho-Oncology*, *19*(7), 756–766.
Melia, E.K.D.A., Putrayasa, I.D.P & Azis, A. (2012). *Hubungan antara frekuensi kemoterapi dengan status fungsional pasien kanker yang menjalani kemoterapi di RSUP Sanglah Denpasar* (Thesis, Universitas Udayana, Denpasar, Indonesia). Retrieved from http://ojs.unud.ac.id/index.php/coping/article/viewFile/6123/4614.

Moorey, S. (2007). Breast cancer and body image. In M. Nasser, K. Baistow & J. Treasure (Eds.), *The female body in mind: The interface between the female body and mental health* (pp. 72–89). New York, NY: Routledge.

Mruk, C.J. (2006). *Self-esteem research, theory, and practice: Toward a positive psychology of self-esteem* (3rd ed.). New York, NY: Springer Publishing.

Münstedt, K., Manthey, N., Sachsse, S. & Vahrson, H. (1997). Changes in self-concept and body image during alopecia-induced cancer chemotherapy. *Support Care Cancer*, *5*(2), 139–143.

Murray, S.L., Griffin, D.W., Rose, P. & Bellavia, G.M. (2003). Calibrating the sociometer: The relational contingencies of self-esteem. *Journal of Personality and Social Psychology*, *85*, 63–84.

Nairn, R.C. (2004). *Improving coping with cancer utilizing mastery enhancement therapy* (Unpublished dissertation, University of Notre Dame, IN).

National Cancer Institute. (2009). *Managing chemotherapy side effect: Skin and nail changes*. Retrieved from https://www.cancer.gov/about-cancer/treatment/side-effects/

Neilson-Clayton, H. & Brownlee, K. (2002). Solution-focused brief therapy with cancer patients and their families. *Journal of Psychosocial Oncology*, *20*(1), 1–13.

Preston, M.M. (2010). *An exploration of appearance-related issues of breast cancer treatment on sense of self, self-esteem, and social functioning in women with breast cancer* (Doctoral dissertation, University of Pennsylvania, Philadelphia).

Robins, R.W. & Trzesniewski, K.H. (2005). Self-esteem development across the lifespan. *Current Direction in Psychology Science*, *14*(3), 158–162. doi:10.1111/j.0963-7214.2005.00353.x

Santrock, J.W. (2007). *Adolescence* (11th ed.). Boston, MA: McGraw-Hill.

Sarafino, E.P. & Smith, T.W. (2011). *Health psychology: Biopsychosocial interactions* (7th ed.). NJ: John Wiley & Sons.

Sebastián, J., Manos, D., Bueno, M.J. & Mateos, N. (2008). Body image and self-esteem in women with breast cancer participating in a psychosocial intervention program. *Psychology in Spain*, *12*(1), 13–25.

Seffrin, J.R., Wilson, J.L. & Black, B.L. (1991). Patient perceptions. *Cancer*, *67*(S6), 1783–1789.

Shaughnessy, J.J., Zechmeister, E.B. & Zechmeister, J.S. (2000). *Research methods in psychology* (5th ed.). Boston, MA: McGraw-Hill Higher Education.

Trzesniewski, K.H., Donnellan, M.B. & Robins, R.W. (2003). Stability of self-esteem across the lifespan. *Journal of Personality and Social Psychology*, *84*, 205–220. doi:10.1037/0022-3514.84.1.205.

Wagner, J., Lang, F.R., Neyer, F.J. & Wagner, G.G. (2013). Self-esteem across adulthood: The role of resources. *European Journal of Aging*, *11*(2), 109–119. doi10.1007/s10433-013-0299-z.

World Health Organization. (2012). *World health statistics 2012*. Retrieved from http://www.who.int/gho/publications/world_health_statistics/EN_WHS2012_Full.pdf.

Diversity in Unity: Perspectives from Psychology and Behavioral Sciences – Ariyanto et al. (Eds)

The contribution of parenting style and theory of mind to the understanding of morally relevant theory of mind in Indonesian children

I.A. Kuntoro, G. Dwiputri & P. Adams
Faculty of Psychology, Universitas Indonesia, Depok, Indonesia

ABSTRACT: This study aims to explore the contribution of parenting style and children's understanding of others' minds in morally relevant situations (morally relevant theory of mind) and moral judgment. To date, existing research has only focused on parenting and children's theory of mind, or children's understanding of theory of mind and moral standards. Therefore, this research aims to establish the connections among parenting, children's theory of mind, and moral at the same time. The understanding of Morally relevant Theory of Mind (MoToM) and children's moral judgment were assessed using MoToM tasks and the Prototypical Moral Transgression Scale developed by Killen et al. (2011), while parenting style was assessed using the Parenting Attitudes Inventory (PAI) of Vinden (2001) and O'Reilly and Peterson (2014). There were 122 participants involved in this study: children ranging from 4 to 6 years old and their parents. The results show that parenting style and children's theory of mind contribute towards the children's MoToM understanding by 35%. Theory of mind understanding also contributes towards children's moral judgment in an intentional moral transgression situation (prototypical transgression). However, no significant correlation between parenting style and children's moral judgment was found.

1 INTRODUCTION

Moral development and empathy are important for children aged from 4 to 6 years to deal with different social situations in school. By learning about moral standards and developing empathy, children become more able to co-operate with other people and follow the rules of society (Carpendale & Lewis, 2006).

The first and foremost place to develop the characters of children, so that they can understand morality and empathy, is within their family. Vinden (2001) found that there are two parenting styles that affect children's ability to understand the thoughts and feelings of others: autonomy—and conformity-oriented parenting styles. Conformity-oriented parenting focuses on rules and parental authority without considering the opinions and thoughts of the children. When children are exposed to situations that require moral judgments, they will only use parental rules as the basis of making judgment (Smetana, 2006). Furthermore, not having a place to exchange thoughts with their parents makes children unable to understand that other people may have different views from them. There are numerous evidences that have shown that children's Theory of Mind (ToM) development will be impeded when they are exposed to a conformity-oriented parenting style (Vinden, 2001; O'Reilly & Peterson, 2014).

An autonomy-oriented parenting style, in contrast, focuses on initiating discussions and negotiations between children and their parents, taking on the children's perspective. Hoffman (1994) and Ruffman et al. (1999) summarised that children whom their parents talk to about the consequences of their actions and reflect on how they and others feel about the actions, are able to have a better understanding of the minds of others, and are able to be more empathetic with others. The findings of O'Reilly and Peterson (2014) support this process and show that there is a positive relationship and association between autonomy-oriented parenting style and children's ToM. Autonomy-oriented parenting style provides space for parents to instil moral values to their children by communicating the consequences of their children's actions (Smetana, 2006). Thus, children will be able to have a better understanding on the consequences of their actions and the consequences of other people's actions. Moreover, children's ability to understand the moral judgment and thoughts of others is supported by their parents who facilitate the reflecting of their views and feelings.

Making a moral judgment is an ability to judge the good and the bad of an action, and it can only be carried out well when children have managed to understand the intention or the willingness of people who did it (Knobe, 2005). Children's ability to see the intention of and make judgment on others is gained through an understanding of one's belief in relation to other people's minds, serving as an input in the process of making moral judgments on the acts of others, and is referred to as morally relevant ToM (Killen et al., 2011). Morally relevant Theory of Mind (MoToM) is a perceiving process of the mental condition (e.g. emotion, belief, and desire) of a person in a morally relevant situation. In a study conducted by Killen et al. (2011), it was found that as of the age of 3.5 years, children who have MoToM are able to perceive the intention or mental conditions underlying people's actions in morally relevant situations. When children were using someone else's perspective in looking at an action, then they would be able to infer the moral judgment performed by others (Lane et al., 2010). Therefore, a child's ability to understand the underlying intention of others influences his or her judgment of other people's actions, which makes an unintended wrongdoing to be perceived as having a lesser degree of fault, and as a result, punishment will not be deemed necessary (Smetana et al., 2012).

To date, however, there has been no research conducted on this topic in Indonesian settings. Moreover, there is no other known study that has used Vinden's parenting measure in the context of children's ToM and moral development at the same time. Yet there is a positive trend that parenting style strongly links to the development of children's understanding of the mind and moral standards. At the same time, children's understanding of the mind will have a strong impact on children's moral and vice versa. To answer this, this study has aimed for two things. Firstly, to explore the contribution of parenting style and children's ToM to their MoToM. Secondly, to investigate the contribution of parenting style and children's ToM to the development of their moral judgment.

2 METHOD

2.1 *Participants*

This study involved 122 children, aged 4, 5, and 6 years, and their parents. The criteria of the children chosen in this study were: 1) they had no language problems/disorders as the participants must possess verbal ability to communicate with others; 2) they were willing to follow the research procedure fully; and 3) their parents were also willing to engage in the research.

2.2 *Instruments and materials*

There were four research instruments applied in this study. The first instrument was the Parenting Attitudes Inventory scale (PAI-scale) developed by Vinden (2001) and O'Reilly and Peterson (2014), which was applied to the assessment of parenting styles. The PAI-scale that was distributed to parents consisted of 12 Likert-like items (six autonomy-related items and six conformity-related items). The second was the ToM scale, which aimed to assess children's understanding of others' mental states. The ToM scale was developed by Wellman and Liu (2004) and has been adapted for Indonesian settings by Kuntoro et al. (2013). It consisted of five stories comprised of subscales: 1) diverse desires; 2) diverse beliefs; 3) knowledge access; 4) false beliefs; and 5) hidden emotions. The third was the MoToM scale that was used to perceive the understanding of a false belief about an event that is associated with moral situations and a child's ability to carry out moral judgments (Killen et al., 2011).

The fourth instrument was the scale of prototypical moral transgression developed by Killen et al. (2011), which consists of a short story about two characters (a moral transgressor and a victim) developed by Smetana (2006).

2.3 *Design*

There were two studies conducted to answer both objectives. The first study aimed to explore the contribution of parenting style and children's ToM to the children's MoToM. The second study was a follow-up study to investigate whether there is a positive link between parenting style as well as children's understanding of the mind and the children's ability to have a judgment on a morally relevant situation. In the second study, we further explored the children's ability to discriminate between an accidental and intentional act.

3 RESULTS

Study 1 aims to explore the influence of parenting style and children's ToM to the understanding of MoToM and children's moral judgment. Based on the analysis, we found that the score of autonomy-oriented parenting style ranges from 11 to 30 ($M = 23.68$, $SD = 3.145$), whereas the score of conformity-oriented parenting style ranges from 6 to 23 ($M = 12.79$, $SD = 3.275$). In terms of false belief ability, in the test of MoToM, there were only 30 children participants able to pass the test of false content ($n = 5$, $n = 9$, $n = 16$) and only 21 children able to pass the test of false location ($n = 1$, $n = 4$, $n = 16$). In order to see the contribution of their parents' parenting style and the understanding of ToM towards false belief in the children's MoToM understanding, the researchers used multiple regression analysis using participants' ages as a covariance.

Table 1 indicates that age, ToM understanding, and conformity-oriented parenting style make a significant contribution to the children's MoToM. Furthermore, we found that ToM contributes significantly to the understanding of MoToM, which indicated that the better the development of ToM, the better MoToM understanding the children have. On the other hand, conformity-oriented parenting style significantly inhibits MoToM. The analysis shows that this latter parenting style contributes significantly high (35%) towards the children's ToM ($F = 17.386$, $p < 0.01$).

Study 2 aims to further investigate the effect of parenting style and ToM towards the moral judgment of children. A moral judgment about an incident that happened by accident (accidental transgression) was measured using a scale of MoToM. The results of moral judgment of children aged from 4 to 6 years old on an accidental transgression incident can be seen in Table 2. When evaluating the accidental moral breaking incident, almost all participants considered it wrong because it causes a loss on others' property. Given the justification, it was concluded that accidents materialised in a form of recklessness should not be done by the transgressor because he/she does not check if there is something inside the plastic bag that he/she tries to throw away.

Table 1. Contribution of parenting style and ToM to MoToM.

Dependent variable/predictor	ΔR^2	F	β	Partial correlation
False Beliefs MoToM				
Step 1:				
Age	0.149**	22.239	0.395**	0.395
Step 2				
Conformity	0.351**	17.386	0.249**	–0.290
Autonomy			–0.100	–0.121
ToM			0.394**	0.427

Note: **p 0.01

Table 2. Mean and deviation standard of children's moral judgment on accidental transgression incident.

	4		5		6	
Age	M	SD	M	SD	M	SD
Evaluation of intention	2.23	1.050	1.98	0.987	2.10	1.091
Evaluation of behaviour	2.15	1.027	1.61	0.862	1.80	1.030

Table 3. Contributions of parenting style and ToM to MoToM and moral judgment.

Dependent variable/predictor	ΔR^2	F	β	Partial correlation
Child's evaluation of intention in *Prototypical Transgression*				
Step 1				
Age	0.007	0.141	–0.34	–0.034
Step 2				
Conformity	0.068*	3.204	–0.120	–0.121
Autonomy			–0.099	–0.100
ToM			–0.303**	–0.290
MoToM			0.000	–0.088
Child's evaluation of behaviour in *Prototypical Transgression*				
Step 1				
Age	0.065**	8.063**	–0.251**	–0.251
Step 2				
Conformity	0.075*	3.441	–0.049	–0.050
Autonomy			0.053	0.057
ToM			–0.207*	–0.203
MoToM			–0.100	–0.145
Child's evaluation of intention in *Accidental Transgression*				
Step 1				
Age	0.006	0.295	–0.050	
Step 2				
Conformity	0.020	0.137	0.046	0.042
Autonomy			0.018	0.007
ToM			–0.011	–0.011
MoToM			–0.156	–0.125
Child's evaluation of behaviour in *Accidental Transgression*				
Step 1				
Age	0.012	2.431	–0.141	
Step 2				
Conformity	0.022	1.675	–0.071	–0.141
Autonomy			0.132	0.070
ToM			–0.113	0.131
MoToM			0.017	–0.103

*$p < 0.05$; **$p < 0.01$

Moral judgment on the prototypical situation was measured using a scale of prototypical moral transgression. The evaluation performed on the children towards the transgressor's intention and the transgression act itself can be seen in Table 2, which shows that the intention of the prototypical moral transgressor and the incident itself are seen as incorrect and something that cannot be done.

A hierarchical regression analysis was subsequently conducted in order to see the contribution of parenting style and understanding of ToM towards moral judgment on the prototypical and accidental incident. In this analysis, the moral judgment from the children in the prototypical and accidental incident is a dependent variable and thus was analysed separately. For each analysis model conducted, the researchers initially included age as a covariance, followed by parenting style and ToM.

Based on the results of multiple hierarchical regression analyses (presented in Table 3), there are several important things to note. The first is the analysis that predicts moral judgment towards the transgressor's intention in the prototypical transgression. It was found that parenting style and ToM have a significant contribution ($F = 3.204$, $p < 0.01$) towards the moral judgment of the children on the transgressor's intention in the prototypical accident. The analyses summarise that only the ToM understanding has a significant contribution ($\beta = -0.303$, $p = 0.01$) to moral judgment towards the transgressor's intention in the prototypical transgression. Simultaneously, all of the variables contributed as much as 7% to the children's ability to carry out moral judgment on the transgressor's intention in the prototypical transgression. Table 3 also reveals that parenting style and ToM have a significant contribution ($F = 3.441$, $p < 0.05$) to moral judgment, in which the children's ToM alone also contributed significantly ($\beta = -0207$, $p = 0:27$). The entire set of variables contributed as much as 13% to the children's moral judgment in the prototypical transgression incident.

4 DISCUSSION AND CONCLUSIONS

The development of MoToM of children aged from 4 to 6 years old was observed by conducting an analysis of the false belief components on a scale of MoToM. In the analysis of the false belief components, it was found that only a few participants could successfully complete the test of MoToM false belief. The result also indicates that the MoToM false belief test with false content is easier to be answered properly by the children than the MoToM false belief test with location change. In addition, it was found that the older the children are, the better they are at passing the test of MoToM false content and false location. This result is in accordance with the development of MoToM false belief in the United States and China, showing that MoToM false belief ability of children will improve as the children get older (Killen et al., 2011; Fu et al., 2014). Judging from the fact that only few participants could answer the MoToM false belief test correctly, it is safe to conclude that at the age of 4–6 years, children's MoToM understanding is not yet well developed.

In terms of moral judgment variables related to the accidental transgression incident, it was found that the children would evaluate the transgressor's intention and then consider it wrong. More specifically, this study finds that the group of five-year-old children considers the accidental moral breaking incident as being more wrong than in the other groups. This finding shows that at the age of five, children still consider that accidental incidents yield a loss, even though at that age, children are already able to perceive the intention of the transgressor. Based on the children's evaluation, it can be concluded that the participants of this study still consider the act of the transgressor a careless action that will result in losses for others, which is in contrast with the findings from the research conducted in the US and mainland China (Killen et al., 2011; Fu et al., 2014). The aforementioned studies found that as children get older, they can see the intention of an accidental transgressor in a more positive light, although most other children perceive them more negatively. Those studies assume that children understand that the transgressor does not have a bad intention which will result in losses for others.

In relation to moral judgment variables, it was found that all children evaluate the intention of the transgressor and judge the prototypical transgression incident as a wrongdoing. The children regard that the moral transgressor has a bad intention to make his/her friend fall, and to cause injury to others. As children grow older, they will consider the prototypical transgression incident a wrongdoing. This finding shows a similar result to the previous research conducted in the US and China (Killen et al., 2011; Fu et al., 2014).

Judging from the given contribution of parenting style and children's ToM to the MoToM of children, it can be concluded that: 1) conformity-oriented parenting style contributes negatively to children's MoToM comprehension. Therefore, a parenting style that emphasises on conformity will significantly impede the development of MoToM of the children; 2) understanding ToM will contribute towards the understanding of MoToM of the children. A good understanding of ToM will enhance the development of MoToM understanding; and 3) 35% of the variance of MoToM is attributed to parenting style. In addition, the better the children's ToM, the better the understanding of MoToM they have. This is in line with the results of previous studies conducted by Killen et al. (2011).

Another illuminating result of this study is that a conformity-oriented parenting style contributes negatively to the understanding of MoToM, while an autonomy-oriented parenting style does not have a significant influence. These findings are in line with the results found in the Anglo-American families but are different from the Korean and Iranian families who do not find the influence of parenting style on the understanding of the mind (Vinden, 2001; Shahaeian et al., 2014). In her research, Vinden (2001) found that, in general, Anglo-American parents have an autonomy-oriented parenting style. However, based on the analysis of the parenting score of Anglo-Americans with false belief ability, only conformity-oriented parenting style was found to have a significant correlation to the children's false belief. Thus, parents who have a strong control over their children without considering the autonomy of the children will inhibit the development of the children's MoToM understanding. On the other hand, autonomy-oriented parenting style has no significant positive influence. The major reason why autonomy-oriented parenting style does not influence the understanding of MoToM is due to the fact that the particular parenting style is contextually too broad. These is in contrast to the previous studies that focus more on seeing the disciplinary patterns applied by the mother and thus successfully prove the effect of the practice of discipline on her child's ability to understand the minds of others (Shahaeian et al., 2014). In addition, these results also indicate that there are other factors at play in developing the children's MoToM ability, such as the educational and work backgrounds of their mothers (Vinden, 2001).

In terms of the contribution of parenting styles and children's ToM to moral judgment, it was concluded that ToM contributes as much as 7% to children's ability to evaluate the intention of others in the prototypical situation and 13% to their ability to evaluate the incident of prototypical moral transgression. As they get older, the understanding of ToM and children's MoToM will mature. Subsequently, children will consider the prototypical transgression incident wrong and they can see that the moral prototypical transgressor has no good intention.

Meanwhile, children's ToM understanding does not have significant contribution towards the children's moral judgment on the accidental incident. This finding indicates that although children are already able to see the intention of other people behind the act, they still consider the intention of the transgressor wrong. This is reflected in the justification of the children who judge that an accident will still be considered wrong if it is caused by careless behaviour that can harm others. It is different from the studies in the US and China where children see accidental incident because of carelessness as not having bad intention behind it (Killen et al., 2011; Fu et al., 2014).

Based on the discussions and conclusions outlined above, two suggestions can be made for future research related to the connections among parenting style, ToM, MoToM, and children's moral judgment. First, in the future, researchers are expected to design a parenting style scale containing moral values to see the connection between parenting style and children's moral judgment. Second, it is essential to see whether there are other factors related to parenting, besides just parenting styles, which contribute more to the improvement of children's understanding of the mind.

REFERENCES

Carpendale, J. (1997). An explication of Piaget's constructivism: Implications for social cognitive development. In S. Hala (Ed.), *The development of social cognition* (pp. 35–64). Hove, UK: Psychology Press.

Carpendale, J. & Lewis, C. (2006). *How children develop social understanding. Understanding children's worlds*. Hoboken, NJ: John Wiley & Sons.
Fu, G., Xiao, W.S., Killen, M. & Lee, K. (2014). Moral judgment and its relation to second-order theory of mind. *Developmental Psychology*, *50*, 1–26.
Hoffman, M.L. (1994). Discipline and internalization. *Developmental Psychology*, *30*(1), 26–28.
Killen, M., Mulvey, K.L., Richardson, C., Jampol, N. & Woodward, A. (2011). The accidental transgressor: Morally relevant theory of mind. *Cognition*, *119*(2), 197–215.
Knobe, J. (2005). Theory of mind and moral cognition: Exploring the connections. *Trends in Cognitive Sciences*, *9*(8), 357–359.
Kuntoro, I.A., Saraswati, L., Peterson, C. & Slaughter, V. (2013). Micro-cultural influences on theory of mind development: A comparative study of middle-class and pemulung children in Jakarta, Indonesia. *International Journal of Behavioral Development*, *37*(3), 266–273.
Lane, J.D., Wellman, H.M., Olson, S.L., LaBounty, J. & Kerr, D.C. (2010). Theory of mind and emotion understanding predict moral development in early childhood. *British Journal of Developmental Psychology*, *28*(4), 871–889.
O'Reilly, J. & Peterson, C.C. (2014). Theory of mind at home: Linking authoritative and authoritarian parenting styles to children's social understanding. *Early Child Development and Care*, *184*(12), 1934–1947.
Ruffman, T., Perner, J. & Parkin, L. (1999). How parenting style affects false belief understanding. *Social Development*, *8*(3), 395–411.
Shahaeian, A., Nielsen, M., Peterson, C.C. & Slaughter, V. (2014). Iranian mothers' disciplinary strategies and theory of mind in children: A focus on belief understanding. *Journal of Cross-Cultural Psychology*, *45*(7), 1110–1123.
Smetana, J.G. (2006). Social-cognitive domain theory: Consistencies and variation in children's moral and social judgments. In M. Killen & J.G. Smetana (Eds.), *Handbook of moral development* (pp. 119–154). Mahwah, NJ: Lawrence Erlbaum Associates.
Smetana, J.G., Jambon, M., Conry-Murray, C. & Sturge-Apple, M.L. (2012). Reciprocal associations between young children's developing moral judgments and theory of mind. *Developmental Psychology*, *48*(4), 1144–1155.
Vinden, P.G. (2001). Parenting attitudes and children's understanding of mind: A comparison of Korean American and Anglo-American families. *Cognitive Development*, *16*(3), 793–809.
Wellman, H.M. & Liu, D. (2004). Scaling of theory-of-mind tasks. *Child Development*, *75*(2), 523–541.

Diversity in Unity: Perspectives from Psychology and Behavioral Sciences – Ariyanto et al. (Eds)
 ISBN 978-1-138-62665-2

The development of mental time travel in Indonesian children

I.A. Kuntoro & E. Risnawati
Faculty of Psychology, Universitas Indonesia, Depok, Indonesia

E. Collier-Baker
School of Psychology, University of Queensland, Brisbane, Australia

ABSTRACT: The link between episodic foresight and episodic memory, known as Mental Time Travel (MTT), has captured scientific imagination in the last decade, with Science listing it as one of the top ten theoretical breakthroughs of 2007. Establishing the timing of MTT could also confer substantial benefits for advancing early childhood education. Surprisingly, until very recently, there has been limited investigation into how MTT develops along childhood time. Moreover, nothing is known about the development of MTT in children outside Western, industrialised cultures. Using a behavioural task administered to Australian children, this study aims to fill this significant gap and provide the data on MTT in young children outside Western countries. We investigated the specific period when MTT first emerged in children from urban areas (Jakarta) and rural areas (Yogyakarta and Riau). In each sample group, we compared the performance of 3- 4- and 5-year-old children. Our analyses indicate that, like Australian children, MTT emerged at the age of four in Indonesian children.

1 INTRODUCTION

Humans are capable of mentally projecting themselves into the past and the future. Planning your next project, reminiscing about last year's travels, or picturing your future retirement years, all involve what Suddendorf and Corballis (1997) have called Mental Time Travel (MTT). We engage in these mental projections into the past and future with little effort and, in fact, very often involuntarily. Buckner and Carroll (2007) suggested that MTT could even be the mind's default 'idle' setting. Certainly, it can prove a considerable challenge to disengage from MTT. Our social, cultural and technological exploits depend heavily on MTT. An ability to foresee possible future scenarios undoubtedly confers a powerful advantage. However, since how this ability develops still largely remains undiscovered, we would like to further investigate it. Episodic memory—the faculty that enables us to mentally relive our personal past experiences—has been the focus of research in decades (Tulving, 1984, 2012). Yet, surprisingly, questions about our ability related to mentally 'prelive' personal futures did not capture scientific imagination until very recently.

Research areas as diverse as clinical psychology, social psychology and neuroscience are now turning their attention to the mechanisms of this profound capacity of foresight. In the last few years, we have witnessed a rapid increase in multidisciplinary interest and debate as the nature of MTT and the means of testing it. The notion that mental travels into the past and future shared neuro-cognitive resources has been persuasively argued. New evidence for the shared bases of episodic memory and foresight has begun to pour in with prominent papers highlighting links in the mind (Suddendorf et al., 2009; Spreng & Levine, 2006) and in the brain (Addis et al., 2007; Lavallee & Persinger, 2010). Indeed, *Science* magazine (2007) declared this evidence as one of the top ten scientific breakthroughs of 2007.

However, despite the fact that MTT originated in children's cognitive development theory (Suddendorf & Corballis, 1997), empirical investigation into the timing and mechanism of MTT in children has been remarkably slow to emerge. This is very surprising considering the potential educational applications for understanding how and when planning develops in children (Meltzer, 2007). Most of the studies that have been conducted on MTT in children have focused on their ability to produce or comprehend verbal references to the past and future instead of finding the emergence of this ability. For example, the study conducted by Atance and Meltzoff (2005) asked 3–4- and 5-year-olds to select one of three items (e.g. a winter coat) they would choose if they were going to walk across a landscape depicted in a photograph (e.g. a wintery scene). Older children selected the appropriate item for the imagined future scenario more often and gave more explanations for their choices with reference to future conditions than younger children.

Notably, very few paradigms have used non-verbal dependent measures. Some data claimed that the evidence for MTT may be explained by associative learning strategies (see Suddendorf et al., 2009) or suffer interpretative problems. For example, do children project themselves into the future when selecting the winter coat in Atance and Meltzoff's (2005) study or have they already formed an association between a coat and the snowy scene in the picture? Do children who fail to see a future situation show less verbal generativity *per se*? These issues limit the conclusions we can draw from some of the previous studies on future planning. Tasks that rely only on verbal dependent measures may underestimate the cognitive skills of younger children since their language skills are still developing at the age of 3 and 4 (Fenson et al., 1994; Suddendorf & Redshaw, 2013).

Only one study by Suddendorf et al. (2011) has directly assessed children's capacity to connect past and future events to solve a problem and used a paradigm that is not wholly dependent on verbal ability. In this 'two-room' test, 3- and 4-year-old children from Australia were given two simple tasks, one involving a physical problem and one a social problem. These tasks were counterbalanced so that each child received one in a delay condition and one in an instant condition. Solving the delay condition required the children to remember a problem from 15 minutes earlier and anticipate returning to that problem in the near future. The instant version was identical except there was no delay period and this served as a control to check that the children understood the basic task requirements. For the physical task, the children first had experience in Room 1 of fitting a shaped key (e.g. square) into a same-shaped lock to open a box and retrieve a reward. A new box was then presented, this box had a different lock (e.g. triangle-shaped) and could not be opened with the square key. In the delay condition, the children were taken to Room 2 to play unrelated activities for 15 minutes before an array of keys (including the target triangle key) was revealed. The children could select one key to be taken with them back to the puzzle in Room 1. In the instant condition, the children made their selection immediately on the other side of Room 1. This task was counterbalanced with a social task using an identical structure but with more verbal components which involved feeding a puppet from a selection of fruit. Each child received one trial of the delayed version of one task and one trial of the instant version so that there were no carry-over effects.

The results revealed that 4- but not 3-year-old children are able to secure a novel solution to a task they have encountered in a different location fifteen minutes earlier. Successful children could also give a verbal explanation for their choice. Learning could be ruled out as an explanation for success because the clever task design means that the solution could only be inferred and is never actually experienced. The successful performance of 3-year-olds in an instant condition shows that they are capable of the 'means-ends' reasoning requirements of the task, implicating the temporal delay as the reason for their failure. This study suggests that—in Australian children at least—the ability to consider the future and remember a past event over this short time delay period does not emerge until the age of four. Our aim in the current study is to extend this behavioural test to children in Indonesia and investigate the timing of emergence of this ability in samples from populations never before tested on MTT.

Henrich et al. (2010) recently urged renewed attention to a pervasive problem in the behavioural sciences of making claims about human psychology based on highly selective, and likely unrepresentative, samples of the population: namely, Western, educated, industrialised, rich, and democratic (W.E.I.R.D.) societies. Assumptions of universality in the development of some core cognitive capacities have been contested by cross-cultural work, for example, for theory of mind (Liu et al., 2008) and spatial cognition (Haun et al., 2006), who reveal the influence of culturally specific, as well as universal, forces on the timing and trajectory of such skills. Uncovering the influence of universal and cultural forces can help us unpack the mechanisms underlying particular cognitive skills.

One recent study directly compared MTT in European, American and Chinese adults and found specificity differences in their descriptions of future events (Wang et al., 2011), adding to previous findings of cultural variation in episodic memory (Han et al., 1998), and in conceptions of time itself (Boroditsky, 2001). It is possible that childhood emergence of MTT is universal in terms of timing. However, considering these cross-cultural differences reported for MTT in adults as well as differences in the developmental trajectory of other cognitive capacities in childhood, such as theory of mind (Kuntoro et al., 2013), emergence of MTT cannot be assumed to be universal. Children now need to be tested on MTT tasks beyond Western cultures. In general, there is limited published data on child cognitive development in Indonesian children (Kuntoro et al., 2013); therefore, this study contributes entirely towards novel data.

The MTT tests of Suddendorf et al. (2011) are readily adaptable to young children across different cultural groups by including behavioural as well as verbal measures. In order to claim with any confidence that the emergence of planning is independent of cultural practices, and is instead a universal capacity, this study included Indonesian children from three groups: (a) urban children living in the south Jakarta metropolitan area, (b) children living in rural areas within 20 km of Yogyakarta, and (c) children living in rural areas in Riau (Sumatera). Within each group, we tested 3-, 4- and 5-year-old children in order to capture the emergence of MTT across this critical period around four years of age, at which Australian children have already been shown to pass the current task (Suddendorf et al., 2011).

2 METHOD

2.1 *Participants*

A sample of 212 Indonesian middle class children (106 boys, 106 girls) from three provinces (DKI, DIY, and Riau) were recruited and tested in their respected playgroups and kindergartens. Each group was divided into three age groups; 3 years-old, 4 years-old, and 5 years-old children. See Table 1 for detailed description of participants' age and gender.

2.2 *Apparatus*

This study used two tasks (i.e. the *Box* and *Food task*) developed by Suddendorf et al. (2011) with some adjustments—for the Food task—based on a pilot study. The *Box* task involved two wooden boxes, each with a large keyhole on the left side of the front panel; one was square, and the other was triangle. Pushing the corresponding square or triangle key through the keyhole rotated a hidden platform within to reveal a reward (brightly coloured sticker) accessible from the right side of the box. Seven keys were used in the task and consisted of wooden dowels with a flat wooden shape securely attached at one end. The target keys were a blue triangle and an orange square and fit the respective triangle and square key holes only. The remaining five keys served as distractors with shapes that could not fit into either keyhole. Materials for the Food task consisted of two commercially

Table 1. Participants' age and gender.

		N	%
Age	3 (36–47 months)	68	32
	4 (48–59 months)	72	34
	5 (60–71 months)	72	34
Gender	Boys	106	50
	Girls	106	50

available hand puppets (frog and elephant) and seven plastic fruits, which varied slightly in each group depending on familiarity with fruits (Jakarta: grapes, strawberry, banana, apple, orange, pear, watermelon; Yogyakarta: grapes, strawberry, banana, apple, orange, guava, mangosteen). The target fruits for both Jakarta and Yogyakarta were grapes and strawberry. The fruits were placed in a randomised order on fixed equidistant positions approximately 4 cm apart in a horizontal linear array. Mobile phone timers or stopwatches were used to record the 15-minute delay period. Sony handycams on tripods were used to record children's selections on the *Box* and *Food tasks*. The boxes and target keys used in this task are presented in Figure 1.

2.3 *Procedure*

Demographic information about the children and their schools was gathered via interviews with schoolteachers. Participants were tested individually in schoolrooms in one 30-minute testing session by Indonesian experimenters. All children were fluent in Indonesian; therefore, they were tested in Indonesian. Written consent to participate in the study was obtained from the parent/village head prior to testing. Parents/guardians were informed that they may withdraw their child without prejudice at any time during the study and that all data would remain confidential. Token rewards were given to the children for participating and books were donated to each participating school. All sessions were videotaped and coded by experimenters and 20% of trials were also scored by independent coders blind to the research hypotheses to assess inter-rater reliability.

Before administering tasks, the experimenter engaged the child in free play with toys and colouring activities to ensure that he/she was relaxed and comfortable with the tester. Modifying the design of Suddendorf et al. (2011), each subject was then presented with one trial on two simple tasks: a physical *Box task* and a social *Food task,* one of which could be solved instantly and the other after a delay of 15 minutes. The temporal condition and order were counterbalanced so that there were four trial types: *Box* delay/*Food* instant, *Food* delay/*Box* instant, *Box* instant/*Food* delay and *Food* instant/*Box* delay. In all conditions, each child was first led into a room identified as the 'flower room' due to a flower poster on the wall. See Figure 2 for the research procedure.

In the *Box task*, the experimenter demonstrated opening a box with the square keyhole using the square key revealing the sticker hidden inside. The participant was then encouraged

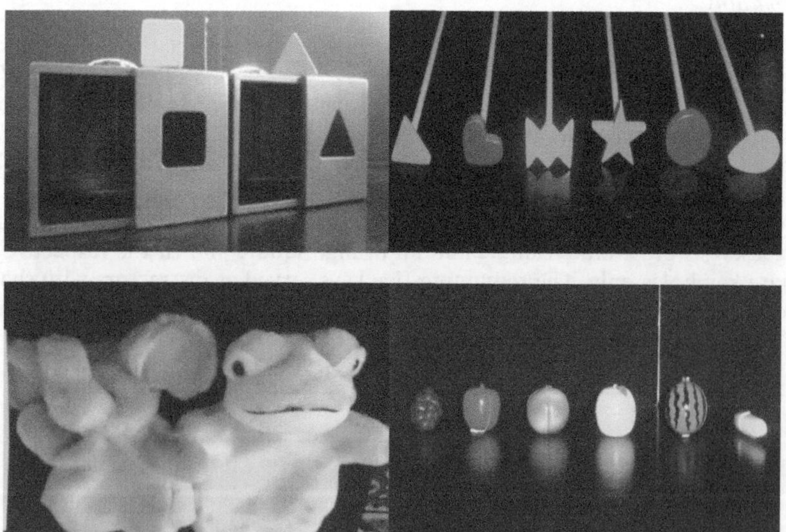

Figure 1. Apparatus used in the box (above) and food tasks (below).

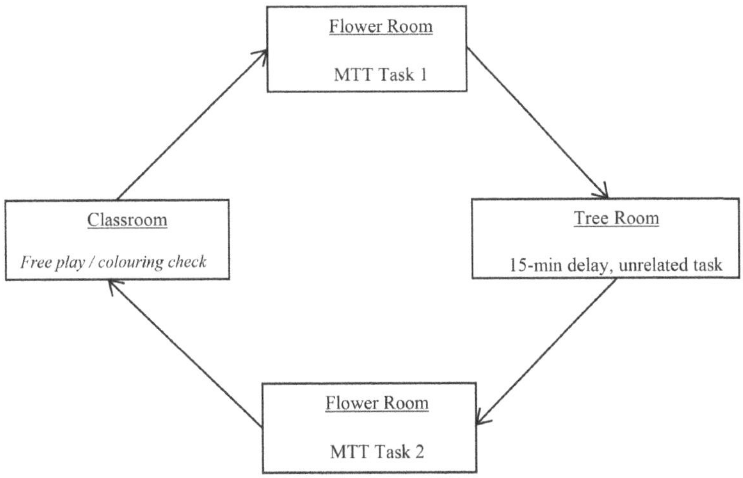

Figure 2. Experiment procedure.

to open the box for themselves to retrieve the reward inside. The triangle box was then presented and the experimenter showed the child that the square key does not work and then allowed the child to try it too. Triangle and square keys and order of box presentation were counterbalanced.

In the delay condition, the children were led into the adjacent 'tree room', identified by a tree poster on the wall and engaged in an unrelated task. After 15 minutes (recorded by timer), the experimenter declared that they would now return to the flower room but first presented the child with an array consisting of the target key and five distracter keys (randomising target position across participants). At no point did the experimenter refer to the puzzle in the other room. Following Suddendorf et al. (2011), the experimenter also obtained verbal measures of future referral by asking the child, 'Why did you choose that one?' and 'what is in the flower room?'. After leading the child back to the flower room, the experimenter allowed the child to use the key on the box. If the children chose the incorrect key, the experimenter simply turned the key around which enabled the children to use it to open the box. This meant that the children received the sticker reward whether they chose the correct key or not in order to minimise carry-over effects when the *Box delay task* was presented first. In the instant versions of either task (counterbalanced with the delay condition) the children were led to a hidden array of keys/fruits on the other side of the flower room. With their back to the box or puppet, the children were asked to choose one to take back to the table. The children were engaged in colouring activities at the end of the session in order to minimise discussion of the MTT task details with their classmates after the session.

The *Food task* had the same structure but involved two puppets and plastic fruit. In the flower room, the experimenter first revealed a plastic strawberry (or grapes) as these were counterbalanced. The experimenter encouraged the child to 'feed' Koko and then put him under the table saying he was now full and needed a sleep, replacing him with the other puppet, 'Gaga the Gadjah' (elephant). The experimenter told the child that Gaga liked grapes (or other target fruit) but pointed out that they did not have that fruit. In the delay condition, the experimenter then led the child into the tree room for 15 minutes following the same procedure and activities described above before asking them to select a fruit from a revealed array of six fruits including the target (thus probability of success by chance was 0.167 in both *Food* and *Box tasks*) to take back to the flower room and asking them the two questions above. Back in the flower room, if the children had chosen the incorrect fruit, the experimenter told them that it was Gaga's second favourite food and let the children feed Gaga with it in order to minimise carry-over effects when the Food task was presented first.

3 RESULTS

3.1 *The onset of MTT*

Since preliminary analyses of performance from boys and girls shows that there were no significant differences for MTT ($R^2 = 0.016$, $p > 0.01$), genders were combined for all the remaining analyses. The analysis summarises the differences in mental time ability among children at the age of three, four, and five. In the delay condition, only 19% of 3 year-old children were able to choose a target and able to give appropriate reasons. This percentage increases up to 29% and 38% in children at the age of 4 and 5 respectively (see Figure 3). However, the increase in percentage is not significant ($p > 0.01$) because of the influence of the age range of participants (see Table 1).

If we compare the performance of the children in the delay and instant conditions, it can be seen that there are significant differences (see Figure 2). Figure 2 shows that the children's success rate was quite high (41–54%) in the instant condition, demonstrating comprehension of the tasks in general. Further, in children at the age of 3, 4 and 5, the percentage differences in the instant and delay conditions were 22%, 21% and 16% respectively. The percentage of success rate in the delay condition increased with the children's age (3 years = 19%; 4 years = 29%; 5 years = 38%). This would indicate that the ability to remember, plan, predict and anticipate future events will develop as the children get older. Whereas, in the instant condition, there were no significant differences between the ages of 3 and 4 (9%), or between the ages of 4 and 5 (4%). This showed that in a short period of time between the problem explanation and issue resolution, there were no significant differences in all three age groups. Nevertheless, the lower score of the 3 year-old group compared to the groups of 4- and 5-year-olds is an indication that the MTT ability of children at the ages of 4 and 5 has developed better than children at the age of three. Therefore, it can be concluded that in Indonesian children MTT appeared at the age of four and develops more as children get older, which is observable from the increased number of 5-year-old participants who were able to complete the MTT task precisely.

Although the onset of MTT in Indonesian participants is no different from children in Australia (Suddendorf et al., 2011), there are subtle differences in the performance of children in the *box* task versus the *food* task. The children in this study generally performed better in the *box* task than the *food* task, both in the instant and delay conditions. In the group of 3 year-olds, in the instant condition, the percentage of children who chose targets correctly in the *box* task was 35% and only 10% for *food* task. Whereas, in the delay condition, the choice of correct targets in the task box was higher (15%) than in the puppet task

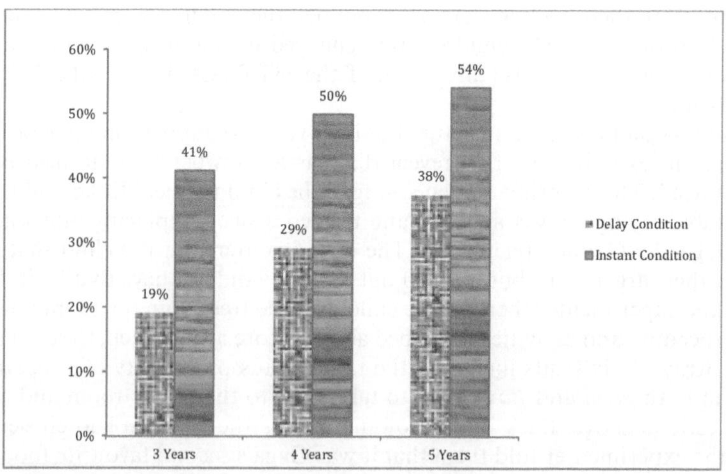

Figure 3. The onset of MTT.

(5%). Furthermore, in the group of 4 year-olds, the percentage of success in the instant box *task* was 63% with the instant *food* task at 17%; while the delay *box* task was 33% and the delay *food* task was 25%. At the age of 5, the percentage of success in the instant *box* task was 75% and the instant *food* task was 17%, while the delay *box* task was 46%, and delay *food* task was 29%. It is speculated that the difference in performance was related to reward. In the *box* task, the children received a sticker as a reward when they successfully opened the *box* with the correct key, whereas in the *food* task, they were not given a reward even if they managed to give the fruit favoured by puppet. It is suspected that rewards were related to the performance of children in both the *box* and *food* tasks. As Cameron and Pierce (1994) suggested, reward increases motivation. The result revealed in this study should be considered for future studies, and rewards should be given for both *box* and *food* tasks.

Another possibility for different performances between the *box* and *food* tasks is, in the *food* task, when the children were asked the reason for choosing a particular fruit they answered with their personal preferences. Supposedly, children should put themselves in puppet situation and feed the puppet with fruit according to puppet's favour. The inability of 5-year-old children to put themselves in a puppet-fruit simulation could be assumed to be a lack of inhibitory control, so they were unable to place themselves in a puppet-fruit situation (Nielsen & Graham, 2009). Another explanation proposed is familiarity differences between the *box* task and *food* task. In the *box* task, when the experimenter showed the key selection, the children immediately took the key and associated it with the hole in the box. This was probably due to the fact that the box task was more novel, and the children had no association with their previous experiences, whereas, in the *food* task, the possibility of choosing fruit which was not the puppet's favourite was bigger. The fruits that were shown and chosen by the children could be associated with other things outside the task and closely related with the children's experience with the fruit. At the end of the research result, it was finally established that the main reason they chose a particular fruit was related to the children's favourite fruit.

4 CONCLUSION

The study aimed to examine the onset of planning behaviour in Indonesian children through a carefully controlled behaviour experiment and comparing children's performance in verbal and non-verbal tasks. This study suggested that like Australian children, planning behaviour emerges at the age of four in Indonesian children. Four-year-old children performed significantly better in the delay task than 3-year-old children. Analysis also indicated that the children participating in this study performed better in non-verbal tasks than verbal tasks in contrast to Australian children.

REFERENCES

Addis, D.R., Wong, A.T. & Schacter, D.L. (2007). Remembering the past and imagining the future: common and distinct neural substrates during event construction and elaboration. *Neuropsychologia, 45*(7), 1363–1377.

Atance, C.M. & Meltzoff, A.N. (2005). My future self: Young children's ability to anticipate and explain future states. *Cognitive Development, 20*(3), 341–361.

Boroditsky, L. (2001). Does language shape thought?: Mandarin and English speakers' conceptions of time. *Cognitive Psychology, 43*(1), 1–22.

Buckner, R.L. & Carroll, D.C. (2007). Self-projection and the brain. *Trends in Cognitive Sciences, 11*(2), 49–57.

Cameron, J. & Pierce, W.D. (1994). Reinforcement, reward, and intrinsic motivation: a meta-analysis. *Review of Educational Research, 64*(3), 363–423.

Fenson, L., Dale, P.S., Reznick, J.S., Bates, E., Thal, D.J., Pethick, S.J., Tomasello, M., Mervis, C.B. & Stiles, J. (1994). Variability in early communicative development. *Monographs of the Society for Research in Child Development, 59*(5), 1–185.

Han, J.J., Leichtman, M.D. & Wang, Q. (1998). Autobiographical memory in Korean, Chinese, and American children. *Developmental Psychology, 34*(4), 701–713.
Henrich, J., Heine, S.J. & Norenzayan, A. (2010). The weirdest people in the world? *The Behavioral and Brain Sciences, 33*(2–3), 61.
Kuntoro, I.A., Saraswati, L., Peterson, C. & Slaughter, V. (2013) Micro-cultural influences on theory of mind development: A comparative study of middle-class and pemulung children in Jakarta, Indonesia. *International Journal of Behavioral Development, 37*(3), 266–273.
Lavallee, C.F. & Persinger, M.A. (2010). A loreta study of mental time travel: Similar and distinct electrophysiological correlates of re-experiencing past events and pre-experiencing future events. *Consciousness and Cognition, 19*(4), 1037–1044.
Liu, D., Wellman, H.M., Tardif, T. & Sabbagh, M.A. (2008). Theory of mind development in Chinese children: A meta-analysis of false-belief understanding across cultures and languages. *Developmental Psychology,* 44(2), 523–531.
Meltzer, L. (2007). *Executive function in education: From theory to practice*. New York, NY: Guilford Press.
Spreng, R.N. & Levine, B (2006). The temporal distribution of past and future autobiographical events across the lifespan. *Memory & Cognition, 34*(8), 1644–1651.
Suddendorf, T. & Corballis, M.C. (1997). Mental time travel and the evolution of the human mind. *Genetic, Social, and General Psychology Monographs, 123*(2), 133–167.
Suddendorf, T. & Redshaw, J. (2013). The development of mental scenario building and episodic foresight. *Annals of the New York Academy of Sciences, 1296*(1), 135–153.
Suddendorf, T., Corballis, M.C. & Collier-Baker, E. (2009). How great is great ape foresight? *Animal Cognition, 12*(5), 751–754.
Suddendorf, T., Nielsen, M. & Von Gehlen, R. (2011). Children's capacity to remember a novel problem and to secure its future solution. *Developmental Science, 14*(1), 26–33.
Tulving, E. (1984). Précis of elements of episodic memory. *The Behavioral and Brain Sciences, 7*, 223–268.
Tulving, E. (2012) Episodic Memory and Autonoesis: Uniquely human? In S. Terrace, S. Herbert & J. Metcalfe (Eds.), *The missing link in cognition: Origins of self-reflective consciousness* (pp. 3–56). Oxford, UK: Oxford University Press.
Wang, Q., Hou, Y., Tang, H. & Wiprovnick, A. (2011). Travelling backwards and forwards in time: Culture and gender in the episodic specificity of past and future events. *Memory, 19*(1), 103–109.

Diversity in Unity: Perspectives from Psychology and Behavioral Sciences – Ariyanto et al. (Eds)

Father involvement and sanctification of parenting in Aceh

N.Z. Amani
Faculty of Psychology, Universitas Indonesia, Depok, Indonesia

L.S.Y. Savitri
Department of Developmental Psychology, Faculty of Psychology, Universitas Indonesia, Depok, Indonesia

D.R. Bintari
Department of Clinical Psychology, Faculty of Psychology, Universitas Indonesia, Depok, Indonesia

ABSTRACT: This study examines the extent to which sanctification of parenting in fathers could predict father involvement in adolescents' upbringing. One hundred and thirty-three pairs of fathers (M_{Age} = 50.44 years) and adolescents (M_{Age} = 16.18) in Aceh participated in this study. This study used the Indonesian version of the Reported Father Involvement Scale (Finley & Schwartz, 2004), Nurturance Fathering Scale (Finley & Schwartz, 2004), and Sanctification of Parenting Scale (Mahoney et al., 2003). Results show that sanctification of parenting contributes significantly to father involvement, both in behaviour domains and in affective quality perceived by the adolescents. The results might be influenced by Acehnese culture and Islamic teaching that should be studied in future research.

1 INTRODUCTION

The increasing number of coffee shops in Aceh after the tsunami hit Aceh in 2004 (Kompas, 2011) has dubbed Aceh as the 'Thousand Coffee Shops State' (Tempo, 2014). The positive impact of the phenomenon can be seen in the areas of economy and tourism, while the city of Banda Aceh now annually organises coffee festivals (Tempo, 2014). However, the phenomenon has also brought a negative impact arising from the considerable amount of time spent by men, especially fathers, in coffee shops. The results show that Acehnese men spend much time at coffee shops while their wives are doing productive work at home (Syukrizal et al., 2009). The considerable amount of time spent by fathers at the coffee shop could actually be used to interact with their wives and children at home (Kahhari, 2015).

Data from Aceh Agency for Women's Empowerment and Child Protection [BP3A Aceh] (2015) show that, in Acehnese culture, parenting is only performed by mothers. Besides that, from 1976 to 2005 Aceh became an Area of Military Operations (Amnesty International, 2014). During this period of prolonged conflict, many children lost their fathers (Amnesty International, 2000). Psychologically, these conditions removed their chance to interact with their fathers and thus took away the figure with which they identify. It is possible that when children who have lost their fathers become fathers themselves, they will not know how to act an involve as a father in the process of child parenting.

According to Kuzucu and Özdemir (2013), Carlson (2006), and Reid and Finley (2010), the involvement of fathers in their children's development has positive benefits because it can help develop those adolescents' mental health, prevent them from being involved in juvenile delinquency and assist in reaching better psychosocial development. A study in South Africa also shows that adolescents who do not have a father figure or biological father have a higher tendency to have alcohol consumption problems due to the absence of authority figure which provides direction and discipline (Ramsoomar et al., 2013).

Furthermore, daughters whose fathers are much involved in their lives have better psychological well-being (Allgood et al., 2012). Byrd-Craven et al. (2012) also found that warm interactions between a father and a daughter make daughters more willing to discuss issues related to interpersonal relationships with their father. However, Byrd-Craven et al. (2012) found that father involvement only reduces aggressive behaviour in boys but not in girls.

Newland et al. (2013) also conducted research regarding fathers' belief of their parenting ability, perception, life context, the involvement of fathers in the attachment, and learning outcomes of children in the United States and Taiwan. One of their findings shows that fathers have different motivations with regard to their involvement as fathers in their children's education. They posited that fathers in the United States have an urge to have closeness with their children in the educational context, while fathers in Taiwan have an urge to be responsible for their children's education. The difference between these findings can be explained by the cultural context of the fathers, in which Asian cultures emphasise the responsibilities of parents towards their children's lives (Newland et al., 2013).

Viewed from an Islamic perspective, a father has a variety of roles in the lives of his children. In general, the task of a father in his family is to fulfil basic needs such as money, clothes and shelter, as well as psychological needs (Istiadah, 1999). More particularly, the father acts as a leader, protector, and educator in his family in order to enhance the development of emotion, cognition, morality and religious views, as well as being an entertaining companion for their children (Amirulloh & Sumantri, 2015). A father is the person who determines and provides direction to the process of parenting, both of which should be performed by the father and the mother. Amirulloh and Sumantri (2015) add that out of the 17 verses in the Qur'an that explain about the teaching process through interaction, 14 are regarding father-child interaction.

According to Weyand et al. (2013), there are several ways to measure the concept of religion in parenting, such as religious coping, biblical conservatism and the sanctification of parenting. Religious coping is the way individuals face stressful events by connecting the self with God and religion (Pargament et al., in Weyand et al., 2013). Biblical conservatism is the religious dimension of individuals regarding literal interpretation of the content of the gospel (Ellison & Sherkat, in O'Laughin et al., 2013). Mahoney et al. (2003) define sanctification of parenting as the extent to which the father perceives his parenting to the child as having a divine character and significance.

Due to different results of research regarding the impact of the religious values towards the process of father parenting and outcomes, the researchers feel that there are still gaps that have not been explained by previous studies. Very little research has been carried out to explain how fathers in Aceh apply religious values to their involvement in their children's parenting. Study by Syukrizal et al. (2009) shows that many fathers in Aceh do not understand Islamic teachings associated with fathers' role in parenting their children. This fact is noteworthy, particularly because Aceh has been one of the Indonesian provinces that have implemented Islamic religious values in the daily lives of its inhabitants since the 19th century (Nurdin, 2013). Moreover, in 2000, these practices of Islamic values officially became Islamic Sharia law which is based on the Regional Regulation No. 5 of 2000 on the Implementation of Islamic Law (Sufi & Wibowo, 2004). Available research was carried out within the context of Western culture and other religions. Every culture and religion certainly has different views concerning how fathers should be involved in their children's lives. Thus, this study is conducted to see whether the sanctification of parenting performed by fathers can predict the extent of father's involvement in adolescents, either in the form of involvement in the behaviour domain or in the quality of affection given, in Acehnese culture. This research also aims to explain qualitatively the extent of Acehnese fathers' knowledge regarding parenting in Islam as well as the attitude towards their children.

2 LITERATURE REVIEW

Finley and Schwartz (2004) describe the involvement of fathers in adolescents as the extent to which he or she perceives that his or her father was involved in various behaviour domains and the quality of affection given. There are several points that Finley and Schwartz (2004) emphasise regarding the involvement of fathers. First, father involvement is a concept that is very diverse. There are various domains of behaviour that allow a father to be involved or not in his child's life. Second, the magnitude of the impact of father's involvement in his child's life is the result of the child's perception. One way to measure the long-term impact of the father's involvement is by asking the child, both as a teenager and as an adult, to assess the involvement and parenting of his or her father in the past. Lastly, any measurement with regard to the child should be conducted by applying the principle of 'the best interests of the child' (Finley, 2002, in Finley & Schwartz, 2004). Meaning that, study about father involvement should focus on child's perspective rather than father's reporton his involvement.

Finley and Schwartz (2004) divide the way to look at perceptions regarding child parenting and father involvement into two forms, which are father involvement and nurturance fathering. Father involvement is a form of father's involvement within the domains of behaviour in a child's life (Finney & Schwartz, 2004). Father involvement is further divided into two forms, namely, reported father involvement and desired father involvement. Reported father involvement reflects the extent to which a child's father has been involved, while desired father involvement reflects the extent to which the child wants their father to be involved (Finley & Schwartz, 2004). Father involvement is seen based on the 20 domains of a child's life, including the spiritual development, as well as the development of responsibility, discipline and counsel. In this study, the researchers only use the reported father involvement.

Nurturance fathering is the quality of affection in a father-child relationship (Finley & Schwartz, 2004). Nurturance fathering has eight indicators, including the enjoyment of parents in parenting, support, allocation of time, emotional closeness and parent-child relationship, as well as an overall assessment (Finley, in Williams & Finley, 1997). In a journal in 2004, Finley and Schwartz refered to this measurement as the Nurturance Fathering Scale.

There are several factors that affect a father's involvement in his child's life. Culture and religion affect how fathers interact with their children. Parenting is a process that takes place in a social context (Juhari et al., 2013; Bronfenbrenner, 1994). Therefore, the conditions and rules that exist in a specific cultural and religious context can affect how people perform their parenting, including father involvement. Level of economy, father's residence, gender of the child and the birth order of the child in the family also affect the father's involvement in his child's life (Gillies, 2009; King et al., 2004; Carlson, 2006; Pleck, 1997).

Pargament and Mahoney (2005) define sanctification of parenting as 'a process through which aspects of life are perceived as having divine character and significance' (p.183). This definition has arisen as a psycho-spiritual concept into which they combine the perspectives of psychology which scientifically study an individual's perception of things that are considered sacred within the spiritual perspective (Mahoney et al., 2003; Pargament & Mahoney, 2005).

Parenting is one of the elements of human life which has a close relationship with religion throughout its process. It is generally believed that relationships in a family have a spiritual value (Mahoney et al., 2003). It is also evident that many religions view having children as the fulfilment of the sacred purpose of a wedding and a blessing for the family (Mahoney et al., 2003). Besides that, Mahoney et al. (2003) also offer several arguments which demonstrate the strong relationship between religion and parenting.

First, every religion considers that a mother's pregnancy and a child's birth are imbued with spiritual significance. For example, Islam teaches that a father has to recite an *adzan* (special prayer) to a newborn child and *tahnik* the child (give food to a newborn child to taste, usually date palms) (Al-Jawziyya, 2001). Second, every religion always considers the process of child parenting as a glorious stage in human life. In Islam, there is a *hadith* which states, 'If a person dies, his or her deeds are cut off except for three: charity, beneficial knowledge, and pious children who pray' (Hadith narated by Muslim). Children who are raised properly by their parents will later become a source of reward for their parents. Third, every religion also teaches children that any interaction with their parents should be in line with religious principles. In the Qur'an, this is mentioned in QS. Luqman Verse 14:

> *And We have enjoined on humans (to be dutiful and kind) to their parents. His mother carried him in difficulty upon difficulty and his weaning is in two years. Be grateful to both your parents! The return is to Me.*

Lastly, every religion gives parents a parenting guide. For example, Islam recommends that parents kiss their children and treat all their children equally (Al-Jawziyya, 2001; Amirulloh & Sumantri, 2015). Islam further encourages fathers to treat their children with fairness, as contained in the *hadith*: Be fair among your children in granting, as you love justice among you in goodness and gentleness (Hadith narated by Abid Dunya). Righteous conduct should also be applied to both daughters and sons.

Sanctification in each individual can occur in two forms, namely the manifestation of God and the sacred qualities. Manifestation of God is the process of sanctification of parenting that happens to fathers who embrace a particular religion (Mahoney et al., 2003). Pargament and Mahoney (2005) explain that every religion encourages its followers to understand that God exists in various aspects of their lives. Their traditions, values and religious teachings allow these individuals to be aware of and to understand the invisible force that governs the world and life (Mahoney et al., 2003). In other words, individuals can actualise their spiritual experience with God (Pargament & Mahoney, 2005).

Sacred qualities are the process of sanctification of parenting that occurs in individuals who do not have any 'specific reference to deity' (Mahoney et al., 2003, p. 221). Thus, it can be inferred that sacred qualities occur in individuals who do not have an attachment to God or any divine figure. According to the results of previous studies, Mahoney et al. (2003) and Pargament and Mahoney (2005) found that individuals who do not adhere to a particular religion still attribute sacred characters to things that they consider important in their lives. Sacred qualities can be attached to any elements of parenting that highlight the parents' perception of parenting as sacred and eternal. Both processes of sanctification can occur in different ways among different people (Pargament & Mahoney, 2005). There are people who perceive that certain events have been caused by God. There are also individuals who attribute divine characteristics to other objects. And then there are also individuals who experience both. Thus, both of these processes may occur in both religious and irreligious individuals (Pargament & Mahoney, 2005).

3 METHODS

3.1 *Measures*

Three scales were used in this study. All scales were tested prior to research to ensure the readability and fulfilment of psychometrics properties.

3.1.1 *Reported father involvement scale*

This scale is intended to measure adolescence's retrospective perception of father' involvement across three behaviour domains: instrumental, expressive, and mentoring/advising (Finley & Schwartz, 2004). Adolescents were asked to give a response to 20 items ranging from (1) 'never involved' to (5) 'always involved'. This measurement has an α of 0.941.

3.1.2 *Nurturance fathering scale*

The purpose of this scale is to measure fathers' affective quality perceived by the adolescents (Finley & Schwartz, 2004). There are nine items, each of which has five possible dissimilar responses across question. One example is 'How emotionally close are you with your father?' with possible responses from 'very close' to 'not close at all'. The alpha Cronbach for this scale is 0.899.

3.1.3 *Sanctification of parenting*

This measurement has two subscales: manifestation of God and sacred qualities, with 14 and 10 items respectively. Manifestation of God is intended to measure the extent to which a father attributes his spiritual experience in parenting to God. This subscale has six possible responses ranging from (1) 'highly disagree' to (6) 'highly agree' (Mahoney et al., 2003). The sacred qualities subscale measures sanctification of parenting in a father who has no 'reference to (a) specific deity' (Mahoney et al., 2003). This part of the measurement has six possible responses ranging from (1) 'very not describing' to (6) 'very describing'. In the last part of the questionnaire, open questions which seek to measure a father's knowledge about parenting in Islam and the extent to which he has practised it were asked. It was expected that the answers would give a concrete example on how the father performs the parenting to the child.

3.2 *Participants*

Participants in this study are pairs of a father and a child from several high schools in Aceh. High school students were selected because higher dynamics between father and child tend to occur at these ages (Duvall & Miller, 1985). Moreover, father involvement measuring tools were designed for adolescents and early adults because it measures the –residue- of the involvement of the father when the child was young (Finley & Schwartz, 2004). Another reason is because the abstract form of the items in the questionnaire is more easily understood by high school students, based on the fact that children at these ages have already acquired the ability to think about abstract concepts (Miller, 2011). The reason for selecting Moslem and Acehnese fathers as participants is because both the Acehnese culture and Islam have a specific set of regulations concerning a father's role in his child's life. Thus, the measurement of sanctification of parenting is applied to those fathers because they are the ones who practise it in their child's parenting.

There were two methods used to perform this research. First, packages of questionnaires were distributed, each of which had a pair of questionnaires for a father, his adolescent child and rdetauled introduction of this research. The packages were distributed to high schools in Banda Aceh city and Aceh Besar district. From each school, researchers recruited students to distribute and collect the packages. This was to ensure the wide distribution of data from various grades. Another method was by using online questionnaires. By means of high school social media groups, announcement about the research and the researchers' contact details were published. If they would like to participate, they could contact the researchers and would receive two links to the questionnaires for the adolescent and the father. All adolescent participants were given a phone credit upon completing and returning the questionnaires.

A great deal of information can be drawn from the adolescents' demographic data. First, 88% of 133 participants were 16–17 years-old with an average age of 16.18 (SD = 1.006). Moreover, 54.6% of the participants were female, and 71.4% of all samples went to non-boarding schools. Almost all participants rated their biological father as their father figure (98.5%). Almost half of the participants (40.9%) are middle children, but one participant did not provide birth order data.

The mean age of fathers who participated in this research was 50.44 with an SD of 5.95. Most fathers did not attend an Islamic boarding school (*pesantren*). Those fathers' education levels were varied, ranging from elementary school to a postgraduate programme, with 34.6% having finished their undergraduate education. The calculation of monthly earnings in this study is divided into two, namely, the father's revenue and the total revenue of both working fathers and mothers. Based on the Ministry of Finance's standard (BPPK Kemenkeu, 2015), 65.4% of fathers were categorised as having a middle income with a mean of 4,501,282 IDR and an SD of 1,868,484 IDR. Meanwhile, based on the total revenue, 47.4% came from middle-income families with an average income of 4,845,588 IDR and an SD of 1,513,909 IDR.

4 ANALYSIS

The researchers used linear regression to ascertain the influence of sanctification of parenting as an independent variable on father involvement as a dependent variable. Statistical results show that 6.3% of father involvement in behaviour domains is attributable to sanctification of parenting, with $F(1.131) = 9.916$, $p = 0.002$. Meanwhile, 5.5% of father affective quality as perceived by children is attributable to sanctification of parenting $F(1.131) = 8.643$, $p = 0.004$.

4.1 *Open question analysis*

On the final part of sanctification of parenting questionnaire, the researchers asked the fathers what they knew about parenting in Islam, as well as examples of behaviours that they have practised. In the previous section, researchers divided the scores of sanctification of parenting variables into two categories, high and low. Qualitative answers were obtained by seeking responses which can adequately represent each category based on the quantitative score. The following are the answers to these questions.

An example of an answer related to the knowledge about parenting in Islam given by fathers who fall into the category of low-score in sanctification of parenting is, 'At the age above ten, if children do not pray, they should be hit.' Meanwhile, an example of answers related to the application of parenting behaviours is, 'The Prophet ordered to hit the child if they do not pray, to practice what is *ma'ruf* (favourable to God) and to avoid what is *munkar* (unfavourable to God)' and 'Forcing my child to do morning prayers and pouring them with water if necessary'.

Examples of answers related to the knowledge about parenting in Islam given by fathers that fall into the category of high-score sanctification of parenting are 'We can see in the Holy Koran on Luqman verses, where The Almighty asks us to take care of our children' and

> In Holy Koran, many verses explain about parenting from Prophet Ibrahim and Ismail, the famous conversation between Luqman and his son, and verses which ask us [father] to guide our family from hell. Islam is a perfect religion. The Almighty arranges everything perfectly.

An example of behaviour that is applied in the parenting process is

> *Mengontrol bacaan dan hafalan anak, mendengar cerita aktivitasnya seharian, memberi contoh yang baik, mengajaknya solat magrib dan isya dan subuh berjama'ah di masjid, makan, berbincang bersama* [Accompanying the children to memorise and recite Holy Koran, listen to his/her daily activity, become a good role model, ask them to pray together in the mosque, have lunch or dinner together and having discussion].

This qualitative answer demonstrates an obvious difference between the father's understanding of Islamic teachings regarding parenting and the examples of behaviour they practised. Fathers who has low-score in sanctification of parenting put more emphasis on discipline, especially in the form of corporal punishment if they do not comply with the rules.

Fathers who have high-score in sanctification of the parenting category tended to give longer answers and showed a broader understanding of parenting practices in Islam. These fathers showed an understanding of Qur'anic verses which discuss examples of good parenting in Islam. These fathers also put emphasis on meaningful interaction with children.

5 DISCUSSION

The results of the linear regression analysis show that sanctification of parenting may predict father involvement perceived by children in terms of both behaviour domain and affective quality. These findings support the argument that father involvement is a behaviour that takes place in a social context. Therefore, culture, as one of the components of a social context, might contribute to a father's involvement in his child's life (Bronfenbrenner, 1994; Juhari et al., 2013). Lamb (1997) also explains that culture supports a father by providing him with a set of parenting rules that he must apply in order to fulfil society's expectation of an ideal father. Therefore, the father will try to nurture his child to meet this expectation.

The answers to the open-ended questions which researchers asked the fathers with regard to parenting in Islam and examples of behaviours known by the fathers provide a description of what is understood by fathers in terms of parenting. Those who earned higher scores in sanctifications of parenting gave more elaborate answers and showed a better understanding of Islamic parenting. Meanwhile, fathers who earned lower scores gave answers that only emphasise disciplinary functions. The results of the qualitative analysis illustrate how the sanctification of parenting score can predict the extent to which fathers truly understand and fulfil their role in the upbringing of children as something that has a divine character and significance (Mahoney et al., 2003).

6 CONCLUSION

This research aimed to find out how far sanctification of parenting predicts father involvement, both in behaviour domains and in affective quality. Based on the analysis, it can be concluded that sanctification of parenting can accurately predict the two forms of father involvement. The findings also shed some light on the dynamics of fathering in Aceh as a cultural context where Islamic teachings and cultural values blend together and influence how individuals behave.

Future studies should seek to answer further questions regarding the dynamics of fathering in Aceh, such as the differences in fathering patterns between a son and a daughter, the influence of a father's prior experience, and a qualitative analysis of fathers' perception of the influence of religious values on fathering. More studies should also be performed in other cultural contexts in Indonesia.

REFERENCES

Al-Jauziyyah, I.Q. (2001). *Tuhfah al-Maudud bi Ahkan al-Maulud* (*Mengantar balita menuju dewasa: Panduan fikih mewujudkan anak saleh*) [Guidance for toodler to adult: Fiqh guidance for virtuous child] (Q. SF, Ed., & Bahreisy, F., Trans.). Jakarta, Indonesia: PT Serambi Ilmu Semesta.

Allgood, S.M., Beckert, T.E. & Peterson, C. (2012). The role of father involvement in the perceived psychological well-being of young adult daughters: A retrospective study. *North American Journal of Psychology*, *14*(1), 95–110.

Amirulloh & Sumantri. (2015). *Ayah Sesungguhnya Seperti Rasulullah Mencontohkannya*. [Real father as The Rasulullah PBUH did] Jakarta, Indonesia: QultumMedia.

Amnesty International. (2000). *Indonesia: Siklus kekerasan bagi anak-anak di Aceh*. [Indonesia: Violence cycle in Acehnese Children] London, UK: Amnesty International.

Amnesty International. (2014). *Indonesia: Lack of truth, justice, and reparation in Aceh for the past abuses undermines peace process*. London, UK: Amnesty International.

Badan Pemberdayaan Perempuan dan Perlindungan Anak Aceh. (2015). *Sosialisasi pola asuh anak, kerja sama dengan badan PP dan PA Aceh dengan Lembaga Rumah Pelangi*. [Socialization of child care pattern, cooperation with PP and PA Aceh with Lembaga Rumah Pelangi.] Retrieved from http://bp3a.acehprov.go.id/ index.php/news/read/2015/05/13/31/sosialsiasi-pola-asuh-anak-kerja-sama-badan-pp-dan-pa-aceh-dengan-lembaga-rumah-pelangi.html.

Badan Pendidikan dan Pelatihan Keuangan Kementerian Keuangan. (2015, April 29). *Penghasilan kelas menengah naik = Potensi pajak?* [Rising income of the middle class: Tax Potential?] Retrieved from http://www.bppk.kemenkeu.go.id/publikasi/artikel/ 167-artikel-pajak/21014-penghasilan-kelas-menengah-naik-potensi-pajak?PageSpeed = noscript.

Bronfenbrenner, U. (1994). Ecological models of human development. *Readings on the Development of Children*, *2*, 37–43.

Byrd-Craven, J., Auer, B.J., Granger, D.A. & Massey, A.R. (2012). The father–daughter dance: The relationship between father–daughter relationship quality and daughters' stress response. *Journal of Family Psychology*, *26*(1), 87–94. doi:10.1037/a0026588.

Carlson, M.J. (2006). Family structure, father involvement, and adolescent behavioral outcomes. *Journal of Marriage and Family*, *68*(1), 137–154.

Duvall, E.R.M. & Miller, B.C. (1985). *Marriage and family development*. New York, NY: Harper & Row.

Finley, G.E. & Schwartz, S.J. (2004). The father involvement and nurturant fathering scales: Retrospective measures for adolescent and adult children. *Educational and Psychological Measurement*, *64*(1), 143–164.

Istiadah. (1999). *Pembagian kerja rumah tangga dalam Islam.* [Division of domestic labor in Islam]. Jakarta, Indonesia: Lembaga Kajian Agama dan Jender.

Kahhari, A. (2015). *Banda Aceh masuk akal dan rencana jam malam.* [A make sense Banda Aceh and curfew plans.] Retrieved from http://www.lintasnasional.com/2015/06/01/banda-aceh-masuk-akal-dan-rencana-jam-malam/.

Kompas. (2011). Aceh, Negeri Seribu Warung Kopi. [Aceh, a Thousand Coffee Shops State] *Kompas.com.* Retrieved from http://travel.kompas.com/read/2011/03/01/08512866/Aceh.Negeri.Seribu.Warung.Kopi.

Kuzucu, Y. & Özdemir, Y. (2013). Predicting adolescent mental health in terms of mother and father involvement, *Egitim Ve Bilim, 38*–168. Retrieved from http://search.proquest.com/docview/1521710442.

Lamb, M.E. (1997). Father and child development: An introductory overview and guide. In M.E. Lamb, (Ed.), *The Role of the father in child development* (pp. 1–18). Canada: John Wiley & Sons.

Mahoney, A., Pargament, K.I., Murray-Swank, A. & Murray-Swank, N. (2003). Religion and the sanctification of family relationships. *Review of Religious Research, 44*(3), 220–236.

Miller, P.H. (2002). *Theories of developmental psychology.* London: Macmillan.

Newland, L.A., Chen, H.H. & Coyl-Shepherd, D.D. (2013). Associations among father beliefs, perceptions, life context, involvement, child attachment and school outcomes in the US and Taiwan. *Fathering, 11*(1), 3–30.

Nurdin. (2013). *Sekilas sejarah Aceh abad ke-16.* [Aceh history in 16th century]. Retrieved from http://kebudayaan.kemdikbud.go.id/bpcbaceh/2013/10/31/sekilas-sejarah-aceh-abad-ke-16-penulis-nurdin-s-sos-staf-pemugaran-bpcb-aceh/.

Pargament, K.I. & Mahoney, A. (2005). Sacred matters: Sanctification as a vital topic for the psychology of religion. *The International Journal for the Psychology of Religion, 15*(3), 179–198.

Pleck, J. (1997). Paternal involvement: Level, sources, and consequences. In M.E. Lamb (Ed.), *The role of the father in child development* (pp. 66–143). Canada: John Wiley & Sons.

Ramsoomar, L., Morojele, N.K. & Norris, S.A. (2013). Alcohol use in early and late adolescence among the birth to twenty cohort in Soweto, South Africa. *Global Health Action, 6*(1), 57–66.

Reid, M. & Finley, G.E. (2010). Trends in African American and Caribbean fathers' nurturance and involvement. *Culture, Society, & Masculinities, 2*(2), 107–119.

Sufi, R. & Wibowo, A.B. (2004). *Kehidupan Sosial Masyarakat Aceh. Budaya Masyarakat Aceh.* [Social Life of Acehnese Community. Culture of Acehnese Community] Banda Aceh, Badan Perpustakaan.

Syukrizal, A., Hafidz, W. & Sauter, G. (2009). *Reconstructing life after the tsunami: The work of Uplink Banda Aceh in Indonesia.* London, UK: International Institute for Environment and Development.

Weyand, C., O'Laughin, L. & Bennett, P. (2013). Dimensions of religiousness that influence parenting. *Psychology of Religion and Spirituality, 5*(3), 182–191.

Williams, S.M. & Finley, G.E. (1997). Father contact and perceived affective quality of fathering in Trinidad. *Revista Interamericana de Psicología, 31*(2), 315–319.

Diversity in Unity: Perspectives from Psychology and Behavioral Sciences – Ariyanto et al. (Eds)

Flat face expression as a typical Sundanese mother's social cue

A.H. Noer
Faculty of Psychology, Universitas Indonesia, Depok, Indonesia
Faculty of Psychology, Universitas Padjadjaran, Jatinangor, Indonesia

S.H.R. Suradijono & T.R. Umar-Ali
Faculty of Psychology, Universitas Indonesia, Depok, Indonesia

ABSTRACT: Mothers use social cues to connect and bond while teaching their children. Western-individualistic culture holds that facial expression and gaze are the most significant cues to revealing a mother's intention, but in Eastern-collectivism culture, such as Sundanese culture, it is forbidden to expresses emotion freely. This study explored Sundanese social cues that were utilised by mothers to state their intention while communicating with their children. Data was collected through naturalistic observation in three *kabupaten* (regencies) in West Java Province. Mother and child interactions in six families were recorded, each for five days. The mothers' instructions were categorised and interpreted based on categories of social cues, from which were identified three vocal intonations, repeating word(s), three facial expressions, two eye gaze(s) and three gesture varieties. The result shows that even though a flat-faced expression does not give any clue about a mother's intention, it is the most frequent cue (82.8%) that Sundanese mothers use, and utilisation of eye contact is restrained. Sundanese mothers use other modalities to indicate their intention, which are high intonation (62.8%) and pointing gestures (71.2%).

1 INTRODUCTION

To interact with one another and build a relationship, humans have to recognise each other's intentions. Besides using language as a tool to understand one another, humans also use cues to emphasise their intention, as well as to express their emotion and intention. Humans could use facial expression, eye gaze, gesture, vocal intonation and posture to express their intention (Keltner & Lerner, 2010; Nummenmaa & Calder, 2009). Human children learn to understand those cues from infancy (see Barry et al., 2015; Fawcett & Gredeback, 2013; Repacholi et al., 2008). Eye gaze and facial expression are the primary cues for others to understand one's intention (see Jack et al., 2009; Langtong et al., 2000; Leekam et al., 2010).

Prior studies of cues reveal that facial expression and eye gaze are the first cues that an infant recognises and processes in order to understand their mother and thus these cues serve as the foundation of social cognition (see Allison et al., 2000; Emery, 2000; Streri et al., 2013).

As an expression of emotional state, cues are not universal, and are dependent on the culture where the emotion is expressed (Markus & Kitayama, 1991). For instance, Jack et al. (2012) found that each culture has specific cultural signs which can be seen as different facial expression. Every culture has its own criterion of appropriate expression, thus each cue is used differently according to the culture (Matsumoto, 2001) and synchronised to culture values. Eastern-collectivism culture is characterised by people submitting and conforming to society values (Triandis et al., 1988).

Sundanese ethnicity is an ethnic group native to the West Java Province in Indonesia which holds collectivism culture (Triandis, 2001) that has a vertical system prioritising seniority and social class to make a social stratification. It is not acceptable to speak frankly in Sundanese culture (Aziz, 2001); neither is it appropriate to reveal their feelings as freely as Western-individualistic people; and they are thus expected to control their expressions (especially emotional expressions). Moreover, it is uncommon for Sundanese to express their true feelings through facial expression. Ekadjati (2014) said that Sundanese people are not allowed to sob when they cry or roar with laughter when they laugh.

As a group member of her community, a mother becomes a child's entrance to the community, and she provides the child with facilities to survive in the world (Buss, 2008). To establish an attachment with its mother, a child has to be able to comprehend her cues. The mother's expression will represent her community norms and values (Jack et al., 2009, 2012; Murata et al., 2013). Facial expression is the most powerful and important cue to human children, because face perception develops on the first day of a newborn's life and is an immediate source of information on the mothers' emotions (Haxby & Ida Gobbini, 2007; Johnson et al., 1991; Leekam et al., 2010). A Sundanese mother also employs facial expression when interacting with her child; however, there are boundaries for the mother to express her emotions within the Sundanese norm, especially in the facial expression and eye contact (Ekadjati, 2014; Rusyana et al., 1988). To overcome it, a Sundanese mother has to adjust and modify her expressions to be implied as social cues but not to violate Sundanese values.

Eye gaze and eye contact are defined as primary cues for infants and children to recognise their mother's facial expression (Emery, 2000). A previous study confirms that an adult's eye gaze could guide infant attention and help infants to comprehend environmental information easily (Hoehl et al., 2008) and recognise others' intention (see de Bordes et al., 2013; Farroni et al., 2004). Eye contact in Sundanese culture is similar to that found in other Eastern-collectivism culture, which is limited (Blais et al., 2008) since it is not polite for youngsters in the community to have eye contact while talking with elders. Therefore, it is reasonable to believe that Sundanese mothers only use eye contact to emphasise important messages.

To be accepted, a Sundanese child has to understand cues to act properly in order to align with community values. Mothers, as agents of the community, should teach their children and provide clues through social cues. In addition to facial expression and eye contact, humans also use vocal intonation, gesture and posture as cues. Intonation is a salient clue of a mother's intention (Keitel & Daum, 2015; Sauter et al., 2013). Furthermore, people often use gestures, especially hand gestures to accentuate their intention (Cook & Tanenhaus, 2009). The use of intonation and gestures gives Sundanese mothers alternative modality to state their intention.

In daily life, humans use not only one expression as cues, but integrate several expressions to indicate their intention (Zaki, 2013). Even though eye contact and facial expression serve as prominent cues and give immediate meaning of a mother's intention, it has limited use in Sundanese culture. Therefore, Sundanese mothers use different ways to combine and integrate cues.

This study aims to describe mothers' utilisation of facial expression, eye contact, and other modalities as cues without violating Sundanese values.

2 METHOD

2.1 *Participants and apparatus*

Participants were six families from three different *kabupaten* or regencies (Garut, Tasikmalaya and Bandung Barat) with children of 31–98 months observed for five days. These families' ethnicity is Sundanese, who live in the Sundanese culture and use Sundanese and Indonesian language in daily conversations. Only mother and child interaction was taken as data. Each observer was given an observation guide sheet and a JVC Everio 30GB Hard Disk Hybrid Camcorder to record the mother–child interactions.

2.2 *Procedure*

Natural observation was chosen to acquire mothers' social cues in a natural setting in their daily life activities. Each family was observed for five days, in at least three daily activities.

The observer could select one of four activities: mealtime, child grooming or play activities, or child assisting mother doing household tasks. The observer recorded the mother–child interactions in their home using a handheld audiovisual camera to afford flexibility in a mobile setting.

Each activity is defined as one of four possible activities with the mother's instruction as the beginning and completion of the activity as the ending. There was no duration or time limit for each activity. For example, mealtime activity began when the mother asked her child to have breakfast and finished when breakfast was done.

Data were analyzed and categorised based on the type of mothers' social cues used during the interaction. Several cue types that mothers used to instruct or respond to child action were recorded as one cue combination. If the cue changed, it was recorded as another cue combination. The children's reaction was recorded every time a different combination of social cues appeared. The modification of a social cues combination was counted as a different cue combination that stimulated a different reaction from the child.

The raw data of mother–child interaction was coded qualitatively by an anthropologist and a psychologist, and then scored quantitatively by trained scorers to describe utilisation of cue in daily life. Training for scorers emphasised the definition and restriction of every type of social cue, the way to code cue combinations, and to specify mother–child activities. There are inter-rater procedures to ensure that the scorers have the same comprehension.

2.3 *Coding*

Data was coded qualitatively and quantitatively. The anthropologist described the context of each cue according to Sundanese values that the mother expressed. The psychologist described the mothers' micro-expressions, interpreted them in context and then formulated a definition for each cue. Coding was based on a social cues framework, which consisted of vocalisations, repeating words, facial expressions, gaze and gestures. Each cue was divided into several categories. The result of qualitative coding was a standardised definition for every cue that appeared in the mother–child interaction and the meaning of cues.

A standardised definition of cues was used to conduct quantitative coding. The scorer was trained to score mother–child interaction data. To validate the data, inter-rater coding was done for 10% of the total data. In order to have standardised quantitative scoring, all video data was rescreened to obtain good quality data with a clear angle and sharp focus. The activity video was truncated into approximately five minutes and dismissed if an activity occurred for less than five minutes.

Cue combinations that consist of 3–5 single cues were coded and recorded in the observation instrument. Children's reactions were also scored in parallel. The scorer tallied every occurrence of the mother's cues, children's reactions and the composition of every cue combination.

3 RESULT

3.1 *Qualitative coding*

Mother's social cues were categorised into five cues which are vocal, repeating word(s), facial expression, gaze and gestures. The vocal cue consists of flat intonation, high intonation, and low intonation. Facial expression was subdivided into flat expression, smile, sullen, and frowning. Gaze includes eye contact and gaze elsewhere. Gestures consist of hand gesture taking over child's tasks, forbid, pointing and idle hand gesture. Children's reactions were subdivided into accept, decline, confused and indifferent reaction. The standardised definition of the mother's social cues is listed in Table 1, and definitions of the children's reactions are listed in Table 2.

3.2 *Quantitative coding*

Quantitative coding consists of single cues, cue combination and children's reaction coding. From the six mothers that were observed for five days, 285 combination cues were recorded (containing 1,161 single cues) and 285 children's reactions.

Table 1. Definition of social cues. article.

Mother's social cues		Definition
Vocal	Higher intonation	Higher pitch intonation, sometimes accompanied by faster speech tempo with additional words such as *'sok', 'sok engal', 'tah', 'ulah'*.
	Lower intonation	Lower pitch intonation while vocalized last syllable in the word. Sometimes accompanied by slower speech tempo.
	Flat	Flat intonation through the sentence. Does not show any emotion.
Repeating word(s)	Repeat	Repeating word(s) at least twice (usually verb or adverb) in one breath. Repetition of word(s) usually occurs in the imperative sentence.
	Not repeat	Not repeat any word.
Facial expression	Sullen/frowning	Lower mouth line, sometimes accompanied with cone mouth, wider or narrowing eyes, not triggered by physiological stimuli.
	Smile	Higher mouth line, sometimes accompanied with appearance of teeth.
	Flat	Neutral expression, no change in face lines (mouth line and/or eye line) due to approval or disapproval of the child's behaviour.
Gaze	Eye gaze	Eye contact between mother and child while mother is talking or giving instruction to the child.
	Gaze elsewhere	Mother's gaze to other place except child's eyes while talking or giving instruction to child. Usually the mother's eyes gaze is on the task or object.
Gesture	Take over	Mother gestures with her hand to take over and do partly/all of the child's task.
	Forbid	Waving hand gesture to inhibit children from doing something.
	Pointing	Mother's hand gestures to direct child to accomplish the task, either by pointing or touching the object or giving an example of the action required to complete the task.
	Idle	No hand gestures movement.

Table 2. Definition of children's reaction.

Children's reaction	Definition
Accept	Do as mother instructed, whether he is doing it correctly or not.
Decline	Refuse to do mother's instruction; usually stated by saying 'no'.
Confused	Not immediately do mother's instruction. Looking back and forth to mother or else uncertain. Inquire mother's instruction.
Indifferent	Does not hear mother's instruction, and do nothing.

The result (Table 3) shows that flat face expression is the most frequent cue (20.70%) used by Sundanese mothers, followed by pointing gesture (17.81%), and high intonation (15.70%). Unlike flat facial expression, high intonation and pointing gesture are meaningful cues and could indicate the mother's intentions.

The most common children's reaction was acceptance of the mother's instruction (73.0%). Or, the child accepts and follows the mother's instructions directed by the mother's cues. It shows that the child understands its mother's intention. If a child does not understand the mother's instruction, the mother repeats her instruction with a different cue combination. It is interesting that child's indifferent reaction (16.8%) is the second most popular reaction, even higher than the decline reaction (8.8%).

Table 3. Frequency occurrences of social cues.

Social cues		Children's reaction: Indifferent	Confused	Accept	Decline	Freq/ cues	% Cues	% Total
Vocal	Flat intonation	6 (2.1%)	0 (0%)	26 (9.1%)	2 (0.7%)	34	11.9	2.98
	Low intonation	10 (3.5%)	0 (0%)	56 (19.6%)	6 (2.1%)	72	25.3	6.32
	High intonation	32 (11.2%)	4 (1.4%)	126 (44.2%)	17 (6%)	179	62.8	15.70
Repeating words	Repeat	19 (6.7%)	3 (1.1%)	45 (15.8%)	10 (3.5%)	77	27	6.63
	Not repeat	29 (10.2%)	1 (0.4%)	163 (57.2%)	15 (5.3%)	208	73	17.91
Facial expression	Sullen	0 (0%)	0 (0%)	5 (1.8%)	3 (1.1%)	8	2.8	0.70
	Smile	2 (0.7%)	2 (0.7%)	35 (12.3%)	2 (0.7%)	41	14.4	3.59
	Flat face	46 (16.1%)	2 (0.7%)	168 (58.9%)	20 (7%)	236	82.8	20.70
Eye Gaze	Eye contact	18 (6.3%)	3 (1.1%)	100 (35.1%)	14 (4.9%)	135	47.4	11.84
	Gaze elsewhere	30 (10.5%)	1 (0.4%)	108 (37.9%)	10 (3.5%)	150	52.3	13.16
Gesture	Taking over	2 (0.7%)	0 (0%)	6 (2.1%)	0 (0%)	8	2.8	0.70
	Forbid	6 (2.1%)	0 (0%)	11 (3.9%)	4 (1.4%)	21	7.4	1.84
	Pointing	27 (9.5%)	4 (1.4%)	152 (53.3%)	20 (7%)	203	71.2	17.81
	Idle	13 (4.6%)	0 (0%)	39 (13.7%)	1 (0.4%)	53	18.6	4.65
		48 (16.8%)	4 (1.4%)	208 (73.0%)	25 (8.8%)			

Table 4. Contingency tables of cue combinations.

Combination	Children's reaction: Indifferent	Confused	Accept	Decline	Total	Rank
HI,FF,GE,PG	7 (2.5%)	0 (0.0%)	28 (9.8%)	2 (0.7%)	37 (13.0%)	1
HI,FF,EC,PG	3 (1.1%)	1 (0.4%)	16 (5.6%)	2 (0.7%)	22 (7.7%)	2
HI,RW,FF,GE,PG	4 (1.4%)	1 (0.4%)	13 (4.6%)	2 (0.7%)	20 (7.0%)	3
HI,RW,FF,EC,PG	3 (1.1%)	0 (0.0%)	12 (4.2%)	3 (1.1%)	18 (6.3%)	4
HI,FF,EC	2 (0.7%)	0 (0.0%)	16 (5.6%)	0 (0.0%)	18 (6.3%)	4
LI,FF,GE,PG	0 (0.0%)	0 (0.0%)	16 (5.6%)	1 (0.4%)	17 (6.0%)	5
LI,FF,EC,PG	2 (0.7%)	0 (0.0%)	13 (4.6%)	0 (0.0%)	15 (5.3%)	6
FI,FF,GE,PG	0 (0.0%)	0 (0.0%)	14 (4.9%)	1 (0.4%)	15 (5.3%)	6

Note: HI = High intonation; LI = Low intonation; FI = Flat intonation; RW = Repeating word(s); FF = Flat face; GE = Gaze elsewhere; EC = Eye contact; PG = Pointing gesture.

Vocal cues, especially high intonation, appears to be a significant cue from mothers since 44.2% of high intonation cues yielded an accepted reaction from children, similar to the pointing gesture which yielded 53.3% accepted reaction from children.

Cue combination. There were 52 cue combinations that appeared but only eight cues combinations frequently appeared (above 5%). All of the most frequent cue combinations employed flat face (FF) regardless of the number of cues that create combinations.

Cue combinations consisted of at least a three-cue combination, but it more often employed a four-cue combination (54.7%). Mothers employed FF in the three-cue combination (HI, FF, EC) but used high intonation (HI) and eye contact (EC). All five four-cue combinations employed pointing gesture (PG) and five out of eight most frequent cue combinations employed HI. Frequency of EC and gaze elsewhere (GE) were similar. FF and gaze did not signify mothers' intention; therefore, utilisation of PG and HI could indicate Sundanese mothers' intention.

Child accepted reaction mostly occurred within HI, FF and PG combinations (HI, FF, GE, PG 9.8%; HI, FF, EC, PG 5.6%; HI, FF, EC 5.6%; LI, FF, EG, PG 5.6%). In the children's accepted reaction, 16 out of 18 times had the occurrence of a three-cue combination (HI, FF, EC). It is the highest percentage of cue combinations that yielded accepted children's reactions.

4 DISCUSSION

FF is a typical facial expression employed by Sundanese mothers while giving instruction to children. In this study, it was found that FF occurred in almost all cue combinations. It implies that FF does not indicate Sundanese mothers' intention. Facial expressions give solid information about an emotional state, thus utilisation of it arouses strong emotional feeling (Chronaki et al., 2015). According to Sundanese values, it is not appropriate to unveil emotional feeling in public (Ekadjati, 2014). Mothers are therefore obligated to maintain a neutral facial expression.

Another robust cue to indicate intention is eye contact. Facial expression together with eye contact are the first strong cues that are perceived and understood by infants, and carried bold emotional and affection content (Leekam et al., 2010; Repacholi et al., 2014, 2016). Meanwhile, the Sundanese gaze rules consider it to be impolite to have eye contact while talking to someone older (Rosidi, 2011; Rusyana et al., 1988). The Sundanese mothers used EC and GE evenly. This could mean that the mother is teaching her child about the values of eye contact in their society. The mother does not provide the child with too many chances to make eye contact; nevertheless, the mother undoubtedly uses eye contact to emphasise her intention.

To signify her intentions, Sundanese mothers use other cues in cue combination. Utilisation of FF and GE lessened emotion and taught children to control their expression. In this study, cue combinations that employed PG and HI obtained the highest acceptance from child reactions; thus, PG and HI can be considered as the prominent cues in stating Sundanese mothers' intentions. HI as vocal cues can be an indication of the mother's emotion (Chronaki et al., 2015; Sauter et al., 2013). Parallel to vocal cues, gestures have an important meaning as a cue, specifically PG, which is a non-verbal communication that emerges before language (Liszkowski, 2011). PG is a salient cue for children.

Zaki (2013) believes that humans use multiple cues and integrate them to understand others due to the complexity of social information. The most powerful cue integration is between face and vocal cues, as newborn infants understand it as a foundation of social cognition (Streri et al., 2013). Cue integration in Sundanese culture combines FF, HI, and PG to fulfil Sundanese values but still gives children clues through modalities other than facial expression.

The ability to recognise intention in vocal cues develops later after facial expression. Hence, older children are more accurate in recognising emotion through facial cues than younger children (Chronaki et al., 2015). A Sundanese mother uses more identical facial expressions, even to a younger child. It is necessary to reveal more about the development of facial cues recognition in Sundanese children for future study.

Another interesting result is that Sundanese children react indifferently to mothers' instruction. By reacting indifferently, children could be (1) not listening to or not understanding the mother's instruction, or, (2) showing a different form of declining the mother's instruction Children's reactions to copy mothers' instructions are an attempt to conform to the values of society, which serves as an adaptation process (Shea, 2009). Children do this in an effort to adapt to the norm of the culture (see Kenward, 2011; Nielsen, 2012; Shea, 2009).

The limitation of this study is that children's age has not been considered as a factor of their ability to comprehend their mother's intention. Therefore, it would be intriguing to see age as a factor being accounted for in the next study.

ACKNOWLEDGEMENTS

This research was supported by DRPM UI. We thank Hendriati Agustiani and Urip Purwono who provided insight and expertise that greatly assisted the research.

REFERENCES

Allison, T., Puce, A. & McCarthy, G. (2000). Social perception from visual cues: Role of the STS region. *Trends in Cognitive Sciences, 4*(7), 267–278.

Aziz, E. & A. Aminudin (2001). *Gaya Ki Sunda Menyatakan 'Tidak: Sebuah Telaan Sosiolinguistik Terhadap Variabel Sosial yang Mempengaruhi Realisasi Kesantunan dalam Pertuturan Menolah oleh Orang Sunda*. [Presentation] Konferensi Internasional Budaya Sunda, Bandung.

Barry, R.A., Graf Estes, K. & Rivera, M.S. (2015). Domain general learning: Infants use social and non-social cues when learning object statistics. *Frontiers in Psychology, 6*, 551.

Blais, C., Jack, E.R., Scheepers, C., Fiset, D. & Caldara, R. (2008). Culture shapes how we look at faces. *PLoS ONE, 3*(8), e3022.

Buss, D.M. (2008). *Evolutionary psychology: The new science of the mind* (3rd ed.). Boston, MA: Pearson Education.

Chronaki, G., Hadwin, A.J., Garner, M., Maurage, P. & Sonuga-Barke, J.S.E. (2015). The development of emotion recognition from facial expressions and non-linguistic vocalizations during childhood. *British Journal of Developmental Psychology, 33*(2), 218–236.

Cook, S.W. & Tanenhaus, K.M. (2009). Embodied communication: Speakers' gestures affect listeners' actions. *Cognition, 113*(1), 98–104.

De Bordes, P.F., Cox, F.A.R., Hasselman, F. & Cillessen, A.H.N. (2013). Toddlers' gaze following through attention modulation: Intention is in the eye of the beholder. *Journal of Experimental Child Psychology, 116*(2), 443–452.

Ekadjati, E.S. (2014). *Kebudayaan Sunda: Suatu Pendekatan Sejarah* (4th ed.). Bandung, Indonesia: Dunia Pustaka Jaya.

Emery, N. (2000). The eyes have it: The neuroethology, function and evolution of social gaze. *Neuroscience and Biobehavioral Reviews, 24*(6), 581–604.

Farroni, T., Johnson, M.H. & Csibra, G. (2004). Mechanisms of eye gaze perception during infancy. *Journal of Cognitive Neuroscience, 16*(8), 1320–1326.

Fawcett, C. & Gredebäck, G. (2013). Infants use social context to bind actions into a collaborative sequence. *Developmental Science, 16*(6), 841–849.

Haxby, J.V. & Ida, G.M. (2007). The perception of emotion and social cues in faces. *Neuropsychologia, 45*(1), 1.

Hoehl, S., Reid, V., Mooney, J. & Striano, T. (2008). What are you looking at? Infants' neural processing of an adult's object-directed eye gaze. *Developmental Science, 11*(1), 10–16.

Jack, R.E., Blais, C., Scheepers, C., Schyns, P.G. & Caldara, R. (2009). Cultural confusions show that facial expressions are not universal. *Current Biology, 19*(18), 1543–1548.

Jack, R.E., Garrod, O.G.B., Yu, H., Caldara, R. & Schyns, P.G. (2012). Facial expressions of emotion are not culturally universal. *Proceedings of the National Academy of Sciences of the United States of America, 109*(19), 7241–7244.

Johnson, M.H., Dziurawiec, S., Ellis, H. & Morton, J. (1991). Newborns' preferential tracking of face-like stimuli and its subsequent decline. *Cognition, 40*(1), 1–19.

Keitel, A. & Moritz M.D. (2015). The use of intonation for turn anticipation in observed conversations without visual signals as source of information. *Frontiers in Psychology,* 6 (FEB). Available on doi:10.3389/fpsyg.2015.00108.

Keltner, D. & Jennifer, S.L. (2010). Emotion. In S.T. Fiske, D.T. Gilbert & G. Lindzey (Eds.), *Handbook of social psychology, 1*, (5th ed.) (pp. 317–352). Hoboken, NJ: John Wiley & Sons.

Kenward, B., Karlsson, M. & Persson, J. (2011). Over-imitation is better explained by norm learning than by distorted causal learning. *Proceedings of The Royal Society B, 278*(1709), 1239–1246. doi:10.1098/rspb.2010.1399.

Langton, S.R.H., Watt, R.J. & Bruce, V. (2000). Do the eyes have it? Cues to the direction of social attention. *Trends in Cognitive Sciences, 4*(2), 50–59.

Leekam, S.R., Solomon, T.L, Teoh, Y-S (2010). Adults' social cues facilitate young children's use of signs and symbols. *Developmental Science, 13*(1), 108–119.

Liszkowski, U. (2011). Three lines in the emergence of prelinguistic communication and social cognition. *Journal of Cognitive Education and Psychology, 10*(1), 32–43.

Markus, H.R. & Kitayama, S. (1991). Culture and the self: Implications for cognition, emotion, and motivation. *Psychological Review, 98*(2), 224–253.

Matsumoto, D. (2001). *The handbook of culture and psychology*. New York, NY: Oxford University Press.

Murata, A., Moser, J.S. & Kitayama, S. (2013). Culture shapes electrocortical responses during emotion suppression. *Social Cognitive and Affective Neuroscience, 8*(5), 595–601.

Nielsen, M. (2012). Imitation, pretend play, and childhood: Essential elements in the evolution of human culture? *Journal of Comparative Psychology*, *126*(2), 170–181.

Nummenmaa, L., & Calder, A.J. (2009). Neural mechanisms of social attention. *Trends in Cognitive Sciences, 13*(3), 135–143.

Repacholi, B.M., Meltzoff, A.N. & Olsen, B. (2008). Infants' understanding of the link between visual perception and emotion: 'If she can't see me doing it, she won't get angry.' *Developmental Psychology, 44*(2), 561–574.

Repacholi, B.M., Meltzoff, A.N., Rowe, H. & Toub, T.S. (2014). Infant, control thyself: Infants' integration of multiple social cues to regulate their imitative behaviour. *Cognitive Development, 32*, 46–57.

Repacholi, B.M., Meltzoff, A.N., Toub, T.S. & Ruba, A.L. (2016). Infants' generalizations about other people's emotions : Foundations for trait-like attributions. *Developmental Psychology, 52*(3), 1–15.

Rosidi, A. (2011) *Kearifan Lokal dalam Pérspéktif Budaya Sunda*. Bandung, Indonesia: Kiblat Buku Utama.

Rusyana, Y., Sariyun, Y., Ekadjati E.S. & Darsz U.A. (1988). *Pandangan Hidup Orang Sunda: Seperti Tercermin Dalam Kehidupan Masyarakat Dewasa ini (Tahap III)*. Bandung, Proyek Penelitian dan Pengkajian Kebudayaan Nusantara Bagian Protek Penelitian dan Pengkajian Kebudayaan Sunda.

Sauter, D.A., Panattoni, C. & Happé, F. (2013). Children's recognition of emotions from vocal cues. *British Journal of Developmental Psychology*, *31*(1), 97–113.

Shea, N. (2009). Imitation as an inheritance system. *Philosophical Transactions of the Royal Society B, 364*(1528), 2429–2443.

Streri, A., Coulon, M. & Guellai, B. (2012). The foundations of social cognition: Studies on face/voice integration in newborn infants. *International Journal of Behavioral Development*, *37*, 79–83.

Triandis, H.C. (2001). Individualism and collectivism: Past, present, and future. In D. Matsumoto (Ed.), *The handbook of culture and psychology* (pp. 35–50). New York, NY: Oxford University Press.

Triandis, H.C., Brislin, R. & Hui, C.H. (1988). Cross-cultural training across the individualism-collectivism divide. *International Journal of Intercultural Relations*, *12*(3), 269–289.

Zaki, J. (2013). Cue integration: A common framework for social cognition and physical perception. *Perspectives on Psychological Science*, *8*(3), 296–312.

Diversity in Unity: Perspectives from Psychology and Behavioral Sciences – Ariyanto et al. (Eds)

Electronic vs non-electronic toys: Which one is better for mother–child interaction?

R. Hildayani, L.S.Y. Savitri, A. Dwyniaputeri, D.V. Tertia,
R. Wukiranuttama & T. Gracia
Faculty of Psychology, Universitas Indonesia, Depok, Indonesia

ABSTRACT: The objective of this study is to find the difference in mother–child interaction between a play situation with electronic toys and a play situation with non-electronic toys. There are four domains of the mother–child interaction as measured by PICCOLO (Parenting Interaction with Children: Checklist of Observations Linked to Outcomes): Affection, maternal responsiveness, encouragement and teaching. Sixty-one pairs of mothers and their preschool-aged children participated in this quasi-experimental study. They were observed while playing with two kinds of toys at different times. The results showed that the mother–child interaction when the mother and child were playing with electronic toys is significantly lower than when they were playing with non-electronic toys. It is shown on all four dimensions of mother–child interaction: affection, maternal responsiveness, encouragement and teaching.

1 INTRODUCTION

There are many benefits of play for young children. Play could increase cognitive, physical and social domains as well as children's emotional well-being (Ginsburg, 2007). In the cognitive domain, play could enhance memory, language, and school adjustment and achievement (Bodrova & Leong, 2005). In the physical domain, play helps children to develop gross and fine motor skills. It also helps children to develop social competence (Hofferth & Sandberg, 2001), autonomy (Goldstein, 2012), and emotional understanding and regulation (Ashiabi, 2007).

There are two types of toys: electronic and non-electronic. Electronic toys are operated by battery, produce light and sounds, and may include programmes that are operated by hardware, such as television, computers and mobile devices (Canadian Toy Association, 2011; Kirriemuir & McFarlene, 2006). On the other hand, non-electronic toys are made from wood, paper, plastic, and other materials that are not operated by electronic equipment (Rachmawati, 2013). They are also related to traditional games, multipurpose, and unstructured (Abdullah et al., 2008).

With the advancement of technology, electronic toys are becoming more popular. Rideout, Vandewater, and Wartella (2003) wrote that thirty percent of children in America below six like to play video games, one type of electronic toy. In Indonesia, data shows that in recent years, children prefer electronic games to non-electronic ones (www.waspada.co.id, 2012). Electronic games are even placed in the first rank of favourite games among children (www.kabartop.com, 2012).

In general, there are similarities between electronic and non-electronic toys. Both of them can include the board games category, provide fun for children (National Association for the Education of Young Children (NAEYC) & Fred Rogers Center, 2011; De Kort & Ijsselsteijn, 2008), and offer some benefits. In appropriate ways, electronic games, as one kind of electronic toy, can even support and increase the function of non-electronic games (NAEYS & Fred Rogers Center, 2011).

Both electronic and non-electronic toys have their own strengths and weaknesses. Non-electronic toys give opportunity for children to directly touch the objects they played.

They can also use any object of play in many ways. However, some non-electronic toys can be hazardous; they might have sharp angles, be easily broken and contain toxic materials (Tedjasaputra, 2001). On the other hand, electronic toys, especially electronic games, give children opportunity to experience things that they could not sense in the real world (NAEYC & Fred Rogers Center, 2011). They also train children to react in a fast and agile manner (Tedjasaputra, 2001). However, electronic toys, especially electronic games, could change the play pattern of children. They could encourage children to only sit in front of a monitor screen and eat snacks while they are playing, which could lead to obesity over time (Abdullah et al., 2008; Darling, 2011). Generally, electronic games are inflexible and not varied, because the games' rules have been programmed (Kirkorian et al., 2008). Another weakness of electronic games is related to social competence (Boccagno, 2012). They are lacking in giving a child the chance to share and cooperate with others, and lacking in developing one's emotional understanding, as well as disrupting direct communication with others. Particularly, they reduce conversation between parent and child (Lavigne et al., 2011). In accordance with this, Wooldrige and Shapka (2012) found that electronic games cause lack of parent-child interaction.

Parent-child interaction, especially interaction between mother and child, is important because mother–child interaction is associated with the optimal development of children in all domains (Landry et al., 2006) and because the mother plays more frequently with the child than the father does (Rothbaum & Weisz, 1994; Borsntein, 1989). Parent-child interaction can be defined as a combination of behaviours, feelings, and expectations that is special for every parent-child dyad (Bayoglu et al., 2013). Roggman et al. (2013) explained four behavioural domains in parent-child interaction: maternal affect, responsiveness, encouragement and teaching. Maternal affect is shown by warmth, physical closeness and positive expression of parents which would be perceived by children as signs of the presence of affection. Responsiveness refers to parent's sensitivity and responsivity to the signs, expressions, needs and interests of a child. Encouragement includes parents' support for playing, exploration, curiosity, initiative and the development of children's skills and creativity. Finally, teaching deals with activities such as sharing stories and play, as well as giving cognitive stimulations, explanations and questions.

Wooldridge and Shapka's (2012) study in British Columbia, Canada, examined the impact of play using electronic and non-electronic toys on mother-toddler interaction. They found that mother–child interaction, especially in the domains of responsiveness, teaching and encouragement, was better when the mother and child played with non-electronic toys than with electronic toys. However, there are some concerns about the results of the study. Cultural differences can raise different patterns of parent-child interaction (Hwa-Froelich & Vigil, 2004). For example, mothers from a Western culture tend to provide a more autonomous support for her child than mothers from an Asian culture. Mothers with a European or American background are also more talkative and give more information to the child in order to increase the child's knowledge as well as stimulate the child's cognitive ability (Hwa-Froelich & Vigil, 2004). Based on the literature review it was also found that research about children's play in Indonesia is limited.

In the present study, the differences of mother–child interaction when the mother and child play with electronic toys and non-electronic toys was examined. This research replicated Wooldridge and Shapka's (2012) study, but with more participants. The hypothesis is that the mother–child interaction is better when the mother and child play with non-electronic toys than with electronic toys.

2 METHODS

2.1 *Participants and design*

Participants of this study consist of 61 mother–child dyads. There were 31 boys and 30 girls aged 3–6 years. Participants were recruited from some playgroups and kindergartens. Fifty-seven mothers are housewives. The remainder work in various occupations, such as private employees, civil servants, artists, physicians, architects and businesswomen. Seventy-five percent of the children have siblings. Family income is highly variable from 1,100,000 IDR to 41,000,000 IDR per month.

This study used the quasi-experimental design. Sequential play sessions of each dyad were recorded with only the toy types being manipulated (electronic vs non-electronic toys).

2.2 *Setting and apparatus*

Data collection through observation was conducted in a natural setting, taking place either in their kindergartens or in their homes. Toys were chosen through a pilot study, in which seven kinds of toys that children often played with were identified: cooking set for pretend play, Barbie doll and her accessories, cars, dress up figure made from paper, plastic balls, puzzle, and Angry Birds' catapult. Thereafter, two sets of toys that were identical were provided; one set was for electronic toys (in particular electronic games) and the other was for non-electronic toys. Electronic games that were chosen and have resemblance to the non-electronic toys were *Angry Birds Rio; Barbie Makeup, Hairstyle, Dress!; Barbie Princess Makeup Dress 2; Cake Now-Cooking Games; Sara's Cooking Class Lite; Crazy Racing 3D; Drag Racing; Kids Jigsaw Puzzle Ocean Free; Kids Preeschool Puzzle Lite; Transport Jigsaw Puzzle Free;* and *Egypt Zuma-Temple of Anubis*. All of the game applications were operated on a computer tablet. Besides a number of non-electronic toys and a computer tablet, a handycam or digital camera and a tripod were used to record each mother–child play activity.

2.3 *Procedures*

In order to prepare the kinds of toys used in this study, 44 children were asked about which electronic and non-electronic toys they often play with and what kind of gadget they use.

Before the observation, mothers were asked to sign an informed consent as a mark of agreement to participate. Subsequently, mothers filled in a questionnaire with the participants' data. Afterwards, the observation began. In the first ten minutes, the mother and her child were instructed to play with non-electronic toys. They could choose to play with any toy they wanted. When the mother and child were playing, researchers were positioned in a different area. This was so that the mother and child could play more comfortably together. After ten minutes had elapsed, the researchers returned to see them, took all the non-electronic toys away and gave them a computer tablet ready for play. In the second ten minutes, they were asked to play with the electronic toys.

2.4 *Measures*

Two instruments were used in this study; the play activity questionnaire and the Parenting Interactions with Children: Checklist of Observations Linked to Outcomes (PICCOLO).

The play activity questionnaire contains open-ended questions which assess information on play activities, including mother and child's identity; child's frequency and duration of play with electronic and non-electronic toys; types of electronic and non-electronic toys owned by the child; mother's definition of playing together; mother's perception of the benefits of playing with electronic and non-electronic toys, mother's preference for one type of toys over another (electronic or non-electronic) and the reasons for such preference; and the child's playmates.

PICCOLO was developed by Roggman et al. (2009). It measures positive interaction between parent and infant, as well as toddler and preschooler through the observational method.

PICCOLO consists of 29 items of behaviour that measure four dimensions of the parent-child interaction; parent affect, responsiveness, encouragement and teaching. The dimension of teaching contains eight items and the rest consist of seven items for each dimension. All items consist of an indicator of behaviour related to each dimension. It was scored using a

Table 1. Index of inter-rater reliability.

Dimension of PICCOLO	Type of toys	Cronbach's alpha coefficient (α)
Parent affect	Electronic toys	.919
	Non-electronic toys	.953
Responsiveness	Electronic toys	.891
	Non-electronic toys	.925
Encouragement	Electronic toys	.920
	Non-electronic toys	.957
Teaching	Electronic toys	.902
	Non-electronic toys	.957

three-point scale ranging from 0 to 2. A score of 0 was given if no behaviour was observed during a ten-minute period of interaction. If a behaviour was observed more than twice in ten minutes, it was scored 2. If the behaviour occurred in between, it was scored 1.

2.5 *Inter-rater reliability*

In order to make the measurement of the mother–child interaction reliable, four coders practised by coding eight videos of mother–child interaction. First, they watched the videos together. Each of them then assigned scores on the PICCOLO form. If there were different scores among coders, they would discuss it and add the criteria to code the behaviour. Afterwards, they went back to check and score the videos based on the new criteria they made.

A test of the inter-rater reliability using Cronbach's alpha showed that the coefficient of reliability among raters is high. Results are displayed in Table 1.

3 RESULTS

3.1 *Description of play activities*

According to the data, the top three non-electronic toys that children have at home consist of equipment for doing dramatic plays such as cars, cooking sets and dolls (55 children); followed by constructive toys, such as blocks, puzzles and play dough (38 children); and functional toys, such as balls, bicycles, swings, and slides (28 children). Whereas, the top three media that children use to play electronic games are tablets (40 children), laptops or computers (25 children) and smartphones (17 children).

The majority of children play with non-electronic toys every day (85.24%) with a play duration of between 1–2 hours (72.13%). On the other hand, 32 children (52.46%) play with electronic toys every day, and 21 children (34.43%) play 1–3 days per week. They spend time playing with electronic toys for approximately 1–2 hours (54.1%) and 3–4 hours (37.7%) each time they play.

Generally for mothers, the meaning of playing together is for enhancing closeness and quality time. By playing together with their children, they can also educate them. Mothers rank first as a child's playmate, followed by siblings and fathers. According to mothers, the benefits of non-electronic toys are the development of gross and fine motor skills and intellectual enhancement, as well as the development of imagination and creativity. On the other hand, the benefits of electronic toys are knowledge improvement, intellectual enhancement and technology awareness. However, mothers like non-electronic plays more than they do electronic plays because of their benefits.

3.2 *Mother–child interaction*

Based on the observation using the PICCOLO form, the differences between the mean values of parent affect, responsiveness, encouragement, and teaching scores in two situations were

Table 2. The mean difference of parent-affects responsiveness, encouragement, and teaching scores when mother and child play with electronic and non-electronic toys.

			M (SD)	
Dimension	T	df	Non-electronic	Electronic
Parent affect	3.666*	60	10.85 (2.227)	10.03 (2.302)
Responsiveness	6.324*		12.08 (2.622)	9.89 (2.811)
Encouragement	6.760*		11.33 (2.959)	9.46 (3.050)
Teaching	8.903*		8.93 (3.936)	4.41 (2.777)

*P<.001

examined, that is, when mothers and children play with electronic toys and when they play with non-electronic toys, using the paired samples two-tailed t-test analysis. The results are presented in Table 2.

Based on Table 2, the mean score of parent affect when the mother and child played with electronic toys is lower than when they played with non-electronic toys. In other words, parent affect is better when the mother and child played with non-electronic toys than with electronic toys. For the dimension of responsiveness, the mean score of maternal responsiveness when the mother and child played with electronic toys is also lower than when they played with non-electronic toys. In other words, the mother is more responsive when she and her child played with non-electronic toys than with electronic toys. A similar result is found for the dimension of encouragement. The mean score of encouragement when the mother and child played with electronic toys is lower than when they played with non-electronic toys. This means that encouragement from the mother is improved when she and her child play with non-electronic toys than with electronic toys. Finally, the mean score of teaching when the mother and child played with electronic toys is lower than when they played with non-electronic toys. In other words, the mother teaches more when she and her child play with non-electronic toys.

4 DISCUSSION

In general, the results of this study support Wooldridge and Shapka's study (2012), which found that electronic games cause lower parent-child interaction when compared to non-electronic games, especially in the domains of responsiveness, encouragement and teaching. One of the indicators of maternal responsiveness is 'looking at the child when the child talks'. Based on observations here, when electronic games were played, mothers' gaze was directed more to the monitor screen than to their children. This supports the finding of Boccagno (2012) that electronic games reduce face-to-face interaction between mother and child. Electronic games also reduced the mothers' opportunity to be involved in play with their children because some electronic games are only played by a single player. Generally, in plays with electronic games, the role of a mother is only as an attendant, and not as her child's playmate.

The low mother–child interaction, especially in the encouragement domain, could be caused by several factors. First, when playing with electronic games, conversation between the mother and child tends to be reduced. Meanwhile, encouragement in general is expressed through conversation. Lavigne et al. (2011) claimed that electronic games have limited parental language, child language output and turn-taking conversation. This is because the child is more focused on the game's instruction than on the mother's explanation. Second, the rules that have been programmed about how to play can restrict children from exploring. Electronic games make it hard for children to play in their own way. When playing electronic games, a mother tends to direct the child to follow the existing rules. They also cannot be involved in the activity in their own way. Third, some mothers do not understand how to play the games, even though their children do. This 'generation gap' causes the mothers to seem passive. In other words, mothers only take the audience role (Johnson et al., 1999). Fourth, electronic games have their own remarks as a reward system, such as 'Great!' or 'Excellent!' which could decrease the role of mothers in giving praise and showing enthusiasm for their children.

As discussed above, a lack of direct interaction when the mother and child are playing with electronic games is also due to the lack of teaching from the mother during play activity. Maternal teaching behaviours are constructed by four components, such as conversation, cognitive stimulation, explanation and questions. Thus, a lack of presence of the four components may lead to a lack of direct interaction during playtime between mother and child. Familiarity of the games was also considered to have influenced mothers' teaching behaviour. Some of the mothers said that they did not understand the games that their children played, and even asked their children how to play. This is in contrast to the non-electronic play, during which the mothers were playing more actively with their children, including having conversations and giving cognitive stimulations, explanations, and questions. In addition, the inflexible nature of electronic games because of the pre-programmed rules causes the games to hold the main control during a play activity. Meanwhile, the control of toys is associated with the amount of knowledge that someone learns. It means that the less control someone has on toys, the less he or she can learn from them (Calvert et al., 2005).

In contrast to the study by Wooldridge and Shapka (2012) which found that there is no difference in maternal affect when the mother and child play with non-electronic toys and electronic toys, this study found that maternal affect is better when the child plays with non-electronic toys. Playing with non-electronic toys gave more opportunities for the mother and child to have face-to-face interactions. In this way, the mother can see and hear her child's tone and social cues, such as facial expressions, and can thus respond to the cues appropriately. A lack of face-to-face interaction when the mother and child play with electronic games leads to a more limited observation of maternal affect. Based on the play activity questionnaire, it was found that mothers are the first playmates of their children and those mothers prefer to play with non-electronic toys. Because of this, the mothers and children are more familiar with non-electronic play activities. Based on this observation, it was found that the more a mother is familiar with the toys, the more she is interested in playing. Furthermore, mothers seem to show more awareness and understanding of their child's social cues and to give appropriate responses. Likewise, the first benefit of playing together that the mothers reported is that it may enhance closeness and quality time, which eventually contributes to the affection that a mother expresses to her child during a play activity.

5 CONCLUSION

There are differences between the mother–child interaction when they play with electronic toys and their interaction when they play with non-electronic toys. The dimensions of maternal affect, responsiveness, encouragement and teaching are improved when the mother and child play with non-electronic toys compared to when they play with electronic toys.

REFERENCES

Abdullah, F., Hoesni, S.M. & Wan A.W.J. (2008). Dari halaman rumah ke hadapan layar: Pola bermain dan fungsinya kepada perkembangan kanak-kanak (From courtyard to front of screen: Play patterns and their function for child development). *Jurnal E-Bangi, 3*(1), 1–14.

Ashiabi, G.S. (2007). Play in the preschool classroom: Its socioemotional significance and the teacher's role in play. *Early Childhood Education Journal, 35*(2), 199–207. DOI: 10.1007/s10643-007-0165-8.

Bayoglu, B., Unal, O., Elibol, F., Karabulut, E. & Innocenti, M.S. (2013). Turkish validation of the PICCOLO (Parenting interaction with children: Checklist of observations linked to outcome). *Infant Mental Health Journal, 3* (4), 330–338. DOI: 10.1002/imhj.21393.

Boccagno, C.E. (2012). *Is screen time ruining our face time?* Retrieved from: http://www.ascd.org/publications/educational-leadership/sept05/vol63/num01/Uniquely-Preschool.aspx.

Bodrova, E., & Leong, D.J. (2005). Uniquely preschool: What research tells us about the ways young children learn. *Educational Leadership, 63*(1), 44–47.

Calvert, S.L., Strong, B.L. & Gallagher, L. (2005). Control as an engagement feature for young children's attention to and learning of computer content. *The American Behavioral Scientist, 48*(5), 578–589. doi:10.1177/0002764204271507.

Canadian Toy Association. (2011). *Canadian Brand Owner Residual Stewardship Corporation Stewardship Plan.* British Columbia, Canada: Canadian Toy Association.

Darling, N. (2011). Is it okay to let your toddler play with the iPad? *Psychology Today.* Retrieved from http://www.psychologytoday.com/blog/thinking-about-kids/201110/is-it-okay-let-your-toddler-play-the-ipad.

De Kort, Y.A.W. & Ijsselsteijn, W.A. (2008). People, places, and play: Player experience in a socio-spatial context. *ACM Computers in Entertainment, 6*(2), 1–11. doi:10.1145/1371216.1371221.

Ginsburg, K.R. (2007). The importance of play in promoting healthy child development and maintaining strong parent-child bonds. *American Academy of Pediatrics, 119*(1), 182–191.

Goldstein, J.H. (2012). *Play in children's development, health, and well-being.* Brussels, Belgium: Toy Industries of Europe.

Hofferth, S.L., & J. Sandberg. (2001). Changes in American children's time 1981–1997. In S. Hofferth & T. Owens (Eds.), *Children at the millenium: Where did we come from, where are we going?* (pp. 193–229). New York, NY: Elsevier Science.

Hwa-Froelich, D.A. & Vigil, D.C. (2004) Three aspects of cultural influence on communication. *Communication Disorders Quarterly, 25*(3), 107–118.

Inilah permainan yang digemari anak perempuan kini (The games which are popular among girls today). (2012). *Kabar Top.* Retrieved from http://kabartop.com/inilah-permainan-yang-digemari-anak-perempuan-kini/.

Johnson, J.E., Christie, J.F. & Yawkey, T.D. (1999), *Play and early childhood development.* New York, NY: Longman.

Kirkorian, H.L., Wartella, E.A. & Anderson, D.R. (2008). Media and young children's learning. *The Future of Children, 18*(1), 39–61. doi:10.1353/foc.0.0002.

Kirriemuir, J. & McFarlene, A. (2006). *Literature review in games and learning (report No. 8).* UK: Futurelab.

Komputer mainan favorit anak (Computer, favourite toys for children). (2012). *Waspada online.* Retrieved from: http://www.waspada.co.id/index.php?option=com_contentandview=articleandid=251506:komputer-mainan-favorit-anakandcatid=204:anakandItemid=197.

Landry, S.H., Smith, K.E. & Swank, P.R. (2006). Responsive parenting: Establishing early foundations for social, communication, and independent problem-solving skills. *Developmental Psychology, 42*(4), 627–642. doi:10.1037/0012–1649.42.4.627.

Lavigne, H.J., Hanson, K.G., Pempek, T.A., Kirkorian, H.L., Demers, L.B. & Anderson, D.R. (2011). *Baby video viewing and the quantity and quality of parent language.* [Presentation] The Biennial Meeting of The Society for Research in Child Development, Montreal, P.Q.

National Association for the Education of Young Children (NAEYC) & Fred Rogers Center. (2011). *Technology in early childhood programs serving children from birth through age 8. Position statement.* Retrieved from http://www.naeyc.org/files/naeyc/file/positions/PS_technology_WEB2.pdf.

Rachmawati, S. (2013). Penerapan media kartu gambar binatang untuk meningkatkan kemampuan gerakan dasar menari pada anak kelompok A di TK Puspita Kecamatan Gunung Anyar Surabaya (Application of animal pictures' cards for improving basic movement skill for dance among students at Puspita Kindergarten, Gunung Anyar, Surabaya). *PAUD Teratai, 2*(1).

Rideout, V.J., Vandewater, E.A. & Wartella, E.A. (2003). *Zero to six: Electronic media in the lives of infants, toddlers, and preschoolers.* Washington, DC: Kaiser Family Foundation.

Roggman, L.A., Cook, G.A., Norman, V.J., Innocenti, M.S. & Christiansen, K. (2009). *Parenting interactions with children: Checklist of observations linked to outcomes (PICCOLO) tool.* Baltimore, MD: Brookes Publishing.

Roggman, L.A., Cook, G.A., Norman, V.J., Innocenti, M.S. & Christiansen, K. (2013). Parenting interaction with children: Checklist of observation linked to outcomes (PICCOLO) in diverse ethnic groups. *Infant Mental Health Journal, 34*(4), 290–306. doi:10.1002/imhj.21389.

Rothbaum, F. & Weisz, J.R. (1994). Parental caregiving and child externalizing behavior in nonclinical samples: A meta-analysis. *Psychological Bulletin APA, 116*(1), 55–74. Retrieved from: http://scholar.harvard.edu/jweisz/files/1994c.pdf.

Tedjasaputra, M.S. (2003). *Bermain, Mainan, dan Permainan: Untuk Pendidikan Usia Dini* (Play. Toys, and Games: For Early Childhood Education). Jakarta, Indonesia: PT Grasindo.

Wooldridge, M.B. & Shapka, J. (2012). Playing with technology: Mother-toddler interaction scores lower during play with electronic toys. *Journal of Applied Developmental Psychology, 33*(5), 211–218. doi:10.1016/j.appdev.2012.05.005.

Diversity in Unity: Perspectives from Psychology and Behavioral Sciences – Ariyanto et al. (Eds)
© 2018 Taylor & Francis Group, London, ISBN 978-1-138-62665-2

The effectiveness of a training programme for kindergarten teachers to teach critical thinking in science learning

J. Suleeman
Faculty of Psychology, Universitas Indonesia, Depok, Indonesia

Y. Widiastuti
Faculty of Psychology, Universitas Indonesia, Depok, Indonesia
Insan Cendekia Madani Kindergarten, Banten, Indonesia

ABSTRACT: This study examines the effectiveness of a training programme to improve kindergarten teachers' ability in teaching critical thinking through science learning. The science learning programme is constructed from the Preschool Pathway to Science (PrePS) programme. Using a post-test only non-equivalent control group design, four teachers acted as an experimental group that received training on critical thinking in science learning consisting of critical thinking concepts, principles of early childhood education, science learning for preschoolers, and teacher's role in teaching critical thinking. Three teachers acted as a control group. Observations using a checklist on the teachers' behaviour in classroom were carried out a week after the training and were repeated three months afterwards. Results show that the experimental group consistently performed better than the control group in using more dialogue and open-ended questions, and when providing constructive feedback. The implication of this study is culturally relevant since dialogue and constructive feedback are not common in teacher–student interaction in an Indonesian setting.

1 INTRODUCTION

Critical thinking is often defined as a way of finding knowledge that is valid, reliable, systematic and logical (Galinsky, 2010). In education, critical thinking is needed to increase understanding, evaluating, developing, and defending arguments and theories, while in daily life critical thinking is needed to prevent unimportant decision-making (such as buying products because of paying attention to advertisements) (Bassham et al., 2011). In short, critical thinking helps us to be more cautious and more logical every time we need to make a decision. An individual cannot automatically become a critical thinker unless he or she is provided with enough opportunities to exercise critical thinking. Dewey (1933) insists that the process to teach critical thinking starts at an early age. The earliest educational institution that can facilitate critical thinking is kindergarten with teachers acting as facilitators (Davis-Seaver, 1994).

The behaviour of teachers in teaching has a major role in supporting student's critical thinking (McBright & Knight, 1993). Teachers should open up opportunities for students to engage in learning by asking about the possibility of solving a problem, encouraging students to work together, giving verbal support, and thinking open-mindedly during the learning process. The classroom's atmosphere should be a pleasant learning environment.

On science learning, there are similarities between the basic skills of critical thinking and skills that should be taught in science lessons. Some of the basic skills of critical thinking that are similar to the skills taught in science are observing, making hypotheses, inferring, reasoning, and evaluating (Ennis, 2002). Through science learning, students are required to continuously carry out the process of critical thinking in understanding scientific concepts (George & Straton, 1999). This condition implies that the learning of science in schools can be used as a preferred way to teach students' critical thinking.

According to Daniel and Gagnon (2011), when four-year olds learn about critical thinking, this will facilitate their science learning later on. Science learning involves a learning process that children should go through. Unfortunately, most of the time, teachers only tell, and then ask the students to recite the laws instead of doing activities on how they work. When the second researcher visited several kindergartens, learning activities only included writing, drawing, counting and singing, and these also happened in science learning. The students appeared to be bored since drawing and writing were the only activities they had for science learning with no activities to explore and no emphasis on processing skills. This created an opportunity for some intervention through the teachers who could be empowered to teach critical thinking in science learning.

A learning programme in science for kindergarteners, *Preschool Pathways to Science* (PrePS), developed by both developmental psychologists and kindergarten teachers, emphasises learning by doing and experiencing a process. The importance of relating students' personal experiences to learning general concepts, mathematicss, and literature in science learning cannot be underestimated if we want children to become active learners (Gelman et al., 2010). This training was presented to kindergarten teachers who teach Level B five-year-olds who will enter first grade after completing kindergarten—as in Indonesia six years-old is the minimum age requirement for first graders. Compared to four-year-olds, five-year-olds are in the transition process from pre-operational level to concrete operational level. Children of this age are ready to receive abstract information stimulated by real objects (Papalia & Feldman, 2012). The contents of the training include development of very young children, conceptualisation of critical thinking, how to teach critical thinking, and science learning, with demonstrations of science experiments that can be used in classroom, using the PrePS programme.

1.1 *Critical thinking in young children*

Even though experts do not agree on one single definition of critical thinking, Lipman's (2003) definition is widely acknowledged and accepted. For Lipman, critical thinking is '*Thinking that facilitates judgement because it relies on criteria, is self-correcting, and is sensitive to context*' (p. 211). Several criteria are used, namely rules, norms, goals to achieve, or methods used. *Self-correcting* means that the individual is able to correct his or her own thinking from his or her learning through the environment. Sensitive to context means the thinking process that one goes through should be appropriate to the topic discussed.

Critical thinking is an ability regarded as too advanced for young children. Research on critical thinking in very young children is considered impossible. However, several studies show that very young children are capable of critical thinking skills (see for instance, Davis-Seaver, 2004; Silva, 2008) or at least, they show some precursors of critical thinking, such as comparing, clarifying, or ordering (Chandra, 2009; Willingham, 2007). Critical thinking is influenced by several factors, namely creativity (Paul & Elder, 2004), metacognition (Halpern, 2003), motivation (Paul & Elder, 2004), external factors (Perkins & Salomon, 1989), skills or abilities (Ennis, 2002), and disposition (Bailin et al., 1999).

1.2 *Facilitating critical thinking development through science learning*

Developing and introducing characters related to critical thinking are regarded as an attitude component (McBright & Knight, 1993). An open attitude of the teachers can provide opportunities for students to research, which can, in turn, create possibilities for children to find something new independently. In providing instruction for activities, teachers should act more as facilitators rather than as deliverers of materials (Meyer, 1986). How teachers manage the learning environment should also be emphasised when teaching for critical thinking. Another behaviour that teachers can apply to develop students' critical thinking is developing experiential learning environments. According to Hickman et al. (2009), an experiment can create a situation that is interactive, participative, applicative, and uncertain. Monitoring critical thinking refers to teachers' activities in observing students. Monitoring will be the teachers' main task after they give instructions to the students and the students start working on the assignment (McBright & Knight,1993).

Teachers can apply all these actions in science learning, which consists of two main components, namely content and process. Plants and animals are examples of contents that can be taught to very young children. Process, usually called *science process skills*, emphasises the methods and attitudes a scientist should have in order to collect information and solve a problem (Henniger, 2013). A programme developed by Gelman and Brenneman (2004, in Gelman et al., 2010) is an application of a learning programme that emphasises science process skills. This programme is called *Preschool Pathways to Science* (PrePS) and combines the pedagogical aspect with cognitive and social development. PrePS emphasises the importance of science process skills and the need to relate experience using general concepts, mathematicss and literature in science learning for very young children (Gelman & Brenneman, 2004).

PrePS programme classifies five main activities considered necessary by every child in science learning (Gelman et al., 2010). These activities are briefly described below.

1. Observe, predict, check
Observe, predict, and check is the first step in PrePS programme, since these three activities are the basis to do more (Gelman et al., 2010). Children should be provided opportunities to look at various geometrical forms, colours, sizes, textures, and other characteristics while they are exposed to various things (Peacock, 2005).
2. Compare, contrast, experiment
Activities such as categorising objects or events by noticing their differences and similarities are helpful to develop children's skills in comparing and contrasting, which are important activities in critical thinking (Gentner, 2005).
3. Vocabulary, discourse and language
Vocabulary in science allows children to discuss their findings and questions deeply in richer language they use daily. Children engaged in this process also start reflecting what they do, think and discuss about their exploration (Gelman et al., 2010).
4. Counting, measurement and maths
Generally speaking, only a small part of mathematics can be taught to very young children (Ginsburg et al., 2008). PrePS encourages mathematical thinking by allowing children to use mathematics as a means to define and describe the world (Gelman et al., 2010).
5. Recording and documenting
Recording and documenting sound like difficult activities for 4–6 year-old children, but children can be asked to draw in detail the activities (events or experiments) in which they are engaged (Gelman et al., 2010).

1.3 *Research question and hypothesis*

One single research question is proposed, 'Is a training programme to teach critical thinking in science learning for kindergarten teachers effective?' The hypothesis is that a training programme to teach critical thinking in science learning for kindergarten teachers is effective.

2 METHOD

This study was conducted in five kindergartens, all of which are located in the South Tangerang area. Five kindergartens were chosen using the accidental sampling method, which is based on availability and willingness. Each of them give Islam religious teachings in addition to the national curriculum for kindergarten (PerMenDikNas No 58 2009). All kindergartens are located in a middle- to low-class socioeconomic environment.

2.1 *Participants*

Eight teachers from Kindergarten Level B (for five-year olds), recruited from five kindergartens in South Tangerang, participated in this study. Their details are described below.

2.2 *Research design*

A post-test for only the non-equivalent control group design was used. This selection of design was considered appropriate since no pretest was able to be administered to control group participants due to their heavy schedule at the end of the semester. Before the training was implemented, each participant was observed by the first researcher in her own classroom, using a checklist on teaching behaviour with the four components mentioned above. Besides this, for post-test, a questionnaire consisting of questions on the training materials was also administered. A post-test was applied on the last day of training. To check whether the results of the training were maintained, three months after the post-test was applied, another observation was carried out both on the experimental as well as the control group. The training was delivered on three different days each lasting for three to four hours. The topics of the training included the development of very young children, critical thinking, teachers' behaviour that is supportive of children's critical thinking, and how to teach science in a classroom setting. Data were analysed quantitatively.

2.3 *Research instruments*

2.3.1 *The training programme*

The overview of the training programme is provided in Table 2. The training lasted for three to four hours per day.

Table 1. Profile of research participants.

Group	Name	Length of education	Age	Teaching experience	Number of students
Experimental	DS	16 years	39 years-old	19 years	17
	MU	16 years	25 years-old	2 years	15
Experimental	NU	16 years	25 years-old	2 years	14
	MA	14 years	43 years-old	23 years	18
Control	YA	16 years	29 years-old	6 years	12
	NR	16 years	40 years-old	15 years	17
	IA	16 years	25 years-old	2 years	15

Table 2. The training content, goals, and method of delivery.

Day	Content	Goals	Method of delivery
1	Overview of the training programme.	To acquire information about participants' prior knowledge of science learning and critical thinking.	Group discussion and lecture
2	The need to teach young children critical thinking Principles of teaching critical thinking. Teachers' behaviour in teaching critical thinking through science learning activities	Participants understand why critical thinking is taught to young children. Participants understand the teacher's role in fostering critical thinking in the classroom. Participants know the strategies that can be applied in teaching critical thinking skills to students through science learning.	Group discussion and lecture
3	Role play	Participants create their own science activity plans that emphasise critical thinking in students.	Role play and group discussion

2.3.2 *List of teachers' behaviour in teaching critical thinking*
A checklist of teachers' behaviours in teaching critical thinking, consisting of 25 items, was used by an observer to check whether the participants' behaviours, both from the experimental and control groups, could be considered as teaching critical thinking in a 30-minute period. The checklist of teachers' behaviours in teaching critical thinking was based on four aspects, (1) developing and introducing characters related to critical thinking, (2) providing instruction for the activities, (3) managing learning environments, and (4) monitoring critical thinking. The behaviour checklist had been assessed by two experts: a kindergarten principal who was also a winner of the Indonesian Best Principal award in 2012, and an academician whose expertise is in young children's learning and psychological test construction. For each item, a score of 2 is given if the behaviour occurs at least twice, while a score of 1 is given if the behaviour occurs only once, and a score of 0 is given if that behaviour does not occur at all. The checklist reliability used inter-rater reliability by two assessors using the index of agreement Cohen's kappa formula (Sattler, 2002). The result is 0.753 and this is considered reliable. The complete list of the indicators can be seen in the appendix.

3 RESULTS

Teachers' behaviours in teaching critical thinking were measured by a checklist as shown below, focusing on four aspects, both at post-test and follow-up stage.

1. Developing and introducing characters related to critical thinking
This aspect consists of five indicators developed into 15 items. For example, Item 2 is 'Teacher encourages students to predict what would happen'.

MA, MU, NU and DS, the experimental group participants, showed higher mean scores compared to NH, IA, and YA from the control group, both on post-test and three-month follow-up measures. Decreasing scores from post-test to follow-up are noted in the indicators 'respecting students' opinion' and 'praising verbally', 'accepting students' opinion without judging or criticising', 'encouraging', and 'doing group activities'.

2. Providing instruction for the activities
This aspect consists of two indicators or four items. The two indicators used are, telling students what is expected and asking open-ended questions.

The experimental group is better than the control group on this second aspect, both on post-test and follow-up measures. Three experimental subjects showed increased scores from post-test to follow-up. Two control group participants did not get any score for this second aspect.

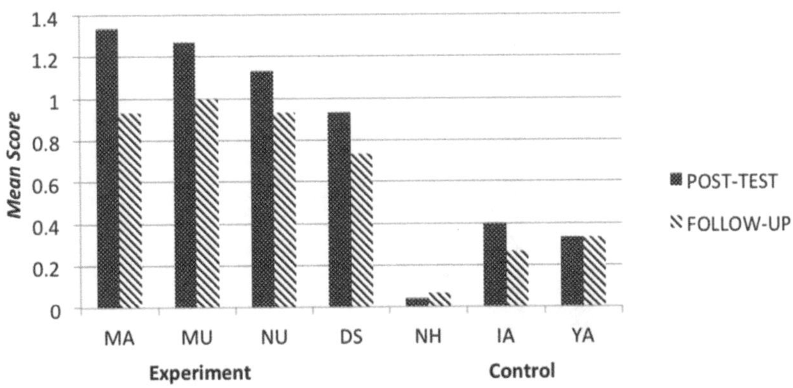

Figure 1. Experimental vs control group differences in developing and introducing characters related to critical thinking.

3. Managing learning environments
This aspect has two indicators or three items. The indicators are, providing demonstration or experiments and developing on-task behaviour.

Compared to the control group, all experimental group participants show higher scores, both on the post-test and follow-up measures and two performed even better on the follow-up measures. One control group participant did not get any score on this aspect.

4. Monitoring critical thinking
This aspect has two indicators or three items.

In this aspect, all experimental group participants performed better than the control group ones who showed none of the checklist behaviours.

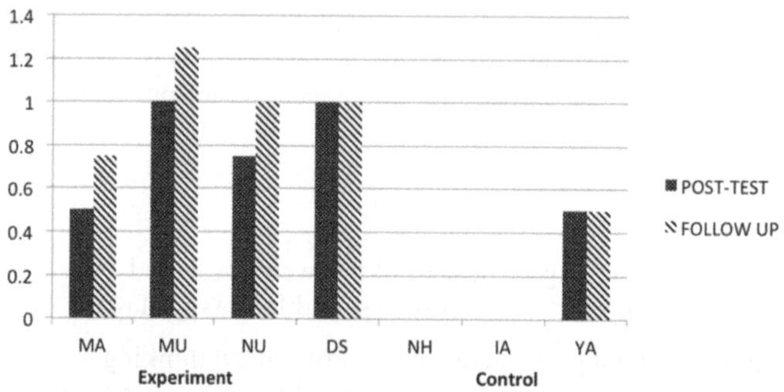

Figure 2. Experimental vs control group differences in providing instruction for the activities.

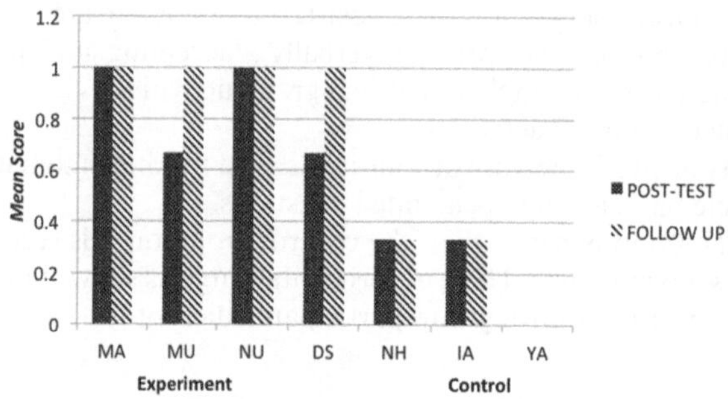

Figure 3. Experimental vs control group differences in managing learning environments.

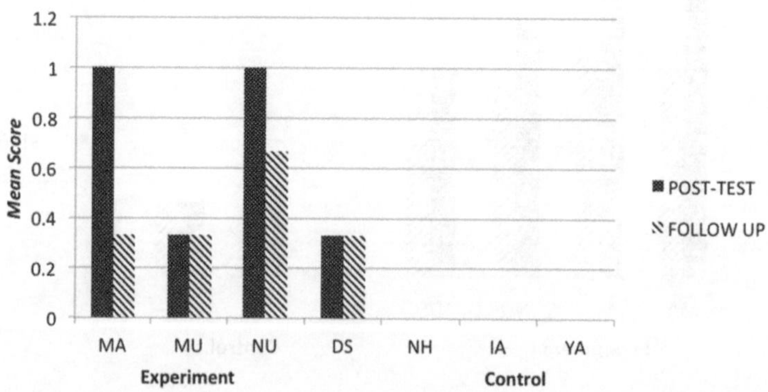

Figure 4. The experimental vs control group differences in monitoring critical thinking.

Based on these results, it can be concluded that the research hypothesis is accepted, that is, the training programme of teaching critical thinking in science learning for kindergartner teachers is effective. The training resulted in changes in teachers' behaviours which indicates that they show teaching behaviours to support critical thinking in their students.

4 DISCUSSION

In general, the programme improved the ability of the experimental group's participants of kindergarten teachers in teaching critical thinking to students. On all four aspects in teaching critical thinking, as suggested by McBright and Knight (1993), the experimental group participants showed higher scores than the control group participants, both at the post-test and follow-up measures. One striking difference between the experimental and the control group participants was regarding the use of questions to facilitate children's understanding of what is being learned. After the training, the experimental group participants often used open-ended questions in their interaction with the students, like 'Tell me what you see there,' and 'Why did you proceed that way?' Asking and answering questions, especially open-ended questions, are important indicators of critical thinking. Teachers can use this technique to engage their students in critical thinking (see for instance, Facione et al., 1995; Tay-Lim, 2012), thus opening up new possibilities for the students to pursue their curiosity. This can create a dialogue between the teacher and the students, which is also important in creating a democratic atmosphere in the classroom (Dewey, 1933).

Experimental group participants also provided examples. Teachers, as facilitators for children's critical thinking (Halpern, 1998), can give instruction to help children understand a problem and find the solution.

The experimental group admitted that they had difficulties on the third aspect; managing the learning environment. As teachers, they focused more on delivery of the course material rather than on how the students could engage in the learning so they could better understand the material discussed. In a critical thinking setting, teachers should play a role as facilitators rather than instructors so that they allow the students to do their own thinking (Daniel & Gagnon, 2011). It is quite a challenge for teachers to make the students become responsible and enthusiastic in their own learning (Terry, 2007).

Another interesting fact is that both the experimental and control groups did not get any score on providing feedback after the students made a mistake. During post-training discussion, the experimental group admitted that they were afraid to let the students make mistakes, so as teachers they would rather prevent the students doing activities that might be done wrongly. As indicated in Chandra's (2009) study, Indonesian mothers tend to prevent their children from doing something 'dangerous' by not giving opportunities to try at all. Without realising it, the mothers inhibit their young children in developing their own thinking.

Demonstrating something so that the students can later do the activity themselves is often done by teachers. However, to be regarded as useful for developing the children's critical thinking, this should be properly done by adjusting the complexities of the demonstration to the children's level of thinking (Papalia & Feldman, 2012). Teachers should be observant of each student's own level of thinking so that the demonstration provided is suitable for each individual.

5 CONCLUSION

Compared to teachers who did not receive training on teaching critical thinking in science learning, those who had received training showed changes in their behaviours, including their knowledge about science learning, critical thinking conceptualisation, and in teaching science learning. These changes appeared both at post-test and follow-up measures. In terms of teaching behaviours, the changes were seen in four aspects, namely in developing and introducing skills related to critical thinking, in delivering instruction for the activities, in managing the learning environment, and in monitoring students' critical thinking.

REFERENCES

Bailin, S., Case, R., Coombs, J.R. & Daniels, L.B. (1999). Conceptualizing critical thinking. *Journal of Curriculum Studies, 31*(3), 285–302. doi:10.1080/002202799183133.

Bassham, G., Irwin, W., Nardone, H. & Wallace, J.M. (2011). *Critical thinking a student's introduction* (4th ed.). Boston, MA: McGraw-Hill.

Chandra, J.S. (2008). *A Vygotskian perspective on promoting critical thinking in young children through mother-child interactions* (Unpublished doctoral dissertation, Murdoch Univerisity, Western Australia).

Daniel, M.F. (2011). Developmental process of dialogical critical thinking in groups of pupils aged 4 to 12 years. *Creative Education, 2*(5), 418–428. doi:10.4236/ce.2011.25061.

Davis-Seaver, J. (1994). *Critical thinking in young children* (Doctoral dissertation, The University of North Carolina at Greensboro, NC).

Dewey, J. (1933). *How we think: A restatement of the relation of reflective thinking to the educative process.* Lexington, MA: D.C. Heath. doi:10.1037/10903-000.

Ennis, R.H. (2002). Goals for a critical thinking curriculum and its assessment. In A. Costa (Ed.), *Developing minds: A resources book for teaching thinking* (3rd ed.). London, UK: Association for Supervision and Curriculum Development.

Facinoe, A., Sanchez, C.A., Facione, N.C. & Gainen, J. (1995). The disposition toward critical thinking. *The Journal of General Education, 44*(1), 1–25. doi:10.2307/27797240.

Fisher, R. (2005). *Teaching children to think* (2nd ed.). Cheltenham, UK: Nelson Thornes.

Galinsky, E. (2010). *Mind in the making.* New York, NY: Harper Collins.

Gelman, R. & Brenneman, K. (2004). Science learning pathways for young children. *Early Childhood Research Quarterly, 19*(1), 150–158. doi:10.1016/j.ecresq.2004.01.009.

Gelman, R., Brenneman, K., Macdonald, G. & Roman, M. (2010). *Preschool pathways to science: Facilitating scientific ways of thinking, talking, doing, and understanding.* Baltimore, MD: Brookes.

Gentner, D. (2005). The development of relational category knowledge. In Gershkoff-Stowe, L. & Rakison, D.H. (Eds.), *Building object categories in developmental time* (pp. 245–275). doi:10.4324/9781410612908.

George, L.A & Straton, J.C. (1999) Approaching critical thinking through science. *The Journal of General Education, 48*(1), 40–47.

Ginsburg, H.P., Lee, J.S. & Boyd, J.S. (2008). Mathematics education for young children: What it is and how to promote it. *Society for Research in Child Development Social Policy Report, 22*(1), 3–22.

Halpern, D.F. (1998). Teaching critical thinking for transfer across domains: Disposition, skills, structure training, and metacognitive monitoring. *American Psychologist, 53*, 449–455. doi:10.1037//0003-066X.53.4.449.

Michael L.H. (2013). *Teaching young children: An introduction.* New York, NY: Pearson.

Hickman, B.T., Mc.Kinney, S.E., Chappel, S. & Berry, R.Q. (2009). An examination of the instructional practices of mathematics teachers in urban school. *Preventing School Failure: Alternative Education for Children and Youth, 53*(2), 278–284.

Kementerian Pendidikan dan Kebudayaan. (2009). *Peraturan Menteri Pendidikan Nasional Republik Indonesia Nomor 58 Tahun 2009 tentang Standar Pendidikan Anak Usia Dini.* Jakarta, Indonesia: Kemdikbud

Lipman, M. (2003). *Thinking in education.* Cambridge, UK: Cambridge University.

McBright, R.E. & S. Knight (1993). Identifying teacher behaviors during critical thinking: A case study. *The Clearing House, 66*, 374–378.

Meyers, C. (1986). *Teaching students to think critically: A guide in all disciplines.* San Francisco, CA: Jossey-Bass.

Papalia, D.E. & Feldman, R. (2012). *A child's world* (12th ed.). New York, NY: McGraw-Hill.

Paul, R. & Elder, L. (2012). *The miniature guide to critical thinking: Concepts and tools.* Tomales, CA: Foundation for Critical Thinking.

Peacock, A. (2005). *Science skills: A problem-solving activities book.* London, UK: Taylor & Francis.

Perkins, D.N. & Salomon, G. (1989). Are cognitive skills context-bound? *Educational Researcher, 18*(1), 16–25. doi:10.3102/0013189X018001016.

Silva, E. (2009). Measuring skills for 21st-century learning. *The Phi Delta Kappan, 90*(9), 630–634. doi:10.1177/003172170909000905.

Tay-Lim, J. (2012). *Developing young children's critical thinking skills through conversation* (Dissertation, National Institute of Education, Singapore).

Terry, D.R. (2007). Do case studies promote critical thinking? *Poster presented at the 2007 Annual Conference on Case Study Teaching in Science, Buffalo, NY.*

Willingham, D.T. (2007). Critical thinking: Why is it so hard to teach? *American Educator, Summer*, 8–19.

APPENDIX

List of indicators of teacher's behaviors in encouraging students' critical thinking.

No	Teacher's behavior in encouraging students' critical thinking	Yes	No	Notes
Aspect 1. Developing and introducing characters related with critical thinking				
1	Providing students' opportunities to observe tools and materials before science learning takes place.			
2	Let students predict what would happen if they carry the activities.			
3	Providing students opportunities to observe the science experiments.			
4	Providing students verbal praises.			
5	Providing positive reward when a student expresses his or her opinion.			
6	Probe a student's statement for a clearer understanding.			
7	Providing opportunities for every student to express his or her opinion.			
8	Identifying a student's answer as wrong without giving a negative label to the student.			
9	Listening to a student's question or statement until he or she finishes.			
10	Encouraging students to express their opinion.			
11	Encouraging students to take notes or other documentation of what happen during science learning.			
12	Encouraging students to make comparisons.			
13	Put students in groups before an experiment takes place.			
14	Encouraging groups to help a member who is not able to finishes an assignment properly.			
15	Asking another student an answer of a question posed by one student before providing explanations.			
Aspect 2. Providing instructions for the activities				
16	Explaining the objective(s) of the lesson before the lesson starts.			
17	Explaining the procedures of an experiment before it starts.			
18	Clarifying the teacher's and the students' roles in running an experiment.			
19	Asking an open-ended question with more than two possible answers.			
Aspect 3. Managing learning environments				
20	Giving an example on how to proceed on difficult task.			
21	Asking students to concentrate on every step of the activities.			
22	Asking students to focus on the activities being carried on.			
Aspect 4. Monitoring students' critical thinking				
23	Let students carry the experiment on their own.			
24	Providing feedback on mistakes after the whole procedure had been finished.			
25	Monitoring around the classroom when students carry the experiment independently.			

Diversity in Unity: Perspectives from Psychology and Behavioral Sciences – Ariyanto et al. (Eds)
© 2018 Taylor & Francis Group, London, ISBN 978-1-138-62665-2

The correlation between young children's emotion regulation and maternal stress in low socioeconomic status families

R.F. Dewi & M.A. Tjakrawiralaksana
Faculty of Psychology, Universitas Indonesia, Depok, Indonesia

ABSTRACT: The objective of this study is to investigate the correlation between young children's emotion regulation and maternal stress in low socioeconomic status (SES) families. The participants (N = 122) were low SES mothers and young children aged 3–6 years living in Jakarta, Depok, and Bekasi (Jadetabek) areas. Maternal stress was measured using Stress Subscale of Depression, Anxiety, and Stress Scale 21-items (DASS-21), while young children's emotion regulation was measured using Emotion Regulation Checklist (ERC) reported by their mothers. The results show that there was a significant, negative correlation between young children's emotion regulation and maternal stress. This indicates that the higher maternal stress is, the lower young children's emotion regulation is. It has also been found that maternal employment status is associated with the level of their stresses.

1 INTRODUCTION

Children are expected to have an understanding of emotion, develop active and well-planned emotion regulation strategies, and increase their awareness of emotion for adaptive skills development since early childhood (Berk, 2012). Poor emotion regulation would impact negatively on some domains of development, i.e. affective, social, and cognitive (Kim-Spoon *et al.*, 2013 in Molina *et al.*, 2014). On the other hand, young children with higher emotion regulation skills would have no difficulties in later psychological adjustments; they also had better social skills and higher academic performances in elementary school (Thompson, 1994).

Emotion regulation becomes more complex with age. At the age of 3–6 years, children develop interpersonal emotion regulation strategies. They are able to understand, express, and modulate emotion independently (Holodynski & Friedlmeier, 2006). Caregivers' involvement, especially mothers, is an important external factor which influences children's emotion regulation development (Calkins, 1994). Bernard and Solchany (2002) have found that the interaction between children and their mothers had a greater impact on children's development than children's interaction with other caregivers.

Mothers have three major roles: to supervise, to nurture, and to response to their children (Winnicott, as cited in Bernard & Solchany, 2002). Mothers with low SES do not perform these roles optimally because they have difficulty in providing developmental facilities for their children, such as health services, education, and quality residences (Scaramella, Neppl, Ontai, & Conger, 2008). Low-income mothers are continually exposed to financial problems and could easily suffer from stress. Maternal stress affects their actions. The mothers tend to be more controlling, give negative feedbacks, and show aggressive behaviors towards their children (Rice, 1999; Blandon, Calkins, Keane & O'Brien, 2008). Maternal stress also correlates with less warmth and inductive discipline techniques during a mother-child interaction (Choe, Olson & Sameroff, 2013).

Prior studies attempted to prove the correlation between maternal negative emotions, such as stress or depression, and children's low emotion regulation (Blandon, Calkins, Keane & O'Brien, 2008; Lengua *et al.*, 2008; Kam *et al.*, 2011). However, a contradictory result is shown by Choe, Olson, and Sameroff (2013) who found that maternal stress is mediated by inductive discipline and maternal warmth and does not correlate directly with children's emotion regulation. Nevertheless, all of those studies only involved low-middle SES families in the US, and most of the participants are of African-American, European-American, and Hispanic origins.

Different from those previous studies, this study was conducted in Indonesia, which is one of the most collectivistic countries, while 10.96% of its 250 million citizens were coming from low SES family (BPS, 2014). Low SES is marked by low income, low household expense, low level of education, and low employment status. Parents with low SES have difficulties in providing a secure environment for stimulating their children's development. Children who live in low SES families are susceptible to various physical, intellectual, and psychological problems (Evans, 2014), such as a low ability to regulate their emotion (Eisenberg, Hoferz & Vaughan, 2007). Therefore, the primary goal of this study is to test the correlation between young children's emotion regulation and maternal stress in low socioeconomic status families.

1.1 *Emotion regulation*

Shields and Cicchetti (1997) define emotion regulation as the capacity to modulate the emotional arousal to reach an optimal level of engagement with the environment. Children's emotion regulation focuses on the capacity to modulate emotions, such as the process of determining, establishing, directing, controlling, managing, and even modifying emotions that are appropriate with specific expectations and conditions. Basic emotion regulation skills began to emerge in babies and toddlers, and then preschoolers and school-age children develop more complex behavior and cognition when they regulate their emotions (Calkins & Hill, 2007).

There are two dimensions of emotion regulation: emotion regulation (ER) and lability/negativity (LN) (Shield & Cicchetti, as cited in Molina *et al.,* 2014). Emotion Regulation (ER) is a set of skills to express an appropriate emotion, empathy, and emotional awareness, while Lability/Negativity (LN) is a dimension to measure rigidity, emotion dysregulation, and mood swing. The two dimensions complement each other in showing emotion regulation processes such as affective lability, flexibility, and compatibility with a particular situation (Molina *et al.,* 2014).

The development of emotion regulation is influenced by internal and external factors. Internal factors consist of age, gender, temperament and language skills, whereas external factors consist of parenting style, coaching, attachment, culture, and socioeconomic status. Santrock (2011) found that children's age affects the complexity of management attention in emotion regulation, the development of emotion modulation capacity, and the change in strategy from interpersonal to intrapersonal. His study also shows differences of emotion understanding and self-control between boys and girls.

Furthermore, harsh parenting and punishment contributed to children emotion dysregulation (Eisenberg *et al.* in Chang, Schwartz, Dodge & McBride-Chang, 2003). Poor emotion regulation was also influenced by parents' coaching techniques. Emotion-coaching parents would help their children to deal with their emotion effectively, whereas emotion-dismissing parents would encourage their children to refuse, ignore, or change the negative emotion (Gottman, as cited in Santrock, 2011). Cassidy (1994) proves that children-mother attachment correlated with children's emotion regulation. Children with a secure attachment are more open to their parents when dealing with negative emotions. On the contrary, children with insecure attachment chose to hide their stress or emotions.

There are also differences between children's emotion regulation developments in individualistic and collectivistic cultures. In an individualistic culture, mothers encourage their children to express their emotions, experience feelings, and conceptualize emotions (Holodynski & Friedlmeier, 2006). Meanwhile, mothers in collectivistic culture directed their children to ignore and modify their emotions.

1.2 *Stress*

Lazarus and Folkman (in Harrington, 2013) define stress as constellation of cognitive, emotional, physiological, and behavioral reactions of people who experience threat and challenge.

Stress cannot be separated from its four constituent components, which are resources, demands, discrepancy, and transaction (Lazarus & Folkman, as cited in Sarafino & Smith, 2011). Stress appears when there is a discrepancy between sources of stress (demands) and potentials (resources) that can be used to overcome the pressure (transaction).

An individual's level of stress is influenced by internal factors, such as intellectual, motivation, self-esteem, and gender. Individuals who have irrational beliefs tend to overstate minor issues, which leads them to experience a higher level of stress than average people do. People who have a greater motivation to achieve a target will also experience a higher level of stress (Patterson & Neufield in Sarafino & Smith, 2011). Moreover, people with high self-esteem tend to see a stressful event as a challenge rather than a threat (Sarafino & Smith, 2011). Research also found that an individual's gender also affects his/her level of stress; Aranda *et al.* (2001) demonstrate that women have a higher level of vulnerabilities and stress than men do.

Stress is also influenced by external factors which consist of stressful situations, social support, and socioeconomic status. Stressful situations have a positive correlation with stress level, whereas social support has a negative correlation with the stress level. Social support helps individuals coping with (Gatchel, Baum & Krantz, 1989) and prevents them from experiencing the negative effects of stress (Jenkins, 1991). Mothers with low SES often experience stressful events, such as divorce, death of a child, dangerous situations, and criminal acts in their residential areas (Rice, 1999).

2 METHOD

2.1 *Participants*

Participants were 122 mothers (age range: 20–40 years old) who had children (64 male, 58 female; age range: 3–6 years old) and lived in Jakarta, Depok, Tangerang, or Bekasi areas (see Table 1 for demographic details and test scores). All of the participants come from low socioeconomic status families (monthly household expenditure ≤ USD 23–54). The mothers'

Table 1. Demographic characteristics of participants.

Characteristics	Frequency	Percentage
Child's age		
3 years	23	18.9
4 years	37	30.3
5 years	28	23
6 years	34	27.9
Child's Gender		
Female	58	47.5
Male	64	52.5
Mother's Level of Education		
Primary or lower	31	25.4
Middle School	28	23
High School	63	51.6
Mother's Employment Status		
Full-time (≥35 hours/week)	15	12.3
Part-time (<35 hours/week)	14	11.5
Freelance	11	9
Unemployed	82	67.2
Household Expense per month		
<USD 23	20	16.4
USD 23–USD 38	33	27
USD 38–USD 54	69	56.6

education levels range from primary school to high school. Sixty-three of the mothers (51.6%) are high school graduates, while the rest are middle school graduates (23%) or lower (25.4%). Most of the mothers are unemployed or housewives (67.2%).

2.2 *Measurement*

2.2.1 *Emotion regulation*

The Emotion Regulation Checklist (ERC) is a 24-item mother-report measure which is used to assess children's emotion regulation (Shileds & Cicchetti, 1997). ERC tested the central processes of emotion regulation, such as affective liability, intensity, valence, flexibility, and situational appropriateness ($\alpha = 0.872$, $r < 0.2$). The Emotion Regulation (ER) subscale (8 items) measures mothers' perceptions of their children's appropriate emotional expression, feeling, empathy, and emotional self-awareness. The Lability/Negativity (LN) is a 15-item mother-report measure which is used to assess children's response of anticipation and strategy to overcome negative emotions. Using a 4-point Likert scale, mothers rated how likely their children are to show certain behaviors. The total score was obtained by adding up all the scores of all dimensions: the higher the total score is, the better the child's emotion regulation is.

2.2.2 *Stress*

Maternal stress level was measured by the stress subscale of Depression, Anxiety, and Stress Scale 21-items (DASS-21)that consists of 7 items, i.e. items number 1, 6, 8, 11, 12, 14, and 18 (Lovibond & Lovibond, 1995). The stress subscale of DASS-Smeasured non-specific aspects of arousal, such as nervous arousal, difficulty in relaxing, impatience, and tendency to be easily annoyed or angry (Lovibond 1998). DASS-S uses a 4-point Likert scale ranging from 0 to 3 ($\alpha = 0.687$, $r < 0.2$). The whole items were added up and multiplied by 2 to obtain a total score. A high total score indicates a high level of stress.

2.3 *Procedure*

We visited slum areas in Jakarta, Depok, Tangerang, and Bekasi (Jadetabek) areas that had been identified in previous surveys carried out by students from Universitas Indonesia. Initially, we asked for permission to conduct research activitiesin those areas and scheduled the date for data collection. Then, data were collected under the supervision and guidance of the researchers. Each participant had to complete two questionnaires (ERC and DASS-S). Most of the mothers completed the questionnaire independently, but we also read the content of questionnaires for those who were illiterate.

2.4 *Data analytic strategy*

Data were analyzed using Pearson product-moment correlation. Two-tailed tests were performed to assess our current hypotheses. For further analysis, the correlation between maternal stress and socioeconomic factors (education level, employment status, and household expenditure) were measured using the analysis of variance (ANOVA).

3 RESULTS

Results show that 43.4% of children had poor emotion regulation, whereas 56.5% of children have good emotion regulation (MERC = 73.11, SDERC = 6.392). Furthermore, 46.7% of mothers have a high level of stress, while 53.3% have a low level of stress (MDASS-S = 10.98, SDDASS-S = 7.378).

There is a significant and negative correlation ($R = -0.255$, LoS.01) between young children's emotion regulation and maternal stress. This finding indicates that the higher the level of maternal stress is, the lower young children's emotion regulation is in low SES families.

Table 2. The correlation between young children's emotion regulation and maternal stress.

N	R	R^2	Significant (2-tailed)	Description
122	−.255**	.065	.005	Significant in LoS.01

**$p < .01$.

Table 3. The association between maternal stress and employment status.

Participant data	N	M	Significant	Description
Employment status				
Full-time	15	10.27	F = 3.094	Significant
Part-time	14	10.71	p = 0.030	
Freelance	11	17.27		
Unemployed	82	10.32		

There is a significant association between maternal stress and mother's employment status ($F = 3.094$ and $p = 0.030$, $p < 0.05$).This finding indicates that there is a significant difference between the stress levelsof mothers who work full-time, work part-time, work freelance, and are unemployed. Mothers with a freelance status have the highest level of stress ($M = 17.27$).

4 DISCUSSION

In line with previous studies, the present study demonstrates that maternal stress correlates with young children's emotion regulation in low socioeconomic status families. Low-income mothers tend to be more stressful and to produce children with poor emotion regulation (Blandon, Calkins, Keane & O'Brien, 2008; Lengua *et al.*, 2008; Kam *et al.*, 2011). This current study shows that mothers' employment status correlates with maternal stress. Working mothers who perform two roles at the same time experience a higher level of stress. They are more exposed to various sources of stress, such as problems in their workplace or house or problems related to mother-child interaction (Lian & Tam, 2014).

However, this study has several limitations that must be taken into consideration. The correlation coefficient ($R = 0.2$) is weak. There is only 6.5% of variance in children's emotion regulation that can be explained by the relationship between maternal stress and children's emotion regulation, whereas 93.5% is explained by other variables that are not measured in this study. One of those variables is that mother and child interaction which could affect children's emotional development, such as mother's expression, mother's reaction, and discussions about emotion between mother and child. Chang, Schwartz, Dodge, and McBride-Chang (2003) mention parenting style as another variable that affects the relationship between maternal stress and children's emotion regulation. Stressful mothers tend to use physical punishment on their children, which would trigger emotion dysregulation in those children. Maternal stress also correlates with inductive discipline and maternal warmth (Choe, Olson & Sameroff, 2013). Lack of maternal warmth may result in maladaptive emotion regulation in children (Bariola, Gullone & Hughes, 2011). Similarly, mothers who rarely give positive responses or assert an excessive level of control on their children will also produce children with poor emotion regulation strategies.

5 CONCLUSION

This study found a significant correlation between young children's emotion regulation and maternal stress in low socioeconomic status families. Mothers with a higher level of stress tend to have young children with a lower emotion regulation level, whereas mothers with a lower level of stress tend to have young children with higher emotion regulation level.

An important note for future research is to reexamine the research methodology. First, we need to revise bad items and retest the reliability and validity of the questionnaires. For sampling, we suggest using the probability sampling method. The whole population data can be retrieved from statistical institutions, such as Tim Nasional Percepatan Penanggulangan Kemiskinan (TNP2K) or Badan Pusat Statistik (BPS). Second, the effectiveness of data collection process can be improved if we provide display media to assist the participants in determining their answers. For example, we can use image or symbol cards which represent "never", "rare", "often", and "almost always". Third, we need to add other methods such as behavioral observation to measure young children's emotion regulation to complement the questionnaire. Further research should also be carried out to include external factors that were not tested in the current study. The correlation between young children's emotion regulation and maternal stress can be tested by using parenting style as the mediator or moderator for both variables.

The results of the current study can be used as a basis for performing intervention and providing information for low-income mothers about maternal stress and its impacts on young children's emotion regulation. We also encourage social workers to empower low-income mothers by conducting psycho-education programs or training about coping stress, parenting, or entrepreneurial skills to improve their welfare.

REFERENCES

Aranda, M.P., Castaneda, I., Lee, P. & Sobel, E. (2001) Stress, social support, and coping as predictors of depressive symptoms: Gender differences among Mexican Americans. *Social Work Research,* 25 (1), 37–48. http://search.proquest.com/docview/212112217?accountid=17242.

Badan Pusat Statistik. (2014) *Jumlah Persentase Penduduk Miskin, Garis Kemiskinan, Indeks Kedalaman Kemiskinan (P1) dan Indeks Keparahan Kemiskinan (P2) Menurut Provinsi, Maret 2014.* http://www.bps.go.id/tab_sub/view.php?tabel=1andid_subyek=23.

Bariola, E., Gullone, E. & Hughes, E.K. (2011) Child and adolescent emotion regulation: The role of parental emotion regulation and expression. *Clinical Child and Family Psychology Review,* 14, 198–212.

Berk, L.E. (2012) *Developmental through Life-span.* 5th edition. Boston, Pearson.

Bernard, K.E., & Solchany, J.E. (2002) Mothering. In Bornstein, M.H. (Ed.), *Handbook of Parenting, Volume 3: Being and Becoming a Parent.* New Jersey, Lawrence Erlbaum Associates, Publishers.

Blandon, A.Y., Calkins, S.D., Keane, S. P. & O'Brien, M. (2008) Individual differences in trajectories of emotion regulation processes: The effects of maternal depressive symptomatology and children's physiological regulation. *Developmental Psychology,* 44(4), 1110–1123. DOI: 10.1037/0012-1649.44.4.1110.

Calkins, S.D. (1994) Origins and outcomes of individual differences in emotion regulation. *Monographs of the Society for Research in Child Development,* 59(2–3), 53–72. http://www.jstor.org/discover/10.2307/1166138.

Calkins, S.D. & Hill, A. (2007) Caregiver influences on emerging emotion regulation. In Gross, J.J. (Ed.), *Handbook Of Emotion Regulation.* USA, The Guilford Press.

Cassidy, J. (1994) Emotion regulation: Influences of attachment relationships. *Monographs of the Society for Research in Child Development,* 59(2/3), 228–249. www.jstor.org/stable/1166148.

Chang, L., Schwartz, Dodge, D.A. & McBride-Chang, C. (2003) Harsh parenting in relation to child emotion regulation and aggression. *Journal Family Psychology,* 17(4), 598–606. DOI: 10.1037/0893-3200.17.4.598.

Choe, D.E., Olson, S.L. & Sameroff, A.J. (2013) Effects of Early Maternal Distress and Parenting on The Development of Children's Self-Regulation and Externalizing Behavior. *Development and Psychopathology,* 25, 437–453. DOI: 10.1017/S0954579412001162.

Cole, P.M., Martin, S.E. & Dennis, T.A. (2004) Emotion regulation as a scientific construct: Methodological challenges and directions for child development research. *Child Development,* 75 (2), 317–333.

Eisenberg, N., Hofer, C. & Vaughan, J. (2007) Effortful control and its socioemotional consequences. In Gross, J.J. (Ed.) *Handbook of Emotion Regulation.* USA, The Guilford Press.
Eisenberg, N., Liew, J. & Pidada, S.U. (2004) The longitudinal relations of regulation and emotionality to quality of indonesian children's socioemotional functioning. *Developmental Psychology,* 40 (5), 790–804. DOI: 10.1037/0012-1649.40.5.790.
Evans, G.W. (2004) The Environment of Childhood Poverty. *American Psychologist* 59 (2), 77–92. DOI: 10.1037/0003-066X.59.2.77.
Gatchel, R.J., Baum, A. & Krantz, D.S. (1989) *An Introduction to Health Psychology*. 2nd edition. New York, McGraw-Hill, Inc.
Harrington, R. (2013) *Stress, Health, and Well-Being: Thriving in The 21st Century.* USA, Wadsworth, Cengage Learning.
Hill, A.L., Degnan, K.A., Calkins, S.D. & Keane, S.P. (2006) Profiles of externalizing behavior problems for boys and girls across preschool: The roles of emotion regulation and inattention. *Developmental Psychology,* 42(5), 913–928. DOI:10.1037/0012-1649.42.5.913.
Hoff, E., Laursen, B. & Tardif, T. (2002) Socioeconomic Status and Parenting. In Bornstein, M.H. (Ed), *Handbook of Parenting,* 2. New Jersey, Lawrence Erlbaum Associates, Publisher.
Holodynski, M., & Friedlmeier, W. (2006) *Development of Emotions and Emotion Regulation.* New York, Springer.
Jenkins, R. (1991) Demographic Aspect of Stress. In Cooper, C.L., & Payne, R. (Eds.). *Personality and Stress: Individual Differences in the Stress Process.* Chichester, John Wiley and Sons Ltd.
Kam, C.M., Greenberg, M.T., Bierman, K.L., Coie, J.D., Dodge, K.A., Foster, M.E. John E., Lochman, McMahon, R.J. & Pinderhughes, E.E. (2011) Maternal Depressive Symptoms and Child Social Preference during the Early School Years: Mediation by Maternal Warmth and Child Emotion Regulation. *Journal of Abnormal Child Psychology,* 39 (3), 365–77. doi:10.1007/s10802-010-9468-0.
Lengua, L.J., Bush, N.R., Long, A.C., Kovacs, E. & Trancik, A.M. (2008) Effortful control as a moderator of the relation between contextual risk factors and growth in adjustment problems. *Development and Psychopathology,* 20 (2), 509–28. doi:10.1017/S0954579408000254.
Lian, S.Y. & Tam, C.L. (2014) Work stress, coping strategies and resilience: A study among working females. *Asian Social Science,* 10 (12), 41–52. doi:10.5539/ass.v10n12p41.
Lovibond, P.F. (1998) Long-Term Stability of Depression, Anxiety, and Stress Syndromes. *Journal of Abnormal Psychology,* 107 (3), 520–26. doi:10.1016/0005-7967(94)00075-U.
Lovibond, P.F. & Lovibond, S.H. (1995) The Structure of negative emotional states: Comparison of the Depression Anxiety Stress Scales (DASS) with the beck depression and anxiety inventories. *Behaviour Research and Therapy,* 33 (3), 335–43. doi:10.1016/0005-7967(94)00075-U.
Lovibond, S.H. & Lovibond, P.F. (1995) Manual for the Depression Anxiety Stress Scales. *Psychology Foundation of Australia*, 56. doi:10.1016/0005-7967(94)00075-U.
Molina, P., Sala, M.N. & Zappulla, C. (2014) The emotion regulation checklistitalian translation. validation of parent and teacher versions. *European Journal of …*,11 (5), 624–34. doi:http://dx.doi.org/10.1080/17405629.2014.898581.
Nolen-Hoeksema, S. (2012) Emotion regulation and psychopathology: The role of gender. *Annual Review of Clinical Psychology,* 8 (1): 161–87. doi:10.1146/annurev-clinpsy-032511-143109.
Rice, P.L. (1999) *Stress and Health.* 3rd edition. USA, Brooks/Cole Publishing Company.
Santrock, J. (2011) *Child Development.* USA, McGraw-Hill.
Sarafino, E.P. & Smith, T.W. (2011) *Health Psychology: Biopsychosocial Interaction.* 7th edition. USA, John Wiley and Sons, Inc.
Scaramella, L.V., Neppl, T.K., Ontai, L.L. & Conger, R.D. (2008) Consequences of socioeconomic disadvantage across three generations: Parenting behavior and child externalizing problems. *Journal of Family Psychology : JFP : Journal of the Division of Family Psychology of the American Psychological Association (Division 43),* 22 (5), 725–33. doi:10.1037/a0013190.
Shields, A., & Cicchetti, D. (1997) Emotion regulation among school-age children: The Development and validation of a new criterion q-sort scale. *Developmental Psychology,* 33 (6), 906–16. doi:10.1037/0012-1649.33.6.906.
Thompson, R.A. (1994) Emotion regulation: A theme in search of definition. *Monographs of the Society for Research in Child Development,* 59 (2–3), 25–52. doi:10.1111/j.1540–5834.1994.tb01276.x.

Diversity in Unity: Perspectives from Psychology and Behavioral Sciences – Ariyanto et al. (Eds)

Optimising executive function in early childhood: The role of maternal depressive symptoms and father involvement in parenting

A.E. Nurilla, D. Hendrawan & N. Arbiyah
Faculty of Psychology, Universitas Indonesia, Depok, Indonesia

ABSTRACT: Depression is the most common psychopathological condition that affects the majority of women, particularly mothers. Not only does this condition predict negative outcomes in maternal lives, but this applies also to their children. Numerous studies have revealed that maternal depressive symptoms could impede the Executive Function (EF), a higher mental function that controls behaviour, cognition, and emotion of their children. Nonetheless, an inconsistency was found in the results of studies about the role of maternal depressive symptoms in predicting the EF of preschoolers because such studies failed to take the role of the father figure into consideration. This study aims to assess the contribution of maternal depression and father involvement to predicting the EF of preschool children. As many as 101 children aged 4–6 and their respective parents were involved. Several EF tests were performed on the children, while a maternal depressive symptoms self-report scale and a father involvement questionnaire were given to the mother and the father respectively. The results show that maternal depressive symptoms was negatively related to the EF performances of children after controlling child's gender and age, maternal work status, and family's socioeconomic level. This study points out the importance of the maternal psychological condition while targeting interventions for improving the EF of preschoolers.

1 INTRODUCTION

Depression, a common yet serious psychological disorder is highly prevalent among mothers of young children (Hall, 1990). The implications of depression are far-reaching. Not only does it lower the well-being of mothers (Merikangas et al., 2007; Marcotte & Wilcox-Gok, 2001), but it also heightens the risk of their children being exposed to a plethora of negative outcomes. Moreover, the risk is higher for children of a younger age (Shonkoff & Phillips, 2000). Indeed, many researchers have extensively studied the relationship between maternal depression and preschool children's poor functioning in emotional, behavioural, and cognitive aspects. Compared to the children of non-depressed mothers, the children of depressed mothers show poor regulation of emotion and emotional expression (Hoffman et al., 2006; Hooper et al., 2015), as well as higher rates of both externalising (e.g. hyperactivity and aggressive behaviour) and internalising problem behaviours (e.g. withdrawal) (Turney, 2012; Conners-Burrow et al., 2014; Hooper et al., 2015). In terms of cognition, they also show lower scores in language tests, intelligence tests, (Brennan et al., 2000; Slykerman et al., 2005), and academic achievement (Claessens et al., 2015). Until recently, however, only little attention has been given to the negative outcomes of maternal depression which affect children's higher cognitive functioning, such as the EF. The current study addresses this gap by focusing on the relationship between maternal depression and children's EF.

The term EF refers to a set of higher mental processes that enables an individual to control his/her thoughts, action, and emotion consciously and purposefully, which is relevant to behavioural adjustment (Zelazo & Carlson, 2012). EF develops most rapidly during preschool years, along with the development of the prefrontal cortex, a brain area that contributes greatly to the execution of EF process. Numerous research studies have revealed that both EF and prefrontal cortex are susceptible to adverse environmental conditions (e.g. Mezzacappa et al., 2001; Odgers & Jaffee, 2013). One condition that is believed to produce adverse outcomes which affect EF and prefrontal cortex is maternal depression.

Previous research has confirmed the relationship between maternal depressive symptoms and a child's performance of EF tasks in the present (Hughes & Ensor, 2009) or a few years later (Hughes et al., 2013; Roman et al., 2016). In their studies, which involved single and non-single mothers as participants, maternal depression was found as one factor that contributes significantly to the poor EF performance of children. However, in another study, a significant relationship was not found (Rhoades et al., 2011). The result from a study conducted by Rhoades et al. (2011), which involved only non-single mothers, suggests that the difference of EF performance between children of depressed mothers and those of non-depressed mothers was not present. This difference in the marital status of the mothers implies that the group of children from non-single mothers has a father figure, while the group of children from single mothers does not. Thus, we argue that such conflicting findings might be due to the existence and non-existence of a father figure.

In the case of maternal depression, the existence of a father figure might be beneficial for both mother and child. For the mother, social support secured by the presence of a partner could lower the intensity of the depression (Kring et al., 2013). Meanwhile for the child, the presence of a father could compensate the ineffective parenting style performed by a depressed mother (Goodman et al., 2014); thus, it could lower the risk of children being exposed to the negative outcomes (Mezulis et al., 2004). Moreover, past research has revealed that a father's involvement in parenting predicted a variety of child outcomes, including the cognitive (Bronte-Tinkew et al., 2008) and social-emotional ones (Baker, 2013). Consequently, when the role of a father figure is taken into account in the studies, different results might be generated.

In addition to producing inconclusive results, previous research also tended to focus on only one domain of EF, the cool EF (domain that covers cognitive problem-solving; Zelazo & Muller, 2010), while less attention has been given to investigate the 'hot' EF, which is another domain covering emotional problem-solving (Zelazo & Muller, 2010). However, maternal depression and children's hot EF might be related, given the fact that both contain emotional elements. Furthermore, there has been considerable research which sought to identify the impacts of maternal depression on children's emotional regulation and expression (Hoffman, et al., 2006; Hooper et al., 2015), one of the functions controlled by the hot EF.

This current study aims to investigate the independent relationship between maternal depressive symptoms, father involvement, and preschool children's cool and hot EF, which take into consideration secondary factors, such as a child's age and gender, a family's socioeconomic level, as well as maternal work status, all of which have previously been found to be related to EF (e.g. Hendrawan et al., 2016; Hewage et al., 2011). Instead of single mothers, this study focuses on non-single mothers because we wanted to examine the role of father involvement in cases of maternal depression. Maternal depressive symptoms and father involvement were assessed using self-report, while child EF was assessed using performance test-tapping cool and hot EF. It was hypothesised that a father who had high involvement with his children while the mother showed depressive symptoms would have children with better cool and hot EF skills, above and beyond the effects of the child's age, the child's gender, the family's socioeconomic level and maternal work status.

2 METHOD

2.1 *Participants*

Participants were recruited from kindergartens in the cities of Jakarta, Bogor, Depok, Tangerang and Bekasi. Out of 110 families, complete data were obtained from 101 families. Nine children did not meet the characteristic requirements; therefore, they were excluded from the analysis. The remaining 101 children (M = 61.75 months, SD = 6.80 months) were relatively proportional in terms of gender (51 girls, 50 boys), but socially diverse in terms of family socioeconomic level. All of the participants included in this study had passed the screening test which was carried out to assess the children's developmental disorder. Low parenting involvement in the participants' lives was also avoided by ensuring that all of their parents are legally married and do not live apart from their respective partners.

2.2 *Measures*

The children were asked to complete four performance tasks on EF: Matahari/Rumput, Backward Word Span (BWS), Dimensional Change Card Sort (DCCS), and Gift Delay, while their fathers and mothers completed the CESD-R and the father involvement questionnaire respectively. The performance tasks were administered by trained research assistants under the supervision of a licensed psychologist.

2.2.1 *Matahari/rumput*

Matahari/Rumput (Hendrawan et al., 2016) is a version of the Grass/Snow task (Carlson et al., 2014) that has been adapted to Indonesian culture. With the objective of assessing a person's inhibitory control, this test consists of 16 trials. Children were asked to point out a green card when a tester said *matahari* (sun) or a yellow card when the tester said *rumput* (grass). Performance was rated by the total number of correct trials (0–16).

2.2.2 *Backward word span*

BWS is a working memory test developed by Carlson et al. (2014) and was adapted to Indonesian culture by Hendrawan et al. (2016). Children were asked to repeat a sequence of words in reverse order. There are five levels, each of which constitutes a different score (0–5).

2.2.3 *Dimensional change card sort*

DCCS was developed by Zelazo (2006) and adapted to Indonesian by Hendrawan et al. (2016) to assess children's set shifting. In this test, children were asked to sort cards according to certain rules. There are three phases of the test: pretest, post-test, and post-test-border. Performance was rated by the total number of correct trials in both post-test (0–6) and post-test-border phase (0–12).

2.2.4 *Gift delay*

Measuring hot EF, Gift Delay was developed by Carlson et al. (2014) and adapted to *Bahasa* (Indonesian) by Hendrawan et al. (2016). Children were asked to wait and resist peeking while the tester wrapped a gift noisily within a two-minute duration. Three scores were derived from this test: latency of peeking, frequency of peeking, and the extent of child peeking.

2.2.5 *CESD-R*

Mothers completed the Indonesian version of the Center for Epidemiologic Studies Depression Scale-Revised (CESD-R), which measures the level of perceived depressive symptoms. CESD-R comprises 20 items that are rated on a 5-point Likert scale. The Cronbach's alpha of CESD-R was 0.84.

2.2.6 *PIQ*

Fathers completed the Indonesian version of Parent Involvement Questionnaire (PIQ) (Meuwissen & Carlson, 2015) that measures the degree of father involvement. The PIQ comprises 13 items that are rated on a 6-point Likert scale, but in the present research one item that was not relevant to the characteristics of the children involved in this study was eliminated. The Cronbach's alpha for the 12 items was 0.86.

2.3 *Procedures*

Meetings in kindergartens were organised to introduce the research to prospective parents of the participants. Parents were asked to complete a family demographic questionnaire as they gave consent to their children's participation in the study. The children then underwent the EF test in an isolated room while their fathers and mothers completed the father involvement questionnaire and CESD-R respectively.

2.4 *Data analysis*

Due to the differences in the scoring method for EF tasks, the raw scores from cool EF tasks (Matahari/Rumput, BWS, and DCCS) and the three scores from Gift Delay were first transformed into standard scores and then combined to form the children's cool and hot EF composite scores, respectively. Data were analysed using Pearson correlation and two-block hierarchical regression analyses. Pearson correlation was conducted to examine the zero-order correlation between variables, while a two-block hierarchical regression analysis was conducted to examine how the predictive power of primary variables would hold after accounting for secondary variables. The secondary variables (i.e. the child's gender and age, the family's socioeconomic level and maternal work status) were entered in the first block, while the predictor variables were entered in the second block.

3 RESULTS

Table 1 shows the descriptive statistics of children's EF and parents' variables. Based on the observed score, the father involvement score could be classified as relatively varied. Most mothers were classified as non-depressed based on their CESD-R scores that fell below the cutoff point of depressed category (≥ 16).

Table 2 presents the zero-order correlations between maternal depressive symptoms, father involvement, and child performance in the two EF domains. Significant correlations were found only in both EF domains and maternal depressive symptoms. While the two EF domains were interrelated, maternal depressive symptoms were negatively correlated with both cool and hot EF, indicating that greater maternal depressive symptoms were related to lower child performance of both cool and hot EF.

Table 3 summarises the result of the hierarchical regression analyses predicting both hot and cool EF. The hierarchical multiple regression revealed that, at step one, secondary variables jointly contributed to the 12.9% variation in hot EF. Adding maternal depression into

Table 1. Means, standard deviations and ranges of children's EF, maternal depression, and father involvement measures.

Variable	N	M	SD	Observed range	Theoretical range
Maternal depressive symptoms	101	10.27	8.25	0–35	0–80
Father involvement	101	28.52	8.11	4–46	0–60
Cool EF	99	9.99	2.01	3.45–13.02	
Hot EF	101	10	2.49	3.24–20.83	

Table 2. Zero-order correlation between maternal depressive symptoms, father involvement and child performance in two EF domains.

	1	2	3
Hot EF	–		
Cool EF	0.462**	–	
Maternal depressive symptoms	–0.308**	–0.202*	1
Father involvement	–0.003	0.091	0.112

Note. *p < 0.05; **p < 0.01 one-tailed.

Table 3. Summary of regression analyses predicting child cool and hot EF.

	EF domains			
	Hot		Cool	
Block	β	ΔR²	B	ΔR²
Covarites		0.129*		0.289**
Gender	0.114		0.053	
Child's age	0.234*		0.453**	
MWS	–0.083		0.093	
SES	0.317**		0.260*	
Parents' variables regression		0074*		0.031
Gender	0.126		0.053	
Child's age	0.256**		0.469*	
MWS	–0.061		0.098	
SES	0.234*		0.205	
MDS	–0.280**		–0.164	
FI	–0.035		0.076	

Note. MWS = maternal work status; SES = family's socioeconomic level; MDS = maternal depressive symptoms; FI = father involvement. $^*p < 0.05$; $^{**}p < 0.01$ one-tailed.

the model further explained the 7.4% of variance in hot EF, thus suggesting that even after controlling the child's age and gender, the family's socioeconomic level and maternal work status, the contribution of maternal depression to children's hot EF was still significant, with $F(6, 94) = 3.979$, $p < 0.05$. A lower level of maternal depressive symptoms was associated with better performance of children's hot EF ($\beta = -0.280$, $p < 0.01$). Unfortunately, this was not the case for cool EF. The significant contribution of maternal depression to predicting cool EF was not found, with $F(6, 92) = 7.215$, $p = 0.127$. However, maternal depression almost reached a significant value ($p = 0.067$). As for the case of father involvement, the analysis failed to find any significant contribution of father involvement to predicting both cool and hot EF.

4 DISCUSSION

This present study serves as an extension of previous works on maternal depression and children EF by focusing on the involvement of a father in parenting. Three major findings were obtained. First, it was demonstrated that maternal depression predicted children's hot EF. This finding was evident even after controlling other factors that might also contribute to the outcomes, that is, the child's age and gender, maternal work status, and the family's socioeconomic level. Second, confirming the previous finding, maternal depression was found to have a correlation with children's cool EF. Greater maternal depressive symptoms were related to a lower performance of children's cool EF. Third, the involvement of fathers was not associated with either cool or hot EF.

To our knowledge, the present study is the first to demonstrate the impact of maternal depression on children's emotional problem-solving, namely hot EF. It appears that a mother's depression may be detrimental to a child's hot EF development. Such relations can be explained through the modelling method which implies that a typical pattern of emotion regulation displayed by parents in daily life may become a learning source for children about what kind of emotion and particular emotional expressions are acceptable and expected in the family (Denham et al., 1997). Depressed mothers tend to show poor emotional regulation marked by uncontrollable tears, hostility, and angry tone. By witnessing their mother, children might unconsciously imitate those emotional regulation and expression, thus internalising them into their own behaviour. Besides the modelling of parents' emotional regulation, children's hot EF may also be affected by maternal depression through the influence of depressed mothers' EF. Research has found that depressed mothers show poor EF performance(Castaneda et al., 2008). The intergenerational transmission of EF by gene-environment mechanism may be one way of explaining how mothers' poor EF is transmitted into their children's poor EF (Cuevas, et al., 2014a, 2014b). Another mechanism is through a mother's poor parenting behaviours, which could negatively affect children's EF (Carlson, 2009).

In terms of the relationship between maternal depression and children's cool EF, these results are consistent with those of Hughes and Ensor (2009) who identified a negative association between maternal depression and children's cool EF. However, when those variables were added into the regression model, which took into consideration other secondary factors, maternal depression was found to have no association with children's cool EF. This result is different from those of Hughes et al. (2013) and Roman et al. (2016) who have identified the independent effects of maternal depression on children's cool EF. These differences in findings can be attributed to two factors. First, Hughes et al. (2013), Roman et al. (2016), and this study have controlled several different secondary variables. While this current study controlled the effects of a child's age, a child's gender, maternal work status, and a family's socioeconomic level on his/her EF, Hughes et al. (2013) controlled the influences of maternal education and parenting of the mother on children's EF. Meanwhile, Roman et al. (2016) controlled the influence of a child's verbal ability on his/her EF. It might be possible that the secondary variables in this study have more predictive power over the outcomes as it was found that a child's age could serve as a significant predictor before and after the primary variables were included. Moreover, numerous studies have confirmed that the development of EF is greatly influenced by a child's age (e.g. Diamond, 2006; Hendrawan et al., 2016). Thus, when the age factor was included in the analysis, the predictive power of maternal depression might be diminished. Second, the difference in the results might be attributed to the small sample size of this study because it was also found that the contribution of maternal depression almost reached the significant value.

In addition to supporting the results of Hughes et al. (2013) and the research of Roman et al. (2016), the current study also contradicts Rhoades et al. (2011), who did not find any differences between the EF performance of the children of depressed mothers and that of the children of non-depressed mothers. Despite the similarity between this study and that of Rhoades et al. (2011), both of which involved only non-single mothers, this study did not generate similar results. It is argued that the differences between the results might be due to the differences in children's ages. This present study involved children aged 4–6, while the study of Rhoades et al. (2011) involved children aged three. It might be possible that the poor EF skills observed in the study of Rhoades et al. was due to the underdeveloped EF skills of 3-year-old children, instead of the influence of their mothers' depression. Future studies should include children at younger ages to get a deeper understanding of how maternal depression could influence children's EF across all ages.

The involvement of the father in parenting did not contribute to either hot or cool EF. These findings are consistent with previous reports that did not find a significant association between father involvement and cool EF (Meuwissen & Carlson, 2015). This confirmatory result may indicate that, instead of the quantity of fathering, the quality of fathering may play a more significant role in the development of children's EF. It might be possible to find fathers who have a high level of involvement, yet show ineffective parenting, and vice versa. It would be beneficial in future studies to include both quantity and quality in the fathering measurement. For example, observations of parenting at home could also be included into the research procedures, in addition to the analysis of father involvement questionnaire scores.

The study presented has several limitations. First, the sample size was narrow. Second, it is still unclear about what the result would be if mothers' parenting performance was controlled. Research has found that depression can disrupt mother's parenting skills (Lovejoy et al.,

2000) which are known to contribute greatly to the development of children's EF (Bernier et al., 2010). Future studies should include mothers' parenting performance to gain a deeper understanding of how maternal depression could influence children's EF.

Finally, in this study, it was also observed that there were differences between the description written in the father involvement questionnaires and the actual behaviour those fathers demonstrated. Only around 25% of fathers who had high scores on the involvement questionnaires showed great interest in research activities and enthusiastically filled in the questionnaires, while the remaining fathers did not. This finding might demonstrate the weakness of using the self-report method to assess an individual's behaviour since it is known that there are many sources of bias in this measurement method. Future studies should include information from other members of the family, such as the mothers, to confirm the accuracy of information provided by the fathers.

5 CONCLUSIONS

Previous research on maternal depression and children's EF has paid little attention to the role of the father figure and tended to focus on cool EF. In order to address this gap, this current study examined the relationship between maternal depression, father involvement, and preschool children's EF and found that maternal depression may be detrimental to the development of hot and cool EF in preschool children. Future interventions to improve children's EF should not only involve observations of children's EF, but also consider maternal psychological conditions when designing an intervention to promote the development of children's EF skills.

ACKNOWLEDGEMENTS

This research was supported by a PITTA research grant from Universitas Indonesia. The authors thank Intan Putri Hertyas, Rizka Nurbatari, Ditta Metta Hestiany, Siti Nurlaila Fiam, Fasya Fauzani, Hanifah Nurul Fatimah, Claudia Carolina, as well as the schools and families who have participated in this study.

REFERENCES

Baker, C.E. (2013). Fathers' and mothers' home literacy involvement and children's cognitive and social emotional development: Implications for family literacy programs. *Applied Developmental Science, 17*(4), 184–197. doi:10.1080/10888691.2013.836034.

Bernier, A., Carlson, S.M. & Whipple, N. (2010). From external regulation to self-regulation: Early parenting precursors of young children's executive functioning. *Child Development, 81*(1), 326–339. doi:10.1111/j.1467–8624.2009.01397.x.

Brennan, P., Hammen, C., Andersen, M.J., Bor, W., Najman, J.M. & Williams, G.M. (2000). Chronicity, severity, and timing of maternal depressive symptoms: Relationships with child outcomes at age 5. *Developmental Psychology, 36*(6), 759–766. doi:10.1037/0012–1649.36.6.759.

Bronte-Tinkew, J., Carrano, J., Horowitz, A. & Kinukawa, A. (2008). Involvement among resident fathers and links to infant cognitive outcomes. *Journal of Family Issues, 29*(9), 1211–1244. doi:10.1177/0192513X08318145.

Carlson, S.M. (2009). Social origins of executive function development. *New Directions for Child and Adolescent Development, 2009*(123), 87–98. doi:10.1002/cd.237.

Carlson, S.M., White, R.E. & Davis-Unger, A.C. (2014). Evidence for a relation between executive function and pretense representation in preschool children. *Cognitive Development, 29* (January), 1–16. doi:10.1016/j.cogdev.2013.09.001.

Castaneda, A.E., Tuulio-Henriksson, A., Marttunen, M., Suvisaari, J., & Jouko L. (2008). A review on cognitive impairments in depressive and anxiety disorders with a focus on young adults. *Journal of Affective Disorders, 106*(1–2), 1–27. doi:10.1016/j.jad.2007.06.006.

Claessens, A., Engel, M. & Curran, F.C. (2015). The effects of maternal depression on child outcomes during the first years of formal schooling. *Early Childhood Research Quarterly, 32*, 80–93. doi:10.1016/j.ecresq.2015.02.003.

Conners-Burrow, A., Swindle, T., McKelvey, L. & Bokony, P. (2015). A little bit of the blues: Low-level symptoms of maternal depression and classroom behavior problems in preschool children. *Early Education and Development, 26*(2), 230–244. doi:10.1080/10409289.2015.979725.
Cuevas, K., Deater-Deckard, K., Kim-Spoon, J., Wang, Z., Morasch, K.C. & Bell, M.A. (2014a). A longitudinal intergenerational analysis of executive functions during early childhood. *British Journal of Developmental Psychology, 32*(1), 50–64. doi:10.1111/bjdp.12021.
Cuevas, K., Deater-Deckard, K., Kim-Spoon, J., Watson, A.J., Morasch, K.C. & Bell, M.A. (2014b). What's mom got to do with it? Contributions of maternal executive function and caregiving to the development of executive function across early childhood. *Developmental Science, 17*(2), 224–238. doi:10.1111/desc.12073.
Denham, S.A., Mitchell-Copeland, J., Strandberg, K., Auerbach, S. & Blair, K. (1997). Parental contributions to preschoolers' emotional competence: Direct and indirect effects. *Motivation and Emotion, 21*(1), 65–86. doi:10.1023/A:1024426431247.
Diamond, A. (2006). The early development of executive functions. In E. Bialystock & F.I.M. Craik (Eds.), *Lifespan cognition: Mechanisms of change* (pp. 70–95). Oxford, UK: Oxford University Press.
Goodman, S.H., Lusby, C.M., Thompson, K., Newport, J. & Stowe, Z.N. (2014). Maternal depression in association with fathers' involvement with their infants: Spillover or compensation/buffering? *Infant Mental Health Journal, 35*(5), 495–508. doi:10.1002/imhj.21469.
Hall, L.A. (1990). Prevalence and correlates of depressive symptoms in mothers of young children. *Public Health Nursing, 7*(2), 71–79. doi:10.1111/j.1525-1446.1990.tb00615.x.
Hendrawan, D., Fauzani, F., Carolina, C., Fatimah, H.N., Wijaya, F.P. & Kurniawati, F. (2016). The construction of executive function instruments for early child ages in Indonesia: A pilot study. In *Proceedings International Conference on Child and Adolescent Mental Health, "Promoting Children's Health, Development and Well-being: Integrating Cultural Diversity", 5–7 November 2015* (pp. 17–28). Banten, Indonesia: Faculty of Psychology, State Islamic University Syarif Hidayatullah Jakarta. Retrieved from http://psikologi.uinjkt.ac.id/wp-content/uploads/2016/08/Prosiding-ICCAMH–2015.pdf.
Hewage, C., Bohlin, G., Wijewardena, K. & Lindmark, G. (2011). Executive functions and child problem behaviors are sensitive to family disruption: A study of children of mothers working overseas. *Developmental Science, 14*(1), 18–25. doi:10.1111/j.1467-7687.2010.00953.x.
Hoffman, C, Crnic, K.A. & Baker, J.K. (2006). Maternal depression and parenting: Implications for children's emergent emotion regulation and behavioral functioning. *Parenting, 6*(4), 271–295. doi:10.1207/s15327922par0604.
Hooper, E., Feng, X., Christian, L. & Slesnick, N. (2015). Emotion expression, emotionality, depressive symptoms, and stress: Maternal profiles related to child outcomes. *Journal of Abnormal Child Psychology, 43*(7), 1319–1331. doi:10.1007/s10802-015-0019-6.
Hughes, C. & Ensor, R. (2009). Independence and interplay between maternal and child risk factors for preschool problem behaviors? *International Journal of Behavioral Development, 33*(4), 312–322. doi:10.1177/0165025408101274.
Hughes, C., Roman, G., Hart, M.J. & Ensor, R. (2013). Does maternal depression predict young children's executive function? A 4-year longitudinal study. *Journal of Child Psychology and Psychiatry, 54*(2), 169–177. doi:10.1111/jcpp.12014.
Kring, A., Davison, G.C., Neale, J.M. & Johnson, S. (2006). *Abnormal psychology* (10th ed.). Hoboken, NJ: John Wiley & Sons.
Lovejoy, M.C, Graczyk, P.A., O'Hare, E. & Neuman, G. (2000). Maternal depression and parenting behavior: A meta-analytic review. *Clinical Psychology Review, 20*(5), 561–592. doi:10.1016/S0272–7358(98)00100-7.
Marcotte, D.E, & Wilcox-Gök, V. (2001). Estimating the employment and earnings costs of mental illness: Recent developments in the United States. *Social Science & Medicine, 53*(1), 21–27. doi:10.1016/S0277–9536(00)00312-9.
Merikangas, K.R., Ames, M., Cui, L., Stang, P.E., Ustun, T.B., Korff, M.V. & Kessler, R.C. (2007). The impact of comorbidity of mental and physical conditions on role disability in the US adult household population. *Archives of General Psychiatry, 64*(10), 1180–1188. doi:10.1001/archpsyc.64.10.1180.
Meuwissen, A.S. & Carlson, S.M. (2015). Fathers matter: The role of father parenting in preschoolers' executive function development. *Journal of Experimental Child Psychology, 140*, 1–15. doi:10.1016/j.jecp.2015.06.010.
Mezulis, A.H., Hyde, J.S. & Clark, R. (2004). Father involvement moderates the effect of maternal depression during a child's infancy on child behavior problems in kindergarten. *Journal of Family Psychology, 18*(4), 575–588. doi:10.1037/0893-3200.18.4.575.

Mezzacappa, E., Kindlon, D. & Earls, F. (2001). Child abuse and performance task assessments of executive functions in boys. *Journal of Psychology and Psychiatry, 42*(8), 1041–1048. doi:10.1111/1469-7610.00803.

Odgers, C.L, & Jaffee, S.R. (2013). Routine versus catastrophic influences on the developing child. *Annual Review of Public Health, 34*(Suppl 2), 29–48. doi:10.1146/annurev-publhealth-031912-114447.

Rhoades, B.L., Greenberg, M.T., Lanza, S.T. & Blair, C. (2011). Demographic and familial predictors of early executive function development: Contribution of a person-centered perspective. *Journal of Experimental Child Psychology, 108*(3), 638–662. doi:10.1016/j.jecp.2010.08.004.

Roman, G.D., Ensor, R. & Hughes, C. (2016). Does executive function mediate the path from mothers' depressive symptoms to young children's problem behaviors? *Journal of Experimental Child Psychology, 142*, 158–170. doi:10.1016/j.jecp.2015.09.022.

Shonkoff, J.P. & Phillips, D.A. (2000). Nurturing relationship. In Shonkoff, J.P. & Phillips, D.A. (Eds.), *From neurons to neighborhoods: The science of early childhood development*. Washington, DC: National Academies Press. Retrieved from http://www.ncbi.nlm.nih.gov/books/NBK225557/.

Slykerman, R.F., Thompson, J.M.D., Pryor, J.E., Becroft, D.M.O.E., Robinson, P.M.C., Wild, C.J. & Mitchell, E.A. (2005). Maternal stress, social support and preschool children's intelligence. *Early Human Development, 81*(10), 815–821. doi:10.1016/j.earlhumdev.2005.05.005.

Turney, K. (2012). Pathways of disadvantage: Explaining the relationship between maternal depression and children's problem behaviors. *Social Science Research, 41*(6), 1546–1564. doi:10.1016/j.ssresearch.2012.06.003.

Zelazo, P.D. (2006). The dimensional change card sort (DCCS): A method of assessing executive function in children. *Nature Protocols, 1*(1), 297–301. doi:10.1038/nprot.2006.46.

Zelazo, P.D. & Carlson, S.M. (2012). Hot and cool executive function in childhood and adolescence: Development and plasticity. *Child Development Perspectives, 6*(4), 354–360. doi:10.1111/j.1750-8606.2012.00246.x.

Zelazo, P.D. & Müller, U. (2010). Executive function in typical and atypical development. In U. Goswami, (Ed.), *The Wiley-Blackwell handbook of childhood cognitive development* (pp. 574–603). Chichester, UK: John Wiley & Sons. Retrieved from http://onlinelibrary.wiley.com/doi/10.1002/9781444325485.ch22/summary.

Diversity in Unity: Perspectives from Psychology and Behavioral Sciences – Ariyanto et al. (Eds)
© 2018 Taylor & Francis Group, London, ISBN 978-1-138-62665-2

The contribution of maternal management language to predicting executive function in early childhood

S.N.F. Putri, D. Hendrawan, D.M. Hestiany & N. Arbiyah
Faculty of Psychology, Universitas Indonesia, Depok, Indonesia

ABSTRACT: Executive Function (EF) is a higher mental process that maintains, manages, controls, and modifies mental processes in goal-directed behavior. The critical development of EF takes place during early childhood. A few studies suggest that mothers' verbal utterances used to guide children's behavior, which are called maternal management language, could influence EF development in their children. Unfortunately, studies covering this issue are still limited in number. This study aims to investigate the contribution of management language to predicting EF in 48 to 72 month-old children. Mother-child interactions (N = 90) were videotaped and categorized into two categories (direction and suggestion) during a structured play. EF was assessed using Executive Function Battery Test which consists of Matahari/Rumput task, Backward Word Span task, Dimensional Change Card Sorting task, and Gift Delay task. Hierarchical multiple regression was used to investigate the predictive pattern of some variables in this study. The results show that direction language could influence EF performance negatively in preschool children, whereas suggestion language could influence EF only if when it was used by mothers with a high SES. This study presents important findings on how to control children's behavior, especially through maternal verbal utterance.

1 INTRODUCTION

Executive function (EF) refers to a set of high level neurocognitive skills that involve controlling mind, emotion, and behavior consciously to achieve goals and solve problems (Zelazo & Müller, 2002). As the definition implies, EF can help individuals to face a situation in which they need to plan, maintain attention, and change their behavior in order to adapt well to their environment (Diamond, 2013). Parenting has a significant influence on children's EF development, especially during early childhood.

During infancy and the preschool period, core compo nents of EF develop, forming a critical foundation that will set the stage for the development of higher cognitive processes into adulthood (Garon, Bryson, & Smith, 2008). This is why children with high EF will be able to compete well and get higher scores than other children (Carlson, Zelazo, & Faja, 2013).

EF dysfunction has been implicated in a number of childhood disorders, in-between attention-deficit/hyperactivity disorder (ADHD), autism, and conduct disorder (see Casey, Tottenham & Fossella, 2002) and is associated with a variety of negative adjustment outcomes. Furthermore, Moffitt, Arseneault *et al.* (2011) and Carlson *et al.* (2013) found that optimizing the development of EF since early childhood could help reduce a number of problems in adulthood, involvement in criminal behavior, alcohol and drug abuse, and lower socioeconomic status (SES). Therefore, it is important for us to pay attention to the development of EF, especially during its rapid growth period in early childhood.

Although the study of parenting and EF in preschool children has long been a field of active research, comparatively little is known about the influence of parents' verbalization on children's EF development. Bindman, Hindman, Bowles, and Marrison (2013) found that mothers' verbal control while regulating children's behavior, defined as management language, could influence children's EF development. Children who had mothers that gave explanation of their advice and did not impose their advice on them (suggestion language) had better EF performance in preschool than children with mothers who simply dictated their children's behavior (direction language)(Bindman *et al.* 2013). Mothers' dictation could hinder their children's ability to regulate themselves since they did not have any opportunity to determine their own choices. These findings are consistent with those of the research conducted by Meuwissen and Carlson (2015) which indicates that father's controlling behavior, both verbal and nonverbal, has a relationship with poor EF performance in children.

However, Weber (2011) obtained a different result which suggests that parents' dictation had no correlation with children's EF performance. This difference in results might be due to the different instruments that they used (Fay-Stammbach, Hawes, and Meredith 2014). In his research, Weber (2011) used self-report methods, while Bindman *et al.* (2013) used the dyadic activity to probe parent's attitude in directing or controlling the child's behavior. Self-report methods could be very vulnerable to bias (Lucassen *et al.,* 2015). Meanwhile, observation methods for maternal control strategy which is applied in a natural setting could depict the whole process of how the mothers interact with or intervene their children's behavior. However, there has been limited research on the use of parental controls, especially in the form of verbal utterance, and the results are also inconsistent (Fay-Stammbach, Hawes, & Meredith, 2014). This condition raises concern about the important role of parents in assisting and guiding their children's behavior on regular and daily basis. Therefore, it is important to carry out further studies about the relationship between EF development in preschool children and parental management, especially those focusing on its language aspect, since children at critical ages could easily internalize their parents' utterances (Papalia & Feldman, 2012; Grolnick, 2002).

The current study aims to investigate the influence of maternal management language on children's EF performance through dyadic activities, such as their attitudes when directing or controlling their children's behavior in a natural setting. We also take into consideration a number of factors, such as a child's age, a child's gender, and family socioeconomic status (SES). Previously, Hendrawan, Fauzani, Carolina, Fatimah, Wijaya, and Kurniawati (2015) found that children's age, gender, and family's SES could predict EF domains (cool, hot, and both domains respectively). This study also investigates the relationship between maternal management language and the domain of hot EF, which had not been studied in the previous research. We hypothesized that maternal management language has a significant relationship with children's EF. Suggestion language could enhance children's EF development, while direction language could decrease it. In addition, we also hypothesized that management language could predict EF in early childhood.

2 METHODS

2.1 *Participants*

As many as 90 mothers and their children, age 48–72 months (M = 60.08 months, SD = 7.12) from various preschools in Jakarta, Bogor, Depok, Tangerang and Bekasi (Jabodetabek) areas volunteered to participate in this study. Children participants had a quite balanced proportion of gender, consisting of 47 boys and 43 girls. The participants ranged from low to high SES with an average monthly expenditure from 600,000.00 IDR to 27,000,000.00. Only participants who had no serious developmental problem and been using Indonesian language in their daily communication were eligible to participate.

2.2 *Measures*

The children were given four subtests, consisting of Matahari/Rumput task, Backward Word Span (BWS) task, Dimensional Card Sorting task (DCCS), and Gift Delay task, all of which had been adapted into Indonesian language by Hendrawan *et al.* (2015). The tests were delivered in a certain sequence by the tester who had been trained under the supervision of an accredited psychologist.

2.2.1 *Matahari/rumput task*

The first subtest is Matahari/Rumput task adapted from 'Grass/Snow' task by Carlson, White, and David-Unger (2014) which was used to measure inhibitory control. Because "Snow" was not a common concept for Indonesians, it was replaced with "Matahari" ("Sun"). In this subtest, children were asked to choose a green card when the tester said "Matahari" and a yellow card when the tester said "Rumput". There were 16 trials in which "Matahari" and "Rumput" were mentioned in a random order. Each correct response was given the score 1, while each incorrect response got the score 0.

2.2.2 *Backward Word Span (BWS) task*

The second subtest is BWS task adapted from Carlson *et al.* (2014) which was used to measure working memory *et al.* In this test children were asked to pronounce a number of words which had been spoken by the tester in reverse order. This test used a hand puppet as a tool. There were five levels of the words in this subtest. Score 1 was given at every level if the child was able to mention the words in reverse order correctly. The maximum score for this test was 5.

2.2.3 *Dimensional Card Sorting Task (DCCS)*

The third subtest was DCCS adapted from Zelazo (2006) which was used to measure flexibility. This test used cards with various combinations of colors (red and blue) and images (flower and house) and two boxes. The test consisted of three stages with three different rules (pre-test, post-test, and border). On the pre-test stage, children were asked to sort the cards only by color. Later on the post-test stage, they were asked to sort the cards based on image only. Lastly, on the border stage, the children were asked to sort bordered cards by color and non-bordered cards by image. Score 1 was given at every stage if the children finished playing with a maximum of 1 error on the first two stages and 3 errors on the last stage. The highest score was 3 and the lowest score was 0.

2.2.4 *Gift delay task*

The last test was Gift Delay task. It was adapted from Carlson *et al.* (2014) and used to measure hot EF. In this test, children were asked to wait and not to peek when the tester wrapped gifts on their back for two minutes. This test assessed the frequency, duration, and intensity of the children's peeking behavior while the tester was wrapping the gift.

2.2.5 *Parent management language*

In this test a mother and her child were given three different toys (fishing, playdough, and bricks) that had been tested previously. Then the mother was instructed to make sure that her child played all of the three toys in 10 minutes. She must also make sure that her children focus on one toy only. This activity was adapted from Schaffer and Crook (1979) and used to measure the intensity of maternal control.

Verbal utterances that the mother made during 10 minutes of the game will be scored based on two categories, i.e. direction language or suggestion language. Sentences such as questions or statements that provided choices for children and took account of the children's perspective were classified as suggestion language. Meanwhile, sentences in the form of commands were categorized as direction language. We used Parental Management Language Code by Bindman *et al.* (2013) to categorize every sentence. The scores were taken by two different raters. Inter-rater reliability was calculated using ICC (Shrout-Fleiss Intraclass Correlation). The ICC coefficient of direction language was 0.86, while that of suggestion language was 0.88.

2.3 *Procedures*

Before data collection, the researchers had conducted an ethical review of the procedures prior to the research. Then the researchers visited several kindergartens in Jabodetabek area and asked for their permission to organize meetings with parents to inform them about the research that would be carried out in the kindergartens. Parents who were interested to participate in this study were asked to complete a screening form containing some demographic variables such as age, the family's monthly expenses, and some questions related to the history of the children's development. The researchers informed them about the research and asked for their consent before administering the tests.

Children who passed the screening test were then invited to take a series of EF tests consisting of Rumput/Matahari, Backward Word Span (BWS), Dimensional Card Sorting Task (DCCS), and Gift Delay tests. The tests were conducted for about 60 minutes in a closed room to avoid distraction. Children's responses were recorded during the test. After the tests were completed, the children were then moved to a different room to do parental management language's activities. Mothers and children were asked to play a number of games for 10 minutes. The mothers' utterances were recorded and scored later using parental management language's code to calculate how many items of each type of management language were used by mothers during the games.

2.4 *Analysis*

First, the EF's scores were standardized and then combined into one composite score. This score was correlated with the scores of maternal management language in terms of both direction and suggestion languages. The capability of maternal management language of predicting EF would be analyzed using hierarchical multiple regression. EF, as the dependent variable, was analyzed together with demographic variables (gender, socioeconomic status, and age) as the first step. Then management language (direction language and suggestion language) as the independent variable was analyzed later in the second step.

3 RESULTS

The correlation between each of the management language types, EF composite score outcome, and some of covariate variables are listed in Table 1. Direction language has a significant and negative correlation with both cool and hot EF. There is no significant correlation between suggestion language and either cool or hot EF.

Table 2 presents a regression statistic analysis for direction and suggestion management language that mothers used. This analysis is used to predict children's executive function (EF) performance after controlling the influence of children's gender, age, and socioeconomic status (SES). According to this statistical analysis, it has been found that only direction language could predict 5.80% of the variance of children's cool EF, with $F(5, 84) = 6.88$, $p < 0.05$. Direction language has a significant negative correlation with children's cool EF ($r = -0.233$, $p < 0.05$) and hot EF ($r = -0.232$, $p < 0.05$). Suggestion language fails to show any significant correlation with children's cool EF ($r = 0.079$, $p > 0.05$) and hot EF ($r = 0.080$, $p > 0.05$).

In order to offer a more comprehensive elucidation, we classified suggestion language into two categories, i.e. low (n = 51) and high (n = 39). We conducted an analysis of t-test to

Table 1. Pearson's correlations between *management language* and child *executive function*.

	Management language	
	Direction	Suggestion
Executive function		
Cool	−0.353**	0.157
Hot	−0.262*	0.079

Note: * $p < 0.05$, ** $p < 0.01$. N = 90.

Table 2. Regression analysis of the influence of management language on children's EF.

	Cool EF		Hot EF	
	B	ΔR^2	B	ΔR^2
Step 1		0.190**		0.038
Child's age	0.281*		0.087	
SES	0.300*		0.012	
Gender	0.091		0.180	
Step 2		0.058*		0.058
Direction language	−0.233*		−0.232*	
Suggestion language	0.079		0.080	

Note: SES = Socioeconomic status.
*$p < 0.05$, **$p < 0.01$.

compare the significance between those two categories. The result showed that there was a significant difference between the cool EF and suggestion language scores (t = 0.405), but there was no significant difference between the hot EF and suggestion language scores of mothers in the two groups (t = 297).

4 DISCUSSION

This study provides an insight into the relationship between maternal management language and EF development in children since previous studies only yielded inconsistent results. This study shows that the use of direction language to control children's behavior could lessen their EF performance. This result is in line with that of Bindman *et al.* (2013), who found that direction language could hinder children's opportunity to make and execute some choices. Children who tend to follow orders will rarely practice self-regulation, so their EF ability was not well stimulated. However, it could not be generalized into younger children since they still need their parents' guidance to behave in an appropriate manner. Landry, Smith, Swank, and Miller-Loncar (2000) found that mothers who use directive language with their two-year-old children had children with high level cognitive skills. However, mothers who used directive language with their older children had children with poor EF. These findings demonstrated the importance of further research on maternal management language in the form of longitudinal studies in order to get more comprehensive information. On the other hand, the influence of suggestion language was found to be different from that found by Bindman *et al.* (2013) since there was no identifiable influence which could predict children's EF performance.

The previous analysis found that high suggestion language was used by mothers with a high SES. Then, the combination of high suggestion language and high SES enhance children's EF performance. This finding is in line with other research which explains that socio-economic status could moderate children's EF development. Children from low SES tend to have cognitive deficits (Carlson *et al.*, 2013). McLoyd (1998) also said that the effect of socio-economic disadvantage on early cognitive functioning were mediated by different levels of learning, academic, and language stimulation that the children received in their home environment. Further, Carlson *et al.* (2013) state that the connection between EF and SES is moderated by distal environmental factors, such as neighborhood (see Caspi, Taylor, Moffitt, & Plomin 2000), and by proximal risk factors, such as maternal education, family chaos, and child stress (see Ardila, Rosselli, Matute, & Guajardo, 2005).

This study also found a pattern of the relationship between management language and hot EF domain which is still rarely examined. The result shows that the more direction language is used by mothers, the lower hot EF scores that the children will get. Consistent with this finding, Kuczynski, Kochanska, Radke-Yarrow, and Girnius-Brown (1987) explain that the more a mother used directive language, the more the children would show non-compliant behaviors as a result of poor emotion regulation. Children who have a poor emotion regulation also have low EF scores (Zelazo & Cunningham, 2007).

The limitation of this current study is the exclusion of maternal education as a control variable. It is possible that maternal education could affect the way parents' utterances are produced to control their children's behavior. Further research should use random sampling to include more heterogeneous participants that could enrich the results. Besides that, it should also involve younger children (2–3 years) to obtain more results which would serve as the basis for developing early intervention programs. Not only mothers, fathers could also influence their children's EF development, especially at older ages (Meuwissen & Carlson, 2015). It would be better if future research also take account of fathers' utterances to produce more comprehensive knowledge which would serve as the basis for improving children's EF development. A longitudinal design would also be a better technique for investigating both fathers' and mothers' management language in early childhood period.

In brief, children's EF could be optimized by taking into account maternal verbal control in parenting practices. An excessively high level of mothers' control in terms of verbal utterance (direction language) could result in poor EF in children, because the children would lose their sense of autonomy in determining their own choices and making their own decisions (Bindman *et al.,* 2013). Therefore, parents, especially mothers, are expected to be able to listen and consider children's opinion in decision-making process while guiding children's behavior at the same time.

ACKNOWLEDGEMENT

This research was fully funded by PITTA research grant from Universitas Indonesia. We would like to thank Hanifah Nurul Fatimah, Fasya Fauzani, Claudya Carolina, Intan Putri Hertyas, Afiyana Eka Nurilla, and Rizka Nurbatari who provided feedback and support in completing this paper, as well as to all of the participants who have participated in this study. We dedicate this study to aiding Indonesian people in improving their EF, starting from early childhood.

REFERENCES

Ardila, A., Rosselli, M., Matute, E. & Guajardo, S., (2005) The influence of the parents' educational level on the development of executive functions. *Developmental Neuropsychology,* 28 (1), 539–60. doi:10.1207/s15326942dn2801_5.

Bindman, S.W., Hindman, A.H., Bowles, R.P. & Morrison, F.J. (2013) The contributions of parental management language to executive function in preschool children. *Early Childhood Research Quarterly,* 28 (3), 529–39. doi:10.1016/j.ecresq.2013.03.003.

Carlson, S.M., White, R.E. & Davis-Unger, A.C. (2014) Evidence for a relation between executive function and pretense representation in preschool children. *Cognitive Development*, 29 (January), 1–16. doi:10.1016/j.cogdev.2013.09.001.

Carlson, S.M., Zelazo, P.D. & Faja, S. (2013) Executive Function. In *Oxford Handbook of Developmental Psychology*, 1, 706–42.

Casey, B.J., Tottenham, N. & Fossella, J. (2002) Clinical, imaging, lesion, and genetic approaches toward a model of cognitive control. *Developmental Psychobiology,* 40 (3), 237–54. doi:10.1002/dev.10030.

Caspi, A., Taylor, A. Moffitt, T.E. & Plomin, R. (2000) Neighborhood deprivation affects children's mental health: Environmental risks identified in a genetic design. *Psychological Science,* 11 (4), 338–42. doi:10.1111/1467-9280.00267.

Diamond, A. (2013) Executive functions. *Annual Review of Psychology,* 64, 135–168. doi:10.1146/annurev-psych-113011-143750.

Fay-Stammbach, T., Hawes, D.J. & Meredith, P. (2014) Parenting influences on executive function in early childhood: A review. *Child Development Perspectives,* 8 (4), 258–64. doi:10.1111/cdep.12095.

Garon, N., Bryson, S.E. & Smith, I.M. (2008) Executive function in preschoolers: A review using an integrative framework. *Psychological Bulletin,* 134 (1), 31–60. doi:10.1037/0033–2909.134.1.31.

Grolnick, W.S. (2002) *The Psychology of Parental Control: How Well-Meant Parenting Backfires.* 1st edition. Mahwah, N.J, Psychology Press.

Hendrawan, D., Fauzani, F., Carolina, C., Fatimah, H.N., Wijaya, F.P. & Kurniawati, F. (2015) The construction of executive function instruments for early child ages in Indonesia: A pilot study. In *Promoting Children's Health, Development, and Well-Being: Integrating Cultural Diversity.* Jakarta, Faculty of Psychology.

Kuczynski, L., Kochanska, G., Radke-Yarrow, M. & Girnius-Brown, O. (1987) A developmental interpretation of young children's noncompliance. *Developmental Psychology,* 23 (6), 799–806. doi:10.1037/0012–1649.23.6.799.

Landry, S.H., Miller-Loncar, C.L., Smith, K.E. & Swank, P.R. (2002) The role of early parenting in children's development of executive processes. *Developmental Neuropsychology,* 21 (1), 15–41. doi:10.1207/S15326942DN2101_2.

Lucassen, N., Kok, R., Bakermans-Kranenburg, M.J., Ijzendoorn, M.H.V., Jaddoe, Vincent,W.V., Hofman, A., Verhulst, F.C., Lambregtse-Van, M.P. & Tiemeier, H. (2015) Executive functions in early childhood: The role of maternal and paternal parenting practices. *British Journal of Developmental Psychology,* 33 (4), 489–505. doi:10.1111/bjdp.12112.

McLoyd, V.C. (1998) Socioeconomic disadvantage and child development. *The American Psychologist,* 53 (2), 185–204. doi:10.1037/0003–066X.53.2.185.

Meuwissen, A.S. & Carlson, S.M. (2015) Fathers matter: The role of father parenting in preschoolers' executive function development. *Journal of Experimental Child Psychology,* 140, 1–15. doi:10.1016/j.jecp.2015.06.010.

Moffitt, T.E., Arseneault, L., Belsky, D., Dickson, N., Hancox, R.J., Harrington, H., Houts, R. *et al.* (2011) A gradient of childhood self-control predicts health, wealth, and public safety. *Proceedings of the National Academy of Sciences of the United States of America,* 108, 2693–98. doi:10.1073/pnas.1010076108.

Papalia, D.E. & Feldman, R.D. (2012) *Experience Human Development, 12th Edition.* New York, McGraw-Hill.

Schaffer, H.R. & Crook, C.K. (1979) Maternal control techniques in a directed play situation. *Child Development,* 50 (4), 989–96. doi:10.2307/1129324.

Weber, R.C. (2011) How hot or cool is it to speak two languages: Executive function advantages in bilingual children. [PhD], Texas A7M University. http://oaktrust.library.tamu.edu/handle/1969.1/ETD-TAMU-2011–08–10028.

Zelazo, P.D. (2006) The Dimensional Change Card Sort (DCCS): A method of assessing executive function in children. *Nature Protocols,* 1 (1), 297–301. doi:10.1038/nprot.2006.46.

Zelazo, P.D. & Cunningham W.A. (2007) Executive function: Mechanisms underlying emotion regulation. *ResearchGate,* January. https://www.researchgate.net/publication/232532328_Executive_Function_Mechanisms_Underlying_Emotion_Regulation.

Zelazo, P.D. & Müller, U. (2002) Executive function in typical and atypical development. In Goswami, U. (ed.) *Blackwell Handbook of Childhood Cognitive Development,* 445–69. Blackwell Publishers Ltd. http://onlinelibrary.wiley.com/doi/10.1002/9780470996652.ch20/summary.

Diversity in Unity: Perspectives from Psychology and Behavioral Sciences – Ariyanto et al. (Eds)
© 2018 Taylor & Francis Group, London, ISBN 978-1-138-62665-2

The effect of literary fiction on school-aged children's Theory of Mind (ToM)

Wulandini, I.A. Kuntoro & E. Handayani
Faculty of Psychology, Universitas Indonesia, Depok, Indonesia

ABSTRACT: This present study examines the effect of reading literary fiction on school-aged children at two levels of Theory of Mind (ToM) understanding, that is, the first-order and the second-order. One hundred and eight children aged from 9 to 10 years (M = 120.5 months, SD = 4.19) participated; they were randomly assigned into experimental and control groups. The experimental group had to read three genres of literature, that is, literary fiction, popular fiction and non-fiction, while the control group did other activities. The result showed that the literary fiction group achieved higher scores, which proved that this genre had a significant influence on the children's first-order ToM [$p = 0.04$, $\chi 2$ (3, N = 108) = 8.55] but did not have a significant influence on their second-order ToM [$p = 0.68$, $\chi 2$ (3, N = 108) = 1.51). Finally, literary fiction had a more significant influence on the children's total ToM understanding [$p = 0.04$, $\chi 2$ (3, N = 108) = 1.51] compared to popular fiction and non-fiction, as well as the activities conducted by the control group. This shows that reading a literary fiction could enhance the ToM of school-aged children (9–10 years-old).

1 INTRODUCTION

In developmental psychology, Theory of Mind (ToM) is defined as the ability to understand the mental state of other people which includes others' thinking, beliefs, desire and emotion (Goldstein & Winner, 2012; Liddle & Nettle, 2006; Wellman, 1990). ToM facilitates the prediction of others' behaviour based on mental state understanding. Researchers have identified two levels of ToM understanding which take place during childhood, that is, the first and the second-orders (Korkmaz, 2011; Moran, 2013). The first-order ToM is defined as a subject's reasoning about another person's thoughts ('I think that he thinks').

Meanwhile, the second-order ToM is defined as a subject's reasoning about what one person thinks about another person's thoughts ('he thinks that she thinks') (O'Reilly et al., 2014). If the first-order only involves inferring one person's mental state, the second-order involves more than one person's mental state. The second-order informs more about a person's understanding of a social interaction condition (Liddle & Nettle, 2006) and is an advanced ToM ability (Moran, 2013).

ToM performance is important for school-aged children to deal with their peers in collaborative social relationships (Lecce, 2014; Liddle & Nettle, 2006). It influences and organises emotion and behaviour during peer and social interactions (Hoglund et al., 2007), empathic responses (Kidd & Castano, 2013; Repacholi & Slaughter, 2003), and conflict resolution (Razza & Blair, 2009; Watson et al., 2001). Children with a better understanding of ToM will be more successful in their social relationships (Razza & Blair, 2009; Watson et al., 2001) and have a more extensive network of friends (Stiller & Dunbar, 2004).

Despite the important role of ToM for social interaction, the ToM performance of school-aged children has been much less studied than that of pre-school children, and such studies have largely been concerned with establishing the chronology and sequence of ToM acquisition (Wellman & Liu, 2004). The limited research on school-aged ToM performance inspired this study, which aims to enhance children's ToM. A prior study with the same aim but which targeted adults suggests that reading literary fiction improves people's ToM (Kidd & Castano, 2013). Reading fiction might serve as a way to enhance and refine interpersonal sensitivity. More research on the effects of literary fiction on school-aged children's ToM should be conducted since the literacy skills of school-aged children have significantly developed. Besides, the increasing level of media exposure in school-aged children, as they read more frequent and diverse reading materials, and concomitant familiarity with fiction will benefit theory of mind understanding.

There are some capabilities that support school-aged children's understanding of first and second-order ToM and their understanding of literary fiction. Their ability to understand sensory inputs and environmental stimulations promotes the development of complex mental abilities such as the ability to think, to believe in something, and to have intentions. Children aged from 9 to 10 have the ability to interpret emotions, which becomes more complex and adequate. This ability stimulates the development of an understanding which distinguishes an individual from others in terms of emotion control and manipulation (Vitulic, 2009). Children at these ages are more able to use reasoning and perspectives that stimulates the development of their social understanding (Banerjee et al., 2011). Kamawar and Olson (2009) explain that the ToM of children aged from nine to ten is also enhanced by mind-reading, which is the ability to map the mind, emotion, and body language, which develops rapidly. This capability can be observed from their use of mental state vocabulary such as 'I know' or 'I feel' which is used to support and monitor ToM processes (de Rosnay, 2004; Louca, 2008; Pons & Harris, 2005). Furthermore, children aged 9–11 have the ability to assess intentional or unintentional actions by understanding the purpose of those actions. They are able to interpret the use of irony or sarcasm (Filippova & Astington, 2008) and to understand *faux pas* (Baron-Cohen et al., 1999). These capabilities can assist school-aged children from 9 to 10 years-old in inferring mental states and social information contained in literary fiction.

It is argued that this study would be beneficial for school-aged children for three important reasons. The first reason is that there is a great need to foster ToM as early as possible, as soon as children are able to read and understand various reading materials independently at school age. Second, children have to rely more on their ToM when they engage in a more complex network of social relationships. The third is the increasing level of media exposure in the form of reading materials at school age. For parents and educators in Indonesia, a good understanding of reading materials will help nurture children's reading interest and improve their level of literacy. At school age, children have acquired an adequate ability to use reasoning for understanding various social contexts. Therefore, the researchers are interested in examining the effect of literary fiction on school-aged children's ToM.

Fiction is a narrative work in the form of text or discourse which is based on imagination (Abrams, 1981), and it needs to be imagined through the context (Walton, 1990). Hodgins (1993) and Stein (1995) describe the characteristic of literary fiction as stories made with caution. This cautiousness is necessary to produce a story which could suit the age and the developmental stage of children (Kurniawan, 2002). Thus, the children can respond to the stories with their emotion and intellectuality (Nurgiyantoro, 2005). Children's literary fiction is characterised by its aim to give pleasure and understanding. This objective is achieved through the use of language, which contains elements of beauty, the exploration of social life, and the disclosure of various characters which gives social information to the reader (Mursini, 2011).

Reading literary fiction stimulates the process of the ToM mechanism which involves knowledge, perception and interpretation, such as the use of characters' perspectives, which may be different from the perspectives of the readers themselves. Its contents provide simulative experiences that lead the readers to feel and think about the social context and various depictions of the mental state. In contrast to literary fiction, non-fiction and popular fiction, for example comics, are less able to stimulate imagination and emotional experience. Popular fiction materials in this research were chosen from the most recent bestseller books sold by one of the largest bookstore chains in Indonesia. Although the difference between literary fiction and popular fiction is often judged to be decisive because both contain social information such as a character's mental state, there is one type of popular fiction that is somewhat distinguishable from others, which is genre of comics. In comics social information is presented through sequences of pictures and callouts with short and direct sentences. Meanwhile, non-fiction only contains academic-related information such as encyclopedic knowledge or biographies that are relied upon for the purpose of studying various fields of inquiry, both natural and social sciences (Dromey, 2010). Therefore, popular fiction and non-fiction cannot stimulate an individual's ToM the way literary fiction can. In this case, the authors seek to prove a hypothesis which states that children who read literary fiction will perform better on tests of the first-order ToM, the second-order ToM, and the total of both orders of ToM, than those who read other literary genres.

2 METHODS

2.1 *Participants*

One hundred and eight children aged from 9 to 10 years (M = 120.5 months, SD = 4.19) took part in this research. These participants were randomly assigned into one of four groups in order to obtain a balanced representation of gender and age; 28 participants read literary fiction, 30 participants read popular fiction, 27 participants read non-fiction, and 23 participants performed non-reading activities as the control group.

2.2 *Instruments*

In this study, the first-order ToM was measured using two scales, while the second-order ToM was measured with one scale. To test the first-order ToM, only five concepts that are suitable for children from a broad range of ages, up to 10 years-old or older were used. The four concepts that were developed by Wellman and Liu (2004) included Diverse Belief or DB (understanding thinking), (2) Knowledge Access or KA (understanding that seeing leads to knowing), (3) False Belief or FB (understanding false belief), and (4) Hidden Emotion or HE (understanding hidden feelings). Meanwhile, the one concept developed by Peterson and Welman (2012) was (5) sarcasm or SARC (understanding sarcasm). To test the second-order ToM, the second-order belief or SOB concept that was developed by Perner and Wimmer (1985) was used.

For the experimental groups three types of reading materials were used, while for the control group drawing materials and puzzles were used. Experimental groups were asked to read literary fiction, popular fiction, and non-fiction. The literary fiction group read the books of Murti Bunanta, a writer of children's literature in Indonesia who has received various prestigious awards (e.g. the International Book Award). Three titles were used, consisting of *Si Bungsu Katak* ('The Youngest Frog'), *Anak Kucing yang Manja* ('The Spoiled Kittens'), and *Legenda Pohon Beringin* ('The Legend of Banyan Tree'). The non-fiction popular literature group read several bestseller children's comics which had been chosen based on a survey recently conducted in Indonesia. The non-fiction group read materials such as biographies, science books, and history books.

2.3 *Procedure*

Each experimental group was asked to read a specific book genre in five consecutive days for duration of 20 minutes per day. Every time any member of an experimental group finished his/her respective reading materials, they would be given a Reading Comprehension Test (RCT) to control his/her reading comprehension abilities. On the fifth day, a ToM test was administrated. A group of testers, who had been previously trained, assessed the participants' first-order ToM in terms of the 'five concepts' with four verbal tasks. Meanwhile, second-order ToM of the participants was assessed with one verbal task. In every task the testers would tell the participants an elaborate story that was accompanied by pictures and materials related to the story. Participants were then requested to predict the mental state of the characters in the story. The ToM test was administered following the sequence from DB,

KA, FB, HE, SARC, and SOB. In order to ensure that the children understood the stories, control questions were provided for each story. One point was given to a child if he/she could answer each target question correctly.

3 RESULTS

Figure 1 shows that, for the first-order tasks, the literary fiction group has the highest percentage of correct answers (DB = 96%, KA and FB = 100%, HE = 93%, SARC = 36%). Meanwhile, the control group has the lowest percentage (DB and FB = 87%, KA = 96%, SARC = 9%), except for the HE concept (78%). For this concept, the popular fiction group obtained the smallest percentage (57%), even smaller than the non-fiction group did (63%).

Figure 2 shows that the sequence of groups based on the results of the second-order tasks is not similar to that based on the results of the first-order tasks. Despite the correct answers, the literary fiction and the control groups have the highest (64%) and the lowest (48%) percentages respectively, which is similar to the results of the first-order tasks, but the non-fiction group obtained a higher percentage than the popular fiction group did.

The Kruskal–Wallis test was performed on the first-order ToM scores, the second-order ToM scores, and total ToM scores obtained by all of the groups. Table 1 presents the comparison of

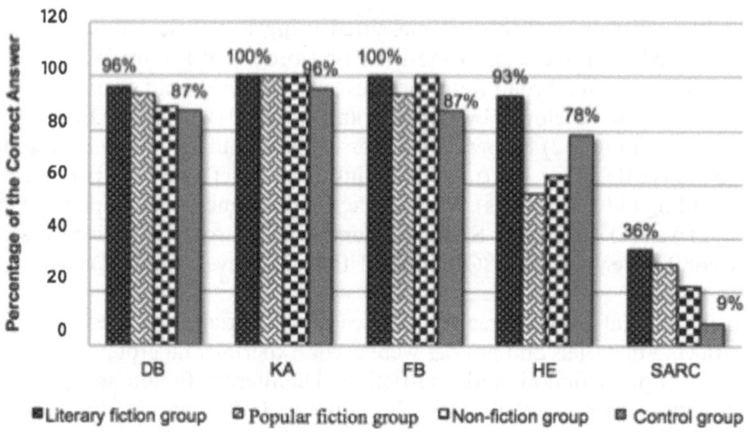

Figure 1. Comparison of the correct answers in first-ToM.

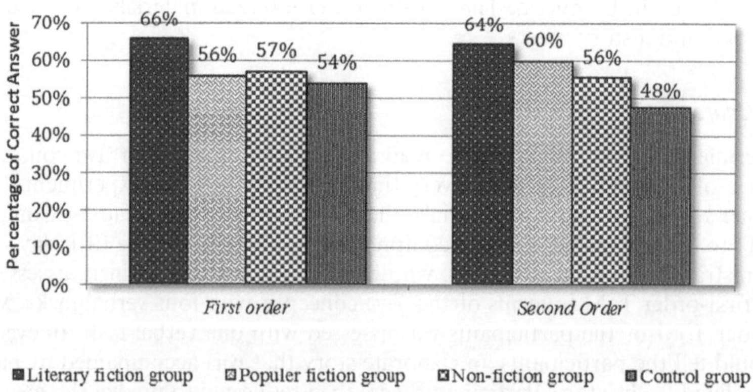

Figure 2. Comparison of the correct answers in first- and second-order ToM.

Table 1. Mean scores of the first-order ToM test scores, the second-order ToM test scores, and the total ToM scores.

	First-order		Second-order		Total ToM (First-order + Second-order)	
Groups	M	SD	M	SD	M	SD
Literary fiction	3.86	0.85	0.64	0.49	4.86	0.85
Popular fiction	3.33	1.27	0.60	0.50	4.33	1.27
Non-fiction	3.30	1.10	0.56	0.50	4.30	1.10
Control group	3.04	0.88	0.48	0.51	4.04	0.88

mean scores of the four groups. It shows that the literary fiction group has the highest percentage, while the control group has the lowest percentage for the first-order ToM, second-order ToM, and total ToM scores. This result shows that the effect of genre of reading material on the participants' first-order ToM is significant at the level of $p < 0.05$, [$\chi 2$ (3, N = 108) = 8.55, p = 0.04]. Further *post hoc* tests show that this significance comes from the difference between the literary fiction group and control group [$\chi 2$ (1, N = 51) = 9.5, p = 0.00].

The results of the Mann–Whitney U test indicate that the literary fiction group obtained the highest score, followed in sequence by the popular fiction group, non-fiction group, and control group. A further analysis of the comparison of the mean scores of the first-order ToM between the literary fiction group and the non-fiction group generates a slightly significant value of U = 269, p = 0.052. For the second-order ToM, the results show that the effect of reading material genre is not significant [$\chi 2$ (3, N = 108) = 1.51, p = 0.68]. However, it was also found that the difference between literary fiction and the control group approximated is significant.

The effect of reading material genre on total ToM scores was significant [$\chi 2$ (3, N = 108) = 1.51, p = 0.04]. From the *post hoc* test, it is known that this significance comes from the difference between the literary fiction group and control group [$\chi 2$ (1, N = 51) = 9.5, p = 0.00]. The results of the Mann–Whitney U test indicate that the literary fiction group obtained the highest scores, followed in sequence by the popular fiction group, non-fiction group, and control group.

4 DISCUSSION

The results show that the literary fiction group obtained the highest percentage of correct answers in the first-order, second-order, and total ToM. This result indicates that literary fiction influences ToM acquisition better than popular fiction and non-fiction. In terms of mean scores, the order from the highest to the lowest is the literary, popular fiction, non-fiction, and control groups. However, an observation of the first-order percentages (see Figure 2) resulted in interesting findings on the differences in the rank order for FB and HE concepts. In terms of the FB concept, the non-fiction group has a higher percentage than the popular fiction group, while, in terms of the HE concept, the control group has a higher percentage than the popular fiction and non-fiction groups. These results of the two concepts show a different pattern than that shown by the results from other concepts. It shows that the content of popular fiction did not exert a better influence on the participants' ToM in terms of FB and HE concepts than that of non-fiction; even the control and the popular fiction groups show more depictions of mental state. The dynamics that these comparisons show demonstrates the influence of other variables on ToM understanding that were not controlled in this study. Even random assignments have been conducted to minimise the effect of these variables, such as security attachment to parents (Pavarini et al., 2012) and social interaction (Stiller & Dunbar, 2006).

The effect of literary fiction on the second-order ToM is not significant. However the result of the mean score analysis has generated values that are slightly significant. Previous research even shows that second-order belief could be conducted on school-aged children. However, the test of this advanced ToM ability is not sensitive enough to show the influence of reading genres on ToM. Moreover, the scale used to assess the second-order ToM only consists of one concept, while that used to assess the first-order ToM consists of five. It is suggested that future research should either add more concepts, or exchange the current scale for another to measure the second-order ToM, such as the second-order desire scale. Furthermore, research with a higher number of participants should be performed to better prove the effect of literary fiction on ToM.

This study shifts the category of participants used in prior research from adults to school-aged children. The similarity between the effect size of this study and that of the research by Kidd and Castano (2013) indicates that the effects of literary fiction on the ToM of adults and school-aged children is quite similar. Even in this study where children were asked to read five times, the short-term results already proved that reading literary fiction regularly could lead to stable improvements of ToM. Literary fiction has facilitated the children's thinking process about mental state. School-aged children from 9 to 10 years-old will be engaged with imagination and emotions while trying to understand the experience of the characters. Without knowing how the characters think, the plot would be difficult to follow (Black & Barnes, 2015).

The limitation of this study lies in the experiment method which was marked by the presence of several controls. This study controlled the form of media consumed by the participants, and ToM was measured simply by answering questions related to ToM tasks. More natural forms of literacy activities require more elaboration in terms of their influence on ToM, which can be done by, for instance, keeping a diary. In their diaries, children usually write about their reflections on social events and express their emotions. In spite of the fact that reading literary fiction has been proven to enhance both ToM and literary capability, which is certainly needed by school-aged children, the availability of children's literature in Indonesia is still limited. A language lesson at school could facilitate their engagement in reading literary fiction more regularly.

If this study reveals that the readers of popular fiction obtained better ToM scores than those obtained by the control group, these results may be more interesting to research which seeks only to compare between the popular and the control groups. Comics were also specifically chosen as one of the reading genres because they are written to enhance readers' imagination and emotion. Given the fact that this genre is the most preferred by children, it is highly recommended in its application as good reading material because it may also potentially enhance their readers' ToM.

In can be concluded that reading literary fiction affects school-aged children's first-order and total ToM understanding. The effect of literary fiction is greater than that of other genres. This suggests that literary fiction supports readers' reasoning process about mental state, possibly by facilitating their cognitive systems, which keep track of the mental states attributed by the readers to the characters of the stories. Although reading has been found to influence the acquisition of ToM test scores at various degrees, this activity, particularly the one which involves whatever genres were applied in this study, has been proven to increase participants' ability to answer ToM tests in a more significant way than non-reading activities (e.g. answering puzzles, solving labyrinths, and colouring). Therefore, in general, it is asserted that reading is a beneficial activity which can be undertaken to improve ToM abilities.

REFERENCES

Abram, M.H.(1981). *Teori pengantar fiksi [Indonesia]*. Yogyakarta, Indonesia: Hanindita Graha Wida.

Banerjee, R., Watling, D. & Caputi, M. (2011). Peer relations and the understanding of faux pas: Longitudinal evidence for bidirectional associations. *Child Development, 82*(6), 1887–1905. doi:10.1111/j.1467-8624.2011.01669.x.

Baron-Cohen, S. (1995). Mindblindness: An essay on autism and theory of mind. *Learning Development and Conceptual Change, 74*. doi:10.1027//0269-8803.13.1.57.
Baron-Cohen, S. (1999). *Mindblindness. An essay on autism and theory of mind*. Cambridge, MA: MIT Press.
Bedi, S. & Babu, N. (2014). Higher order theory of mind and social competence in school age children. *Indian Journal of Positive Psychology, 5*(1), 72.
Bellagamba, F., Addessi, E., Focaroli, V., Pecora, G., Maggiorelli, V., Pace, B., & Paglieri, F.(2015). False belief understanding and "cool" inhibitory control in 3-and4-years-old Italian children, *Front Psychology, 6,678*.
Black, J. & Barnes, J.L. (2015). Fiction and social cognition: The effect of viewing award-winning television dramas on theory of mind. *Psychology of Aesthetics, Creativity, and the Arts, 9*(4), 423–429. doi:10.1037/aca0000031.
Bunanta, M. (1998). *Problematika Penulisan Cerita Rakyat Untuk Anak Di Indonesia*. Jakarta, Indonesia: Balai Pustaka.
Byom, L.J. & Mutlu, B. (2013). Theory of mind: Mechanisms, methods, and new directions. *Frontiers in Human Neuroscience, 7*, 413. doi:10.3389/fnhum.2013.00413.
Caputi, M., Lecce, S., Pagnin, A. & Banerjee, R. (2012). Longitudinal effects of theory of mind on later peer relations: The role of prosocial behavior. *Developmental Psychology, 48*(1), 257–270. Retrieved from http://ovidsp.ovid.com/ovidweb.cgi?T=JS&PAGE=reference&D=paovftm&NEWS=N&AN=00063061-201201000-00024.
Dunn, J., Cutting, A.L. & Fisher, N. (2002). Old friends, new friends: Predictors of children's perspective on their friends at school. *Child Development, 73*, 621–635. doi:10.2307/3696378.
Filippova, E. & Astington, J.W. (2008). Further development in social reasoning revealed in discourse irony understanding. *Child Development, 79*(1), 126–138. doi:10.1111/j.1467-8624.2007.01115.x.
Flavell, J.H. (2004). Theory-of-mind development: Retrospect and prospect. *Merrill-Palmer Quarterly, 50*(3), 274–290. doi:10.1353/mpq.2004.0018.
Goldstein, T.R. & Winner, E. (2012). Enhancing empathy and theory of mind. *Journal of Cognition and Development, 13*(1), 19–37. doi:10.1080/15248372.2011.573514.
Gopnik, A., Slaughter, V. & Meltzoff, A. (1994). Changing your views: How understanding visual perception can lead to a new theory of the mind. In Lewis, C. & Mitchell, P. (Eds.), *Children's early understanding of mind: Origins and development* (pp. 157–181). Hove, UK: Lawrence Erlbaum Associates. Retrieved from http://ilabs.washington.edu/meltzoff/pdf/94Gopnik_Meltzoff_etal_View.pdf.
Grazzani, I., Ornaghi, V., Cherubin, E. & Piralli, P. (2013). Promoting children's social cognition through story-based conversation: An intervention study. Retrieved from https://www.researchgate.net/publication/257307299_Promoting_children%27s_social_cognition_through_story-based_conversation_An_intervention_study.
Happe, F. (1999). Autism: Cognitive deficit or cogntive style? *Trends in Cogntive Science, 3*(6), 216–222. doi:10.1016/S1364-6613(99)01318-2.
Hodgins, J. (1993). *A passion for narrative: A guide for writing fiction*. Toronto, Canada: McClelland & Stewart.
Kamawar, D. & Olson, D.R. (2009). Children's understanding of referentially opaque contexts: The role of metarepresentational and metalinguistic ability. *Journal of Cognition & Development, 10*(4), 285–305. doi:10.1080/15248370903389499.
Kidd, D.C. & Castano, E. (2013). Reading literary fiction improves theory of mind. *Science (New York), 342*(6156), 377–380. doi:10.1126/science.1239918.
Korkmaz, B. (2011). Theory of mind and neurodevelopmental disorders of childhood. *Pediatric Research, 69*(5), 101R–108R. doi:10.1203/PDR.0b013e318212c177.
Kuntoro, I.A., Saraswati, L., Peterson, C. & Slaughter, V. (2013). Micro-cultural influences on theory of mind development: A comparative study of middle-class and pemulung children in Jakarta, Indonesia. *International Journal of Behavioral Development, 37*(3), 266–273. doi:10.1177/0165025413478258.
Kurniawan, H. (2009). Sastra Anak (dalam Kajian Strukturalisme, Sosiologi, Semiotika, hingga Penulisan Kreatif). Yogyakarta, Indonesia: Graha Ilmu.
Lecce, S., Bianco, F., Devine, R.T., Hughes, C. & Banerjee, R. (2014). Promoting theory of mind during middle childhood: A training program. *Journal of Experimental Child Psychology, 126*, 52–67. doi:10.1016/j.jecp.2014.03.002.
Lerner, M.D. & Lillard, A.S. (2015). From false belief to friendship: Commentary on Fink, Begeer, Peterson, Slaughter, and de Rosnay. *British Journal of Developmental Psychology, 33*(1), 18–20. doi:10.1111/bjdp.12070.
Liddle, B. & Nettle, D. (2006). Higher-order theory of mind and social competence in school-252 age children. *Journal of Cultural and Evolutionary Psychology, 4*(3–4), 231–246. doi:10.1556/JCEP.4.2006.3.

Louca, E.P. (2008). *Metacognition and theory of mind*. Newcastle upon Tyne, UK: Cambridge Scholars Publishing.
Mar, R.A., Oatley, K. & Peterson, J.B. (2009). Exploring the link between reading fiction and empathy: Ruling out individual differences and examining outcomes. *Communications, 34*(4), 407–428. doi:10.1515/COMM.2009.025.
Mar, R.A., Tackett J.L. & Moore, C. (2010). Exposure to media and theory-of-mind development in preschoolers. *Cognitive Development, 25*(1), 69–78. doi:10.1016/j.cogdev.2009.11.002.
McIntyre, C.W., Watson, D. & Cunningham, A.C. (1990). The effects of social interaction, exercise, and test stress on positive and negative affect. *Bulletin of the Psychonomic Society, 28*, 141–143.
Miller, S.A. (2012). *Theory of mind: Beyond the preschool years*. New York, NY: Psychology Press.
Moran, J.M. (2013). Lifespan development: The effects of typical aging on theory of mind. *Behavioural Brain Research*. doi:10.1016/j.bbr.2012.09.020.
Mursini, B, & Amran, S. (2011). *Pembelajaran Apresiasi Prosa Fiksi Dan Puisi Anak-Anak*. Medan, Indonesia: Libraries Unlimited.
Nurgiantoro, B. (2005). *Sastra Anak Pengantar Pemahaman Dunia Anak*. Yogyakarta, Indonesia: Gajah Mada Press.
O'Reilly, K., Peterson, C.C. & Wellman H.M. (2014). Sarcasm and advanced theory of mind understanding in children and adults with prelingual deafness. *Developmental Psychology, 50*(7), 1862–1877. doi:10.1037/a0036654.
Ornaghi, V., Brockmeier, J. & Grazzani, I. (2014). Enhancing social cognition by training children in emotion understanding: A primary school study. *Journal of Experimental Child Psychology, 119*, 26–39. doi:10.1016/j.jecp.2013.10.005.
Paal, T. & Bereczkei T. (2007). Adult theory of mind, cooperation, Machiavellianism: The effect of mindreading on social relations. *Personality and Individual Differences, 43*(3), 541–551. doi:10.1016/j.paid.2006.12.021.
Pace, B. & Paglieri, F. (2015). False belief understanding and 'cool' inhibitory control in 3-and 4-year-old Italian children. *Frontiers in Psychology, 6*, 872. doi:10.3389/fpsyg.2015.00872.
Perner, J. (1988). Developing semantics for theories of mind: From proportional attitudes to mental representation. In J. Astington, P. Harris, & D. Olson (Eds.), *Developing theories of mind*. Cambridge MA: Bradford Books/MIT Press.
Perner, J. & Wimmer, H. (1985). 'John thinks that Mary thinks that...': Attribution of second-order beliefs by 5- to 10-year-old children. *Journal of Experimental Child Psychology, 39*(3), 437–471. doi:10.1016/0022-0965(85)90051-7.
Peterson, C.C., Wellman, H.M. & Slaughter, V. (2012). The mind behind the message: Advancing theory-of-mind scales for typically developing children, and those with deafness, autism, or asperger syndrome. *Child Development, 83*(2), 469–485. doi:10.1111/j.1467-8624.2011.01728.x.
Pons, F., Harris, P. & de Rosnay, M. (2004). Emotion comprehension between 3–11 years: Developmental periods and hierarchical organizations. *European Journal of Developmental Psychology, 192*, 127–152.
Prosen, S., Škraban, O.P. & Smrtnik, H. (2011). *Teachers' emotional expression in interaction with students of different ages*. University of Ljubljana, Faculty of Education. Retrieved from https://repozitorij.uni-lj.si/IzpisGradiva.php?id=68130&lang=eng.
Razza, R.A., & Blair, C. (2009). Associations among false-belief understanding, executive function, and social competence: A longitudinal analysis. *Journal of Applied Developmental Psychology, 30*, 332–343.
Repacholi, H. & Slaughter, V. (2003). Individual differences in theory of mind: Implications for typical and atypical development. New York, NY: Psychology Press.
Shahaeian, A. (2013). *Developing an understanding of mind: A view across cultures*. Retrieved from http://espace.library.uq.edu.au/view/UQ:314880.
Smrtnik Vitulić, H. (2009). The development of understanding of basic emotions from middle c hildhood to adolescence. *Studia Psychologica, 51*, 3–20.
Stein, S. (1995). *Stein on writing*. New York, NY: St. Martin's Press.
Stiller, J. & Dunbar, R. (2007). Perspective-taking and social network size in humans. *Social Networks, 29*, 93–104.
Watson, A.C., Nixon, C.L., Wilson, A. & Capage, L. (1999). Social interaction skills and theory of mind in preschoolers. *Developmental Psychology, 35*, 386–391.
Wellman, H.M. (1990). *The child's theory of mind*. Cambridge, MA: MIT Press.
Wellman, H.M. & Liu, D. (2004). Scaling of theory-of-mind tasks. *Child Development, 75*(2), 523–541. doi:10.1111/j.1467-8624.2004.00691.x.

Diversity in Unity: Perspectives from Psychology and Behavioral Sciences – Ariyanto et al. (Eds)
© 2018 Taylor & Francis Group, London, ISBN 978-1-138-62665-2

Analyzing the influence of parent involvement and co-parenting on parenting self-efficacy

S.Y. Indrasari & M. Dewi
Faculty of Psychology, Universitas Indonesia, Depok, Indonesia

ABSTRACT: Parenting self-efficacy is parents' belief about their roles as parents or their perceptions of their ability to determine their children's behavior and development. Factors influencing parenting self-efficacy include parents' childhood-rearing experience, parents' experience with children, parents' cognitive and behavioral levels, social support, and children's characteristics. The purpose of this study is to examine whether or not parent involvement (as part of parents' cognitive and behavioral level) and co-parenting (as part of social support) predicts parenting self-efficacy. As research sample, we selected 306 parents which consisted of 152 fathers and 154 mothers aged between 25 and 45 years who had middle childhood children aged between 5 and 12 years. These parents were residents of Greater Jakarta. Parents were required to fulfill three questionnaires about parenting self-efficacy which had been adapted from the Self-Efficacy for Parenting Task Index (SEPTI). The questionnaire used to measure parent involvement was the Father Involvement Scale-Reported/Mother Involvement Scale-Reported (FIS-R/MIS-R), while the questionnaire used to measure co-parenting was the Co-parenting Relationship Scale. Results indicate that parent involvement ($\beta = 0.363$, $p < 0.01$) and co-parenting ($\beta = 0.434$, $p < 0.01$) significantly influence parenting self-efficacy ($R^2 = 0.440$. $F(2, 303) = 120.702$, $p < 0.01$). The findings and implications of this study are also discussed.

1 INTRODUCTION

Parenting is the process of interaction between parents and children in which both parties affect each other until the children grew older (Brooks, 2011). In parenting, parents served as figures that play a significant role in supporting and nurturing their children in all aspects of their lives, including their physical, emotional, and psychological states (Brooks, 2011).

Parenting is a complex process influenced by some factors. Belsky (in Sevigny & Loutzenhiser, 2009) stated that parenting behavior is determined by the interplay of three distinct components: the parents' personal characteristics (e.g. personality traits and psychological functioning), social and contextual influences of stress and support (e.g. marital relationship, employment, and social networks), and the children's personal characteristics. There are many ways children's characteristics can determine parenting behavior. For instance, middle childhood children (5–12 years old) have developed a more complex cognitive ability to understand about their own selves, to elaborate their emotions, and to practice self-control, so they tend to spend more time at school and socialize with their peers (Collins, Madsen, & Bornstein, 2002; Papalia & Feldman, 2012). Therefore, parents with middle childhood children would monitor and give direction to their children at some distance when they are at school or spending time with their friends (Brooks, 2008, 2011). Papalia and Feldman (2012) state that, in terms of parenting middle childhood children, some parents were worried about sharing responsibility as parents, fulfilling commitments with regard to time allocation, and energy consumption. Problems found among middle childhood children were fighting (Bond, Pragerm, Tiggemann, & Taoi, 2001; Cutrona & Troutman, 1986 in Altman, 2005), and this particular problem might cause feelings of fatigue, guilt, and failure in parents (Altman,

2005). These conditions might disrupt parenting activities. Hence, parents were expected to acquire certain competencies to ensure their children's positive development.

1.1 *Parenting self-efficacy*

Cognitive competencies are regarded as some of the most important competencies because they could influence parents' behavior and emotion while parenting their children (Jones & Prinz, 2005). Moreover, Coleman and Karraker (1998, 2000) also stated that one of the most important cognitive elements of parenting competence is parenting self-efficacy (PSE). Parenting self-efficacy is defined as parents' self-referent estimations of competence in the parental role or as parents' perceptions of their ability to positively influence their children's behavior and development (Coleman & Karraker, 2000). In general, parents with strong beliefs in themselves, especially in their PSE, also engage in positive parenting behaviors, and they might also have greater satisfaction in parenting their children, even in hard times.

Previous studies have shown links between parenting self-efficacy, parenting practices, and children's behavior (Jonas & Prinz, 2005). Research has shown that parents with low parenting self-efficacy tend to produce negative outcomes in the parenting process (Bandura in Coleman & Karraker, 1998). As an advocacy institution for children in Indonesia (Setyawan, 2015), the Indonesian Child Protection Commission (Komisi Perlindungan Anak Indonesia) has gathered facts related to child abuses which are committed by parents. Furthermore, Al-Hakim (2015) asserts that the main reason for such abusive parenting practices is because, when children make a mistake, those parents usually take it as their failure as parents. This condition leads to anger and other abusive actions toward the children. In contrast, parents with high parenting self-efficacy perceive parenting problems (especially during hard times) as challenges, instead of threats that might hurt their children (Donovan in Coleman & Karraker, 1998). Furthermore, it is generally accepted that high parenting self-efficacy is associated with a larger capacity to provide an adaptive, stimulating, and nurturing child-rearing environment.

Previous studies found that parenting self-efficacy is influenced by several factors. Those factors were childhood experience, culture and community, parents' experience in nurturing children, parental readiness in terms of cognition and behavior (Coleman & Karraker, 1998), and social support from family and friends (Holloway, Suzuki, Yamamoto, & Behrens, 2005). Most studies which found a positive correlation between parenting self-efficacy and parenting only involved mothers. Because of this, Coleman and Karraker (1998) suggest that future research on parenting self-efficacy should also involve fathers because parenting is the responsibility of both parents, not only of one of them. This present study explores the influence of parent involvement and co-parenting on parenting self-efficacy of both fathers and mothers.

1.2 *Parent involvement*

Parent involvement has been found as a factor which influences parenting self-efficacy (Glatz 7 Buchanan, 2015; Shumow & Lomax, 2002; Coleman & Karraker, 2000). Lean and Karraker (in Coleman & Karraker, 2005) stated that parenting self-efficacy is related to parents' readiness in cognitive and behavioral aspects of parenting. These would include the parents' own decision to get involved in parent-child relationship. In contrast, Peiffer (2015) achieved different results that suggest that parent involvement has no correlation with parenting self-efficacy. Thus, the current study seeks to explore whether or not parenting involvement has a correlation with parenting self-efficacy.

According to Finley, Mira, and Schwartz (2008), parent involvement can be defined as the extent to which parents participate in various aspects of their children's lives. Finley, Mira, and Schwartz (2008) have also identified twenty domains of parental involvement, including providing income, caregiving, monitoring schoolwork, encouraging independence, discipline, and companionship. A previous study conducted by Finley and Schwartz (2004) regarding father involvement explains that the two primary dimensions of parenting, as outlined by Parsons and Bales (1955), are the instrumental dimension (e.g. discipline, protection, and income provision) and the expressive dimension (e.g. companionship, caregiving, and sharing leisure activities). Furthermore, Finley and Schwartz (2004) added a third subscale called "mentoring/advising". Taken together, these subscales can serve as a measuring instrument which could be applied to both fathers and mothers as parents.

1.3 *Co-parenting*

Parenting self-efficacy could be influenced by various environmental aspects, especially by other people with whom an individual share a close relationship (Bandura in Merrifield and Gamble 2012). Moreover, co-parenting is one type of social support that a parent could have in raising children in a family. Bandura has reported that a person who feels satisfied with his or her social support will also have more confidence and have better emotional well-being. This can happen because he or she knows that there will always be friends and family who are willing to support him or her (Cutrona & Troutman, 1986; MacPhee *et al.,* 1996; Simons & Johnson, 1996; Teti & Gelfand, 1991 in Holloway *et al.,* 2005). In contrast, a person with less social support tends to be less competent in terms of self-development and have a higher level of stress (Holloway *et al.,* 2005). Previous studies found that support from one's partner makes the most meaningful contribution to one's parenting self-efficacy (Elder *et al.,*1995; Ozer, 1995; Simons *et al.,* 1993; Williams *et al.,* 1987 in Holloway *et al.,* 2005).

Merrifield and Gamble (2012) explain that co-parenting construct consists of both parents' ability to coordinate their activities and to support one another, as well as their non-supportive responses or efforts to interfere with one another. In addition, the ways partners share their responsibilities as parents could be an almost inexhaustible source of information, either supportive or undermining, regarding one's performance as a parent. Little is known about the effects of co-parenting on parental cognition, and this present study seeks to address this gap by specifically investigating the effects of co-parenting on PSE.

1.4 *The present study*

This study examines PSE as a product of a variety of contextual factors using parent involvement and co-parenting as the predictors. It is expected that parent involvement and co-parenting will be found to promote or positively influence PSE. In addition, this study seeks to identify which, between the two variables of parent involvement and co-parenting, is the strongest predictor of PSE.

2 METHOD

2.1 *Participants*

The current study involved 306 parents who consist of 152 fathers and 154 mothers as respondents. The inclusion criteria were (a) parents who had a child aged between 5 to 12 years and (b) parents aged between 25 to 45 years. The majority of the fathers and mothers had a child with an average age of 8.77 years, while the average age of the participants was 38.06 years. The participants resided in Greater Jakarta area that includes Jakarta, Depok, Bogor, Bekasi, and Cilegon. All participants were married (n = 306, 100%).

2.2 *Procedures*

Two data collectors were employed in this current study. They were students of the Faculty of Psychology, Universitas Indonesia. They went to several kindergartens and elementary schools in Greater Jakarta, particularly those within the areas of Jakarta, Depok, Bogor, Bekasi, and Cilegon.

The questionnaires were distributed to all students (from kindergarten to sixth-grade elementary school level) who studied at public or private schools. These students then delivered the questionnaires to their parents. Only parents who are in the age range of 25 to 45 years could voluntarily participate in this current study. They were then instructed to read the informed consent form before they agreed to participate in this study; therefore, they could choose whether or not to participate in this study. The fathers or mothers were asked to fill out the questionnaires as if they had only one child, in spite of the fact that the fathers or mothers might have had more than one child in middle childhood years.

The questionnaires were submitted by the students at school after 1–2 days. All data were collected from all schools within a certain time limit, averaging 14 days, depending on the agreement between each school and our data collectors. 350 questionnaires were brought home by the children, but only 306 questionnaires were submitted to our data collectors.

2.3 *Measurements*

2.3.1 *Parenting self-efficacy*

Parenting self-efficacy was assessed using the Self-Efficacy for Parenting Task Index (Coleman & Karraker, 2000) which had been adapted to Indonesian language by Erniza Miranda Madjidin (2011). This instrument assesses parenting self-efficacy across five dimensions of parenting: achievement, recreation, discipline, nurturance, and health.

Coleman and Karraker (2000) construct the five dimensions which make up the scale and they are designed to assess parents' sense of competence pertaining to these distinct categories of parenting task: (1) facilitating children's achievement in school ("achievement"); (2) supporting children's need for recreation, including socializing with peers ("recreation"); (3) providing structure and discipline ("discipline"), (4) providing emotional nurturance ("nurturance"), and (5) maintaining children's physical health ("health"). This instrument consists of 36 items with a Likert scale ranging from 1 ("strongly disagree") to 6 ("strongly agree"). Higher scores were indicative of higher self-efficacy on all dimensions after several items were reversely scored. For the current study, the internal reliability was found to be at $\alpha = 0.889$.

2.3.2 *Parent involvement*

Parent involvement was assessed using the Father/Mother Involvement Scale–Reported (Finley & Schwartz, 2004; Finley *et al.*, 2008). The instrument was originally constructed by Finley *et al.* (2008), and was adapted to Indonesia language by Mitranti (2005) according to children's perspective of parent involvement. Mitranti used children's perspective of parent involvement, while this study used parents' own perspectives of parent involvement. Thus, the instrument was re-adapted for the purpose of this study. To distinguish between the Father Involvement Scale–Reported and the Mother Involvement Scale–Reported as in the original version, the word "ayah" (father) and "ibu" (mother) were also used in the instructions to represent parent involvement.

The instrument consists of 20 items which assess three dimensions: expressive, instrumental, and mentoring/advising. First, the expressive dimension consists of eight domains which include companionship, caregiving, sharing activities/interests, emotional development, social development, leisure/fun/play, physical development, and spiritual development. Second, the instrumental dimension consists of eight domains which include developing responsibilities, ethical/moral development, career development, developing independence, being protective, school/homework, discipline, and providing income. Third, the mentoring/advising dimension consists of four domains which include mentoring, developing competence, advising, and intellectual development. The Likert scale is used in five alternative options ranging from 1 ("never involved") to 5 ("highly involved"). For the current study, the internal reliability of FIS-R was found to be at $\alpha = 0.941$, while the MIS-R was found to be at $\alpha = 0.894$.

2.3.3 *Co-parenting*

Co-parenting was assessed using the Co-parenting Relationship Scale (Feinberg, 2003). For the purpose of this study, this instrument was adapted to Indonesian language. This instrument consists of 35 items which assess various dimensions of co-parenting: (1) child-rearing agreement which consists of one sub-scale: co-parenting agreement; (2) co-parenting support/undermining which consists of three sub-scales: co-parenting support, endorsement of partner's parenting, and co-parenting undermining; (3) division of labor which consists of one sub-scale: division of labor; (4) family management which consists of one sub-scale: exposure to conflict. The Likert scale was used in six alternative options ranging from 1 ("strongly disagree") to 6 ("strongly agree"). For the current study, the internal reliability of the co-parenting relation scale was found to be at $\alpha = 0.861$.

2.4 *Data analysis*

The analyses for this study were performed using Microsoft Excel for data input and SPSS version 20.0 for Windows for data analysis. The first step was to analyze descriptive statistics to obtain general information about participants' characteristics. The descriptive statistics involved frequency analysis, mean, standard deviation, and range of total scores. This analysis generates information and percentages related to participants' age, educational background, marriage status, child's age, and child's gender. The second step was to examine the influence of parent involvement and co-parenting on PSE. Multiple linear regression analyses were conducted in the second step.

3 RESULTS

The demographic information about research participants could be found on Table 1 below.

Our research participants consist of 306 parents (152 fathers and 154 mothers). In terms of cultural background, most participants are Javanese (39.5%), while the rest arew Sundanese (27.2%) and 33.3% were classified as "other". In terms of educational background, half of all participants (53%) graduated from Elementary to High Schools, while 39% had earned graduate degrees and 8% had earned post-graduate degrees. In terms of occupation, most fathers worked as employees in private companies, while most mothers were housewives. In terms of family expenditure, most parents (38.6%) spent less than IDR 2.5 million per month, 38.2% spent IDR 2.5 to 7.5 million per month, while the remainder (23.2%) spent more than IDR 7.5 million per month.

Multiple Regression Analyses of the influence of Parent Involvement and Co-Parenting on Parenting Self-Efficacy.

In order to analyze whether parent involvement and co-parenting predict parenting self-efficacy among parents with middle childhood children, multiple linear regression analyses were conducted.

Table 2 shows that parent involvement and co-parenting significantly influence parenting self-efficacy, with $F\,(2{,}303) = 120.702$, $p < 0.01$. These results show that parent involvement ($\beta = 0.363$, $p < 0.01$) and co-parenting ($\beta = 0.434$, $p < 0.01$) significantly influence parenting self-efficacy. In addition to that, the strongest predictor of parenting self-efficacy is co-

Table 1. Parenting Self-Efficacy Task Index (SEPTI).

Dimension	Total item	Item number	Item example
Discipline	8	1, 2*, 3*, 4*, 5, 6*, 7, 8*	I am pretty good in disciplining my child.
Achievement	7	9, 10, 11, 12*, 13, 14*, 15	I am sure my child knows I am interested in his/her life at school.
Recreation	7	16, 17*, 18*, 19, 20, 21, 22*	I don't do enough to make sure my child has fun.
Nurturance	7	23, 24, 25*, 26, 27, 28, 29	I have trouble expressing my affection to my child.
Health	7	30, 31*, 32, 33, 34, 35, 36*	I work hard to encourage healthy habits in my child.

*Unfavorable items.

Table 2. Father/Mother involvement scale—reported.

Dimension	Total item	Item number	Item example
Expressive	8	4, 7, 8, 9, 12, 16, 18, 19	How involved am I in my child's moral development or ethics?
Instrumental	8	2, 3, 5, 6, 11, 13, 15, 20	How involved am I in my child's emotional development?
Mentoring/advising	4	1, 10, 14, 17	How involved am I in my child's intellectual development?

Table 3. The co-parenting relationship scale.

Dimension	Subscale	Total item	Item number	Item example
Co-parenting Agreement	Co-parenting Agreement	4	6, 9*, 11*, 15*	My partner and I have the same target for our child.
Division of Labor	Division of Labor	2	5*, 20*	I feel my partner and I didn't have an equal share in the performance of our parenting tasks.
Co-parenting Support and Undermining	Co-parenting support	6	3, 10, 19, 25, 26, 27	My partner made me feel that I am the best parent for our child.
	Endorsement of Partner's Parenting	8	1, 4, 7*, 14, 18, 23, 29*, 32	I believe that my partner is a good parent.
	Co-parenting Undermining	7	8*, 12*, 13*, 16*, 21*, 22*, 31*	My partner didn't trust my parenting skills.
Joint Family Management	Exposure to Conflict	5	33*, 34*, 35*, 36*, 37*	Did I argue with my partner about our child in front of the child?
	Co-parenting Closeness	5	2, 17, 24, 28*, 30	Parenting made us focus on our future.

*Unfavorable items.

Table 4. Demographic data of participants.

	Parents	
Data of participants	Frequency	Percentage
Gender		
Male	152	49.7%
Female	154	50.3%
Cultural Background		
Javanese	121	39.5%
Sundanese	83	27.2%
Other	102	33.3%
Educational Background		
Elementary—High School	162	53.0%
College	119	39.0%
Post Graduated	25	8%
Family Expenditure		
2,500,000	118	38.6%
2,500,000–7,500,000	117	38.2%
>7,500,000	71	23.2%

Table 5. The results of multiple regression analyses of the influence of parent involvement and co-parenting on parenting self-efficacy (N = 306).

	Parenting self-efficacy			
Predictor	R^2	F	Sig	B
	0.440	120.702	0.000	
Parent Involvement				0.363**
Co-parenting				0.434**

**p < 0.01, (one-tailed).

parenting. The results show that 44.0% of parenting self-efficacy variance is predicted by parent involvement. Therefore, it can be concluded that parent involvement and co-parenting are strong predictors of PSE, with $R^2 = 0.440$ (Gravetter & Wallnau, 2013).

4 DISCUSSION

The aim of this present study is to assess the influence of parent involvement and co-parenting on parenting self-efficacy. The results show that parent involvement and co-parenting can significantly predict parenting self-efficacy. It means that high levels of parent involvement and co-parenting can increase the level of parenting self-efficacy. Previous studies have also found that there is a correlation between parent involvement and PSE (Coleman & Karraker, 1997; Shumow & Lomax, 2002). Jones and Prinz (2005, in Glatz & Buchanan, 2015) also explain that parent involvement may influence an individual's parenting self-efficacy. This is also in line with Bandura who found that parent involvement as part of promotive parenting practices also had a significant influence on parenting self-efficacy. Thus, a higher level of parent involvement will lead to a higher level of parenting self-efficacy (Bandura in Glatz & Buchanan, 2015).

In terms of co-parenting, the results of this present study are also in line with Baker's findings (2007). A high level of parenting self-efficacy is indicative of a high level of co-parenting relations, which suggests a lower level of conflict and a higher level of support given by one's partner. Merrifield and Gamble (2012) state that the correlation between co-parenting and parenting self-efficacy is attributable to positive qualities of co-parenting, such as supports given by one's partner which can increase one's parenting self-efficacy, especially when dealing with various difficulties in life (e.g. conflicts in marriage or the undermining of one's co-parenting). This means that an individual with good co-parenting (i.e. having a supportive partner) tends to have a stronger belief in his or her success in parenting.

In this study, we also seek to identify the strongest predictor of parenting self-efficacy, whether it is parent involvement or co-parenting. Our findings show that co-parenting is the strongest predictor of parenting self-efficacy. Merrifield and Gamble (2012) have further proposed that the strength of co-parenting is determined by parents' own perception of the quality of their own marriage relationship, which will in turn influence the extent of their engagement in parenting their children (parenting self-efficacy).

5 LIMITATIONS AND CONCLUSION

There are some limitations in this present study. First, the instruments were distributed to parents through institutions (schools), and this resulted in several incomplete questionnaires which could not be used for a further analysis. Second, similar to other research on this topic, we used only self-report measures which tend to have a high level of social desirability. In order to broaden our understanding of parenting self-efficacy, future studies should make use of information which are gathered through various methods and obtained from multiple sources. Third, the sample should include participants from more varied cultural, educational, and economic backgrounds in order to produce more comprehensive results.

This study has provided a significant contribution to the body of knowledge about parenting self-efficacy. This study is also one of the few studies that try to investigate the influence of parent involvement and co-parenting on parenting self-efficacy. Therefore, this study could serve as a basis for other local research and comparative studies in the future. Most previous research on parenting self-efficacy had focused on mothers, so the inclusion of fathers as a determinant factor can be considered as an important strength of this study. Second, the instruments used in this study (i.e. the Self-Efficacy for Parenting Task Index (SEPTI), Father/Mother Involvement Scale–Reported (F/MIS-R), and the Co-parenting Relationship Scale) have shown a high level of reliability. Third, the provision of questionnaires closed-ended questions has facilitated not only the scoring process, but also the interpretation of the data.

REFERENCES

Al, H., Fauziah, R. (2015). *Mendidik Anak Tanpa Kekerasan*. http://www.mirajnews.com/id/mendidik-anak-tanpa-kekerasan/71782.

Altman, M.D. (2005). Vicarious Experience and The Development of Parenting Self-Efficacy. [PhD diss.], George Mason University.

Bornstein, M.H. (2002). *Handbook of Parenting*. 2nd edition. New Jersey, Lawrence Erlbaum.

Brooks, J.B. (2008). *The Process of Parenting.* New York, McGraw-Hill.

Brooks, J.B. (2011). *The Process of Parenting*. 8th edition. Boston, McGraw Hill.

Coleman, P.K. & Karraker K.H. (1998). Self-efficacy and parenting quality: Findings and future applications. *Developmental Review,* 18(1), 47–85. http://dx.doi.org/10.1006/drev.1997.0448.

Coleman, P.K. & Karraker K.H. (2000). Parenting self-efficacy among mothers of school-age children: conceptualization, measurement, and correlates. *Family Relations*, 49(1), 13–24. http://search.proquest.com/docview/213935105?accountid=17242.

Finley, G.E., Mira S.D. & Seth J. (2008). Perceived paternal and maternal involvement: Factor structures, mean differences, and parental roles. *Fathering*, 6(1), 62–68.

Glatz, T. & Buchanan C.M. (2015). Over-time associations among parental self-efficacy, promotive parenting practices, and adolescents' externalizing behaviors. *Journal of Family Psychology,* 29(3), 427–437.

Holloway, S.D., Suzuki, S., Yamamoto Y. & Behrens K.Y. (2005). Parenting self-efficacy among japanese mothers. *Journal of Comparative Family Studies,* 36(1), 61–76.

Jones, T.L. & Prinz R.J. (2005). Potential roles of parental self-efficacy in parent and child adjustment: A review. *Clinical Pyschological Revie,* 25(3), 341–363.

Merrifield, K.A. & Gamble W.C. (2012). Associations among marital qualities, supportive and undermining co-parenting, and parenting self-efficacy: Testing spillover and stress-buffering processes. *Journal of Family Issues,* 34(4), 510–533.

Papalia, D., Olds, S., Feldman, R., Martorell, G., & Papalia, D. (2012). *Experience human development*. 12th edition. New York, McGraw-Hill.

Setyawan, D. (2015). KPAI: Pelaku kekerasan terhadap anak tiap tahun meningkat. Komisi Perlindungan Anak Indonesia. http://www.kpai.go.id/berita/kpai-pelaku-kekerasan-terhadap-anak-tiap-tahun-meningkat/

Sevigny, P.R. & Loutzenhiser, L. (2009). Predictors of parenting self-efficacy in mothers and fathers of toddlers. *In Child: Care, Health, and Development,* 36(2), 179–189.

Shumow, L. & Lomax R. (2002). Parental efficacy: Predictor of parenting behavior and adolescent outcomes. *Parenting: Science and Practice,* 2(2), 127–150.

Diversity in Unity: Perspectives from Psychology and Behavioral Sciences – Ariyanto et al. (Eds)
© 2018 Taylor & Francis Group, London, ISBN 978-1-138-62665-2

The relationship between social expectation and self-identity among adolescents

J. Suleeman & N. Saputra
Faculty of Psychology, Universitas Indonesia, Depok, Indonesia

ABSTRACT: The objective of this research is to explore the relationship between social expectation and self-identity among adolescents in Jakarta. Self-identity is measured by the Extended Objective Measure of Ego-Identity Status (EOM-EIS II) Indonesian version, originally developed by Adams (1998), while social expectation is measured by Social Expectation Scale, developed specifically for this study. One hundred and ninety adolescents from six districts in Jakarta participated. Results show that there is a relationship between social expectation and self-identity. Differences in achievement and diffusion status identity are found between early adolescents and late adolescents, and between middle adolescents and late adolescents, but not between early and middle adolescents, nor on moratorium and foreclosure status identity. Differences are also found on family social expectation and peer social expectation across three age groups. Further research could look at gender differences on these variables, and how parenting and significant others influence self-identity formation among adolescents.

1 INTRODUCTION

Adolescence is a period from childhood to adulthood where changes in biological, cognitive, emotional and social aspects take place. Erikson (1968) recognises the internal conflict adolescents go through. In one sense, they want to be independent enough to take care of themselves, but on the other hand they still need help from significant adults. 'Who am I' is a question often asked by adolescents that indicates they are still in the process of forming their identities (Cremers, 1989). The answer to this simple question is not that simple though. Identity vs identity confusion is theorised by Erikson (1968) to characterise what adolescents go through in their development for self-identity.

Identity confusion, a term referring to identity crisis in adolescents, can cause a withdrawal, namely isolating oneself from family and peers, or being among peers without having a personal identity or self-identity. Adolescents who know what they are good at, what their characters are, and what they want to accomplish and pursue in the future, are referred to as having formed their identity. Adolescents who are not sure about themselves in terms of character, potentials and other matters are referred to as having identity diffusion.

1.1 *Self-identity*

Self-identity is an answer an individual gives when he or she is asked 'Who are you?' (Cremers, 1989). Adolescence is a period during which one looks for self-identity; characterised as a life where one lives according to the commitment to attaining objectives, goals, values and beliefs that are important to oneself (Papalia & Feldman, 2012). In essence, there are two pathways how an individual can have his/her own self-identity (Marcia et al., 1993). One is through the experiences received by significant others in one's life. The other is through one's own experiences. For those with identities that have been assigned by other people, the activities revolve around fulfilling other people's expectation; also referred to as social expectation. On the other hand, those who have affirmed their own self-identity would rather build their future based on meanings they set aside for themselves, based on the experiences they regard as valuable and relevant. Two important elements in the formation of self-identity are exploration and commitment (Marcia et al., 1993). Exploration refers to the various alternatives one goes through before selecting one suitable for his- or herself. Commitment refers to the effort and strategy used to accomplish what has been decided as objectives or goals.

Based on exploration and commitment elements, Marcia et al. (1993) distinguish four identity statuses: diffusion, foreclosure, moratorium and achievement. Diffusion is when both exploration and commitment are low. Foreclosure is when one already has some commitment but no exploration. Moratorium is when one has already started the exploration but is still low in commitment. Achievement is for those who have already explored and made commitments.

1.2 *Social expectation as one factor influencing self-identity*

Of all factors influencing self-identity, the factors associated with one's family and environment are the most important, and these include child-rearing practices and behaviour by parents from when a child is very young, modelling towards significant others, social expectation, successful attempts to reveal oneself through various ways and media, and developmental tasks one has even before entering adolescence (Marcia et al., 1993). Social expectations are values (good and/or bad), expected in the environment, and for adolescents, environment can mean family, peers, and school (Marcia et al., 1993). Fulfilling these expectations is what adolescents are trying to do so that they can be accepted in their environment.

Marcia et al. (1993) label a 'destructive period' for early adolescents (13–15 years-old) where cognitive, psychosexual and physiological aspects are in transition from an earlier age. The 'restructuration period' for middle adolescents (16–18 years-old) is where old and new knowledge is combined. Late adolescents (19–22 years-old) have a 'consolidation period' where the identity status is more clearly formed before being finally established.

Since Indonesia is identified as a collectivist society (Triandis, 2001) with some personality characteristics different from those in an individualist society, it would be interesting to find out how adolescents in Indonesia generally, and Jakarta metropolitan area (as the national capital city) specifically, would score in their self-identity. Jakarta is also the centre for governmental, educational, business and industrial activities. Jakarta has long been an attractive city for those who wish for a better life. It would not be surprising if Jakarta were to become multicultural because as a modern city it offers more freedom compared to other cities in Indonesia. Jakarta has six districts, Central, North, East, South, West and Seribu Islands (this last district was only included a few years ago and the location is separated further north of Jakarta beyond the Jakarta Gulf.) Generally, adolescents in Jakarta compared to other cities have more direct access to modern life, and not only through social media. Nightlife is quite often looked after by adolescents especially during weekends, not to mention alcohol and free sex. This condition might influence the formation of identity among Jakarta adolescents. There are indications that peers are very influential in for Indonesian adolescents (Naibaho, 2013. In Naibaho's study she had 100 adolescents, aged 12 to 15 years-old, all were Junior High Schoolers. Fifty of them came from families with full time working mothers and the other 50 came from families with no working mothers. Her study showed that the majority of adolescents whose mothers were working and whose mothers did not work had a secure relationship with their peers, that was 40% and 68%, respectively. The other type of relationship with peers were ambivalent, avoidant, and disorganized and the instrument used to measure peer and mother closeness was Inventory of parental and peer attachment. Similar results of the closeness between adolescents and their peers are also shown in a study by Ningrum (2013) who had 721 Grade 10 students as her participants.

This particular study aims at identifying the type of self-identity, the patterns of social expectation, and the relationship between these two variables in early- compared to middle- and late adolescents as this idea has not been explored in Indonesia. Previous studies on self-identity among Indonesian adolescents have focused on adolescents from poor families

(Mardikoesno, 1999), athletes (Arsendy, 2013), and adolescent club goers (Yunita, 2002), but none have taken social expectation as the research variable.

1.3 *Research questions and hypotheses*

The questions this study has tried to answer are: 1) What is the type of self-identity in early, middle and late adolescents? 2) What is the social expectation element of early, middle and late adolescents? 3) Is there any relationship between identity status and social expectation among adolescents?

Only the third question requires a hypothesis, namely that a positive relationship exists between identity status and social expectation among adolescents. Adolescents in this instance are those who live in the Jakarta metropolitan area.

2 METHOD

2.1 *Participants*

The participants were 190 adolescents who were further categorised into early, middle, and late adolescents according to their chronological age, and live in Jakarta metropolitan area (consisting of Central Jakarta, North Jakarta, East Jakarta, South Jakarta, West Jakarta, and Seribu Islands districts), who were recruited using accidental sampling, participated. Each signed the informed consent before being asked to fill out the research instruments.

2.2 *Research instruments*

2.2.1 *EOM-EIS II*

To measure self-identity, the Extended Objective Measure of Ego-Identity Status (EOM-EIS II), with a 4-point Likert-scale (1 = Strongly disagree; 2 = Disagree; 3 = Agree; and 4 = Strongly agree) in the Indonesian language, adapted from Adams (1998), was used, with the usual procedure of back translation. Among all measures on identity status, EOM-EIS II is the most commonly used (Schwartz et al., 2006).

However, through prior elicitation with 49 adolescents aged 18–21 years-old, it was found that two domains, politics and recreation, were not popular or relevant issues. Therefore they were dropped which thus left only six domains (career, lifestyle, friendship, dating, religion and gender) with 47 items used from the original 64 items. From here, four identity statuses are identified: diffusion, foreclosure, moratorium and achievement. Since each item is targeted for a specific identity status, the total score for each identity status can be summed up.

Adams (1998) suggests two approaches in scoring and interpreting identity status with EOM-EIS II. The first is by taking the total raw score taken from all identity status' items. The identity status that has the highest total score is regarded as the strongest identity status that one has. The second approach is by calculating the *z-score* for each identity status. Thus, each individual has four identity statuses. The highest *z-score* among these four identity statuses is regarded as the dominant identity status. For this study, the second approach was

Table 1. Item specifications for EOM-EIS II.

Identity status	Number of items	Example of item
Diffusion	12 items	I don't think about dating much. I just kind of take it as it comes.
Foreclosure	11 items	I might have thought about a lot of different jobs, but there's never really been any question since my parents said what they wanted.
Moratorium	12 items	There are so many ways to divide responsibilities in marriage; I am trying to decide what will work for me.
Achievement	12 items	A person's faith is unique to each individual; I consider and reconsider it myself and know what I can believe.

used, assuming that *z-score* is more standardised to indicate the position one has in a distribution rather than total score.

Previous reliability analyses with Cronbach Alpha for items in each identity status from 154 participants who were university students resulted in 0.703 (for diffusion), 0.823 (for foreclosure), 0.706 (for moratorium) and 0.732 (for achievement). External validity using the Rosenberg Self-esteem Scale as the external criterion resulted in $r = 0.374$ ($p = < 0.05$). Another study has also found positive correlation between self-esteem and achievement identity status (Basak & Ghosh, 2008).

2.2.2 *Social expectation scale*

This scale was specifically constructed for the purpose of this research. It consists of 12 items with six items each for the family and peers elements. Each item represents a specific domain, namely religion, lifestyle, career, friendship, spouse and gender role. Each item is measured using a 4-point Likert-scale (1 = Not fulfilled at all; 2 = Not fulfilled; 3 = Fulfilled; and 4 = Very much fulfilled). The total score of these six items correspondingly constitutes the score for each of the elements family and peers. A higher score among these two elements is regarded as the preferred social expectation element for that individual. Exploratory Factor Analyses reveal two factors for this scale, and they are separate factors for the family element (eigenvalue 9.814, 34.967%), and for the peers element (eigenvalue 2.185, 17.057%), indicating that this scale is targeted for different constructs and they are family and peers. The total variance explained was 52.024%.

2.3 *Research design*

A correlational research design is considered to be appropriate since the relationship between social expectation and identity status is assessed through a correlational technique.

3 RESULTS

Each participant lives in one of the six districts of Jakarta metropolitan area. Below is the demographic data of all participants.

There were more female than male participants even though the number of participants in each age group was comparable.

3.1 *Results on identity status and social expectation*

Table 3 below shows the results from the EOM-EIS II and Social Expectation Scale. For EOM-EIS II, the number of participants who achieved each identity status is recorded. Only one identity status was identified for each individual, depending on which identity status had the highest *z-score* in comparison with the other three identity statuses. For the Social Expectation Scale, the number of participants for each element (family or peers) was also

Table 2. Participants' demographic data.

Age category	Frequency (percentage)	Male	Female
Early adolescents (13–15 years-old)	63 (33.2%)	26 (41.27%)	37 (58.73%)
Middle adolescents (16–18 years-old)	63 (33.2%)	25 (39.68%)	38 (60.32%)
Late adolescents (19–22 years-old)	64 (33.7%)	32 (50%)	32 (50%)
	N = 190	83 (43.7%)	107 (56.3%)

Table 3. Results on EOM-EIS II and social expectation scale.

	Identity status				Social expectation	
Category	Diffusion	Foreclosure	Moratorium	Achievement	Family	Peers
Early ($n_1 = 63$)	22 (34.92%)	17 (26.99%)	18 (28.57%)	6 (9.52%)	34 (53.97%)	29 (46.03%)
Middle ($n_2 = 63$)	15 (23.81%)	18 (28.57%)	15 (23.81%)	15 (23.81%)	27 (42.86%)	36 (57.14%)
Late ($n_3 = 64$)	13 (20.31%)	9 (14.06%)	15 (23.44%)	27 (42.19%)	45 (70.31%)	19 (29.69%)

Table 4. Differences in identity status among each age category.

Diffusion achievement				
	Middle	Late	Middle	Late
Early	0.107	0.001***	0.099	0.001***
Middle	—	0.026*	—	0.044*

Table 5. Differences in social expectation elements across age categories.

	Family		Peer	
	Middle *p*	Late *p*	Middle *p*	Late *p*
Early	0.012	.001	0.003	0.001
Middle	—	0.002	—	0.023

recorded. Each individual was only entitled to one element of social expectation, family or peers, depending on which element had bigger value for him or her.

There are more participants with achievement identity status among the late adolescents than among the early and middle ones. On the opposite side, there are more participants with diffusion identity status among early rather than middle and late adolescents. As for social expectation, early and late adolescents are more associated with family than peer expectations, while for middle adolescents, the opposite is true.

Further analyses with Analysis of Variance (ANOVA) were conducted to identify whether across age groups there were any differences on identity status. Only diffusion and achievement statuses reveal differences across age group ($F_{2.187} = 10.904$; $p = < 0.001$ for diffusion identity and $F_{2.187} = 10.128$; $p = <0.001$ for achievement identity status). *Post hoc* analyses using Tukey's Honest Significant Difference (HSD) test reveal further significant differences among early, middle and late adolescents on diffusion and achievement identity statuses as summarised in Table 4.

For diffusion and achievement identity statuses, differences were found between early- and late adolescents, and between middle- and late adolescents, thus similar patterns of differences characterise these cross-age comparisons.

Accordingly, further analyses were conducted to identify whether there were any differences in social expectation elements in the cross-age groups. The results are shown in Table 5.

Across age groups, differences were found in family and peers social expectations, namely between early and middle adolescents, between early and late adolescents, and between middle and late adolescents.

Table 6. Relationships between social expectation and identity status.

	Social expectation elements			
	Family		Peers	
Identity status	*r*	*p*	*r*	*p*
Diffusion	–0.924	<0.001	–0.734	0.001
Foreclosure	0.818	<0.001	0.541	0.001
Moratorium	–0.333	<0.001	–0.143	0.049
Achievement	0.929	<0.001	0.727	0.001

3.2 *Relationship between social expectation elements and identity status*

Using Pearson product-moment correlation technique, it was found that each of these identity statuses is correlated with both family and peers' expectation elements.

All relationships are significant with family social expectation having stronger relationships than peers' social expectation.

4 DISCUSSION

The research results show that social expectation, both from family and peers, is related with identity status among adolescents from three age categories, early, middle and late, with family having a stronger relationship in middle, compared to early and late, adolescents. Following Erikson (1968), receiving recognition both from family and peers is important in identity development and setting a future life as fulfilling the social expectation is also some of adolescents' developmental tasks. And since this is found among adolescents who live in Jakarta, these findings are quite encouraging, indicating that family and peers are still inseparable in adolescents' daily life.

Across age groups, differences were found in identity status. Since exploration and commitment are two bases on how identity status develops, and each age category provides different exploration and comment, it is not surprising that differences across age groups are found. For Marcia et al. (1993), one's personal experiences of exploration and commitment for each of the domains will characterise one's identity status. The personal experiences themselves are influenced by factors such as child-rearing practices and other family and environmental backgrounds. It is quite possible to have an identity status for one domain that is different for an identity status for another domain.

Therefore, it is not surprising that this study shows differences in identity status and social expectation among age groups. Early adolescents who are still in diffusion identity status have more family than peers' social expectation. Middle adolescents who are in the foreclosure identity status have more peers than family social expectation. Late adolescents who are in achievement identity status have more family than peers' social expectation. According to Whitmire (2000), early adolescents face a period where they need to solve concrete problems and peers are seemingly becoming more realistic to hang out with than parents. Middle adolescents, on the other hand, are more interested in issues outside familial ones, for instance, physical appearance, while they are also more able to do abstract thinking and become more idealistic. Late adolescents have more self-confidence than early and middle adolescents; they are also able to solve more complex issues regarding interpersonal relationships, and generally they are more mature and independent in making decisions. That different age categories—early, middle and late adolescents—show different results in both identity status and social expectation is not surprising as each age category has a different function in the identity development (Marcia et al., 1993). Fulfilling both family and peer social expectation might give different influence towards adolescents' identity development (Whitmire, 2000). For instance, family is regarde as important for Unfortunately, since no personal interviews took place, the researchers were unable to establish how adolescents balance family and peers' social expectation.

While these results are also in line with what Marcia et al. (1993) had suggested, there is a notable difference in the school social expectation. Marcia et al. (1993) stated that school social expectation also influences identity development, but for the participants in this study it is unknown since school social expectation is not included as part of social expectation element as recommended by prior result from a pilot study. This pilot study found only two out of eight domains, namely religion and career domains, which are significant from school social expectation. However, future research might need to account for school social expectation to find out whether different results can be collected.

5 CONCLUSION

This study shows that early, middle and late adolescents have different identity statuses. Early adolescents are more associated with diffusion identity status; middle adolescents with foreclosure identity status; and late adolescents with achievement identity status. Each age category also has a specific social expectation element associated with it. Early and late adolescents preferred family social expectation while middle adolescents prefer peers' social expectation. This study also supports the idea that among adolescents, family social expectation and peers' social expectation are related significantly with each of the four identity statuses, namely diffusion, for closure, moratorium and achievement. However, there are some differences in the pattern of the relationship across early, middle, and late adolescents. These findings have to be interpreted in the dynamics of identity development among adolescents.

REFERENCES

Adams, G.R. (1998). *The objective measure of ego identity status: A reference manual*. Canada: University of Guelph.

Arnet, J.J. (2013). *Adolescence and emerging adulthood: A cultural approach* (5th ed.). Hoboken, NJ: Pearson.

Arsendy, S. (2013). *Hubungan antara Status Identitas dengan Kematangan Karir pada Pelajar-Atlet di Sekolah Atlet Ragunan*. Depok, Indonesia: Fakultas Psikologi, Universitas Indonesia.

Basak, R. & Ghosh A. (2008). Ego-identity status and its relationship with self-esteem in a group of late adolescents. *Journal of the Indian Academy of Applied Psychology, 34*(2), 337–344.

Beyers, W. & Luyckx K. (2016). Ruminative exploration and reconsideration of commitment as risk factors for suboptimal identity development in adolescence and emerging adulthood. *Journal of Adolescence, 47*, 169–78. doi:10.1016/j.adolescence.2015.10.018

Cremers, A. (1989). *Identitas dan Siklus Hidup Manusia*. Jakarta, Indonesia: Gramedia.

Erikson, E.H. (1968). *Identity: Youth and crisis*. New York, NY: Norton.

Gravetter, F.J. & Forzano L.B. (2012). *Research methods for the behavioral sciences* (4th ed.). Belmont, CA: Wadsworth Cengage Learning.

Gulman, A. (1999). Hubungan antara Parental dan Peer Attachment dengan Perkembangan Identitas pada Remaja (Bachelor's thesis, Fakultas Psikologi, Universitas Indonesia).

Hoffman, L., Paris S. & Hall, E. (1994). *Developmental psychology today* (6th ed.). New York, NY: McGraw-Hill.

Kaplan, R.M. & Saccuzzo D.P. (2001). Psychological testing: Principles, applications, and issues (5th ed.). *44*, 1–11. doi:10.1017/CBO9781107415324.004.

Kerlinger, F.N. & Lee, H.B. (2000). *Foundations of behavioral research*. Tokyo, Japan: Harcourt College Publishers.

Marcia, J.E., Waterman A.S., Matteson, D.P., Archer S.L. & Orlofsky J.L. (1993). *Ego identity: A handbook for psychological research*. New York, NY: Springer-Verlag.

Mardikoesno, B.Y. (1999). *Gambaran identitas diri remaja miskin yang putus sekolah* (Description of self identity among drop out adolescents) Bachelor's thesis, Fakultas Psikologi, Universitas Indonesia).

Naibaho, D.F. (2013). *Gaya pengasuhan ibu, kelekatan dengan teman sebaya, dan konsep diri remaja pada keluarga ibu bekerja dan tidak bekerja* (Relationship between mother's parenting style, closeness to peers, and self esteem among adolescents' from families with working mothers and families with no working mothers). (Bachelor's thesis, Fakultas Ekologi Manusia, Institut Pertanian Bogor).

Ningrum, L.R. (2013). *Hubungan Dukungan Sosial Teman Sebaya dengan Konsep Diri Remaja pada Siswa Kelas X di SMKN 2 Malang*. (The relationship between social support from peers and self-concept among Grade X SMKN 2 Malang students). Bachelor's thesis, Fakultas Psikologi, Universitas Islam Negeri Maliki Malang.).

Papalia, D.E. & Feldman, R.D. (2012). *Experience human development* (12th ed.). New York, NY: McGraw-Hill.

Papalia, D.E. & Olds, S.W. (2001). *Human development.* New York, NY: McGraw-Hill.

Purwadi, P. (2004). Proses pembentukan identitas remaja. *Humanitas: Jurnal Psikologi Indonesia, 1*(1), 43–52.

Sarwono, S.W. (2011). *Psikologi Remaja.* Jakarta, Indonesia: Rajawali Pers.

Schwartz, S.J., Adamson, L., Ferrer-Wreder, L., Dillon F.R. & Berman, S.L. (2006). Identity status measurement across contexts: Variations in measurement structure and mean levels among white American, hispanic American, and Swedish emerging adults. *Journal of Personality Assessment, 86*(1), 61–76.

Steinberg, L. (1993). *Adolescence* (3rd ed.). New York, NY: McGraw-Hill.

Thornburg, H.D. (1982). *Development in adolescents* (3rd ed.). Belmont, CA: Wadsworth.

Triandis, H.C. (2001). Individualism-collectivism and personality. *Journal of Personality, 6*(6), 907–924. doi:10.111/1467-6494.696169.

Whitmire, K.A. (2000). Adolescence as a developmental phase: A tutorial. *Topics in Language Disorders, 20*(2), 1–14. doi:10.1097/00011363-200020020-00003.

Yunita, F. (2002). *Gambaran Identits Diri Remaja yang Melakukan Aktivitas Clubbing* (Bachelor's thesis, Fakultas Psikologi, Universitas Indonesia).

Diversity in Unity: Perspectives from Psychology and Behavioral Sciences – Ariyanto et al. (Eds)
© 2018 Taylor & Francis Group, London, ISBN 978-1-138-62665-2

The correlation between parenting style of working mothers and mothers' perception of their school-aged children's academic achievement

G.A.F. Tinihada & F.M. Mangunsong
Faculty of Psychology, Universitas Indonesia, Depok, Indonesia

ABSTRACT: The objective of this present study is to investigate the correlation between the parenting style of working mothers and the mothers' perception of their school-aged children's academic achievement. The instruments used in this study are the Parenting Styles and Dimensions Questionnaire (PSDQ—Short Version) and the Scale of Perceived Academic Achievement (SPAA). One hundred and fifty full-time working mothers with at least one school-aged child living in the Jabodetabek region participated in this study. The result of this study shows that authoritative and authoritarian parenting style are significantly correlated with the mothers' perception of their school-aged children's academic achievement. The result of this study also shows that the age, educational level, socioeconomic status, culture of origin of the mothers and the children's gender are not significantly correlated with the parenting style of the working mothers. Moreover, the result of this study shows that the educational level, socioeconomic status of the mothers and the children's gender are not significantly correlated with the mothers' perception of their children's academic achievement.

1 INTRODUCTION

Parenting is a process of action and interaction between parent and child, involving the processes of change towards both the parent and the child as the child grows up into adulthood (Brooks, 2008). As cited in Brooks (2008), Belsky *et al.* maintains that the parent's behavior and efforts may significantly influence a child's development and competence. According to Baumrind (as cited in Baumrind, 1991), the behavior of the parents that describes the efforts of parents in fulfilling the needs for their children in rearing and supervising their children is referred to as parenting style.

Baumrind (1966; 1991) suggested three prototypes of parenting style: (1) authoritative, (2) authoritarian and (3) permissive. Authoritative parenting style is the prototype in which parents direct the child's behavior in a rational manner and problem-oriented. Authoritative parents are parents who give demands but are still responsive to the child. Authoritarian parenting style is the prototype in which parents seek to establish, monitor and evaluate the child's behavior based on a set of absolute standards of behavior. Moreover, authoritarian parents tend to demand and direct, but are not responsive to the child. Permissive parenting style is the prototype which parents behave in a way not to give punishment, tend to accept and approve the child's desires and behaviors. Permissive parents are more responsive to children and do not provide demand

Today in Indonesia, especially in major cities, there has been a shift in the role of the mother in the family. Many housewives are now also working. Based on data from the Central Bureau of Statistics of Indonesia (BPS) in 2012, the number of female workers in Indonesia in February 2012 reached 46,509,689 women which suggest an increase of 1 million female workers when compared to the total female work force in 2011 that reached 45,118,964 women. Improvement of educational levels and rising living costs are the reasons for the increase in the number of working mothers (Tjiptoherijanto in Tjaja, 2000). Dwijanti (in

Suryadi & Damayanti, 2003) defines a working mother as a mother who earns a salary from someone to perform certain tasks as a worker or as an employee with a specific work schedule, and is rarely at home, so the working mothers have less time to meet with their children.

Based on various earlier studies, Greenberger and Goldberg (1989) concluded that working parents can affect the lives of children in various aspects, such as: (1) affecting the investments made by both parents in parenting, (2) adjusting the expectations of parents on the children's behavior, (3) encourage parents to modify the type of parenting style or discipline and control strategies applied to the children, and (4) affects the perception and evaluation of the parent to the child. In addition, there are several other factors that can affect the parenting style, such as: (1) parents' age (Meggiolaro& Ongaro, 2013), (2) parents' educational level (Klebanov *et al.* in Davis—Kean, 2005), (3) parents' socioeconomic status (Berns, 2013), (4) parents' culture of origin (Berns, 2013), (5) child's gender (Bornstein as cited in Berns, 2013) and (6) child's age (Bornstein as cited in Berns, 2013).

School-aged children are children aged 6–11 years that are in the middle phase of childhood (Papalia & Feldman, 2012). Based on the theory of Erikson's Psychosocial stages (in Santrock, 2011), school-aged children are entering the fourth stage, namely the Industry versus Inferiority stage, in which children are encouraged to master the knowledge and skills that are taught in schools to attain good academic achievement. School-aged children that fail to achieve good academic performance tend to feel inferior, unproductive and unable to perform.

Children's academic achievement can be assessed objectively and subjectively. Objective measurement of a child's academic performance can be assessed by reviewing the scores of a child's school report cards or test scores. Farkas *et al.* (in Carbonaro, 2005) concluded that the study habits of students can be assessed from the teacher's reports on completion of homework assignments, the student's participation in class, the students' efforts and orderliness which are positively correlated with the students' mastery of the subject matters taught in class and are reflected in the grades achieved by the students. Meanwhile, subjective assessment on a child's academic performance is based on the parents', teachers' and the child's perception towards the child's ability. According to Parsons, Adler and Kaczala (in Arbreton, Eccles, & Harold, 1994), the parents' perceptions of their children's academic abilities have greater impact than the feedback received through the school report. The parents' perceptions of the ability of their children contribute to the child's confidence in his/her ability, their attitude towards school and academic achievement (Parsons, Adler & Kaczala; Galper, Wigfield & Seefeldt; Frome & Eccles; Pomerantz & Dong; Chamorro-Premuzic; Artecbe; Furnham & Trickot in Raty, 2014). Children—with parents who perceive their children as having high academic ability—receive higher scores on standardized tests and they are more likely to survive in school compared to children with parents who perceive their child as having low academic ability. There are several factors that influence the perception of parents, especially factors affecting the perception of working mothers towards their children's academic achievement. The factors are: (1) parenting style (Kordi & Baharudin, 2010), (2) parents' educational level (Dizon-Ross 2014), (3) parents' socioeconomic status (Dizon-Ross, 2014) and (4) child's gender (McGrath & Repetti, 2000).

Hence, based on the above mentioned findings, this study is designed to examine the correlation between parenting style of working mothers and the mothers' perception of the academic achievement of school-age children. This study is also designed to examine the correlation between the mothers' age, the mothers' educational level, the mothers' socioeconomic status, the mothers' culture of origin, the children's gender and the authoritative, authoritarian and permissive parenting style of working mothers. Moreover, this study is designed to examine the correlation between the mothers' educational level, the mothers' socioeconomic status and the children's gender and the mothers' perception of their school-aged children's academic achievement.

2 METHOD

This study is part of a joint study on parenting that involved Universitas Indonesia, Depok, Widya Mandala Catholic University, Surabaya and the University of Queensland, Australia.

The variables and instruments used in this the study on parenting were determined, adapted and provided by a team of researchers from the University of Queensland. Therefore, the researchers could not replace nor revise the instruments even though the results from the reliability and the validity testing of the instruments in Indonesia were different from the results in Australia.

2.1 *Population and sample*

Participants of this study consist of 150 working mothers with at least one school-aged child and living in the Jabodetabek region. The sampling technique used in this study is a non-probability sampling in the form of quota sampling.

2.2 *Instruments*

The Parenting Styles and Dimensions Questionnaire—Short Version (PSDQ—Short Version). PSDQ—Short Version was developed by Robinson, Mandleco, Susanne and Hart (2001). PSDQ—Short Version consists of 32 items and divided into three factors, (1) authoritative, (2) authoritarian and (3) permissive. Authoritative factor consists of 15 items divided into three sub-factors: (1) connection dimension (five items), (2) regulation dimension (five items) and (3) granting autonomy dimension (five items). The authoritarian factor consists of 12 items divided into three sub-factors: (1) physical coercion dimension (four items), (2) verbal hostility dimension (four items) and (3) non-reasoning or punitive dimension (four items). Permissive factor consists of only one sub-factor, indulgent dimension (five items). The PSDQ—Short Version was adapted to an Indonesian version by Dr. Agnes Sumargi in 2014. The reliability of PSDQ—Short Version is measured using Cronbach's Alpha. The α value is 0.871. Therefore, PSDQ—Short Version is considered as reliable to measure parenting style. The validity of PSDQ—Short Version was tested by assessing the correlation between the items contained therein. According to Aiken and Groth-Marnat (2006), the minimum limit of validity coefficient is 0.20. There were three items of the PSDQ—Short Version which had coefficient validity under 0.20. However, these items were maintained within the PSDQ—Short Version for the benefit of a larger study.

2.3 *Scale of Perceived Academic Achievement (SPAA)*

The SPAA was developed by Sumargi, Haslam and Filus (2014) and adapted for the use in Indonesia by Dr. Agnes Sumargi in 2014. The SPAA consists of eight items, four items for academic outcomes dimension and four items for effort to achieve dimension. The reliability of SPAA is measured using Cronbach's Alpha. The α value is 0.892. The SPAA is considered as a reliable tool to assess the academic achievement of children based on parents' perceptions. The validity of SPAA was tested by measuring the correlation between the items contained therein. The validity test resulted in a coefficient ranging from 0.440 to 0.837 for each item. Therefore, this instrument can produce an accurate measurement of the construct.

A Pearson correlation test was conducted to examine the correlation between the parenting style of working mothers and the mothers' perception on the academic achievement of school-aged children. A multiple regression was applied (1) to examine the correlation between various potential predictors, such as the mothers' age, mothers' educational level, mothers' socioeconomic status, mothers' culture of origin, children's gender and the parenting style adopted by the working mothers and (2) to examine the correlation between various potential predictors, such as the mothers' educational level, mothers' socioeconomic status, children's gender and mothers' perception of their school-aged children's academic achievement.

3 RESULTS

The result in Table 1 indicates that authoritative parenting style of working mothers and the mothers' perception of their school-aged children's academic achievement were significantly

Table 1. The correlation between parenting style of working mothers and mothers' perception of their school-aged children's academic achievement.

Parenting style		Perception of academic achievement
Authoritative	*R*	0.375**
	Sig. (2-tailed)	0.000
Authoritarian	*R*	–0.199*
	Sig. (2-tailed)	0.015
Permissive	*R*	–0.375
	Sig. (2-tailed)	0.814

$\alpha = 0.01$**; $\alpha = 0.05$*.

Table 2. The correlation between demographic factors and authoritative parenting style.

Demographic factors		Authoritative
Mothers' Age	*R*	–0.025
	Sig. (2-tailed)	0.762
Mothers' Educational Level	*R*	0.134
	Sig. (2-tailed)	0.102
Mothers' Socioeconomic	*R*	0.088
	Sig. (2-tailed)	0.287
Mothers' Culture of Origin	*R*	–0.025
	Sig. (2-tailed)	0.758
Children's Gender	*R*	0.043
	Sig. (2-tailed)	0.605

correlated, r(149) = 0.375, $p < 0.01$. This means that when working mothers use authoritative parenting style, then the mothers' perception of their school-aged children's academic achievement will be more positive. Similarly, the authoritarian parenting style of working mothers and the mothers' perception of their school-aged children's academic achievement were also significantly correlated, with r(149) = –0.199, $p < 0.05$. This suggests that the more working mothers use authoritarian parenting style, then the mothers' perception of their school-aged children's academic achievement will be more negative. However, there was a non-significant correlation between permissive parenting style of working mothers and the mothers' perception of their school-aged children's academic ability, with r(149) = –0.375, $p> 0.05$.

The results in Table 2 show that all of the demographic factors of working mothers and authoritative parenting style were not significantly correlated. The multiple regression model with five predictors produced $R^2 = 0.029$, F (5, 144) = 0.875, $p > 0.05$ (Table 3). It is apparent that none of the predictors significantly identified the authoritative parenting style of working mothers.

The results in Table 4 show that all of the demographic factors of working mothers and authoritarian parenting style were not significantly correlated. The multiple regression model with five predictors produced $R^2 = 0.043$, $F(5, 144) = 1.301$, $p>0.05$ (Table 5). It is apparent that none of the predictors significantly predicted the authoritarian parenting style of working mothers.

The results in Table 6 show that all of the demographic factors of working mothers and permissive parenting style were not significantly correlated. The multiple regression model with five predictors produced $R^2 = 0.771$, $F(5, 144) = 0.026$, $p>0.05$ (Table 7). It is apparent that none of the predictors significantly predicted the permissive parenting style of working mothers.

Table 3. Predicting authoritative parenting style.

Predictor	b	B	*T*	*P*
Constant	3.558		11.321	0.000
Mothers' Age	0.016	–0.013	–0.162	0.872
Mothers' Educational Level	0.091	0.139	1.648	0.102
Mothers' Socioeconomic	–0.011	–0.050	–0.570	0.569
Mothers' Culture of Origin	0.139	0.080	0.958	0.340
Children's Gender	0.061	0.052	0.618	0.538

$F = 0.875$, $R^2 = 0.029$.

Table 4. The correlation between demographic factors and authoritarian parenting style.

Demographic factors		Authoritarian
Mothers' Age	*R*	–0.074
	Sig. (2-tailed)	0.185
Mothers' Educational Level	*R*	–0.028
	Sig. (2-tailed)	0.366
Mothers' Socioeconomic	*R*	0.042
	Sig. (2-tailed)	0.306
Mothers' Culture of Origin	*R*	0.003
	Sig. (2-tailed)	0.485
Children's Gender	*R*	–0.185
	Sig. (2-tailed)	0.012

Table 5. Predicting authoritarian parenting style.

Predictor	B	β	*t*	*P*
Constant	2.227		8.406	0.000
Mothers' Age	–0.072	–0.072	–0.874	0.384
Mothers' Educational Level	–0.019	–0.035	–0.419	0.102
Mothers' Socioeconomic	0.050	0.051	–0.589	0.569
Mothers' Culture of Origin	0.010	0.034	0.034	0.340
Children's Gender	–0.192	–0.191	–0.191	0.538

$F = 1.301$, $R^2 = 0.043$.

Table 6. The correlation between demographic factors and permissive parenting style.

Demographic factors		Permissive
Mothers' Age	*R*	–0.119
	Sig. (2-tailed)	0.074
Mothers' Educational Level	R	–0.023
	Sig. (2-tailed)	0.389
Mothers' Socioeconomic	R	0.061
	Sig. (2-tailed)	0.229
Mothers' Culture of Origin	R	–0.074
	Sig. (2-tailed)	0.186
Children's Gender	R	0.031
	Sig. (2-tailed)	0.352

Table 7. Predicting permissive parenting style.

Predictor	B	β	*t*	*P*
Constant	2.561		7.857	0.000
Mothers' Age	–0.155	–0.126	–1.525	0.130
Mothers' Educational Level	–0.016	–0.023	–0.278	0.782
Mothers' Socioeconomic	0.085	0.048	0.568	0.571
Mothers' Culture of Origin	–0.019	–0.081	–0.923	0.358
Children's Gender	0.066	0.054	–0.191	0.521

$F = 0.771$, $R^2 = 0.026$.

Table 8. The correlation between demographic factors and mothers' perception of their school-aged children's academic achievement.

Demographic factors		Perception academic achievement
Mothers' Educational Level	*R*	0.030
	Sig. (2-tailed)	0.359
Mothers' Socioeconomic	*r*	0.113
	Sig. (2-tailed)	0.085
Children's Gender	*r*	0.052
	Sig. (2-tailed)	0.265

Table 9. Predicting the mothers' perception of their school-aged children's academic achievement.

Predictors	b	B	*t*	*P*
Constant	40.838		12.167	0.000
Mothers' Educational Level	0.202	0.025	0.301	0.764
Mothers' Socioeconomic	2.541	0.118	1.430	0.155
Children's Gender	0.894	0.060	0.733	0.465

$F = 0.849$, $R^2 = 0.017$.

The results in Table 8 show that all of the demographic factors of working mothers and mothers' perception of their children's academic achievement were not significantly correlated. The multiple regression model with three predictors produced $R^2 = 0.017$, $F(3, 144) = 0.026$, $p > 0.05$ (Table 9). It is apparent that none of the predictors significantly predicted the mothers' perception of their school-aged children's academic achievement.

4 DISCUSSION

According to Hoffman (1998), working mothers tend to apply authoritative parenting style to their children rather than applying other types of parenting styles. This is confirmed by the results from this study that working mothers have a tendency to adopt the authoritative parenting style rather than the authoritarian and permissive parenting style. Kohn (in Talib, Mohamad & Mamat, 2011) describes professional workers as adhering to the values associated with self-direction such as freedom, individualization, initiative, creativity and self-actualization. The values held by working mothers who are professional workers will affect their parenting style which applies the authoritative parenting style.

As proposed by Greenberger and Goldberg (1989), parenting style can influence the perception of parents towards their children. Working mothers who apply the authoritative parenting style will create more positive interactions with school-aged children compared to mothers who do not work outside their household. Positive interactions created by working mothers who apply the authoritative parenting style associated with positive behaviors were demonstrated in their parenting style such as reception, support, nurturing and giving praise of school-aged children.

The positive behaviors of working mothers will certainly have an impact on the improvement of the academic achievement of school-aged children, which will be perceived positively by the mothers. Positive perception by the parents towards the children's academic achievement will lead to higher academic scores of the children compared with the scores of children whose parents perceive their children's academic achievement negatively. This is because the perception of the mother would affect the child's own assessment of their academic ability. Working mothers who apply authoritarian parenting style will tend to create negative interactions with school-aged children. Negative behaviors exhibited by working mothers who apply authoritarian parenting style will affect their perception of the academic achievement of school-aged children. Low academic achievement of school-aged children will also be perceived negatively by working mothers.

The permissive parenting style of working mothers did not show any significant correlation with the mother's perception of their school-aged children's academic achievement. The researchers assume that the number of items in the indulgent dimension were lacking in proportion if compared to the number of items in the authoritative and authoritarian dimension. This may have influenced the significance of the correlation between the permissive parenting style of working mothers and the mothers' perception of their school-aged children's academic achievement. In addition, the results from the Pearson correlation technique that was conducted by the researchers, showed that there was a correlation between the authoritarian parenting style and the permissive parenting style. The researchers also assumed that these correlations may have a significant influence on the correlation between the permissive parenting style of working mothers and the mothers' perception of their school-aged children's academic achievement.

The results indicated that there were no significant correlations between the mothers' age and the parenting styles (authoritative, authoritarian and permissive) of the working mothers. This showed that there were no differences in the application of parenting style despite the differences in the age group of the working mothers in this study that consisted of young adults and middle-aged adults. Many studies show that mothers' educational level and mothers' socioeconomic status are correlated with parenting style of working mothers and conclude that socioeconomic status is the most powerful influence to form the parenting style, but this study did not find significant correlations between mothers' educational level and the mothers' socioeconomic status with parenting style of the working mothers. Brooks (2008) stated that the influence of socioeconomic status and educational level of parents towards parenting style are not fixed. Parents are more adaptable to change their view of parenting as they obtain updated information on parenting rather than from the improvement of the parents' education level.

Each cultural group has specific attention and specific objectives related to the development of children, parents' behavior towards children have the same impact regardless of cultural groups, family structure, social status, parent gender and child gender (Amato & Fowler as cited in Brooks, 2008). This can explain cultural origin was not correlated with the parenting style of working mothers. In this study it was found that children's gender is not correlated to parenting style of working mothers. This result is consistent with the results of research conducted by Hoffman (as cited in Hoffman, 1998) which stated that there was no difference in the application of parenting style based on child gender adopted by working mother. This is because not only boys who required to be independent, girls were also demanded by their mother to become more independent rather than to become obedient or feminine.

The result shows that there were no significant correlations between the participants' demographic factors—such as the mothers' educational level, the mothers' socioeconomic status and the children's gender—with the mothers' perception of their school-aged children's academic achievement. In terms of the demographics of the participants in which the working mothers' educational backgrounds are mostly from bachelor and master degree and the socioeconomic status of the working mothers are predominantly from upper middle class, it apparent that the working mothers do not have a difference in perceiving the academic achievement of their school-aged children, regardless of the children's gender.

5 CONCLUSION

There are three results from this study: (1) there were significant correlations only between the authoritative parenting style and the authoritarian parenting style of working mothers and the mothers' perception of their school-aged children's academic achievement, while permissive parenting style of working mothers had a non-significant correlation with the mothers' perception of their school-aged children's academic achievement, (2) there were non-significant correlations between the mothers' age, the mothers' educational level, the mothers' socioeconomic status, the mothers' culture of origin and the children's gender with all three types of parenting styles (authoritative, authoritarian and permissive) of working mothers, and (3) there were non-significant correlations between the mothers' educational level, the mothers' socioeconomic status and the children's gender and the mothers' perception of their school-aged children's academic achievement.

For further research, it is suggested that the results of this study are used as a basis for preliminary research on topics of academic achievement perceived by working mothers and as a reference for mothers, especially working mothers to implement authoritative parenting style and avoid authoritarian parenting style and permissive parenting style, since the authoritative parenting style affects the mothers' perception of the academic achievement of their children to be positive and eventually will also have positive impact on the academic achievement of their school-aged children.

REFERENCES

Aiken, L.R., & Groth-Marnat G. (2006). *Psychological Testing and Assessment 12th ed.* Boston, MA, Pearson Education.

Arbreton, A.J., Eccles J.S. & Harold R.D. (1994). Parent's perceptions of their children's competence: the role of parent attributions. *Paper presented in Society for Research on Adolescence, San Diego, USA.* http://www.rcgd.isr.umich.edu/garp/presentations/eccles94.pdf.

Badan Pusat Statistik (BPS). (2012). *Perkembangan Beberapa Indikator Utama Sosial-Ekonomi Indonesia.* Jakarta, Badan Pusat Statistik.

Baumrind, D. (1966). Effects of authoritative parental control on child behavior. *Child Development,* 37 (4), 887–907. doi:10.2307/1126611.

Baumrind, D. (1991). The Influence of parenting style on adolescent competence and substance use. *The Journal of Early Adolescence,* 11 (1), 56–95. doi:10.1177/0272431691111004.

Berns, R.M. (2013). Child, Family, School, Community: Socialization and Support (Ninth Edition). United States of America, Wadsworth Cengage Learning.

Brooks, J.B. (2008). *The Process of Parenting: Seventh Edition.* New York, USA, McGraw-Hill.

Carbonaro, W. (2005). Tracking, Students' Effort, and Academic Achievement. *Sociology of Education,* 78 (1), 27–49. doi:10.1177/003804070507800102.

Davis-Kean, P.E. (2005). The influence of parent education and family income on academic achievement: The indirect role of parental expectations and the home environment. *Journal of Family Psychology,* 19 (2), 205–304.

Dizon-ross, R. (2014). Parents' perceptions and children's education: Experimental evidence from malawi. *Working Paper.*

Greenberger, E. & Goldberg W.A. (1989). Work, Parenting, and the socialization of children. *Developmental Psychology,* 25 (1), 22–35. doi:10.1037/0012-1649.25.1.22.

Hoffman, L.W. (1998). The effects of the mother's employment on the family and the child. *Parenthood in America.* http://parenthood.library.wisc.edu/Hoffman/Hoffman.html
Kordi, A. & Baharudin R. (2010). Parenting attitude and style and its effect on children's school achievement. *International Journal of Psychological Studies,* 2 (2), 217–222.
McGrath, E.P. & Repetti R.L. (2000). Mothers' and fathers' attitudes toward their children's academic performance and children's perceptions of their academic competenece. *Journal of Youth and Adolescence,* 29 (6), 713–23. doi:10.1023/A:1026460007421.
Meggiolaro, S. & Ongaro F. (2013). Maternal age and parenting strategies. *Genus,* 49 (3), 1–24.
Papalia, D.E. & Feldman, R.D. (2012). *Experience Human Development 12th ed.* New York, McGraw-Hill.
Räty, Hannu. (2014). A 9-year study of academically and vocationally educated parent's perceptions of their children's general abilities. *Journal for Educational Research Online,* 6 (2), 3–20.
Robinson, C.C., Mandleco, B., Roper S.O. & Hart C.H. (2001). The parenting styles and dimensions questionnaire (psdQ). In B.F. Perlmutter, J. Touliatos, and G.W. Holden (Eds.), *Handbook of Family Measurement Techniques: Vol. 3.* Thousand Oaks, Sage. pp. 319–321.
Santrock, J.W. (2011). *Educational Psychology (2nd Ed.).* New York, USA, McGraw-Hill
Sumargi, A, Haslam D. & Filus A. (2014). *Scale of Perceived Academic Achievement.* Brisbane, Parenting and Family Support Centre.
Suryadi, D. & Damayanti C. (2003). Perbedaan tingkat kemandirian remaja puteri yang ibunya bekerja dan yang tidak bekerja. *Jurnal Psikologi,* 1 (1), 1–28.
Talib, J., Mohamad Z. & Mamat M. (2011). Effects of parenting style on children development. *World Journal of Social Sciences,* 1 (2), 14–35.
Tjaja, R.P. (2000). Wanita bekerja dan implikasi sosial. *Kementerian PPN/ Bappenas.* http://www.bappenas.go.id/files/6513/5228/3053/ratna__20091015151137__2386__0.pdf.

Hoffman, L.W. (1989). The effects of maternal employment on the family and the child. [illegible]
[illegible]
[illegible] (2016). Parenting [illegible]
[illegible] Mothers and father [illegible]
[illegible]
[illegible]
[illegible]
[illegible]
[illegible]
[illegible]
Santrock, J.W. (2011). [illegible] McGraw-Hill.
[illegible]
[illegible]
[illegible]
[illegible]
[illegible] Wanita [illegible] .pdf

Diversity in Unity: Perspectives from Psychology and Behavioral Sciences – Ariyanto et al. (Eds)
 ISBN 978-1-138-62665-2

Parental support and achievement motivation differences between adolescents whose parents work as migrant workers and those who work as non-migrant workers

Q. Masturoh, W. Prasetyawati & S.S. Turnip
Faculty of Psychology, Universitas Indonesia, Depok, Indonesia

ABSTRACT: People in rural areas work as migrant workers to overcome poverty. They work abroad and leave their families behind. This condition gives disadvantaged consequences to the adolescents left in the rural areas, especially in their academic life. Several studies have found that there is a positive relationship between parental support and achievement motivation. This study aimed to compare the parental support and achievement motivation between adolescents whose parents work as migrant workers and adolescents whose parents work as non-migrant workers in rural areas. The samples of study were 171 adolescents whose parents work as migrant workers and 257 adolescents whose parents work as non-migrant workers from rural areas in Karawang. Children and Adolescents Social Support Scale (CASS) and Achievement Motivation Inventory (AMI) were used to measure parental support and achievement motivation. The results showed that there are significant differences in parental support and achievement motivation between adolescents whose parents work as migrant workers and adolescents whose parents work as non-migrant workers. It was found that adolescents from the non-migrant worker parents group have higher parental support and achievement motivation.

1 INTRODUCTION

The rural areas in Indonesia are larger than urban areas and are known for their low economic level (Griffith, 1982). Abrar (2012) stated that Indonesia has 30 million citizens living with poverty issues, 19 million of whom live in rural areas. In an attempt to meet their needs, people in rural areas try to find jobs with a high salary. Unfortunately, it's a great challenge for rural people to find jobs with a high salary within the rural areas. According to the International Labour Conference (2008), low payment becomes the main problem for rural people that urged them to find proper job outside their living area. Low payment is the main problem for rural people to find proper jobs within rural area. Rural people with limited skills and level of education were urged to find a job outside their area because of the difficulty in finding work within their own area. Working as migrant workers in other countries is one solution for them.

Level of education and skills are not important requirements in finding work as a migrant worker. According to the government institution for migrant workers (known as *Badan Nasional Penempatan dan Perlindungan Tenaga Kerja Indonesia* or BNP2TKI), the majority of migrant workers work as caregivers (23,288 people) and domestic workers (16,362 people) in several countries. Further data show that migrant workers who were placed have the following education: junior high school (40.49%), senior high school (29.23%), primary school (28.57%), diploma (1.16%), bachelor (0.54%), and postgraduate (0.01%). The data prove that most migrant workers are senior high school graduates, while only a few of them have a university degree. In addition, migrant working also generates a better income for rural people. Furthermore, migrant workers also offer a better salary for rural people.

The migrant workers' remittances has improved their financial condition. Remittances or salary improve migrant worker's financial situation (Fajriah, 2015). One factor that encourages people to work as migrant workers is the salary. The migrant worker's salary is higher than work as, for example, farmers, the most common work in rural area. In 2014, a migrant worker could earn as much as USD 1,700/year (Sitepu, 2015) which is a higher salary than that obtained if they worked as farmers in rural areas (USD 461–646/year). Migrant workers' salaries have a positive impact on their financial situation, such as better access to education and health facilities (De La Garza, in Nuraini, 2015). Financial improvement also decreases the number of underage workers trying to help fulfil their family's needs.

Although working as migrant workers has a positive impact on their financial aspect, there are some negative impacts. Migrant workers who work overseas must leave their family for a long period of time. This has an unfavourable impact on the family left behind, particularly the children. Some research has shown that there are negative consequences on both the mental health and academic performance of the adolescents left behind.

Research conducted by Umami (2015) shows that 73.7% of adolescents who are left behind by one or both parents to work as migrant workers overseas experience loneliness, and 81.9% of them have tendencies towards developing mental health problems. Parents are significant figures for adolescents; thus their absence increases loneliness in adolescents, as they do not have any significant figures with whom they can share, ask and discuss their changes or problems. Gursoy and Bicakci (in Umami, 2015) state adolescents who do not spend enough time with their parents tend to have higher levels of loneliness. Besides loneliness, these adolescents tend to have higher behavioural problems, such as externalising behaviour (stealing, vandalism, underage drinking, etc.) than their peers who have not been left behind by their parents (Nordhani, 2016).

Another negative impact of parents' departure to work as migrant workers also can be found in adolescent's academic life, one of which is their achievement motivation. Nuraini's (2015) research found that adolescents who are left behind by one or both parents to work as migrant workers have lower achievement motivation scores than adolescents who have both parents staying with them (Nuraini, 2015). Jeynes and Thomas (2009) found students who live with their own parents have more stable emotions because the parents watch and support them in the achievement of their goals. Meanwhile, for students who do not live with their own parents there is an unfavourable impact on their life and academic motivation.

Human behaviour is driven by motivation, which involves processes that guide, support and sustain behaviour. Motivation is a critical aspect in learning activities. Thus, students with no motivation will not make the effort to learn (Santrock, 2011). Achievement motivation is an attitude to reach a goal, included action planning and the desire to fulfil particular internal standards (Chetri, 2014). Achievement motivation is known as an important key to success in academic performance (Bridgeman & Shipman, 1978). Someone with achievement motivation has a strong will to strive for something important and gains gratification from finishing challenging tasks, which makes them work hard over a long-term period in order to reach their goals (Beuke, 2011).

Academic demands and academic environment changes can impact student's achievement motivation. A student's achievement motivation changes in every developmental period. Pickhardt (2009) states the early stage of adolescents is a crucial moment for academic achievement. From an adolescent's point of view, socialising and being popular among other teens is more important than academic achievement. Peer group gives greatest influence on adolescent. This view leads them to spending less effort to achieve something than building relationships with their peers (Pickhardt, 2009).

Schonert-Reichl et al. (in Bajema et al., 2002) stated that adolescents in rural areas have less career and education aspiration than adolescents in urban areas. Researchers argue that traditional values in rural areas, limited available schools and psychosocial aspects limit an adolescent's aspiration in rural areas. Hawley (2006) states that emphasizing of farming and working activities for rural people may make their adolescents has limited their views of education. Hawley (2006) states that values in rural areas which emphasise farming activities or looking for jobs also limiting adolescents' knowledge about education that is available them. Adolescents in rural areas have narrow perspectives about the relationships between their education and their choices of work (Reid, in Hawley, 2006).

Achievement motivation can be affected by internal and external factors. Internal factors guide students to do activities or to behave because of the insight they can get for doing it (Rothstein, 1990). For example, students do painting because they enjoy and feel proud of painting. Internal factors involve students' own internal factors such as self-esteem (Awan et al., 2011; Nwanko et al., 2013) and their attribution interpretation of for success-failure (Rothstein, 1990). Students with high self-esteem have higher expectations of success, which encourage them to reach higher success possibilities (Nwanko et al., 2013). Besides internal factors, achievement motivation can also be influenced by external factors, for example, both the school environment and family play the role of achievement motivation external factors.

The rural educational environment, as an external factor of achievement motivation among adolescents in rural areas, is inferior to the educational environment in urban areas. Schools in rural areas provide less extracurricular activities (Schonert et al., in Guiffrida, 2008). There are also different education qualities between schools in rural and urban areas. There is also a gap in education qualities between schools in rural and urban areas. According to *Neraca Pendidikan Daerah* (NPD) for 2015, Karawang, as a city with many rural areas within it, has 2,031 classes in good condition; this number is lower than the nearby urban city Bekasi which has 3,616 classes in good condition. The difference in quality is even higher if we compare it with Jakarta's urban areas. Besides the quality of education facilities, equality problems and the quality of teachers also pose great challenges in rural education. Karawang has 16,369 teachers, while Bekasi has 21,199 teachers. In the remote area of Riau, schools often face insufficient human resources in a lack of teachers, and because of this, they often have to borrow teachers from other schools (Virdhani, 2014). The quality of teachers' knowledge about what they teach can be seen from teacher evaluation test results (known as *Uji Kompetensi Guru* or UKG). In the 2015 results nationwide, Karawang was positioned as the 4th lowest. Education problems in rural areas create a gap which results in negative consequences for rural adolescents. Research conducted by Chetri (2014) proves that there are significant differences in academic achievements between rural and urban schools. Students in rural schools tend to have lower academic achievements.

Parent-adolescent relationships also have an impact on adolescents' academic life. Although someone might spend more time with their peers than their parents during the adolescent period (Papalia, 2012), parents still play an important role in their life. Parents give secure spaces for their adolescents to explore new things in their attempts to become adults (Papalia, 2012). Research conducted by Igbo et al. (2015) found that there is a significant effect of parental relationship and children's academic achievement. Juvenon and Wentzel (in Wentzel, 1998) state that social support from parents influences achievement motivation.

One important role of parents is as social support provider for their adolescents. Parents motivate children's education by facilitating and creating learning environments, giving approval, appreciating, and helping children to overcome their problems (Acharya & Joshi, 2011). According to Acharya and Joshi (2011), parental support includes guidance, communication and interest in their children's development in school. Parents' communication with their children shapes the children's perception about the world in which they live (Maximo et al., 2011). Parent-child communication does not only build attachment but also determines the depth of their relationship. Maximo et al. (2011) also found that the parent-child communication methods influence an adolescent's achievement motivation. Parents evaluate their own achievement and share the evaluation with their children. Parents who show a supportive attitude to their children's ideas increase children's aspiration to achievement, while parents who are not involved and give less feedback to their children decrease their children's achievement aspiration.

Parental support has been proved as a positive predictor for students' interest and goal orientation in school (Wentzel, 1998). Parental support has a great effect on adolescent achievement motivation (Acharya & Joshi, 2011). Nuraeni and Supraningsih's research found that there is a positive correlation between parental support and male adolescents in junior high school. Putri's research (2014) also found similar results of a positive correlation between parental support and achievement motivation in both academic and sports activities.

Living with other people while their parents work abroad decreases adolescents' chances to communicate and interact directly with their parents. The relationship between adolescents and their parents can have a great impact on adolescents' life, especially in their academic achievement (Santrock, 2012). Furthermore, Santrock (2012) states if parents spend less time with their children and doing other things, adolescents achievement motivation can be troubled. If parents spend more time doing other things than take care of their adolescent children, it will give negative impact to their children's achievement motivation. Parents who give frequent verbal support and praise, regular feedback for schoolwork and talk directly about schoolwork and activities may facilitate the adolescent's progress in school (Acharya & Joshi, 2011). Thus, parents' absence in their adolescent's daily life will result in adolescent's low achievement motivation.

Parents play an important role in an adolescent's academic life. Poverty drives parents to earn money abroad by working as migrant workers, and as a consequence they have to leave their children in the care of another person. The absence of parents in children's life will impact their parental support and achievement motivation. This research aims to find out the differences between parental support and achievement motivation among adolescents who are left behind by their migrant worker parents and adolescents who have their parents in their daily life in the rural areas of Karawang, which is billed as the biggest migrant worker supplier in Indonesia.

2 METHODS

This research used quantitative research method and two kinds of data, primary and secondary data. The primary data was collected from participants whose parents do not work as migrant worker and the secondary was collected from participants whose parents work as migrant workers. The primary data was collected from three junior high schools in Karawang rural area. The secondary data was collected from the same schools as primary data in the previous year.

2.1 *Participants*

The participants were adolescents aged 11–16 years who live in Karawang. There are two groups in this research; the first is a sample from primary data (children of non-migrant workers) and the second group is from secondary data (children of migrant workers). After the data collection, the researcher selected data based on the completed responses given by the participants. The researcher was able to proceed finally with 428 participants (Migrant worker = 171; Non-migrant = 257).

2.2 *Instruments*

The Achievement Motivation Inventory (AMI) developed by Muthee and Thomas (2009) was used to measure achievement motivation in this study. It consists of 32 items with a 4-point Likert scale. 14 items in AMI are negative ('I think I am lazy'), while 18 items are written in positive sentences ('I want to be the best student in classroom'). AMI has good reliability (Cronbach alpha = 0.64) and validity (0.18–0.47) to measure achievement motivation. Adolescents' perceived parental support was measured by the Children and Adolescents Social Support Scale (CASS) created by Malecki et al. (1999, in Wilendari, 2015). CASS consists of three sub-scales (parents, peers, teachers and classmates) and has four dimensions of social support; emotional, instrumental, information and appraisal. For this research only the parental support sub-scale was used. Each sub-scale has 12 items with four Likert answer choices. The original version of CASS had six answer choices, whereas this research reduced the choices into four to help adolescents give an exact condition for each item (Nuraini, 2015). All items are summed up and multiplied by six before they are divided by four. Thus, the score gained by participants in this research will be equal to the score they would get from the original version. CASS has good reliability (Cronbach alpha = 0.77) and validity (0.284–0.479).

2.3 *Procedures*

The researchers distributed the questionnaire to the target respondents in their classrooms. The principal of each school assigned some classes where the researcher could distribute the questionnaires directly. This research used some statistical methods, such as descriptive statistics, the independent sample t-test and Pearson Product Moment, to analyse data.

3 RESULTS

The demographic information for 448 participants in this study is presented in Table 1. As can be seen, the majority of participants in both sample groups is composed of females

Table 1. Participant demographic information.

		Migrant worker		Non-migrant worker	
Demography		Total	Percentage	Total	Percentage
Gender	Male	82	48	114	44.4
	Female	89	52	143	55.6
Born	First	56	32.7	96	37.4
	Second	68	39.8	72	28.0
	After second	47	27.4	88	34
	Didn't answer	0	0	1	0.4
Age	<12 y.o	2	1.2	7	2.7
	12–15 y.o	148	86.5	248	96.4
	15 y.o	20	11.7	2	0.8
Father's last education	Primary	108	63.2	170	66.1
	Junior high	35	20.5	47	18.3
	Senior high	19	11.1	25	9.7
	College	1	0.6	10	3.9
	Didn't go to school	7	4.1	4	1.6
	Didn't answer	1	0.6	1	0.4
Mother's last education	Primary	109	63.7	191	74.3
	Junior high	37	21.6	33	12.8
	Senior high	14	8.2	12	4.7
	College	1	0.6	2	0.8
	Didn't go to school	10	5.8	12	4.7
	Didn't answer	0	0	7	2.7
Parents who work	Father	9	5.3	130	50.6
	Mother	156	91.2	23	8.9
	Both	6	3.5	103	40.1
Father's job	Migrant worker	15	8.7	0	0
	Farmers	0	0	109	42.4
	Labour	0	0	21	8.2
	Entrepreneur	0	0	68	26.5
	Others	0	0	26	10.1
	Unemployed	0	0	23	8.9
	Didn't Answer	0	0	10	3.9
Mother's job	Migrant Worker	15	8.7	0	0
	Farmer	0	0	109	42.4
	Labourer	0	0	21	8.2
	Entrepreneur	0	0	68	26.5
	Others	0	0	26	10.1
	Unemployed	0	0	23	8.9
	Didn't answer	0	0	10	3.9

(Non-migrant = 55.6%; Migrant = 52%). The majority of participants' parents level of education is elementary school ($Father_{Non\text{-}Migrant}$ = 66.1%; $Father_{Migrant}$ = 63.2%; $Mother_{Non\text{-}Migrant}$ = 74.3%; $Mother_{Migrant}$ = 63.7%). The majority of mothers in the migrant group work (91.2%), while in the non-migrant group the majority of breadwinners are the fathers (50.6%). The majority of mothers in the migrant group work as migrant workers (94.7%), while in the non-migrant groups most of the participants' mothers are unemployed (53%). The majority of fathers who work in the non-migrant worker group are farmers (42.3).

From Table 2 it can be seen that the majority of participants whose parents work as migrant workers communicate with their parents once a month (33.9%), while participants whose parents work as non-migrant workers communicate every day (70%). The majority of participants from the migrant worker group were not taken care of by their parents as caregivers (74.3%), while the majority of the non-migrant group had their parents as caregivers (92.2%).

The comparison of parental support and achievement motivation variables will be examined next. Table 3 shows the results of parental support and achievement motivation comparisons. It is revealed that there are significant differences between parental support ($t = 4.3^{**}$, $p = 0.000$) and achievement motivation ($t = 14.57^{**}$, $p = 0.000$) among adolescents who are left behind by their parents and those who are not. Adolescents whose parents work as non-migrant workers have higher mean score of parental support ($M_{Non\text{-}migrant}$ = 55.11; $M_{Migrant}$ = 51.7) and achievement motivation ($M_{Non\text{-}migrant}$ = 115.01; $M_{Migrant}$ = 103.6) than adolescents whose parents work as migrant workers.

Table 4 shows that there are also significant differences in each dimension of parental support: emotional support ($t = -3.03$, $p < 0.01$), informational support ($t = -3.43$, $p < 0.01$), appraisal support ($t = -2.3$, $p < 0.05$), and instrumental support ($t = -3.96$, $p < 0.01$).

The results (Table 5) also show that there is a significant association between parental support and achievement motivation ($r = 0.283$, $p < 0.05$).

Table 2. Participant's frequency of communication with their parents.

		Migrant worker		Non-migrant worker	
		Total	Percentage	Total	Percentage
Frequency of communication	Everyday	25	14.6	180	70
	Once a week	52	30.4	45	17.5
	Twice a week	21	12.3	11	4.3
Frequency of communication	Once a month	58	33.9	7	2.7
	Once a year	10	5.8	4	1.6
	Never	5	2.9	10	3.9
Caregiver	Father/Mother	67	39.2	137	92.2
	Non-parent	127	74.3	20	7.8

Table 3. Parental support and achievement motivation comparison among adolescents whose parents work as migrant workers and adolescents whose parents work as non-migrant workers.

	Mean			
Variable	Migrant worker	Non-migrant worker	Sig.	Description
Parental support	51.7 (SD = 8.56)	55.11 (SD = 7.62)	P = 0.000; t = –4.30	Significant
Achievement motivation	103.6 (SD = 8.24)	115.01 (SD = 7.68)	P =0.000; t = –14.57	Significant

Table 4. Parental support dimensions comparison among adolescents whose parents work as migrant workers and adolescents whose parents work as non-migrant workers.

	Mean				
Parental support dimension	Migrant worker (n = 171)	Non-migrant worker (n = 257)	t	*P*	Description
Instrumental support	11.92 (SD = 3.03)	13.06 (SD = 2.79)	−3.96	0.000	Significant
Informational support	13.93 (SD = 2.62)	14.81 (SD = 2.54)	−3.43	0.001	Significant
Emotional support	13.28 (SD = 2.92)	14.06 (SD = 2.39)	−3.03	0.003	Significant
Appraisal support	12.56 (SD = 2.65)	13.17 (SD = 2.58)	−2.3	0.018	Significant

Table 5. Association of parental support and achievement motivation.

	Achievement motivation	
Parental support	R	P
	0.283**	0.000

4 DISCUSSION

Based on the results, it can be seen that there are significant differences in parental support and achievement motivation among adolescents whose parents work as migrant workers and adolescents whose parents work as non-migrant workers. Adolescents whose parents work as non-migrant workers have a higher score in both parental support and achievement motivation. Further analysis also shows that there is a significant positive association between parental support and achievement motivation; it means that if adolescents have high parental support, their achievement motivation will also increase.

Instrumental support was found as the most different among other dimensions. Instrumental support consists of financial and time support (Malecki & Demaray, 2002). In the parental sub-scale, instrumental support has three items concerning parent's frequencies in spending their time helping the children: giving time, helping children to decide something, and supporting the children. Two of the items are related to the time that parents spend together with their children. Adolescents whose parents work as migrant workers might not have spent much time with their parents, thus, they had lower scores than the other group. From the results, it was found that adolescents whose parents work as migrant workers have lower emotional support. Emotional support is shown by the affection and care given by parents, which makes adolescents feel loved. Adolescents who are left behind often experience depression, anxiety and have behavioural problems, which results in poor academic performance (Feng, 2016). Most of the adolescents who are left by their parents are taken care of by their relatives. Muthee (2011) states that children who do not live with their own parents tend to lack love and care; they also tend to develop undisciplined behaviour which can disrupt their academic performance.

The association between parental support and achievement motivation can provide the answer for the achievement motivation differences between the two groups. Adolescents whose parents work as migrant workers have lower parental support, since parental support has a positive association with achievement motivation; thus adolescents whose parents work as migrant workers also have lower achievement motivation. Parental support enables children to be more confident and braver in their efforts to try and achieve something (Santrock, 2012).

Most adolescents whose parents work as migrant workers are taken care of by their relatives or one of their parents. During the times migrant worker parents work abroad, they leave their children to be taken care of by other family members, such as the children's grandparents (Nuraini, 2015). Non-parent's parenting often brings other consequences. Lu (2011) states that the older generation tend to pamper the children left behind and focus more on fulfilling their non-emotional needs rather than moral aspects or the children's spirituality. Children who live with their own parents tend to have higher cognitive abilities, emotion and good behaviour than children who live with only one parental figure or with relatives (Kim, 2008).

Parents working as migrant worker also have limited opportunities for adolescents to communicate with them. Students' relationship with their parents influences their achievement (Santrock, 2012). Parent-children communication enables adolescents to share their problems so parents can help their children. Parents who are involved in their children's education could help their children face academic problems in school and facilitate them to raise their achievement in school (Acharya & Joshi, 2011).

This research has several limitations. First, data for the research was only collected from schools in one rural area. Secondly, the information about educational levels (such as drop-out, no schooling, etc.) are not included in the sample of this research. Thus, this research can only be generated to a group of students. There is also the possibility of high social desirability when participants filled in the questionnaire. High social desirability might have led participants to choose 'good' answers not 'reality-based answers' when they completed the questionnaire. This is because achievement motivation is constructed related to their academic performance, so they tend to think that this will affect their scores in school. Second, the participants in this research are still in the early stage of adolescence. According to Santrock (2013) the adolescent period has sub-stages; the early and late stages, both of which have their own characteristics. The early stages of adolescence are the critical moments for achievement motivation and parents are needed in this stage. Thirdly, the measurement of parental support was done by measuring participant's perceived parental support. This research did not measure the perceived attachment from the parents' point of view.

REFERENCES

Abrar. (2012). Letter: Boosting education in rural areas. *The Jakarta Post*. Retrieved from http://www.thejakartapost.com/news/2012/06/02/boosting-education-rural-areas.html.

Acharya, N. & Joshi, S. (2011). Achievement motivation and parental support to adolescents. *Journal of the Indian Academy of Applied Psychology, 37*(1), 132–139.

Awan, R.N., Noureen, G. & Naz, A. (2011). A study of relationship between achievement motivation, self-concept, and achievement in English and mathematics at secondary level. *International Education Study, 4*(3), 72–79.

Bajema, D., Miller, W.W. & Williams, D.L. (2002). Aspiration of rural youth. *Journal of Agricultural Education, 3*(43), 61–67.

Beuke, C. (2011). How do high achievers really think? *Psychology Today*. Retrieved from https://www.psychologytoday.com/blog/youre-hired/201110/how-do-high-achievers-really-think.

Bridgeman, B. & Shipman, V.C. (1978). Preschool measures of self-esteem and achievement motivation as predictors of third-grade achievement. *Journal of Educational Psychology, 70*(1), 17–28.

Chetri, S. (2014). Achievement motivation of adolescent and its relationship with academic achievement. *International Journal of Humanities and Social Science Invention, 3*(6), 8–15.

Fajriah, L.R. (2015). Poverty as the reason of being migrant worker Kemiskinan Penyebab Masyarakat Pilih Jadi TKI. *Sindonews.com*. Retrieved from http://ekbis.sindonews.com/read/968471/34/kemiskinan-penyebab-masyarakat-pilih-jadi-tki-1424772483.

Feng, E. (2016). China to survey children left behind by migrant workers. *The New York Times*. Retrieved from http://www.nytimes.com/2016/03/30/world/asia/china-left-behind-children-survey.html?_r = 0

Griffiths, V.L. (1982). Educational problems in rural areas. *Masalah Pendidikan di Daerah Pedesaan*. Retrieved from http://unesdoc.unesco.org/images/0007/000764/076492 indb.pdf.

Guiffrida, D. (2008). Preparing rural students for large colleges and universities. New York: University of Rochester.

Hawley, C.W. (2006). Remote possibilities: Rural children's educational aspirations. *Peabody Journal of Education, 81*(2), 62–88.

Igbo, J.N., Odo, A.S., Onu, V.C. & Meiziobi, D. (2015). Parent-child relationship motivation to learn and students academic achievement in mathematics. *International Journal of Research in Applied, Natural and Social Science, 3*(9), 87–108.
Kim, C.C. (2008). Academic success begins at home: How children can succeed in school. Washington: The Heritage Foundation, 22185.
Malecki, C.K. & Demaray, M.K. (2002). Measuring perceived social support: Development of the child and adolescent social support scale. *Psychology in the Schools, 39*(1), 1–18. doi:10.1002/pits.10004
Muthee, J.M. & Thomas, I. (2009). *Achievement motivation inventory.* Trivandrum: Department of Psychology, University of Kerala.
Maximo, S.I., Tayaban, H.S., Cacdac, G.B., Cacanindin, M.J.A., Pugat, R.J.S., Rivera M.F. & Lingbawan, M.C. (2011). Parents' communication styles and their influence on the adolescents' attachment, intimacy and achievement motivation. *International Journal of Behavioral Science, 6*(1), 59–72.
Nuraini, F. (2015). Achievement motivation level differences between girl and boy adolescent whose migrant worker parents (Unpublished manuscript, Universitas Indonesia, Depok, Jawa Barat).
Nuraeni, Y. & Supraptiningsih, E. (2014). Parental support and academic motivation among 7 grade girl students at Mts Misbahumur Cimahi City. Proceeding in Psychology: Universitas Islam Bandung.
Nordhani, M. (2016). Association between self-esteem and externalizing problem among adolescent whose parents as migrant worker and non-migrant worker in Karawang. (Unpublished manuscript, Universitas Indonesia, Depok, Jawa Barat).
Nwanko, B.E., Obi, T.C. & Agu, S.A. (2013). Relationship between self-esteem and achievement motivation among undergraduates in south eastern Nigeria. *Journal of Humanities and Social Science, 13*(5), 102–106.
Papalia, D.E. & Feldman, R.D. (2012). *Experience human development.* New York, NY: McGraw-Hill.
Putri, E.P. (2014). Association of parental social support, coach and peer with academic achievement motivation and sport achievement motivation among athlete college students at Universitas Surabaya. Surabaya college student scientific journal, 3 (1).
Pickhardt, C.E. (2009). Early adolescent achievement drop: Falling effort and grades. *Psychology Today*. Retrieved from https://www.psychologytoday.com/blog/surviving-your-childs-adolescence/200903/early-adolescent-achievement-drop-falling-effort-and
Rothstein, P.R. (1990). McGraw-Hill's college review books: Educational psychology. Singapore: McGraw-Hill.
Santrock, J. (2011). *Educational psychology*. New York, NY: McGraw-Hill.
Santrock, J. (2012). *Adolescence*. New York, NY: McGraw-Hill.
Sepfitri, N. (2011). The effect of social support to achievement motivation among high school students at MAN 16 Jakarta. Fakultas Psikologi Universitas Islam Negeri Syarief Hidayatullah.
Sia, N. (2014). Perceived social support, achievement motivation and academic of rural adolescents. *Indian Journal of Health and Wellbeing, 5*(4), 452–456.
Singh, K. (2011). Study of achievement motivation in relation to academic achievement of students. *International Journal of Educational Planning and Administration, 1*(2), 161–171.
Sitepu, A.D. (2015). Nusron: Migrant worker's remmitance 2014 reach Rp 107.15 billion *Sindonews.com*. http://ekbis.sindonews.com/read/965081/34/nusron-remiten-tki-2014-tembus-rp107-15-t-1424077667.
Umami, R. (2015). Loneliness and psychotic-like symptom among adolescent who left by migrant worker parents. (Unpublished manuscript, Universitas Indonesia, Depok, Jawa Barat).
Virdhani, M.H. (2014). Teacher deficit is a fatal mistake. *Okezone.com*. Retrieved from http://news.okezone.com/read/2014/01/24/560/930975/kekurangan-tenaga-pendidik-jadi-kesalahan-fatal
Wilendari, S.F. (2014). Association of parental social support and self-esteem among adolescent who left by migrant worker parents. *Unpublished Manuscript* Universitas Indonesia, Depok, Jawa Barat.
Wigfield, A. & Eccles, J.S. (2002). The development of competence beliefs, expectancies for success, and achievement values from childhood through adolescence. In *Development of achievement motivation* (pp. 91–120). San Diego, CA: Academic Press.
Wigfield, A., Ho, A. & Mason-Singh, A. (2011). Achievement motivation. In B. Brown & M. Prinstein (Eds.), *Encyclopedia of adolescence: Vol. 1. Normative processes in development* (pp. 10–19). London, UK: Elsevier.

Diversity in Unity: Perspectives from Psychology and Behavioral Sciences – Ariyanto et al. (Eds)
© 2018 Taylor & Francis Group, London, ISBN 978-1-138-62665-2

The relationship between interest differentiation, interest consistency and career maturity in Grade 10 school students

W. Indianti & N. Sinaga
Faculty of Psychology, Universitas Indonesia, Depok, Indonesia

ABSTRACT: The main purpose of this research was to find the relationship between interest differentiation and interest consistency with career maturity. This research used a quantitative approach with 222 participants from Grade 10 school students in Jakarta. The instruments for data collection were career development inventories (Super, 1975 in Hinkelman & Kivlighan, 1998) and self-directed search questionnaires (Holland, 1997b in Larson & Borgen, 2002). The result of this research shows that interest differentiation has a positive and significant relationship with career maturity. Furthermore, interest consistency has no significant relationship with career maturity. This research also makes suggestions for further research and implications for practical purposes.

1 INTRODUCTION

After completing junior high school, students have to choose their major area of interest. In senior high school three of the major areas of interest are, social science (IPS), natural science (IPA), and language and literature. When deciding on their major, students choose based on teacher's recommendation, general evaluation scores, psychological report, and placement test. Actually, when students have to make a decision about their choice, the real condition is not that simple. According to Jawa Post National Network, 90% students choose natural sciences but then move to social sciences after three months. The reason is they are not sure about their competency. This situation shows that many students are confused about their career choices. They do not understand yet about their capacities and their real interests; and they do not have any plans for the future.

Seligman (1994) proposed that to make career choices, students have to identify their strengths, weaknesses, and interests. The knowledge may help students explore all the information about careers, understand career requirements, and make plans for their career choices.

Otherwise students are unable to make a correct choice, and as a consequence, they cannot build strong motivation and tend to give up easily. They also do not have high commitment and unsatisfied with what they do for the rest of their life. According to Seligman (1994), the situation above deals with career maturity (the maturity to make a career choice). Students who cannot make a career choice face a big problem because they have difficulty deciding the optimal career choice and then maintaining it, and thus finding career satisfaction (Crites & Semler, 1967). Therefore, career maturity is an important aspect in students' life and development.

Some researchers have found that career interest is related positively to career exploration. A clear career interest makes it easier for an individual to determine career decisions and career plans (Donohue, 2006; Hirschi & Läge, 2008; Hirschi & Läge, 2007; Hirschi, 2009; Nauta, 2005; Davis, 2007). Some other researchers have reported that students who have clear career interests obviously also have a high level of confidence in achieving a higher level of education (Nauta, 2007; Hirschi & Läge, 2007; Davis, 2007). The assumption is that when high school students recognise or have an interest about their career, they are more able to make career plans and explore career options. They can also make the right career decisions and thus have a high level of satisfaction in their career life. High school students who knows their career interest, will be able to make career decision more easily on their future area competencies when they finished junior high school (grade 9). Moreover, these career mature students have high task commitment and are able to choose their university major more easily when they are in Grade 12.

Super (in Greenhaus & Callanan, 2006) suggested that individuals who have career maturity are those who can make the right career decisions when they are involved in the exploration of career planning. In this case, individuals can make the right career decisions based on adequate career knowledge, self-knowledge, and skills in making the career decision. Alvarez (2008), Crites (1978) and Wu (2009) revealed that career maturity is the behaviour displayed during the fulfilment of tasks. Savickas (1994) and Ahlgren and Etzel (2001) defined career maturity as the readiness of an individual to make career decisions, along with the individual's ability to go through the stages of his/her development. High career maturity in an individual shows that he/she is ready to obtain information, make career decisions, and is able to overcome obstacles in his/her development duties.

In Super's theory, career maturity has two dimensions: cognitive and attitude. The cognitive dimension has two sub-dimensions: career planning and career exploration. The career planning focuses on how an individual thinks about working; knows about work conditions; what education to take in accordance with the goal; and collects information about the work, such as talking to adults about work and career plan, taking courses, extracurricular activities, or internship. Whereas in career exploration, the individual shows a willingness to obtain information about specific job using his/her own resources, such as asking professionals or adults about the job, reading a book or watching a movie, and doing well in education. Furthermore, the attitude dimension has two sub-dimensions: decision-making and world-of-work information. The decision-making sub-dimension illustrates the ability of an individual to use his/her knowledge and thoughts to make a decision or career plan. The world-of-work information sub-dimension is related to the knowledge about finding one's interests, capabilities and knowledge on how to do certain jobs.

In Holland's RIASEC (Realistic, Investigative, Artistic, Social, Enterprising and Conventional, 1985), individuals with real interest have a high orientation towards activities that involve physical skills and masculinity. They tend to choose jobs that are oriented to practical matters and outdoor activities. They do not like to interact with other people and prefer simple things. Individuals with investigative interest are generally oriented to reasoning tasks and using deep thinking in problem-solving. They can troubleshoot and analyse problems well because they tend to think mathematically, such as observation and evaluation; are cautious, critical, and highly curious. Individuals with artistic interest tend to think about abstract matters, such as beauty and creativity. They also like complex, emotional, intuitive and imaginative matters. Individuals with social interest tend to be communicative, and they are friendly and sociable, helpful, and tolerant. Moreover, they have verbal skills and good interpersonal relations. Individuals with enterprising interest tend to be confident, adaptable, assertive and ambitious. They are full of energy, extroverted, optimistic, risk-takers and are spontaneous; they have good verbal skills and can lead others. Individuals with conventional interest generally have good abilities in clerical and numerical skills. They like doing things that are detailed and systematic; and are reasonably stable and traditional in nature. Some examples of jobs related to this orientation are cashiers, statisticians, bank employees.

Each of Holland's six basic personality types (Realistic, Investigative, Artistic, Social, Enterprising and Conventional) (1997b in Nauta, 2005) has a unique constellation of preferred activities, self-beliefs, abilities, and values. Holland (in Nauta, 2005) suggested that there are four basic important concepts about career interest, namely congruence, differentiation, consistency and identity. Congruence is described as how the person's interest fits the environment. When the individuals are in an environment that is congruent with their career interest, they will be more satisfied and survive better.

The concept of differentiation illustrates the difference between the highest score and the lowest score of the six personality types in career interest measured by Self-Directed Search

(SDS), (Holland, 1990, in Brown & Lent 2013). Individuals with high interest differentiation have a clear distinction between the types of interest in the RIASEC. Individuals with low level of interest differentiation tend to have clear boundaries in choosing their career. Hirschi and Läge (2007) explained that students' level of differentiation is the difference between the three highest scores (predominant interest) and the three lowest scores in RIASEC. If students have big differences between the three highest scores (predominant interest) and the three lowest scores, it means that they would have a high degree of interest differentiation. On the other hand, if the differences between the three highest scores (predominant interest) and the three lowest scores were small, then they would have a low level of interest differentiation. Hence, individuals with a high level of differentiation have a clear career interest and that makes them more prepared and easily able to make a career decision and choice (Davis, 2007; Hirschi, 2010). This research will therefore see students' level of differentiation in accordance with their career maturity.

Several pairs of interest have similarities with other interests. Bullock (2006) and Lehberger (1988) described the concept of interest consistency as the proximity of two dominant types owned by a person in the hexagonal model proposed by Holland. Theoretically, Holland (in Nauta, 2005) suggested that individuals who have two dominant interests are adjacent to the hexagonal model, whereas individuals with a low level of consistency have two dominant interests that are opposed to the model. Other researchers found that career interest consistency is associated with the stability of career choice (Villwock et al., 1976), the persistence and academic achievement in the higher education stage (Wiley & Magoon, 1982), the stability in employment (Gottfredson & Lipstein, 1975), and the ability to choose and perform in the career (Holland et al., 1975). It means that individuals with high consistency will be more stable in career choice, and have high academic achievement and a good career life. The concept of identity is a situation in which a person has an idea about career goals with clear and stable interests and talents.

Based on the exposure and the findings of research described above, this research seeks to find the relationship between the level of differentiation and consistency and career maturity. The two variables (differentiation and consistency) have received less attention than the congruence. In fact, interest differentiation and interest consistency are predictive in career decision-making, career choice stability, and work satisfaction (Holland, in Brown & Lent, 2013). It also shows career maturity It means the interest differentiation and consistency give many advantages in building career maturity in adolescence but few researches found, especially in Indonesia. This research uses only two concepts of Holland's theory (differentiation and consistency) because the other two concepts (congruence and identity) are usually used qualitatively in a clinical setting, and as such these two concepts could not be included in this research.

The research question of this study is whether or not there is a relationship between interest differentiation and interest consistency and career maturity in Grade 10 students in Jakarta. The goal of this study is to find a correlation between interest differentiation and interest consistency and career maturity in Grade 10 students in Jakarta.

2 METHODS

This research is a quantitative study that uses correlation techniques in data analysis. The sampling method is non-probability (non-random sampling). The respondents were Grade 10 students in Senior High School (SMA) in Jakarta of 11–19 years age range, and majoring both in natural science and social science. Three high schools were involved as respondents. To measure career maturity, Career Development Inventory (CDI) by Super (1975, in Hinkelman & Kivlighan, 1998) was used after being translated and adapted into an Indonesian language and situation by Larasati (2010). After performing a trial, the CDI found the coefficient of Cronbach's alpha score of 0.824 for cognitive dimension and Cronbach alpha score of 0.710 for attitude dimension. To measure the interest consistency and interest differentiation, the researchers used a measuring instrument of SDS (Holland), which had been translated and adapted by Suzanna et al. (1990, in Wijaksana, 1992). For SDS Holland, researcher did not do any trial because this inventory is usually used as an assessment test which reliability and validity proved.

3 RESULTS

Respondents in this research were Grade 10 high school students in Jakarta. This research used 222 respondents.

3.1 *Career maturity*

A significant mean difference was found in both the natural science group (IPA) and the social science group (IPS) in their career maturity ($t = 2{,}136$; $p = 0.034 < 0.05$). Students in the natural science group (IPA) had a higher mean score of career maturity than students in the social science group (IPS). Nevertheless, different results showed that mean differences were not found between male and female students in career maturity. This indicates that there are no differences between males and females among the respondents in their career maturity.

3.2 *Interest differentiation and consistency*

A significant mean difference was found in interest differentiation of student genders ($t = -2.252$; $p = 0.025 < 0.05$). Female students had a higher differentiation mean score than male students. This indicates that female students had clearer interests than male students. Regarding the other result, there was no significant mean difference of interest differentiation among the students of both the natural science group and the social science group.

Moreover, there were no differences in interest consistency both in gender and in groups of interest. In other words, males and females did not differ in interest consistency. In addition, there were no differences in interest consistency between the natural science group and the social science group.

Correlation among variables:

Table 1. Correlation result among variables.

Variable	N	Career maturity	Interest differentiation	Interest consistency
Career maturity	222	1		
Interest differentiation	222	0.160*	1	
Interest consistency	222	–0.047	0.145*	1

Table 1 shows that the differentiation variable has a significant correlation with career maturity ($r = 0.160$; sign. los.005). It indicates that higher differentiation will be followed by higher career maturity. In contrast to the differentiation variable, the consistency variable is not significantly correlated with career maturity. Consistency has a significant correlation only with the differentiation variable. In other words, a higher level of interest differentiation will be followed by a higher level of interest consistency.

4 DISCUSSION

The positive correlation between differentiation and career maturity indicates that the higher the students' interest differentiation is, the higher the students' career maturity will be. In other words, when students can identify a deep and specific career interest they will have better preparation in the planning and making of a career decision, conducting exploration of career fields, and making career decisions based on their interest and information about their career. This result is consistent with previous studies conducted by Holland et al. (1975) and Lunneborg (1975). On the contrary, students with low level interest differentiation cannot identify deep and specific interest, and this vagueness of interest eventually makes it more difficult for them to make career decisions.

This research also found a significant difference of interest differentiation between male and female students. The result is in line with previous studies (Fouad & Mohler, 2004; Hirschi, 2009; Stark, 2001) which found that female students had a higher level of differentiation career interest than male students. According to Hirschi and Läge (2008), female students have a higher level of interest differentiation because they are generally more knowledgeable and have more information about the career world than male students. Bullock and Reardon (2005) suggested that students with higher sensitivity level in career interest will be open to various options and opportunities related to existing career fields, and therefore they are prepared to make better career decisions. According to Crites (1978, in Hirschi & Läge, 2008), students' level of sensitivity in career interest is an important preparation aspect in making career decisions. Fuller et al. (1999) also suggested that students who cannot identify their career interests tend to have low willingness to explore ideas and this limits their information about career interest and fields.

In contrast to differentiation, consistency is not related to career maturity. According to Sharf (2006), an individual may have low level interest consistency, but has good ability and readiness to choose a career field. For example, a student who has interest in a career field associated with clerical and numerical ability and accounting, but who also has interest in music, may only have readiness in making career decision by choosing to become a bank employee. In this case, the career decision to become a bank clerk has been influenced by the interest depth and the extent of information he/she has related to the career field. Thus, consistency can be said to be influenced by many variables, and therefore could be another research topic to be investigated.

This research has implications for practical matters. The research result can predict that students with high career interest differentiation are able to identify their deep and specific interest so it helps them in making a correct decision and to be more mature in career. Adults can intervene in developing student career maturity by giving them more opportunity to explore, to obtain information, and to make plans in their career activities. This would increase the level of student interest differentiation. Parents, teachers and other adults can assist students in achieving these skills to make proper preparation for their career development.

REFERENCES

Ahlgren, R.L. & Etzel, E.F. (2001). An investigation of demographic, psychosocial and self-reported behavioral influences on career maturity levels of college student-athletes. Ann Arbor West Virginia University Libraries.

Alvarez, G.M. (2008). Career Maturity: a priority for secondary education. *Electronic Journal of Research in Educational Psychology, 6*(3).

Brown, Steven,D., Lent,Robert,W.(2005). *Career Development and Counseling Putting Theory and Research to Work.* John Willey & Sons, Inc,. Canada.

Brown, Steven,D., Lent,Robert,W.(2013). *Career Development and Counseling Putting Theory and Research to Work.* John Willey & Sons, Inc,. Canada.

Bullock, E.E. (2006). Self-directed search interest profile elevation, big five personality factors, and interest secondary constructs in a college career course. (3232370 Ph.D.), The Florida State University, Ann Arbor. Retrieved from http://search.proquest.com/docview/305332009?accountid = 17242 ProQuest Dissertations & Theses Global database.

Bullock, E E., & Reardon, R C. (2005). Using Profile elevation to Increase the Usefulness of the Self-Directed Search and Other Inventories. *The Career Development Quarterly, 54*(2), 175–183.

Crites, J.O. & Semler I.J. (1967). Adjustment, educational achievement, and vocational maturity as dimensions of development in adolescence. *Journal of Counseling Psychology, 14*(6), 489–496. doi:10.1037/h0025227.

Davis, G.A. (2007). *Interest differentiation and profile elevation: investigating correlates of depression, confidence, and vocational identity* (Doctoral dissertation, University of North Texas, TX). Retrieved from http://remote-lib.ui.ac.id:2073/docview/304814905/abstract/D0F7CC87C86343EFPQ/1.

Donohue, R. (2006). Person-environment congruence in relation to career change and career persistence. *Journal of Vocational Behavior*, *68*(3), 504–515.

Fouad, N A., & Mohler, C J. (2004). Cultural validity of Holland's theory and the strong interest inventory for five racial/ethnic groups. *Journal of Career Assessment, 12*, 423–439. doi: 10.1177/1069072704267736.

Fuller, B E., Holland, J L., & Johnston, J A. (1999). The Relation of Profile elevation in the Self-Directed Search to Personality Variables. *Journal of Career Assessment, 7*, 111–123.

Gottfredson, G D., & Lipstein, D.J. (1975). Using personal characteristic to predict parolee and probationer employment stability. *Journal of Applied Psychology, 60*, 644–648.

Greenhaus, J H., & Callanan, G A. (2006). *Encyclopedia of career development*: Sage Publications.

Hinkelman, J M., & Kivlighan, D M. (1998). The effects of DISCOVER on the career maturity and career indecision of rural high school students: A randomized field experiment (Vol. 9904847, pp. 305–305 p.). Ann Arbor: University of Missouri–Columbia.

Hirschi, A. (2009). Development and criterion validity of differentiated and elevated vocational interests in adolescence. *Journal of Career Assessment, 17*(4), 384–401. doi:10.1177/1069072709334237.

Hirschi, A. & Läge, D. (2007). Holland's Secondary constructs of vocational interests and career choice readiness of secondary students: Measures for related but different constructs. *Journal of Individual Differences*, *28*(4), 205–218. doi:10.1027/1614-0001.28.4.205.

Hirschi, A. & Läge, D. (2008). Using accuracy of self-estimated interest type as a sign of career choice readiness in career assessment of secondary students. *Journal of Career Assessment, 16*(3), 310–325. doi:10.1177/1069072708317372.

Hirschi, A. (2010). Vocational interests and career goals: Development and relations to personality in middle adolescence. *Journal of Career Assessment, 18*, 223–238. Doi: 10.1177/106907271036478.

Larasati, W P. (2010). Goal-Orientation and Career Maturity on Senior High School. *Scription*. Universitas Indonesia. Depok.

Larson, L M., Rottinghaus, P J., & Borgen, F H. (2002). Meta-analyses of Big Six interests and Big Five personality factors. *Journal of Vocational Behavior, 61*, 217–239. doi: 10.1006/jvbe.2001.1854.

Lehberger, P.H. (1989). Factors related to a congruent choice of college major. Dissertasion. Texas Tech University.

Lunneborg, P W. (1975). Interest differentiation in high school and vocational indecision in college. *Journal of Vocational Behavior, 7*, 297–303.

Nauta, M.M. (2005). Holland's theory of vocational choice and adjustment. In S.D. Brown & R.W. Lent (Eds.), *Career development and counseling: Putting theory and research to work* (2nd ed.). 36–41 Hoboken, NJ: John Wiley & Sons.

Nauta, M.M. (2007). Career interests, self-efficacy, and personality as antecedents of career exploration. *Journal of Career Assessment, 15*(2), 162–180. doi:10.1177/1069072706298018.

Sharf, R. (2006). *Applying Career Development Theory to Counseling* (4th ed.). USA: Thomson Higher Education.

Savickas, M.L. (1997). Career adaptability: An integrative construct for life-span, life-space theory. *The Career Development Quarterly; Mar 1997;* 45; 3; ProQuest pg.247

Stark, R M. (2001). The Application of Holland's Theory to African-American College Students: A Test of Consistency, Differentiation, and Calculus: University of Wisconsin–Milwaukee.

Seligman, L., (1994). Developmental Career Counseling (2nd edition). California: Sage.

Villwock, J.D., Schnitzen, J.P., & Carbonari, J.P. (1976). Holland's personality constructs as predictor of stability of choice. *Journal of Vocational Behavior, 9*, 77–85.

Wijaksana, A. (1992). Adolecent's interest in career decision making on family with high, moderate and low economic status. *Scription*. Universitas Indonesia. Depok.

Wiley, M. O, & Magoon, T.M. (1982). Holland high point social types: Is consistency related to persistence and achievement? *Journal of Vocational Behavior, 20*, 14–21

Wu, M. (2009). The relationship between parenting styles, career decision self -efficacy, and career maturity of Asian-American College Students (Thesis, University of Southern California, Los Angeles, CA). Retrieved from http://remote-lib.ui.ac.id:2073/docview/304998440/abstract/8DB02DDC20F7405CPQ/1.

Diversity in Unity: Perspectives from Psychology and Behavioral Sciences – Ariyanto et al. (Eds)
© 2018 Taylor & Francis Group, London, ISBN 978-1-138-62665-2

Enhancing reading motivation through the teaching of RAP (Read, Ask, Put) reading strategy and writing reading diaries for an underachieving student

F. Febriani, S.Y. Indrasari & W. Prasetyawati
Faculty of Psychology, Universitas Indonesia, Depok, Indonesia

ABSTRACT: Underachievement is defined as a discrepancy between ability and performance. Many students with underachievement problems in primary school have concurrent reading problems. Moreover, one of the causes of underachievement in students is the lack of motivation in reading. In order to enhance reading motivation, many methods can be applied. Teaching reading strategies and applying reading diaries, for example, have been implemented. Using a single-case study design, an elementary school student with reading problems is trained to use the Read-Ask-Put (RAP) reading strategy and to make reading diaries after the reading session. The result shows that using RAP reading strategy and writing reading diaries enhance the reading motivation of the student, although the difference is not significant. However, after intervention, the participant claims to experience increased reading capability and reports higher confidence in dealing with more difficult reading material.

1 INTRODUCTION

Academic achievement is one of the indicators of learning success in school (Scheerens, 1990). A student with good academic ability has the potential to excel in their academic achievement. Yet in reality, many students with good academic ability may not necessarily perform as well as they should (Coil, 2010). This phenomenon is known as academic underachievement. Conceptually, academic underachievement refers to a significant gap between ability and performance (Coil, 2010). According to Coil (2010), academic underachievement is usually degenerative, which means that it begins in the early grades and the effects accumulate as the child grows older. To help underachievers, there are several strategies that can be applied (Coil, 2010); but in choosing the appropriate strategies, the teachers and parents should first consider the students' characteristics (Rahal, 2010). Mandel and Marcus (as cited in Chukwu-Etu, 2009) categorized underachiever students into six categories, i.e., anxious underachiever, defiant underachiever, wheeler-dealer underachiever, identity-search underachiever, sad or depressed underachiever, and coasting underachiever. The type of underachiever that can be found in most schools is the coasting underachiever (Bond, n.d.; Rahal, 2010). Students of this type are usually considered as lazy and unmotivated, which may not always be true (Bond, n.d.; Chere & Hlalele, 2014). They may actually have high motivation, but only in activities that interest them (Bond, n.d.). They tend to only do fun activities and typically do not show a strong and consistent effort in learning (Rahal, 2010).

One example of a coasting underachiever is O, an 11-year-old male student in 6th grade. O's intellectual ability was above average. Having such ability, he has the potential to obtain good marks in school. However, both his teacher and his parents reported that his performance was not satisfactory. As a coasting underachiever, O showed high motivation only in subjects that he liked, such as sports, computer science, arts, and other subjects related to practical activities. O did not like reading activities. O seemed disinterested in subjects or activities related to reading activities. He would frequently lose his concentration and often chatted with friends during study sessions. As a result, he was often unable to finish his tasks on time and would likely receive poor grades. O's condition was consistent with the research findings of Ryan (as cited in Chukwu-Etu, 2009) and Stroet (2009). Ryan found that 75% of students with underachievement problems in primary school also had reading problems. In addition, Stroet (2009) found that the lack of reading motivation, especially in male students, is one contributing factor of underachievement. Therefore, reading activity is an important issue in the topic of underachievement (Chukwu-Etu, 2009).

In many literatures, reading activity is considered as an important element of student achievement, both in academic and in other aspects of life (Ryan, as cited in Chukwu-Etu, 2009; Pečjak & Košir, 2004). According to Baier (2011), reading activity is a fundamental factor in academic achievement because this activity is found in every school subject. Wigfield and Guthrie (1997) explained that the volume and variety of readings are affected by a student's reading motivation. Students with high reading motivation tend to read more and show better results in reading tests or in their academic performance (Pečjak & Košir, 2004; Applegate & Applegate, as cited in Saw, 2014).

Reading motivation is defined as the likelihood to engage in reading or choosing to read (Gambrell, 2011). It is a construction that consists of three aspects, namely self-efficacy belief, achievement belief, and social aspect (Wigfield & Guthrie, 1997). Self-efficacy belief refers to the students' beliefs about their competence and ability in terms of reading. This aspect consists of three dimensions, which include reading efficacy, reading challenge, and reading work avoidance. On the other hand, the achievement belief refers to factors related to reading orientation, an individual's perception of the usefulness of reading, and whether the encouragement for reading comes from within or outside the student. This aspect consists of six dimensions, including reading curiosity, reading involvement, importance of reading, recognition for reading, reading for grades, and competition in reading. Lastly, the social aspect comprises of two dimensions. The first dimension is social reasons for reading, which refers to the desire to share the meaning of the reading material to other people, while the second is compliance, which involves reading due to external demands.

Saw (2014) explained that reading motivation is an important issue in primary school. A primary school student who is disinterested or inactive in reading activities usually does not develop the reading motivation needed in the subsequent level of education (Pečjak & Košir, 2004). Therefore, in this research, the intervention was focused on the enhancement of reading motivation in a primary school underachiever. To enhance reading motivation, many strategies can be used. Some researchers have devised various strategies that can be used to improve students' reading motivation, such as using a wide range of reading material and providing opportunities to discuss, offering the chance to read in the classroom and to make choices about what and how they read, sharing reading experiences, providing a challenging environment for reading as well as appreciation and positive feedback, and teaching reading strategies and writing reading diaries (Pečjak & Košir, 2004; Gambrell, 2011).

The strategies implemented in this research included teaching the use of reading strategies and writing reading diaries. In teaching reading strategies, the participant was trained to use reading strategies to enhance reading comprehension. According to Afflerbach, Pearson, and Paris (2008), teaching reading strategies can help the participant become fluent readers, which in turn can make the student more confident and motivated to read more. Harvey (n.d.) proposed four reading strategies commonly used to enhance reading comprehension among primary school students, including Rainbow Dots, Story-Map, Manipulative Object, and RAP (Read, Ask, and Put). Rainbow Dots is a strategy that combines four general reading strategies: visualization, summarization, making inferences, and making connection. This strategy is suitable for third grade elementary school students. A Story-Map is a strategy for helping students understand the elements of a story through the visual pathway and is suitable for narrative readings. Manipulative Object is a strategy that utilizes real objects associated with stories or readings and is generally used with students who are in the early grades of elementary school. While in RAP (Read, Ask, and Put) strategy, a student is directed to read the paragraph, ask questions about the reading, and summarize the information in their own words. This strategy is mostly used with primary school students or secondary school students with learning difficulties.

Among these four strategies, the strategy that was selected for this research is the RAP (Read, Ask, and Put) strategy. This strategy was chosen because it was deemed suitable for the participant who at the time of study was in 6th grade. In contrast, Rainbow Dots and Manipulative Object are more suitable for younger students. The RAP strategy is also well suited to better understand various types of reading, such as narrative and expository readings. Moreover, the RAP strategy, that involves a metacognitive process (El-Koumy, 2004; Hagaman, Luschen, & Reid, 2010), is also appropriate for the participant, who has above average intelligence. In metacognitive processes, students are expected to be aware of what they know and what they do not know, as well as to regulate the learning process by means of planning, monitoring, and evaluating (Livingston, 1997). A student with a low or average intellectual ability is likely to encounter difficulties engaging in such a process.

In writing reading diaries, students are trained to reflect what they have read and to then relate the materials they read with their life experiences. This strategy was adapted from the intervention made by Hulleman, Godes, Hendricks, and Harackiewicz (2010). Hulleman *et al.* (2010) believed that by perceiving a connection between a lesson and its usefulness in life, students' interest in the subject will increase. In the intervention, the researchers asked high school students to write down the relationship between the materials and their lives. This intervention principle was then applied in the context of reading by Gambrell (2011). By writing down the relationship between what they read and their lives, students realize that what they read have meanings in their lives. As a result, they will begin to perceive reading as a beneficial activity and will be more motivated to read.

Both teaching reading strategies and writing reading diaries have different targets. The target of teaching reading strategies is the cognitive aspect of reading. By learning strategies, the participant can improve his ability to understand and remember the content of reading material. On the other hand, the target of writing reading diaries pertains more to the affective aspect because it raises the participant's awareness about the benefits of reading. By combining the two strategies, it is expected that the participant's reading motivation will improve. Therefore, the problem that the current research is addressing is, "Can teaching RAP reading strategies and writing reading diaries enhance the reading motivation of an underachiever?"

2 METHOD

The study design applied in this study is the A-B single case design, which was developed from a within-subjects design, involving two measurements of the same participant (Bordens & Abbott, 2011). The study design with a single participant generally has three components: the repeated measures, the baseline phase, and the treatment phase (Engel & Schutt, 2008). The research participant was selected through purposive sampling, wherein investigators intentionally attempted to find a subject with characteristics consistent with the objectives of the study (Kumar, 2005). In this study, the participant was O (11 years old), a 6th grade student possessing the characteristics of a coasting underachiever.

The general objective of the intervention program was to increase reading motivation in an underachieving student through the teaching of RAP (Read, Ask, Put) reading strategies and writing reading diaries. The specific objectives of this intervention were to teach the use of RAP (Read, Ask, Put) reading strategies and to train the student to write reading diaries. The measurement tool used to measure reading motivation was the Motivations for Reading Questionnaire (MRQ) developed by Wigfield and Guthrie (1997). The measurement was chosen because it is intended for determining reading motivation in students from grades 3 through grade 8. The instrument consists of 54 items that are divided into eleven dimensions. Some examples of the MRQ items are presented in Table 1.

For each item of the MRQ, the subject was asked to respond by choosing among one of four possible answers: (1) very different from me, (2) a little different from me, (3) a little like me, and (4) a lot like me. Scores obtained from this measure are generally classified into

Table 1. The examples of MRQ items.

Dimension	Indicator	Items example
Reading Efficacy	Student's feeling that he has good competence in reading activities	I am a good reader
Reading Challenge	Student's strong willingness to read difficult reading materials and looking at it as a challenge	I like hard, challenging books
Reading Work Avoidance	Student's desire to avoid reading activities	I don't like vocabulary questions
Reading Curiosity	Students read for their desire to learn about specific topics of interest	I like to read about new things
Reading Involvement	Students read for their urge for pleasure or enjoyment through reading passages like	I make pictures in my mind when I read
Importance of Reading	Students read because they consider that it is important to do and beneficial for him	It is important to me to be a good reader
Recognition for Reading	Students read because they feel happy to get recognition or appreciation from others was significant for him	I like having the teacher say I read well
Reading for Grades	Students read to pursue a satisfactory achievement	I read to improve my grades
Competition in Reading	Students read for their urge to compete or trying to exceed the other students.	I like being the best at reading
Social Reasons for Reading	Students read because they have a desire to share a story or meaning of the text they have read	My friends and I like to trade things to read
Compliance in Reading	Students read for commands, demands, or task.	I read because I have to

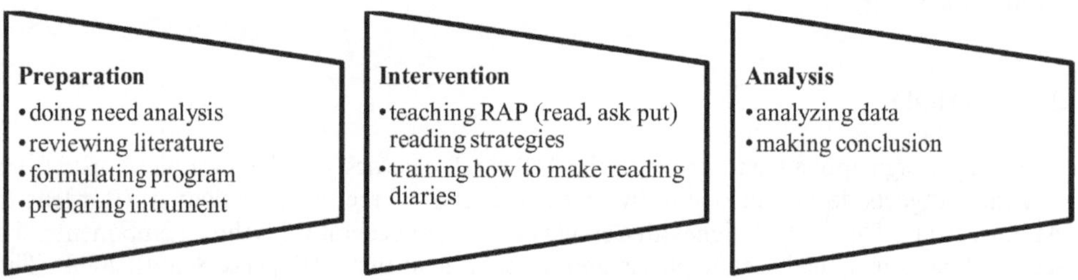

Figure 1. Research procedure.

three categories based on hypothetical means, namely low (54–108), medium (109–162), and high (163–216). The indicators of success of this intervention included the increase of the participant's MRQ score after intervention and positive changes in the participant's daily evaluation.

The research procedure was divided into three phases, i.e. the preparation phase, the intervention phase, and the analysis phase. The details of each phase are shown in Figure 1.

The intervention program was designed to be implemented across 7 sessions, with each session lasting 45–60 minutes. The duration of teaching RAP strategies was 20–30 minutes. Afterwards, the participant was asked to find the relationship between the content of the reading material and his real life experiences. Results from these reflections were then written in reading diaries worksheets. After the intervention program was completed, the data obtained during the intervention were then analyzed.

3 RESULTS

The intervention was implemented in 7 sessions within a 3-week timeframe. Each session lasted about 45–60 minutes. Each meeting started at 5 p.m. In general, the participant was very cooperative throughout the intervention. He noted the instructions carefully and followed each activity properly. He also consistently followed the intervention, starting from the very first session to the last session, although his health had declined on the fourth day of the intervention. Nevertheless, the intervention was implemented smoothly and managed to meet the targets that had originally been set.

Based on the evaluations of the participant's comprehension of the reading material, the participant demonstrated that he was able to capture the essence of the reading and to explain it again using his own words. He also showed an ability to apply the RAP strategy independently and to find the relationship between the content of reading and his life. By comparing the MRQ scores he obtained before and after the intervention, the total score of the participant's reading motivation was found to increase by 8 points from 137 to 145. While this suggested some positive changes, the changes were not significant as both scores were in the same category (i.e., medium). Analyses of the changes occurring in various dimensions of reading motivation are illustrated in two parts, separated based on whether the dimensions are internally or externally motivated. The first part pertains to the dimensions of internal motivation, which is depicted in Figure 2. Part two, on the other hand, is the dimensions of external motivation and is illustrated in Figure 3.

Figure 2 shows the increase in three dimensions, i.e. reading work avoidance, reading curiosity, and reading involvement. The levels of two other dimensions, i.e. reading efficacy and importance for reading, had decreased. One other dimension, namely reading challenge, did not change.

In Figure 3, increases in two dimensions were noted, which include the increase in recognition for reading and in social reasons for reading. One dimension (i.e., compliance in reading) declined, while two other dimensions (i.e., reading for grades and competition in reading) did not change.

Using daily evaluation worksheets, the participant's affective aspects were assessed. The results revealed that the participant showed positive emotional changes over time. It is

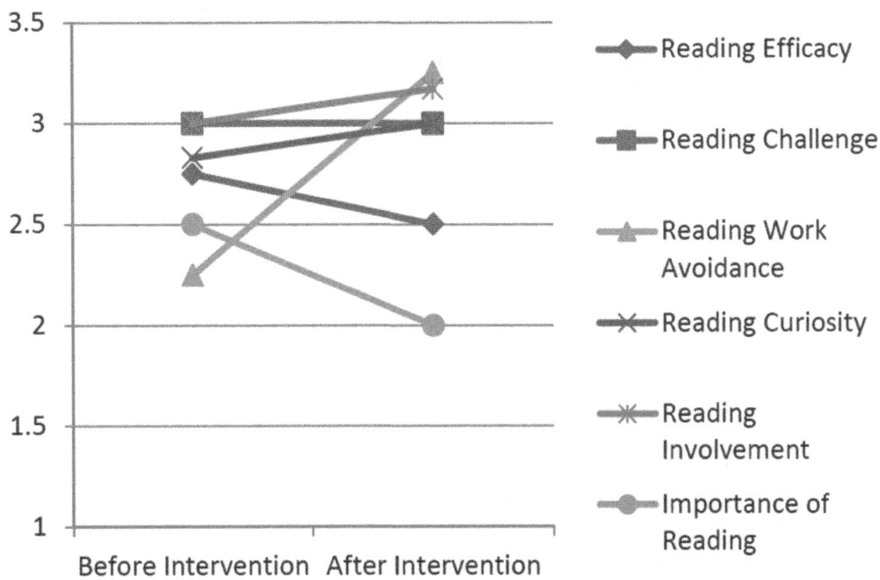

Figure 2. Comparisons of MRQ scores in the dimensions of internal motivation.

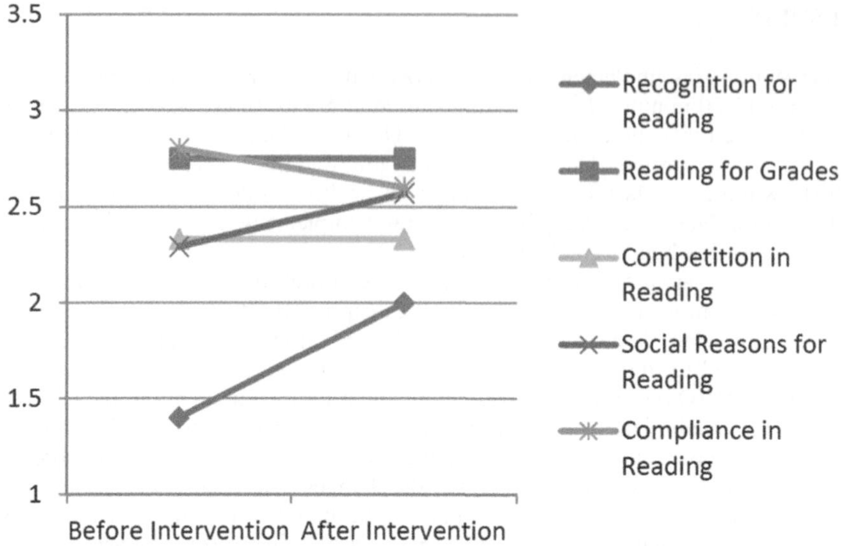

Figure 3. Comparisons of MRQ scores in the dimensions of external motivation.

important to note that in the first three sessions, the participant's emotions tended to be neutral, but during the fourth session, there was a change in which the participant was unhappy. This negative emotion was apparently due to his physical condition as he was not feeling well that day. Finally, in the last three sessions, the participant reported the feeling of excitement.

At the end of the intervention, the participant stated that he was happy to follow the series of intervention activities because he was able to learn new things. He also claimed that the teaching of reading strategies was beneficial because he became more skilled in finding the main idea and supporting ideas in a paragraph. In addition, the participant admitted that by practicing to retell the reading content in his own words, he was able to find the finer points of the reading more quickly. This ability is useful for working on essay examination questions. When answering questions on an exam, answers usually cannot be written down in verbatim as they are expressed in textbooks. Therefore, learning to find the essence of reading through this strategy can prove to be extremely helpful. The participant also reported to have applied this strategy while taking his final exams, which happened to take place in the middle of the implementation of the intervention.

4 DISCUSSION

The purpose of the intervention in the present study was to improve the reading motivation of the participant. In general, the intervention program was implemented without trouble. However, the improvement of the participant's reading motivation was not optimal. The total scores on the MRQ obtained by the participant both before and after the intervention were in the same medium category. Moreover, three dimensions of motivation had decreased while three other dimensions remained unchanged. Several factors that possibly contributed to the results will be discussed further in this section.

First, as mentioned earlier, the RAP reading strategy consists of three stages, i.e. reading the paragraph, asking questions about the readings, and summarizing the readings using the participant's own words. Also as discussed earlier, these stages involve a metacognitive process (El-Koumy, 2004; Hagaman, Luschen, & Reid, 2010). During these stages, an individual is trained to monitor the student's reading process so as to gain awareness of what he understands and what he does not understand. This awareness can in turn influence the student's self-efficacy. More particularly, as long as he understands what he is reading, his self-efficacy will increase. On the other hand, if he believes that he does not understand the reading, his self-efficacy decreases. In this intervention, the score of reading efficacy was found to decline, which presumably occurred because the participant was aware of the difficulties he was facing while trying to understand the reading using the RAP reading strategy. The same reason may also help explain the lack of change in the reading challenge score. In other words, because the participant did not feel confident in his reading abilities, he did not feel challenged to read.

Second, the purpose of writing reading diaries was to develop the participant's awareness about the importance and benefits of reading activities. In this intervention, however, the results seemed to indicate the absence of such awareness. Based on the results of the MRQ, the change in the dimension of importance of reading tended to be negative. This probably occurred due to limitations in the implementation of the strategy of writing reading diaries in this intervention. As stated earlier, this strategy was adapted from a strategy introduced by Hulleman *et al.* (2010). However, in the study conducted by Hulleman *et al.*, this strategy was implemented for the course of one semester and focused on one particular subject. This gave students the chance to explore the benefits of reading in greater depth and in turn created greater motivation to learn. In contrast, in the intervention of the current study, writing reading diaries was implemented in only 7 sessions, and the reading theme was changed every day. As a result, the internalization process of the importance of reading did not happen.

Third, except for the reading efficacy, reading challenge, and the importance of reading, the levels of the other three dimensions (i.e. reading for grades, competition in reading, and compliance in reading) did not increase. This was probably due to the external factors of motivation for these other three dimensions. On the other hand, both teaching reading strategies and the writing reading diaries that was applied in the current study are perhaps more effective for improving internal motivation.

5 CONCLUSION

The result of this study shows that teaching RAP reading strategy and writing reading diaries can enhance the reading motivation of an underachiever, but this enhancement was not significant. The participant's motivation score in the Motivations for Reading Questionnaire (MRQ) only increased by 8 points after the intervention (from 137 to 145), and these pre-test and post-test scores were in the same category range. Even though the result was not significant, however, based on the qualitative evaluation, the participant felt more confident in dealing with more difficult reading material after the intervention.

REFERENCES

Afflerbach, P., Pearson, P.D. & Paris S.G. (2008). Clarifying differences between reading skills and reading strategies. *The Reading Teacher,* 61 (5), 364–73. doi:10.1598/RT.61.5.1.

Baier, K. (2011). *The Effects of SQ3R on Fifth Grade Students' Comprehension Levels.* [Master's Thesis], College of Bowling Green, State University. https://etd.ohiolink.edu/!etd.send_file?accession=bgsu1300677596anddisposition=inline.

Bond, B. (n.d.). *Types of Underachievers and Strategies to Help Them.* http://www.flemingclt.ca/ccei/documents/CA/PMS_underachievers.pdf.

Bordens, K.S. & Abbott, B.B. (2011). *Research Design and Methods: A Process Approach (8th ed).* New York, USA, McGraw-Hill Companies.

Chere, N.E. & Hlalele, D. (2014). Academic underachievement of learners at school: A literature review. *Mediterranean Journal of Social Sciences,* 5 (23), 827–39. doi:10.5901/mjss.2014.v5n23p827.

Chukwu-Etu, O. (2009). Underachieving learners: Can they learn at all? *ARECLS,* 6, 84–102. research.ncl.ac.uk/ARECLS/volume_6/ogbonnia_vol6.pdf.

Coil, C. (2010). *Motivating Underachiever: 220 Strategies for Successes Refined and Expanded Edition.* Saline, MI, USA, McNaughton and Gunn, Inc. https://www.piecesoflearning.com/UserFiles/File/FreePreviews/clc0256.pdf.

El-Koumy, A.S.A.K. (2004). Metacognition and Reading Comprehension: Current Trends in Theory and Research. Cairo, Egypt, Anglo Egyptian Bookshop.

Engel, R.J. & Schutt, R.K. (2008). *The Practice of Research of Social Work (2nd ed).* New York, USA, McGraw Hill.

Gambrell, Linda B. (2011). seven rules of engagement: what's Most Important to Know about Motivation to Read. *Reading Teacher,* 65 (3), 172–78. doi:10.1002/TRTR.01024.

Hagaman, Jessica L.Luschen. (2010). The 'RAP' on reading comprehension. *Teaching Exceptional Children,* 43, 22–29.

Harvey, M.n.d. (2013). *Reading Comprehension: Strategies for Elementary and Secondary School Students.* http://www.lynchburg.edu/wp-content/uploads/volume-8–2013/HarveyM-Reading-Comprehension-Elementary-Secondary.pdf.

Hulleman, C.S., Godes, O., Hendricks, B.L. & Harackiewicz, J.M. (2010). Enhancing interest and performance with a utility value intervention. *Journal of Educational Psychology,* 102 (4), 880–895. http://psych.wisc.edu/cmsdocuments/HullemanGodesH2010.pdf.

Kumar, R. (2005). Research Methodology, A Step by Step Guide for Beginners (2nd ed). Malaysia, SAGE Publication.

Livingston, J.A. (1997). *Metacognition: An overview.* http://gse.buffalo.edu/fas/shuell/cep564/metacog.htm.

Pečjak, S. & Košir, K. (2004). Pupils' reading motivation and teacher's activities for enhancing it. *Review of Psychology.* http://search.ebscohost.com/login.aspx?direct=true&db=psyh&AN=2005-04723-002&loginpage=Login.asp&site=ehost-live&scope=site.

Scheerens, J. (1990). School effectiveness research and the development of process indicators of school functioning. *School Effectiveness and School Improvement,* 1 (1), 61–80. doi:10.1080/0924345900010106.

Rahal, M.L. (2010). *Identifying and Motivating Underachievers.* www.edweek.org/media/fo-motivation-resources.pdf.

Saw, V.A. (2014). Creating classroom environments that foster students' reading motivation. [Master's Thesis], University of Toronto. https://tspace.library.utoronto.ca/.../1/Saw_ValerieAnne_T_201406_MT_MTRP.pdf

Stroet, S. (2009). Mad about The Boy: Research Project to Develop Whole School and Cross Curricular Strategies for Raising Boys' Achievement, With Particular Focus on Lesson Planning, and Teaching and Learning Materials And Approaches. http://www.farlingaye.suffolk.sch.uk/Information/LeadingEdge/Developing%20Strategies%20For%20Raising%20Boys'%20 Achievement.doc.

Wigfield, A. & Guthrie, J.T. (1997). Relations of children's motivation for reading to the amount and breadth or their reading. *Journal of Educational Psychology,* 89 (3), 420–32. doi:10.1037/0022-0663.89.3.420.

Diversity in Unity: Perspectives from Psychology and Behavioral Sciences – Ariyanto et al. (Eds)
© 2018 Taylor & Francis Group, London, ISBN 978-1-138-62665-2

Applying a sex education programme in elementary schools in Indonesia: Theory, application, and best practices

S. Safitri
Faculty of Psychology, Universitas Indonesia, Depok, Indonesia

ABSTRACT: Sexuality is a part of life. Therefore, receiving accurate information about sexuality—known as sex education—is one of the rights of children. UNESCO (2009) stated that sex education aims to equip children with knowledge, skills, and values regarding sexual and social relationships to make them better decision makers in their adulthood. It is specifically geared to remove gender inequalities as one of the Sustainable Development Goals set by the United Nations (UN). Furthermore, enabling sex education is closely linked to the local culture. In Indonesia, discussing sexuality is commonly considered taboo. Thus, as a UN member, Indonesia does not yet have national programmes related to sex education for students, even though it is mandated by the constitution. In contrast, in response to the current school climate, the need for sex education is recognised by the school community as essential. This study describes a pilot programme of sex education that is delivered to fourth grade students at one elementary school in Jakarta, Indonesia. It is found that their knowledge of sexuality is generally below the requisite level and is significantly different between boys and girls. Both genders are also different in terms of their success in achieving the programme's goals.

1 INTRODUCTION

Sexuality is a natural part of life. According to the Indonesian Ministry of Health (2009), sexuality refers to the total aspect of human life associated with the genitals. Therefore, it is not surprising that receiving accurate information about sexuality—known as sex education—is one of the rights of children (UNFPA, 2014). Sex education aims to equip children with the knowledge, skills, and values regarding sexual and social relationships to make them better decision makers in their adulthood (UNESCO, 2009).

The importance of sex education for students' own advantage lies in the main premise of the United Nations (UN) mandate that the governments of each of its member nations will provide the service (UNESCO, 2009).

Sex education is known to be the key for preventing risky sexual behaviour by teenagers. It is also a means to encourage teenagers to get involved in maintaining their own health as individuals (Bearinger et al., 2007). The need for sex education is forced by the global urgency to combat human immunodeficiency virus infection.

UNFPA (2014) specifically mentioned that sex education should take place both in school and in community-based settings. Schools provide opportunities for large numbers of children to be recipients of sex education (Gordon, 2008, in UNESCO, 2009). Through the medium of the school, sex education can also be formally given to children just before they start to become mature adults and sexually active. Furthermore, sexuality as a subject is known to be a component of a comprehensive educational curriculum. By equipping the curriculum with the subject of sex education, students can obtain accurate information for developing their own personal attitudes related to sexuality (UNESCO, 2009).

The need for sex education is crucial for students in elementary schools. Elementary school students are in the late childhood period that ranges from 6 to 12 years of age. It is the time when students start to become sexually mature and will progress to adolescence. Therefore, this period is marked as the starting point of bodily maturation (Hurlock, 1991). Physical and sexual development happens fast, but it varies from one child to another. Some children enter the stage of puberty earlier than others. Girls also tend to mature earlier than boys. This difference of time makes some children feel awkward about the changes in their body, as their friends do not experience it at the same (Sukadji, 2000). Many of them also begin to feel shy or worry about their body. It can be concluded that change of the body affects children's personal and social adjustment (Hurlock, 1991).

Children in late childhood generally also have high curiosity (Hurlock, 1991). In terms of sexuality, children aged 6–9 years start to show a desire to learn about the body of the opposite sex. They begin to question the basic nature of sex. Older children aged 10–12 years start to have a romantic interest towards the opposite sex. They also have a deeper understanding of sexuality, compared with younger children, and are eager to know more about it by themselves. However, the more children learn about sexuality by themselves, the more they become anxious about it (Berman, 2009). Thus, it appears that children in late childhood have a thirst for knowledge about sexuality, but they need adequate assistance to understand it properly.

From a developmental point of view, developmental tasks of late childhood require children to be able to develop their capacity as social beings. They need to be able to get along and work with others. They are also expected to learn gender roles, understand themselves as growing human beings, and learn to be independent and responsible for themselves (Sukadji, 2000). All of these developmental tasks, especially on the aspect of human sexuality, are consistent with the purpose of and topics covered in sex education.

Contrary to the UN mandate for delivering a comprehensive sex education for children, in Indonesia sex education is still viewed as a controversial topic. The programme has met strong criticism from conservative officials who argued that it should not be made compulsory (IPPF, 2013). On the other hand, there are also parties who support and strongly urge the government to include sex education as a part of school curriculum. The Indonesian National Child Protection Commission (Komisi Perlindungan Anak Indonesia/KPAI) is one of the parties to push the government to implement sex education in school. The reason is that the majority of Indonesian parents do not understand the topic in order to inform their own children adequately. KPAI has been proposing the inclusion of sex education in the school curriculum since 1999, but the government has not yet responded to the proposal (Yosephine, 2016). Thus, Indonesia still does not have a national programme of sex education delivered to its children. The current implementation of sex education for children in Indonesia is, therefore, run sporadically by those who think it is necessary.

This study describes a pilot programme of sex education that is delivered to fourth grade students at one elementary school in Jakarta, Indonesia. The implementation of the sex education programme at this school is encouraged by the current school climate where students tease each other by touching others' private parts. The initial assessment revealed that students had no idea regarding the appropriateness of touching others' body parts. Thus, they touched others' private parts to tease or in jest. This finding is in line with the broader context of Indonesia where the need and delivery of sex education are still considered taboo (IPPF, 2013).

2 THEORETICAL BACKGROUND

Sex education aims to equip children with the knowledge, skills, and values regarding sexual and social relationships to make them better decision makers in their adulthood. In general, sex education has the following objectives (UNESCO, 2009):

1. Improving knowledge and understanding about sexuality
2. Explaining and clarifying personal feelings and values about sexuality
3. Developing skills of decision making around issues of sexuality
4. Promoting behaviour that avoids risks related to the sexual act

Sex education, especially for students in kindergarten and elementary schools, is mainly aimed to understand issues related to the body and health. This understanding is achieved within the framework of promoting the value of health and socio cultural value of the body, along with the effort towards raising child awareness about sexual harassment (UNFPA, 2015).

Four components are ideally included in sex education service. These are as follows (UNESCO, 2009):

1. Knowledge: Sex education should provide accurate information about human sexuality.
2. Values, attitudes, and social norms: Sex education should facilitate students to explore values, attitudes, and social norms that are applied regarding sexual behaviour, including mutual respect, human rights, tolerance, and gender equality.
3. Interpersonal skills: Sex education should encourage acquisition of decision making, assertiveness, negotiation, and refusal skills in relation to the process of forming and maintaining relationships with others, such as family members, friends, and spouse.
4. Responsibility: Sex education should encourage students to take responsibility for their behaviour. Responsibility can be defined in terms of caring for themselves and avoiding sexual harassment.

Furthermore, UNFPA (2012) provides an overview of the topics to be discussed in sex education, which is as follows:

1. Awareness of self and personal relationship with others
2. Human development, including puberty, bodily functions, and reproduction
3. Sexual behaviour
4. Sexual health, including pregnancy, prevention of sexually transmitted diseases, contraception, and abortion
5. Communication skills, negotiation, and decision making

In accordance with UNFPA, Berman (2009) also stated that these five topics are the main topics of sex education. In addition, the sequence of topic delivery starts with a discussion about the human body, its anatomy, and the concept of puberty. It then continues with a discussion about gender roles and sexually related behaviour, followed by sexual health and self-protection.

The wide range of objectives and topics of sex education makes it a long-term learning process, and it is strongly associated with the age and stage of development of students (UNFPA, 2015). Apart from the range of topics covered in sex education, the basic premise of effective sex education content is that it has to be suitable for the participants' age (i.e. age appropriate) (UNESCO, 2009). Moreover, sex education should be given as early as possible. It is known that children as young as in kindergarten already can be the recipients of sex education services (UNFPA, 2015). In addition, the content also has to be relevant to the local culture, without ignoring the need for scientifically validated accuracy (UNESCO, 2009).

3 METHOD

3.1 *Participants*

Participants in this study were all students, both boys and girls, in the fourth grade of elementary schools. There were 31 male students and 34 female students who participated.

3.2 *Programme development*

The activities were constructed based on the initial assessment of the students' baseline knowledge about sexuality. This assessment was done by asking all fourth grade students in the targeted school to answer a set number of questions related to the topic of sexuality covered in the sex education programme, as stated by UNFPA (2014). From the answers given, we found that the majority of those fourth grade students have no basic knowledge of sexuality, ranging from naming body parts (including private parts) to understanding how to protect one's body from unwanted touch. Therefore, all the unknown information, as revealed

Table 1. The sex education programme: Topic, session, and goals.

Topics	Session no. (duration in minutes)	Programme goals
Awareness of self and personal relationship with others	1(45)	Identify/name the body part correctly (including genitals)
Human development	2 (25)	Understand development of the human body and puberty
Sexual behaviour	3 (30)	Understand the differences of gender roles
Sexual health	4 (60)	Understand how to maintain personal hygiene
Communication skills, negotiation, and decision making	5 (60)	Understand how to protect one's body from unwanted touch

by the majority of students, was made part of sex education programme topics and goals, as listed in Table 1.

Compared with the Ontario curriculum of the Canadian Ministry of Education and Training (2015), the participants' baseline knowledge of sexuality was generally below the requisite level with respect to their grade and age of development. Some of the programme goals stated earlier are materials for the first to third grade students. These goals are: (1) to identify/name the body part correctly, (2) to understand the differences of gender roles, and (3) to understand how to maintain personal hygiene. Furthermore, the Ontario curriculum content for sex education for fourth grade students is already expanded to cover the comprehension of sexual exploitation prevention, intimate human relationships, and healthy sexual conduct and its related preventable diseases.

The sex education programme for this study was held as a one-day training programme. To make sure the learning processes were conducive, participants were grouped by gender while taking into account the number of students within one group. Thus, the activities were carried out by four groups from the whole class, two groups of boys and two groups of girls. The five sessions listed in Table 1 lasted for 220 minutes.

3.3 *Programme implementation*

Right before the sex education class started, all participants were asked to write down on a piece of paper what they wanted to know about the topic of sexuality. This was done as a way to assess their aspirations about sexuality and to customise the content of the class discussion to address what the participants wanted to know. However, all the participants still received the same basic information of sexuality related to each of the topic's goals. Table 2 shows the percentage of responses gathered.

Table 2 shows that the majority of male and female participants asked a specific question about sexuality. Examples of these questions included: 'Why do people need to wear pants?', 'I want to know about my genitals', and 'What is puberty?'. This is in line with what Hurlock (1991) stated that children in their late childhood generally have high curiosity, especially about the topic of sexuality. These children start to show a desire to learn about their own body and that of the opposite sex.

The second leading response from all participants included questions unrelated to sexuality, such as 'Does God forgive the terrorist?' and 'I want to be a great man'. However, focusing on the gender of the participants reveals that the majority of girls (80%) asked a question related to sexuality whereas only 50% of boys asked the same kind of questions. Half the number of male participants responded with irrelevant questions or responses. Also, only male participants declared that they did not have any questions related to sexuality or they did not want to know anything. This might be because children are more anxious to know more about sexuality (Berman, 2009). The dynamics between boys and girls might represent

Table 2. Participant responses and questions about sexuality.

Type of response	Boys (%)	Girls (%)
Nothing	15	—
Irrelevant questions	16	—
Irrelevant responses	19	20
Asked one to two question(s) related to sexuality	50	80

the differences among children in approaching sexuality and entering the stage of puberty. Girls also tend to mature earlier than boys (Sukadji, 2000).

3.4 *Measurements*

The difference in the participants' knowledge of sexuality was examined through a pre-test and post-test. The pre-test also functioned as an assessment of the participants' baseline knowledge. In other words, a set number of questions related to the content of the programme were given before and after the participants completed the programme. The frequency of correct answers for each question was compared between the pre-test and post-test assessments. The significant difference in the number of correct responses was determined by chi-square goodness of fit. Using this technique, the difference in the proportion distribution of the sample can be compared with the general population (Gravetter &Wallnau, 2008).

3.5 *Hypotheses*

For the five sessions delivered to the participants, there were five hypotheses regarding the effectiveness of the programme. These hypotheses were as follows:

1. There is no significant difference in the frequency of correct answers between the pre-test and post-test for the topic of awareness of human body.
2. There is no significant difference in the frequency of correct answers between the pre-test and post-test for the topic of puberty.
3. There is no significant difference in the frequency of correct answers between the pre-test and post-test for the topic of gender roles.
4. There is no significant difference in the frequency of correct answers between the pre-test and post-test for the topic of sexual health.
5. There is no significant difference in the frequency of correct answers between the pre-test and post-test for the topic of self-protection.

4 RESULTS

Table 3 provides details of the rates of correct responses recorded for the questionnaire given to the participants before and after the completion of the programme.

Table 3 shows that boys and girls are different in terms of their success in achieving the programme's goals. For the five goals targeted in the programme, boys only succeeded to achieve three goals. These goals were to identify/name the body part correctly, to understand how to maintain personal hygiene, and to understand how to protect the body from unwanted touch. On the other hand, girls succeeded in achieving all the five goals. With respect to the previous finding of gender differences regarding raising questions about sexuality, majority of girls raised more specific questions on sexuality compared with boys; thus, these findings add to the dynamics of variability between genders regarding sexuality. The several findings of this study re-emphasise the differences among children in approaching sexuality, where girls also tend to mature earlier than boys (Sukadji, 2000).

Table 3. Participant rate of correct responses in pre-test and post-test.

Programme goals	Boys Pre-test (%)	Boys Post-test (%)	Boys Significance	Girls Pre-test (%)	Girls Post-test (%)	Girls Significance
Identify/name the body parts correctly (including genitals)	3	81	$p < 0.05$*	29	91	$p < 0.05$*
Understand the development of the human body	0	26	$p > 0.05$	0	36	$p < 0.05$*
Understand gender roles	0	9	$p > 0.05$	3	23	$p < 0.05$*
Understand how to maintain personal hygiene	56	77	$p < 0.05$*	89	97	$p < 0.05$*
Understand how to protect the body from unwanted touch	0	69	$p < 0.05$*	57	83	$p < 0.05$*

*Significant at $p = 0.05$.

Table 3 also shows that boys failed to achieve the programme's goals of understanding the development of the human body and gender roles, whereas girls succeeded. From a practical viewpoint, boys tended to give a concrete but incomplete answer in the post-test. For example, they mentioned body parts that transform (e.g. muscle) to answer a question about the definition of puberty. Such answers cannot be counted as correct answers to substitute the expected response because the boys already participated in the programme and received the related learning materials. Further, the number of boys participating in the programme gives another perspective from a practical viewpoint. Boys appeared to be in a class situation that was less conducive for learning as it was too noisy. This fact is somewhat related to the first finding that boys were not very curious about the material or the topic of sexuality included in the sex education programme. Therefore, they appeared to be less attracted to the programme and found it easier to make noise. They needed more enforcement to engage in the programme during the sessions given.

5 CONCLUSION

The UN has given a mandate to each of its members to deliver sex education as a school-based learning programme. Furthermore, the sex education programme in Indonesia is also mandated by law. Nonetheless, Indonesia still has no national curriculum related to sex education, and a formal school-based sex education programme for children is still commonly considered taboo or controversial (IPPF, 2013). On the other hand, in response to the current school climate, the need for sex education is recognised by the grassroots school community as essential. In one elementary school in Jakarta, it was found that students tease each other by touching others' private parts because they have no knowledge about the appropriateness of such conduct. Therefore, a pilot programme of sex education was delivered to fourth grade students at that school.

From the implementation of the pilot sex education programme, a number of interested findings were made. The students' baseline knowledge of sexuality is generally below the requisite level with respect to their age and school grade. It was found that they lacked the knowledge that is necessarily known by first to third graders. Moreover, there was also a gender difference in the students' curiosity about sexually related issues. When they were asked to write down what they wanted to know about sexuality, girls tended to ask more specific questions than boys. The latter tended to ask more irrelevant questions or give irrelevant responses related to sexuality. Therefore, there was a difference between both genders in approaching sexuality.

Both genders also showed a difference in their success at achieving the goals of the sex education programme. From the five goal sessions delivered to the participants, the boys only succeeded in achieving three goals. In contrast, the girls succeeded in achieving all five goals of the programme. This result re-emphasises the difference between the two genders in approaching sexuality. Girls tend to mature early and are more receptive than boys. From a practical point of view, several factors that lead to this difference have been highlighted. One factor is the way in which the boys answered the question on pre-test and post-test. They tended to respond with a concrete but incomplete answer; thus, it cannot be counted as a correct answer. They remembered only a fragment of the materials given without comprehensive understanding. Another factor is that the class was too noisy. The boys appeared to be less attracted to the programme, made more noise easily, and needed more enforcement during the sessions given.

Based on the conclusions of this research, some suggestions for the successful implementation of sex education programmes for elementary school children are as follows:

1. The setting of the class during the programme is one of the crucial factors for the success of the programme. Furthermore, it needs more facilitators in each class to facilitate the learning process whenever the number of students increases.
2. The programme needs to consider further about the content in terms of providing a concrete one. Moreover, the programme should also assess the kind of concrete information from participants.
3. Sex education for boys seems to require a more attractive packaging to make them more enthusiastic about participating in the programme. This is related to the characteristics of boys who show a less curiosity about sexuality than girls.

REFERENCES

Bearinger, L.H., Sieving, R.E., Ferguson, J. & Sharma, V. (2007). Global perspective on the sexual and reproductive health of adolescents: Patterns, preventions, and potentials. *Lancet: Adolescents Health, 2*(369), 1220–1231.

Berman, L. (2009). *Sex ed: How to talk to your kids about sex.* London: Dorling Kindersley.

Canadian Ministry of Education and Training. (2015). *The Ontario curriculum grade 1–8: Health and physical education.* Ontario: Ministry of Education and Training.

Gravetter, F.J. & Wallnau, L.B. (2008). *Statistic for behavioural sciences* (7th Ed.). New York: Thomson & Wadsworth.

Hurlock, E.B. (1991). *Developmental psychology: A life span approach* (5th Ed.). Jakarta: Erlangga.

Indonesian Ministry of Health.(2009). *Behaviour change intervention training module (B-3): Sex, sexuality, and gender*. Jakarta: Indonesian Ministry of Health.

IPPF. (2013, January 29). Indonesia's young people demand comprehensive sexuality education.London, UK: International Planned Parenthood Federation. Retrieved from http://www.ippf.org/news/indonesias-young-people-demand-comprehensive-sexuality-education.

Sukadji, S. (2000). Educational psychology and school psychology. Depok: LPSP3.

UNESCO. (2009). *International technical guidance on sexuality education: An evidence-informed approach for schools, teachers, and health educators*. Paris: UNESCO.

UNFPA. (2012). *Sexuality education: A ten-country review of school curricula in East and Southern Africa*. New York, NY: United Nations Population Fund.

UNFPA. (2014). *UNFPA operational guidance for comprehensive sexuality education: A focus on human right and gender*. New York: United Nations Population Fund.

UNFPA. (2015). *The evaluation of comprehensive sex education programs: A focus on gender and empowerment outcomes.* New York: United Nations Population Fund.

Yosephine, L. (2016, May 26). Sex education must be taught in schools: Child protection commission. *The Jakarta Post*. Retrieved from http://www.thejakartapost.com/news/2016/05/26/sex-education-must-be-taught-in-schools-child-protection-commission.html.

Diversity in Unity: Perspectives from Psychology and Behavioral Sciences – Ariyanto et al. (Eds)
© 2018 Taylor & Francis Group, London, ISBN 978-1-138-62665-2

Training a father to better use prompt and reinforcement: Effects on the initiation of joint attention in a child with pervasive developmental disorder, not otherwise specified

H. Ekapraja, F. Kurniawati & S.Y. Indrasari
Faculty of Psychology, Universitas Indonesia, Depok, Indonesia

ABSTRACT: Initiation of Joint Attention (IJA) has been considered essential in the establishment of human social interaction for children with autism spectrum disorder. Three aspects are involved in such skill, namely, communication, language and social interaction. The goals of this study are, first, to increase the rate of prompting and reinforcement a father delivers to his child with pervasive developmental disorder, not otherwise specified and, second, to determine the effects of increased prompt and reinforcement on the child's IJA. An A–B design study was set up, and the effects of the 23-session programme were measured before and after the intervention. The results reveal that, following the training, the father slightly increased use of prompt and reinforcement, but was unsuccessful in improving the child's IJA. The length of programme, child's perseverative interest in a certain object, child's ability in perceiving eye gaze, and measurement of programme effectiveness were among the factors considered to contribute to the result of the study.

1 INTRODUCTION

The Diagnostic and Statistical Manual of Mental Disorders, Fourth Edition, Text Revision (DSM-IV-TR) (American Psychiatric Association, 2000) explains the symptoms of autistic disorder with qualitative impairments in social interaction and communication, and restricted repetitive and stereotyped patterns of behaviour, interests, and activities. As cited in Mash and Wolfe (2010), autism is a spectrum disorder. This means that 'its symptoms, abilities, and characteristics are expressed in many different combinations and in any degree of severity' (Lord et al., 2000, in Mash & Wolfe, 2010). Under the same category of Pervasive Developmental Disorder (PDD), DSM-IV-TR classifies a disorder that shares the main characteristics of autistic disorder, but does not meet the full criteria, as PDD-Not Otherwise Specified (PDD-NOS). This disorder is referred to as 'a severe and pervasive impairment in the development of reciprocal social interaction associated with impairment in either verbal or non-verbal communication skills or with the presence of stereotyped behavior, interests, and activities, but the criteria are not met for a specific Pervasive Developmental Disorder, Schizophrenia, Schizotypal Personality Disorder, or Avoidant Personality Disorder' (American Psychiatric Association, 2000). As cited in Mash and Wolfe (2010), the primary problems of individuals with Autism Spectrum Disorder (ASD) stem from a deficit in their Theory of Mind (ToM) mechanism, a capability to predict others' intention, goal, and desires behind their behaviours (Johnson & de Haan, 2011). ToM deficit seems to be most specific to children with autism (Mash & Wolfe, 2010).

Researchers have considered that Joint Attention (JA) is the underlying skill of ToM's development (Sebanz et al., 2005, in Ma, 2009). According to Batherton (1991) and Tomasello (1995), as cited in Schietecatte, et al. (2011), JA is closely related to one's social–cognitive development in understanding others' mental condition, such as feelings, thoughts, and intentions. Baron-Cohen et al. (1985, in Ma, 2009) concluded that ToM and JA are theoretically related as they share similar underlying cognitive capacity.

As cited in Vismara and Lyons (2007), JA refers to a child's ability to coordinate attention between themselves and another individual and an object/event, in order to share interest or experience. There are two types of JA: 1) Response to Joint Attention (RJA), in which a child *responds* to another's initiation to share attention towards an event/object, and 2) Initiation of Joint Attention (IJA), in which a child *initiates* sharing of attention towards an event/object with another (Ma, 2009; Schertz & Odom, 2007; Taylor & Hoch, 2008).

Studies have suggested that children with ASD can be trained in JA, and its success may lead to a progress in other aspects, such as language and communication, social behaviour, and positive affect (Koeger et al., 1999, in Jones & Carr, 2004; Jones et al., 2006; Whalen et al., 2006). Meindl and Cannella-Malone's (2011) literature review concludes that most researchers trained RJA and IJA separately, with specific target behaviour and different instructions. According to Whalen and Schreibman (2003), although both forms of JA have similar social function, each skill has a different motivational parameter. As cited in Jones and Carr (2004), Corkum and Moore (1995, 1998) proposed that a child's attempt of RJA is not necessarily socially motivated. The behaviour may be produced as learnt behaviour of 'looking where someone else is looking', and maintained by an external reward (i.e. a toy/object) (Butterworth & Jarret, 1991, as cited in Jones & Carr, 2004). In contrast, a child's IJA is maintained by intrinsic rewards (i.e. social sharing) (Mundy, 1995, in Whalen & Schreibman, 2003). Ma (2009) explains that low JA is an early indicator of autism in a child. This low ability is not noted in the behaviour of eye gaze or pointing, but in the child's lack of interest to interact (Jones & Carr, 2004). These findings suggest that training a child with ASD for IJA is more relevant to the functional aspect of JA, namely to share attention, than as a means to obtain an object.

The purposes of this study are, first, to increase the rate of prompting and reinforcement a father delivers to his child with PDD-NOS and, second, to determine the effects of increased prompt and reinforcement on the child's IJA. To achieve these purposes, the father of the child is trained to implement prompting and reinforcement during allocated playtime with the child.

2 METHODS

This was a single-subject study (Gravetter & Forzano, 2009) with an A–B design (Stocks, 2000). The programme consisted of two time periods, before (A) and after (B) the intervention. There were five sessions during the baseline period, 15 sessions of intervention, and three sessions after the intervention. From the baseline to intervention session 5, the programme was set for 30 minutes per session. Given that the opportunity for the father to prompt the child, and to anticipate the child's boredom, was in the first minutes of each session, the rest of the programme was then conducted for 10 minutes per session.

Participants of this study were a father and son duo. The father acted as the programme implementer. The son was aged 10 years and had been diagnosed with PDD-NOS when he was 2.5 years. At the age of 10 years, the child's autistic characteristics were still obvious: he had limited verbal expression, his articulation was not clear, his communication type was mostly proto-imperative, he occasionally showed echolalia, and he was preoccupied with cigarette packages and Lego.

The variable measured was the child's IJA, defined as his initiation to conduct an activity, by means of non-verbal behaviour with or without verbal behaviour (i.e. comment), in order to gain another's attention. Non-verbal behaviour included 1) coordinated gaze shifting, 2) proto-declarative pointing, 3) showing, and/or 4) giving. Verbal behaviour was defined as making comment: expressing one or more word(s) about the object while making eye contact with the father.

The experiment was conducted in a 2.5 m × 3 m room. At one corner some objects, such as toys (Lego, cars), storybooks, coloured pencils, and some pictures, were placed on a table. The father and the son were then requested to enter the room. When the child showed IJA the moment he saw the objects, the father was required to reinforce this behaviour by expressing comments that were related to the objects. On the contrary, when the child did not show IJA, the father was required to show physical prompts to assist the child to first shift his gaze to him, then to the object, and to point towards the object he touched or stared at.

Event-recording (Sattler, 1988) was used to record the frequencies of three behaviours: 1) child's IJA, 2) father's prompt (either physical or verbal), and 3) father's reinforcement when the child showed IJA, through either a relevant comment or social engagement with the child's activity.

To measure the success of the programme, data analysis was conducted.

1. Child's IJA: Score 1 was given on every correct attempt at IJA followed by the child's gaze shifting towards the father and the object. The child's attempts at IJA that were prompted by the father were not scored.
2. Father's prompts: Score 1 was given on the father's attempt to physically and verbally assist the child to shift his gaze and point.
3. Father's reinforcement: Score 1 was given on the father's attempt to reinforce the child's IJA by expressing a relevant comment or by joining in the child's play activity. The score was given to the father's reinforcement of the child's correct IJA, not of the child's attempt at IJA.

For each measurement, the same procedure of data analysis was conducted. Total frequency and mean were calculated for comparison at different moments of the programme (before and after intervention) to analyse its effectiveness. Intervention to increase the child's IJA was considered effective when there was an increase in the child's attempt at IJA during final assessment. The father's training was effective when he implemented prompting and reinforcement procedures correctly, with a minimum score of 85% for fidelity of implementation (Taylor & Hoch, 2008).

Inter-observer reliability (Kerlinger & Lee, 2000) was measured by calculating the inter-observer agreement (Gravetter & Forzano, 2009; Taylor & Hoch, 2008). A minimum rate of 80% was required (Sulzer & Mayer, 1972, as cited in Sarafino, 2012).

3 RESULTS

During five sessions of the baseline period, the child showed 13 attempts at IJA; four attempts could be scored correct, resulting in a mean score of 0.8. During 15 sessions of intervention, the child's attempt at IJA increased almost two times the frequency during the baseline period. However, over 24 attempts, only four could be scored correct, resulting in a mean score of 0.27. In the three sessions after the intervention, the child showed two correct attempts out of eight attempts, resulting in a mean score of 0.67. Figures 1 shows the mean of the child's IJA.

The materials used in this study were toys. One of them was Lego, which was the child's favourite. Throughout the whole programme, as long as there was Lego, the child's attention was preoccupied. The moment he saw Lego, the child did not attempt IJA. However, once he finished building something with Lego, he showed an attempt at IJA. For the first eight sessions of intervention, Lego was provided. The child attempted IJA after he finished building something, such as a robot, helicopter, or tree. From the ninth session, Lego was removed from the room and was replaced by other constructive materials, such as mini building bricks and puzzles. Unlike Lego, these building bricks and puzzles were pre-determined constructive materials. As a result, during the second half of the intervention period, the child focused on

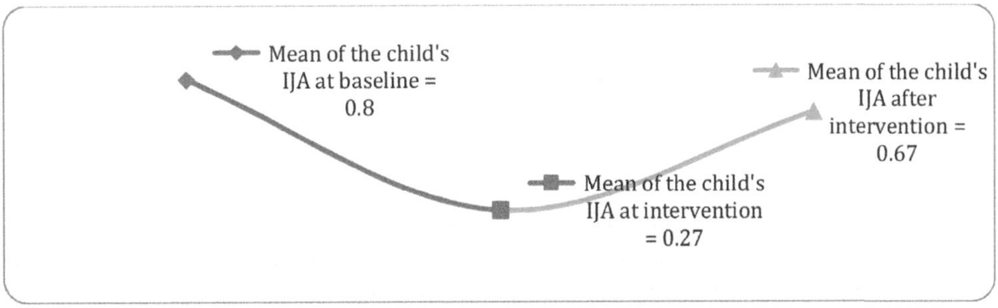

Figure 1. Mean of the child's initiation of joint attention (IJA).

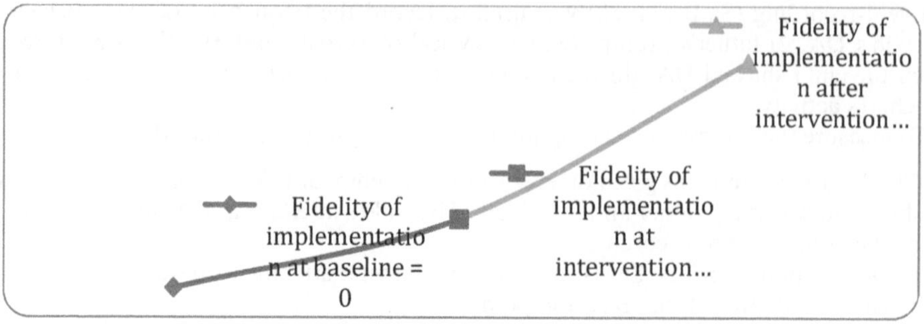

Figure 2. Mean of the father's fidelity of implementation on prompt.

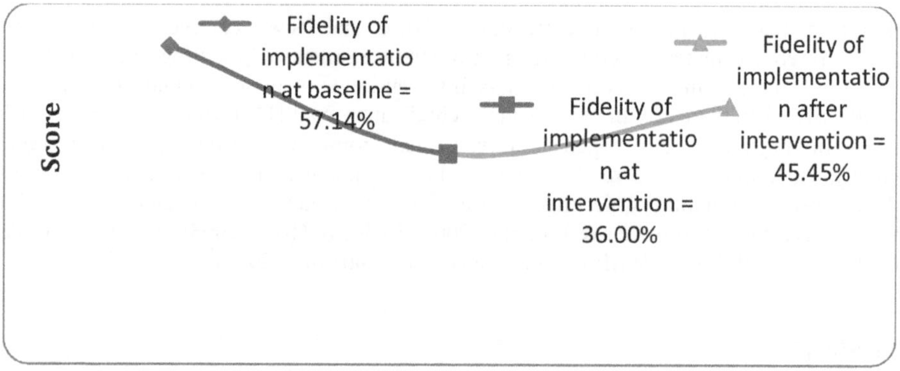

Figure 3. Mean of the father's fidelity of implementation on reinforcement.

constructing the puzzle or mini building bricks according to its pre-determined result/form and showed only one attempt at IJA. This material replacement was believed to be related to the lower rate of the child's IJA during the second half of the intervention programme.

With regard to the father's attempt at prompting the child, results showed that the father did not make any attempt to prompt the child during the baseline period. During and after the intervention, he made several attempts in most of the sessions. However, only a few attempts could be scored correct, resulting in a score of fidelity of implementation below the minimum rate of 85% (shown in Figure 2).

During the baseline period, the father gave 57.15% correct reinforcement for the child's IJA. This rate decreased to 36% during intervention. Although it increased to 45.45% after intervention, the rate was still below baseline and did not meet the minimum requirement rate of 85%. Figure 3 shows mean values of the father's fidelity of implementation on reinforcement.

During the baseline the father gave 57.15% correct reinforcement on child's IJA. This rate 137 decreased to 36% during intervention. Although it increased to the rate of 45.45% during post- 138 intervention period, the rate was still below baseline, and did not meet the minimum require- 139 ment rate of 85%. Figure 3 below shows mean of father's fidelity of implementation on rein- 140 forcement.

4 DISCUSSION

This study aimed to train a father to better use prompt and reinforcement in order to increase a child's IJA. Some factors were considered to contribute to the results. Time-wise, the programme was conducted for 23 sessions from baseline to final assessment after the intervention, with the correct IJA frequency between 0 and 2. In comparison, previous research that proved to be effective was conducted over a longer period of time. Taylor and Hoch's (2008) study was conducted for 90 sessions with different increases in IJA: 100% in two subjects and 40% in one subject, with a frequency between 0 and 6 per session. Therefore, although this study does not show an effective result, it is aligned with that of Taylor and Hoch (2008). To have better results, a longer period of time is required.

Another factor that contributed to the result of this study was parental involvement. In this study, the father of the child played a role as the implementer of the programme through the use of prompt and reinforcement. The score of fidelity of implementation was below 85%, for both prompt and reinforcement. This might lead to the conclusion that the training programme was not effective. However, when referring to the formula, fidelity of implementation score was obtained by calculating the correct prompt/reinforcement attempts divided by the total number of prompt/reinforcement attempts (Taylor & Hoch, 2008). This may lead future studies to consider different perspectives. With regard to the father's attempt to prompt the child, fidelity of implementation limited us to see only the father's correct and incorrect prompting. This formula does not allow us to see the success of the father's prompt in eliciting the child's IJA. A future study may consider recording and analysing occurrence of the child's IJA following the father's correct prompting in order to measure the success of the prompts. In this study, a correct reinforcement attempt was scored for the father's reinforcement only of the child's independent IJA, and not for the child's prompted IJA. For a better description of the father's reinforcement, a future study may also consider analysing the father's correct reinforcement of the child's prompted IJA.

Many studies have proved to be successful when involving parents as agents. Among these were Vismara and Lyons (2007), and Ma (2009). Continued assistance to parents is required, and may be gradually decreased when parents' skills increase. To evaluate parents' skills, fidelity of implementation may be measured regularly.

This study shows that the rate of the child's IJA was low. This low rate did not simply indicate the fewer attempts of the child at IJA. The child's attempts during baseline actually doubled during the intervention. However, these attempts were mostly either pointing, showing, giving, or verbalising, without gaze shifting, whereas in this study we scored only the attempt that was followed by gaze shifting from the object to the father. Taylor and Hoch (2008) found that gaze shifting 'was challenging to teach and remained a fairly inconsistent response'. This is related to an impairment in the brain, specifically in the superior temporal sulcus; because of this, individuals with ASD have difficulties in responding to dynamic social stimulus, such as eye gaze (cited in Johnson & de Haan, 2011), as shown in the study of Speer et al. (2007). The child's effort, although it was not followed by gaze shifting, could be an indication of the child's awareness of the other's presence and his desire to share experience or interest with the other person (in this study, the father).

Considering children with ASD and their perseverative interest on certain objects, the materials used in an intervention need to be selected carefully. In their study, Taylor and Hoch (2008) replaced and relocated the objects in order to preserve the novelty and maintain the child's interest. As proved in the Vismara and Lyon study (2007), objects of the child's perseverative interest may function as a motivating operator that helps to elicit the child's IJA. In this study, Lego was replaced with other constructive toys all at once in the second half of the intervention. This was believed to be the reason for the non-occurrence of the child's IJA.

5 CONCLUSION

This study aimed to train a father to better use prompt and reinforcement in order to increase a child's IJA. Although it does not show the intended result, this programme shows a positive impact. First, the child showed attempts at IJA through the behaviour of pointing, showing, giving, and verbalising. Second, compared with baseline, the father's increased attempt to prompt and reinforce the child's behaviour during and after the intervention suggests that the father made some effort to coordinate attention and be involved in the child's play activity. Consistently implemented, this form of communication may lead to success of the child's IJA, as shown in previous studies (i.e. Ma, 2009; Taylor & Hoch, 2008).

To increase its effectiveness, future research needs to train parents as intervention agents more intensively, both for prompting and for reinforcement. The scoring formula needs to be reconsidered to measure efforts of by parents and positive outcomes. A longer period of intervention with continuous evaluation is required in the design of the programme. Intensive communications with the family are suggested to guarantee the novelty and variety of the materials. This study shows that on his attempts at IJA, the child rarely shifts his gaze towards his father and the object. This suggests the need for future research to investigate whether training gaze shifting prior to intervention leads to better results.

REFERENCES

American Psychiatric Association. (2000). *Diagnostic and statistical manual of mental disorders* (4th. Ed.). Text Revision. doi:10.1016/B978-1-4377-2242-0.00016-X.

Gravetter, F.J. & Forzano, L.B. (2009). *Research methods for the behavioral sciences.* Belmont, CA: Wadsworth.

Johnson, M.H. & de Haan, M. (2011). *Developmental cognitive neuroscience* (3rd. Eds.). Chichester, UK: Wiley-Blackwell.

Jones, E.A. & Carr, E.G. (2004). Joint attention in children with autism: Theory and intervention. *Focus on Autism and Other Developmental Disabilities*, *19*(1), 13–26. Retrieved from https://search.proquest.com/docview/205057507?accountid=17242.

Jones, E.A., Carr, E.G. & Feeley, K.M. (2006). Multiple effects of joint attention intervention for children with autism. *Behavior Modification*, *30*, 782–834. doi:10.1177/0145445506289392.

Kerlinger, F.N. & Lee, H.B. (2000). *Foundations of behavioral research.* Belmont, CA: Wadsworth.

Ma, C.Q. (2009). *Effects of a parent-implemented intervention on initiating joint attention in children with autism* (Doctoral dissertation, Department of Psychology, University of South Carolina). Available from ProQuest Dissertations and Theses database (UMI No. 3367523).

Mash, E.J. & Wolfe, D.A. (2010). *Abnormal child psychology.* Belmont, CA: Wadsworth.

Meindl, J.N. & Cannella-Malone, H.I. (2011). Initiating and responding to joint attention bids in children with autism: A review of the literature. *Research in Developmental Disabilities*, *32*, 1441–1454. doi:10.1016/j.ridd.2011.02.013.

Rocha, M.I., Schreibman, L. & Stahmer, A.C. (2007). Effectiveness of training parents to teach joint attention in children with autism. *Journal of Early Intervention*, *29*(2), 154–172. doi:10.1177/105381510702900207.

Sarafino, E.P. (2012). *Applied behavior analysis: Principles and procedures of modifying behavior.* Hoboken, NJ: John Wiley & Sons.

Sattler, J.M. (1988). *Assessment of children* (3rd ed.). San Diego, CA: Jerome M. Sattler.

Schertz, H.H. & Odom, S.L. (2007). Promoting joint attention in toddlers with autism: A parent-mediated developmental model. *Journal of Autism and Developmental Disorders*, *37*(8), 1562–1575. doi:10.1007/s10803-006-0290-z.

Schietecatte, I., Roeyers, H. & Warreyn, P. (2012). Exploring the nature of joint attention impairments in young children with autism spectrum disorder: Associated social and cognitive skills. *Journal of Autism and Developmental Disorders*, *42*(1), 1–12. doi:10.1007/s10803-011-1209-x.

Speer, L.L, Cook, A.E., McMahon, W.M. & Clark, E. (2007). Face processing in children with autism: Effects of stimulus contents and type. *Autism: The International Journal of Research and Practice*, *11*(3), 265–277. doi:10.1177/1362361307076925.

Stocks, J.T. (2000). Introduction to single subject designs. Retrieved from https://www.msu.edu/user/sw/ssd/issd01.htm.

Taylor, B.A., & Hoch, H. (2008). Teaching children with autism to respond to and initiate bids for joint attention. *Journal of Applied Behavior Analysis*, *41*(3), 377–391.

Vismara, L.A. & Lyons, G. (2007). Using perseverative interests to elicit joint attention behaviors in young children with autism: Theoretical and clinical implications for understanding motivation. *Journal of Positive Behavior Interventions*, *9*(4), 214–228.

Whalen, C., Schreibman, L. & Ingersoll, B. (2006). The collateral effects of joint attention training on social initiations, positive affect, imitation, and spontaneous speech for young children with autism. *Journal of Autism and Developmental Disorders*, *36*(5), 655–664. doi:10.1007/s10803-006-0108-z.

Diversity in Unity: Perspectives from Psychology and Behavioral Sciences – Ariyanto et al. (Eds)
© 2018 Taylor & Francis Group, London, ISBN 978-1-138-62665-2

Self-directed learning as a mediator of the relationship between contextual support and career decision self-efficacy

P.L. Suharso, F.M. Mangunsong & L.R.M. Royanto
Faculty of Psychology, Universitas Indonesia, Depok, Indonesia

ABSTRACT: The difficulty of career planning can become a serious problem if a person experiences career uncertainty or career indecision, resulting in their inability to survive, to persist, and to maintain stability in their study and their work. This study aims to examine the role of self-directed learning as a mediator in the relationship between contextual support and career decision self-efficacy. This quantitative research used 496 4th semester Universitas Indonesia students as participants. The participants completed the Contextual Support Self-report, Career Decision Self-efficacy Short Form (CDSE-SF), and Student Self-directed Learning Questionnaire (SSDLQ). The result of this study shows self-directed learning as a partial mediator in the relationship between contextual support and career decision self-efficacy. The implications of this research is that factors in the higher education environment, especially those pertaining to the role of lecturers, must be optimized to increase students' self-directed learning to improve their career decision self-efficacy.

1 INTRODUCTION

In the era of globalization and the ASEAN Economic Community (AEC), undergraduate students need to be ready to compete, because the population of workers in Indonesia will soon be dominated by workers from ASEAN countries. From the needs analysis of Universitas Indonesia (UI) Career and Scholarship EXPO XIX held in the year 2015, which was attended by students from various universities, it was found that almost 70% of the attendees were UI students who experienced difficulties in career planning and who therefore required the services of career counseling and workshops/seminars on career preparation (Need Assessment Data UI Career and Scholarship EXPO XIX by the Faculty of Psychology UI with CDC UI, April 10, 2015).

The difficulty of career planning can become a serious problem, as the issue highlights the effects of career uncertainty or career indecision in a person's career development. This situation will further threaten one's persistence and stability in pursuing their chosen field of study (Restubog, Florentino, & Gracia, 2010), eventually resulting in the person's lack of career decision self-efficacy (CDSE). CDSE is defined as an individual's belief that he or she can successfully complete the tasks necessary in his or her career decision making process (Hackett and Betz 1981; Whiston & Keller, 2004). It follows that if someone is confident with the field of study he or she has chosen, and then the person will pursue this choice and strive to complete his or her education in the relevant area of interest. The person's success in the chosen field of study will in turn result in the person's persistence to adhere to his or her own decision (Wright, Jenkins-Guarnieri, & Murdock, 2012).

In recent years, career decision self-efficacy (CDSE) has become a popular issue of research in career development, due to its significant impact on the career decision making process of college students. Research on CDSE has been linked to several areas of the career development process, such as career planning and exploration (Gushue *et al.* 2006; Rogers, Creed, & Glendon, 2008) and career choice commitment (Jin, Watkins, & Yuen, 2009).

However, most studies on CDSE have been conducted and addressed within the context of career development literature (Hui-Hsien & Jie-Tsuen, 2014), while little has been done to examine the causes of CDSE in terms of the social and individual contexts (e.g., the context of learning in universities).

Using the social cognitive career theory (SCCT) proposed by Lent, Brown, and Hackett (2000), the current research chose to focus on the role of contextual support and self-directed learning in CDSE. Although many researchers have investigated the importance of contextual support in career development, it is believed that social support may be associated indirectly with career decision self-efficacy, especially considering that career development is not only determined by external factors but also by individual factors that can be developed through learning situations.

1.1 *Contextual support and CDSE*

According to SCCT, career development is influenced by objective and perceived environmental factors (Lent, Brown & Hackett, 2000). SCCT proposes that environmental factors as contextual affordances may affect the development of an individual's self-efficacy beliefs through learning experiences. Environmental factors are further associated with such factors as family or parents, relationships with lecturers and peers, and a variety of other factors that are commonly found in the environment and that can form a person's perception, such as socioeconomic status and social discrimination.

Among the factors that contribute to career decision self-efficacy is contextual support, which is defined as the condition of a person or the person's environment that supports the development of his or her sources. Several studies have indicated that career decision self-efficacy is influenced by contextual support factors, including social support from parents (Rogers, Creed, & Glendon, 2008; Guay *et al.,* 2003; Garriott, Flores, & Martens, 2013) and peers (Kracke, 2002; Kiuru *et al.* 2011; Nawaz & Gilani, 2011). Singaravelu, White, and Bringaze (2005) mentioned that students' confidence and persistence in their chosen majors not only depend on their interaction with peers, but also on their relationship with the faculty and staff, as well as on their participation in campus activities that include their commitment in completing academic assignments that would affect the students' existence, belief, and persistence in their chosen majors (Beal & Noel, 1980; Gim, 1992). Based on such considerations, the present study hypothesized that contextual support would be directly and positively associated with career decision self-efficacy.

Hypothesis 1: Contextual support is directly and positively associated with career decision self-efficacy.

1.2 *Contextual support and Self-directed learning*

The commitment in academic assignment that shows individual (personal) responsibility is an important aspect in self-directed learning (Brockett & Heimstra, 1991), because self-directed learning is an activity that involves the individual's responsibility for planning, implementing, and evaluating their learning effort (Brockett & Heimstra, 1991). The concept of self-directed learning becomes important in college, particularly since such active learning approaches as collaborative learning and problem-based learning are frequently used as a common learning system in university-level courses (Prabjandee & Inthachot, 2013). Self-directed learning is required when prior knowledge of the topic discussed in the assignment is limited, such that students are expected to independently find and formulate the issues associated with the assignment's topic (Loyens, Magda, & Rikers, 2008).

Past research has demonstrated that self-directed learning can be influenced by environmental conditions and personality traits (Raemdonck, 2006). Environmental conditions are comprised of the characteristics of the experience gained from the teaching-learning environment. Environmental conditions that affect self-directed learning are those conditions of the environment that provide opportunities for individuals to learn and develop their career

(Raemdonck, 2006). Based on previous findings, the current research hypothesized that contextual support is related to self-directed learning.

Hypothesis 2: Contextual support is positively associated with self-directed learning.

With regard to career development, Bruin and Cornelius (2011) suggested that self-directed students would also manage to self-direct in their career development, as they tend to be better prepared to make decisions in their career. Such a suggestion is reasonable, considering that people who take a greater role in their career development also need to develop self-directed learning abilities to determine the direction of and changes in their career (Bruin & Cornelius, 2011). Thus, Bruin and Cornelius's (2011) study suggested a relationship between self-directed learning and career decision self-efficacy, whereby an individual's posession of self-directed learning ability would naturally affect his or her career.

Hypothesis 3: Self-directed learning is positively associated with career decision self-efficacy.

In the current research, it is assumed that if contextual support and self-directed learning are shown to be associated with career decision self-efficacy, while self-directed learning is influenced by environmental factors, then there should be an indirect relationship between contextual support and career decision self-efficacy that is mediated by self-directed learning. Based on this assumption, the researchers of the current study proposed the following model that was examined in this research.

The aim of the current study was to examine the role of self-directed learning in career decision self-efficacy and to demonstrate whether the obtained data fit into the proposed theoretical model of the relationships among contextual support, self-directed learning, and career decision self-efficacy. The study involved 4th semester undergraduate students enrolled at Universitas Indonesia (UI). University students were chosen based on the consideration that the majority of people who pursue university education are preparing for a career (Wright, 1982, in Singgih & Sukadji, 2006). Undergraduate students are typically aged 18–25 years and have the following primary tasks for development: (a) to explore themselves and the world of work, (b) to enforce their interests and establish their vocational identity, (c) to develop their educational and work aspirations, (d) to engage in career planning, and (e) to specify their initial job of choice (Lent and Brown 2013). An additional reason for choosing UI undergraduate students as research participants was the fact that all UI students are familiar with student-centered learning approaches, as they all have been exposed to collaborative learning (CL) and problem-based learning (PBL) strategies, both of which encourage students to learn actively, to work in teams as well as independently, to have good communication skills, to think holistically, and to care for the environment (Pedoman Penjaminan Mutu Akademik UI, 2007). Through the independent learning promoted by student-centered learning strategies, UI students are expected to be ready to apply self-direction in their development.

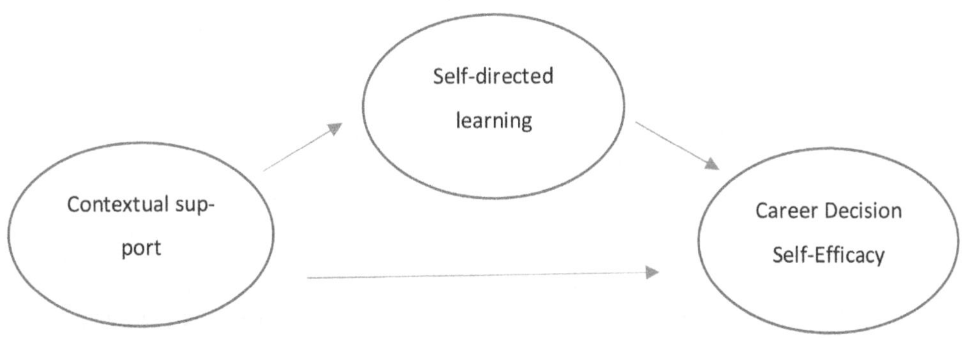

Figure 1. The theoretical model of variables.
Notes: 1 = contextual support, 2 = self-directed learning, 3 = career decision self-efficacy.

2 METHOD

2.1 *Participants*

The participants consisted of 496 4th semester undergraduate students from 13 faculties in Universitas Indonesia. Participants consisted of students majoring in the clusters of Social and Humanities (37.7%), Health Sciences (38.5%), and Science and Technology (23.8%). Their ages ranged from 18 to 24 years, although most were aged between 18 and 20 years (Mean = 19.64).

2.2 *Procedure*

The current study consisted of a preliminary study and a main (primary) study. The preliminary study began with focus group discussions involving the participation of ten students, with the purpose of exploring possible factors that influence career decision self-efficacy and followed by reconstitution of the instruments. Results from the preliminary study were eventually used as considerations for strengthening the research variables previously constructed based on literature review, as well as for preparing the instruments. On the other hand, the main (primary) study was a quantitative study intended to test the structural equation model of career decision self-efficacy.

2.3 *Instruments*

2.3.1 *Career Decision Self-efficacy Short Form (CDSE-SF)*

Consisting of 25 items, the CDSE-SF was used to assess undergraduate students' self-efficacy in making career-related decisions (Betz & Taylor, 2006). The CDSE-SF contained five items for each of the five subscales: Self-Appraisal, Gathering Occupational Information, Goal Selection, Planning, and Problem Solving. To avoid a neutral or midpoint response from the participants, response choices for the CDSE-SF that originally used a 5-point scale were adapted into 6-point scale response choices ranging from 1 (no confidence at all) to 6 (complete confidence). Higher scores indicated a higher degree of CDSE. Before the instrument was used in the main (primary) study, the instrument was tested on 52 UI undergraduate students that matched the characteristics of the main study's sample. The Cronbach alpha for the total of 25 items of CDSE-SF was 0.815.

2.3.2 *Contextual support self-report*

A modified version of the Contextual Supports and Barriers form originally developed by Lent *et al.* (2001) was used to examine perceived environmental support. The instrument contained 15 items regarding contextual support provided by parents (family) and institutions (lecturers and peers), as well as financial support (social economics) and non-discriminative support. To avoid a neutral or midpoint answer from the participants, the Contextual Support Self-report's original 5-point Likert scale responses were adapted into 6-point Likert scale response choices ranging from 1 (not at all likely) to 6 (extremely likely). Higher scores on the instrument reflected stronger contextual support. Coefficient alpha value for the Contextual Support scale was 0.78.

2.3.3 *Student Self-Directed Learning Questionnaire (SSDLQ)*

The SSDLQ was developed by Bruin (2008, in Bruin & Cornelius, 2011) and contained 22 unidimensional scale items that indicated self-directed learning behavior. Researchers of the current study modified the SSDLQ, which amounted to a final of 13 items out of 22, because some items were found to have the same meaning as other items. The objective of modifying the items was to equalize the self-directed learning items pertaining to planning, implementation, and evaluation. To avoid a neutral or midpoint response from the participants, the SSDLQ's items that originally consisted of 5 Likert-like response choices were adapted into 6-point choices ranging from 1 (strongly disagree) to 6 (very strongly agree). Higher scores were interpreted to indicate a higher degree of self-directed learning. The Cronbach alpha for the 13 final items of the modified SSDLQ was 0.69.

2.3.4 *Translation of instrument*

As the measurement scales were originally written in English, the instruments had to first be translated into Indonesian. A back-translation procedure was used to ensure the accuracy and semantic similarity between the English and Indonesian versions of the instruments. The back-translation process involved having one expert initially translate the English versions of the instruments into Indonesian, before then asking a second independent expert to translate the Indonesian versions back into English. The original and the back-translated English versions were then compared and contrasted to identify any discrepancies between them. The discrepancies were discussed among the researchers and the two experts until a consensus concerning the accuracy and semantic similarity between the original versions and the translations was reached.

2.4 *Data analysis*

The data analysis was followed by tests of the reliability and validity of the measuring instruments. After all the instruments had demonstrated acceptable reliability coefficients, a linear structural relations (LISREL) analysis was conducted to test the measurement model and structural equation model. This method of structural equation modeling can help explain the causal relationship proposed in a theoretical model of research.

3 RESULT

Table 1 presents the descriptive statistics and correlations among the study variables. As can be seen, the results indicate positive and significant relationships among the variables of interest.

Before getting into results from the structural equation modelling of career decision self-efficacy, it is necessary to first look at the analyses of the measurement model. The factor loadings for the indicator constructs of contextual support and career decision-making were found to be greater than 0.60. Only contextual support from parents, financial support, and problem solving competencies had factor loadings less than 0.60. This means that contextual support from institution (lecturers and peers) and non-discriminative environment both made large contributions to contextual support.

Figure 2 depicts the results of the structural equation modeling of career decision self-efficacy. The results show that contextual support was positively associated with career decision self-efficacy ($\beta = 0.29^*$, $p \leq 0.05$), implying a direct effect between contextual support and career decision self-efficacy. The result also shows that contextual support was positively associated with self-directed learning ($\beta = 0.44^*$, $p \leq 0.05$), while self-directed learning was positively associated with career decision self-efficacy ($\beta = 0.49^*$, $p \leq 0.05$). In other words, self-directed learning mediated (as a partial mediator) the relationship between contextual support and career decision self-efficacy.

As seen from the contributions made toward career decision self-efficacy, even though contextual support and self-directed learning were both found to contribute to career decision

Table 1. Means, standard deviations, correlations among the measured variables.

Variable	Mean	SD	1	2	3
CDSE	4.47	0.43	–		
SDL	4.41	0.52	0.62**	–	
CSupp	4.08	0.61	0.51**	0.44**	–

Note: N = 496. CDSE: career decision self-efficacy, SDL: self-directed learning, CSupp: Contextual support.
** Significant p < 0.01.

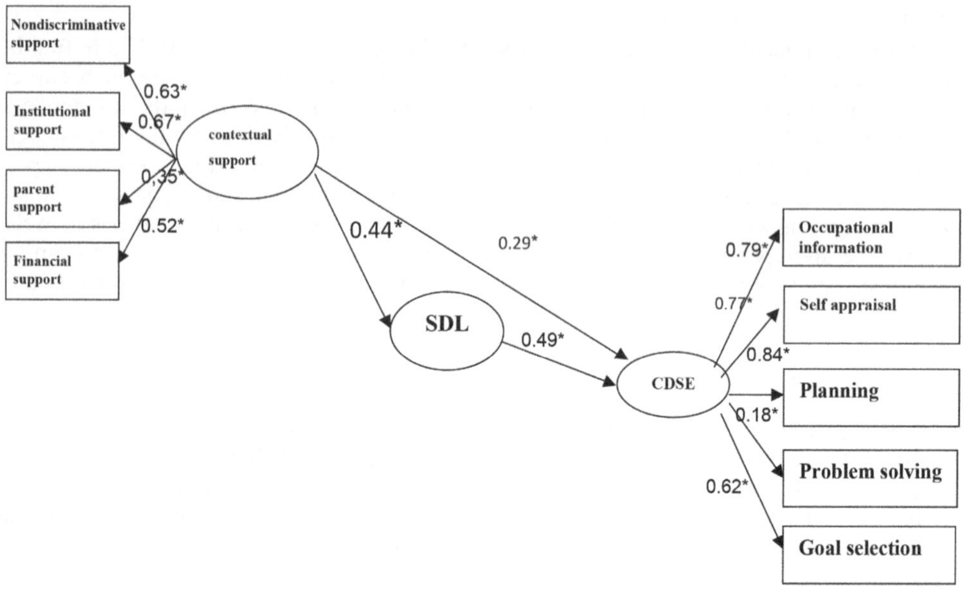

Figure 2. Structural equation model of career decision self-efficacy.
Note: CDSE: career decision self-efficacy, SDL: self-directed learning, Chi Square = 39.30, df = 29, p Value = 0.09599, RMSEA = 0.027.

self-efficacy ($R^2 = 0.457$), contextual support provided greater contribution ($R^2 = 0.262$) than self-directed learning ($R^2 = 0.195$).

4 DISCUSSION

There is direct relationship between contextual support and career decision self-efficacy. The undergraduate students are more efficacious in their major choice decision if they are given contextual support by either from parents, peers, lecturers, also financial support and a non-discriminative environment. This means that the more support they get, the more convinced the students will be of their major choice. Moreover, contextual support can be indirectly related to the career decision self-efficacy through self-directed learning. The higher support from parents, peers, lecturers will develop students' responsibility to make planning, implementation and evaluation related to learning and especially their career choices. This will further develop their efficacy in their major choice.

Self-directed learning becomes an important concept to be examined by researchers studying issues in the higher education context, especially since student-centered learning (i.e., active learning) has become a customary learning approach in higher education (Bruin & Cornelius, 2011; Prabjandee & Inthachot, 2013) as it requires students to be responsible for their own learning process. The current research has provided evidence to support the notion that self-directed learning acts as a mediator in the structural equation modeling of career decision self-efficacy of undergraduate students. However, the participation of self-directed learning is limited to the role of a partial mediator in the relationship between contextual support and career decision self-efficacy, as contextual support was additionally found to be directly associated with career decision self-efficacy. It has also been suggested that although students may be self-directed in learning, they still need contextual support for their career decision self-efficacy because the instability that characterizes undergraduate students (who belong to the emerging adulthood age group classification) still requires them to seek the support and guidance of an older person (Arnett, 2000). If we traced from the contextual support instruments, based on the statement suggests more connected to career. That is why contextual support contributed greater than self-directed learning.

Beginning from their first semester, all UI undergraduate students are introduced to various student-centered learning approaches that include collaborative learning, problem-based learning, and experiential learning through the Integrated Personal Development Courses (*Mata Kuliah Pengembangan Kepribadian Terintegrasi*) that all first-year students are required to take. Evaluations of the courses' results have revealed that following exposure to the student-centered learning approaches, students come to develop a positive emotion, an optimistic attitude, and a focused regulation of their learning. Furthermore, students also develop the courage to ask questions or express their opinions because the lecturers encourage students to be active in their learning processes. Students additionally demonstrate initiative and enthusiasm whenever their lecturers stimulate them with the introduction of a particular topic (Singgih, Diponegoro, Takwin, & Prasetyawati, 2014). As such, it follows that a lecturer's role is still very much needed to provide stimulation and support for students. This argument is in line with Primana's (2015) finding that even when students are provided with the opportunity to study independently and learn from their social interactions with other students, in reality students still need their lecturers to be present as a source of information and source of support who can help clarify the meaning of what the students have learned. In other words, a lecturer's task is not limited to stimulating students to be active learners, but also demonstrating the initiative to embody a contextual support role that enhances self-directed learning (Raemdonck, 2006).

Moreover, the quality of college life can affect students vocationally. How students perceive their college environment, the curriculum, and achievements during the time they pursue their education will affect their efficacy beliefs. For instance, students who perceive their lecturers as supportive demonstrate higher career decision self-efficacy (Gushue & Whiston, 2006). Results from the current study indicate that for UI students, lecturers are expected to provide contextual support by offering students advice on courses prior to the start of each semester, supporting students when they have a problem, providing explanations that are easy to understand, and providing assignment feedbacks. Such a finding is consistent with Gloria and Ho's (2003) claim that a strong relationship between the college environment and students' persistence in their chosen major is characterized by the guidance or mentoring from faculty members. Peers were additionally found to be an important stakeholder in an individual's career decision self-efficacy (Nawaz & Gilani, 2011) because peers can influence the emotional and social development of the individual, ultimately affecting their career decision self-efficacy (Guay *et al.*, 2003). Peers who can work together in a group, are committed to the group task, and are non-discriminative, are some of the characteristics of peers who are perceived by students as those who are able to offer contextual support.

Student-centered learning allows students to develop self-directed learning, because when prior knowledge of the problem to be solved is limited, then students must independently find and formulate the issues associated with the problem. In student-centered learning, personal responsibility, in addition to intrinsic motivation to learn and find information, is needed to enrich one's knowledge (Loyens, Magda, & Rikers, 2008). Thus, students with better self-directed learning abilities tend to be more successful both academically and in the workplace than individuals who do not have self-directed learning skills (Thompson & Wulffi, 2004; Gerber *et al.*, 2005 in Bruin & Cornelius, 2011). Therefore, by introducing student-centered learning early in their academic career, students are expected to be able to engage in self-directed learning and eventually display the same level of independence in directing their own career development.

It is important to note, however, that this study is not without its limitations. The first limitation concerns the limited recruitment of participants who were in their 4th semester of undergraduate studies in Universitas Indonesia, despite the fact that career problems are also experienced by other undergraduate students who are at different stages of their academic careers. Therefore, the results of this study may not necessarily be applicable to other students in other stages of their undergraduate studies. The second limitation, according to the competencies on career decision self-efficacy, there is small contribution from problem solving competencies on career decision self-efficacy. The researchers believe that the instrument used to measure problem solving competencies need to be revised for better contributions to career decision self-efficacy because there are much needed problem-solving competence in the career decision self-efficacy process. So in the further studies, the items in the problem solving competencies should be revised.

5 CONCLUSION

From the current research, it can be concluded that while self-directed learning can enhance career decision self-efficacy, the effect would be amplified by the provision of contextual support in academic tasks by lecturers and peers. Therefore, the results of the present study imply that in higher education environments, the role of lecturers must especially be optimized to increase students' self-directed learning with the ultimate goal of improving students' career decision self-efficacy.

REFERENCES

Arnett, J.J. (2000). Emerging adulthood: A theory of development from the late teens through the twenties. *American Psychologist,* 55 (5), 469–80. doi:10.1037//0003-066X.55.5.469.

Betz, N.E. & Taylor, K.M. (2006). *Manual Of The Career Decision Self-Efficacy Scale And CDSE-Short Form.* Unpublished Material. http://www.mindgarden.com/79-career-decision-self-efficacy-scale.

Brockett, R.G. & Hiemstra, R. (1991). A Conceptual Framework for Understanding Self-Direction in Adult Learning' in Self-Direction in Adult Learning: Perspectives on Theory, Research, and Practice. London and New York, Routledge. http://www.infed.org/archives/e-texts/hiemstra_self_direction.htm.

Bruin, K.D. & Cornelius, E. (2011). Self-directed learning and career decision-making. *Acta Academia 2011,* 43 (2): 214–35.

CDC UI and Faculty of Psychology UI. (2015). *Need Assessment UI Career and Scholarship EXPO XIX.* Depok, Universitas Indonesia.

Garriott, P.O., Flores, L.Y. & Martens, M.P. (2013). Predicting the math/science career goals of low-income prospective first-generation college students. *Journal of Counseling Psychology,* 60 (2), 206–209. DOI: 10.1037/a0032074.

Gloria, A.M. & Ho, T.A. (2003). Environmental, Social, and psychological experiences of asian american undergraduates: examining issues of academic persistence. *Journal of Counseling and Development : JCD,* 81 (1), 93–105. doi:10.1002/j.1556-6678.2003.tb00230.x.

Guay, F., Senecal, C., Gauthier, L. & Fernet, C. (2003). Predicting career indecision: A self-determination theory perspective. *J Couns Psychol,* 50 (2), 165–77. doi:10.1037/0022-0167.50.2.165.

Gushue, G.V. & Whitson, M.L. (2006). The relationship among support, ethnic identity, career decision self-efficacy, and outcome expectations in African American High school students: Applying social cognitive career theory. *Journal of Career Development,* 33 (2), 112–24. doi:10.1177/0894845306293416.

Gushue, G.V., Clarke, C.P., Pantzer, K.M., & Scanlan, K.R.L. (2006). Self-Efficacy, perceptions of barriers, vocational identity, and the career exploration behavior of latino/a high school students. *The Career Development Quarterly,* 54 (4), 307–17. doi:10.1002/j.2161-0045.2006.tb00196.x.

Hackett, G. & Betz, N.E. (1981). A self-efficacy approach to the career development of women. *Journal of Vocational Behavior,* 18 (3), 326–39. doi:10.1016/0001-8791(81)90019-1.

Hsieh, H.H. & Huang, J.T. (2014). The effects of socioeconomic status and proactive personality on career decision self-efficacy. *The Career Development Quarterly,* 62 (1), 29–43. doi:10.1002/j.2161-0045.2014.00068.x.

Jin, L., Watkins, D. & Yuen, M. (2009). Personality, career decision self-efficacy and commitment to the career choices process among chinese graduate students. *Journal of Vocational Behavior,* 74 (1), 47–52. doi:10.1016/j.jvb.2008.10.002.

Kiuru, N., Salmela-Aro, K., Nurmi, J.E., Zettergren, P., Andersson, H. & Bergman, L. (2012). Best friends in adolescence show similar educational careers in early adulthood. *Journal of Applied Developmental Psychology,* 33 (2), 102–11. doi:10.1016/j.appdev.2011.12.001.

Kracke, B. (2002). The role of personality, parents and peers in adolescents career exploration. *Journal of Adolescence*, 25, 19–30. DOI: 10.1006/jado.2001.0446. (Kracke 2002) Davis, Alanah, and Deepak Khazanchi. (2008) An empirical study of online word of mouth as a predictor for multi-product category e-commerce sales. *Electronic Markets,* 18 (2), 130–41. doi:10.1080/10196780802044776.

Lent, R.W., Brown, S.D., Brenner, B., Chopra, S.B., Davis, T., Talleyrand, R. & Suthakaran, V. (2001). The role of contextual supports and barriers in the choices of math/science educational options: A test of social cognitive hypotheses. *Journal of Counselling Psychology,* 48, 474–483. doi:10.1037/0022-0167.48.4.474.

Lent, R., Brown, S. & Hackett, G. (2000). Contextual supports and barriers to career choice: A social cognitive analysis. *Journal of Counseling Psychology,* 47 (1), 36–49. doi:http://dx.doi.org/10.1037/0022-0167.47.1.36.

Lent, R.W. & Brown, S.D. (2013). Social cognitive model of career self-management toward a unifying view of adaptive career behavior across the life span. *Journal of Counseling Psychology,* 60 (4), 557–568. doi: 10.1037/a0033446.

Loyens, S.M.M., Magda, J. & Rikers, R.M.J.P. (2008). Self-directed learning in problem-based learning and its relationships with self-regulated learning. *Educational Psychology Review,* 20 (4), 411–27. doi:10.1007/s10648-008-9082-7.

Nawaz, S. & Gilani, N. (2011). Relationship of parental and peer attachment bonds with career decision-making self-efficacy among adolescents and post- adolescents. *Journal of Behavioural Sciences,* 21 (1).

Prabjandee, D. & Inthachot, M. (2013). Self-directed learning readiness of college students in Thailand. *Journal of Educational Research and Innovation,* 2 (1), 1–11.

Primana, L. (2015). Pengaruh Dukungan Makna Belajar dari Dosen, Motivasi Intrinsik, Self-Efficacy, dan Pandangan Otoritas Sumber Informasi Terhadap Keterlibatan Belajar Mahasiswa Universitas Indonesia. [PhD Diss.], Universitas Indonesia.

Raemdonck, I. (2006). Self-Directedness in learning and career processes. a study in lower-qualified employees in Flanders. University of Gent. [Online] Available from: http://users.ugent.be/~mvalcke/CV/raemdonck%20definitief.pdf.

Restubog, S.L.D., Florentino, A.R. & Garcia, P.R.J.M. (2010). The mediating roles of career self-efficacy and career decidedness in the relationship between contextual support and persistence. *Journal of Vocational Behavior,* 77 (2), 186–195. DOI: 10.1016/j.jvb.2010.06.005.

Rogers, M.E., Creed, P.A. & Glendon, A.I. (2008). The role of personality in adolescent career planning and exploration: A social cognitive perspective. *Journal of Vocational Behavior,* 73 (1), 132–42. doi:10.1016/j.jvb.2008.02.002.

Singaravelu, H.D, White, L.J. & Bringaze, T.B. (2005). Factors Influencing international students' career choice. a comparative study. *Journal of Career Development,* 32 (1), 46–59. doi:http://dx.doi.org/10.1177/0894845305277043.

Singgih, E.E., Diponegoro, M., Takwin, B. & Prasetyawati, W. (2013). Laporan Evaluasi Program Pengembangan Kepribadian Pendidikan Tinggi (PPKT) Tahap 1. Depok.

Singgih, E.E., & Sukadji, S. (2006). *Sukses Belajar di Perguruan Tinggi.* Yogyakarta, Panduan

Whiston, S.C. & Keller, B.K. (2004). The influences of the family of origin on career development: A review and analysis. *The Counseling Psychologist,* 32 (4), 493–568. doi:10.1177/0011000004265660.

Wright, S.L., Jenkins-Guarnieri, M.A. & Murdock, J.L. (2012). Career development among first-year college students: College self-efficacy, student persistence, and academic success. *Journal of Career Development,* 40 (4), 292–310. doi:10.1177/0894845312455509.

Diversity in Unity: Perspectives from Psychology and Behavioral Sciences – Ariyanto et al. (Eds)
© 2018 Taylor & Francis Group, London, ISBN 978-1-138-62665-2

The correlation between shame and moral identity among undergraduate students

H.R. Kautsar, E. Septiana & R.M.A. Salim
Faculty of Psychology, Universitas Indonesia, Depok, Indonesia

ABSTRACT: Individual moral behaviour is motivated by shame and moral identity. As a part of self-identity, moral identity is assumed to be influenced by shame, but there is no research proving the assumption. This study aims to evaluate the correlation between shame and moral identity using a quantitative approach. The participants of the study were 520 undergraduate students, 379 females and 141 males, from several regions in Indonesia, mainly Jakarta, West Java, East Java, Central Java, Bali, Aceh and Banten. Pearson product moment was used to test correlation. The result showed a small but significant correlation between shame and moral identity ($r = 0.149$; $p < 0.01$). The result of the study encourages the use of shame-based moral education in schools and other education institutions.

1 INTRODUCTION

According to Indonesian Law No. 20 of 2003, the aim of the educational system in Indonesia is to shape learners into noble and responsible human beings. In reality, however, there are many learners' behaviours that are not in line with the terms of nobility and responsibility, such as academic dishonesty and bullying.

Moral issues have become the main focus in the discussion of such behaviours. Goles et al. (2006) stated that morality is closely related to individual decision making in terms of whether an individual will or will not perform the behaviour. According to Nather (2013), education contributes to an individual's moral reasoning. Further, an individual's moral reasoning affects their decision making in producing moral behaviour.

Undergraduate students as higher education learners have been exposed to morals from the community and education system over long periods of time. The age range of undergraduate students (around 18–22 years) is considered an advanced stage of moral reasoning (Papalia, Olds & Feldman, 2012). Based on Nather (2013), undergraduate students have higher scores in moral reasoning compared with non-students and other individuals in lower grades of education.

Although individuals, especially undergraduate students, have advanced moral reasoning, their behaviour is not always consistent with morals. For example, although a student knows that cheating on an examination is morally wrong, on certain occasions s/he might still do so. Based on that, a question arises about an individual's morality if s/he still performs morally inappropriate behaviour.

According to Blasi (1983, in Hardy & Carlo, 2005) every individual has his/her own belief in the importance of morals for him/herself. This belief is called moral identity; that is, the belief of how important it is to act according to one's morals and how important these morals are for oneself that they become a characteristic of oneself (Blasi, 1983, in Hardy & Carlo, 2005). Moral identity motivates an individual's moral behaviour (Hardy & Carlo 2005). An individual having a high score in moral identity regards morals as an important part of the self, so that is followed by a consistent appearance of moral behaviour (Aquino & Reed 2002; Blasi, 1983, in Hardy & Carlo, 2005).

Moral identity is a part of self-identity (Hardy, 2006). Shame, which is associated with collectivist cultures such as that of Indonesia, affects the development of self-identity (Czub, 2013; Su, 2011). The role of shame in affecting self-identity is assumed to be very strong among individuals in a collectivist culture. Shame renders aversive feedback to individuals when performing behaviour that does not conform to morals. Then, the individual values the importance of morals and regards morals as an important part of self-identity.

Tangney et al. (2007) claim shame to be a trait variable; every individual has his/her own predisposition or tendency to feel shame compared with other emotions. Shame is described as an individual's tendency to experience negative emotions when s/he fails to behave according to social norms and morals. This leads the individual to focus on how people evaluate him/her and the behaviour performed. In turn, this leads the individual to globally evaluate him/herself, which then generates denying, hiding, and/or avoiding behaviours (Tangney, 1999, in Barlian, 2013; Tangney et al., 2007).

The role of shame is important to discuss since it is associated with collectivist cultures like that of Indonesia (Su, 2011), and affects self-identity (Czub, 2013), including moral identity as a part of self-identity. Although Czub (2013) concludes that shame affects self-identity, there is still no specific research discussing the correlation of shame and moral identity. Then, the question arises of whether there is any correlation between shame and moral identity.

Shame and moral identity are assumed to be related based on Czub's (2013) statement that shame affects self-identity. Moral identity, as part of self-identity, is also assumed to be affected by shame. Another assumption that underlies the correlation of shame and moral identity is that an individual with a high score in shame and an individual with a high score in moral identity have the same consistency in producing moral behaviours. It is because shame and moral identity have the same role in motivating moral behaviour (Hardy & Carlo, 2005; Hardy, 2006). Based on the assumptions above, it can be presumed that individuals with high scores in shame also possess high scores in moral identity.

The undergraduate student population is important as the population of the study because undergraduate students are assumed to have more awareness of morals than non-students or individuals with lower grades of education. The assumption is based on the fact that undergraduate students have been exposed to morals in the community and education system for long periods of time. Additionally, Hardy and Carlo (2005) conclude that Blasi's view on the role of identity as the motivator of moral behaviour can only be observed after an individual has exited the adolescent age range (i.e. more than 18 years), which is equivalent to the undergraduate students' age range. Thus, evaluation of the correlation between shame and moral identity especially among undergraduate students is important.

From the explanation above, it can be hypothesised that there is significant correlation between shame and moral identity among undergraduate students in Indonesia.

1.1 *Shame*

Shame is defined as an individual's tendency to experience negative emotions when s/he fails to behave according to social norms and morals. This leads the individual to focus on how people evaluate him/her and the behaviour performed and to globally evaluate him/herself, which then generates denying, hiding, and/or avoiding behaviour (Tangney, 1999, in Barlian, 2013; Tangney et al., 2007). Based on the definition, shame is considered to be strongly related to the presence of others when an individual performs a particular behaviour.

Svensson et al. (2013) state that shame is associated with reduced rule-violating behaviours. Rahel (2014) also found that shame is negatively related to misconduct behaviours among adolescents. On the other hand, Tangney et al. (2007) claim that shame is associated with morally inappropriate behaviours.

Every individual has different tendencies towards experiencing shame. Individuals in a collectivist culture like Indonesia (Ardi & Maison, 2014) are associated with shame (Su, 2011); this is because individuals in a collectivist culture focus on the causality of individual behaviour on a group's particular aspects and performances, and perceives individual failures as group failures. Lewis et al. (1992, in McInerney, 1995) found that females have more tendency towards experiencing shame than males. Furthermore, Breugelmans and Poortinga (2006) also found that individuals from Javanese ethnic groups are associated with the tendency to experience shame.

1.2 *Moral identity*

According to Blasi (1983, in Hardy & Carlo, 2005), moral identity is defined as the belief of how important it is to act according to one's morals and the how important these morals are for oneself so that they become a characteristic of oneself. Blasi (1983, in Hardy & Carlo, 2005) and Hardy (2006) state that moral identity motivates the individual's moral behaviour.

Moral identity does not change as quickly as the environment around an individual changes, rather it is constructed from time to time following the maturity of an individual's subjective identity as one of the structures of self-identity (Hardy & Carlo, 2005). Subjective identity is described as an individual's interpretation of the aspects that adhere to self-identity (Hardy & Carlo, 2005). According to Blasi (1988, in Hardy & Carlo, 2005) the maturity of subjective identity is followed by more internally focused individual interpretation, such as moral and individual goals, thus making morals an important part of an individual's self-identity (Blasi, 1988, in Hardy & Carlo, 2005). The maturity of subjective identity is also followed by the individual's need to protect his/her self-consistency, thus making the individual consistently act morally (Blasi, 1988, in Hardy & Carlo, 2005).

Black and Reynolds (2016) have proposed moral self and moral integrity as the components of moral identity. Moral self describes how strongly an individual values the importance of morality and how closely an individual identifies him/herself with morality. Moral integrity describes how consistently an individual acts morally, indicated by his/her desire to make his/her intentions become observed behaviour and by his/her belief on the importance of producing moral behaviour for him/herself (Black & Reynolds, 2016).

1.3 *Dynamics of the relationship of shame and moral identity*

Aside from having a same role to motivate moral behaviour, shame and moral identity are assumed to be related. As previously explained, shame affects the individual's self-identity. As a part of self-identity, moral identity is also affected by shame. Moral identity is constructed from time to time following the maturity of an individual's subjective identity. More internally focused interpretation inherited from the maturity of subjective identity results in an interpretation of the importance of morality. Thus, this makes an individual identify him/herself close to morality (moral self-component) and consistent in producing moral behaviour (moral integrity component).

The internally focused interpretation of subjective identity producing the importance of morality for oneself is motivated by numerous situations. One of the situations is an individual's experience of shame. An individual experiences shame when receiving negative evaluation from another individual following the individual's failure in acting according to certain morals. Shame provides aversive feedback following the situation. Then, the individual's subjective identity interprets the situation by concluding that it is important to act according to morals, thus making subjective identity regard morality as important.

Placing importance on morality leads to the emergence of moral identity in oneself. The more frequently an individual is stimulated and tends to experience shame, the higher his/her score in moral identity. The relationship is supported by the fact that both an individual with a high score in shame and an individual with a high score in moral identity consistently produce moral behaviours.

2 METHODS

2.1 *Participants of the study*

Participants of the study were undergraduate students taking courses in Indonesia, with an age range of around 18–22 years. Participants were recruited using the convenience sampling method, choosing easily accessed individuals as participants of the study (Gravetter & Forzano, 2011). Aside from a printed questionnaire, an online questionnaire was also used in order to reach participants from various regions in Indonesia.

There were 520 participants in the study, with 379 (72.84%) females and 141 (27.16%) males. The majority of participants were 21 years old (33.27%) and came from Javanese ethnic groups (38.85%). Based on their academic institution, the majority of participants were from public universities (76.15%), from social and humanities studies (75.38), and were attending university in 2012 (35%).

2.2 *Design and statistical analysis*

The design of this study is a correlational study. In evaluating the correlation between shame and moral identity, Pearson product moment was used for statistical analysis.

2.3 *Instruments of the study*

The Test of Self-Conscious Affect-3 (TOSCA-3) from Tangney and Dearing (2002) was used to measure shame. TOSCA-3 that has been adapted to Bahasa (Indonesian) by Barlian (2013) consists of 17 scenarios categorised into four categories: family scenario, friendship scenario, occupational scenario, and self-scenario. The scenarios are followed by 34 items measuring shame and guilt. However, in this study only 16 items measuring shame were used. The choice of answers is presented in a five-point Likert scale ranging from 'tidak sesuai' (Disagree) to 'sangat sesuai' (Strongly agree). The possible score range that can be obtained is 16–80, with 48 as the middle score.

The Moral Identity Questionnaire (MIQ) from Black and Reynolds (2016) was used to measure moral identity. MIQ consists of two subscales each corresponding to moral self and moral integrity. There are a total of 20 MIQ items with a six-point Likert scale with the choices of answers ranging from 'sangat tidak sesuai' (Strongly disagree) to 'sangat sesuai' (Strongly agree). The possible score range that can be obtained is 20–120, with 70 as the middle score.

3 RESULTS

The average score of the participants on TOSCA-3 (shame) was 51.17 and the average score of the participants from MIQ (moral identity) was 94.45. Based on the descriptive statistics, 260 participants (50%) scored below average on TOSCA-3 (shame) and the rest scored above average. On the other hand, 252 participants (48.46%) scored below average on MIQ (moral identity) and another 268 participants (51.54%) scored above average.

The Pearson product moment with SPSS was used to evaluate the correlation of the participants' total score on TOSCA-3 (shame) and on MIQ (moral identity). The result showed a correlation index of 0.149 and a level of significance of 0.01; $r(518) = 0.149$, $p < 0.01$, with the effect size of $r^2 = 0.02$.

4 DISCUSSION

The positive correlation of shame and moral identity observed in the study supports findings of Svensson et al. (2013) and Rahel (2014) who state that shame is associated with the reduction of morally inappropriate behaviours. It is because individuals possessing a high tendency to experience shame also believe that morals and behaving accordingly is important for oneself, thus making individuals consistently produce moral behaviours.

The background of the study was the collectivist culture of Indonesia and this showed a significant result because, according to the characteristics of a collectivist culture, the individual always focuses on the causality of his/her behaviour on the group's particular aspects and performances (Chen & West, 2008). On the other hand, in a collectivist culture, individual failure is considered as the group's failure (Chen & West, 2008). Thus, this makes the individual evaluate all one's actions based on how s/he will be judged by others in the group. This kind of evaluation is one of the characteristics of shame. Based on that, the role of shame in affecting an individual's moral identity is very strong in a collectivist culture, thus making the correlation of shame and moral identity significant in a collectivist culture, especially in Indonesia.

Since it was presumed that a collectivist culture encourages shame to control a person's behaviour, the correlation index between shame and moral identity is expected to be large. However, the results of the study showed a small but significant correlation between shame and moral identity. The results showed that only a small variance of moral identity can be explained by shame. The small correlation index is assumed to be caused by the limitation of TOSCA-3 used to measure shame in the undergraduate student population. In TOSCA-3, there is no scenario representing an academic setting, even though undergraduate students spend most of their time in academic activities. Thus, the expression of shame in the study population was not optimally observed or captured. Aside from this limitation, TOSCA-3 is highly valid and reliable in capturing the expression of shame in the target population.

Another explanation of the small correlation result is that the effect of shame on moral identity only applies when an individual evaluates how others judge his/her behaviour. According to the characteristics of shame proposed by Tangney et al. (2007), an individual experiencing shame focuses on an evaluation of how others judge him/her; thus, the presence of others around the individual plays an important role. On the other hand, moral identity plays an important role in motivating moral behaviour, but not all moral behaviour is necessarily linked to the presence of others.

Furthermore, demographical data with females and individuals from Javanese ethnic groups constituting the majority of samples showed an uneven distribution of the samples, which is another limitation of the study. This uneven distribution of samples was because only 196 questionnaires were printed and these were distributed mainly in the Jakarta region. The culture in the big city is influenced by industrialisation and westernisation that is associated with individualistic culture (Ram, 2010). This condition is assumed to occur in Jakarta too. The implications of the condition can be found in the parenting practices of the urban community in Jakarta. Shame is not necessarily encouraged among children, so it does not significantly contribute to moral identity in the urban community.

The implication of the study can be drawn from the educational context. The findings of the study encourage the use of shame-based education materials, especially when it comes to moral education. Although the results showed a small correlation index, the use of shame to educate learners can support the development of learners' self-identity and moral identity and contributes to their consistently producing moral behaviour in the future.

5 CONCLUSION

There is a small but significant correlation between shame and moral identity among undergraduate students in Indonesia. The results encourage the use of shame-based education materials to develop learners' self-identity and moral identity. The small correlation index can be explained by the fact that the instrument does not specifically include an academic context; moral behaviour is not necessarily linked to the presence of others as a characteristic of shame or to the trend of an individualist culture in an urban community.

REFERENCES

Aquino, K. & Reed, A. (2002). The self-importance of moral identity. *Journal of Personality and Social Psychology, 83*(6), 1423–1440. doi:10.1037/0022-3514.83.6.1423

Ardi, R. & Maison, D. (2014). How do Polish and Indonesian disclose in Facebook? *Journal of Information, Communication and Ethics in Society, 12*(3), 195–218. doi:10.1108/JICES-01-2014-0006.

Barlian, I.Y. (2013). Perbedaan Emosi Malu dan Emosi Bersalah pada Generasi Tua dan Generasi Muda (The Difference of Shame Emotion and Guilt Emotion in Older and Younger Generation) (Thesis, Universitas Indonesia, Depok, Indonesia).

Black, J.E. & Reynolds, W.M. (2016). Development, reliability, and validity of the moral identity questionnaire. *Personality and Individual Differences, 97*. 120-129. doi:10.1016/j.paid.2016.03.041.

Breugelmans, S.M. & Poortinga, Y.H. (2006). Emotion without a word: Shame and guilt among Rarámuri Indians and rural Javanese. *Journal of Personality and Social Psychology, 91*(6), 1111–1122. doi:10.1037/0022-3514.91.6.1111.

Chen, F.F. & West, S.G. (2008). Measuring individualism and collectivism: The importance of considering differential components, reference groups, and measurement invariance. *Journal of Research in Personality, 42*(2), 259–294. doi:10.1016/j.jrp.2007.05.006.

Czub, T. (2013). Shame as a self-conscious emotion and its role in identity formation. *Polish Psychological Bulletin, 44*(3), 245–253. doi:10.2478/ppb-2013-0028.

Goles, T., White, G.B., Beebe, N., Dorantes, C.A. & Hewitt, B. (2006). Moral intensity and ethical decision-making: A contextual extension. *Advances in Information Systems, 37*(2&3), 86–95. doi:10.1145/1161345.1161357.

Gravetter, F.J. & Forzano, L.B. (2011). *Research methods for the behavioral sciences* (4th ed.). Stamford, CT: Wadsworth Publishing.

Hardy, S.A. (2006). Identity, reasoning, and emotion: An empirical comparison of three sources of moral motivation. *Motivation and Emotion, 30*(3), 205–213. doi:10.1007/s11031-006-9034-9.

Hardy, S.A. & Carlo, G. (2005). Identity as a source of moral motivation. *Human Development.*, 48, 232–256. Doi:10.1159/000086859.

McInerney, F.R. (1995). Shame And Guilt: Their Role In The Relation Between Moral Reasoning And Behavior In Early Adolescent Girls. (Doctoral Thesis, University of Delaware).

Nather, F. (2013). Exploring the impact of formal education on the moral reasoning abilities of college students. *College Student Journal, 47*(3), 470–477.

Rahel. (2014). Asosiasi antara Tekanan Teman Sebaya, Emosi Malu, dan Emosi Bersalah, pada Remaja *(The Association between Peer Pressure with Shame Emotion and Guilt Emotion Among Adolescence)* (Thesis, Universitas Indonesia, Depok, Indonesia).

Papalia, D.E., Olds, S. W, & Feldman, R.D. (2012). *Human Development.*, New York: McGraw Hill, Inc.

Ram, S.I. (2010). Exposure to western culture in relation to individualism, collectivism and subjective well-being, in India (Doctoral dissertation, Illinois Institute of Technology, Chicago, IL).

Su, C. (2011). A cross-cultural study on the experience and self-regulation of shame and guilt (Doctoral dissertation, University of York, UK).

Svensson, R., Weerman, F.M., Pauwels, L.J.R., Bruinsma, G.J.N. & Bernasco, W. (2013). Moral emotions and offending: Do feelings of anticipated shame and guilt mediate the effect of socialization on offending? *European Journal of Criminology, 10*(1), 22–39. doi:10.1177/1477370812454393.

Tangney, J.P. & Dearing, R.L. (2002). *Shame and guilt.* New York, NY: The Guilford Press.

Tangney, J.P., Stuewig, J. & Mashek, D.J. (2007). Moral emotions and moral behavior. *Annual Review of Psychology, 58*, 345–372. doi:10.1146/annurev.psych.56.091103.070145.

Diversity in Unity: Perspectives from Psychology and Behavioral Sciences – Ariyanto et al. (Eds)
 ISBN 978-1-138-62665-2

Challenges to facilitating social interaction among students in the inclusive classroom: Relationship between teachers' attitudes and their strategies

Y. Candraresmi & F. Kurniawati
Faculty of Psychology, Universitas Indonesia, Depok, Indonesia

ABSTRACT: Research has argued that inclusive education may benefit students with Special Educational Needs (SENs) in their academic and social development. This implies the critical role of teachers in accommodating learning needs and mediating social interaction of such students in the classroom. This study aims to measure teachers' attitudes towards students with SENs and their use of effective strategies in supporting positive social interaction through the role of peers. Participants of this study were 40 classroom teachers, drawn from inclusive primary schools in Depok, Indonesia. Teachers responded to two questionnaires, indicating their attitudes towards students with SENs on cognitive, affective, and behavioural components; the questionnaire focused on the teachers' strategies supporting positive interaction through peers in inclusive classrooms. The results reveal that teachers held positive attitudes towards inclusive education and had high scores on using their strategies; they showed acceptance and enthusiasm. The results also show the strong correlation between teachers' attitudes and their strategies in supporting positive social interaction through the role of peers. Other methods to enrich data could be added for future research. Teachers' training covering knowledge about students with SENs and skills for teaching strategies is recommended to support teachers for better practices in an inclusive educational setting.

1 INTRODUCTION

Inclusive education was first formally introduced to the Indonesian educational system in 2002. In that year, the government formally conducted preliminary testing in nine provinces that have resource centres and, at the time, there were more than 1,500 students with Special Educational Needs (SENs) going to regular schools (Sunardi et al., 2011). Inclusive education here is defined as formal education that involves students with SENs participating in classroom activities together with regular students in a regular school. According to Hallahan and Kauffman (2006), students with SENs are those who need educational services for their needs, to cope with problems such as physical disabilities, cognitive and sensory problems, learning difficulties, emotional problems, behavioural problems, or combinations of them.

Inclusive education itself gives many academic and social advantages to the students, especially those with SENs. From an academic aspect, students with SENs have more opportunities to receive a better education due to the development of these inclusive educational schools by the government.

From a social aspect, the benefits obtained are not limited only to students with SENs. In inclusive education, it is hoped that not only students with SENs but also regular students can develop optimally. Booth (2000) explains that the core of the approach to inclusive education includes the commitment to increase the sense of belonging of the students, to increase student participation in learning, to develop culture, to make and implement policies towards diversity and respect for others, and also to put emphasis on values in developing a positive school community, such as achievement.

The role of peers is very important in developing social interaction in an inclusive classroom between students with SENs and other regular students. This is also a challenge for teachers to ensure that the development of all students' interactions, communication, and language skills are functional. In this situation, students with SENs need support and help from the teachers and that is what makes the teacher's role in developing and maintaining this social interaction very important. Developing and maintaining conducive class atmosphere for positive interaction is one of the most critical parts of inclusive education.

In developing positive social interaction in the classroom, the teacher's role in the inclusive classroom must be optimal. A journal published by the National Association of Special Education Teachers (NASET, n.d.) states that for the inclusive classroom to be effective, there must be a strategy to increase positive social interaction in the classroom by the teacher. The teacher's strategies in the inclusive classroom must be consistently related with the successful implementation to inclusive education (Florian, 2009; Florin & Chambers, 2011). If related to the function of peers, as explained previously, it will have a big influence on acceptance and social interaction in the inclusive classroom; so, the teacher must make a plan or strategy to increase positive social interaction in the classroom by involving peers.

Besides the teacher's strategies, several studies also claim that the positive attitude of the teacher towards inclusive education is the most important factor for the success of inclusive education (see McGhie-Richmond et al., 2013; Jordan & Stanovich, 2003; Moberg et al., 1997; Murphy 1996; Sharma et al., 2008). Through the positive attitudes of the teacher, students with SENs will receive more opportunities in education to learn with their peers and will get the most educational benefits (Olson, 2003). The negative attitudes of the teacher will give less benefit to students with SENs in the inclusive classroom (Elliott, 2008).

Previous studies about teachers' attitudes towards inclusive education for students with SENs tend to vary in outcomes. Some studies show that teachers have positive attitudes (Leatherman & Niemeyer, 2005; Kurniawati et al., 2012) whereas other studies show neutral attitudes (de Boer et al., 2010. Some studies also show that teachers indicate negative attitudes towards inclusive education (Avramidis & Norwich, 2002). Teachers have evaluated themselves as not having enough skills and self-confidence to teach students with SENs in an inclusive classroom (Kurniawati et al., 2012). Previous studies have revealed that teachers' attitudes towards inclusive education might be related to their teaching strategies in the classroom (see Kurniawati, 2017; Bransford et al., 2008; Hastings & Oakford, 2003). It can be said that the more positive a teacher's attitude, the more willingly the teachers uses effective strategies in the classroom. Kochar et al. (2000) also state that there is a possibility that the most important influence on the positive relationship or interaction and social attitudes in the classroom is the attitude of the teacher to the degree to which the teacher shows acceptance towards students with SENs. The inclusive classroom teacher must have positive acceptance and show support in the classroom towards students with SENs.

Studies on teachers' attitudes have also been conducted in Indonesia; however, to the best of the authors' knowledge, a specific study that explains the correlation between teachers' attitudes and their strategies in supporting positive social interaction has not yet been conducted. Therefore, the main purpose of this research is to measure teachers' attitudes and strategies in supporting positive social interaction through the role of peers. The study has three objectives:

1. to measure teachers' attitudes in inclusive classrooms;
2. to measure teachers' strategies supporting positive social interaction through the role of peers in inclusive classrooms; and
3. to evaluate the strength of the relationship between teachers' attitudes and strategies in supporting positive social interaction through the role of peers in inclusive classrooms.

2 METHODS

2.1 *Procedure*

This study was focused in Depok, Indonesia. Depok was chosen as the location for acquiring the research data for several reasons. The city of Depok has pioneered and applied inclusive education in schools since 2009, according to Permendiknas No. 70/2009. In addition, Depok has proclaimed itself as '*Kota Layak Anak*' or, loosely translated, the 'City Worthy of Children' since 2010. In this programme, the city has a system to develop an administration area that will integrate the commitment and resources of the government, community, and the business world in order to meet children's rights through a holistic and sustainable plan in which these rights are the priority. The declaration of *Kota Layak Anak* has made children's education one of the priorities of the city of Depok. The meaning of education for all children is broad; it is not limited to regular students but extends to students with SENs. Therefore, inclusive education gets much attention in the policies of the government of Depok. This research was conducted by taking the population of all teachers in inclusive schools in Depok.

This study uses data from the Inclusive Working Group (*Kelompok Kerja Inklusif/Pokjasif*) in Depok and the Depok Inclusive School Community (*Paguyuban Sekolah Inklusif Depok*). The teachers who participated in this study were from public elementary schools that were appointed by both institutions mentioned above as schools applying the inclusive programme and active in the organisation of both institutions.

2.2 *Participants*

Forty teachers from inclusive schools that met the criteria of this study became study participants. The participants consisted of seven male teachers and 33 female teachers, who teach in grade 2 through grade 6, from inclusive schools in Depok. The demographic characteristics of the participants are shown in Table 1.

2.3 *Measures*

In order to understand teachers' attitudes and strategies in supporting positive interaction through the role of peers in inclusive classrooms, two questionnaires were used in this study.

Table 1. Demographic characteristics of inclusive classroom teachers in Depok (N = 40).

Characteristics category	Amount	Percentage
School		
A	8	20.0
B	3	7.5
C	4	10.0
D	5	12.5
E	8	20.0
F	3	7.7
G	1	2.5
H	8	20
Gender		
Males	7	17.5
Females	33	82.5
Age (years)		
<40	15	37.5
40–50	9	22.5
>50	16	40.0
Education		
Diploma	2	5.0
Strata 1	38	95.0
Teaching experience (years)		
<10	14	35.0
10–20	9	22.5
>20	17	42.5
Join inclusive education training		
Yes	14	35.0
No	26	65.0

The first questionnaire focused on teachers' attitudes using the Multidimensional Attitudes Toward Inclusive Educational Scale (MATIES), which was originally developed by Mahat (2008). After developing and testing the measuring instrument, the Indonesian version of MATIES went through some adaptations regarding not only the language but also the items themselves. The Indonesian version of MATIES has 18 items that measure three dimensions: cognitive, affective, and behavioural. It uses a Likert type scale, with scores ranging from 'Strongly disagree' (1) to 'Strongly agree' (6). The second questionnaire, which focuses on teachers' strategies in supporting positive social interaction through peers in inclusive classrooms, was developed based on the 13 strategies in supporting social interactions suggested by Yang (2005). It consisted of 38 items to measure two dimensions: direct strategies and indirect strategies. Direct strategies consisted of nine strategies used by teachers directly to support positive social interaction, such as prompting for respect, inviting participation, and helping with movement. Indirect strategies consisted of four strategies used by teachers indirectly to support positive social interaction, such as providing sensory input and fading from the interaction. This second questionnaire also uses a Likert type scale, with scores ranging from 'Strongly disagree' (1) to 'Strongly agree' (4).

2.4 *Analysis*

All the data were analysed to determine the research variables. All the completed questionnaires were obtained from the participants and processed to determine the validity and reliability of results. Statistical analysis was conducted using SPSS software (IBM Corporation).

The descriptive statistical analysis was used to obtain the frequency, mean, maximum and minimum scores, and standard deviation (SD) in order to estimate the levels of teachers' attitudes. The Pearson correlation calculation was run to find out the relationship between teachers' attitude and teachers' strategies supporting the positive social interaction.

3 RESULTS

Based on the demographic data, teachers were predominantly female (85%, n = 34) and 15% were male (n = 6). The age range of participants was <40 years (37.5%) to >40 years (62.5%), and only 14 out of 40 teachers (35%) had training in special education.

3.1 *Teachers' attitudes towards inclusive education*

Table 2 shows an overview of the respondents' attitudes towards inclusive education. The questionnaire consisted of 18 items. The possible score range was between 18 and 108, with higher scores reflecting more positive attitudes. The mean score of 86.58 (SD = 8.482) showed that teachers held positive attitudes. The results for the component cognitive, affective, and behavioural attitudes were also above the mid-point (21) of the response scale score, indicating positive attitudes towards inclusive education. Teachers were positive in their beliefs, held positive feelings, and showed positive behavioural intentions towards inclusive education.

Table 2. Teachers' attitudes towards inclusive education (N = 40).

Variable	Maximum	Minimum	Mean	SD
Attitudes	103	69	86.58	8.482
Cognitive	36	19	27.30	4.244
Affective	36	18	28.52	4.267
Behaviour	36	21	30.75	3.418

SD: Standard deviation.

Table 3. Means and SDs for teachers' strategies (N = 40).

Variables	Maximum	Minimum	Mean	SD
Teacher's strategies	152	109	125.90	12.448
Direct strategies in supporting social interaction	108	76	90.15	9.057
Indirect strategies in supporting social interaction	56	40	45.85	4.886

Table 4. Relationship between teachers' attitudes and strategies (N = 40).

	Teachers' strategies		Direct strategies		Indirect strategies	
Variable	*r*	*p*	*R*	*p*	*r*	*p*
Attitudes	0.635*	0.00	0.572*	0.00	0.722*	0.00
Cognitive attitudes	0.473*	0.00	0.454*	0.00	0.519*	0.00
Affective attitudes	0.339**	0.03	0.238	0.14	0.448*	0.00
Behavioural attitudes	0.556*	0.00	0.559*	0.00	0.537*	0.00

*Significant correlation of level of significance 0.01 (two-tailed); **Significant correlation of level of significance 0.05.

3.2 *Teacher's strategies supporting positive social interaction through the role of peers*

Table 3 shows an overview of teachers' strategies supporting positive social interaction. The questionnaire consisted of 38 items. The possible score range was between 38 and 152, with higher scores indicating that more strategies were used. The mean score of 125.90 (SD = 12.448) showed high use of teachers' strategies supporting positive social interaction. The results for the direct and indirect strategies were above the mid-points (direct = 65, indirect = 30) of the response scale score, also indicating high use by teachers' of strategies supporting positive social interaction.

3.3 *Relationship between teachers' attitudes and strategies supporting positive social interaction through the role of peers*

Table 4 shows that the relationship between teachers' attitudes and strategies is strongly significant (r = 0.63, p = 0.01). This means that the more positive the attitudes of teachers towards inclusive education, the higher the use of strategies in supporting positive social interaction through the role of peers in inclusive classrooms. The relationship between the components of attitudes (cognitive, affective, and behaviour) and teachers' strategies was also significant.

Related to the direct and indirect strategies, the results showed that the correlation values between attitudes and the direct and indirect strategies were significant. The results also indicated that almost all components of attitudes were significantly related to teachers' strategies supporting social interaction, both direct and indirect. The only component that was not related to the direct strategies was the affective attitudes.

4 DISCUSSION

The results of the study show that, in general, teachers of inclusive schools in Depok held positive attitudes towards inclusive education in the three components of attitudes tested. Similar results are also found in previous studies (Leatherman & Niemeyer, 2005; Kurniawati et al., 2012). If related to demographic factors of teachers, such as gender, age, teaching experience, training, and educational environment, this study confirmed previous study results about teachers' attitudes. Avramidis and Norwich (2002) found that factors influencing a teacher's attitude included gender, age, teaching experience training, and educational environment. In this study, the respondents were mainly women (85%), 62.5% were aged above 40 years, and 65% had teaching experience of more than ten years.

Another finding of the study shows high use of teachers' strategies supporting positive social interaction, both direct and indirect. This is surprising considering that only 35% of the teachers had been trained in inclusive education. Similar contradiction was also found in the findings on teachers' attitudes. Previous studies show that teachers who received training in special needs education held more positive attitudes towards inclusive education compared with teachers who did not receive this training (Avramidis & Norwich, 2002; de Boer et al., 2010). However, findings on teachers' attitudes reveals some other factors that might have a role in shaping the attitudes of the respondents and affecting the use of teachers' strategies, both direct and indirect.

A number of studies have examined some environmental factors and their influences in the formation of teachers' attitudes towards inclusive education. Support as an environmental factor can be regarded as physical and non-physical, such as the enthusiastic support from head teachers and co-workers (see Avramidis & Norwich, 2002; Janney et al., 1995). Environmental factors have also been found in the present study. Socialisation, informing the school members about inclusive education, and the availability of work groups and *paguyuban* (community) seem to have an important role in shaping positive attitudes and in the use of interaction strategies. The work group and the community (*paguyuban*) could function as a source of knowledge to share teachers' problems related to inclusive education.

The results of this study showed a close relationship between attitudes and strategies supporting positive social interaction through the role of peers, proving that the respondents have positive cognitive, affective, and behavioural attitudes and that they are able to use resources and knowledge to implement their strategies in supporting positive social interaction between regular students and students with SENs. This confirms previous study results that teachers' attitudes in inclusive education are related to teaching strategies in the classroom (see Kurniawati et al., 2012 Bransford et al., 2008; Hastings & Oakford, 2003). Only affective attitude is not related to the direct strategies of supporting social interaction. A possible explanation for this finding might come from the environmental factors. The learning methods in Indonesia are dominated by the teacher speaking in front of the class. This results in the teachers rarely giving a role to the peer to support positive social interaction. However, the data from the measurement of teachers' attitudes and strategies supporting social interaction are only based on their responses to questionnaires. Therefore, the data from the questionnaires must be incorporated with other data that show what is really happening in the field of inclusive education.

5 CONCLUSION

With regard to the first aim of the research, it can be concluded that teachers hold positive attitudes towards inclusive education in the three components of attitudes tested. Regarding the second research question, this study reveals that there is high use of teachers' strategies in supporting positive social interaction, both direct and indirect.

Another finding of the study is that there is a strongly significant relationship between attitudes and teachers' strategies supporting positive social interaction through the role of peers. A strongly significant relationship exists between each attitude component (cognitive, affective, and behavioural) and teachers' strategies. The results also indicate that almost all of the components of attitudes are significantly related to teachers' strategies, both direct and indirect. Affective attitudes are the only component that is not related to the direct strategies in supporting social interaction.

6 FUTURE RECOMMENDATIONS

Based on the results of this research, it is suggested that future research should find additional data using other research methods, such as interviews and classroom observations by video recording. The purpose is to enrich the data available so the overall data are not based on only the questionnaires. Although the results show positive teachers' attitudes and high use of teachers' strategies, the related teacher training programmes covering knowledge about students with SENs and skills for teaching strategies are recommended to support teachers for better practices in an inclusive educational setting, since many teachers have not received the proper training yet.

REFERENCES

Ajzen, I. (2005). *Attitudes, personality and behaviour* (2nd ed.). Maidenhead, UK: Open University Press.

Anderson, J.R. (1981). *Cognitive skills and their acquisition*. Hillsdale, NJ: Lawrence Erlbaum Associates.

Avramidis, E. & Norwich, B. (2002). Teachers' attitudes towards integration/inclusion: A review of the literature. *European Journal of Special Needs Education, 17*(2), 129–147.

Booth, A., Ainscow, M., Black-Hawkins, K., Vaughn, M. & Shaw, L. (2000). *Index for inclusion: Developing learning and participation in schools*. Bristol, UK: Centre for Studies on Inclusive Education.

Bransford, J.D., Jacobs, J., Eitoljorg, E. & Pittman, M.E. (2008). Introduction. In L. Darling-Hammond, J. Bransford, P. LePage, K. Hammerness & H. Duffy (Eds.), *Preparing teachers for a changing world: What teachers should learn and be able to do* (pp. 1–39). San Francisco, CA: Jossey-Bass.

Cozby, P. & Bates, S. (2011). *Methods in behavioral research* (11th ed.). New York, NY: McGraw-Hill.

de Boer, A., Pijl, S.J. & Minnaert, A. (2010). Regular primary school teachers' inclusive education: A review of the literature. *European Journal of Special Needs Education, 25*(2), 165–181.

Depdikbud. (2003). *Undang Undang Republik Indonesia Nomor 20 tahun 2003 Tentang Sistem Pendidikan Nasional*. Jakarta, Indonesia. (Depdikbud. (2003). *Indonesian Republic Laws Number 20 year of 2003 about National Educational System*. Jakarta, Indonesia)

Elliott, S. (2008). The effect of teachers' attitude toward inclusion on the practice and success levels of children with and without disabilities in physical education. *International Journal of Special Education, 23*(3), 48–55.

Florian, L. (2009). *Psychology for inclusive education: New directions in theory and practice*. London, UK: Routledge/Falmer.

Forlin, C. & Chambers, D. (2011). Teacher preparation for inclusive education: Increasing knowledge but raising concerns. *Asia-Pacific Journal of Teacher Education, 13*(2), 195–209.

Gresham, F.M. & MacMillan, D.L. (1997). Social competence and affective characteristics of students with mild disabilities. *Review of Educational Research, 67*(4), 377–415. doi:10.3102/00346543067004377.

Hallahan, D.P. & Kauffman, J.M. (2006). *Exceptional learners: An introduction to special education* (10th ed.). Upper Saddle River, NJ: Pearson Education.

Hastings, R.P. & Oakford, S. (2003). Student teachers' attitudes toward the inclusion of children with special needs. *Educational Psychology, 23*(1), 87–94.

Hurlock, E.B. (1998). *Psikologi Perkembangan*. Jakarta, Indonesia: Erlangga.

Janney, R.E., Snell, M.E., Beers, M.K. & Raynes, M. (1995). Integrating students with moderate and severe disabilities into general education classes. *Exceptional Children, 61*(5), 425–439. doi:10.1177/001440299506100503.

Jordan, A & Stanovitch, P. (2003). Teahers' personal epistemological beliefs about students with disabilities as indicators of effective teaching practices. *Journal of Research in Special Educational Needs, Vol 3 (1)*. doi:10.1111/j.1471-3802.2003.00184.x

Kochar, C.A., West, L.L., Taymans, J.M., (2000). *Successfull inclusion*, Columbus: Merill.

Koster, M., Pijl, S.J., Nakken, H. & Van Houten, E. (2010). Social participation of students with special needs in regular primary education in the Netherlands. *International Journal of Disability, Development and Education, 57*(1), 59–75. doi:10.1080/10349120903537905.

Kurniawati, F., Minnaert, A., Mangunsong, F. & Ahmed, W. (2012). Empirical study on primary school teachers' attitudes towards inclusive education in Jakarta, Indonesia. *Procedia – Social and Behavioral Science, 69*, 1430–1436. doi:10.1016/j.sbspro.2012.12.082.

Kurniawati, F. (2017). Indonesian teachers' attitudes and their observed teaching strategies in primary inclusive classrooms. Submitted for publication. Teachers and Teaching: Theory and Practice.

Leatherman, J. & Niemeyer, J. (2005). Teachers' attitudes toward inclusion: Factors influencing classroom practice. *Journal of Early Childhood Teacher Education, 26*(1), 23–36. doi:10.1080/10901020590918979.

Mahat, M. (2008). The development of a psychometrically sound instrument to measure teachers' multidimensional attitudes toward inclusive education. *International Journal of Special Education, 23*(1), 82–92.

McGhie-Richmond, D., Irvine, A., Loreman, T., Cizman, J.L. & Lupart, J. (2013). Teacher perspectives on inclusive education in rural Alberta, Canada. *Canadian Journal of Education, 36*(1), 195–239.
Moberg, S., Zumberg, M. & Reinmaa, A. (1997). Inclusive education as perceived by prospective special education teachers in Estonia, Finland, and the United States. SAGE journals, *22*(1). doi: 10.117/154079699702200105.
Moberg, S. (2003). Education for all in the North and the South: Teachers' attitudes towards inclusive education in Finland and Zambia. *Education and Training in Developmental Disabilities, 38*(4), 417–428.
Murphy, D.M. (1996). Implications of Inclusion for General and Special Education. *The University of Chicago Press Journals, Vol. 96 (5),* 469–493.
Mangunsong, F. (2011). *Psikologi dan Pendidikan Anak Berkebutuhan Khusus. Jilid Kedua.* Depok, Indonesia: LPSP3.
Olson, J.M. (2003). *Special education and general education teacher attitudes toward inclusion* (Master's thesis, University of Wisconsin–Stout, Menomonie, WI).
Saifuddin, A. (1998). *Sikap Manusia Teori dan Pengukurannya.* Yogyakarta, Indonesia: Pustaka Pelajar.
Santrock, J.W. (2011). *Life span development* (13th ed.). New York, NY: McGraw-Hill.
Sharma, U., Forlin, C. & Loreman, T. (2008). Impact of training on pre-service teachers' attitudes, concerns and sentiments about inclusive education: An international comparison of the novice pre-service teacher. *International Journal of Special Education, 21*(2), 80–93.
Sunardi, Yusuf, M., Gunarhadi, Priyono & Yeager, J.L. (2011). The implementation of inclusive education for students with special needs in Indonesia. *Excellence in Higher Education, 2*(1), 1–10. doi:10.5195/ehe.2011.27.
UNESCO. (1994). *The Salamanca statement and the framework for action on special needs education.* Paris, France: UNESCO.
United Nations. (2006). *Convention on the rights of persons with disabilities and optional protocol.* New York, NY: United Nations.
Yang, C. (2005). *Training teachers to use strategies supporting social interaction for children with moderate to severe disabilities in inclusive preschool classroom* (Thesis, University of Kansas, Lawrence, KS).
Yang, C. & Rusli, E. (2012). Teacher training in using effective strategies for preschool children with disabilities in inclusive classrooms. *Journal of College Teaching & Learning, 9*(1), 53–64.

Diversity in Unity: Perspectives from Psychology and Behavioral Sciences – Ariyanto et al. (Eds)

The role of parental involvement in student's academic achievement through basic needs satisfaction and school engagement: Construct development

J. Savitri
Faculty of Psychology, Universitas Padjadjaran, Sumedang, Indonesia
Faculty of Psychology, Maranatha Christian University, Bandung, Indonesia

I.L. Setyono, S. Cahyadi & W. Srisayekti
Faculty of Psychology, Universitas Padjadjaran, Sumedang, Indonesia

ABSTRACT: Parents play an important role in supporting the education of students in the school. This study aims to develop a conceptual constructs on the effect of parental involvement on student's academic achievement through basic needs satisfaction and school engagement. The study was based on two preliminary studies in grades 4–6 elementary school students in Bandung, Indonesia. Study 1 ($N = 350$) was conducted to examine the effect of parental involvement on school engagement and academic achievement. Using structural equation modelling, the result indicated that parental involvement affected academic achievement, with school engagement as a mediator fit with the data (chi-square = 21.52, degree of freedom = 14, $p = 0.089$, root mean square error of approximation = 0.039). There was a positive and significant effect of parental involvement in academic achievement through school engagement, but parental involvement had a negative effect on academic achievement. Study 2 ($N = 231$) tested the effect of parental involvement on basic needs satisfaction which consist of a need for autonomy, need for competence, and the need for relatedness. This research indicated that parental involvement significantly influenced basic needs satisfaction ($F = 3.422$, $p = 0.018$). The model constructed was based on preliminary studies and will be examined in future research.

1 INTRODUCTION

Parents are the people most responsible for their children's development. Issues concerning parents' significant roles are still very topical in the global postmodern era. The ways in which parents treat their children have strong impacts on many aspects of children's lives, including their education. In every phase of a child's development, parents are required to perform specific roles (Brooks, 2001). For example, when children are at the elementary school level, they need parental involvement that can enable them to adapt themselves to the academic requirements at school. Parents who show involvement in their children's education help them to reach their expected educational outcomes.

In Indonesia, demand for high academic achievement is quite high. Therefore, there is tight competition among schools. Schools determine a standard academic achievement known as *Kriteria Ketuntasan Minimum*, which refers to the criteria for minimum requirements in the form of scores. Children spend most of their time at school preparing for academic competence, besides being required to do extracurricular activities.

Teachers considered parental roles as significant in enabling children to meet the demands of the schools. This statement came from 31 teachers from three elementary schools in Bandung, West Java (Savitri et al., 2015). The Indonesian government has acknowledged the importance of parental roles in supporting children's education and aims to promote it by establishing a *direktorat keayah-bundaan*, which refers to specific department that have responsibility to develop parenthood in Indonesia. Presently, the government is still looking for a model of parents' roles as positive impacts on their children's education.

1.1 *Parental involvement and student academic achievement*

Pomerantz et al. (2005) assert that parents are the central figures in their children's lives; therefore, they have the potentials to sharpen their children's approach to achievements. In the framework of parents as partners in education, Berger (1995) exposes the importance of parental involvement in children's education. Parental involvement is required to bring about children's school performance/ achievement (Grolnick & Slowiaczek, 1994). Academic achievement, in general, can be understood as grades that students obtain in certain school subjects.

Research on the impact of parental involvement in academic achievement shows varying results. Hoover-Dempsey et al. (2001) note that testing the influence of parental involvement results in various findings, with both positive and negative relationships. Likewise, Domina (2005) mentions that policy makers and theorists have assumed that parental involvement has positive consequences, but many studies have shown that it is negatively associated with some children's outcomes in educational context. Furthermore, there is indication that types of parental involvement on achievement tests determine the results.

1.2 *Self-determination theory as a resource of engagement*

An engagement construct is developing nowadays. Appleton et al. (2008) stated that an engagement construct is relevant for all students, not only for those who are at risk of drop-out. In addition, an engagement is seen to be very suitable for giving explanations about motivation and other constructs affecting school-related outcomes/significant results (Appleton et al., 2008).

Previous research focused more on how parents influence their children's motivation, resulting in high academic achievements. However, the research was insufficient to explain how the motivational component could make children achieve high academic performance. The action component by children is worth studying to gain insight in this area. Experts use 'engagement' as a term to illustrate children's action component at school.

Motivation is central to understanding how social context influences engagement. Grolnick et al. developed research connecting a context of family/parents to children's achievement based on a motivational framework (Grolnick et al., 1991, 2009; Grolnick & Slowiaczek, 1994; Pomerantz et al., 2005). Furrer et al. (2006, in Appleton et al., 2008) also emphasised the importance of examining engagement through the motivational framework because engagement could change through its interactions with contextual variables, which would finally influence academic, behavioural, and social outcomes.

Basic needs satisfaction is part of the Self-Determination Theory (SDT) used in the self-system model of motivational development (Connell & Wellborn, 1991; Deci & Ryan, 2000; Skinner & Pitzer, 2012). SDT states that every individual has three basic psychological needs that are universal across different cultures, age-groups, and gender (Deci & Ryan, 2000).

The fulfilment of students' need for autonomy, competence, and relatedness is crucial because the three basic needs influence formation of students' motivation. Motivation involves energy, persistence, direction, and an ultimate goal (Deci & Ryan, 2000). According to Ryan and Connell (1989), students' motivation is the most important determinant of students' success or failure at school; therefore, motivation is something worth highlighting.

Deci and Ryan (2000) explain that when the three basic needs are met, individuals will be interested in doing activities focused on goals. In terms of education, students' expected motivation is intrinsic. When they have the intrinsic motivation, they will be interested and enjoy themselves in following lessons. Thus, they will be involved in learning activities (Ryan & Deci, 2009).

Appleton et al. (2008) wrote an article related to various concepts and methodology issues dealing with the engagement construct. They tried to integrate influences from various social

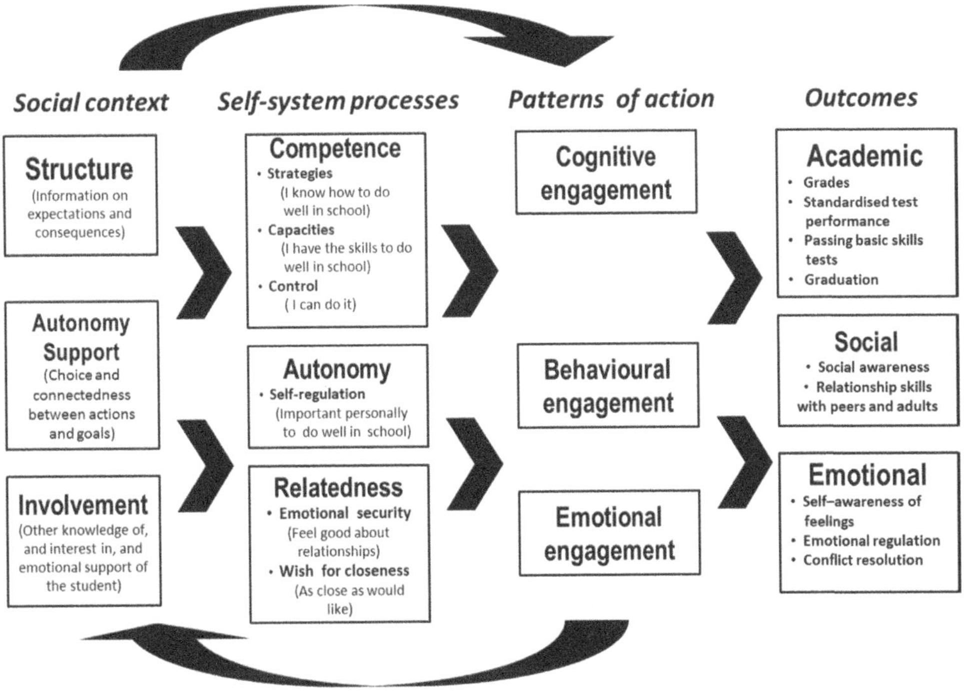

Figure 1. Self-processes model applied to educational setting (Appleton et al., 2008, p. 380).

contexts, self-system processes, action patterns, and outcomes. The naturalness of this integration offered the promise of a design intervention. Their theoretical framework is illustrated in Figure 1.

Preliminary studies were conducted to obtain a description of the relationship among the variables of parental involvement, basic needs satisfaction, school engagement, and academic achievement. Based on the result of the preliminary studies, a conceptual construct was built for further research. This research conducted two studies at three elementary schools in Bandung.

2 THE PRESENT STUDY

Study 1 was conducted to investigate the effect of parental involvement on school engagement and academic achievement. The constructs of parental involvement in previous research vary. This research, referred to the theory of Grolnick and Slowiaczek (1994), which regarded parental involvement as parents' self-initiated behaviours. They built a parental involvement construct based on psychology, which was formulated as parents' resources for children in an educational context.

On the other hand, the school engagement construct in this research referred to the theory of Fredricks et al. (2004), which illustrated school engagement as a multi-dimensional construct consisting of behavioural, emotional, and cognitive components. Behavioural engagement is students' active involvement in academic and non-academic activities at school. Emotional engagement is students' affective reaction towards teachers, peers, tasks, and school in general. Cognitive engagement is students' investment or commitment (i.e. efforts) and cognitive strategies to master the lessons.

Study 2 was conducted to examine the influence of parental involvement on basic needs satisfaction of the students of grades 4–6 at elementary school "Z" in Bandung. Self-determination theory (SDT) is part of human motivation that consider innate psychological needs for autonomy, competence, and relatedness (Deci & Ryan, 2000).

2.1 *Methods*

2.1.1 *Participants*

The participants of study 1 were 350 students (167 boys and 183 girls) of grades 4–6. The students' ages ranged from 9 to 12 years (mean = 10.33 years). The majority of them were of middle to high-class socioeconomic status, of Chinese descent and from two urban elementary schools in Bandung.

The participants of study 2 were 231 students (118 girls and 113 boys) of grades 4–6. The majority of the participants were of middle socioeconomic status, in an urban elementary school in Bandung.

2.1.2 *Measures*

The measurement of construct was done by referring to conceptualisation and operationalization based on theories or previous research.

2.1.2.1 *Parental involvement*

A questionnaire about parental involvement was developed, based on the theory of Grolnick and Slowiaczek (1994). Students rated the 21 given items, five items represented school involvement, eight personal involvement, and eight cognitive involvement. Students indicated the extent to which each of the statements of student perception is true (4 = Very true, 3 = True, 2 = Less true, 1 = Not true). Confirmatory Factor Analysis (CFA) showed that parent involvement as a construct fits with the empirical data: chi-square = 4.41, Degree of Freedom (df) = 6, $p = 0.622$, Root Mean Square Error of Approximation (RMSEA) = 0.000.

2.1.2.2 *Basic needs satisfaction*

The questionnaire for the Basic Needs Satisfaction in General Scale (BNSG-S), entitled "Feeling I Have" from Deci and Ryan (2000), measured students' basic needs satisfaction. The English version of the questionnaire was translated into *Bahasa* (Indonesian) by reference to the steps of a translation process as described by Purwono (2010) that consists of reviewing the co-existence of the construct, language translation, and empirical testing, which according to the "International Test Commission Guidelines for translating and adapting tests" (2010). The number on the scale was adjusted from 7 to 4, given that the cognitive development of elementary school children was at the level of concrete operational. The choices of answers were 'Not True', 'Less True', 'True', and 'Very True'. The results of the validity test from the Pearson correlation coefficient was 0.412–0.692. One item to assess the needs for autonomy ("In my daily life, I frequently have to do what I am told") was invalid according to the criteria of Friedenberg (1995), with a score of 0.187, but it was significant ($p < 0.01$). Meanwhile, the result of the reliability test using the calculation of Cronbach Alpha was 0.736.

2.1.2.3 *School engagement*

A questionnaire was constructed based on the theory of Fredricks et al. (2004). The participants were asked to choose one of four options (4 = Very often; 3 = Often; 2 = Less often; 1 = Not often) for every statement. The questionnaire consisted of 28 items, with 11 items representing behavioural engagement, nine representing emotional engagement, and eight cognitive engagement. CFA showed that parental involvement as a construct from the three types of school engagement with the indicators fits with the empirical data (chi-square = 30.78, df = 20, $p = 0.058$, RMSEA = 0.039).

2.1.2.4 *Academic achievement*

The assessment of academic achievement was based on average scores of daily quizzes, worksheets, homework, and performance assessment for eight subjects, namely *Agama* (Religion), *Sosial* (Social Sciences), *PPKn* (Civics Education), *Bahasa Indonesia* (Indonesian Language), *Matematika* (Mathematics), *Sains* (Natural Sciences), *Ketrampilan* (Handicrafts), and *Pendidikan Jasmani* (Physical Education). Based on CFA calculation of academic achievement, the academic achievement model as a construct integrating the eight subjects fits with the empirical data (chi-square = 21.52, df = 14, $p = 0.089$, RMSEA = 0.039).

2.1.3 *Procedures*

2.1.3.1 *Study 1*

The data were collected by distributing questionnaires among students of grades 4–6 at elementary schools "X" and "Y" in Bandung. The survey was done at the beginning of the second half of the 2014–2015 school years. Students filled out two sets of questionnaires about parental involvement and school engagement in the form of a self-report. Academic scores/grades were obtained from their homeroom teachers.

2.1.3.2 *Study 2*

Data collection was carried out by distributing questionnaires to all students in grades 4–6 at elementary school "Z" in Bandung. The survey was administered in the mid-period of the first half of the 2015–2016 school year. The students filled out two sets of self-reporting questionnaires related to parental involvement and basic needs satisfaction.

2.1.4 *Data analyses*

2.1.4.1 *Study 1*

Structural equation modelling analysis with LISREL 8.8. software was used to test the hypothesis that parental involvement played a role in academic achievement mediated by school engagement (see Figure 2). First, CFA was used to test the measurement model, particularly to measure whether the indicators were associated adequately with latent variables. Then, the influences among the latent variables were compared.

The results of CFA calculation of the measurement model of parental involvement revealed that parental involvement as the construct of school involvement, personal involvement, and cognitive involvement fits with the empirical data [$\lambda = 4.41$, df = 6, p = 0.62, RMSEA = 0.0, Comparative Fit Index (CFI) = 1.00, Normed Fit Index (NFI) = 0.99, Goodness of Fit Index (GFI) = 1.00].

Furthermore, the result of CFA calculation of the measurement model of school involvement showed that school engagement as the construct of behavioural, emotional, and cognitive engagements fits with the empirical data ($\lambda = 30.78$, df = 20, p = 0.058, RMSEA = 0.039, CFI = 0.99, NFI = 0.97, GFI = 0.98).

The results of CFA calculation of the measurement model of academic achievement showed that academic achievement as the construct, which was the integration of school subjects, such as Religion, Social Studies, Civil Studies, Indonesian, Mathematics, Natural Sciences,

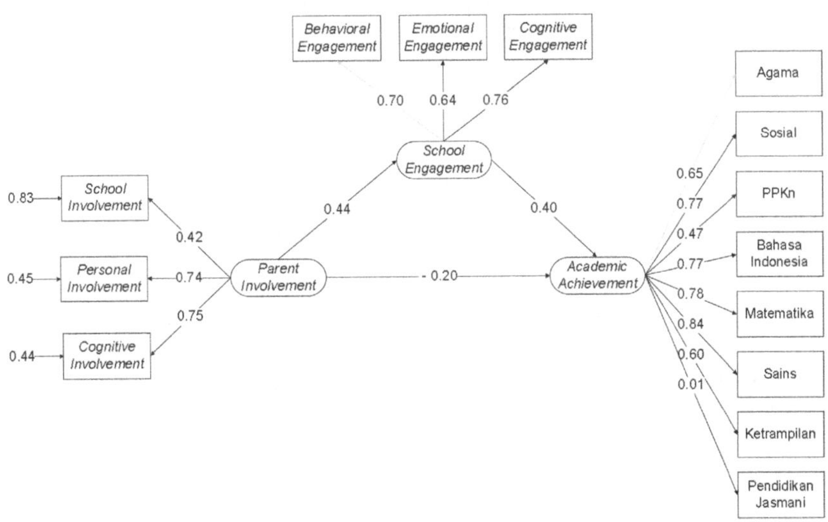

Figure 2. The influence of parental involvement on school engagement and academic achievement.

and Life Skills, fits with the empirical data ($\lambda = 21.52$, df = 14, p = 0.89, RMSEA = 0.039, CFI = 1.00, NFI = 0.99, GFI = 0.98).

2.1.4.2 *Study 2*

Regression analysis was used to test the hypothesis, that parental involvement was influential on basic needs satisfaction. First, the score of parental involvement was regressed with the score of basic needs satisfaction. Next, the score of parental involvement was regressed with each score of the three basic needs of autonomy, competence, and relatedness. The entire processing of the test was carried out with the SPSS statistical software program (IBM Corporation) for Windows.

3 RESULTS AND DISCUSSION

Study 1

Statistical analysis showed that the data supported the model of the influence of parental involvement on school engagement and academic achievement. The results of the calculation can be summarised as follows: chi-square = 168.11 (65), p = 0.00, RMSEA = 0.067, CFI = 1.00, NFI = 0.99, GFI = 1.00, which indicated a good model fit.

The analysis of the influence of parental involvement on school engagement and academic achievement confirmed that the hypothesis of the research was accepted. Parental involvement had a significant influence on school engagement ($\gamma = 0.44$), school engagement had a significant influence on academic achievement ($\beta = 0.40$), and parental involvement had a significant influence on academic achievement ($\alpha = 0.20$). Considering the direct influence coefficient (0.20) of parental involvement on academic achievements and the indirect influence through school engagement (0.20), the level of influence was relatively similar; however, the direction was different.

The results of the data analysis are described in Figure 2.

The research proved that parental involvement influenced the academic achievement of elementary school students in grades 4–6, with school engagement as the mediator. These results were in line with the theoretical framework of Appleton et al. (2008), who stated that parental involvement as a part of a social context was influential on behaviour patterns, such as student engagement at school that resulted in academic achievement/performance.

Figure 2 shows that parental involvement had a more positive impact on school engagement than on academic achievement. Moreover, parental involvement had an impact on student academic achievement in a negative direction. This means that examples of parental involvement, such as coming to school, paying attention to children's school life, and having discussions with the children, as well as providing facilities for children's cognitive development could increase the school engagement. Student engagement shown by participating in academic and non-academic activities, showing positive emotion in social interactions at school, and showing commitment to study lesson would enable students to accomplish better academic achievement. This finding supported the research of Mo and Singh (2008) on high school students, which found that parental involvement was influential on school engagement; furthermore, it had an impact on children's school performance.

However, a gap still remains to be explained about how parental involvement can influence students' school engagement. Meanwhile, parents had little interaction with their children at school. Therefore, further research was conducted to test the influence of parental involvement on the satisfaction of basic needs in the motivational framework. Grolnick et al. (2009) explained that the motivation theory considered children as active individuals who interpret inputs from social contexts, which further describes why they engage at school and how they behave, perform activities, and feel emotions.

Study 2

The result of this study shows that parental involvement had significant influences on basic need satisfaction of students of grades 4–6 at elementary school "Z" in Bandung (F = 3.422, p = 0.018). This finding explained that parental involvement construct in the form of students' perception of their parents' behaviours, related to school involvement, personal involvement, and cognitive involvement, was able to meet the basic psychological needs of the students. Parental involvement in providing necessary resources for their children's education had a significant role in fulfilling the students' basic psychological needs. The satisfaction of basic needs becoming the source of motivation for students to make achievements, as mentioned by Grolnick et al. (2009), showed that parental involvement had an impact on the facilitation of students' motivation at school. Based on SDT, students are motivated when their basic psychological needs are met. This research specifically revealed that parental involvement makes a contribution in fulfilling the basic needs of students of grades 4–6.

If it was analysed independently, parental involvement was influential merely on the fulfilment of the needs for competence ($F = 3.201$, $p = 0.024$), but was not influential on the fulfilment of the needs for autonomy ($F = 2.146$, $p = 0.095$) and relatedness ($F = 2.544$, $p = 0.056$). It means that parents who attend school events or parent gatherings at school (school involvement), communicate with their children about their school activities (personal involvement), and discuss current events or visit bookstores with their children (cognitive development) could fulfil children's needs to feel confident, be effective in their social interactions, and develop their potentials. Pomerantz et al. (2005) explained that one of the effects of parental involvement was that children have a sense of accomplishment. Lack of influence of parental involvement on the needs for autonomy and competence was actually possible due to other dimensions of parenting. Grolnick (2009) pointed out that autonomy support and structure from the parents had the potential to be sources of motivation for the students. Nonetheless, it would require further empirical research to find accurate correlations between other dimensions of parenting and basic needs satisfaction of students.

In addition, referring to this research, we look into consideration of the validity of the BNSG questionnaire and the meaning of the statements as perceived by the students of grades 4–6 at the elementary schools. Even though the internal consistency reliability is 0.736, there were still some invalid items based on the calculation of Pearson correlation. This research suggested that it was necessary to adapt the BNSG-S questionnaire based on the cognitive development of upper-elementary students in their concrete operational level. Another test could then be carried out by using factor analyses.

4 CONCLUSION

Pomerantz et al. (2005) stated that since parents are the central figures in children's lives, they have the potentials to sharpen their children's approach to achievements. However, the result of the first study (study 1) in this research showed that parental involvement did not directly influence the children to attain higher academic achievement. On the other hand, there was an indication that parental involvement could influence children's motivation to engage at school. Most of the parental involvement-related research on schooling has focused more on children's academic achievement than on how the children attain that achievement (Pomerantz et al., 2005).

Referring to the framework of Appleton et al. (2008), there are unclear mechanism on how parental involvement influence school engagement of the students. Fredricks et al. (2004) stated that school contexts had correlation with behavioural engagement, emotional engagement, and cognitive engagement of the students. They also mentions that fulfilment of needs for autonomy, needs for competent, and needs for relatedness could influence student's school engagement. Nevertheless, in their article, Fredricks et al. (2004) do not explain how parental involvement can influence the students' school engagement. Therefore, there is an opportunity for the conduct of further research to develop the theory of Fredricks et al. (2004) with parenting studies to support student engagement.

Grolnick (2009) suggests that parental involvement does not have any direct influences on the development of academic competence, such as in mathematics, but it can facilitate the sources of motivation for children to reach academic goals. Accordingly, it is necessary to conduct research to establish a structural model of the influence of parental involvement on school engagement on basic needs satisfaction, as the theory of motivation connects with individual social contexts. Fan and Williams (2010) underline the importance of the intrinsic motivation for students to engage at school. With reference to SDT from Deci and Ryan (2000), it is expected that further research can give more comprehensive explanations about the influence of parental involvement on students' school engagement. Raftery et al. (2012) stated that the engagement from the SDT framework is the outward manifestation of motivation and occurs in a context that fulfils the children's needs for autonomy, competence, and relatedness.

In studies 1 and 2, only parental involvement was investigated as one of the parenting dimensions. Meanwhile, previous studies (Grolnick, 2009; Grolnick et al, 2009; Farkas & Grolnick, 2010) suggest that there are three dimensions of parenting: parental involvement, parent autonomy support, and parental structure. The other two dimensions of parenting can be influential in the research on the effects of parental involvement related to the constructs. Thus, it can give recommendations for supporting parental involvement in their children's education.

Study 2 revealed that parental involvement influenced the basic needs satisfaction in general. If investigated partially, parental involvement has an impact only on the fulfilment of the need for competence. It does not have an impact on the fulfilment of the needs for autonomy and relatedness. In their article, Grolnick et al. (2009) described the correlation between each parenting dimension and various variables in the educational contexts of students, particularly in relation to motivation and school outcomes. The school engagement variables are rarely discussed in that article.

Consequently, the engagement theoretical framework (Figure 1) suggested by Appleton et al. (2008), can be used to answer the research questions related to how parental involvement influences school engagement through basic needs satisfaction of students. The model that has been constructed based on preliminary studies will be examined for future research in an Indonesian parenting context.

REFERENCES

Appleton, J.J., Christenson, S.L. & Furlong, M.J. (2008) Student engagement with school: Critical conceptual and methodological issues of the construct. *Psychology in the Schools*, *45*(5), 369–386. doi:10.1002/pits.20303.

Berger, E.H. (1995) *Parents as partners in education* (4th ed.). Upper Saddle River, NJ: Prentice Hall.

Brooks, J.B. (2001) *Parenting* (3rd ed.). Mountain View, CA: Mayfield Publishing.

Connell, J.P. & Wellborn, J.G. (1991) Competence, autonomy, and relatedness: A motivational analysis of self-system processes. *Self-processes and development. The Minnesota symposia on child psychology*, *23*, 1954, 43–77. Retrieved from http://www.jamesgwellborn.com/pdf/ConnellWellbornChapter.pdf.

Deci, E.L. & Ryan, R.M. (2000) The 'what' and 'why' of goal pursuits: Human needs and self-determination of behaviour. *Psychological Inquary*, *11*(4), 227–268. doi:10.1207/S15327965PLI1104.

Deci, E.L. & Vansteenkiste, M. (2004) Self-determination theory and basic need satisfaction: Understanding human development in positive psychology. *Ricerche Di Psicologia*, *27*, 23–40. Retrieved from http://psycnet.apa.org/psycinfo/2004-19493-002.

Domina, T. (2005) Leveling the home advantage: Assessing the effectiveness of parental involvement in elementary school. *Sociology of Education*, *78*(3), 233–249. doi:10.1177/003804070507800303.

Fan, W. & Williams, C.M. (2010) The effects of parental involvement on students' academic self-efficacy, engagement and intrinsic motivation. *Educational Psychology*, *30*(1), 53–74. doi:10.1080/01443410903353302.

Farkas, M. & Grolnick, W.S. (2010) Examining the components and concomitants of parental structure in the academic domain. *Motivation and Emotion*, *34*(3), 266–279. doi:10.1007/s11031-010-9176-7.

Fredricks, J.A., Blumenfeld, P. & Paris, A. (2004) School engagement: Potential of the concept, state of the evidence. *Review of Educational Research*, *74*(1), 59–109. doi:10.3102/00346543074001059.

Fredricks, J.A., Blumenfeld, P., Friedel, J. & Paris, A. (2005) School engagement. In K.A. Moore & L. Lippman (Eds.), *What do children need to flourish? Conceptualizing and measuring indicators of positive development* (pp. 305–321). New York, NY: Springer Science and Business Media.

Friedenberg, L. (1995) *Psychological testing: Design, analysis, and use*. Boston, MA: Allyn and Bacon.

Furrer, C. & Skinner, E. (2003) Sense of relatedness as a factor in children's academic engagement and performance. *Journal of Educational Psychology*, *95*(1), 148–162.

Grolnick, W.S. (2009) The role of parents in facilitating autonomous self-regulation for education. *Theory and Research in Education*, *7*(2), 164–173. doi:10.1177/1477878509104321.

Grolnick, W.S. & Pomerantz, E.M. (2009) Issues and challenges in studying parental control: Toward a new conceptualization. *Journal Compilation, Society for Research in Child Development*, *3*(3), 165–170. doi:10.1111/j.1750-8606.2009.00099.x.

Grolnick, W.S. & Ryan, R.M. (1989) Parent styles associated with children's self-regulation and competence in school. *Journal of Educational Psychology*, *81*(2), 143–154. doi:10.1037/0022-0663.81.2.143.

Grolnick, W.S. & Slowiaczek, M.L. (1994) Parents' involvement in children' s schooling: A multidimensional conceptualization and motivational model. *Child Development*, *65*(1), 237–252. doi:10.1111/j.1467-8624.1994.tb00747.x.

Grolnick, W.S., Ryan, R.M. & Deci, E.L. (1991) Inner resources for school achievement: motivational mediators of children's perceptions of their parents. *Journal of Educational Psychology, 83*(4), 508–517.

Grolnick, W.S., Kurowski, C.O., Dunlap, K.G. & Hevey, C. (2000) Parental resources and the transition to junior high. Journal of Research on Adolescence, 10(4), 465–488. doi:10.1207/SJRA100405.

Grolnick, W.S., Friendly, R. & Bellas, V. (2009) Parenting and children's motivation at school. In *Handbook of motivation at school, 1966* (pp.279–300). Retrieved from http://books.google.com/books?hl+nl&lr=&id=P5GOAgAAQBAJ&pgis=1.

Hoover-Dempsey, K.V., Battiato, A.C., Walker, J.M.T., Reed, R.P., DeJong, J.M. & Jones, K.P. (2001) Parental involvement in homework. *Educational Psychologist, 36*(3), 195–209. doi:10.1207/S15326985EP36035.

Jang, H., Reeve, J., Ryan, R.M. & Kim, A. (2009) Can self-determination theory explain what underlies the productive, satisfying learning experiences of collectivistically oriented Korean students? *Journal of Educational Psychology*, *101*(3), 644–661. doi:10.1037/a0014241.

Jeffery, L., Kwok, O., Chang, Y., Chang, B.W. & Yeh, Y. (2014) Parental autonomy support predicts academic achievement through emotion-related self-regulation and adaptive skills in Chinese American adolescents. *Asean American Journal of Psychology, 5*(3), 214–222. doi:10.1037/a0034787.

Jiang, Y.H., Yau, J., Bonner, P. & Chiang, L. (2011) The role of perceived parental autonomy support in academic achievement of Asian and Latino American adolescents. *Electronic Journal of Research in Educational Psychology*, *9*(2), 497–522.

Johnston, M.M. & Finney, S.J. (2010) Measuring basic needs satisfaction: Evaluating previous research and conducting new psychometric evaluations of the basic needs satisfaction in general scale. *Contemporary Educational Psychology*, *35*(4), 280–296. doi:10.1016/j.cedpsych.2010.04.003.

Mo, Y. & Singh, K. (2008) Parents' relationships and involvement: Effects on students' school engagement and performance. *Research in Middle Level Education Online*, *31*(10), 1–11. doi:10.1080/19404476.2008.11462053.

Pomerantz, E.M., Grolnick, W.S. & Price, C.E. (2005) The role of parents in how children approach achievement: A dynamic process perspective. In A.J. Elliot & C.S. Dweck (Eds.), *Handbook of competence and motivation* (pp. 259–278). New York, NY: The Guilford Press.

Purwono, U. (2010) Metode dan prosedur adaptasi tes psikologi. In Supratiknya A. & S. Tjipto (Eds.). *Redefinisi Psikologi Indonesia Dalam Keberagaman* (pp. 347–373). Indonesia: Penerbit HIMPSI.

Raftery, J.N., Grolnick, W.S. & Flamm, E.S. (2012) Families as facilitators of student engagement: toward a home-school partnership model. In S.L. Christenson, A.L. Reschly, & C. Wylie (Eds.), *Handbook of research on student engagement* (pp. 343–364). New York, NY: Springer Science and Business Media.

Reeve, J. (2002) Self-determination theory applied to educational settings. In E.L. Deci & R.M. Ryan (Eds.), *Handbook of self-determination research* (pp.183–203). Rochester, NY: University of Rochester Press.

Ryan, R.M. & Connell, C.P. (1989) Perceived locus of causality and internalization: Examining reasons of acting in two domains. *Journal of Personality and Social Psychology*, *57*(5), 749–761.

Ryan, R.M. & Deci, E.L. (2000) Intrinsic and extrinsic motivations classic definitions and new directions. *Journal of Educational Psychology*, *25*, 54–67. doi:10.1006/ceps.1999.1020.

Ryan, R.M. & Deci, E.L. (2009) Promoting self-determined school engagement: Motivation, learning, and well-being. In K.R. Wentzel & A. Wigfield (Eds.), *Handbook of motivation at school* (pp. 171–195). New York, NY: Routledge publishing.

Savitri, J., Setyono, I.L., Cahyadi, S. & Srisayekti, W. (2015) Teachers' views about parent involvement and school engagement. Paper presented at the Seventh International AAICP Conference. Sumedang, Indonesia: Faculty of Psychology, Padjadjaran University.

Skinner, E.A. & Pitzer, J.R. (2012). Developmental dynamics of student engagement, coping, and everyday resilience. In S.L. Christenson, A.L. Reschly, & C. Wylie (Eds.), *Handbook of research on student engagement* (pp. 21–44). New York, NY: Springer Science and Business Media.

Diversity in Unity: Perspectives from Psychology and Behavioral Sciences – Ariyanto et al. (Eds)

Applied behaviour analysis and video modelling programme to enhance receptive and expressive abilities in children with mild autism

F. Putra & F.M. Mangunsong
Faculty of Psychology, Universitas Indonesia, Depok, Indonesia

ABSTRACT: The objective of this research is to examine whether a programme of Applied Behaviour Analysis (ABA) and video modelling can enhance receptive and expressive abilities in children with mild autism. Receptive ability is defined as the ability to match, point to, and name basic emotions on facial expression cards, whereas expressive ability is the ability to express inconvenient feelings to others. This research uses a single-subject design in relation to a child with mild autism. The programme was administered for two weeks. After that, the generalisation phase was introduced for one week. The result of this research shows that receptive and expressive abilities improved after the programme was administered. Even though the programme was stopped for a week, the participant still mastered the receptive and expressive abilities well. According to this research, parents can teach receptive and expressive abilities to their children by using the ABA method in a child's natural setting.

1 INTRODUCTION

Language is a basic skill that children need to communicate with their social environment. Children can express their ideas and feelings through language (Papalia et al., 2008). Language skills not only have a role in children's communication skills, but they also determine children's social skills (Maurice, 1996). Language is an important tool for children to interact with other people (Helland et al., 2014). Children can control their emotion and behaviour on a daily basis with language. Language skills have a significant role in determining children's social skills; however, some children with special needs, especially autism, experience language deficit from the time they are toddlers (Sundheim & Voeller, 2004; Hallahan & Kauffman, 2006; Paelt et al., 2014; Pinborough-Zimmerman et al., 2007).

Research by Loveland and Kelly (1991) found that language skills increase with age in children with autism, but another study found that language skills do not automatically develop as chronological age increases (Kasari et al., 2008). Kasari et al. (2008) found that children with autism who are given a special intervention programme to increase their language skills show better receptive and expressive language skills than children who do not get the intervention. This indicates that language skills need to be taught to children with autism who have deficits in receptive and expressive language. The common receptive language problem faced by children with autism is comprehending emotional expression of the face (Wright & Poulin-Dubois, 2012), whereas the common expressive language problem faced by children with autism is the ability to express one's own feelings (Rapin, 1999, in Pry et al., 2005).

A few studies show that children who are given a special intervention programme to increase language skills show better behaviour in other areas (Carpenter & Tomasello, 2000; Anderson et al., 2009; Mawhood et al., 2000). Research by Carpenter and Tomasello (2000) found that by giving intervention to children with autism, not only do the children's language skills increase, but also their social skills are enhanced. After receiving the intervention programme, the children become more interested in social interaction.

Intervention in language is important for children with autism (Anderson et al., 2009) because language skills at an early age are a strong predictor of social skills when these children grow up. There are four intervention techniques commonly used to develop the language skills of children with autism: direct instruction, computer-assisted instruction, Applied Behaviour Analysis (ABA), and Video Modelling (VM). First, the direct instruction method is preferred for children who have enough receptive language skills. Children with limited receptive language skills tend to have difficulties when guided by the direct instruction method that consists of more verbal instructions (Cole & Chan, 1990). Second, the computer-assisted instruction programme (Chen & Bernard-Opitz, 1993) does not significantly increase children's language skills, making them dependent on devices and less dependent on communication in social interaction (Chen & Bernard-Opitz, 1993). Third, ABA is a method that can be applied to children with autism at early age (Maurice, 1996). It applies a behavioural approach that uses reinforcement to display or increase certain behaviour. ABA is suitable for children with autism who have receptive language deficit because this method emphasies firm, clear, and brief instructions (Maurice, 1996). Lastly, VM is a new method considered effective for increasing language skills of children with autism (Delano, 2007; Ganz et al., 2014; Shukla-Mehta et al., 2009, in Laarhoven et al., 2010). Usually, children with autism can absorb information better if the information is delivered visually (Ganz et al., 2013). VM is an intervention technique in which the materials are delivered through visual aids. This method is highly interesting for children with autism because videos can grab their attention more easily as they require minimum social interaction with other people.

ABA and VM seem to be appropriate intervention methods for children with autism who have a language deficit. According to Shukla-Mehta et al. (2009), VM is an intervention method that can be paired with other methods. This video priming method can be used to demonstrate skills that will be taught and a prompt or facility that can correct the child's behaviour during the intervention. Therefore, this study examines the use of ABA and VM intervention methods to enhance receptive and expressive abilities in a child with mild autism who has difficulties in communication and social interaction. Based on systematic observation using the Childhood Autism Rating Scale, the study subject is assessed to have limited language skills. Therefore, in this study, an intervention programme using ABA and VM is administered to increase language skills of a child with mild autism. The subject will be taught receptive and expressive language skills (Maurice, 1996). This study hypothesises that ABA and VM intervention methods will effectively increase receptive and expressive abilities in children with mild autism.

2 METHODS

2.1 *Design*

This study used a single-subject design and an A–B–A assessment method. A–B–A assessment is a method where receptive and expressive language skills are assessed in three stages: baseline (A), intervention (B), and post-test (A). Apart from the three stages of assessment, receptive and expressive language skills were assessed at the maintenance stage.

2.2 *Subject*

The subject, named G, was a five-year-old boy, enrolled in a kindergarten private school at Jakarta Selatan. He was an only child and had mild autism. G was chosen by purposive sampling. According to the psychological test administered, the subject had average intelligence, especially in concrete-practical problem solving (intelligence quotient = 103, Wechsler Scale), and enough abilities to absorb information, but G's language development was not on a par with that of other children his age. He could not comprehend sentences with long instructions. It was so difficult for him to do tasks correctly that he rarely finished his tasks at school.

G did not have enough emotional maturity at the time of the study. He was still unable to express feelings of sadness and anger in an appropriate way through language. This affected his behaviour such that he tended to express his anger through smashing and hitting objects, shouting loudly. G's language deficit made it difficult for him to express unpleasant emotions that he felt, so he was often silent when he experienced unpleasant feelings. During the examination, G often wet his pants because he could not express his need to urinate. Also, when he could not comprehend the teacher's instructions in the class, G looked like he was thinking but he did not do the task.

2.3 *Intervention design*

Based on the interview, we found that G could not recognise basic emotions, so he had difficulties in expressing his emotions. G was unable to express verbally whether he was sad or angry. He usually hit the door and smashed toys when he was hungry or sleepy. Moreover, when he was at school, he stayed silent when he did not understand the task set by the teacher. He was unable to express that he did not understand the instructions and needed help.

The skills taught in this study were sequential. The subject needed to master the most basic skill first, and then he would be taught a more advanced skill. Once the subject has successfully mastered the ability to identify basic emotions, by the end of the programme the subject should have learned how to express unpleasant feelings verbally to other people. Based on results of the assessment that we did in baseline phase, we chose the target behaviour to be taught to the subject. This target behaviour was based on material constructed by Maurice (1996).

1. Receptive language is the ability to comprehend meaning from non-verbal cues and also basic emotions that are shown through facial expressions. In this study, the subject was taught the ability to comprehend only basic emotions. The skill of comprehending basic emotions was chosen based on the assessment that refers to Maurice's (1996) curriculum. The assessment shows that G had mastered materials at the beginner level, so we chose materials for the intermediate level. At the intermediate level, G had already mastered skills of matching but not receptive language, so receptive language was chosen as the material in the intervention. Receptive language is the ability to comprehend basic emotions, such as sadness, anger, surprise, and fear.
2. Expressive language is the ability to express intention and feelings, such as pleasant or unpleasant emotions. In this study, the subject was taught the ability to recognise only unpleasant feelings. Materials in this intervention programme were constructed based on the ABA curriculum by Maurice (1996) at the intermediate level. Materials at this level consist of the ability to express unpleasant and pleasant feelings; however, in this study, we only chose the ability to express unpleasant feelings. The materials were chosen based on results of the assessment at baseline. The unpleasant feelings that G was taught to express are feeling sleepy, hungry, confused, and the urge to urinate and defecate.

2.4 *Intervention module*

The intervention module was designed according to the results of an assessment through observation and interview and Maurice's (1996) ABA materials. The module materials were adapted in consultation with the child's parents and the teachers' needs at school, because a good intervention programme is one that has practical benefits for the child's daily life (Luiselli et al., 2008).

The classic ABA method that was developed by Lovas (1987, in Luiselli et al., 2008) emphasises that intensive intervention must be conducted for 40 hours in a week. However, several studies state that intervening for 12–27 hours is also effective in developing the language skills of a child with autism (Sheinkopf & Siegel, 1998, in Luiselli et al., 2008). Luiselli et al. (2008) found that giving intervention for 6–20 hours in a week is effective enough in developing six developmental domains of a child with autism. The one domain developed through the intervention programme in this study was receptive and expressive language.

Table 1. Intervention schedule design.

Week	Session	Location
I	1st session: Matching cards skill	Subject's home
	2nd session: Showing expression cards skill	
	3rd session: Naming emotions cards skill	
	4th session: Repeating session 1–3 materials	
	5th session: Receptive language post-test	
II	Receptive language generalisation	Subject's school
	6th session: Expressing unpleasant feelings	Subject's home
	7th session: Repeating Session 6 materials	
III	8th session: Expressive language post–test	Subject's home
	Expressive language generalisation	Subject's school
IV	Maintenance period: Generalise skills into day-to-day environment	Subject's school

The module used ABA and VM. Receptive and expressive language was taught through a Discrete Trial Teaching (DTT) technique and VM was an additional technique. After G learned the two skills in a therapy setting, they were generalised in a natural setting. Generalisation is a process that aims to transfer the skills acquired by the subject in a therapy setting to his day-to-day environment. Generalisation was conducted in the school setting because G was more likely to interact with other people when at school. Intervention was held for three weeks at G's home and school. At the home, intervention was held in a room with adequate lighting and few stickers or objects, so the child would not be distracted. Every session was held for 60–120 minutes, depending on speed and the child's feelings during the session. Intervention was given everyday from Monday to Friday for two weeks. The intervention schedule design is given in Table 1.

2.5 *Scoring method and success indicator*

The scoring method and success criteria followed the principles of the ABA method (Maurice, 1996). Assessment of conditions before and after the intervention used a systematic observation method with an intra-individual focus, so that the observation results become more comprehensive on one individual (Sattler, 2002). Scoring was based on the rate of success, which is the success gained in ten trials in every material (%). For example, in ten trials if the child succeeds four times without prompt, the child gains the score 4/10 (40%). Score criterion considered successful is 80%, so this programme was considered successful if the subject obtained a minimum score of 80% out of ten trials on receptive and expressive language intervention (Maurice, 1996). In every trial, the child received 1 point if he succeeded in responding correctly without prompt, 0 point for responding incorrectly, and 0.5–0.75 point if he succeeded with prompt. The subject received 0.5 point if he responded correctly with the help of a physical or modelling prompt, whereas he received 0.75 point if he responded correctly with the help of a verbal prompt.

3 RESULTS

3.1 *Receptive language*

Receptive language materials were given through three activities: matching, pointing, and naming emotions on expression cards. Matching and pointing were assessed at the baseline, intervention 1, intervention 2, post-test, and maintenance stages. Both the aspects were not assessed at the generalisation stage because they were just requirements for the naming emotions skill. At the generalisation stage, receptive language taught was only for the naming emotions skill because it is the behavioural aspect that is expected to come out in a natural setting. There was a positive increase in the subject's receptive language. The subject's receptive language based on the pre-test was low. He could only match, point to, and name emotions with <80% success. At the intervention stage, the subject's receptive language increased to >80%. This progress continued until the generalisation stage. Details are given below.

1. Matching expression cards

As seen in Figure 1, there was an increase of 60–80% in receptive language at intervention 1stage. This positive trend continued till the post-test stage where the subject's skill to match expression cards for all basic emotions reached 100% and this success rate remained until the maintenance stage.

2. Pointing to emotions

Accurately pointing to emotions on expression cards at baseline was low; the success rate was ≤30% for all the basic emotions. After the subject received intervention through the ABA technique, his skill of pointing to cards increased by ≥80%. Even at the material repetition stage or intervention 2, G's ability to point to emotions on expression cards reached 100% for every basic emotion. G was still able to point to emotions on expression cards of surprise, fear, anger, and sadness with a success rate of 100% at the maintenance stage (Figure 2).

3. Naming emotions on expression cards

Similar to the two aspects of receptive language already discussed, at baseline, G could name basic emotions with a low success rate of <30%. After receiving the intervention programme with the ABA method, G's ability to name basic emotions increased to 95% for the emotion

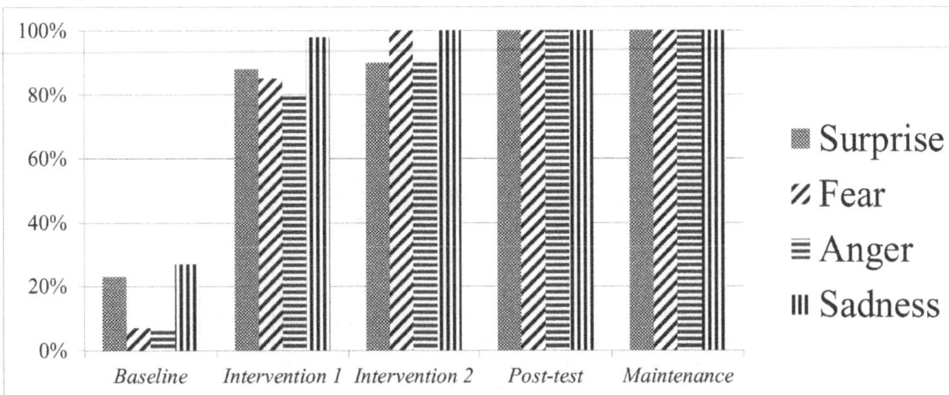

Figure 1. Matching emotions on expression cards from baseline to the maintenance stage.

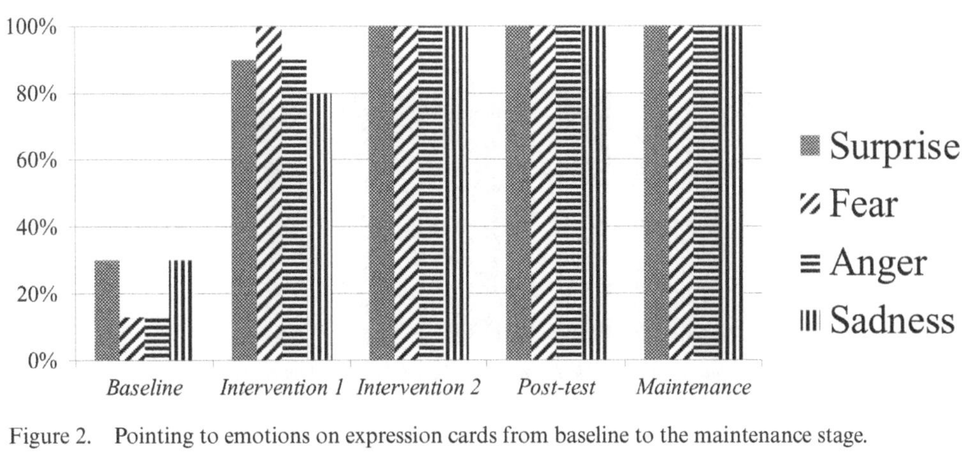

Figure 2. Pointing to emotions on expression cards from baseline to the maintenance stage.

of anger; his success rate for the emotions of surprise, fear, and sadness remained <80%. On the next day, after repeating the materials for naming emotions, G gained a success rate of >80% for every basic emotion. This positive increase continued until the post-test stage where the subject was able to name basic emotions with a success rate of 100%. The success rate of this skill stayed at 100% until the maintenance stage.

3.2 *Expressive language*

Expressive language materials were administered through two intervention methods of ABA and VM. The result of observation at baseline showed that the subject still could not express his unpleasant feelings. G's expressive language started to increase when he received the intervention. On the first day of intervention, G succeeded in expressing all of the unpleasant feelings that he was trained to identify, with a success rate of ≥ 80%. The rate displayed by G until the maintenance stage.

Figure 4 shows that there was a high increase in expressive language at the intervention stage. After video priming by the VM method and using ABA, G's expressive language increased to 70–100%. At the first intervention stage, G could express unpleasant feelings—feeling confused and the need to defecate—with a success rate of >80%. It continued to increase until the post-test stage, from 80% to 90%. Although G's ability to express confused feelings decreased by 10% at the generalisation stage, G once asked about the meaning of confused at this stage. After a verbal prompt, G finally responded correctly. At the final or maintenance stage, G mastered expressive language with a success rate of 100% for feeling sleepy, hungry, and the need to urinate or defecate. Expressive language for feeling confused still had a success rate of 80%.

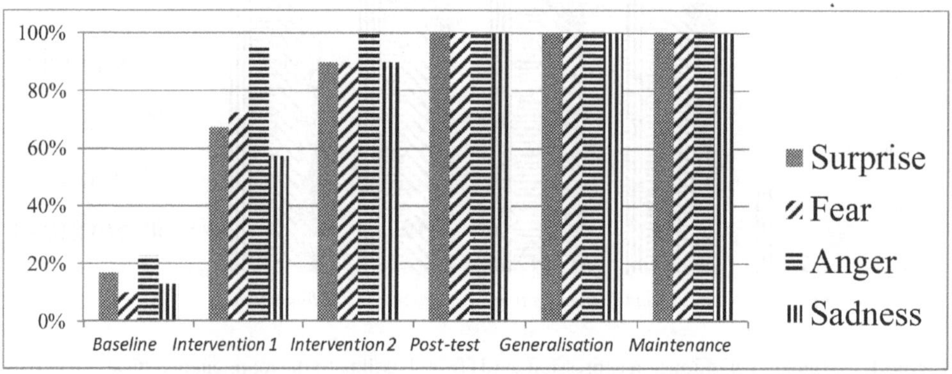

Figure 3. Naming emotions on expression cards from baseline to the maintenance stage.

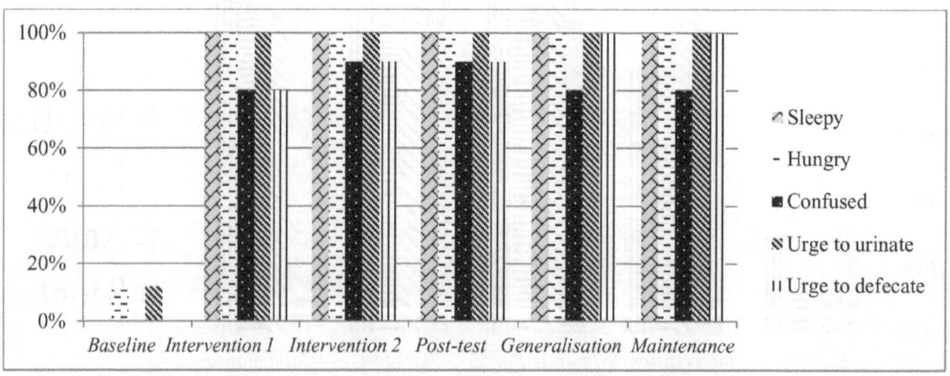

Figure 4. Expressive language from baseline to the maintenance stage.

4 DISCUSSION

In this research, it has been proved that ABA successfully increases receptive and expressive language in children with mild autism. DTT is one of the main principles in the ABA method and helps the subject learn to gradually discriminate instructions because the instructions were delivered clearly, explicitly, and consistently. Slowly, the participant could respond accurately from the stimuli presented.

Children with autism generally have some similarities (Hallahan & Kauffman, 2006). They can easily get disturbed if there is an environmental or instructional change that they had learned. Giving instructions with the DTT principle definitely supports this characteristic. In this study, there was one instruction that was not understood by G. It was the instruction to express feeling confused. Confusion is a condition where the context is very broad. Unlike feeling hungry, sleepy, and the need to urinate and defecate, confusion can appear in various conditions. This complicated the intervention, especially at the generalisation stage. At the generalisation stage, we tried to introduce confusion with new contexts, but G needed a long time to comprehended the condition, so the subject's success rate on expressing feeling confused decreased by 10%.

Applying reinforcement strategy is effective in increasing children's interests in participating in an intervention. Unfortunately, we did not get a chance to administer the reinforcement trial before the intervention was held. The trial aims to determine the reinforcement degree at which the child will get his most liked reinforcement if he succeeds at responding without any prompt. The impact of the lack of a reinforcement test was that the subject was not interested on the first day of intervention. However, from the second day onwards after we successfully identified the type of reinforcement G likes the most, he began to be interested. Laarhoven et al. (2010) found that using the VM method itself is effective enough to teach receptive materials to children with autism. The VM method is usually combined with the behaviouristic approach. Generally, reinforcement is used to increase children's interests in learning. In the VM method, using the video itself is already a reinforcement for children. Children with autism really like visual stimuli, so they are more likely to put their attention on studying when the materials are presented through visual tools (Ganz et al., 2013).

It has been proved that the ABA method increases receptive and expressive language skills of children with autism, but criticism of the method says that the ability that is trained cannot be generalised in real life (Luiselli et al., 2008). However, the current study found the opposite result of this critique. In this study, most of the ability that was taught increased to a success rate of 100% at the generalisation stage. The subject could display the right behaviour quickly although he received different stimuli. The VM method helped G understand the social context through video because it uses less verbal instruction. Delano (2007) found that the ABA method is more effective if it is combined with the VM method. When we used VM as a supporting method with the ABA method, the ability that was taught could be more generalised in daily life.

In this research, we found that the ABA and VM methods effectively increased receptive and expressive abilities in children with mild autism. However, a notable limitation in the current study is that we cannot generalise this to other subjects or conditions even with the same characteristics because it was only specific to one subject. Nevertheless, that it focuses on only one subject is also an advantage, as the effectiveness of the intervention can be investigated (Gravetter & Forzano, 2009). The sustainability of the behaviour that is formed by the intervention may be a problem. Maurice (1996) said that receptive and expressive language of children with autism need to be trained continuously even after the children have mastered them. We suggest that the programme is continued with more advanced materials. As a child's age increases, the skills to do academic activities at school also increase. Therefore, ABA and VM methods need to be applied in children's daily life, so they can master new skills in order to adapt to the environment.

5 CONCLUSION

The result shows that there was an increase in the success rate of the subject's receptive and expressive language skills after receiving the intervention. G's ability to accurately match and point expression cards and name basic emotions increased to >80%. G's ability to accurately express feeling sleepy, hungry, confused, and the need to urinate and defecate increased to ≥ 80%. Receptive and expressive language skills that were taught stayed at a success rate of ≥ 80%, even when G no longer received repetition of materials for one week. In short, ABA and VM can increase receptive and expressive language skills of children with autism.

REFERENCES

Anderson, D.K., Oti, R.S., Lord, C. & Welch, K. (2009). Patterns of growth in adaptive social abilities among children with autism spectrum disorders. *Journal of Abnormal Child Psychology*, *37*(7), 1019–1034. doi:10.1007/s10802-009-9326-0.

Carpenter, M. & Tomasello, M. (2000). Joint attention, cultural learning, and language acquisition: Implications for children with autism. In A.M. Wetherby & B.M. Prizant (Eds.), *Autism spectrum disorders: A transactional developmental perspective* (pp. 31–54). Baltimore, MD: Paul H. Brookes Publishing.

Chen, S.H.A. & Bernard-Opitz, V. (1993). Comparison of personal and computer-assisted instruction for children with autism. *Mental Retardation*, *31*(6), 368–376. doi:10.1017/CBO9781107415324.004.

Chiang, H.M. (2009). Differences between spontaneous and elicited expressive communication in children with autism. *Research in Autism Spectrum Disorders*, *3*(1), 214–222. doi:10.1016/j.rasd.2008.06.002.

Cole, P.G. & Chan, L.K.S. (1990). *Methods and strategies for special education*. Victoria, Australia: Prentice Hall.

Delano, M.E. (2007) Video modeling interventions for individuals with autism. *Remedial and Special Education*, *28*(1), 33–42. doi:10.1177/07419325070280010401.

Ganz, J.B., Boles, M.B., Goodwyn, F.D. & Flores, M.M. (2014). Efficacy of handheld electronic visual supports to enhance vocabulary in children with ASD. *Focus on Autism and Other Developmental Disabilities*, *29*(1), 3–12. doi:10.1177/1088357613504991.

Gravetter, F.J. & Forzano, L.B. (2009). *Research methods for the behavioral sciences.* Belmont, CA: Wadsworth Cengage Learning.

Hallahan, D.P. & Kauffman, J.M. (2006). *Exceptional learners: Introduction to special education* (10th ed.). Upper Saddle River, NJ: Pearson Education.

Helland, W.A., Lundervold, A.J., Heimann, M. & Posserud, M.B. (2014). Stable associations between behavioral problems and language impairments across childhood—The importance of pragmatic language problems. *Research in Developmental Disabilities*, *35*(5), 943–951. doi:10.1016/j.ridd.2014.02.016.

Howlin, P. (1998). *Children with autism and Asperger syndrome: A guide for practitioners and carers.* Chichester, UK: John Wiley.

Kasari, C., Paparella, T., Freeman, S. & Jahromi, L.B. (2008). Language outcome in autism: Randomized comparison of joint attention and play interventions. *Journal of Consulting and Clinical Psychology*, *76*(1), 125–137.

Kumar, R. (2012). *Research methodology: A step by step guide for beginners* (3rd ed.). London, UK: SAGE Publications.

Laarhoven, T.V., Kraus, E., Karpman, K., Nizzi, R. & Valentino, J. (2010). A comparison of picture and video prompts to teach daily living skills to individual with autism. *Focus on Autism and Other Developmental Disabilities*, *25*(4), 195–208.

Lewis, F.M., Woodyatt, G.C. & Murdoch, B.E. (2008). Linguistic and pragmatic language skills in adults with autism spectrum disorder: A pilot study. *Research in Autism Spectrum Disorders*, *2*(1), 176–187. doi:10.1016/j.rasd.2007.05.002.

Loveland, K.A. & Kelly, M.L. (1991). Development of adaptive behaviour in preschoolers with autism or down's syndrome. *American Journal on Mental Retardation*, *96*(1), 13–20.

Luiselli, J.K., Russo, D.C., Christian, W.P. & Wilczynski, S.M. (2008). *Effective practices for children with autism: Educational and behavioral support interventions that work.* New York, NY: Oxford University Press.

Maurice, C. (Ed.). (1996). *Behavioral intervention for young children with autism*. Austin, TX: Pro-Ed.

Mawhood, L., Howlin, P. & Rutter, M. (2000). Autism and developmental receptive language disorder—A comparative follow-up in early adult life. I: Cognitive and language outcomes. *Journal of Child Psychology and Psychiatry and Allied Disciplines*, *41*(5), 547–559. doi:10.1111/1469-7610.00642.

Papalia, D.E., Olds, S.W. & Feldman, R.D. (2008). *Human development* (10th ed.). New York, NY: McGraw-Hill.

Paelt, S.V.D., Warreyn, P. & Roeyers, H. (2014). Social-communicative abilities and language in preschoolers with autism spectrum disorders: Associations differ depending on language age. *Research in Autism Spectrum Disorders*, *8*(5), 518–528.

Pinborough-Zimmerman, J., Satterfield, R., Miller, J., Bilder, D., Hossain, S. & McMahon, W. (2007). Communication disorders: Prevalence and comorbid intellectual disability, autism, and emotional/behavioral disorders. *American Journal of Speech-Language Pathology*, *16*(4), 359–367. doi:10.1044/1058-0360(2007/039).

Pry, R., Petersen, A. & Baghdadli, A. (2005). The relationship between expressive language level and psychological development in children with autism 5 years of age. *Autism*, *9*(2), 179–189. doi:10.1177/1362361305047222.

Sattler, J.M. (2002). *Assessment of children: Behavioral and clinical applications* (4th ed.). San Diego, CA: Jerome M. Sattler.

Shukla-Mehta, S., Miller, T. & Callahan, K.J. (2009). Evaluating the effectiveness of video instruction on social and communication skills training for children with autism spectrum disorders: A review of the literature. *Focus on Autism and Other Developmental Disabilities*, *25*(1), 23–36. doi:10.1177/1088357609352901.

Sundheim, S.T.P.V. & Voeller, K.K.S. (2004). Psychiatric implications of language disorders and learning disabilities: Risks and management. *Journal of Child Neurology*, *19*, 814–826. doi:10.1177/08830738040190101001.

Wright, K. & Poulin-Dubois, D. (2012). Modified checklist for autism in toddlers (M-CHAT) screening at 18 months of age predicts concurrent understanding of desires, word learning and expressive vocabulary. *Research in Autism Spectrum Disorders*, *6*(1), 184–192.

[illegible], D., [illegible], S. W., & [illegible], R. D. (2004). [illegible] (10th ed.). New York, NY: [illegible].

[illegible], S. D., [illegible], & [illegible], H. (2014). [illegible] in preschoolers with [illegible] developing use of language. [illegible], 8, 518–[illegible].

[illegible], L., [illegible], S., [illegible] (2007). [illegible]. doi: [illegible]/2007[illegible]

[illegible], R., [illegible], A., & [illegible], A. (2008). [illegible] language [illegible]. [illegible]

[illegible], M. (1957). [illegible]

[illegible] (2014). [illegible]

[illegible], R. V., [illegible] (2016). [illegible]

[illegible] (2017). [illegible]

Diversity in Unity: Perspectives from Psychology and Behavioral Sciences – Ariyanto et al. (Eds)
© 2018 Taylor & Francis Group, London, ISBN 978-1-138-62665-2

The use of the pajares principles to increase mathematics self-efficacy in a middle childhood student

A. Selandia, W. Prasetyawati & R.M.A. Prianto
Faculty of Psychology, Universitas Indonesia, Depok, Indonesia

ABSTRACT: Self-efficacy is an important factor that determines a student's ability to achieve the intended learning outcomes (Bandura, 2010 in Santrock, 2012). In this research, E is a middle childhood student who displays low mathematics self-efficacy. The intervention given to increase E's math self-efficacy proposed in the research followed the principles of Pajares, which consist of modeling, attributional feedback, and goal setting methods. The research is conducted using a single subject design. E's mathematics self-efficacy is measured using a Likert scale given on the pre-test and post-test. The program was administered for 6 days, and the result of this research shows that there was an increase of E's mathematics self-efficacy as it can be seen from the comparison of the Likert scale in the pre-test and post-test.

1 INTRODUCTION

Among the various factors that affect a student's achievement in school is self-efficacy (Papalia, Olds & Feldman, 2005). Self-efficacy is defined as an individual's subjective perception of their ability to perform in a given situation to achieve their desired result (APA Dictionary of Psychology, 2015). People perceive information to judge their self-efficacy from their actual performance, indirect experience (observation), a variety of persuasion, and psychological symptoms (Usher, 2009; Usher & Pajares, 2008 in Schunk, Meece & Pintrinch, 2014). Moreover, Bandura (1997 in Jackson, 2012) states that people form their self-efficacy by choosing and interpreting information from at least four sources: past performance or expertise experience, observation of others' actions, social persuasion based on others' judgments and feedbacks (i.e., from parents, teachers, and peers), and their own psychological states that include anxiety, passion*, mood, or fatigue. Among them, performance has the strongest impact.

Individuals who believe that they have the understanding and the ability to apply the strategies to support their learning process effectively are more likely to display a better sense of control on their learning results, which will promote the self-efficacy and motivation to apply those strategies (Schunk, Meece & Pintrich, 2014). Students with high self-efficacy (i.e., students who believe that they can accomplish their school tasks and manage their learning) are more likely to try to perform and also more likely to succeed compared to students who do not believe in their own abilities (Bandura, Barbaranelli, Caprara & Pastorelli, 1996 in Papalia, Olds & Feldman, 2007). Moreover, individuals with high self-efficacy will most likely put in more effort when faced with challenges and will show endurance to complete a task as long as they have the skills required (Schunk, Meece & Pintrich, 2014). On the contrary, students with low self-efficacy and expectations are more likely to display resigned attitude, ignorance, and reluctance or inability to exert effort (Schunk, Meec & Pintrich, 2014). The characteristics of such students are similar to the characteristics of students with a learned helplessness attitude (Peterson, *et al.,* 1993 in Schunk, Meece & Pintrinch, 2014). A few attitudes that suggest learned helplessness, according to Santrock (2012), include saying "*I cannot do this,*" not paying attention to instructions from teachers, not asking ask for help even when needed, doing nothing (e.g., looking out the window), guessing or giving random answers without trying, not showing pride in success, looking bored and uninterested, not responding to teachers' advice to try, easily giving up, not answering teachers' questions voluntarily, and being more likely to avoid work (e.g., going to the sick bay).

A student's self-efficacy is an individual's judgment of their ability to fulfill the assigned performance and achieve specific results (Pajares, 1996). In this research, the method of intervention follows the Pajares principle. These principles from Pajares are based on Schunk's (1981 in Pajares, 1996) theory, which consists of methods, such as modeling, attributional feedback, and goal setting. Modeling is a show-and-tell method (i.e., showing and explaining) commonly conducted by teachers to explain arithmetic operations. Teachers explain how to solve a problem and demonstrate it on the white board. This approach is reasonable and most students can learn using this method (Sulzer-Azaroff & Mayer, 1986). Providing explanations and examples through modeling is more effective than giving only the explanation (Rosenthal & Zimmerman, 1978 in Schunk, 1986). Observing the strategy effectively through modeling should increase the observer's self-efficacy to learn because modeling implicitly delivers a message to students that they have the ability to learn and apply strategies (Schunk, 1984, 1985 in Schunk, 1986).

In school, the most important result in modeling is feedback from the teacher. For example, feedback from the teacher to a student can inform other students of the success (*"That's right"*), the progress (*"You've shown progress"*), the types of strategy used (*"You've implemented the steps in the right order"*), and what is expected in a task (*"Don't get distracted with the details in the story"*). Such statements can increase the student's self-efficacy to the limit where they believe that they are giving a similar performance or they are able to give a similar performance (Schunk, 1986). The involvement of parents has been found to have a positive effect on student's achievement, including on their grades and test results (Deslandes, *et al.*, 1999 in Hoover-Dempsey, *et al.*, 2005). At home, a child needs feedback from their parents regarding their work. Parents can support their learning activity and increase their children's self-efficacy through appropriate reinforcement of their academic behavior (Hoover-Dempsey & Sandler, 1995 in Peiffer, 2015). The reinforcement mentioned above can be in the forms of praises, support, and rewards. Reinforcement helps build self-efficacy by persuading children to complete a task. When children are given reinforcements they enjoy, they will be more likely to perform the expected behavior in order to get additional reinforcements (Lysakowski & Walberg, 1981 in Peiffer, 2015). If the behavior that increases their academic skills continues, consequently the student's academic self-efficacy will also increase (Peiffer, 2015).

According to the research conducted by Foote (1999), students who receive positive feedback regarding their ability and effort generally have higher self-efficacy and mathematical skills compared to those who do not receive attributional feedback. Successful students generally plan the steps they need to take to achieve their goal, and then they consciously master the steps to where they can achieve their goal naturally and easily (Kline, *et al.*, 1992 in Barron, 2000). Usually, a difficult task can become easier when it is broken down into a few steps (Barron, 2000). Without a goal, students are more likely to be less sure about their ability because they do not have a standard as a benchmark of their progress (Schunk, 1983a, 1983c in Barron, 2000).

Social cognitive theorists hypothesize that believing in one's own ability affects academic achievement, as students' confidence regarding their ability determines what they will do with their skills and knowledge (Pajares & Miller, 1997). Thus, social cognitive theorists believe that academic performance is determined largely by students' confidence in performing academic tasks (Bandura, 1997; Schunk, 1991 in Pajares & Miller, 1997).

The term mathematics self-efficacy in this research refers to an individual's self-efficacy to solve mathematical problems and to demonstrate academic success in subjects related to mathematics (Betz & Hackett, 1983; Pajares & Miller, 1995 in Ogden, 2012). The students' confidence regarding their ability additionally helps determine what they will do with their knowledge and skills (Pajares & Miller, 1997). For example, when a student is performing a mathematical test, the confidence they have when reading and analyzing a problem specifically will help determine the effort that they will contribute to solve the problem. Students with higher self-confidence will more likely work harder, longer, and less anxiously. As a result, the chance of succeeding in their academic performance increases (Pajares & Miller, 1997).

In an academic setting, self-efficacy instruments may ask the students to rate their confidence in solving specific mathematical problems (Hackett & Betz, 1989 in Pajares, 1996). Researchers usually assess mathematics self-efficacy by asking students to indicate their level of confidence in solving various mathematical problems using a Likert scale. Specifically, students are given a few numbers of mathematical problems and are then asked to rate their confidence in completing each problem correctly (Pajares & Miller, 1997). Pajares and Miller (1997) in their research gave two types of assessment, which consist of an assessment on mathematics performance and a self-efficacy assessment. A sample of mathematics performance assessment was a mathematical problem, while the self-efficacy assessment used instructions like, "How confident are you in answering the question below correctly without using a calculator?" The sample problems were given either in open-ended or multiple choice formats. In the self-efficacy instrument, students were asked to respond to a six-point Likert scale ranging from 1 (not confident at all) to 6 (very confident) to rate their confidence in answering each question correctly.

The subject in this research is a female student (E) who was 11 years and 6 months of age. E's numeric ability was classified as average (based on the WISC scale) so she should have had sufficient ability to follow studies involving numbers in school. Her performance in math was not optimal, as shown by her many below average scores in math. The low scores that she constantly obtained in math contributed to her low self-efficacy in math. E usually thought that she did not have adequate ability and was reluctant to put in efforts. She always said "*I cannot do it*" and tended to guess or give random answers without much effort, in addition to being easily discouraged. The researchers felt that E would benefit from an intervention to overcome her low self-efficacy in math because mathematics is used in daily life (Proctor, 2005). In everyday life, people commonly use numbers, describe shapes, or use measuring units to communicate ideas. Therefore, whether we realize it or not, people frequently require the use of numbers (Proctor, 2005). Another reason that necessitated an intervention is the fact that E was able to increase her scores on other subjects in school, except for mathematics. She consistently obtained a below average score in math, which affected her confidence and led her to believe she was unable to perform mathematical tasks.

The intervention to increase math self-efficacy proposed in this research followed the principles of Pajares (1996). The researchers felt that this particular method of intervention was ideal for E because based on the information provided by her teachers, E needed to enhance her understanding of the concepts of multiplication and division, as well as learning strategies to answer mathematical problems involving multiplications and divisions. Modeling should help E improve her ability to apply strategies to complete division and multiplication tasks. Furthermore, E needed to be given attributional feedback so that she could obtain more information about the strategies she could use and the measures she could take to increase her ability to complete a mathematical task. The goal setting method, on the other hand, was intended to help E focus more on her current performance, which was expected to increase and ultimately contribute to her higher self-efficacy in math.

2 RESEARCH METHOD

This research is categorized as a small N design, which according to Myers and Hensen (2006) examines only one or just a few participants, due to specific participant characteristics and for the sake of learning the participants' behaviors in depth. Small N design research is also known as a single case design. According to Gravetter and Forzano (2009), using a single case design, researchers can recognize the impact of an intervention by measuring the condition twice, once before intervention (i.e., pre-test) and once after the intervention (i.e., post-test).

As mentioned earlier, the subject in the current research was E. Due to her low self-efficacy in math, E needed to receive an intervention to increase her self-efficacy in mathematics. By increasing her mathematics self-efficacy, the researchers expected that E would perform more optimally in school.

The researchers conducted several intervention sessions. Before starting with the intervention, the researchers conducted a pre-intervention (baseline) phase, which was followed by the intervention phase, and finally the evaluation phase. In their research, Pajares and Miller (1997) gave two types of assessment, including the math performance assessment and the self-efficacy assessment.

In the pre-intervention phase, E was given 45 minutes to complete 15 numbers of mathematical problems based on the competency standards for grade 2 to grade 5. The mathematical problems were obtained randomly from the Mathematics e-Book published on the Ministry of Culture and Education website. The questions were arranged from the easiest to the most difficult. In the intervention phase, the researchers used the Pajares principle of modeling, attributional feedback, and goal setting. The researchers modeled how to simplify the task, ensuring that E created her own targets by motivating her and also giving her guidance to solve mathematical problems similar to those given in the pre-intervention phase. During the evaluation phase, E was given the same questions as the ones in the pre-intervention phase (baseline), without being given modeling and attributional feedback. Goal setting, on the other hand, was provided in this phase.

3 RESULT

Based on the observation of the pre-intervention (pre-test) phase, E displayed the traits of students who have low mathematics self-efficacy, such as guessing or giving random answers, looking bored and uninterested, and easily giving up. E skipped a few questions because she thought she could not do it, and most of her answers were incorrect. Furthermore, from the Likert scale measurements, it was found that E tended to be unconfident in answering the questions given. Her self-efficacy scale was generally low, and it can be seen in the table below that E mostly chose 1 or 2 in the self-efficacy scale.

By analyzing the scores from the Likert scale obtained in the pre-test and post-test phases, there was an increase of confidence when solving problems in the post-test (i.e., after receiving the intervention) compared with the subject's confidence during the pre-test (i.e., prior to the intervention). In the evaluation or the post-test phase, the participant did not skip any of the questions given and seemed to put in more effort in solving the mathematical

Table 1. Mathematics self-efficacy pre-test and post-test results.

			Pre-test		Post-test	
Question number	Question category	Math grade questions	Math Self-efficacy scale	Answer result	Math Self-efficacy scale	Answer result
1	Multiple Choice (MC)	2	2	Correct	4	Correct
2	MC	2	1	Correct	5	Correct
3	MC	2	6	Incorrect	4	Correct
4	MC	3	2	Incorrect	4	Incorrect
5	MC	3	1	Incorrect	5	Correct
6	MC	3	2	No answer	5	Correct
7	MC	3	2	No answer	4	Correct
8	MC	4	1	Correct	4	Incorrect
9	MC	4	3	Correct	4	Correct
10	MC	4	2	Incorrect	4	Correct
11	Open ended	4	1	Correct	4	Correct
12	Open ended	5	1	No answer	4	Incorrect
13	Open ended	5	1	Incorrect	4	Correct
14	Open ended	5	1	Incorrect	4	Incorrect
15	Open ended	5	1	Incorrect	5	Incorrect

problems. She also used the strategies modeled in the intervention phase. The measurement in the evaluation phase (Post-test) was conducted in a day within a 45-minute period. The mathematical problems given were the same as the ones given in the pre-intervention phase (Pre-test), consisting of 10 multiple choice questions and 5 open-ended questions printed in a booklet. The target or goal was for E to be able to correctly answer at least 10 out of 15 questions. The specified target was reached, with E answering 10 questions correctly. Her self-efficacy also increased, as it can be seen in the table below that E mostly chose 4 or 5 in the self-efficacy scale.

Overall, the results show an increase in the participant's mathematics self-efficacy after being given the Pajares principles that consist of modeling, attributional feedback, and goal setting methods. The participant's ability to complete mathematical tasks increased from 33% to 66%. The participant's mathematical self-efficacy also increased, as can be seen from comparisons of the scores from the baseline phase (pre-test) and the evaluation phase (post-test).

4 DISCUSSION

The results of the intervention show an increase in the participant's mathematics performance and her mathematics self-efficacy. However, there are a few things that need to be considered in this research.

It is important to note that when a strategy is introduced by modeling, students may believe that they can master the strategy and that they can achieve success in various tasks, which should increase their sense of self-efficacy (Schunk, 1984, 1985 in Schunk, 1986). Moreover, students who are given feedback on their ability are more likely to have higher self-efficacy and mathematical skills compared to those who receive feedback only on their effort (Foote, 1999). Throughout the intervention sessions, the participant received attributional feedback when solving a mathematical task. This made her more confident in completing the task given.

The participant was asked to create her own targets throughout the program. This was considered necessary because students who make their own goals are more likely to own them and subsequently be more committed to them (Owings & Follo, 1992; Schunk, 1985 in Barron, 2000). In the intervention phase, the participant asked for an additional division task, and this also added to memorizing multiplications in her target. The participant believed that she was able to solve the division problems when she performed it carefully. This shows that she possessed the willingness to reach her target and also suggested her extra effort to reach her goal. Therefore, the participant appeared to demonstrate an increase in self-efficacy.

The learning contract is considered effective in this research. The contract should be signed by the participant and also the teacher (in this case, the researchers). Both participant and the researchers should respect the essential agreement and apply them until the intervention program ends. Generally, the participant is cooperative and also able to stick with the agreement.

Unexpected situation may happen during the intervention, for example the participant can be sick and unable to conduct the intervention. During the intervention period, the researchers involved the mother of E (S) in the program. However, S was unable to attend the intervention twice due to her work obligations as a preschool teacher. To make up for her absence, S was asked to give E guidance in completing the task given by the researchers at home. S was initially given a list of attributional feedback example sentences that she would use to guide E when completing tasks at home. Unfortunately, S admitted that she did not give any guidance because she thought that E did not need any guidance from her. Basically, a child needs feedback from their parents regarding their work. Parents can support their learning activity and increase their children's self-efficacy through appropriate reinforcement of their academic behavior (Hoover-Dempsey & Sandler, 1995 in Peiffer, 2015). The reinforcement can be in the forms of praises, support, and rewards. Reinforcement helps build self-efficacy by persuading children to complete a task. When children are given reinforcements they enjoy, they will be more likely to perform the expected behavior in order to get additional reinforcements (Lysakowski & Walberg, 1981 in Peiffer, 2015). When S was able to be involved in the intervention program, sometimes she tended to be impatient when giving guidance to E so she gave E the answer directly. According to Tuft (2005), teachers should be a facilitator in the learning process and guide their students. Researchers needed to remind S to guide E to find the answer rather than giving E the answer directly. When involving parents, it would be beneficial to provide a written agreement to minimize the absence and ask for the parents' willingness to provide time to be involved in the program.

5 CONCLUSION

According to the research results, it can be concluded that the Pajares principles which consist of methods, such as modeling, attributional feedback, and goal setting can increase the participants' mathematics self-efficacy. Furthermore, after given the intervention the participant's ability to complete the task given also increased. E was able to answer more questions correctly and confidently compared to the time before she was given the intervention.

There should be other researchers who would like to use a similar intervention, in this case using methods such as modeling, attributional feedback, and goal setting. It is best to conduct an assessment beforehand in order to understand which mathematical concepts that the participants need to work on so that the researchers can give modeling based on the participants' needs. Besides the researchers, the peer of the participants can also provide modeling. It would also be beneficial to share attributional feedback sentence examples to parents and teachers so that they can also be involved to increase the participant's math self-efficacy.

REFERENCES

APA Dictionary of Psychology. (2015) Second Edition. Washington (DC), American Psychological Association. Retrieved from: http://www.apa.org/pubs/books/4311022.aspx.

Barron, T.L. (2000) *Mathematics Learning and Goal Setting*. [Ph.D.], United States, Georgia, University of Georgia. Retrieved from: http://remote-lib.ui.ac.id:2073/docview/304592520/abstract/DDAF94 A91E3D4051PQ/1.

Foote, C.J. (1999) Attribution feedback in the elementary classroom. *Journal of Research in Childhood Education,* 13 (2), 155–66. Available on doi:10.1080/02568549909594736.

Gravetter, F.J. & Forzano, L.B. (2009) *Research Methods for the Behavioral Sciences*. Stamford, CT, Wadsworth Publishing.

Hoover-Dempsey, K.V., Walker, J.M.T., Sandler, H.M., Whetsel, D., Green, C.L., Wilkins, A.S. & Closson, K. (2005) Why do parents become involved? Research findings and implications. *The Elementary School Journal,* 106 (2), 105–30. Available on doi:10.1086/499194.

Jackson, S.F. (2012) *Self-Regulated and Communal Learning Contexts as They Relate to Math Achievement and Math Self Efficacy among African American Elementary Level Students*. [Ph.D.], United States, District of Columbia, Howard University. Retrieved from: http://remote-lib.ui.ac.id:2073/docview/1435635242/abstract/6CA915E1086442CPQ/1.

Myers, A. & Hansen, C. (2006) *Experimental Psychology.* 6th edition. USA, Thomson Wadsworth.

Ogden, M.A. (2012) *Effects of course experiences on self-efficacy in teaching math: A case study of preservice elementary teachers* (Order No. 3514279). Available from ProQuest Dissertations & Theses Global. (1037797905). Retrieved from http://search.proquest.com/docview/1037797905?accountid = 17242.

Pajares, F. (1996) Self-efficacy beliefs in academic settings. *Review of Educational Research,* 66 (4), 543–78. Available on doi:10.3102/00346543066004543.

Pajares, F. & Miller, M.D. (1997) Mathematics Self-efficacy and mathematical problem solving: implications of using different forms of assessment. *The Journal of Experimental Education,* 65 (3), 213–28.

Papalia, D.E., Olds, S.W. & Feldman, R.D. (2005) *Human Development – 10th Edition*. 10th edition. Boston, McGraw-Hill.

Peiffer, G.D. (2015) *The effect of self-efficacy on parental involvement at the secondary school level* (Order No. 3725690). Available from ProQuest Dissertations & Theses Global. (1717301609). Retrieved from http://search.proquest.com/docview/1717301609?accountid=17242.

Proctor, D. (2005) The importance of math in our vocational world. *Water Environment & Technology, 17* (10), 103–104. Retrieved from http://search.proquest.com/docview/205340561?accountid=17242 . [Accessed on 7 September 2015].

Santrock, J.W. (2012) *Educational psychology.* 5th edition. New York, McGraw-Hill.

Schunk, D.H. (1986) Vicarious influences on self-efficacy for cognitive skill learning. *Journal of Social and Clinical Psychology,* 4 (3), 316–27. Available on doi:10.1521/jscp.1986.4.3.316.

Schunk, D.H., Meece, J.R. & Pintrich, P.R. (2014) *Motivation in Education: Theory, Research, and Applications*. 4th edition. Edinburgh Gate, Pearson.

Sulzer-Azaroff, B. & Mayer, G.R. (1986) *Achieving Educational Excellence: Using Behavioral Strategies*. New York, Holt, Rinehart & Winston.

Tuft, E.A. (2005) *What Is Mathematics? Stability and Change in Prospective Teachers' Conceptions of and Attitudes toward Mathematics and Teaching Mathematics*. [Ph.D.], United States, Michigan, Michigan State University. Retrieved from: http://remote-lib.ui.ac.id:2073/docview/305456718/abstract/23B3812E1E1C4FFEPQ/1.

Diversity in Unity: Perspectives from Psychology and Behavioral Sciences – Ariyanto et al. (Eds)
© 2018 Taylor & Francis Group, London, ISBN 978-1-138-62665-2

Teacher's perception of school climate and social-emotional learning, job satisfaction, teaching efficacy, and stress among teachers in special schools for the mentally disabled

M.S. Natalia & F.M. Mangunsong
Faculty of Psychology, Universitas Indonesia, Depok, Indonesia

ABSTRACT: The focus of this study was to investigate whether teachers' perception of school climate and perception of Social-Emotional Learning (SEL) correlated with their job satisfaction, teaching efficacy, and sense of stress. The samples included 99 teachers from five schools that provide special education for mentally disabled children (SLB C/C1) in Jakarta, Indonesia. Participants completed a self-report questionnaire about job satisfaction, teaching efficacy, teachers' stress, perception of school climate, and perception of SEL. Pearson correlation coefficient was used to examine the relationships among the variables. The results showed significant correlations between pairs of variables. Teachers' perceptions of school climate and SEL significantly predicted job satisfaction, teaching efficacy, and stress among teachers in special schools for the mentally disabled.

1 INTRODUCTION

In the world of education, a teacher is a figure who holds an important role in learning activities. A teacher's profession demands many obligations to provide good education for students. These obligations lead to teachers facing obstacles and problems at schools, especially teachers in special needs schools. Teachers in special needs schools are often confronted with problems such as the recruitment of teachers who lack competency in special education, difficulties accessing the school's facilities, difficulties handling students' disabilities, and poor collaboration among teachers, school, and parents (Afifah, 2012; Harsojo, 2014; Rozali, 2015). Moreover, teachers' educational backgrounds and lack of competencies in special education often become another issue affecting their teaching efficacy. Special needs school administrators often complain that not all of the teachers employed in their schools have sufficient competencies to manage class rooms of special needs students (Afifah, 2015; Savitri, 1998). These problems become fascinating to study because the profession of special education teacher requires a person to be an ideal figure of an educator, but on the other hand the person is frequently hindered by self and environmental constraints. Therefore, special education teachers, as well as various aspects related to their working experiences, are an interesting area of study and research.

In order to understand teachers' working experiences, several variables pertaining to their teaching job need to be examined, including job satisfaction, teaching efficacy, and stress among teachers (Collie, Shapka & Perry, 2012). Previous studies have shown that teaching efficacy is a determinant of teachers' job satisfaction (Caprara, *et al.*, 2003; Caprara, *et al.*, 2006), and both stress and teaching efficacy contribute to teachers' job satisfaction (Klassen & Chiu, 2010). Job satisfaction refers to the sense of fulfillment, gratification, and satisfaction from working in an occupation (Collie, *et al.*, 2012), and such satisfaction is considered as a general feeling towards a job (e.g., global job satisfaction) or as a set of interrelated attitudes towards several aspects of the job (e.g., job facet satisfaction) (Spector, 1997). Special education teachers, as well as many other professions, need to have a high level of job satisfaction in order to maintain their commitment to the profession and to reduce the probability of quitting their jobs (DeSio, 2014; Johnson, 2010; Mattingly, 2007; McArthur, 2008).

Other studies also have shown that teaching efficacy influences special education teachers' job satisfaction (McArthur, 2008). Teachers who lack confidence in their capabilities of teaching are not satisfied with their job, conversely, teachers who are not satisfied with their profession question their competence as teachers. Teaching efficacy refers to a teacher's "judgment of his or her capabilities to bring about desired outcomes of student engagement and learning, even among those students who may be difficult or unmotivated" (Tschannen-Moran & Hoy, 2001). The three factors of teaching efficacy examined in the current study include efficacy for student engagement, efficacy for classroom management, and efficacy for instructional strategies (Tschannen-Moran & Hoy, 2001). Positive teaching efficacy has been found to be correlated with positive teachers' outcomes, such as effective use of teaching strategies, better class management, and teachers' psychological well-being (Tschannen-Moran & Hoy, 2001). Teaching efficacy and job satisfaction are also negatively correlated with stress among teachers in special schools (McArthur, 2008; Mattingly, 2007; Johnson, 2010; Kerr, 2013; DeSio, 2014).

Teachers' work stress refers to the experience of unpleasant emotions as a result of working as teachers (Kyriacou, 2001). Work stress can be defined based on three approaches: the characteristics of external environmental stimulus (engineering model), individual emotional state (physiological model), or interaction between individual and environment (transactional model) (Boyle, *et al.*, 1995). Two types of stress that consistently appear in the teaching profession are stress related to students' behaviors and discipline, as well as stress related to workload (Borg & Riding, 1991; Boyle, *et al.*, 1995; Klassen & Chiu, 2010). Stress among teachers has been associated with teachers' negative outcomes, such as lower teaching efficacy, lower job satisfaction, and lower commitment to their jobs (Klassen & Chiu, 2010; 2011). These three variables do not only affect teachers' outcomes, but also students' achievements. Teachers with lower stress levels, more positive teaching efficacy, and higher job satisfaction tend to encourage greater achievement and self-efficacy among the students (Caprara *et al.*, 2006; Colie *et al.*, 2012).

Teachers' job satisfaction, teaching efficacy, and level of stress can be explored through teachers' perceptions of the climate of the school where they teach (Collie et al. 2012). School climate is the quality and character of a school, which is reflected in four essential dimensions of school climate: physical-emotional safety, quality of teaching and learning, relationships and collaboration, and structural environment (Cohen, *et al.*, 2009; Collie, *et al.*, 2012). Every teacher perceives the school climate where they work differently from another. How a teacher perceives the school climate is based on four factors, namely collaboration, teacher-student relationship, school facilities, and decision-making (Johnson, Stevens & Zvoch, 2007). Perception of school climate has previously been associated with teachers' burnout and work commitment, as well as students' achievement and school connectivity (Collie, *et al.*, 2012). Previous studies have additionally shown that teachers' perception of school climate plays a role in determining teachers' outcomes, which include their level of stress, teaching efficacy, and job satisfaction (Collie, *et al.*, 2012). Moreover, the three outcomes have also been shown to interact with and influence one another (Klassen & Chiu, 2010).

Along with their perception of school climate, teachers' outcomes (i.e., job satisfaction, teaching efficacy, and level of stress) can also be explored through teachers' perceptions of Social-Emotional Learning (SEL) (Collie, *et al.*, 2012). Teachers who perceive SEL positively were reported feeling greater sense of accomplishment in teaching, greater adaptive self-efficacy, and lower sense of stress (Brackett, *et al.*, 2011). SEL is defined as the nurturing of the social and emotional awareness and skills of students (Collaborative for Academic, Social, and Emotional Learning (CASEL, 2003; Weissberg, *et al.*, 2008). There are five core social and emotional competencies that SEL should address, which consist of self-awareness, self-management, social awareness, relationship skills, and responsible decision-making (Weissberg, *et al.*, 2008). A teacher's belief of SEL is the key indicator of a teacher's perception of SEL. Teacher's belief of SEL influence the type of learning environments created by the teacher, as well as students' academic performance and beliefs about their own abilities that later influence their teaching efficacy (Brackett, *et al.*, 2011). Three factors contribute to a teacher's SEL belief, namely comfort, commitment, and culture (Brackett, *et al.*, 2011). In the current study, the SEL culture factor is excluded due to its similarity to school climate. Furthermore, teachers' perceptions of SEL have been found to influence the implementation of SEL in the classroom (Brackett, *et al.*, 2011), which correlates positively with job satisfaction and teaching efficacy, while at the same time being negatively correlated with stress among teachers (Collie, *et al.*, 2012). In special education, for example in the teaching of mentally disabled students, SEL is already integrated as part of the curriculum that the students must be able to learn (Dewi, 2016; Mustam, 2015). Teachers in special schools for the mentally disabled (called SLB C/C1 in the Indonesian educational setting), in their effort to implement SEL usually come to develop their own perceptions of SEL. How teachers perceive SEL then represents their belief of SEL. Teachers might develop different beliefs of SEL depending on the support, training, and experience they receive in social-emotional learning (Collie, *et al.*, 2015b).

The focus of the current study was job satisfaction, teaching efficacy, and stress among teachers in special schools for the mentally disabled (SLB C/C1). Through special education in SLB C/C1, mentally disabled students might be able to show more self-esteem, create and maintain social relations, find a productive job and support the economy, and show responsibilities (Mangunsong, 2009). Competencies, such as having self-esteem, social relations, and responsibilities, reflect some of the five core competencies of SEL. These competencies also show that social and emotional skills are some of the focus of special education in SLB C/C1. Therefore, the aim of the current study was to examine the relationship of teachers' perception of school climate and social-emotional learning (SEL) with job satisfaction, teaching efficacy, and stress among teachers in special schools for the mentally disabled.

Perceptions of school climate and SEL are categorized as school-based variables, which are the independent variables of this study. The outcome variables (job satisfaction, teaching efficacy, and teacher stress) are the dependent variables of the current study. In this study, we hypothesized that the school-based variables would impact the three outcomes variables. Therefore, there are two hypotheses to be examined in the current study:

H1: Teachers' perception of school climate shows significant positive correlation with job satisfaction and teaching efficacy, and significant negative correlation with stress among teachers in SLB C/C1.

H2: Teachers' perception of social-emotional learning (SEL) shows significant positive correlation with job satisfaction and teaching efficacy, and significant negative correlation with stress among teachers in SLB C/C1.

2 METHOD

2.1 *Samples*

Participants were recruited from five special schools for mentally disabled children (SLB C/C1) in South Jakarta. There were eight SLB C/C1 contacted to recruit the participants, and five of the schools gave permission to administer the study. All the schools were chosen based on the region (all schools were in South Jakarta). There were 101 participants, but only 99 teachers were included in the current study. Participants were chosen using non-probability sampling. In particular, the participants were recruited using convenience sampling based on two criteria: (1) must be active as a teacher in the current academic year, and (2) must have a minimum of five years working experience in SLB C/C1. Based on the second criteria, there were two participants that had less than five years working experience in SLB C/C1 and they were not included in the current study.

2.2 *Measurement*

All the variables in the current study was measured using *Skala Guru* (namely Indonesian version of Teacher's Scale) which was adapted by the researchers of this study from Collie, Shapka, and Perry (2012) Teacher's Scale. *Skala Guru* was a self-report questionnaire containing 37 items using Liker-like scale that were grouped into five sub-scales. Specifically, sub-scales I (teacher's perception of school climate) contained 11 items, II (teacher's perception of social-emotional learning) had seven items, subscales III (job satisfaction) had six items, IV (teacher's stress) had three items, and V (teaching efficacy) had 10 items. All five sub-scales in *Skala Guru* showed high reliability, as measured by calculating the Cronbach's Alpha coefficients: 0.76 for perception of school climate, 0.94 for perception of social-emotional learning (SEL), 0.95 for job satisfaction, 0.93 for teacher stress, and 0.97 for teaching efficacy.

2.3 *Procedures*

The questionnaires were distributed to five participating SLB C/C1 in South Jakarta. The participants were give informed consents which they filled before continuing the study. Prior to data collection, the principals of the schools were given briefings regarding the questionnaire. These school principals then assisted the researcher to administer the questionnaires. Teachers who participated in the study were given 15–20 minutes to complete the questionnaires. After the questionnaires had been returned, each was checked to make sure all the required data were completed.

2.4 *Data analysis*

All the collected data were processed for hypothesis testing using Pearson's r analyses using IBM SPSS version 24.0 as the statistical software. Pearson correlation coefficient was used to examine the relationships between the dependent variables (job satisfaction, teaching efficacy, and teacher stress) and the independent variables (perception of school climate and perception of social-emotional learning).

3 RESULTS

The results of the hypothesis tests indicated significant relationships between teachers' perception of school climate and job satisfaction, teachers' perception of school climate and teaching efficacy, teachers' perception of school climate and teacher stress, teachers' perception of SEL and job satisfaction, teachers' perception of SEL and teaching efficacy, as well as teachers' perception of SEL and teacher stress. In other words, each of the two independent variables was found to correlate with each of the three dependent variables.

The data in Table 1 indicates a significant positive relationship between teachers' perception of school climate and teachers' job satisfaction ($r = 0.39$, $p < 0.01$), suggesting that the higher a teacher's perception of school climate, the higher the teachers' job satisfaction. Table 1 also indicates a significant positive relationship between teachers' perceptionof school climate and teacher stress, Table 1 indicates a significant negative correlation ($r = -0.30$, $p < 0.01$), whereby a higher perception of school climate tends to be accompanied by a lower teacher stress score. Therefore, the results provided support for the first hypothesis of the study.

Table 1. Correlation between perception of school climate and job satisfaction, teaching efficacy, teacher stress.

Variables	Teacher stress	Teaching efficacy	Job satisfaction	Perception of school climate
Teacher stress	1			
Teaching efficacy	–0.32**	1		
Job satisfaction	–0.27**	0.58**	1	
Perception of school climate	–0.30**	0.48**	0.39**	1

** $p < 0.01$. n = 99.

Table 2. Correlation between perception of social-emotional learning and job satisfaction, teaching efficacy, teacher stress.

Variables	Teacher stress	Teaching efficacy	Job satisfaction	Perception of SEL
Teacher stress	1			
Teaching efficacy	–0.32**	1		
Job satisfaction	–0.27**	0.58**	1	
Perception of SEL	–0.26**	0.71**	0.54**	1

** $p < 0.01$. n = 99.

Furthermore, data in Table 2 indicates a significant positive relationship between teachers' perception of SEL and teachers' job satisfaction ($r = 0.54$, $p < 0.01$), showing that the higher a teacher's perception of SEL, the higher the teachers' job satisfaction. Table 2 also indicates a significant positive relationship between teachers' perception of SEL and teaching efficacy ($r = 0.71$, $p =< 0.01$), wherein a higher perception of SEL is accompanied by a higher teaching efficacy score. Consistent with results in Table 1, results in Table 2 also indicates a significant negative relationship between teachers' perception of SEL and teacher stress ($r = -0.26$, $p < 0.01$), therefore suggesting that the higher a teacher's perception of SEL, the lower the teacher stress score. Therefore, as with the first hypothesis, the second hypothesis of the study was also supported by the results.

4 DISCUSSION

Results from the current study suggest that teachers' perceptions of school climate and social-emotional learning (SEL) have significant positive correlations with job satisfaction among teachers in SLB C/C1. This implies that the more positive teachers' perceptions of school climate and SEL, the higher the job satisfaction that teachers would experience. This supports findings from previous studies which suggest that the way teachers perceive the school climate and SEL influence their job satisfaction (Collie, *et al.*, 2012) and their level of stress (Pas & Bradshaw, 2014). Teachers who perceived the school climate as unsupportive were found to experience higher levels of stress or even burnout (Pas & Bradshaw, 2014). In contrast, teachers who found suitability with the school climate showed lower levels of stress and higher levels of job satisfaction. Job satisfaction is comprised of feelings of satisfaction obtained from teaching and is associated with teachers' perception of school climate and belief in the implementation of SEL (Skaalvik & Skaalvik, 2011). Teachers' job satisfaction also had a significant relationship with comfort, commitment, and culture for implementing SEL at school (Collie, *et al.*, 2015a). Evidently, the results of the current study support the findings of previous research. Therefore, it can probably be assumed that teachers' perception of school climate and SEL are crucial for increasing teachers' job satisfaction in special needs schools.

The results of the current study also suggest that teachers' perceptions of school climate and SEL have significant positive correlations with teaching efficacy among teachers in SLB C/C1. In other words, the more positive teachers' perception of school climate and SEL, the higher the teaching efficacy reported by teachers. Teaching efficacy is an important aspect in a teacher's work, as it is related to the use of effective teaching strategies, better class management, and greater teacher well-being (Collie, *et al.*, 2012). From the current study, we found that positive school climate and positive SEL contributed to better teaching efficacy among teachers in special needs schools. It is likely that teachers who feel comfortable with the school climate and the implementation of SEL would consequently develop better understandings of themselves, resulting in more positive feelings toward their own teaching efficacy.

Results from the current study also indicate that teachers' perceptions of school climate and SEL are significantly negatively correlated with the level of stress experienced by teachers in SLB

C/C1. This is to say that the more positive teachers' perceptions of school climate and SEL, the lower the level of stress they experience. This particular finding is consistent with the previous finding that teachers' perceptions of school climate and SEL are related to teacher stress (Collie, *et al.,* 2012; Collie, *et al.,* 2015a), as well as with the discovery that teachers who have negative perception of school climate tend to experience higher level of stress, lower job satisfaction, and lower sense of teaching efficacy (Tran, 2015). Teachers who have positive perceptions of SEL tend to display more positive results and better performance at work; their implementation of SEL is also relevant to their daily experiences at work (Collie, *et al.*, 2015a). This positive perception does not only affect how teachers deliver SEL, but also influences their psychological experiences such as their capability to cope with stress at work (Brackett, *et al.*, 2012).

Teachers are influenced by their perceptions of their working environment, which further influence their working outcomes that include job satisfaction, teaching efficacy, and level of stress (Collie, *et al.*, 2012). In order to maintain a high level of job satisfaction, teaching efficacy, and a low level of stress for their teachers, schools must be able to provide an overall positive school climate for the teachers. Positive school climate could be obtained through physical-emotional safety, improvements in the quality of teaching and learning, positive relationships and collaboration between teachers, students, and parents, as well as maintenance of the quality of the structural environment (Cohen, *et al.*, 2009; Collie, *et al.*, 2012). Also, teachers' perception of SEL might vary from one teacher to another, depending on the overall school climate. Therefore, schools need to provide a positive school climate to support the positive perception of SEL among teachers in special needs schools. Teachers who find compatibility between their profession, school climate, and social-emotional learning are expected to display more positive results at work, lower stress levels, higher job satisfaction, and more positive teaching efficacy (Collie, *et al.,* 2012; Collie, *et al.,* 2105a).

5 CONCLUSION

The overall results of the current study provide evidence for the relationships of perceptions of school climate and social-emotional learning with job satisfaction, teaching efficacy, and stress among teachers in special schools for the mentally disabled. However, considering that the instrument was adapted from a questionnaire developed by Collie, Shapka, and Perry (2012) based on a regular education context, future studies could perhaps focus on the instrument as an area for improvement. For example, items from *Skala Guru* can be added or revised to obtain a more sensitive measurement of the research variables that would be more appropriate for use in a special education context. Therefore, we might be able to develop a scale which able to measure the research variables in special education with various disabilities context, or even in inclusive education context.

It is necessary to recall that the current study not only found a relationship between the predicting variables (teachers' perception of school climate and SEL) and the outcome variables (job satisfaction, teaching efficacy, and teacher stress), but also significant correlations among the outcome variables. In other words, a teacher's experience of one variable influences the experience of one of the other variables (Klassen & Chiu, 2010). When a teacher experiences a high level of stress, he or she would feel incapable as a teacher, resulting in lower teaching efficacy that in turn could lower job satisfaction. Likewise, when a teacher feels confident of his or her capability as a teacher, this would result in the increase of their teaching efficacy, which leads to higher job satisfaction and lower stress level (Collie, *et al.*, 2012). These findings might be put into consideration for policy making at school level. Special schools could give more attention to maintain teacher's job satisfaction, teaching efficacy, and level of stress.

The current study was a preliminary study to examine the relationships between the five research variables. From this study, we can conclude there were significant relationships between all the variables. Further work is needed to examine the causal or long-term relationship among the variables investigated in this study. By examining these relationships, we might be able to obtain better understanding of the relationships between the five research variables.

REFERENCES

Afifah, R. (2012) Empat Masalah Guru Yang Tak Kunjung Selesai. *Kompas.com*. Retrieved from: http://edukasi.kompas.com/read/2012/11/26/1337430/4. Masalah. Utama. Guru. yang. Tak. Kunjung. Selesai [Accessed on March 11th 2016].

Boyle, G.J., Borg, M.G., Falzon, J.M. & Baglioni, A.J. (1995) A structural model of the dimensions of teacher stress. *British Journal of Educational Psychology,* 65 (1), 49–67. Available on doi:10.1111/j.2044-8279.1995.tb01130.x.

Brackett, M. A., Reyes, M.R., Rivers, S.E., Elbertson, N.A. & Salovey, P. (2011) Assessing teachers' beliefs about social and emotional learning. *Journal of Psychoeducational Assessment,* 30 (3), 219–36. Available on doi:10.1177/0734282911424879.

Caprara, G., Barbaranelli, C., Steca, P. & Malone, P. (2006) Teachers' self-efficacy beliefs as determinants of job satisfaction and students' academic achievement: A Study at the school level. *Journal of School Psychology,* 44 (6), 473–90. Available on doi:10.1016/j.jsp.2006.09.001.

Caprara, G.V., Barbaranelli, C., Borgogni, L. & Steca, P. (2003) Efficacy beliefs as determinants of teachers' job satisfaction. *Journal of Educational Psychology,* 95 (4), 821–32. Available on doi:10.1037/0022-0663.95.4.821.

Cohen, J., Mccabe, E.M. & Michelli, N.M. (2009) School climate: Research, policy, practice, and teacher education. *Teachers College Record,* 111 (1), 180–213. Retrieved from: http://www.schoolclimate.org/climate/documents/policy/School-Climate-Paper-TC-Record.pdf.

Collaborative for Academic Social and Emotional Learning. (2005). Safe and sound: An educational leader's guide to Evidence-Based Social and Emotional Learning (SEL) Programs. *Collaborative for Academic Social and Emotional Learning*.

Collie, R.J, Shapka, J.D., Perry, N.E. & Martin, A.J. (2015a) Teachers' Psychological functioning in the workplace: Exploring the roles of contextual beliefs, need satisfaction, and personal characteristics. *Journal of Educational Psychology,* 107 (4), 1–12. Available on doi:10.1037/edu0000088.

Collie, R.J., Shapka, J.D. & Perry, N.E. (2012) School climate and social–emotional learning: Predicting teacher stress, job satisfaction, and teaching efficacy. *Journal of Educational Psychology,* 104 (4), 1189–204. Available on doi:10.1037/a0029356.

Collie, R.J., Shapka, J.D., Perry, N.E. & Martin, A.J. (2015b) Teachers' Beliefs about social-emotional learning: identifying teacher profiles and their relations with job stress and satisfaction. *Learning and Instruction*, 39, 148–57. Available on doi:10.1016/j.learninstruc.2015.06.002.

DeSio, M.J.A. (2014) *The Impact of special education litigation upon the key factors of job satisfaction, level of stress, and self-esteem that lead to teacher attrition and retention*. [PhD Dissertation.], Fielding Graduate University. ProQuest Dissertations and Theses database, UMI 3479462.

Harsojo, A. (2014) *Profesionalisme Guru dan Calon Guru Menghadapi Tantangan Dunia Pendidikan." Sekolah Tinggi Keguruan dan Ilmu Pendidikan*. Retrieved from: http://www.stkippgrismp.ac.id/profesionalisme-guru-dan-calon-guru-menghadapi-tantangan-dunia-pendidikan/, on March 11th 2016.

Johnson, B.W. (2010) Job satisfaction, self-efficacy, burnout, and path of teacher certification: predictors of attrition in special education teachers. *Dissertation Abstracts International Section A: Humanities and Social Sciences,* 71, 1534. Retrieved from: http://ovidsp.ovid.com/ovidweb.cgi?T=JS&CSC=Y&NEWS=N&PAGE=fulltext&D=psyc6&AN=2010–99210–218\n. http://ezproxy.lib.monash.edu.au/login?url=http://sfx.monash.edu.au:9003/monash2?sid=OVID:psycdb&id=pmid:&id=doi:&issn=0419-4209&isbn=9781109765038&volume=71&issue=5-A&spage=1534&pages=1534&date=2010&title=Dissertation+Abstracts+International+Section+A:+Humanities+and+Social+Sciences&atitle=Job+satisfaction,+self-efficacy,+burnout,+and+path+of+teacher+certification:+Predictors+of+attrition+in+special+ed.

Johnson, B., Stevens, J.J. & Zvoch, K. (2007) Teachers' perceptions of school climate: A validity study of scores from the revised school level environment questionnaire. *Educational and Psychological Measurement,* 67 (5), 833–44. Available on doi:10.1177/0013164406299102.

Kerr, M.B. (2013) *Teacher Stress and Administrative Support as Predictors of Teachers' Elf-Efficacy for Special Education Teachers in California's Central Valley*. [PhD Diss.], California State University. ProQuest Dissertations and Theses Database, UMI 3567289.

Klassen, R.M. & Chiu, M.M. (2010) Effects on teachers' self-efficacy and job satisfaction: Teacher gender, years of experience, and job stress. *Journal of Educational Psychology,* 102 (3), 741–56. Available on doi:10.1037/a0019237.

Klassen, R.M. & Chiu, M.M. (2011) The occupational commitment and intention to quit of practicing and pre-service teachers: influence of self-efficacy, job stress, and teaching context. *Contemporary Educational Psychology,* 36 (2), 114–29. Available on doi:10.1016/j.cedpsych.2011.01.002.

Kyriacou, C. (2001) Teacher stress: Directions for future research. *Educational Review,* 53 (1), 27–35. Available on doi:10.1080/00131910120033628.

Mangunsong, F. (2009) *Psikologi dan Pendidikan Anak Berkebutuhan Khusus*. Universitas Indonesia: Lembaga Sarana Pengukuran dan Pendidikan Psikologi (LPSP3).

Mattingly, J.W. (2008). A Study of Relationships of School Climate, School Culture, Teacher Efficacy, Collective Efficacy, Teacher Job Satisfaction and Intent to Turnover in the Context of Year-Round Education Calendars. *Dissertation Abstracts International Section A: Humanities and Social Sciences*. Retrieved from: http://ovidsp.ovid.com/ovidweb.cgi?T=JS&PAGE=reference&D=psyc5&NEWS=N&AN=2008-99070-584.

McArthur, C.L. (2008) *Teacher Retention in Special Education: Efficacy, Job Satisfaction, and Retention of Teachers in Private Schools Serving Students with Emotional/Behavioral Disabilities*. [PhD Diss.], The State University of New Jersey. ProQuest Dissertations and Theses database, UMI Microform 3319533.

Pas, E.T. & Bradshaw, C.P. (2014) What affects teacher ratings of student behaviors? The Potential influence of teachers' perceptions of the school environment and experiences. *Prevention Science*, 15(6), 940–950. Available on DOI 10.1007/s11121-013-0432-4.

Rozali, A. (2015) Permasalahan dan Tantangan yang Dihadapi Guru. *Balikpapan.co.id*. Retrieved from: http://www.balikpapanpos.co.id/index.php?mib=berita.detailandid=61067, on March 11th 2016.

Savitri, I. (1998) *Burnout pada Guru Sekolah Luar Biasa Tuna Ganda (Studi Kualitatif Mengenai Gambaran, Sumber, dan Proses Burnout pada Guru Sekolah Luar Biasa Tuna Ganda di Sebuah Sekolah Luar Biasa Tuna Ganda di Jakarta)*. [Master's Theses], Universitas Indonesia.

Skaalvik, E.M. & Skaalvik, S. (2010) Teacher self-ef fi cacy and teacher burnout : A Study of relations. *Teaching and Teacher Education,* 26 (4), 1059–69. Available on doi:10.1016/j.tate.2009.11.001.

Spector, P.E. (1997) *Job Satisfaction: Application, Assessment, Cause, and Consequences. Sage Publications*, 35. Available on doi:10.5860/CHOICE.35-0383.

Tschannen-Moran, M. & Hoy, A.W. (2001) Teacher efficacy: Capturing an elusive construct. *Teaching and Teacher Education,* 17 (7), 783–805. Available on doi:10.1016/S0742-051X(01)00036-1.

Tran, V.D. (2015) Effects of gender on teachers' perceptions of school environment, teaching efficacy, stress and job satisfaction. *International Journal of Higher Education,* 4 (4), 147–57. Available on doi:10.5430/ijhe.v4n4p147.

Weissberg, R.P., Durlak, J.A., Dymnicki, A.B., Taylor, R.D. & Schellinger, K.B. (2008) The Positive impact of social and emotional learning for kindergarten to eighth-grade students findings from three scientific reviews. *Learning*, 1–15. Retrieved from: http://www.casel.org/sel/meta.php.

Diversity in Unity: Perspectives from Psychology and Behavioral Sciences – Ariyanto et al. (Eds)
© 2018 Taylor & Francis Group, London, ISBN 978-1-138-62665-2

Teachers' attitude and instructional support for students with special educational needs in inclusive primary schools

A. Marhamah, F. Kurniawati & F.M. Mangunsong
Faculty of Psychology, Universitas Indonesia, Depok, Indonesia

ABSTRACT: The teachers' attitude and instructional support to students with Special Educational Needs (SEN) have positive effects in the success of inclusive education. The purpose of this research is to examine the correlation between teachers' attitude and their instructional support in inclusive primary school. Forty primary school teachers were given Multidimensional Attitudes toward Inclusive Education Scale (MATIES) and instructional support questionnaires. The results have revealed that teachers have positive attitudes towards SEN students and they provide strong instructional support when they interact with such students, thus there is a significant correlation between the two variables.

1 INTRODUCTION

The teachers' role in the success of inclusive education is quite important, because the teachers must be able to teach in accordance with the needs of the regular and Special Educational Needs (SEN) students. The preamble of the 1945 Constitution of Indonesia regarding the national education system, in article 5 paragraph (1), states that all citizens of Indonesia have the right to access education. UNESCO (1994) affirms that all students, regardless of their capability, ethnicity, culture, religion, language, gender and other conditions, have the right to access education, to enhance the students' knowledge and skills, including children with SEN. Children with SEN are students who are significantly different from regular students that have intellectual, physical, social or emotional disparities compared to the SEN students in the classroom (Hallahan & Kauffman, 2011; Mangunsong, 2014). The SEN students become major challenges for teachers as these teachers have been trained for special education and inclusive education (Kurniawati, De Boer, Minnaert & Mangunsong, 2016).

The teachers' attitude is one of the important factors in enhancing the success of inclusive education (Leatherman & Niemeyer, 2005). Attitude is a thought or idea that reflects positive or negative feelings and it influences behavior towards certain objects (Azwar 1995; Fishbein & Ajzen, 1975; Leatherman & Niemeyer, 2005). The teachers' attitude refers to their positive or negative feelings in developing the students' knowledge and skills, and providing the opportunity to such students to acquire equal education as for regular students (Ajzen, 2005). In inclusive settings, the teachers' attitudes towards SEN students refer to the teachers' positive or negative feelings in developing students' knowledge and skills, and providing the opportunity to SEN students to acquire equal education as for regular students. The research shows that teachers will show a positive attitude towards SEN students if the teachers have experience and prior knowledge in teaching SEN students in an inclusive setting (Leatherman & Niemeyer, 2005; Kurniawati, Minnaert, Mangunsong & Ahmed, 2012).

Triandis (1971, in Leatherman & Niemeyer, 2005) stated that attitude consists of three components (cognitive, affective and behavioral) which is known as the "triadic scheme", or the "tripartite model", or "tricomponent" (Azwar, 1995; Fishbein & Ajzen, 1975). In inclusive education, the cognitive component is related to the teachers' knowledge and perception of the SEN student's behavior in inclusive schools. Affective component is based on the teachers' feelings towards the students' disability that will motivate the teachers to interact with SEN students. While the third component, behavioral component is related to the teachers' behavior or the responses tendency towards SEN students when interacting with them. These three components of attitudes will determine the teachers' success in implementing the learning process for SEN students in inclusive schools.

Teachers with positive attitudes may significantly contribute to enhance the success of inclusive practices in schools which is reflected from their interaction with the SEN students (Leatherman & Niemeyer, 2005; Taylor & Ringlaben, 2012). The interaction between the teacher and student is regarded as instructional support (Allen, Gregory, Mikami, Hamre & Pianta, 2013; Curby, Kauffman, Arby, 2013; LaParo, Pianta & Stuhiman, 2004). Instructional support is reflected by the way the teacher delivers the material to the students, the questions posed to the students so they can develop their thinking skills through class discussion, and by the case analysis and problem solving introduced in the classroom, as well as the feedback from the teacher to strengthen the students understanding of the learning materials (Allen, Gregory, Mikami, Hamre & Pianta, 2013; Curby, Kauffman & Arby, 2013; LaParo, Pianta & Stuhiman, 2004).

Instructional support consists of three components: i.e., 1) content understanding, 2) analysis and problem solving, and 3) feedback quality (Allen, Gregory, Mikami, Hamre & Pianta, 2013; Curby, Kauffman & Arby, 2013; LaParo, Pianta & Stuhiman, 2004). Content understanding is closely related to the teachers' ability in providing intellectual support and delivering knowledge in a systematic and structured way. The analysis and problem solving component can be measured by assessing to what extent the teacher provides opportunities to SEN students in developing their critical thinking skills on the material. On the other hand, the quality of the feedback component is subject to the teachers' feedback in responding to the students' questions and giving answers that is aimed to enhance the students understanding (Allen, Gregory, Mikami, Hamre & Pianta, 2013; Curby, Kauffman & Arby, 2013).

Research shows that the teachers' instructional support towards SEN students is influenced by the culture of the country in which the school is located (Helgesen & Brown, 1994). As examples, Japanese teachers give their SEN students the opportunity to listen to their teachers without asking any questions or giving any answers, while teachers in Arabic culture allow their students to interact with their teacher by raising their hand and calling their teacher before giving or answering the questions (Dukmak, 2010; Helgesen & Brown, 1994). Therefore, in designing the most effective instructional support, cultural factors must be considered.

Earlier researches in several countries have noted the correlation between teachers' positive attitudes and the success of inclusive education. A similar result was also observed in primary schools in Indonesia, showing the teachers positive attitudes towards inclusive education (Kurniawati, Minnaert, Mangunsong & Ahmed, 2012). The purpose of this research is to answer the question "Is there a correlation between the teachers' attitude and the instructional support provided to SEN students?"

2 METHOD

The respondents were 40 classroom teachers from eight public inclusive primary schools in the Depok municipality. The respondents were chosen using purposive sampling from the following schools: SDN A, SDN B, SDN C, SDN D, SDN E, SDN F, SDN G, and SDN H. The Depok Inclusive Task Force stated that these schools were actively implementing inclusive education.

The teachers' attitude towards SEN students is measured using the Multidimensional Attitude towards Inclusive Education Scale—Indonesian Version (MATIES-IV) (Marantika, 2015). This is an adaptation from MATIES (Mahat, 2008), which consists of questionnaires with 18 items that measure 3 of attitude, i.e., cognitive (6 items), affective (6 items) and behavioral (6 items). These 18 items applies the Likert-type scale answers, which are: strongly agree (1), disagree (2), somewhat disagree (3), somewhat agree (4), agree (5), strongly agree (6). The validity test shows that the validity coefficient of MATIES_IV is good, ranging from

0.26–0.80 (Nunnaly & Bernstein, 1994). While the reliability test of the three components of MATIES_IV resulted in a good cronbach alpha score (Kapplan & Saccuzo, 2004), cognitive ($\alpha = 0.77$), affective ($\alpha = 0.80$), and behavioral ($\alpha = 0.81$).

To measure the teachers' instructional support, the researcher developed a Teacher Student Interaction instrument. The results from the confirmatory factor analysis (CFA) showed that the items used to measure instructional support are valid with a *p-value* = 0.05098 (*p-value* > 0.05), RMSEA = 0.061 (RMSEA < 0.08), and GFI = 0.90 (GFI > 0.90). The instructional support questionnaire has 11 items that consists of 3 components: i.e., content understanding (5 items), analysis and problem solving (2 items), and feedback quality (4 items). This instrument uses the Likert-type scale with 4 options for the answers ranging from very inappropriate (1) to very appropriate (4).

The teachers' attitudes and the teachers' instructional support towards SEN students are analyzed using descriptive statistics, by analyzing the mean and standard deviations of the collected data. The Pearson Correlation is used to answer the correlation between the two variables.

3 RESULTS

3.1 *Respondents characteristics*

From the 40 teachers involved in this research, 82.5% of the respondents were female, 40% of them were 50 years of age or older. Most of respondents have a bachelor degree (S1) and 42.5% of them have more than 20 years of teaching experience. Thirty five percent of the teachers have participated in training on inclusive education (see Table 1).

The results showed that the teachers have positive attitude in all three components, cognitive (M = 4.55, SD = 0.707), affective (M = 4.75, SD = 0.711), and behavioral (M = 5.12, SD = 0.570), by analyzing the mean score of the 6-point Likert-type scale to measure positive or negative attitudes. The mean score of the 6-point Likert-type of teachers' attitudes was 3.5. The instructional support is also reflected in the teachers' answers, showing high scores in all three components: i.e., content understanding (M = 3.34, SD = 0.351), analysis and problem solving (M = 3.39, SD = 0.431), and feedback quality (M = 3.35, SD = 0.375), which was done by analyzing the mean score of the 4-point Likert-type scale to measure higher or lower teachers' instructional support. The mean score of the 4-point Likert-type of teachers' instructional support was 2.5.

Table 1. Respondents characteristics ($N = 40$).

Respondents characteristics	N	%
Sex		
Male	7	17.5
Female	33	82.5
Age		
<40 years	15	37.5
40–50 years	9	22.5
>50 years	16	40
Education		
D3	2	5
S1	38	95
Teaching experience		
<10 years	14	35
10–20 years	9	22.5
>20 years	17	42.5
Inclusive training		
Trained	14	35
Never	26	65

Table 2. Correlation between teachers' attitude and instructional support ($N = 40$).

Attitude components	Components of instructional support					
	Content understanding		Analysis & Problem solving		Quality of feedback	
	R	*Sig (p)*	*R*	*Sig (p)*	*R*	*Sig (p)*
Cognitive	0.149	0.358	0.152	0.349	0.239	0.138
Affective	0.287	0.073	–0.002	0.991	–0.086	0.599
Behavioral	0.502**	0.001	0.442**	0.004	0.416**	0.008

*Correlation is significant with *level of significant* 0.05 (2-*tailed*).
**Correlation is significant with *level of significant* 0.01 (2-*tailed*).

There is a significant relationship between the teachers' attitude and the teachers' instructional support to SEN students (r = 0.361*, p = 0.022). Further analysis showed (see Table 2) that only the behavioral component of attitude have influenced the teachers' instructional support to SEN students (r = 0.514**, p = 0.001).

4 DISCUSSION

The research aims to measure the teachers' attitude and instructional support towards SEN students, and examine the correlation between the attitude and instructional support. The research results revealed that teachers have positive attitude towards SEN students. Kurniawati *et al,* (2012) also stated that teachers in Indonesia have strong positive attitudes towards SEN student and show strong willingness to include SEN student in their classes. Another research, however, showed different results, that most teachers have negative or neutral attitude towards SEN students (De Boer, Pijl & Minnaert, 2011).

The positive teachers' attitude found in the study could be linked to several issues. First, from the data, 35% teachers have participated in training on inclusive education. The knowledge from the training has equipped the teachers with the ability to understand the learning challenges of the SEN students (Leatherman & Niemeyer, 2005). Second, their experience in interacting and teaching SEN students may have strengthened the teachers' acceptance and confidence that SEN students are able to learn in regular classes, as stated by De Boer et al (2011).

The teachers showed strong instructional support when interacting with SEN students. The strong instructional support implies that the teachers are able to create fun learning processes which could enhance the students' knowledge (Allen, Gregory, Mikami, Hamre & Pianta, 2013; Curby, Kauffman & Arby, 2013). By applying strong instructional support, the teachers are able to deliver the learning process in a systematic way and in a slower pace with a more process-oriented manner, especially when giving feedback. However, these research results still have room for doubt because only 35% of the teachers have been trained for teaching in inclusive education settings. The training itself emphasizes more on basic matters in inclusive education. Relevant trainings on inclusive education for the teachers is very crucial in preparing the teachers to be more effective in the class.

A significant correlation between teachers' attitude and instructional support given by the teachers to SEN students in inclusive primary schools, shows that the more positive the teachers' attitude towards SEN students, the higher the instructional support is. If we observe more closely the components of attitude and instructional support, we can see that the behavioral component of teachers' attitude has a significant correlation in strengthening the teachers' instructional support to the SEN students. The teachers' experience and teachers' prior knowledge working with SEN students present a positive impact to the teachers, thus the successful teachers' experience in interacting with SEN students in inclusive settings create a positive experience, as stated by Leatherman and Niemeyer (2005). Therefore, as the teachers have more experience in working with SEN students and have successful interaction with SEN students, the greater the teachers' effort to give instructional support, i.e., giving intellectual support, teaching knowledge systematically, facilitating opportunities to SEN students in enhancing their knowledge and skills.

However, the results from the cognitive and affective components were quite different. The cognitive component and the affective component of the teachers' attitudes do not show any correlation with the teachers' instructional support. Although the teachers have sufficient knowledge on inclusive education, and also believe that the SEN students are able to learn, and that the teachers feel confident to be able to educate such students, however, this does not help the teachers to be willing and to be able to give instructional support needed by the SEN students. Various factors such as poor learning facilities (Ferguson, 2008; Leatherman & Niemeyer, 2005), the size of the classroom (Leatherman, 2007), limited knowledge and time (Fakolade, Adeniyi & Tella, 2009; Kurniawati, De Boer, Minnaert & Mangunsong, 2016; Stella, Forlin & Lan, 2007) may be related to these factors.

Research on inclusive education in Indonesia, in terms of teachers' attitude and instructional support is rare, that is why this research is aimed to contribute to the success of inclusive education. In this research, special attention is necessary in the following aspects: i.e., the sample size that only consists of 40 teachers, and the self-report for data collection by the teachers tend to produce subjective answers. For further research, it is highly recommended to involve more teachers, i.e., broaden the area of research, and use observational methods.

5 CONCLUSION

The conclusion of this research is that there is a significant correlation between behavioral component of the teachers' attitudes and instructional support in teaching SEN students. The results show that the more experienced the teacher is in working and interacting with SEN students, the greater the teachers' effort to give instructional support to SEN students. It is suggested that the teachers' positive attitude towards SEN students should be enhanced by reinforcing their willingness and belief that the SEN students are able to learn in regular classes, and by interacting postively with the SEN students in the classroom, the teachers can develop stronger instructional support.

REFERENCES

Allen, J.P., Gregory, A., Mikami, A., Lun, J., Hamre, B. & Pianta, R.C. (2013) Observations of effective teacher-student interactions in secondary school classrooms: predicting student achievement with the classroom assessment scoring system-secondary. *School Psychology Review,* 42: 76–97.

Ajzen, I. (2005) *Attitudes, personality, and behavior.* 4th edition. England, McGraw-Hill.

Azwar, S. (1995) *Sikap manusia: Teori dan pengukurannya.* Yogyakarta, Pustaka Pelajar.

Curby, T.W., Rimm-Kaufman, S.E. & Abry, T. (2013) Do emotional support and classroom organization earlier in the year set the stage for higher quality instruction? *Journal of School Psychology,* 51 (5), 557–69. Available on doi:10.1016/j.jsp.2013.06.001.

De Boer, A., Pijl, S.J., Minnaert, A. & Tied, F. (2011) Regular primary schoolteachers' attitudes towards inclusive education: A review of the literature. *International Journal of Inclusive Education,* 15 (3), 331–53. Available on doi:10.1080/13603110903030089.

Dukmak, S. (2010) Classroom interaction in regular and special education middle primary classrooms in the United Arab Emirates. *British Journal of Special Education,* 37 (1), 39–48. Available on doi:10.1111/j.1467-8578.2009.00448.x.

Fakolade, O. A., Adeniyi, S.O. & Tella, A. (2009) Attitude of Teachers Toward the Inclusion of Children with Special Needs in the General Education Classroom: The Case of Teachers in Selected Schools in Nigeria. *Education,* 1 (3), 155–169. Retrieved from: http://search.ebscohost.com/login.aspx?direct=true&db=ehh&AN=42513483&site=ehost-live.

Ferguson, D.L. (2008) International trends in inclusive education: the continuing challenge to teach each one and everyone. *European Journal of Special Needs Education,* 23 (2), 109–20. Available on doi:10.1080/08856250801946236.

Fishbein, M. & Ajzen, I. (1975) *Belief, attitude, intention and behavior: An introduction to theory and research*. USA, Addison-Wesley Publishing Company.
Hallahan, D.P. & Kauffman, J.M. (2011) *Handbook of special education*. New York & London, Routledge.
Helgesen, M. & Brown, S. (1994) *Active listening: Building skills for understanding*. Melbourne, Cambridge University Press.
Kapplan, R.M. & Saccuzzo, D.P. (2004) *Psychologycal testing: Principles, applications, and issues*. USA, Thomson Wadsworth.
Kurniawati, F., De Boer, A.A., Minnaert, A.E.M.G. & Mangunsong, F. (2016) *Teachers' attitudes and their teaching strategies in inclusive classrooms*.
Kurniawati, F., Minnaert, A., Mangunsong, F. & Ahmed, W. (2012) Empirical study on primary school teachers' attitudes towards inclusive education in Jakarta, Indonesia. *Procedia—Social and Behavioral Sciences*, 69, 1430–36. Available on doi:10.1016/j.sbspro.2012.12.082.
La Paro, K.M., Pianta, R.C. & Stuhlman, M. (2004) The classroom assessment scoring system: findings from the prekindergarten year. *The Elementary School Journal*, 104 (5), 409. Available on doi:10.1086/499760.
Leatherman, J.M. (2007) I just see all children as children: Teachers' perceptions about inclusion. *The Qualitative Report*, 12 (4), 594–611. Retrieved from: http://nsuworks.nova.edu/tqr%5Cnhttp://www.nova.edu/ssss/QR/QR12-4/leatherman.pdf.
Leatherman, J. & Niemeyer, J. (2005) Teachers' attitudes toward inclusion: factors influencing classroom practice. *Journal of Early Childhood Teacher Education*, 26 (1), 23–36. Available on doi:10.1080/10901020590918979.
Mahat, M. (2008) The development of a psychometrically-sound instrument to measure teachers' multidimensional attitudes toward inclusive education. *International Journal of Special Education*, 23 (1), 1–11.
Mangunsong, F. (2014) *Psikologi dan Pendidikan Anak Berkebutuhan Khusus*. 1th edition. Depok, Lembaga Pengembangan Sarana Pengukuran dan Pendidikan Psikologi.
Marantika, D. (2015) *Hubungan antara Sikap terhadap Pendidikan Inklusif dan Strategi Pengajaran pada Guru SMA Negeri Inklusif dan SMA Swasta Inklusif*. [Undergraduate's thesis], Universitas Indonesia.
Nunnaly, J.C. & Bernstein, I.H. (1994) *Psychometric Theory*. 3th edition. New York, McGraw-Hill.
Stella, C.S.C, Forlin, C. & Lan, A.M. (2007) The Influence of an inclusive education course on attitude change of pre-service secondary teachers in Hong Kong. *Asia-Pacific Journal of Teacher Education*, 35 (2), 161–79. Available on doi:10.1080/13598660701268585.
Taylor, R.W. & Ringlaben, R.P. (2012) Impacting pre-service teachers' attitudes toward inclusion. *Higher Education Studies*, 2 (3), 16–23. Available on doi:10.5539/hes.v2n3p16.
UNESCO. (1994) Final Report: World Conference on Special Needs Education: Access and Quality. Paris, UNESCO.

Diversity in Unity: Perspectives from Psychology and Behavioral Sciences – Ariyanto et al. (Eds)
© 2018 Taylor & Francis Group, London, ISBN 978-1-138-62665-2

Relationship between parental attachment and career adaptability in grade 12 senior high school students

U.J. Khusna & W. Indianti
Faculty of Psychology, Universitas Indonesia, Depok, Indonesia

ABSTRACT: The research aims to find the correlation between parental attachment and career adaptability among grade 12 students. The data were collected from 272 grade 12 students from public and private senior high schools in Jakarta, Indonesia. Parental attachment was measured by the Inventory of Parents and Peer Attachment—Revised (father–mother version) (Armsden & Greenberg, 2009). Career adaptability was measured by the Career Adaptability Scale (Indianti, 2015). The results indicate that there is a positive significant relationship between parental attachment and career adaptability, which implies that the stronger the parental attachment, the greater the career adaptability. Research also found that attachment to the mother contributes more to career adaptability than attachment to the father. For future studies, an equal number of participants from public and private senior high schools should be recruited to ensure comparability.

1 INTRODUCTION

Grade 12 students who want to continue their studies after graduating from high school must choose their college major, including the type of college they want to apply to. However, according to the Susenas (National Social Economic Survey) in Harian Sinar Harapan (28 May 2010), 61% of high school students in Indonesia did not know where they should pursue higher education (Setiyowati, 2015). Susilowati (2008) explains that the school and its staff (i.e. teachers and guidance counsellors), parents, and friends play an important role in assuring that students choose the right college majors. Most importantly, the role of parents includes having a discussion about career options with their children and providing support when they choose a certain college major (Susilowati, 2008). The preparation for college should be done as early as possible since choosing the wrong college major has psychological, academic, and relational impacts (Susilowati, 2008). To avoid such impacts, it is important for students to plan their career transition before choosing their college major. Choosing majors should be based on interests, talents, beliefs, and values. The majors are closely linked to careers that will be pursued in the future.

Career is a sequence of work experiences that occurs throughout a person's life (Arthur et al., 1989, in Seligman, 1994). Career development is defined as the process experienced by the individual to obtain knowledge, interests, beliefs, and values concerning the selection of work (Blaustein, 1996). According to Super (1954, in Seligman, 1994), the development of a career consists of five stages: growth stage (0–14 years), exploration stage (15–24 years), determination stage (25–44 years), maintenance stage (45–64 years), and decline stage (65 years onwards). To successfully complete the task at each stage of career development, career adaptability is required. Savickas (1997) states that career adaptability involves the ability to cope with situations that are not predictable because of the changes and working conditions. Later, he added that career adaptability consists of four dimensions: concern, control, curiosity and confidence (Savickas & Porfeli, 2012).

Concern about the future helps individuals to look ahead and prepare for what might come next. Career concern can be increased by forming an optimistic attitude towards the future to see the connection between actions and future plans (Indianti, 2015). Control enables individuals to become responsible for shaping themselves and their environments to meet what comes next. Career control also enables individuals to become responsible for building a career and making career choices (Indianti, 2015). Possible selves and alternative scenarios that individuals might shape are explored when curiosity prompts a person to think about him/herself in various situations and roles. These exploration experiences and information-seeking activities produce aspirations and build confidence so that individuals can actualise choices to implement their life design (Savickas & Porfeli, 2012). Individuals need to be responsible for future choices and decisions, open to new experiences, and confident in their choices.

The development of career adaptability is influenced by internal and external factors. Internal factors include personality variables such as emotional stability, anxiety, internal control, and vocational identity. External factors include support from the environment or social support, such as support from the school, peers, and family. For an adolescent, family or parents play an important role in career adaptability as they are the major figures who have attachments and emotional bonds throughout his/her life. The role of family or parents in an adolescent's career adaptability is to provide information and advice, help identify his/her talents, provide him/her freedom to make decisions, and support his/her activities that are connected to future plans.

Armsden and Greenberg (1987) define a parental attachment as an enduring bond of affection from parents who convey a feeling of safety to adolescents. Nawaz and Gilani (2011) found that the relationship between career decision-making and parental attachment is stronger than that with peer attachment. Support and emotional closeness that comes from parents can help adolescents to explore themselves and their environment. As expressed by Ryan et al. (1996), parental attachment is conceptualised as a feeling of safety that encourages individuals to actively explore and gather experience. Good parental attachment can create a feeling of security that underlies adolescent exploring and engaging in the development of behaviour. Parental attachment consists of three dimensions: communication, trust, and alienation.

Communication helps to maintain a strong bond of affection between parents and adolescents throughout life. Trust is a feeling of safety and belief in other people for the fulfilment of all needs (Armsden & Greenberg, 1987). Trust makes individuals feel confident that their parents will respect and understand what they want and need (Hellenthal, 2006 in Sudiarty, 2010). On the other hand, alienation is associated with avoidance and denial makes an adolescent feel uncomfortable with their parents. Hellenthal (2006; Sudiarty, 2010) described the feeling of alienation as feeling embarrassed to discuss personal issues, not being comfortable close to one's parents, feeling angry, and not getting enough attention. Strong parental attachment forms a high sense of trust and communication and a low sense of alienation that gives adolescents a feeling of security to explore their career options.

Career adaptability is important during the transitional period, such as from senior high school to college. In the transition to college, individuals need someone who can give support and advice as well as be a source of reliable information. According to Sawitri et al. (2012), people who have grown up in a collectivist culture have a tendency to act as a collectivist culture implies. For example, in the process of selecting a career, a person tends to make career decisions in order to follow their parents (Leong et al., 2001, in Sawitri et al., 2012). The role of parents is very important to fulfil the needs of intimacy and warmth, to build confidence leading to a feeling of safety, and to independently make decisions and take action (Gunarsa, 2001). Adolescents who perceive secure attachment with their parents develop efficacy to make career decisions. On the other hand, adolescents who do not feel attached to their parentsface career indecision (Emmanuelle, 2009).

There are several studies aiming to find the relationship between parental attachment and career adaptability. For instance, research conducted by Salami and Aremu (2007) reveals that attachment to the mother is significantly associated with career information-seeking behaviour, whereas attachment to the father is not. In their research about parental attachment and career maturity, Choi et al. (2012) show that changes in the quality of parental attachment are positively correlated with changes in career maturity. Another study, by Blustein et al. (1995), explains that parental attachment is related to career exploration among adolescents. Moreover, Sudiarty (2010) found that parental attachment is significantly associated with high school students' career maturity in Indonesia. In her research, Sudiarty (2010) measured attachment to the mother and the father separately, and the result shows that attachment to the father has a significant relationship with career maturity where as attachment to the mother does not. However, research conducted by Gemeay et al. (2015) shows that parental attachment is not significantly related to the academic achievement of students of nursing in Egypt and Saudi Arabia.

The results of previous studies discussed above only involved one dimension of career adaptability (curiosity) and could not be applied to grade 12 high school students who are transitioning from high school to college. As explained earlier, grade 12 high school students are in the exploration stage of career development where they actively explore themselves and their environment to choose the right college major. It is a fairly complex challenge because they have to choose just one college major from a broad set and need to consider their interests, talents, beliefs, and values to make sure that the chosen college major aligns with the career they want to pursue in the future. They need support from their closest circle of family (including parents) and friends during this important stage. Therefore, this study aims to determine whether there is a relationship between parental attachment and career adaptability in grade 12 high school students.

2 METHODS

In this study, researchers tested the alternative hypothesis (H_A),'there is a relationship between parental attachment and career adaptability in grade 12 high school students' and the null hypothesis (H_0),'there is no relationship between parental attachment and career adaptability in grade 12 high school students', and that parental attachment is positively related with the career adaptability of grade 12 senior high school students. This research used the non-experimental study that has no control of the independent variable because the variable has occurred or cannot be manipulated (Kerlinger & Lee, 2000). This research was a cross-sectional study design conducted only once, gathering its data from individuals who have similar characteristics (Kerlinger & Lee, 2000). The non-probability sampling technique was chosen because researchers did not know the exact number of individuals in the population.

The participants in this research were 272 grade 12 students from public and private senior high schools in Jakarta, Indonesia. They were selected based on their availability and willingness to participate in this research (convenience sampling) (Gravetter & Forzano, 2012). The characteristics of the respondents were: 1) adolescents aged 14–18 years; 2) students in grade 12 senior high schools in Jakarta; 3) students who had a parent or a parent substitute.

2.1 *Instrument*

2.1.1 *Career adaptability scale*

In this study, researchers used the *Skala Adaptabilitas Karir* measurement tool (Indianti, 2015) modified from the Career Adapt-Ability Scale-International Form (CAAS-IF) by Savickas and Porfeli (2012). It consists of 24 items that are divided into four subscales: concern (six items), control (six items), curiosity (seven items), and confidence (five items). Career adaptability is measured from the total score of the 24 items. Participants responded to each item employing a scale from 1 (Not at all like me) to 4 (Very much like me).

2.1.2 *Inventory of Parents and Peer Attachment-Revised (IPPA-R)*

This research used the father–mother version of the IPPA-R to measure parental attachment (Armsden & Greenberg, 2009). Although the instrument that measures attachment to the father is separated from the instrument that measures attachment to the mother, it is still able to measure parental attachment as a single unit by adding the total score from both versions. The IPPA instrument was designed to measure positive and negative perceptions of affective and cognitive aspects of adolescents in conjunction with parents and peers, and also to see how parents and peers give a sense of security in adolescents. Both IPPA-R versions of attachment to the father and the mother consist of 22 items that measure three dimensions: trust, communication, and alienation. All items are combined to form a total score that indicates parental attachment. Participants responding to each item employed a scale from 1 (Not at all like me) to 4 (Very much like me).

The data were collected by distributing questionnaires that contained an instrument for measuring peer attachment and career adaptability. The collected data were analysed using SPSS Statistics for Windows through multiple measures of statistical techniques: Cronbach's α, descriptive statistics, Pearson correlation, independent sample t-test, one-way analysis of variance, and multiple regression.

3 RESULTS

The statistical technique used to determine the relationship between parental attachment and career adaptability is the Pearson product moment correlation. Based on the results of statistical calculations, it was found that parental attachment—mean (M) = 128.39, standard deviation (SD) = 22.582 – and career adaptability (M = 73.22, SD = 8.602) had a significant positive correlation coefficient ($r = 0.281$, $p < 0.01$, two-tailed). From this result, we can see that the null hypothesis (H_0) is rejected and the alternative hypothesis (H_A) is accepted, implying that there is a significant relationship between parental attachment and career adaptability in grade 12 high school students. By the coefficient of determination (r^2), it is known that the 7.9% variance score for career adaptability of a student can be explained by parental attachment, whereas the remaining 92.1% is explained by other factors.

The statistical calculation gives a coefficient score of IPPA-R for attachment to the mother ($\beta = 0.168$, $p < 0.01$), which means that each additional point of attachment to the mother increases career adaptability to 0.168, and a coefficient score of IPPA-R for attachment to the father ($\beta = 0.136$, $p < 0.01$), which means that each additional point of attachment to the father increases career adaptability to 0.136. Thus, it can be seen that attachment to the mother has a higher score to the adaptability of a career than attachment to the father.

4 DISCUSSION

The existence of a significant relationship between parental attachment and career adaptability is aligned with the research conducted by Choi et al. (2012), which states that parental attachment is positively related to career maturity. In addition, the present study is also consistent with research conducted by Salami and Aremu (2007), which revealed that parental attachment is significantly related to career information-seeking behaviour, but, if measured separately, only attachment to the mother has a significant relationship. Participants in this study were grade 12 high school students who were faced with a career transition to college. Based on the developmental stages of a career, adolescents are included in the exploration stage in which they actively explore themselves and their environment. The presence of an attachment figure is important for adolescents. During the process of exploration, they need their parents to provide information and advice, to help identify their interests and talents, and support their activities connected to future plans. A high sense of trust and communication and a low sense of alienation form a good attachment that gives adolescents a feeling of security to explore their career options.

In this study, attachment to the mother as a bigger influence on career adaptability than attachment to the father. This is aligned with research by Salami and Aremu (2007), revealing that attachment to the mother is significantly associated with career information-seeking behaviour. In Indonesia, there are different roles for the father and the mother and their contributions to adolescent career development. The role of the father as the breadwinner for the family causes him to spend a lot of time outside, so the father does not have as much time as the mother. In contrast to the role of the father, the mother's role is usually associated as a caregiver to the household. Parenting duties are delegated to mothers more than fathers (Andayani & Koentjoro, 2004). Another opinion comes from Parsons and Bales (1955, in Finley et al., 2008) who argue that in Eastern cultures, the father is more often represented as a figure who works and earns for the family; they also act as the head of the family rather than as a discussion partner for their children.

5 CONCLUSION

Results of the data processing show that parental attachment is correlated positively and significantly with career adaptability in grade 12 high school students. This indicates that a person with higher parental attachment will have higher career adaptability. Conversely, the lower the parental attachment the lower the career adaptability. Good attachment to parents creates a feeling of security and builds confidence to explore oneself and the environment during the transition stage. Hence, adolescents can be responsible for future choices and decisions, open to new experiences, and confident in their choices. In addition, the results also found that attachment to the mother had a higher score for career adaptability than attachment to the father. This implies that attachment to the mother has a bigger influence on career adaptability compared with attachment to the father.

REFERENCES

Andayani, B. & Koentjoro. (2004). *Family psychology: Father's role toward coparenting*. Surabaya, Indonesia: Citramedia.

Armsden, G.C. & Greenberg, M.T. (1987). The inventory of parent and peer attachment: Individual differences and their relationship to psychological well-being in adolescence. *Journal of Youth & Adolescence*, *16*(5), 427–454. doi:10.1007/BF02202939.

Armsden, G., and Greenberg, M.T. (2009). Inventory of Parent and Peer Attachment (IPPA). (online).

Blaustein, B.S. (1996). *Effects of attachment and gender on college students'career development* (Master's thesis, Kean University, NJ). Available from ProQuest Dissertations and Theses database (UMI No. EP15226).

Blustein, D.L., Prezioso, M.S. & Schultheiss, D.P. (1995). Attachment theory and career development: Current status and future directions. *The Counseling Psychologist*, *23*(3), 416–432. doi:10.1177/0011000095233002.

Bronfenbrenner, U. (1994). Ecological models of human development. In T. Husen & T.N. Postlethwaite (Eds.), *International Encyclopedia of Education* (Vol. 3, 2nd ed.). Oxford, UK: Elsevier.

Choi, S., Hutchison, B., Lemberger, M.E. & Pope, M. (2012). A longitudinal study of the developmental trajectories of parent attachment and career maturity of South Korean adolescents. *The Career Development Quarterly*, *60*(2), 163–177. doi:10.1002/j.2161-0045.2012.00014.x.

Emmanuelle, V. (2009). Inter-relationships among attachment to mother and father, self-esteem, and career indecision. *Journal of Vocational Behavior*, *75*, 91–99. doi:10.1016/j.jvb.2009.04.007.

Finley, G. E., Mira, S.D. & Schwartz, S.J. (2008). Perceived paternal and maternal involvement: Factor structures, mean differences, and paternal roles. *Fathering*, *6*, 62–82. doi:10.3149/fth.0601.62.

Gemeay, E.M., Ahmed, E.S., Ahmad, E.R. & Al-Mahmoud, S.A. (2015). Effect of parents and peer attachment on academic achievement of late adolescent nursing students—A comparative study. *Journal of Nursing Education and Practice*, *5*(6), 96. doi:10.5430/jnep.v5n6p96

Gravetter, F.J. & Forzano, L.B. (2012). *Research methods for the behavioral sciences*. Belmont, CA: Wadsworth Cengage Learning.

Gunarsa, Y.S. (2001). *Adoloscent psychology*. Jakarta, Indonesia: Gunung Mulia.

Indianti, W. (2015). *Social support and self regulation in learning to build a career adaptability in new students of the university of Indonesia* (Doctoral dissertation, University of Indonesia, Depok, Indonesia).

Kerlinger, F.N. & Lee, H.B. (2000). *Foundations of behavioral research*. San Diego, CA: Harcourt College Publishers.

Marliyah, L., Dewi, F.I.R. & Suyasa, P.T.Y.S. (2004). Perceptions of parental support and making career decisions for adolescent. *Jounal Provitae*, *1*(1), 59.

Nawaz, S. & Gilani, N. (2011). Relationship of parental and peer attachment bonds with career decision-making self-efficacy among adolescents and post-adolescents. *Journal of Behavioural Sciences*, *21*(1) 33–47.
Ryan, N.E., Solberg, V.S. & Brown, S.D. (1996). Family dysfunction, parent attachment, and career search self-efficacy among community college students. *Journal of Counseling Psychology*, *43*(1), 84–89. doi:10.1037/0022-0167.43.1.84.
Salami, S.O. & Aremu, A.A. (2007). Impact of parent-child relationship on career development process of high school student in Ibadan Nigeria. *Career Development International*, *12*(7), 596–616.
Savickas, M.L. (1997). Career adaptability: An integrative construct for life-span, life-space theory. *The Career Development Quarterly*, *45*(3), 247–259. doi:10.1002/j.2161-0045.1997.tb00469.x.
Savickas, M.L. & Porfeli, E.J. (2012). Career adapt-abilities scale: Construction, reliability, and measurement equivalence across 13 countries. *Journal of Vocational Behavior*, *80*(3), 661–673. doi:10.1016/j.jvb.2012.01.011.
Sawitri, D.R., Creed, P.A. & Zimmer-Gembeck, M.J. (2012). The adolescent–parent career congruence scale: Development and initial validation. *Journal of Career Assessment*, *21*(2), 210–226. doi:10.1177/1069072712466723.
Seligman, L. (1994). *Developmental career counseling and assessment* (2nd ed.). Thousand Oaks, CA: SAGE Publications.
Setiyowati, E. (2015). *Relationship the effectiveness of career guidance and future orientation with adolescents career decision.* (Thesis, University of Muhammadiyah Surakarta, Surakarta, Indonesia). Retrieved from http://eprints.ums.ac.id/33872/1/NASKAH%20PUBLIKASI.pdf.
Sudiarty, F. (2010). *Relationship between parental and career maturity in senior high schools students* (Thesis, University of Indonesia, Depok, Indonesia).
Susilowati, P. (2008). Choosing major in university. Journal of Psychology (online) Retrieved from http://www.academia.edu/6218532/Memilih_Jurusan_di_Perguruan_Tinggi.

Diversity in Unity: Perspectives from Psychology and Behavioral Sciences – Ariyanto et al. (Eds)
© 2018 Taylor & Francis Group, London, ISBN 978-1-138-62665-2

Effectiveness of a self-regulated strategy development programme based on metacognition in improving story-writing skills of elementary school students

A.K. Banuwa, D. Maulina & P. Widyasari
Faculty of Psychology, Universitas Indonesia, Depok, Indonesia

ABSTRACT: Previous studies have found that most elementary school students have difficulties in the writing process. This study examines the effectiveness of a Self-Regulated Strategy Development (SRSD) programme based on metacognition in improving the story-writing skills of elementary school students. This single-subject study was conducted using one elementary school student. The participant was an 11-year-old girl with above average intelligence (Wechsler Scale), and who had problems in story-writing skills, especially in independently regulating the writing process. The intervention was conducted across six phases: 1) develop background knowledge to ensure that the student successfully understands, learns, and applies the strategy; 2) discuss the student's writing skills and strategy; 3) model the strategy by using a 'think aloud' process; 4) help the student memorise the strategy; 5) give feedback and encouragement; and 6) let the student perform the strategy independently. These phases were given in ten sessions, with each session lasting 60 minutes. Qualitative analysis was applied to measure changes in writing score before and after the intervention. The result shows that the SRSD model is effective in developing metacognition, not only in planning but also in reviewing and improving story-writing skills. Therefore, the SRSD programme is a promising intervention to improve story-writing skills, especially in elementary school students.

1 INTRODUCTION

The story-writing skill is particularly important because a certain level of writing skill is required in many aspects of life, such as at school, in the work environment, and in social life (Graham, 2006; Hacker, Dunlosky & Grasser., 2009a). Through writing, students learn to develop strategies for planning, evaluating, and reviewing texts for specific purposes, such as writing a story, composing a report, or delivering arguments. Writing also broadens and improves students' knowledge (Keys, 2000). However, most students find writing difficult. Previous studies have found that most elementary and junior high school students face difficulty in the writing process, especially in writing narration, exposition, and persuasion (Harris, Graham & Mason, 2011). They also experience difficulty in finding ideas, defining the purpose of writing, representing ideas logically and orderly, and regulating the writing process (De La Paz & Graham, 1997). In addition, De La Paz and Graham (1997) state that most students tend to write randomly from their mind without paying attention to the organisation of the story. Sandler et al. (1992) found that students with story-writing difficulties also have problems with the speed of writing, resulting in a story that is structured simply and with limited vocabulary.

Writing involves a complex cognitive process. Generally, writing consists of three main cognitive processes: planning, translating, and reviewing (Glaser & Brunstein, 2007). Firstly, planning involves three components: producing ideas about writing and how to write them, organising ideas to build a full concept, and setting the goal. Secondly, translating consists of two kinds of transformations: idea transformation into sentence or text generation, and transformation into symbols or transcriptions. Thirdly, reviewing is when the writer evaluates the story s/he writes and then modifies it. Reviewing requires coordination of several basic skills, such as reading, checking, and correcting, as well as the skill to rewrite the text that has been evaluated.

The process of writing as stated above shows that story writing requires metacognitive skills. Metacognition refers to awareness and management of one's own thoughts (Kuhn & Dean, 2004). Metacognition enables a student who has been taught a particular problem-solving strategy in a particular context to retrieve and deploy that strategy in a similar but new context (Kuhn & Dean, 2004). Additionally, metacognition engages conscious awareness of the mind, which relates to memory capabilities and how someone uses knowledge effectively. The basic elements of metacognition are planning, monitoring, and reviewing, all of which play an important role in every process involved in writing a story (McCutchen, 1988; Wong, 1994, in Santangelo, Harris & Graham, 2008; Graham, 2006). Students must have independent skills to plan the writing process, to monitor the story already written, as well as to revise mistakes in writing, spelling, and the content to ensure a good story (i.e. reviewing). Students with difficulty in writing show that they are unable to activate their metacognitive skills in writing. Some studies (Graham & Harris, 2003; McCutchen, 1995) demonstrate that students who face problems in planning and reviewing suffer from a lack of knowledge in writing strategies and mostly fail in the monitoring process of their writing.

So far, several interventions have been designed to overcome writing difficulties. Scaffolding and cognitive behaviours are some general strategies used to improve writing skills (Graham, McKeown, Kiuhara, & Harris, 2012). The use of computer-assisted tools has also become one of the strategies that help students improve their writing skills (Esperet, 1991). Additionally, there is a metacognitive development programme that can help students improve both their writing skills as well as their knowledge and motivation pertaining to writing (Graham, 2006). Metacognition also takes an active part in the writing process. Hacker, Kirner, and Kircher (2009b) state that writing, including story writing, involves an individual's metacognition. Thus, good metacognitive abilities enable students to compose well-written stories. Using metacognition, students can learn how to monitor and regulate their writing process. Students' knowledge about their own thought processes not only enables them to integrate what they have learned into the theme or context of the story, but also allows them to apply their knowledge in a variety of story types they set out to write. Therefore, the goal of the intervention in the current research is to improve a student's metacognition needed in the writing process.

An intervention programme that focuses on metacognitive ability during writing is the Self-Regulated Strategy Development (SRSD) programme. The SRSD programme views learning as a complex process that relies on changes that occur in the learners' skills and abilities, including many aspects of metacognition such as self-regulation, strategic knowledge, domain-specific knowledge and abilities, self-efficacy, and motivation (Harris, Graham, Brindle & Sandmel, 2009). In addition, because the SRSD programme is based on an instructional approach, it is designed to help students master the metacognitive processes in writing that include composing, improving independence, reflecting, using self-regulation to write effectively, developing knowledge about the characteristics of good writing, and setting a positive attitude in writing (Graham, Harris & Troia, 1998). Specifically, this programme should enable students to use various assignment-oriented strategies, such as planning and reviewing strategies, to collaboratively and explicitly compose a story (Graham & Harris, 2003; Santangelo & Olinghouse, 2009).

The SRSD programme has been shown to consistently and significantly improve students' writing skills (Graham et al., 1998; Graham, 2006; Schnee, 2010). However, SRSD has not been applied to the process of revision, although it has demonstrated some effectiveness when used for teaching students to plan and write their stories. Meanwhile, in order to produce a good story, a planning strategy has to be paired with a reviewing strategy. Therefore, in this study we developed an SRSD programme to improve metacognitive skills in writing, using planning and reviewing strategies, for elementary school students who face story-writing difficulties. This study aims to examine the following question: 'Is the SRSD programme based on metacognition effective for improving story-writing skills of elementary school students?'

2 METHODS

2.1 *Participant*

The participant was an 11-year-old girl (known as S) with above average intelligence as measured on the Wechsler Scale. S experienced problems in story writing, especially in regulating the writing process independently. This limited writing skill was believed to contribute to the participant's low academic performance in school. She needed extra time to finish assignments that required translating an idea into words or sentences, resulting in her frequent tardiness in submitting assignments. Moreover, her writing contained relatively few word choices and simple sentence structures written in poor handwriting.

2.2 *Research design*

The current research used the A–B single-subject design. In this design, A refers to the pre-test/baseline phase prior to intervention, whereas B refers to the condition after the treatment. The effects of the programme were obtained by comparing data from the baseline (pre-test), during treatment, and after the completion of treatment (post-test).

2.3 *Measures*

Observation checklists, writing worksheets, interview guidelines, and intervention modules were used to measure story-writing skills. Furthermore, a story reminder was provided to help the participant memorise the metacognitive strategy taught in the intervention programme. The effectiveness of the programme was indicated by the behavioural progress obtained by comparing the pre-test (baseline) and post-test data. As proposed by Nurgiyantoro (2009), the quality of the story was evaluated according to its content, organisation, vocabulary, grammar, and spelling. Successful intervention was assumed following a 25% increase in the quality of the story between pre-test and post-test. Four raters were used in order to maintain the interscorer reliability in scoring the quality of the story.

2.4 *Procedures*

In the first meeting, the facilitator asked the participant and her parents to fill out an informed consent form. The intervention programme took place at the participant's house. The whole intervention was planned to be completed in 10–12 sessions and was estimated to take approximately three weeks to finish. Each session was carried out in 45–60 minutes, including one or two five-minute rest periods. The programme was divided into three stages: baseline, intervention, and post-test. At the baseline stage, an interview was conducted to collect early data about the difficulties encountered and strategies used by the participant during the writing process. The participant's handwriting was also assessed during this stage. Next, the intervention stage was divided into five steps (Santangelo et al., 2008; Schnee, 2010). In the first step of the intervention, the participant was provided with an explanation about story writing and the importance of story writing in education, after which she was asked to discuss the difficulties she faced in writing. The second step involved teaching the participant some planning and reviewing strategies to help her develop metacognitive activities in writing a story. In the third step, the facilitator provided an example of how one would implement the taught strategies. The fourth step, the participant was asked to apply the appropriate strategy under the guidance of the facilitator, who employed a scaffolding procedure (i.e. giving instructions, questioning, and providing feedback). The facilitator's guidance was reduced gradually until the last step, where the participant was asked to write independently. In the independent exercise, she was asked to write a story using the 'think aloud' method. Finally, at the end of the intervention, there was the post-test stage, during which the facilitator evaluated the participant's performance on a writing task. The post-test was conducted five days after the intervention ended.

3 RESULTS

Overall, the intervention was found to be an effective method for developing metacognition when used in the context of planning and reviewing strategies in story writing. The effectiveness of the programme can be seen from the 50% increase in writing quality between the baseline score and the post-test score. Such an increase shows that the programme helped improve the participant's metacognitive skill in writing even without the facilitator's help.

Table 1 shows that planning, as a metacognitive strategy, started to appear after the intervention was introduced. Baseline data showed that the participant spent a lot of time before starting to write. Moreover, she seemed confused when she had to continue the story. She tended to write anything that crossed her mind without paying attention to the organisation of the story. Thus, some parts of the story were not logically related. From the interview, it was discovered that the participant did not use any strategy in writing the story.

In the practice session for tutoring (session 3), the participant was able to organise ideas and apply them to the story-writing process. The participant was becoming more fluent in conveying her ideas into the story. The development of her metacognitive strategy also continued to progress in the following sessions. In the self-exercise sessions (sessions 4 and 5), the participant used a planning strategy before writing the story. She applied self-statement in choosing ideas and defining the storyline (organisation) independently without any help from the facilitator.

The development of a metacognitive strategy also occurred in the post-test session, in which the participant managed to spend more time trying to find an idea and constructing a storyline. She then started to write and develop the story based on the storyline. Every sentence and paragraph of her story was logically related. Interview data show that the participant had begun to enjoy the process of writing a story. She claimed it was easier to find ideas and to translate them into words.

The baseline data additionally indicate that the participant did not monitor her writing process at all. She never checked or re-read her writing. Therefore, the quality of the story, as well as her handwriting, was relatively poor in the pre-test session. From the interview, the participant claimed to not realise her mistakes in writing and did not have any strategy to correct the mistakes. On the other hand, in the self-practice sessions (sessions 9 and 10), the participant started to use the reviewing strategy she had learned from the intervention. The participant also re-read the story and used the story reminder to revise her writing. At session 9, the participant noticed her own mistakes in writing. She continued to carefully rewrite and fix the story based on the correction checkmarks she made beforehand. However, the participant could not identify all the mistakes by herself, as she missed several punctuation and spelling errors. The result shows an improvement in the participant's use of metacognitive strategy in story writing, although its use was not yet optimal. In the post-test session, the participant was able to use the reviewing strategy independently. As a result, the participant seemed more careful throughout the writing process to avoid any mistakes. She monitored her writing process, re-read her story voluntarily, and identified and corrected her mistakes immediately.

Table 1. Metacognition development during intervention.

Metacognition	Baseline	During intervention	Post-test
Planning strategy	The participant did not have a planning strategy before writing a story.	At session 2, the participant was able to mention the purpose of the programme and characteristics of a good story. She also tried to make a story reminder before writing. At session 2, the participant was able to explain the planning strategy with the facilitator's help. However, at session 4, she was able to do it independently. At the beginning of session 3, the participant practised planning strategies, such as choosing ideas, constructing a storyline (organisation), and writing it logically. She managed to do it by herself at sessions 4 and 5.	The participant had a planning strategy and was capable of using it when writing a story, which included choosing ideas, constructing a storyline (organisation), and writing it logically.
Reviewing: Monitoring and correcting story elements (SKD/ Siapa-Kapan-Dimana, Apa = 2, Bagaimana = 2)	The participant did not re-read her writing. When the participant was asked to re-read her writing, she did not realise her mistakes related to the story elements. When the participant was asked to re-read her writing, she did not read entire sentences.	At session 8, the participant was able to explain the reviewing strategy with the facilitator's help. However, at session 10, she managed to do it independently. The participant started to use reviewing strategies at sessions 9 and 10: Re-reading her story to find a mistake related to story elements; realising her mistakes, then fixing them immediately.	The participant re-read her story voluntarily. During writing, she realised her mistakes related to the story elements without being reminded. She corrected her mistakes immediately.
Reviewing: Monitoring and correcting mistakes (TIK/Temukan-Identifikasi-laKukan+TEKA/Tanda baca-Ejaan-Kapital-tAmpilan)	The participant did not re-read her writing. When the participant was asked to re-read her writing, she did not realise her mistakes. When the participant was asked to re-read her writing, she did not read entire sentences.	At session 8, the participant could explain the reviewing strategy with the facilitator's help. However, at session 10, she was able to do it independently. The participant started to use reviewing strategies independently at sessions 9 and 10: Re-reading her story to find a mistake related to grammar, such as sentence structure, punctuation, and spelling; realising her mistakes, then fixing them immediately.	The participant re-read her story voluntarily. During writing, she realised her mistakes related to grammar without being reminded. She corrected her mistakes immediately.
Reviewing strategy	The participant did not evaluate her writing at all.	After the participant finished her writing, she re-read the entire story.	Evaluation was applied by re-reading and correcting mistakes.
Story-writing score	52		78 (50% increase)
Number of words	81		315 (3.88 times more)

4 DISCUSSION

The result of the present study shows that the instructional planning and reviewing strategies based on metacognition is effective in improving a student's story-writing skills. The result also supports data from previous studies (Schnee, 2010; Santangelo et al., 2008). Schnee (2010) states that the teaching of instructional planning and reviewing strategies is more effective for improving story-writing skills of elementary school students in grades 3 and 4 compared with the teaching of only an instructional planning strategy. In corroboration with Schnee's (2010) study, the current study found instructional planning and reviewing strategies to also be highly effective for grade 6 elementary school students.

In line with research by Schnee (2010) in which a modified SRSD was used to improve the writing skills of elementary school students, the present study also modified the SRSD programme to try to improve metacognitive strategy (i.e. planning and reviewing) in story writing among elementary school students. Across various themes, the participant in this study managed to plan a storyline. She was also able to find the mistakes in her writing and revise them as necessary. The development in the participant's ability to detect and revise the mistakes in writing was the result of the monitoring strategy used during the writing process.

The characteristics of the SRSD programme are an important factor that supports the effectiveness of this programme (Schnee, 2010). The intervention programme was carried out in two phases and divided into several stages. Each phase consisted of the same stages. In the first stage, the participant was asked to analyse the writing problem and to set a goal before commencing the writing process. Through this analysis, the participant was expected to comprehend and be more aware of the weaknesses in her writing. The subsequent stages included teaching planning or reviewing strategy and monitoring. Generally, information provided by the facilitator included how to use a strategy, when to use it, and why the strategy is useful (Schraw, 2006 in Lai, 2011). Following the completion of these stages, the participant was asked to implement the strategy with a story-writing task.

Other research has found scaffolding techniques, such as modelling, questioning, instructing, and giving feedback (Gallimore & Tharp, 1990), to be effective for building participants' metacognitive strategy (Danli, 2008). In the current research, a scaffolding technique was subsequently applied from sessions 1 to 10, while ensuring that the technique used was relevant to the purpose of every session. For instance, the facilitator gave a modelling technique before the participant was asked to practice the use of the taught metacognitive strategy. Moreover, in the early sessions, the facilitator provided plenty of instructions about what the participant had to do during writing. The facilitator gradually reduced the number of instructions and instead began asking the participant more questions to help her memorise the taught strategy. As a consequence, the metacognitive strategy in story writing was gradually mastered, as observable from the participant's ability to use the strategy eventually without the facilitator's help. Furthermore, self-statement also played an important role in the development of metacognition, enabling the participant to apply the metacognitive strategy independently.

Even though the current study demonstrated the effectiveness of the SRSD programme in improving story-writing skills, the study is not without limitations. Firstly, the research design was a single-subject design, suggesting that the result needs to be replicated in future research to establish consistency. Secondly, the use of only one participant and the A–B design (pre-test, intervention, and post-test) further reduced the ability to generalise the obtained results to other students and populations. Thirdly, all of the instructions in the intervention programme were given by means of a one-on-one method, which is exhaustive and time-consuming. Although the method is customary in Indonesia, where the teaching and learning processes are mostly conducted in large classrooms through conservative methods, it may not necessarily be the most efficient method for teaching writing strategy. Finally, prior research (Tracy et al., 2009) has only provided evidence for the effectiveness of the planning strategy in a conservative teaching and learning process, but not for the effectiveness of combined strategies of planning and reviewing in conservative teaching environments.

5 CONCLUSION

The result of this study shows that the SRSD model is effective for developing a student's metacognition, which can be effectively applied not only for planning a story but also for reviewing and improving story-writing skills. The participant was able to produce and organise her ideas, write them down carefully, and order them in a logical sequence. She was also able to monitor her writing process, re-read and review her story, and correct any mistakes. Evidence for the effectiveness of the programme was found in the post-test evaluation, which showed that planning and reviewing strategies were successfully applied by the participant both during intervention and after the intervention had ended.

REFERENCES

Abbott, R.D., Berninger, V.W. & Fayol, M. (2010). Longitudinal relationships of levels of language in writing and between writing and reading in grades 1 to 7. *Journal of Educational Psychology*, *102*(2), 281–298.

Danli, L. (2008). *Scaffolding and its impact on learning grammatical forms in tertiary Chinese EFL classroom* (Master's thesis, Hong Kong Baptist University, Hong Kong). Available from ProQuest Dissertations and Theses database.

De La Paz, S. & Graham, S. (1997). Strategy instruction in planning: Effects on the writing performance and behavior of students with learning disabilities. *Exceptional Children*, *63*, 67–187.

Esperet, E. (1991). Improving writing skills: Which approaches and what target skills? *European Journal of Psychology of Education*, *6*(2), 215–224.

Gallimore, R. & Tharp, R. (1990). Teaching mind in society: Teaching, schooling, and literate discourse. In L.C. Moll (Ed.), *Vygotsky and education: Instructional implications and applications of sociohistorical psychology* (pp. 175–205). Cambridge, UK: Cambridge University Press.

Glaser, C. & Brunstein, J.C. (2007). Improving fourth-grade students' composition skills: Effects of strategy instruction and self-regulation procedures. *Journal of Educational Psychology*, *99*(2), 297–310.

Graham, S. (2006). Writing. In P.A. Alexander & P.H. Winne (Eds.), *Handbook of educational psychology* (pp. 457–478). Mahwah, NJ: Erbaum.

Graham, S. & Harris, K.R. (2003). Students with learning disabilities and the process of writing: A meta-analysis of SRSD studies. In H.L. Swanson, K.R. Harris & S. Graham (Eds.), *Handbook of learning disabilities* (pp. 323–344). New York, NY: Guilford Press.

Graham, S., Harris, K.R. & Troia, G.A. (1998). Writing and self-regulation: Cases from the self-regulated strategy development model. In D.H. Schunk & B.J. Zimmerman (Eds.), *Self-regulated learning: From teaching to self-reflective practice* (pp. 20–41). New York, NY: Guilford Press.

Graham, S., McKeown, D., Kiuhara, S., & Harris, K.R. (2012). A Meta-Analysis of Writing Instruction for Students in the Elementary Grades. *Journal of Educational Psychology. Advance online Publication*. doi: 10.1037/a0029185.

Graham, S., Santangelo, T. & Harris, K.R. (2010). Metacognition and strategies instruction in writing. In H.S. Waters & W. Schneider (Eds.), *Metacognition, strategy use, and instruction*. New York, NY: Guilford Press.

Hacker, D.J., Dunlosky, J. & Graesser, A.C. (Eds.). (2009a). *Handbook of metacognition in education*. New York, NY: Routledge.

Hacker, D.J., Kirner, M.C. & Kircher, J.C. (2009b). Writing is applied metacognition. In D.J. Hacker, J. Dunlosky & A.C. Graesser (Eds.), *Handbook of metacognition in education* (pg. 154–172). New York, NY: Routledge.

Harris, K.R., Graham, S., Brindle, M. & Sandmel, K. (2009). Metacognition and children's writing. In D.J. Hacker, J. Dunlosky & A.C. Graesser (Eds.), *Handbook of metacognition in education* (pg. 131–153). New York, NY: Routledge.

Harris, K.R., Graham, S. Mason, L.H. (2011). Self-regulated strategy development for students with writing difficulties. *Theory into Practice*, *50*, 20–27.

Keys, C.W. (2000). Investigating the thinking processes of eighth grade writers during the composition of a scientific laboratory report. *Journal of Research in Science Teaching*, *37*, 676–690.

Kuhn, D. & Dean, D., Jr. (2004). A bridge between cognitive psychology and educational practice. *Theory into Practice*, *43*(4), 268–273.

Lai, E.R. (2011). *Metacognition: A literature review*. Pearson Research Report. Retrieved from http://images.pearsonassessments.com/images/tmrs/Metacognition_Literature_Review_Final.pdf

McCutchen, D. (1995). Cognitive processes in children's writing: Developmental and individual differences. *Issues in Education, 1*, 123–160.
Nurgiyantoro, B. (2009). *Penilaian pengajaran bahasa dan sastra* (*Assessment of language and literature teaching*). Yogyakarta, Indonesia: BPFE.
Sandler, A.D, Watson, T., Foto, M., Levine, M., Coleman, W. & Hooper, S. (1992). Neurodevelopmental study of writing disorders in middle childhood. *Developmental and Behavioral Pediatrics, 13*(1), 17–23.
Santangelo, T. & Olinghouse, N.G. (2009). Effective writing instruction for students who have writing difficulties. *Focus on Exceptional Children, 42*(4), 1–18.
Santangelo, T., Harris, K.R. & Graham, S. (2008). Using self-regulated strategy development to support students who have 'trouble getting things into words'. *Remedial and Special Education, 29*(2), 78–89.
Schnee, A.K. (2010). *Student writing performance: Identifying the effects when combining planning and revising instructional strategies* (Doctoral dissertation, University of Nebraska, Lincoln, NE).
Tracy, B., Reid, R. & Graham, S. (2009). Teaching young students strategies for planning and drafting stories: The impact of self-regulated strategy development. *The Journal of Educational Research, 702*, 323–331.

Diversity in Unity: Perspectives from Psychology and Behavioral Sciences – Ariyanto et al. (Eds)
© 2018 Taylor & Francis Group, London, ISBN 978-1-138-62665-2

The relationship between proactive personality and self-directed learning among undergraduate students

S.M. Sari & P.L. Suharso
Faculty of Psychology, Universitas Indonesia, Depok, Indonesia

ABSTRACT: The era of globalization is characterized by the rapid growth of information discovery potentially impacting on the learning process of students. Students need self-directed learning, which is the internal drive and motivation to drive themselves autonomously to solve various problems. Self-directed learning can improve opportunities for learners in distance and open education to review their learning success. It is essentially influenced by conscientiousness and extraversion, which are personality traits. Although there have been many studies on self-directed learning, its relationship with personality traits, especially proactive personality remains largely under investigated. This study aims to investigate the relationship between proactive personality and self-directed learning among college students. This quantitative research involved 520 undergraduate students as the participants. The results of this cross-sectional study showed correlation, with $r = 0.546$ significant at $p < 0.01$. This study used self-report, with Proactive Personality Scale (PPS) to measure proactive personality and Student Self-Directed Learning Questionnaire (SSDLQ) to measure self-directed learning. The implication of this study is college students realize the importance of knowing the role of proactive personality and student-directed learning in their learning success.

1 INTRODUCTION

University students should be able to independently manage, monitor, and complete their academic process (Hall, 2011). During the learning process, students are required to actively seek and independently understand the knowledge (Brookfield, 1986) and should implement part of adult learning process (Knowles, 1975). Adult learners have independently learned knowledge, skills, and experiences (Brookfield, 1986), and this ability enables them to solve academic and non-academic problems during learning activities. Entering the second year, students start to specialize in their specific major related to their career plan. While the university facilitates them in their studies and reports their academic results, students are expected to do their tasks and responsibilities and overcome challenges effectively in order for them to finish their study successfully.

University students are expected to be able to understand, remember, arrange, and practice what they learn from their environment, and they have individual differences in learning. The major assumption in this study is that personality as one of individual differences has influence on understanding university learning process. This research aims to investigate the relationship between personality and learning, proactive personality and self-directed learning in particular.

1.1 *Self-directed learning*

Knowles (1975) defines self-directed learning as "a process in which individuals take the initiative, with or without the help of others, in diagnosing their learning needs, formulating learning goals, identifying human and material resources for learning, choosing and implementing appropriate learning strategies, and evaluating learning outcomes" (p. 18). Sze-yeng and Hussain (2010) argue that self-directed learners are more responsible in completing tasks because of their self-reflection. They are willing to try hard and make more effort to complete challenging tasks. Furthermore, they actively search and solve tasks without others' assistance (Din, Haron & Rashid, 2016).

Self-directed learners will take initiative and control to complete tasks (Loyens, Magda & Rikers, 2008) and have awareness to develop assessment strategies in their learning process (Warburton & Volet, 2012). Moreover, they are engaged in personal learning environment to implement self-directed learning (Haworth, 2016). Self-directed learning engages internal motivation and self-esteem, curiosity, achievement motivation, and comfort (Hall, 2011). Internal locus of control, motivation, performance, support, self-efficacy also relate to self-directed learning during learning process (Boyer, *et al.,* 2014). Students still need guidance and feedback to maintain their performance and tasks towards self-directed learning (Regan, 2003) although they independently develop their capabilities and assess their method of learning in self-directed learning (Macaskill & Denovan, 2013).

1.2 *Proactive personality*

Proactive personality is defined as "a disposition relating to individual differences in people's proclivity to take personal initiative in acting to influence their environments in a broad range of activities and situations" (Bateman & Crant, 1993). Seibert, Kraimer, and Crant (2001) define proactive personality as "a stable disposition to take the personal initiative in a broad range of activities and situations" (p. 3). Proactive individuals take part and engage in activities to improve opportunities and challenges (Crant, 2000). Individuals examine social interactions towards proactive personality (Thompson, 2005). Individuals with proactive personality show initiative to take more effort to achieve learning goal orientation in their environment (Brown & O'Donnel, 2011).

Proactive personality is related to conscientiousness and extraversion as personality traits. Conscientiousness is the level of striving and persistance to achieve goals, and extraversion refers to the need for social interaction in seeking new experiences and activities (Batemant & Crant, 1993; Crant & Batemant, 2000). Proactive behavior is a manifestation of proactive personality which is a predictable factor (Marler, 2008).

1.3 *Proactive personality and self-directed learning*

Self-directed learning is influenced by personality traits (Lounsbury, *et al.,* 2009), which are different in each person (Long, 2007). This has been known from pyschological variables related to self-directed learning (Oliveira & Simoes, 2006). Lounsbury *et al.* (2009) argue that self-directed learning is related with conscientiousness and extraversion. Previous studies found different results regarding the correlation between self-directed learning and conscientiousness and extraversion. Some suggest that self-directed learning is significantly related to conscientiousness (Kirwan, Lounsbury & Gibson, 2010; Lounsbury, *et al.*, 2009; Oliveira & Simoes, 2006). Furthermore, some theorists report inconsistent relations between self-directed learning and extraversion. Some found a significant relationship (Lounsbury *et al.* 2009; Johnson 2001), while some others found no significant relationship (Kirwan Lounsbury & Gibson, 2010; Oliveira & Simoes, 2006).

In previous research, conscientiousness and extraversion were measured by the Big Five Personality Traits (Batemant & Crant, 1993; Crant, 2000; Crant & Batemant, 2000; Kirwan, Lounsbury & Gibson, 2010; Lounsbury, *et al.*, 2009; Oliveira & Simoes 2006; Johnson, 2001). Because personality traits of conscientiousness and extraversion were considered as personality constructs of proactive personality (Batement & Crant, 1993), proactive personality can be examined using the Big Five Personality (Brown & O'Donnell, 2011). Although there has been little research relating proactive personality with self-directed learning, self-directed learning can actually be related with conscientiousness and extraversion as part of proactive personality. The present study aimed to address the gap by examining both conscientiousness and extraversion as personality domains of proactive personality in relation to self-directed learning. Based on previous research, it was hypothesized that proactive personality might be related to self-directed learning.

2 METHODS

The participants in this study were 520 second year undergraduate students at Universitas Indonesia. However, responses from four participants were excluded from the statistical analyses, as these students did not complete the data collection process. Therefore, only 516 participants whose data were analyzed. The participants were 33% males and 67% females, aged between 18 and 25 years.

Proactive Personality Scale (PPS) by (Bateman & Crant, 1983) was used to measure proactive personality and consisted of 17 six-point likert scale items, ranging from 1 (Strongly Disagree) to 6 (Strongly Agree). The Cronbach Alpha reliablity of PPS was 0.82.

Student Self-Directed Learning Questionnaire (SSLDQ) by Bruin (2008; Bruin & Corenelius, 2011) was used to measure self-directed learning and consisted of 15 items. The result of Cronbach Alpha reliability measure was 0.594. Two items, item 1 and item 13, were eliminated due to their negative value. Therefore, the total number of items was 13.

Participation in the study was voluntary. During the registration period, participants were invited to fill in a questionnaire, which contained items measuring their proactive personality and self-directed learning and items asking about the demographic and background information of the students (e.g., gender, age, faculty, and activity engagements).

3 RESULTS

The correlation analysis show a correlation between proactive personality and self-directed learning, r (516) = 0.546, $p < 0.01$. The result of $r^2 = 0.29$ indicates that 29% variation self-directed learning scores can be explained by proactive personality scores. The results indicate that self-directed learning is positively correlated with conscientiousness, r (516) = 0.484, $p < 0.01$, and extraversion, r (516) = 0.518, $p < 0.01$. The results of $r^2 = 0.23$ suggest that 23% variation self-directed learning scores can be explained by conscientiousness scores. The results of $r^2 = 0.26$ imply that 26% variation self-directed learning scores can be explained by extraversion scores. Proactive Personality Scale shows that 53.2% of the participants have scores above the mean score of 68.52, while 46.7% of participants have scores below the mean. Self-Directed Learning Questionnaire shows that 53.5% of the participants have scores above the mean score of 56.84, while 46.5% have scores below the mean.

These results indicate a relationship between proactive personality and self-directed learning among the students of Universitas Indonesia. The more students with proactive personality, the more self-directed learners. Furthermore, conscientiousness and extraversion show a consistent relation to self-directed learning.

Table 1. Correlation between proactive personality and self-directed learning.

	Self-directed learning
Variable	Sig
Proactive personality	0.546**
Conscientiousness	0.484**
Extraversion	0.518**

Note: **p < 0.01.

4 DISCUSSION

This study attempts to investigate conscientiousness and extraversion as proactive personality in relation to self-directed learning. The findings of this study show that proactive personality is positively related to self-directed learning. The more students with proactive personality, the more self-directed learners. Furthermore, proactive personality should be reported through conscientiousness and extraversion (Bateman & Crant, 1993; Crant & Bateman, 2000).

Some literature suggests that self-directed learning is only related to conscientiousness (Kirwan, Lounsbury & Gibson, 2010; Lounsbury, *et al.*, 2009; Oliveira & Simoes, 2006). Others have reported inconsistent relations between self-directed learning and extraversion. Some found a significant relationship (Lounsbury, *et al.*, 2009; Johnson, 2001), while some others found no significant relationship (Kirwan Lounsbury & Gibson 2010; Oliveira & Simoes, 2006). The findings of this study support Lounsbury *et al.* (2009) revealing that conscientiousness and extraversion are related to self-directed learning. Furthermore, proactive personality consisting of conscientiousness and extraversion is consistently related to self-directed learning.

Conscientiousness is a predisposition to organizing tasks correctly and understanding what should be done in one's learning (Brothen & Wambach, 2001). Extraversion is assertiveness and desire on development activity (Major, Turner & Fletcher, 2006). Students who are more responsible in organizing and completing tasks have internal motivation to achieve goals (Komarraju, Karau & Schmek, 2009).

University students, entering adulthood age (18–25 years old), often face a variety of options (Arnett, 2000). Age differences have been examined in relation to self-directed learning and proactive personality. Jones (1993) found that age contributes to self-directed learning, while Lee (2004) reports that age and gender do not have influence on self-directed learning. Others also reveal that self-directed learning does not have any relation with gender and age differences (Brockett *et al.* 2005; Evelyn 2010). Furthermore, males and females show relatively similar levels of self-directed learning. Teng (2005) argues that age is a factor in self-directed learning. Evelyn (2010) reports that self-directed learning in males is higher if the females have more learning experiences. Grove and Miller (2014) further argue gender can influence self-directed learning.

Learning programs have examined how students become independent learners in self-directed learning (Jiusto & Dibiasio, 2006). Self-directed learning enables teachers and institutions to appropriately develop students with individual differences (Lee, 2004). Instructors or teachers encourage students to take initiative by adding their responsibilities (Hiemstra, 2013). One of the learning programs at Universitas Indonesia is problem-based learning, which should contribute to the development of self-directed learning among students. Because of the different learning programs, student might have different levels self-directed learning as described by their proactive personality.

According to Johnson (2001), teachers or instructors develop self-directed learning based on a personality type, which is proactive personality. This finding supports Cazan and Schiopca (2014) who suggest that if students have more conscientiousness and extraversion, they will be more self-directed in learning even though proactive personality might include other traits. Proactive people will be more adaptable and persistent in facing their challenges (Harvey, Blouin & Stout 2006). Proactive personality engages self-consciousness in managing and seeking outcomes (Chiaburu, Baker & Pitariu, 2006).

This research has several limitations. First, the male and female participants are unequal in numbers. Second, it does not include additional indicators such as Grade Point Average (GPA). It is recommended that future research includes this indicator and othres. Third, the participants are limited to one cohort (second year students) at Universitas Indonesia.

This study focuses on the internal condition of self-directed learners that contributes to describe how they are willing to know and solve their problems. This study has confirmed that self-directed learning has a relationship with extraversion and conscientiousness. Future research might include external factors besides internal factors such as individual differences. Considering samples from various educational institutions is also highly suggested.

5 CONCLUSION

The findings of this research have contributed to a deeper understanding of the relation between proactive personality and self-directed learning. Conscientiousness and extraversion as proactive personality have been shown to be consistently related to self-directed learning. Teachers or institutions should describe self-directed learners through proactive personality. The more students with proactive personality, the more self-directed learners. Proactive personality and self-directed learning are important in the success of students' learning process.

REFERENCES

Arnett, J.J. (2000) Emerging adulthood: A Theory of development from the late teens through the twenties. *American Psychologist,* 55 (5), 469–80. Available on doi:10.1037//0003-066X.55.5.469.

Bateman, T.S. & Crant, J.M. (1993) The proactive component of organizational-behavior—a measure and correlates. *Journal of Organizational Behavior,* 14 (2), 103–18. Available on doi:10.1002/job.4030140202.

Brothen, T. & Wambach, C.A. (2001) Refocusing Developmental education. *Journal of Research and Teaching in Developmental Education,* 18 (1), 25–31.

Boyer, S.L., Edmondson, D.R., Artis, A.B. & Fleming, D. (2014) Self-Directed learning: A tool for lifelong learning. *Journal of Marketing Education,* 36 (1), 20–32. Available on doi:10.1177/0273475313494010.

Brockett, R.G., Carré, P., Washington, G., Guglielmino, P.J., Ludwig, G.D., & Maher, P.A. (2005) Age and gender differences in self-directed learning readiness: A developmental perspective. *International Journal of Self-Directed Learning,* 2 (1), 40–9.

Brookfield, S.D. (1986) *Understanding and Facilitating Adult Learning.* San Francisco, Jossey-Bass.

Brown, S. & O'Donnell, E. (2011) Proactive Personality and goal orientation: A Model of directed effort. *Journal of Organizational Culture, Communications & Conflict,* 15 (1), 103–19. Retrieved from: http://libweb.ben.edu/login?url=http://search.ebscohost.com/login.aspx?direct=true&db=bth&AN=64876560&site=ehost-live.

Bruin, K.D. & Cornelius, E. (2011) Self-directed learning and career decision-making. *Acta Academia 2011,* 43 (2), 214–35.

Cazan, A. & Schiopca, B. (2014) Self-directed learning, personality traits and academic achievement. *Procedia-Social and Behavioral Sciences,* 127, 640–44. Available on doi:10.1016/j.sbspro.2014.03.327.

Chiaburu, D.S., Baker, V.L. & Pitariu, A.H. (2006) Beyond being proactive: What (else) matters for career self-management behaviors? *Career Development International,* 11 (7), 619–32. Available on doi:10.1108/13620430610713481.

Crant, J.M. (2000) Proactive behavior in organizations. *Journal of Management,* 26 (3), 435–62. Available on doi:10.1177/014920630002600304.

Crant, J.M. & Bateman, T.S. (2000) Charismatic leadership viewed from above: The impact of proactive personality. *Journal of Organizational Behavior,* 21 (1), 63–75.

Din, N., Haron, S. & Rashid, R.M. (2016) Can self-directed learning environment improve quality of life?. *Journal of Social and Behavioral Sciences,* 222, 219–227.

Evelyn, M.N. (2010) *The Effect of Gender, Age, Learning Preferences, and Environment on Self-Directed Learning: An Exploratory Case Study of Physician Learner Preferences.* [PhD Diss.], Northern Illinois University. Retrieved from: http://gradworks.umi.com/34/57/3457780.html.

Grover, K.S. & Miller, M.T. (2014) Gender differences in self-directed learning practices among community members. *Journal of Lifelong Learning,* 23, 19–31.

Hall, J.D. (2011) *Self-Directed Learning Characteristics of First-Generation, First-Year College Students Participating in a Summer Bridge Program.* [Graduate Theses and Dissertations], University of South Florida. Retrieved from: http://scholarcommons.usf.edu/etd/3140.

Harvey, S., Blouin, C. & Stout, D. (2014) Proactive Personality as a moderator of outcomes for young workers experiencing conflict at work. *Journal of Personality and Individual Differences,* 40, 1063–1074. Available on doi:10.1016/j.paid.2005.09.021.

Hiemstra, R. (2013) Self-directed learning: Why do must instructors still do it wrong? *International Journal of Self-Directed Learning,* 10 (1), 23–34.
Haworth, R. (2016) Personal learning environments: A solution for self-directed learners. *TechTrends,* 60 (4), 359–64. Available on doi:10.1007/s11528-016-0074-z.
Jiusto, S. & DiBiasio, D. (2006) Experiential learning environments: Do they prepare our students to be self-directed, life-long learners? *Journal of Engineering Education,* 95 (3), 195–204. Available on doi:10.1002/j.2168-9830.2006.tb00892.x.
Johnson, A.H. (2001) Predicting Self-directing learning from personality type. *International Journal of Self-Directed Learning,* 3(2), 208–34.
Jones, J.E. (1993) The influence of age on self-directed learning in university and community adult art students. *Studies in Art Education,* 34 (3), 158–66.
Kirwan, J.R., Lounsbury, J.W. & Gibson, L.W. (2010) Self-directed learning and personality: The Big five and narrow personality traits in relation to learner self-direction. *International Journal of Self-Directed Learning,* 7 (2), 21–34.
Knowles, M.S. (1975) Self-Directed learning: A guide for learners and teachers. *Selfdirected Learning a Guide for Learners and Teachers.*
Komarraju, M., Karau, S.J. & Schmeck, R.R. (2009) Role of the big five personality traits in predicting college students' academic motivation and achievement. *Learning and Individual Differences,* 19 (1) 47–52. Available on doi:10.1016/j.lindif.2008.07.001.
Lee, I.H. (2004) *Readiness for self-directed learning and the cultural values of individualism/collectivism among American and South Korean college students seeking teacher certification in agriculture.* [Thesis]. Texas A&M University. Retrieved from http://hdl.handle.net /1969.1 /3281
Long, H.B. (2007) Themes and theses in self-directed learning. *International Journal of Self-Directed Learning,* 4 (2), 1–18. Retrieved from: http://www.sdlglobal.com/journals.php.
Lounsbury, J.W., Levy, L.J., Park, S.H., Gibson, L.W. & Smith, R. (2009) An investigation of the construct validity of the personality trait of self-directed learning. *Learning and Individual Differences,* 19, 411–418. Available on doi:10.1016/j.lindif.2009.03.001.
Loyens, S.M.M., Magda, J. & Rikers, R.M.J.P. (2008) Self-directed learning in problem-based learning and its relationships with self-regulated learning. *Educational Psychology Review,* 20 (4), 411–27. Available on doi:10.1007/s10648-008-9082-7.
Macaskill, A. & Denovan, A. (2013) Developing autonomous learning in first year university students using perspectives from positive psychology. *Studies in Higher Education,* 38 (1), 124–42. Available on doi:10.1080/03075079.2011.566325.
Major, D.A., Turner, J.E. & Fletcher, T.D. (2006) Linking proactive personality and the big five to motivation to learn and development activity. *Journal of Applied Psychology,* 91 (4), 927–35. Available on doi:10.1037/0021-9010.91.4.927.
Marler, L.E. (2008) *Proactive Behavior: A Selection Perspective.* [PhD Dissertation.], Lousiana Tech University.
Oliveira, A.L. & Simoes, A. (2006) Impact of Socio-demographic and psychological variables on the self-directedness of higher education students. *International Journal of Self-Directed Learning,* 3, 1–12.
Regan, J.A. (2003) Motivating students towards self-directed learning. *Nurse Education Today,* 23 (8), 593–99. Available on doi:10.1016/S0260-6917(03)00099-6.
Seibert, S.E., Kraimer, M.L. & Crant, J.M. (2001) What do proactive people do? A longitudinal model linking proactive personality and career success. *Personnel Psychology,* 54, 845–74. Available on doi:10.1111/j.1744-6570.2001.tb00234.x.
Sze-yeng, F. & Hussain, R.M.R. (2010) Self-Directed learning in a socioconstructivist learning environment students. *International Journal of Self-Directed Learning,* 3 (1), 1–12. Available on doi:10.1016/j.sbspro.2010.12.423.
Teng, K.H. (2005) *Perceptions of Taiwanese Students to English Learning as Functions of Self-Efficacy, Motivation, Learning Activities, and Self-Directed Learning.* [PhD Dissertation.], University of Idaho.
Thompson, J.A. (2005) Proactive personality and job performance: A social capital perspective. *Journal of Applied Psychology,* 90 (5), 1011–17. Available on doi:10.1037/0021-9010.90.5.1011.
Warburton, N. & Volet, S. (2012) Enhancing self-directed learning through a content quiz group learning assignment. *Active Learning in Higher Education,* 14, 9–22. Available on doi:10.1177/1469787412467126.

Diversity in Unity: Perspectives from Psychology and Behavioral Sciences – Ariyanto et al. (Eds)
 ISBN 978-1-138-62665-2

The relationship between teacher efficacy and teaching strategies in inclusive private primary schools

A.A. Novara, F.M. Mangunsong & P. Widyasari
Faculty of Psychology, Universitas Indonesia, Depok, Indonesia

ABSTRACT: The objective of this research is to examine the relationship between teacher efficacy and teaching strategies in inclusive private primary schools. This research was conducted in 11 inclusive private primary schools in Jakarta and Depok, Indonesia, with a total sample of 70 teachers. The Teacher's Sense of Efficacy Scale and the Bender Classroom Structure Questionnaire were administered to measure teacher efficacy and inclusive teaching strategies, respectively. Significant positive correlation was found between teacher efficacy and teaching strategies in inclusive private primary schools ($r(70) = 0.247$, $p < 0.05$). This showed that teachers with high efficacy were more often employing different teaching strategies. Differences in the correlation between teacher efficacy and teaching strategies were observed from whether the teacher had had training before. No differences were found in the correlation between teacher efficacy ($t(68) = -0.026$, $p > 0.05$) and teaching strategies ($t(68) = 0.188$, $p > 0.05$) on teachers who had and had not received training. This showed that both teachers with and without training had the same efficacy in teaching and employing inclusive teaching strategies in the classroom.

1 INTRODUCTION

The development of inclusive education in Indonesia is still facing many challenges. Regular schools that are assigned as inclusive schools have constraints in adjusting their perspectives towards inclusive education, which may lead to ineffective implementation of inclusive education. Also, regular schools are not yet fully equipped to provide special services for students with special needs (Napitupulu, 2011). Apparently, it is necessary to closely examine the preparation of certain aspects in inclusive education, which includes changing the attitude of teachers and school staff, changing teaching methods and classroom management, adapting the school environment, adjusting the role of teachers and parents, and modifying the educational system (Mangunsong, 2014). This preparation is aimed at ensuring successful and smooth implementation of inclusive education.

According to Avramidis and Norwich (2002), the roles of teachers are important in implementing inclusive education. Teachers determine what is needed or what is best for students (Eslami & Fatahi, 2008). However, teachers in inclusive schools in Indonesia have limited understanding of the needs of students with special needs. Teachers also consider that the effort to meet the needs of students with special needs is burdensome and demands extra work (Juwono & Kumara, 2011).

Based on a study by Soodak et al. (1998), there are several factors that determine effective teaching in inclusive education. Effective teaching strategies are one of the consistent factors associated with successful inclusive programmes (Soodak et al., 1998). If teachers apply effective inclusive teaching strategies, the implementation of inclusive education will be better.

Teaching strategies are the learning activities chosen by teachers in implementing the teaching and learning process (Mangunsong, 2014). Bender (Bender, Vail, and Scott, 1995) states that a teaching strategy is the practice of teaching to facilitate the provision of learning materials in inclusive education. In the context of inclusive education, students with special needs require an adaptive instructional strategy so that it can meet students' needs (Shippen et al., 2011). There are two dimensions of teaching strategies: individual and cognitive strategies. Individualised strategies are variations of the learning content, materials, and processes of teaching and learning that are suitable for each unique individual (Choate, 2000). Cognitive strategy can facilitate the process of learning and understanding of knowledge (Novak, 1990, in Pate, 2009). This strategy is the result of mental representation that can guide the thinking process to solve problems more effectively (Resnick, 1985, in Pate, 2009).

The effectiveness of teaching strategies is influenced by the teachers' character, the teachers' training, and the school's support (Scott et al., 1998). In another study, it was found that teacher efficacy could affect instructional practices (Soodak et al., 1998; David & Kuyini, 2012; Chester & Beaudin, 1996). Teacher efficacy has the potential to affect the environment around the teacher, as well as a variety of instructional teaching practices used in the classroom (Tschannen-Moran & Hoy, 2001). Teachers who have confidence in their ability have the potential to perform variations of teaching practices.

Bandura identified teachers' efficacy as a type of self-efficacy (1977, in Tschannen-Moran et al., 1998). Armor et al. (1976, in Tschannen-Moran & Hoy, 2001) define teacher efficacy as the assessment of teachers' ability to get the level of involvement and expected student achievement, including students who have difficulties or students who are not motivated. Tschannen-Moran et al. (1998) state that teacher efficacy is the belief in the ability of the teachers to organise and implement actions necessary to achieve success in specific teaching assignments.

There are three dimensions of teacher efficacy: efficacy in instructional strategies, efficacy in classroom management, and efficacy in student engagement (Tschannen-Moran et al., 1998). Efficacy in instructional strategies refers to a teacher's confidence in delivering learning materials using instruction and evaluation methods. Efficacy in classroom management refers to a teacher's belief in self-efficacy to control and prevent misbehaviours of students. Efficacy in student engagement refers to a teacher's self-confidence in handling matters related to students, such as providing motivation to students in the learning process and involving all students in the learning process.

One of the factors that affect teacher efficacy and teaching strategies is the training of teachers. In the study by Sunanto (2008, in Juwono & Kumara, 2011), the teachers who attended training showed a successful rate of implementation of inclusive education. Another study, conducted by Schumm and Vaughn (1991, in Scott et al., 1998), concludes that a lack of training will lead to less confident teachers as they lack the knowledge and ability to adapt and work with students with special needs. The problem is that many teachers in inclusive settings rarely receive training on dealing with students with special needs (Scott et al., 1998). In Indonesia, there are many teachers who have not received training in inclusive education (Rudiyati, 2011). From the results of the research conducted by Juwono and Kumara (2011), teachers of inclusive education in Indonesia still find it difficult to design suitable teaching strategies to meet the needs of students due to this lack of training.

In Indonesia, there are two types of inclusive schools. Based on Act No. 17 of 2010, education is divided into two groups: public and private schools. The Act also explains that inclusive public and private schools differ in character, management, organisational structure and authority, as well as responsibility. In inclusive private schools, students must pay a tuition fee; therefore, facilities in inclusive private schools are better than those in inclusive public schools (Azka, 2011). Private school teachers tend to use a student-centred approach, such as working in small groups and discussions (McKinnon, Barza, & Moussa-Inaty, 2013). Teachers in inclusive private schools are also considered to have a more positive attitude towards inclusive education (Kurniawati et al., 2012). Nonetheless, Herdiana (2010) found that implementation of inclusive education in inclusive private schools still has some challenges. These include individual and group problems, such as the relationship between teachers and students and between regular students and those with special needs. This shows that, in spite of the various advantages in inclusive private schools, there are still problems arising from the implementation of inclusive education in inclusive private schools.

This research focuses on teacher efficacy, teaching strategies, and training in inclusive private primary schools in Depok and Jakarta, Indonesia. This research aims to examine the relationship between teacher efficacy and teaching strategies and to compare trained teachers with those who never received any training.

2 METHODS

The total sample was 70 teachers from inclusive private primary schools in Jakarta and Depok. Participants were teachers who have direct contact with students with special needs in the classroom. Non-probability sampling with a purposive sampling method was used.

In this study, the Teacher's Sense of Efficacy Scale (TSES) by Tschannen-Moran et al. (2001) was used to measure teacher efficacy and the Bender Classroom Structure Questionnaire (BCSQ) by Bender (1992) was used to measure teaching strategies. Both of the instruments were adapted to the Indonesian language. Expert judgement was used on both instruments. The reliability of TSES was 0.814 and that of BCSQ was 0.73. The validity of TSES was 0.037–0.771 and of BCSQ was –0.171–0.675. The items with validity below 0.2 were deleted or revised. The questionnaire was given to teachers teaching students with special needs.

3 RESULTS

First, the relationship between teacher efficacy and teaching strategies in inclusive private primary schools was tested using correlational analyses. Second, correlational analyses were conducted to examine the relationship between the dimensions of teacher efficacy and teaching strategies. Finally, the scores of trained and untrained teachers were examined for differences.

The result showed that there is a significant positive relationship between teacher efficacy and teaching strategies in inclusive private primary schools ($r(70) = 0.247$, $p < 0.05$). This shows that the more confident the teacher, the better they teach and use variation in teaching strategies.

No significant relationship was found between the dimensions of efficacy in instructional and individual strategies (Table 1). This shows that teachers' beliefs in their ability to implement the strategy instruction are not related to the variations in individual strategy applies in the classroom. On the other hand, a significant positive relationship was found between the dimensions of efficacy in instructional and cognitive strategies ($r(70) = 0.364$, $p < 0.01$) (Table 1). However, no significant relationship was evident between the dimensions of efficacy in classroom management and individual strategy. The self-confidence of teachers to organise the students is not related to the variations in individual strategy applied in the classroom. There was a significant positive relationship between the dimensions of efficacy in classroom management and cognitive strategies ($r(70) = 0.365$, $p < 0.01$) (Table 1). On the dimension of efficacy in student engagement, there is no significant relationship between individual and cognitive strategies.

Table 1. The relationship between dimensions of teacher efficacy and teaching strategies.

	Teaching strategies	
Teacher efficacy	Individual	Cognitive
Efficacy in instructional strategies	0.171	0.364*
Efficacy in classroom management	0.222	0.365*
Efficacy in student engagement	0.017	0.210

*Level of significance: p = 0.01.

Table 2. Comparison between teachers with and without training.

	Trained		Untrained			
Variable	N	M (SD)	N	M (SD)	t	p
Teacher efficacy	40	101.88 (7.046)	30	101.83 (6.259)	–0.026	0.980
Teaching strategies	40	120.50 (8.143)	30	120.90 (9.618)	0.188	0.851

There was no significant difference in teacher efficacy among a group of teachers who attended training (M = 101.88, SD = 7.046) and teachers who have never attended training (M = 101.83, SD = 6.259; $t(68) = -0026$, $p > 0.05$; Table 2).

4 DISCUSSION

The results of this study indicate that in general there is a significant positive relationship between teacher efficacy and teaching strategies in inclusive private primary schools. This finding supports the research of Tschannen-Moran and Hoy (2001), which states that teacher efficacy is a strong predictor of teaching strategies in inclusive teaching. Teachers who have confidence in their ability to teach a class are more likely to use a variety of teaching strategies. It also concurs with a previous study of Emmer and Hickman (1991, in Shaukat et al., 2013), which states that teachers who have a high efficacy are more willing to try to use a variety of teaching methods suitable for the needs of their students. It can be said that the results of this study support the findings of previous studies.

There are dynamic relationships between the dimensions of teacher efficacy and teaching strategies. Based on results of the correlations between the three dimensions of teacher efficacy and individual strategy, it was found that there was no significant relationship between the three dimensions of teacher efficacy and individual strategy. This is not consistent with the research by Bender et al. (1995). The teachers in this study seem to execute a cognitive learning strategy rather than an individual learning strategy. Thus, if they must carry out an individual learning strategy, they do not show efficacy in the three dimensions.

The relationship between the dimensions of teacher efficacy and cognitive strategy apparently is not consistent. The reason is that the dimensions of efficacy in instructional strategies and classroom management are influenced by teachers' belief in their ability to implement cognitive strategies in the classroom. These cognitive strategies can facilitate the process of learning and understanding in students (Novak, 1990, in Pate, 2009).

Meanwhile, there is no relationship between efficacy in student engagement and cognitive and individual strategies. Apparently, teachers are not concerned with the engagement of students. Tschannen-Moran and Hoy (2007) stated that it was only recently that student engagement became a focus in the field of teaching, as well as strategies to cultivate student involvement. For this reason, teachers apparently are not ready to engage students in more in-depth learning. This is also similar for the case of applying individual and cognitive strategies.

The results show that there is no significant difference in either teacher efficacy or teaching strategies in both groups of trained and untrained teachers. This is inconsistent with the research conducted by Scott et al. (1998) that states that the training of teachers can enhance teaching strategies. The intervention was also not in line with the research conducted by Schumm and Vaughn (in Scott et al., 1998), which explains that the training may be associated with teachers' belief in having the knowledge and the ability to adapt and work with children with special needs. The second study was conducted in the United States. This may indicate that there are differences in the research conducted in Indonesia and in the United States. The researchers have tried to determine why the results of the research in Indonesia become insignificant.

According to Nguyet and Ha (2010), ideal training in inclusive education should include several important components, such as the explanation of inclusive education, an overview of students with special needs, an individual education plan, how to design and adapt activities for students with special needs, and how to assess the work of students with special needs. Apparently, the content of training is different from one place to another. The different types, frequency, and intensity of training, may lead to different results in efficacy and strategies in both group of teachers.

5 CONCLUSION

From these results, it can be concluded that there is a significant positive relationship between teacher efficacy and teaching strategies in inclusive private primary schools. This means the more confident the teacher, the more likely they are to use a variety of teaching strategies. In terms of the relationship between the dimensions of teacher efficacy and teaching strategies, there is a significant positive correlation between efficacy in instructional and cognitive strategies; also, there is a significant positive relationship between the efficacy in classroom management and cognitive strategies. This means that teachers believe in their ability to carry out instructional strategies that increase the variety of cognitive strategies.

To enrich this research, it is suggested that interviews are conducted and observations are made in future work. Training in inclusive education can also be examined for further improvement.

REFERENCES

Avramidis, E. & Norwich, B. (2002) Teachers' attitudes towards integration/inclusion: A review of the literature. *European Journal of Special Needs Education, 17*(2), 129–147. doi:10.1080/08856250210129056

Azka, N. (2011, August 28). Sekolah Negeri Vs Sekolah Swasta [Public school vs private school]. *Kompasiana.com*. Retrieved from http://www.kompasiana.com/nadiaazka/sekolah-negeri-vs-sekolah-swasta_55090b3ca333114a442e3af5

Bender, W.N. (1992). The Bender classroom structure questionnaire: A tool for placement decisions and evaluation of mainstream learning environment. *Intervention in School and Clinic, 27*(5), 307–312.

Bender, W.N, Vail, C.O. & Scott, K. (1995). Teachers' attitudes toward increased mainstreaming: implementing effective instruction for students with learning disabilities. *Journal of Learning Disabilities, 28*(2), 87. doi:10.1177/002221949502800203

Choate, J.S. (2000). *Successful inclusive teaching: Proven ways to detect and correct special needs.* Boston, MA: Allyn and Bacon.

David, R. & Kuyini, A.B. (2012). Social inclusion: Teachers as facilitators in peer acceptance of students with disabilities in regular classrooms in Tamil Nadu, India. *International Journal of Special Education,* 27 (2), pp. 157–168.

Eslami, Z.R. & Fatahi, A. (2008). Teachers' sense of self-efficacy, English proficiency, and instructional strategies: A study of nonnative EFL teachers in Iran. *Teaching English as a Second or Foreign Language, 11*(4). Retrieved from http://tesl-ej.org/ej44/a1.html

Herdiana, W.A. (2010). *Perbedaan Pengelolaan Kelas Inklusi di SD Negeri dan SD Swasta se-Kota Malang* [Differences in the management of inclusive classroom between private's and public's primary schools in Malang city] (Thesis, Universitas Negeri Malang, Malang, Indonesia).

Juwono, I.D. & Kumara, A. (2011). Pelatihan Penyusunan Rancangan Pembelajaran pada Guru Sekolah Inklusi: Studi Kasus pada SD X di Yogyakarta [Training on formulation of teaching plan for teachers in inclusive school: case study of X primary school in Yogyakarta]. In *Proceedings PESAT Vol. 4, Universitas Gunadarma, Depok, 18–19 October 2011*. ISSN: 1858-2559.

Kurniawati, F., Minnaert, A., Mangunsong, F. & Ahmed, W. (2012). Empirical study on primary school teachers' attitudes towards inclusive education in Jakarta, Indonesia. *Procedia – Social and Behavioral Sciences, 69*, 1430–1436. doi:10.1016/j.sbspro.2012.12.082

Mangunsong, F. (2014). *Psikologi and Pendidikan Anak Berkebutuhan Khusus Jilid 1* [Psychology and Child with Special Needs Education Chapter 1]. Depok, Indonesia: LPSP3-UI.

McKinnon, M., Barza, L., & Moussa-Inaty, J. (2013). Public versus private education in primary science: The case of Abu Dhabi schools. *International Journal of Educational Research*, *62*, 51–61.

Napitupulu, E.L. (2011, November 9). Pendidikan Inklusif Hadapi Tantangan [Inclusive education face challenge]. *Kompas.com*. Retrieved from http://edukasi.kompas.com/read/2011/11/09/2341052/Pendidikan.Inklusif.Hadapi.Tantangan

Nguyet, D.T. & Ha, L.T. (2010). *How-to guide: Preparing teachers for inclusive education.* Baltimore, MD: Catholic Relief Services Vietnam.

Pate, M.L. (2009). Effects of metacognitive instructional strategies in secondary career and technical education courses (Dissertation, Iowa State University, Ames, IA).

Rudiyati, S. (2011). Potret Sekolah Inklusif di Indonesia [Potrait of inclusive school in Indonesia]. *Pertemuan Nasional Asosiasi Kesehatan Jiwa dan Remaja: Memilih Sekolah yang Tepat Bagi Anak Berkebutuhan Khusus*, Yogyakarta, 5 May 2011.

Scott, B. J., Vitale, M.R. & Masten, W.G. (1998). Implementing instructional adaptations for students with disabilities in inclusive classrooms: A literature review. *Remedial and Special Education*, *19*(2), 106–19. doi:10.1177/074193259801900205

Shaukat, S., Sharma, U. & Furlonger, B. (2013). Pakistani and Australian pre-service teachers' attitudes and self-efficacy towards inclusive education. *Journal of Behavioural Sciences*, *23*(2), 1–16. Available at https://www.researchgate.net/publication/262344457_Attitudes_and_self-efficacy_of_pre-service_teachers_towards_inclusion_in_Pakistan

Shippen, M.E., Flores, M.M., Crites, S.A., Patterson, D., Ramsey, M.L., Houchins, D.E. & Jolivette, K. (2011). Classroom structure and teacher efficacy in serving students with disabilities: Differences in elementary and secondary teachers. *International Journal of Special Education*, *26*(3), 36–44.

Soodak, L.C., Podell, D.M. & Lehman, L.R. (1998). Teachers, student, and school attributes as predictors of teachers' responses to inclusion. *Journal of Special Education*, *31*(4), 480–497. doi:10.1177/002246699803100405

Tschannen-Moran, M. & Hoy, A.W. (2001). Teacher efficacy: Capturing an elusive construct. *Teaching and Teacher Education*, *17*(7), 783–805. doi:10.1016/S0742-051X(01)00036–1

Tschannen-Moran, M. & Hoy, A.W. (2007). The differential antecedents of self-efficacy beliefs of novice and experienced teachers. *Teaching and Teacher Education*, *23*(6), 944–956. doi:10.1016/j.tate.2006.05.003

Tschannen-Moran, M., Hoy, A.W. & Hoy, W.K. (1998). Teacher efficacy: Its meaning and measure. *Review of Educational Research*, *68*(2), 202–248. doi:10.3102/00346543068002202

Diversity in Unity: Perspectives from Psychology and Behavioral Sciences – Ariyanto et al. (Eds)
© 2018 Taylor & Francis Group, London, ISBN 978-1-138-62665-2

Inclusive education in primary school: Do teachers' attitudes relate to their classroom management?

M. Maulia & F. Kurniawati
Faculty of Psychology, Universitas Indonesia, Depok, Indonesia

ABSTRACT: International research has shown that successful implementation of inclusive education is highly dependent on teachers' positive perception of students with Special Educational Needs (SENs) and their willingness to accommodate the learning needs of such students in the classroom. It has been argued that teachers' attitudes towards students with SENs have an impact on their classroom management. The aims of this study, therefore, are: 1) to measure teachers' attitudes towards the inclusion of students with SENs, 2) to identify classroom management in the inclusive classroom, and 3) to investigate the relationship between teachers' attitudes and their classroom management. Forty teachers from eight inclusive public primary schools in Depok, Indonesia, voluntarily got involved in the study. Teachers' attitudes and classroom management at an inclusive public primary school, grades 2–6, were measured by a questionnaire. The findings of the study revealed that teachers were positive in their attitudes towards inclusion of students with SENs and were providing a high level of classroom management in the classroom. It was also found that only the behavioural component of teachers' attitudes towards students with SENs significantly correlated with their classroom management in the inclusive classroom. Future research and implications of the study are also described.

1 INTRODUCTION

Discussion about inclusive schools cannot be separated from the subject of inclusive education, namely, students with Special Educational Needs (SENs). Students with SENs are those who are different from average children in several ways, such as mental characteristics, sensory abilities, communication skills, emotional development and behaviour, and/or physical characteristics (Kirk et al., 2015). Inclusive education is defined as an educational reformation that accommodates students with SENs to learn together with their typically developing peers and to participate in all classroom activities in a regular school, with the aim that students with SENs participate fully in social life (Leatherman & Niemeyer, 2005; Leung & Mak, 2010; de Boer et al., 2011).

Inclusive education in Indonesia was formally introduced in 2003 (Direktorat Pembinaan Sekolah Luar Biasa, 2007). According to the Direction Letter of the Directorate General of Primary and Secondary Education No. 380/C.66/MN/2003 on special education in regular schools, every district must operate at least four inclusive schools, consisting of primary, secondary, general high, and higher vocational schools (Sunardi et al., 2011).

There are some important things supporting the success of inclusive education. Mangunsong (2014) stated that the process towards inclusive education requires a change of heart and attitude; reorientation with regard to assessment, teaching methods, and classroom management including environmental adjustment; redefinition of the role of the teacher; reallocation of human resources; redefinition of the role of special schools; the provision of professional assistance for teachers in the form of training; itinerant teacher services; establishment and development of partnerships between teachers and parents; and a flexible education system. Several international research (Avramidis & Norwich, 2015; Forlin, 2010; Leatherman & Niemeyer, 2005; Leung & Mak, 2010) stated that teachers play an important role for the success of students with SENs in inclusive schools because they are the ones who directly interact and spend more time with the students, especially in the classroom (Pianta et al., 2012).

To ensure good interaction between teachers and students, the teacher must have a positive attitude towards students (Leyser & Tappendorf, 2001; Atta et al., 2009; Ahsan et al., 2012). Students who are taught by teachers who demonstrate positive attitudes towards inclusive education have higher satisfaction in learning and lower anxiety levels (Monsen & Frederickson, 2004). Teachers' attitudes are integral in ensuring the success of inclusive practices, as teachers' acceptance and support (or lack thereof) for inclusion is likely to affect their commitment to implementing it (Norwich, 1994).

Ajzen (2005) defined attitudes as an individual's viewpoint to respond favourably or unfavourably to an object, person, institution, or event. Attitude consists of three components: cognitive, affective, and behavioural (Ajzen, 2005). In the context of inclusive education, Triandis (1971, as cited by Leatherman & Niemeyer, 2005) said that the cognitive component is the individual's knowledge about inclusive education and students with SENs, the affective component reflects the individual's feeling about students with SENs, and the behavioural component refers to the predisposition to act in a particular way when dealing with students with SENs.

Attitudes need to be followed by the establishment of a good classroom environment (Hannah, 2013), because classroom is an important place for the growth of a child, so it is important to understand the ways in which to affect this environment in order to receive maximum effectiveness in instruction. In this case, teachers should be able to control the class, handle problematic behaviours, and use instructional time efficiently. The way in which teachers organise their class or how they control it yields positive or negative consequences for their students that lead to a better learning environment (Hannah, 2013).

All the things done by the teacher to organise students, time, and materials so learning can take place properly refers to classroom management (Wong & Wong, 1998, as cited by Timor, 2011). Classroom management also refers to the ability of teachers to use a variety of methods to prevent and redirect undesirable behaviour, the ability of teachers to manage the efficiency of time and learning routine, and to the ability of teachers to maximise involvement and students' ability to learn by preparing an interesting lesson (Pianta et al., 2012).

A previous study by Marzano and Marzano (2003) states that classroom management is the most important factor in influencing school learning. Ben (2006, as cited by Yasar, 2008) states that effective classroom management is significant for successful delivery of instruction. These statements explain why classroom management is important. Effective classroom management prepares the classroom for effective instruction, which is crucial for the progress of learning.

As stated above, teachers' positive attitudes and their classroom management skills are critical in the implementation of inclusive education. Few studies on teachers' attitudes within Indonesian inclusive settings have been carried out (Kurniawati et al., 2012, 2015), but there have not yet been any empirical data on how teachers' attitudes correlate with teachers' classroom management. Therefore, the aim of the study is to: 1) measure teachers' attitudes towards the inclusion of students with SENs, 2) identify classroom management in inclusive classroom, and 3) investigate the relationship between teachers' attitudes and their classroom management.

2 METHODS

Participants were chosen according to purposive sampling and were classroom teachers who teach children with special needs in inclusive public primary schools. A total of 40 teachers from eight public primary inclusive schools in Depok, Indonesia, were involved in the study. Teachers' attitudes towards inclusive education were measured with the Multidimensional Attitudes towards Inclusive Educational Scale (MATIES) (Mahat, 2008). The instrument was adapted in order to suit Indonesian teachers; thus, the name changed to the MATIES Indonesian Version (MATIES-IV). A validity test result showed that MATIES-IV has a quite good validity coefficient, ranging from 0.26 to 0.80 (Nunnaly & Bernstein, 1994). Meanwhile, based on the three components of the reliability test results, MATIES-IV had good Cronbach α (Kapplan & Saccuzo, 2004): cognitive $\alpha = 0.77$, affective $\alpha = 0.80$, and psychomotor $\alpha = 0.81$. This MATIES-IV consisted of 18 items representing the cognitive (six items), affective (six items) and behavioural (six items) components of attitude. The questionnaire used a six-point Likert type scale that ranged from 1 to 6 (1 = Strong disagreement, and 6 = Strong agreement). MATIES-IV has been used to measured teachers' attitudes towards inclusive education in Indonesia for teachers in primary, middle and high schools (Soviana, 2014; Marantika, 2014; Sihombing, 2014).

To measure classroom management, teachers were asked to fill out classroom management questionnaires that had been developed by researchers based on the theory of classroom management by Pianta et al. (2012). The questionnaire used a four-point Likert type scale that ranged from 1 to 4 (1 = Not appropriate, and 4 = Appropriate). This instrument had been through the process of focus group discussions, expert judgement, and readability. For reliability and validity using confirmatory factor analysis, the instrument had acceptable goodness-of-fit indices ($p = 0.16338$, root mean square error of approximation = 0.041, goodness-of-fit index = 0.90). The measuring instrument had 12 items representing the ability of teachers to use a variety of methods to prevent and redirect undesirable behaviour (six items), the ability of teachers to manage the efficiency of time and learning routine (four items), and the ability of teachers to maximise involvement and students' ability to learn by preparing an interesting lesson (two items).

Regarding the first research question, the scale's mid-point or mean on a six-point Likert type scale was used to determine positive or negative attitudes. To answer the second research question, the scale's mid-point or mean of scale on a four-point Likert type scale was used to determine a high or low level of classroom management. With regard to the third research question, the Pearson product moment was used to examine the relationship between teachers' attitudes and their classroom management.

3 RESULTS

Based on the total sample of the study, the majority of female participants (85%) were aged -less than 40 and more than 40 years. Most of the participants were undergraduates (95%) and had experience in teaching for less than 10 years and more than 20 years. Fourteen participants had received training related to inclusive education (see Table 1).

With regard to the first research question, teachers' attitudes towards the inclusion of students with SENs in public primary inclusive schools are presented in Table 2. Table 2 shows that the overall mean of teachers' attitudes is 4.81 (Standard Deviation (SD) = 0.47). Results indicate that teachers' attitudes that include the cognitive, affective and behaviour components are above the mean score of a six-point Likert type scale (3.5), indicating a positive attitude towards students with SENs.

Regarding the second research question, the results indicate that the overall mean of teachers' classroom management is 3.46 (SD = 0.34). Results indicate that teachers' classroom management is above the mean score of a four-point Likert type scale (2.5), indicating a high level of classroom management in inclusive classrooms.

The third research question of this study investigated the correlation between teachers' attitudes and their classroom management. The results are presented in Table 3.

Table 3 shows that teachers' attitudes were significantly and positively correlated with teachers' classroom management in public primary inclusive schools ($r = 0.475$, $p < 0.01$).

Table 1. Characteristics of participants (N = 40).

Demographic variables	N	%
Gender		
Males	6	15
Females	34	85
Age range (years)		
<40	15	37.5
40–50	9	22.5
>50	16	40
Education level		
Diploma	2	5
Bachelor's degree	38	95
Teaching experience (years)		
<10	14	35
10–20	9	22.5
>20	17	42.5
Training in special needs		
Yes	14	35
No	26	65

Table 2. Means and SDs of teachers' attitudes (N = 40).

	Mean	SD
Attitudes	4.81	0.47
Cognitive component	4.55	0.71
Affective component	4.75	0.71
Behavioural component	5.12	0.57

Table 3. Correlation between teachers' attitudes and classroom management scores (N = 40).

	Classroom management	
	R	p
Attitudes	0.475**	0.002
Cognitive component	0.254	0.114
Affective component	0.247	0.125
Behavioural component	0.554**	0.000

Further analysis found that only the behavioural component was significantly and positively correlated with teachers' classroom management.

4 DISCUSSION

This research covers some important points: the majority of teachers in public primary schools in Depok have positive attitudes towards inclusive education in all three components of attitude (cognitive, affective, and behavioural). There are a few reasons for this finding: first, since its start in 2003, the implementation of inclusive education in Indonesia has entered its fourteenth year and all public schools are encouraged to accept all children including students with SENs. Although not all public schools were ready to implement it, the socialisation of inclusive education and students with SENs has been done. Second, Depok announced itself as a city that is friendly to all children regardless of their ability and need, in accordance with Depok local regulation No. 15, in 2013. The local government has formed an inclusive task force to support teachers. Third, the participants' demography showed that 85% teachers were women. Avramidis (2000) state that female teachers have more positive attitudes towards students with SENs than male teachers. Fourth, more than 40% of the participants had more than 20 years of teaching experience. Based on the research conducted by Leung and Mak (2010), teachers with more than 10 years of teaching experience (i.e. more experienced teachers) are more sympathetic towards such students than those with less than 10 years of teaching experience. Fifth, 35% of the participants stated that they had training in inclusive education. Avramidis and Norwich (2002) and Kurniawati et al. (2015) state that teachers' attitudes towards inclusive education are influenced by how much training they have had in inclusive education; the more training they have, the more positive their attitude.

There is no significant correlation between the cognitive component of attitude and classroom management. This means that although teachers have the confidence and knowledge of inclusive education, they may not be willing or able to perform better classroom management in inclusive classrooms. One of the factors that might affect this result is the large class size. The schools participating in this study have at least 35 students in a classroom. Research by Blatchford et al. (2011) explains that if a primary teacher teaches a large number of students, it will affect their capability of managing. When teachers have confidence that inclusive schools facilitate the emergence of acceptable behaviour by all students, but they are faced with a large number of students with diverse behaviours, including students with SENs, the teachers may not be capable of organising students, time and materials in order to make the learning process run properly.

There is no significant correlation between the affective component of attitude and classroom management. This means that although teachers have a positive feeling about inclusive education, they may not be willing or able to perform better classroom management in inclusive classrooms. One of the factors that might affect this result is the target to complete the curriculum in a limited time. Research by Akin et al. (2016) explains that when teachers are required to finish teaching all their material for the curriculum in a very limited time, their focus is more on completing the materials than on managing their classroom. When teachers are given responsibility to share their attention between regular students and those with SENs while also having the responsibility of finishing teaching the materials, the teacher may not be capable of organising students, time, and materials in order to make the learning process run properly.

There is a positive relationship between the behavioural component of attitude and classroom management. When teachers are willing to deliver supporting action to students with SENs in the classroom, they provide greater effort to regulate students, time, and material so the learning process might happen well in the classroom. Research by Sharma et al. (2008) also explains that teachers with a more positive attitude towards students with SENs tend to more readily accept and accommodate their students with SENs. However, from the research interview, most teachers involve students with SENs into regular classes while not having any individual educational programme that relates to such students. This might mean that they take action to support students with SENs, but without any relation to knowledge about the students. On the other hand, background knowledge of students, particularly students with SENs, is an important aspect to know in order to be able to improve classroom management skills, for example, in planning how to deliver learning materials to engage students.

5 CONCLUSION

The study has revealed that teachers hold positive attitudes towards students with SENs for all components of attitude and provide a high level of classroom management. With respect to the third aim of this study, the results show that teachers' attitudes on the behavioural component were significantly and positively correlated with teachers' classroom management. However, further observation is necessary to prove conformity of the teachers' recognition with suitability in the field. This is also to see whether they teach not just to perform their obligations.

This research may contribute in the successful implementation of inclusive education in Indonesia. However, a few limitations should be taken into account, including the use of the self-report and the number of participating teachers. Also, more representative numbers of teachers must be considered in further research.

REFERENCES

Ahsan, M.T., Sharma, U. & Deppeler, J.M. (2012). Exploring pre-service teachers' perceived teaching-efficacy, attitudes and concerns about inclusive education in Bangladesh. *International Journal of Whole Schooling*, *8*(2), 1–20.

Ajzen, I. (2005). *Attitudes, personality and behaviour* (2nd ed.). Maidenhead, UK: Open University Press. doi:10.1037/e418632008–001

Akin, S., Yıldırım, A. & Goodwin, A.L. (2016). Classroom management through the eyes of elementary teachers in Turkey: A phenomenological study. *Educational Sciences: Theory and Practice*, *16*(3), 771–797. doi:10.12738/estp.2016.3.0376

Atta, M.A., Shah, M. & Khan, M.M. (2009). Inclusive school and inclusive teacher. *The Dialogue*, *4*(2), 272–283.

Avramidis, E., Bayliss, P., & Burden, R. (2000). Student teachers' attitudes toward the inclusion of children with special educational needs in the ordinary school. *Teaching and Teacher Education, 16,* 277–293.

Avramidis, E. & Norwich, B. (2002). Teachers' attitudes towards integration/inclusion: A review of the literature. *European Journal of Special Needs Education*, *17*(2), 129–147. doi: 10.1080/08856250210129056

Blatchford, P., Bassett, P. & Brown, P. (2011). Examining the effect of class size on classroom engagement and teacher-pupil interaction: Differences in relation to pupil prior attainment and primary vs. secondary schools. *Learning and Instruction*, *21*(6), 715–730. doi:10.1016/j.learninstruc.2011.04.001

Cameron, C.E., Connor, C.M., Morrison, F.J. & Jewkes, A.M. (2008). Effects of classroom organization on letter-word reading in first grade. *Journal of School Psychology*, *46*(2), 173–192. Doi: 10.1016/j.jsp.2007.03.002

de Boer, A., Pijl, S.J. & Minnaert, A. (2011). Regular primary schoolteachers' attitudes towards inclusive education: A review of the literature. *International Journal of Inclusive Education*, *15*(3), 331–353. doi:10.1080/13603110903030089

Direktorat Pembinaan Sekolah Luar Biasa (Directorate of Special Education). (2007). *Pedoman khusus penyelenggaraan pendidikan inklusif: Identifikasi anak berkebutuhan khusus.* Jakarta: Direktorat Pembinaan Sekolah Luar Biasa.

Emmer, E.T. & Stough, L.M. (2001). Classroom management: A critical part of educational psychology, with implications for teacher education. *Educational Psychologist*, *36*(2), 103–112. doi:10.1207/S15326985EP3602_5

Evertson, C.M. & Harris, A.H. (1999). Support for managing learning-centered classrooms: The classroom organization and management program. In J. Freiberg (Ed.), *Beyond behaviorism: Changing the classroom management paradigm*. Boston, MA: Allyn and Bacon.

Forlin, C. (2010). Developing and implementing quality inclusive education in Hong Kong: Implications for teacher education. *Journal of Research in Special Educational Needs*, *10*(Suppl. 1), 177–184. doi:10.1111/j.1471–3802.2010.01162.x

Hannah, R. (2013). *The effect of classroom environment on student learning* (Thesis, Western Michigan University, Kalamazoo, MI). Retrieved from http://scholarworks.wmich.edu/honors_theses/2375

Kapplan, R.M.m & Sazzuzzo, D.P. (2004). *Psychologycal testing: Principles, applications, and issues*. USA: Thomson Wadsworth. Kirk, S., Gallagher, J.J. & Coleman, M.R. (2015).—*Educating exceptional children* (14th ed.). Stamford, CT: Cengage Learning.

Kurniawati, F., Minnaert, A., Mangunsong, F. & Ahmed, W. (2012). Empirical study on primary school teachers' attitudes towards inclusive education in Jakarta, Indonesia. *Procedia—Social and Behavioral Sciences*, *69*, 1430–1436. doi:10.1016/j.sbspro.2012.12.082

Kurniawati, F., De Boer, A.A., Minnaert, A.E.M.G. & Mangunsong, F. (2015). Teachers' attitudes and their teaching strategies in inclusive classrooms. *Journal of Teachers and Teaching: Theory and Practice*, 1–30.

Leatherman, J. & Niemeyer, J. (2005). Teachers' attitudes toward inclusion: Factors influencing classroom practice. *Journal of Early Childhood Teacher Education*, *26*(1), 23–36. doi:10.1080/10901020590918979

Leung, C. & Mak, K. (2010). Training, understanding, and the attitudes of primary school teachers regarding inclusive education in Hong Kong. *International Journal of Inclusive Education, 14*(8), 829–842. doi:10.1080/13603110902748947

Leyser, Y. & Tappendorf, K. (2001). Are attitudes and practices regarding mainstreaming changing? A case of teachers in two rural school districts. *Education, 121*(4), 751–760.

Mahat, M. (2008). The development of a psychometrically-sound instrument to measure teachers' multidimensional attitudes toward inclusive education. *International Journal of Special Education, 23*(1), 82–92.

Mangunsong, F. (2014). *Psikologi & Pendidikan Anak Berkebutuhan Khusus* (1st ed.). Depok, Indonesia: LPSP3 Fakultas Psikologi UI. Retrieved from http://www.lpsp3.com/product/2/187/Psikologi-Pendidikan-Anak-Berkebutuhan-Khusus-Jilid-1

Marantika, D. (2014). *The relationship between attitudes towards inclusive education and teaching strategies of teachers in inclusive public high school and inclusive private high school* (Thesis, Universitas Indonesia, Depok, Indonesia).

Marks, H.M. (2000). Student engagement in instructional activity: Patterns in the elementary, middle, and high school years. *American Educational Research Journal, 37*(1), 153–184. doi:10.3102/00028312037001153

Marzano, R.J. & Marzano, J.S. (2003). The key to classroom management. *Educational Leadership, 61*(1), 6–18.

Monsen, J.J. & Frederickson, N. (2004). Teachers' attitudes towards mainstreaming and their pupils' perceptions of their classroom learning environment. *Learning Environments Research, 7*(2), 129–142. doi:10.1023/B:LERI.0000037196.62475.32

Norwich, B. (1994). The relationship between attitudes to the integration of children with special educational needs and wider socio-political views: A US-English comparison. *European Journal of Special Needs Education, 9*(March 2015), 91–106. doi:10.1080/0885625940090108

Nunnaly, J.C., & Bernstein, I.H. (1994). *Psychometric theory* (3th ed.). New York: McGraw-Hill.

Pianta, R.C., Hamre, B.K., & Allen, J.P. (2012). Teacher-student relationships and engagement: Conceptualizing, measuring, and improving the capacity of classroom interactions. On S.L. Christenson, et al. (eds.), *Handbook of Research on Student Engagement* (365–386). doi:10.1007/978-1-4614-2018-7

Sharma, U., Forlin, C. & Loreman, T. (2008). Impact of training on pre - service teachers' attitudes and concerns about inclusive education and sentiments about persons with disabilities. *Disability & Society, 23*(7), 773–785. doi:10.1080/09687590802469271

Sihombing, C.M. (2014). *The relationship between attitudes towards inclusive education and teaching strategies of teachers in inclusive public junior high school teacher based on national exam and non-national exam subject group* (Thesis, Universitas Indonesia, Depok, Indonesia).

Soviana, D. (2014). *The relationship between attitudes towards inclusive education and teaching strategies of teachers in inclusive public high school and inclusive private high school* (Thesis, Universitas Indonesia, Depok, Indonesia).

Sunardi, M.Y., Gunarhadi, P. & Yeager, J.L. (2011). The implementation of inclusive education for students with special needs in Indonesia. *Excellence in Higher Education, 2*(1), 1–10. doi:10.5195/ehe.2011.27

Timor, T. (2011). Attitudes of beginner teachers of special education to classroom management: Who's the boss here? *Electronic Journal for Inclusive Education, 2*(7). Retrieved from http://corescholar.libraries.wright.edu/ejie/vol2/iss7/2

Yasar, S. (2008). *Classroom management approaches of primary school teachers* (Master's thesis, Middle East Technical University, Ankara, Turkey).

Diversity in Unity: Perspectives from Psychology and Behavioral Sciences – Ariyanto et al. (Eds)
 ISBN 978-1-138-62665-2

Successful implementation of inclusive education on primary school: Roles of teachers' attitudes and their emotional support for students with special educational needs

A. Virgina & F. Kurniawati
Faculty of Psychology, Universitas Indonesia, Depok, Indonesia

ABSTRACT: Research has consistently shown the critical role of teachers' attitudes in the successful implementation of inclusive education, but how it relates to teachers willingness to emotionally support students with Special Educational Needs (SEN) is less known. The aims of this study, therefore, were 1) to measure teachers' attitudes towards the inclusion of students with SEN, 2) to identify emotional support teachers provide to such students and 3) to investigate the relationship between teachers' attitudes and their emotional support. The attitudes and the emotional support of forty primary school teachers (mean age = over 50 years old) were collected by the MATIES IV and the emotional support scale. The teachers were purposively selected from schools that are identified as actively implementing an inclusive program in Depok. The findings of the study revealed that teachers were positive in their attitudes towards the inclusion of students with SEN and were providing high emotional support in the classroom. It was also found that teachers' attitudes were related to their emotional support towards students with SEN in the classroom. These findings suggest that this study might contribute to the successful implementation of inclusive education in Indonesia.

1 INTRODUCTION

Inclusive education in Indonesia was formally introduced in 2003 (Directorate of Special Education, 2009). According to the Direction Letter of the Directorate General for Primary and Secondary Education No. 380/C.66/MN/2003 on Special Education in Regular Schools, every district must operate at least four inclusive schools, consisting of a primary, a secondary, a general high, and a high vocational school (Sunardi, *et al.*, 2011).

The inclusive education system is implemented by considering students' needs as well as adapting the environment and activities that could be done by all students, both students with and without SEN (UNESCO, 1994). Meanwhile, inclusive education is defined as an educational reform that accommodates students with SEN to learn together with their typically developing peers and to participate in all classroom activities in regular school with the aim that students with SEN participate fully in social life (de Boer, *et al.*, 2011; Leatherman & Niemeyer, 2005; Leung & Mak, 2010).

The successful implementation of inclusive education has been consistently associated with teachers' attitudes (Ahsan, Sharma & Deppeler, 2012; Atta, Shah & Khan, 2009; Leyser & Tappendorf, 2001). Ajzen (2005) defines attitude as an individual's viewpoint to respond favorably or unfavorably to an object, person, institution, or event. Attitude consists of three components: cognitive, affective, and behavioral (Ajzen, 2005). In the context of inclusive education, Triandis (1971) argues that the cognitive component is the individual's knowledge about inclusive education and students with SEN, the affective component reflects the individual's feeling about students with SEN, and the behavioral component refers to the predisposition to act in a particular way when dealing with students with SEN (Leatherman & Niemeyer, 2005).

Previous research on teachers' attitudes towards the inclusion of students with SEN reveals that teachers hold positive attitudes towards inclusive education (Kurniawati, *et al.*, 2012; Letherman & Niemeyer, 2005), whereas other research reported that teachers indicate negative attitudes towards inclusive education (see Avramidis, Bayliss & Burden, 2000). Besides positive and negative attitudes, other researchers found that teachers are neutral in their attitudes towards inclusive education (see de Boer, *et al.*, 2011). The research concludes that teachers have varied attitudes towards inclusive education.

Besides attitudes, teachers' emotional support was also important in determining the success of implementing inclusive education. Emotional support is defined as teachers' efforts to support students in the classroom in social and emotional functioning through teacher student interaction (Pianta, Hamre & Allen, 2012). Emotional support is comprised of three dimensions: positive climate, teacher sensitivity, and regard for students' perspectives (Ruzek, *et al.*, 2016). The first dimension, positive climate, is characterized by warm caring relationships between teacher and students, which includes the presence of shared positive affect, an interactive peer environment, communication of positive expectations, and the use of respectful language and cooperation. The second dimension, teacher sensitivity, refers to teachers' awareness and responsiveness to students' cues and needs in the classroom, which includes their consistent, timely provision, and responsive interactions to help students when they ask for it. The final dimension, regard for students' perspectives, encompasses the degree to which classrooms and interactions are structured around students' ideas and opinions. Teachers create positive climates respond to students' needs and incorporate students' interests into lessons in order to promote students' motivation and engagement (Madill, Gest & Rodkin, 2014; Ruzek, *et al.*, 2016).

Research by Merritt *et al.* (2012) concluded that higher teachers' emotional support reduces student aggression and increase students' behavioral self-control. Moreover, Hamre and Pianta (2005) found that students with high functional risks (a disability combination of behavioral, attentional, social, or academic problems) had higher achievement when teachers provide high emotional support in the classroom. Hence, teachers' emotional support are essential for students' social and academic development (Pianta, Hamre & Allen, 2012).

In order to implement inclusive education successfully, teachers are required to fully support students socially and academically (Booth & Ainscow, 1998). It is argued that research on teachers' attitudes and teachers' emotional support in inclusive education is important as these might be associated with students' motivation for learning and students' engagement in the classroom (Madill, Gest, Rodkin, 2014). Research about attitudes and emotional support of teachers in inclusive education setting in Indonesia is less known. Therefore, the aim of this study was to set up three objectives; 1) to measure teacher's attitudes towards the inclusion of students with SEN, 2) to identify emotional support teachers provide to such students and 3) to investigate the relationship between teachers' attitudes and their emotional support.

2 METHODS

A total of 40 teachers from eight public primary inclusive schools in Depok; School A (n = 8), School B (n = 5), School C (n = 4), School D (n = 3), School E (n = 3), School F (n = 1), School G (n = 8), and School H (n = 8) participated in this study based on purposive sampling.

Out of 40 teachers, fourteen teachers had special education training (35%). Participants were predominantly female (85%, n = 34). Participants' age were split across under 40 years old (37.5%), between 40–50 years old (22.5%), and over 50 years old (40%) (see Table 1 for details).

In this study, teachers' attitude was measured using the Multidimensional Attitudes Towards Inclusive Education Scale (MATIES) by Mahat (2008). The instrument had been adapted to measure teachers' attitudes in Indonesia, then named the MATIES Indonesian Version (MATIES IV). The instrument posed good reliability ($\alpha > 0.70$) (Soviana, 2014).

Table 1. Characteristics of teachers (N = 40).

Demographic variables	N	%	
Gender	Male	6	15
	Female	34	85
Age range	<40 years old	15	37.5
	40–50 years old	9	22.5
	>50 years old	16	40
Education level	Diploma	2	5
	Bachelor's degree	38	95
Teaching experiences	<10 years	14	35
	10–20 years	9	22.5
	>20 years	17	42.5
Training in special needs	Yes	14	35
	No	26	65

The instrument consists of 18 items representing the three components of attitudes, which are the cognitive (6 items), affective (6 items), and behavioral (6 items) components. The questionnaire used a six-point Likert scale that ranged from 1 to 6 (1 = strong disagreement and 6 = strong agreement).

The emotional support of teachers was measured using the instrument constructed by researchers based on theory of emotional support by Ruzek *et al.* (2016). The instrument had acceptable goodness-of-fit indices ($\chi^2 = 55.57$, df = 44, p = 0.11333, RMSEA = 0.051, GFI = 0.91) based on a pilot study to public primary inclusive school teachers in Jakarta. The instrument consists of 11 items representing the three dimensions of emotional support, which are the positive climate (5 items), teacher sensitivity (4 items), and regard for students' perspectives (2 items). Ratings were made on a four-point Likert scale, ranging from very inappropriate (1) to very appropriate (4).

In order to answer the first and second research questions, we used the mid-point of MATIES IV and emotional support scale in categorizing the score. With regard to the third research question, the Pearson Product Moment was used to investigate the relationship between teachers' attitudes and their emotional support.

3 RESULTS

With regard to the first research question, teachers' attitudes towards the inclusion of students with SEN in public primary inclusive school are presented in Table 2.

Table 2 shows that teachers' attitudes on the cognitive component (M = 4.55, SD = 0.70), the affective component (M = 4.75, SD = 0.71), and the behavioral component (M = 5.12, SD = 0.56) are above the mid-point (3.5) of the response scale score. These indicate that teachers have positive attitudes towards inclusive education. It means, teachers have knowledge about inclusive education and students with SEN, teachers are willing to involve students with SEN in all classroom activities and teachers have a strong drive in doing all necessary actions to support students with SEN.

Regarding the second research question, it was found that the overall mean of teachers' emotional support was 3.56 (SD = 0.30). Further analysis, Table 3 shows that teachers' emotional support on the positive climate dimension (M = 3.57, SD = 0.33), the teacher sensitivity dimension (M = 3.51, SD = 0.34), and the regard for students' perspectives dimension (M = 3.62, SD = 0.41). The results indicate that teachers' emotional support are above the mid-point (2.5) of the response scale score, indicating high emotional support for students with SEN. It means, teachers might create a warm caring classroom climate, pay more attention to students according to their needs and abilities, and provide students a wide opportunity to share their thoughts and ideas in the classroom.

Table 2. Means and SDs on teachers' attitudes (N = 40).

Components of attitudes	*M*	*SD*	Score minimum	Score maximum
Cognitive	4.55	0.70	3.17	6.00
Affective	4.75	0.71	3.00	6.00
Behavioral	5.12	0.56	3.50	6.00

Table 3. Means and SDs on teachers' emotional supports (N = 40).

Dimensions of emotional support	*M*	*SD*	Score minimum	Score maximum
Positive climate	3.57	0.33	3.00	4.00
Teacher sensitivity	3.51	0.34	2.75	4.00
Regard for students' perspectives	3.62	0.41	3.00	4.00

With regard to the third research question, Table 4 shows that teachers' attitudes on the cognitive component was significantly and positively correlated with teachers' emotional support in the teacher sensitivity component ($r = 0.34$, $p < 0.05$). It means that teachers' knowledge about inclusion correlated with teachers' paying more attention to students according their needs and abilities. Table 4 also showed that teachers' attitudes in the behavioral component was significantly and positively correlated with teachers' emotional support in the positive climate component ($r = 0.40$, $p < 0.01$), the teacher sensitivity component ($r = 0.70$, $p < 0.01$), and the regard for students' perspectives component ($r = 0.44$, $p < 0.01$). It means that teachers' actions to support SEN students was showed by creating a warm caring classroom climate, paying more attention to students according their needs and abilities, and providing students wide opportunities to share their thoughts and ideas in the classroom.

4 DISCUSSION

The results reveal that teachers in public primary inclusive schools have positive attitudes towards students with SEN. These results are in accordance with the research that was done by Kurniawati *et al.* (2012) that teachers in primary inclusive schools in Jakarta had positive attitudes towards inclusion and showed a strong will to involve SEN students in all classroom activities. The teachers' positive attitudes and their high emotional support found in this study might be related to several factors. Firstly, the majority of participants were women, and it is argued that female teachers have more tolerance and sympathy to SEN students compared with male teachers (Avramidis & Norwich, 2002). Secondly, although only 35% of the teachers have had training in inclusion, the teachers' overall mean on attitudes was positive. Having daily interaction with SEN students at school appears to play a role in the formation of positive attitudes in teachers, as also found by Eagly and Chaiken (1993).

The results also reveal that behavioral components of attitudes have positive and significant relationship with teachers' emotional support on three dimensions (i.e. positive climate, teacher sensitivity, and regard for students' perspectives). It is not surprising as in order to provide emotional support teachers are required to show a strong willingness to accommodate such students in the classroom (Avramidis & Norwich, 2002). The more active teachers are in to accommodating SEN students, the greater their efforts to provide emotional support to such students.

Meanwhile, the results revealed that the cognitive component of attitudes have no relationship with the positive climate dimension and regard for students' perspectives dimension of emotional support. It might be related to the fact that the majority of participants have no educational background in special education. The affective component of attitudes has no relationship with teachers' emotional support in the three dimensions (i.e. positive climate,

Table 4. Correlations between attitudes and emotional supports scores (N = 40).

	Components of emotional supports					
	Positive climate		Teacher sensitivity		Regard for students' perspectives	
Components of attitudes	*r*	*P*	*R*	*p*	*r*	*p*
Cognitive	0.24	0.12	0.34*	0.03	0.21	0.18
Affective	0.27	0.08	0.07	0.65	0.26	0.10
Behavioral	0.40**	0.00	0.70**	0.00	0.44**	0.00

Note: $*p < 0.05$. $**p < 0.01$.

teacher sensitivity, and regard for students' perspectives). It might be related to the limited time teachers have to accommodate students with SEN and lots of other students in the classroom.

As a preliminary study, this study might contribute to the successful implementation of inclusive education in Indonesia, though there are a few limitations such as the use of self-report and the number of participants. For further research, observation and a higher number of participants are recommended in order to gain a better view of the learning process within the inclusive school setting.

5 CONCLUSION

Correlation between teachers' attitudes and their emotional support to students with SEN in this study might be related to the Government Decree on implementing inclusive education in Indonesia. Actually, public primary inclusive schools' teachers in Depok have no educational background in special needs and lack raining in special needs. It seems that they are only willing to educate students with SEN because it is a legal requirement, implying that they are resistant to the concept of inclusive education. Therefore, training programs in special education should be provided for teachers because it might enhance teachers' knowledge as well as their attitudes and emotional support in relation to including students with SEN in regular education.

REFERENCES

Ahsan, M.T., Sharma, U. & Deppeler., J.M. (2012) Exploring pre-service teachers' perceived teaching-efficacy, attitudes and concerns about inclusive education in Bangladesh. *International Journal of Whole Schooling,* 8 (2), 1–20. Available on doi:EJ975715.

Ajzen, I. (2005) *Attitudes, personality and behavior.* 2nd edition. New York, Open University Press.

Atta, M.A., Shah, M. & Khan, M.M. (2009) Inclusive school and inclusive teacher. *The Dialogue,* 4 (2), 272–283. Retrieved from: http://www.qurtuba.edu.pk/thedialogue/The%20Dialogue/4_2/06_mumtaz_khan.pdf

Avramidis, E., Bayliss, P. & Burden, R. (2000) Student teachers' attitudes towards the inclusion of children with special educational needs in the ordinary school. *Teaching and Teacher Education,* 16 (3), 277–293. Available on doi:10.1016/S0742–051X(99)00062-1.

Avramidis, E. & Norwich, B. (2002) Teachers' attitudes towards integration/inclusion: A review of the literature. *European Journal of Special Needs Education,* 17 (2), 129–147. Available on http://dx.doi.org/10.1080/08856250210129056.

De Boer, A., Pijl, S.J. & Minnaert, A. (2011) Regular primary school teachers' attitudes towards inclusive education: A review of the literature. *International Journal of Inclusive Education,* 15 (3), 331–353. Available on doi:10.1080/13603110903030089.

Directorate of Special Education. (2009) *General Guidelines of the Implementation of Inclusive Education.* Jakarta, Author.

Eagly, A.H. & Chaiken, S. (1993) *The psychology of attitudes*. Fort Worth, TX, Harcourt Brace, Jovanovich.
Hamre, B.K. & Pianta, R.C. (2005) Can instructional and emotional support in the first-grade classroom make a difference for children at risk of school failure? *Child Development,* 76 (5), 949–967. Available on doi:10.1111/j.1467–8624.2005.00889.x.
Kurniawati, F., Minnaert, A. Mangunsong, F. & Ahmed, W. (2012) Empirical study on primary school teachers' attitudes towards inclusive education in Jakarta, Indonesia. *Procedia—Social and Behavioral Sciences,* 69, 1430–1436. Available on doi:10.1016/j.sbspro.2012.12.082.
Leatherman, J.M. & Niemeyer, J.A. (2005) Teachers' attitudes toward inclusion: Factors influencing classroom practice. *Journal of Early Childhood Teacher Education,* 26 (1), 23–36. Available on doi:10.1080/10901020590918979.
Leung, C. & Mak, K. (2010) Training, understanding, and the attitudes of primary school teachers regarding inclusive education in Hong Kong. *International Journal of Inclusive Education,* 14 (8), 829–842. Available on doi:10.1080/13603110902748947.
Leyser, Y. & Tappendorf, K. (2001) Are attitudes and practices regarding mainstreaming changing? A case of teachers in two rural school districts. *Education,* 121 (4), 751–760. Retrieved from: http://search.proquest.com/docview/196438486?accountid=14375%5Cnhttp://rh4hh8nr6k.search.serialssolutions.com/?ctx_ver=Z39.88–2004&ctx_enc=info:ofi/enc:UTF-8&rfr_id=info:sid/ProQ%3 Aeducation&rft_val_fmt=info:ofi/fmt:kev:mtx:journal&rft.genre=article&rft.j
Madill, R.A., Gest, S.D. & Rodkin, P.C. (2014) Students' perceptions of relatedness in the classroom: The role of emotionally supportive teacher-child interaction, children's aggressive disruptive behaviors, and peer social preference. *School Psychology Review,* 43 (1), 86–105.
Mahat, M. (2008) The development of a psychometrically-sound instrument to measure teachers' multidimensional attitudes toward inclusive education. *International Journal of Special Education*, 23(1), 82–92.
Merritt, E.G., Wanless, S.B., Rimm-Kaufman, S.E., Cameron, C.E. & Peugh, J.L. (2012) The contribution of teachers' emotional support to children's social behaviors and self-regulatory skills in first grade. *School Psychology Review,* 41 (2), 141–159. Available on http://search.ebscohost.com/login.aspx?direct=true&db=ehh&AN=77346814&site=ehost-live.
Ruzek, E.A., Hafen, C.A., Allen, J.P., Gregory, A., Mikami, A.Y. & Pianta, R.C. (2016) How teacher emotional support motivates students: The mediating roles of perceived peer relatedness, autonomy support, and competence. *Learning and Instruction,* 42, 95–103. Available on doi:10.1016/j.learninstruc.2016.01.004.
Soviana, D. (2014) *Hubungan antara Sikap terhadap Pendidikan Inklusif dan Strategi Pengajaran Guru SMP Negeri Inklusif Ditinjau dari Kelompok Mata Pelajaran Ujian Nasional dan Non-Ujian Nasional.* [Thesis] Depok, Universitas Indonesia.
Sunardi, Y.M., Gunarhadi, P. & Yeager, J.L. (2011) The implementation of inclusive education for students with special needs in Indonesia. *Excellence in Higher Education,* 2, 1–10. Available on doi: 10.5195/ehe.2011.27.
UNESCO. (1994) *Salamanca Statement and Framework for Action on Special Education Needs*. Paris, Author. Retrieved from: http://www.unesco.org/education/pdf/SALAMA_E.PDF

Diversity in Unity: Perspectives from Psychology and Behavioral Sciences – Ariyanto et al. (Eds)
© 2018 Taylor & Francis Group, London, ISBN 978-1-138-62665-2

Relationship between parental involvement and student self-regulation in music practice

W.G.Y. Kesawa & L. Primana
Faculty of Psychology, Universitas Indonesia, Depok, Indonesia

ABSTRACT: Previous studies have explained the relationship between parental involvement and self-regulated learning in many contexts, such as academic learning. However, recently, there have been a few studies that have explored the specific relationship between parental involvement and self-regulated learning in other contexts such as music practice. Previous studies have shown that there are many positive impacts on the intellectual, social and personal development of individuals while actively practising music. Therefore, this study aimed to examine the relationship between parental involvement and self-regulation in music practice. Parental involvement focused on two dimensions, behavioural and cognitive involvement. The respondents in this study consisted of 103 students from two music senior high schools in the Jabodetabek (Greater Jakarta) area. The data were collected using questionnaires for Parental Involvement Measure and the Self-Regulated Practice Behavior Scale. The results showed that there was a significant relationship between parental involvement and self-regulated music practice and that the behavioural involvement dimension of parental involvement has a significant relationship with self-regulation in music practice.

1 INTRODUCTION

There are many positive effects when individuals actively engage in music practice, such as intellectual, social and personal development. Practising music provides children the opportunity to develop their creativity, language and numerical skills. Children also benefit from positive social and personal development, in terms of self-esteem, self-discipline, motivation, self-efficacy, well-being, and many more (Hallam, 2010). It is little wonder, therefore, that millions of children around the world begin learning to play a musical instrument each year. However, very few continue their learning for a longer period of time (McPherson & Davidson, 2006, in Zimmerman & Schunk, 2011).

In regards to the processes of music learning, McPherson (2005) found that development in music learning among individuals is different. For example, there are individuals who are fast in mastering their instrument, but others needed significantly more effort.

Basically, there are complex motoric, sensory, and cognitive skills involved when individuals practice music (Lehmann et al., 2007). In music practice, individuals should be able to focus their attention for a long time, handle a competitive learning environment, and be able to accomplish their goals (Martin, 2008). To be able to play music, there are three domains of music learning: cognitive, affective, and psychomotor skills (Benton, 2014). Individuals should be able to integrate them all at once during music practice. In the cognitive domain, individuals should know and understand the correct rules for playing an instrument. In the affective domain, individuals should be able to communicate the message or meaning of songs through music. In the psychomotor domain, they should be able to master the skills that are required to play the instrument. Therefore, self-regulation is needed to help individuals perform better in the process of mastering musical instruments (McPherson & Renwick, 2011, in Zimmerman & Schunk, 2011).

To succeed in practising music, practice must be made more structured. Ericsson (1997) found that high-intensity practice increased monitoring and evaluation skills to achieve good outcomes. Moreover, more practice would help form self-regulation of music practice. Previous research conducted by McPherson and McCormick (1999) showed that high intensities of practice enabled music students to organise their practice in more effective ways, such as allocating more time to practice difficult parts.

In the music practice context, having self-regulation (Zimmerman, 1986) while practising can assist individuals in attaining optimal performance and achievement through strategies that they use during practice. Moreover, students who are self-regulated in learning also have an intrinsic motivation because they are no longer affected by external factors such as rewards. McPherson et al. (2012) found that individuals successful in learning music had strong resilience even when they got stressed or distracted; something which can affect their motivation and desire to continue their musical learning.

According to McPherson and Zimmerman (2002), there are six dimensions of self-regulation that are relevant to music learning. These are motive, method, behaviour, time management, physical environment and social factors. *Motive* refers to individuals' beliefs about their capacity, which may or may not influence learning, such as self-efficacy. Self-efficacy is considered as the most important dimension, and it relates to goal setting, effort, and persistence in educational settings (Zimmerman et al., 1992), and in music education too (McPherson & McCormick, 2006). *Method* refers to task-oriented learning strategies, mental strategies, and any other strategies that individuals use during learning. The *behaviour* dimension denotes reflective thinking, metacognition, and learners' capabilities to conduct self-evaluation, or how they monitor their own learning processes. *Time management* is about how individuals concentrate, focus on assignments, and make plans of how to use their time during learning processes. The *environment* dimension represents the physical structures surrounding learning, such as where the activities take place. The last dimension, *social factors*, refers to learners' initiative in looking for assistance from others, such as teachers, parents, peers, or siblings, during their learning.

Zimmerman and Martinez-Pons (1986) have identified that there are two sources that can be used to support individuals as self-directed learners, that is, direct help from the teacher, other students, or parents, and from books or symbols (pictures, diagrams or formulas). In the music practice context, involvement from others such as parents will help the formation of self-regulation of music practice in individuals. In addition, support from parents in the learning process can help children become more independent and responsible (Epstein, 1988).

Schunk and Zimmerman (1997) state that parental involvement can help individuals develop self-regulated learning. Through such involvement, parents dedicate their resources to their children within a particular domain (Grolnick & Slowiaczek, 1994). This study focuses on parental involvement in individuals' music practice. There are three types of parental involvement, these being behavioural, cognitive, and personal involvement. Behavioural involvement refers to parents' participation in activities at school and home, such as attending parent–teacher conferences. For example, in the music practice context, a parent asking about his or her child's progress in music practice can be considered as parental behavioural involvement. Cognitive involvement includes stimulating the child intellectually through materials, such as giving them some books or taking them to the library. Similarly, parents might take their child to a music concert. Personal involvement relates to parents' knowledge about their children's environment in school, for example, how their children interact with their friends in music school.

Grolnick and Slowiaczek (1994) have found that behavioural involvement and cognitive involvement are significantly correlated with self-regulated learning. They suggested that parents' behaviours, such as engaging in school activities, teach their child about the importance of school and how to manage situations. Moreover, parents' behaviour may affect their child's motivation. Individuals' exposure to intellectual/cognitive activities may enable them to master such activities in school. However, parental personal involvement might have influence on both the child's behaviour and cognitive music practice. There is an overlapping variance between the personal factors and the other two factors involved (Grolnick & Slowiaczek, 1994). Therefore, in this study, parental involvement focuses on behavioural and cognitive involvement.

The results of previous studies and literature reviews show that parental involvement is important in the development of self-regulation in individuals' learning. Furthermore, there are only a few studies that focus on the relationship between parental involvement and self-regulation in other learning contexts such as music learning. Therefore, this study focused on the music practice context, with the research question of the study being: is there a relationship between parental involvement and self-regulation in music practice?

2 METHODS

2.1 *Respondents*

The respondents (N = 103) were students in grades 10 and 11 from *Sekolah Menengah Kejuruan* (SMK) *Musik* (music vocational school) in Jakarta and Cibinong. Their ages ranged from 15 to 18 years old, and they consisted of 61 females (59%) and 42 males (41%). Most of the respondents' music practice spanned from 1 to 3 years (56.3%). A majority (64.7%) of the respondents played string instruments, such as the violin, viola, cello and double bass, with others playing wind instruments (19.41%), guitar and harp (3.88%), percussion (6.8%), piano (3.88%), and singing (1.94%).

2.2 *Instruments*

The Parental Involvement Measure (PIM) of Zdzinski (1996) was used to measure parental involvement. The scale coefficient Alpha Cronbach was 0.88, therefore the scale reliability was considered good. (α 0.88). This incorporates 19 items about parents' frequency of engagement in individuals' music activities, including two dimensions of parental involvement, namely behavioural involvement (11 items) and cognitive involvement (eight items). An example of a behavioural involvement item is "My parents ask about my progress in music practice". "My parents invite me to music concerts" is an example of a cognitive involvement item. These measures used a four-point Likert style, ranging from *Never* to *Very often* (1 = Never; 2 = Infrequently; 3 = Often; 4 = Very often).

The Self-Regulated Practice Behavior Scale (SRPB) of Ersozlu and Miksza (2014) was used to measure self-regulation in music practice. The scale coefficient Alpha Cronbach was 0.85, therefore the scale reliability was considered good. (α 0.85). It incorporates 25 items, including six items about the dimension of time management (e.g. "It is easy for me to remain focused on my daily musical goals when practising alone"), five items about the metacognition dimension (e.g. "I am focused on monitoring my improvement when I practice"), four items about the dimension of social influence (e.g. "I listen carefully to my lesson teacher's practice advice"), and ten items about the self-efficacy dimension (e.g. "I feel I can solve any musical problem I encounter"). These measures use a six-point Likert style, ranging from *Strongly disagree* to *Strongly agree* (1 = Strongly disagree; 6 = Strongly agree).

2.3 *Control variables*

Demographic variables, such as gender, practice routines, total of practice duration each day, music practice experience, and practice sessions for the self-regulation variable, were controlled in this study. A study by Zimmerman and Martinez-Pons (1990) showed that there were differences between males and females in planning, monitoring, and goal orientation skills. Moreover, Miksza (2012) states that self-regulation in learning will be formed throughout the learning experience. In addition, demographic variables relating to parents, such as their education, occupation, income, instrument mastery, and which parent is involved in children's musical practice, were controlled in the parental involvement variable.

2.4 *Procedures*

The PIM and SRPB questionnaires were conducted in a classical fashion in each class by the researchers. The respondents were asked to signify their informed consent before filling out the questionnaire. Then, the researchers provided instructions about how to fill out each section of the questionnaire. After completing all of the items in the questionnaire, respondents were given rewards.

3 RESULTS

Table 1 displays the descriptive statistics of the questionnaire responses. Based on the four-point Likert scale ranging from 1 (Never) to 4 (Very often), the mean of parental involvement was 2.33. This shows that, on average, the parents in this study were infrequently involved in their children's music practice.

Based on a six-point Likert scale, the mean of self-regulated music practice behaviour was 4.46. This shows that, on average, when it comes to practising music, SMK music students in Jakarta and Cibinong were showed above average self-regulation in music practice good enough in regards to being active in the aspects of metacognition, motivation, and behaviour.

In Table 2, it can be seen that there was a positive and significant correlation between parental involvement and their child's self-regulated music practice behaviour ($r = 0.279$; $n = 103$; $p < 0.01$). Higher parental involvement also indicated higher self-regulated music practice behaviour in music students. The calculation of R^2 (0.079) shows that 7.9% of the variance score for self-regulation in music practice can be explained by means of the parental involvement score, whereas 92.1% of the variance is attributable to other factors not measured in this study.

Table 1. Descriptive statistics of SRPB and PIM.

Variables	Mean	SD
PIM	2.33	0.541
Behavioural	2.41	0.541
Cognitive	2.23	0.567
SRPB	4.46	0.454
Social Influence	4.87	0.642
Self-Efficacy	5.10	0.569
Metacognition	4.77	0.477
Time	3.71	0.850

Table 2. Correlation between parental involvement and self-regulated practice behaviour.

	SRPB	Social influence	Self-efficacy	Meta-cognition	Time
PIM	0.279**	0.260**	0.363**	0.091	0.076
Behavioural involvement	0.342**	0.279**	0.431**	0.161	0.101
Cognitive involvement	0.137	0.179	0.198*	–0.019	0.027
Control variables					
Practice routines	0.426**				
Practice durations	0.447**				
Gender	–0.021				
Student music experience	–0.052				

*Significant in Level of Significance 0.05; **Significant in Level of Significance 0.01.

The other results show that there was a quite strong relationship between parental behavioural involvement and their child's self-regulated music practice behaviour ($r = 0.342$; $n = 103$; $p < 0.01$). This means that the higher the behavioural involvement, the higher the self-regulated music practice behaviour in music students. The calculation of R^2 (0.117) indicates that 11.7% of the variance score in self-regulation in music practice can be explained by means of the behavioural involvement score, whereas 88.3% of other variance is attributable to factors that were not measured in this study. These findings show that there was no relationship found between parental cognitive involvement and self-regulated music practice behaviour.

The relationship between parental involvement and the dimensions of self-regulated music practice behaviour show that only social influence and self-efficacy correlated positively and significantly with parental involvement. Other findings show there were positive and significant correlations between parental behavioural and cognitive involvement and their child's self-efficacy in music practice behaviour. Of the two parental involvement dimensions, only behavioural involvement had a positive and significant correlation with the social influence dimension of self-regulated music practice behaviour.

Based on the results displayed in Table 2, we also discovered that there were positive and significant relationships between the control variables, such as practice routines and practice duration, and self-regulated music practice behaviour.

4 DISCUSSION

This study found that there was a positive and significant correlation between parental involvement and self-regulated practice behaviour in music students. This result supports Schunk and Zimmerman (1997), who state that students' self-regulation can be formed through parental involvement. Further, the study also found that there was a positive and significant relationship between parental behavioural involvement and self-regulated practice behaviour. This result supports Grolnick and Slowiazcek (1994), who argue that behavioural involvement can significantly predict self-regulation. We also believe that when students are at home, parents can give them direct support that can increase their motivation, and can give them feedback on their performance to improve their skills. On the other hand, we found that there was no significant relationship between parental cognitive involvement and student self-regulation in music practice. We assume that there were some parents who had little interest in music and could not give optimum input when it came to cognitively stimulating their child's musical activities.

In addition, the results of this study also show that there was a significant relationship between behavioural and cognitive involvement and the self-efficacy dimension of self-regulated practice behaviour. This result supports Grolnick and Slowiazcek (1994), who found that behavioural and cognitive involvement are associated with perceived competence. When the teenagers have competency, then self-efficacy can be developed indirectly. In addition, there was a significant relationship between parental behavioural involvement and the social influence dimension of self-regulated practice behaviour. This might be because the sentence in each item of the questionnaire always begins with "My parents ...", leading to parental involvement being experienced more directly. As a result, parental behavioural involvement was more visible and directly felt by children than parental cognitive involvement. The other results show that behavioural and cognitive involvements were insignificantly correlated with the two other dimensions of self-regulated practice behaviour, time management and metacognition. This might be due to the fact that the instrument of parental involvement did not have items that measured time management and metacognition aspects, so additional items that can measure those aspects are needed.

This study also found that there were positive and significant relationships between self-regulation in music practice and practice routines and the total practice duration for each day. This result supports the findings of McPherson and McCormick (1999), in which they found that high intensity in practice enables individuals to organise their practice pattern more effectively, such as practising some parts that need more time and being more capable of doing some difficult tasks. This is related to self-regulated practice behaviour.

Although this study has revealed some interesting findings, we realise there are some limitations. Firstly, according to Turner (2006), self-regulation is a dynamic process, which can change from time to time. Therefore, the questionnaire data collection method should be followed by an interview or direct observation, which will provide clearer and more comprehensive descriptions of students' self-regulated learning in music practice. Secondly, this study did not consider the respondents who took a music course or were autodidactic in playing music. We believe different outcomes may result from such scenarios. Thirdly, the respondents in this study were limited to adolescents. It is recommended that future studies include other age groups because this might yield different outcomes. In addition, one type of parental involvement was not included in the measurement of this study, namely personal involvement. Including this in future studies is highly recommended, as it will provide more comprehensive information and enrich parents' knowledge about the relationship between parental involvement and students' self-regulation in music practice. Moreover, the respondents in this study did not adequately represent all music students in Indonesia because the data were only collected from two music schools, SMK Musik Jakarta and Cibinong. Thus, future studies should involve more samples from other areas of Indonesia in order to better represent the population.

5 CONCLUSION

The purpose of this study was to better understand the relationship between parental involvement and student self-regulation in music practice. The results show that there is a positive and significant relationship between them, and that the greater the involvement of parents, the greater the self-regulation by students in their music practice.

5.1 *Future studies*

For future studies, the researchers suggest analysis of the specific dimensions of both the variables involved. In addition, studies should also be conducted in other contexts besides music practice.

REFERENCES

Benton, C.W. (2014). *Thinking about thinking: Metacognition for music learning.* Plymouth, UK: Rowman & Littlefield Education.

Epstein, J.L. (1988). How do we improve programs for parent involvement? *Educational Horizons*, Winter, 58–59.

Ericsson, K.A. (1997). Deliberate practice and the acquisition of expert performance: An overview. In Jérgensen, H. & Lehmann, A.C. (Eds.), *Does practice make perfect? Current theory and research on instrumental music practice*. Oslo, Norway: Norges Musikkhøgskole.

Ersozlu, Z.N. & Miksza, P. (2014). A Turkish adaptation of a self-regulated practice behavior scale for collegiate music students. *Psychology of Music*, *43*(6), 855–869.

Grolnick, W.S. & Slowiaczek, M.L. (1994). Parents' involvement in children's schooling: A multidimensional conceptualization and motivational model. *Child Development*, *65*(1), 237–252. doi:10.1111/j.1467-8624.1994.tb00747.x

Hallam, S. (2010). The power of music: Its impact on the intellectual, social and personal development of children and young people. *International Journal of Music Education*, *28*(3), 269–289. doi:10.1177/0255761410370658

Lehmann, A.C., Sloboda, J.A. & Woody, R.H. (2007). *Psychology for Musicians: Understanding and Acquiring the Skills*. doi:10.1093/acprof:oso/9780195146103.001.0001

Martin, A.J. (2008). How domain specific is motivation and engagement across school, sport, and music? A substantive–methodological synergy assessing young sportspeople and musicians. *Contemporary Educational Psychology*, *33*(4), 785–813. doi:10.1016/j.cedpsych.2008.01.002
McPherson, G.E. (2005). From child to musician: Skill development during the beginning stages of learning an instrument. *Psychology of Music*, *33*(1), 5–35. doi:10.1177/0305735605048012
McPherson, G.E. & McCormick, J. (2014). Motivational and self-regulated learning components of musical practice. *Bulletin of the Council for Research in Music Education*, *141*, 98–102. doi:10.13140/RG.2.1.4263.4080.
McPherson, G.E. & Zimmerman, B.J. (2002). Self-regulation of musical learning: A socialcognitive perspective. In Colwell, R. & Richardson, C. (Eds.), *The new handbook of research on music teaching and learning* (pp. 327–347). New York, NY: Oxford University Press.
McPherson, G.E., Davidson, J.W. & Faulkner, R. (2012). *Music in our lives: Rethinking musical ability, development and identity*. Oxford, UK: Oxford University Press. doi:10.1093/acprof:oso/9780199579297.001.0001
Miksza, P. (2011). The development of a measure of self-regulated beginning and intermediate instrumental music students. *Journal of Research in Music Education*, *59*(4), 321–338.
Schunk, D.H. & Zimmerman, B.J. (1997). Social origins of self-regulatory competence. *Educational Psychologist*, *32*, 195–208. doi:10.1207/s15326985ep3204_1
Turner, J.C. (2006). Measuring Self-Regulation: A Focus on Activity. *Educational Psychological Review*, *18*(3), 293–296. doi:10.1007/s10648–006–9022–3dzinski, S.F. (1996). Parental involvement, selected student attributes, and learning outcomes in instrumental music. *Journal of Research in Music Education*, *44*(1), 34–48. doi:10.2307/3345412
Zimmerman, B.J. (1986). Becoming a self-regulated learner: Which are the key subprocesses? *Contemporary Educational Psychology*, *11*(4), 307–313. doi:10.1016/0361-476X(86)90027-5
Zimmerman, B.J. & Martinez-Pons, M. (1986). Development of a structured interview for assessing student use of self-regulated learning strategies. *American Educational Research Journal*, *23*(4), 614–628. doi:10.3102/00028312023004614
Zimmerman, B.J. & Martinez-Pons, M. (1990). Student differences in self-regulated learning: Relating grade, sex, and giftedness to self-efficacy and strategy use. *Journal of Educational Psychology*, *82*(1), 51–59. doi:10.1037/0022–0663.82.1.51
Zimmerman, B.J. & Schunk, D.H. (2011). *Handbook of self-regulation of learning and performance.* Abingdon, UK: Routledge.
Zimmerman, B.J., Bandura. A. & Martinez-Pons, M. (1992). Self-motivation for academic attainment: The role of self-efficacy beliefs and personal goal setting. *American Educational Research Journal*, *29*, 663–676.

Diversity in Unity: Perspectives from Psychology and Behavioral Sciences – Ariyanto et al. (Eds)
© 2018 Taylor & Francis Group, London, ISBN 978-1-138-62665-2

Me or us? How values (power and benevolence) influence helping behaviour at work

N. Grasiaswaty, D.E. Purba & E. Parahyanti
Faculty of Psychology, Universitas Indonesia, Depok, Indonesia

ABSTRACT: As one of the extra-role behaviours, helping is known as a construct to enable the increase of productivity in an organisation. As previous research was lacking in the contextual role of helping behaviours, especially the group role, this study aimed to investigate the group context in understanding the underlying mechanism of the relationship between personal values, that is, power and benevolence, and helping behaviours. Data was collected from 99 participants and their supervisors, in which supervisors were asked to rate the helping behaviours of their subordinates. The results showed that there was a significant effect of benevolence on helping behaviours (*effect* = 1.82; *SE* = 0.92; *p* = 0.05; *CI* [0.0026, 3.6454]), whereas power was not found to be significantly correlated with helping behaviours (*effect* = –0.142; *SE* = 0.52; *p* > 0.05; *CI* [–1.1631, 0.8796]). Perceived group power also affected helping behaviours (*effect* = –0.37; *SE* = 0.16; *p* < 0.05; *CI* [–0.69, –0.0544]), but it did not moderate the relationship between values and helping behaviours (*effect* = 0.36; *SE* = 0.137; *p* > 0.05; *CI* [–0.472, 0072]). On the other hand, perceived group benevolence did not affect helping behaviours (*effect* = –0.01; *SE* = 0.2; *p* > 0.05; *CI* [–0.472, 0.072]), and it did not moderate the relationship between values and helping behaviours either (*effect* = –0.20; *SE* = 0.5058; *p* > 0.05; *CI* [–0.6524, 1.3664]). The theoretical and practical implications are further discussed.

1 INTRODUCTION

Helping behaviour refers to any action taken by a person voluntarily with the purpose of relieving the problems of others (Dovidio, 1984). Research on this behaviour originated from social psychology (Dovidio, 1984). This behaviour is known to give advantages for an organisation, such as reducing intrigues between individuals in a group, increasing performance, and improving organisational efficiency (Podsakoff et al., 2009; Rotundo & Sackett, 2002). Thus, many scientists from industrial psychology are interested in learning about this behaviour.

Helping behaviour at work has received more attention recently as team-based organisations have also become a trend (Ilgen & Pulakos, 1999, in Ng & Van Dyne, 2005). LePine et al. (2002) pointed out the same reality; that, in this era, organisations focus more on people and networks. This reality has made human interaction, such as helping behaviour, more important than before. If it is brought into the context of organisations, helping behaviour can be directed at helping private interests or organisations in which individuals are located, such as helping co-workers when encountering a new job or helping employees adjust. For the organisation itself, prosocial behaviour can improve efficiency and increase job satisfaction among employees (Podsakoff et al., 2000).

Helping behaviour at work could be considered as an extra-role behaviour. Extra-role behaviour refers to every behaviour which a worker demonstrates in addition to his or her formal role in the workplace that aims to benefit the organisation. Contrary to in-role behaviour (or behaviour that is part of a worker's formal duty within an organisation and considered in performance appraisal), extra-role behaviour does not enter the reward system of an organisation, is done voluntarily for the benefit of the organisation, and the worker will not be penalised when he or she does not display such behaviour (Van Dyne & LePine, 1998). Van Dyne and LePine (1998) combined the Organizational Citizenship Behavior constructed by Organ (1988) with the contextual performance proposed by Borman and Motowidlo (1993) and produced four types of extra-role behaviour. Helping behaviour was considered as falling into the promotive–affiliative category. This is less 'mundane' than other kinds of extra-role behaviour (e.g. whistle-blowing and voice), and is thus more easily detected in daily life. In addition, this behaviour consistently has a more positive effect on organisations than other types of extra-role behaviour, such as voice (speaking out/up), which is also associated with discomfort and tendency to build a public image (Milliken et al., 2003).

Many constructs have been proposed to provide the reason why people engage in helping behaviour at work. Highly charismatic leaders could encourage their subordinates to develop higher levels of helpfulness (Hayibor et al., 2011). Good relations with colleagues (Bowler & Brass, 2006), perception of organisational support, procedural fairness (Lavelle et al., 2009), and confidence in the company (Chiaburu & Byrne, 2009) have all been identified as external factors which increase the frequency of helping behaviour.

Bateman and Organ (1983) believe that helping behaviour, as one dimension of organisation citizenship behaviour, is influenced by a person's level of altruism. Another study reveals weak correlation between conscientiousness and a person's tendency to help other colleagues (Organ & Ryan, 1995). Various aspects of personality are also known to affect helping behaviour in Indonesia; not only conscientiousness, but also extraversion, agreeableness, and emotional abilities (Purba et al., 2015). Other internal factors which are also known to have links with helping behaviour are job satisfaction (Bateman & Organ, 1983; Williams & Anderson, 1991), an individual's self-efficacy (McAllister et al., 2007), and personal values (Liu & Cohen, 2010; Arthaud-Day et al., 2012).

One of the internal constructs that could be an antecedent of helping behaviour is personal values. Values are a set of guidelines that are used as a benchmark on what is important in life (Schwartz, 2006). An individual assesses objects, behaviours, and other individuals based on the values which he or she holds. Schwartz (1992) defined values as a series of beliefs that are used as a reference for evaluating a specific situation and guiding people in making or evaluating a behaviour. Schwartz (2006) later proposed ten types of values that are relatively consistent across cultures. These ten types can be seen as having two dimensions. The first dimension shows the extent to which a person focuses either on improving him or herself (self-enhancement) or on the welfare of others (self-transcendence). Meanwhile, the second dimension shows the extent of a person's focus on either seeking a challenge (openness to change) or on seeking security and comfort (conservation). Values can be a predictor of a person's behaviour based on two aspects (Bardi & Schwartz, 2003): (1) there is a need for equalisation between a person's beliefs and their own behaviour, and (2) if the person displays a behaviour that is consistent with their values, the person will get what they want. People with values in which power figures highly behave consistently with this, like assuming the position of a leader or taking a role in every task.

Different from attitude or personality, personal values are known for their ability to reveal the motives behind someone's behaviour (Schwartz, 2006). They answer the question of 'why' people display certain patterns of behaviour and have personal beliefs about what should and should not be done (Bardi & Schwartz, 2003). Moreover, different values might lead to the same behaviour. Arthaud-Day et al. (2012) found that the values of power and benevolence have a significant correlation with the way a person helps other people. A person with a high value in relation to power tends to help people because they want to bolster their self-image, while a person with a high benevolence value tends to help people because they really care about other people.

This research also examines how contextual factors moderate the correlation between personal values and helping behaviour. Group value is considered as a form of contextual value because it forms the 'guidelines' or norm, that is, what 'must' and 'should not' be done in groups. The importance of groups in shaping the attitudes and behaviour of all employees in a workplace has been known for years (Cohen et al., 2012; Ehrhart & Naumann, 2004).

Moreover, most research on helping behaviour in Indonesia is more concerned with personal factors, such as organisational commitment and personality (Purba et al., 2015), and trust and satisfaction (Pekerti & Sendjaya, 2011). Meanwhile, another study shows that, for Southeast Asian people, individual performance is more determined by the groups in which he or she is working than by the organisation or the job itself (Oh et al., 2014). An individual's actions are often based on what they consider important in their group (Cable & Edwards, 2004). Based on Schwartz's personal values theory, Cable and Edwards (2004) coin the term organisational value, which in this research is modified as group value. Schwartz (2004) emphasises that Indonesia is a country with high levels of embeddedness, which strengthens the individual's sense of belonging to their group. Thus, the group and group values play an important role in the behaviour of Indonesian individuals.

Two values highlighted in this study are power and benevolence. Power is a kind of value prominently held by an individual who prioritises social status and dominance, while benevolence is a kind of value strongly held by an individual who prioritises the welfare of a group (Schwartz, 1992). Schwartz (2013) suggests that these two values generate different responses when an individual is facing a situation that involves cooperative behaviour, such as helping behaviour. These two values produce opposing views about a group. People with a high benevolence value tend to believe that a group's interest should be prioritised, while people with a high power value tend to believe that their personal interests should be prioritised. People with a high power value would regard their groups as an opportunity to gain status or authority (Schwartz, 1992). Because of these differences, both values would produce unique forms of group dynamic in terms of helping behaviour.

Bardi and Schwartz (2003) found that a high level of benevolence is associated with a high level of helping behaviour. Schwartz (2013) reveals that, among the various forms of personal values, benevolence is the strongest predictor of cooperative behaviours, one of which is helping behaviour. People with high benevolence emphasise the harmony or balance of a group, such that they will attempt to further the interests of groups or organisations in which they are placed (Schwartz, 2010). Van Dyne and LePine (1998) suggest that helping behaviour is intended to maintain group harmony. Thus, people who have a high level of benevolence would help their co-workers in order to maintain harmony within the group. Furthermore, Schwartz (2010) reveals that people with a high level of benevolence would assume that they are responsible for the weaknesses of others and thus are more likely to take prosocial actions or, in the context of this study, to display helping behaviour.

Schwartz (1992) reveals that, in contrast to people with high benevolence, those with high power have the need to seek social status and prestige, as well as devising ways to control other people or resources. When an individual holds the value of power strongly, they will perform activities that allow them to dominate others (Boer & Fischer, 2013), such as taking a leading role in an organisation or even intimidating others. Although some empirical evidence indicates that helping behaviour can be induced by power through the need for impression management (Bolino et al., 2012), another study showed that in China a high-power value does not have any effect on helping behaviour (Liu & Cohen, 2010).

Schwartz (2010, 2013) stresses that power could have a negative relationship with prosocial behaviours such as helping. Because this value is based on self-interested motives, an individual holding it would only display behaviours which are considered beneficial to him or herself. In a multinational study, Schwartz (2004) found that power has a positive correlation with anxiety, while benevolence has a negative one. This anxiety leads an individual with high levels of power value to be less prosocial and to hold resources only for him or herself (Schwartz, 2010).

In a group context, group values can influence the behaviour of individuals through at least three mechanisms (Sagiv & Schwartz, 2000): (1) environment affordance, in which people regard the environment outside them as a set of opportunities for, or obstacles to, achieving their goals; (2) social sanctions, in which a group places a range of expectations on a person's behaviour; and (3) group reinforcement, in which being in a group which holds the same value also leads an individual to demonstrate helping behaviour through the belief that he or she will be appreciated or get positive feedback, either directly or indirectly.

Positive feedback in a group is known to strengthen the attachment between people in terms of helping behaviour (Bachrach et al., 2001). Theoretically, a person will feel more comfortable and be less likely to experience dissonance when he or she is in a working group which is valuable to them (Cable & Edwards, 2004), and this condition will ultimately increase their helping behaviour.

People with a high level of power have the need to obtain a high social status in their group (Schwartz, 1992, 2007). When such a person perceives their group as holding a high level of power, he or she would believe that features such as prestige, the desire to be acknowledged by other groups, and status are important to the group. Even though, through impression management, helping behaviour could be seen as one way to acquire social status (Bolino, 1999), Arthaud-Day et al. (2012) reveal that helping behaviour that derives from impression management might be less beneficial for the group and is more likely to trigger negative reactions. Thus, Farrel and Finkelstein (2011) found that, if an individual sees his or her co display helping behaviour and perceives it as part of their self-image management, he or she would view it in a less favourable light. Moreover, Bardi and Schwartz (2003) reveal that, in a collectivist society, the value of power is not appreciated when expressed. Thus, in such societies, individuals who perceive their group as a high power group would tend to restrain their helping behaviour.

2 METHODS

2.1 *Participants*

There were 99 pairs of supervisors and subordinate employees who participated fully in this research. Participants consisted of employees and their direct supervisors in a government-owned factory which deals with electricity distribution. Based on a recommendation from a previous study (Arthaud-Day et al., 2012), the participants should be within working age (22 to 40 years) and have worked for at least three months with their respective supervisors, who would measure their helping behaviour. Questionnaires were distributed to 160 such pairs but only 129 made returns. Based on Schwartz's (2010) suggestion, participants who marked "7" ('Very important') for more than 35 items should be excluded, after which the number of eligible pairs remaining was 99.

2.2 *Measuring instrument*

Research data were gathered from two sources, consisting of a questionnaire for measuring helping behaviour, completed by an employee's direct supervisor, and a self-reporting questionnaire for measuring personal and group values of the employees. The researchers assigned a specific code to each respondent (e.g. a supervisor who fills out questionnaire A-1 is required to measure the helping behaviour of employee A-1). The helping behaviour questionnaire was adapted from helping behaviour questionnaire items developed by Van Dyne & LePine (1998). A careful adaptation and legibility test were performed on this instrument in accordance with the recommendations of Beaton et al. (2000). Feedback from an expert committee was regularly provided throughout the processes of translation and synthesis. This instrument consisted of six items with a five-point Likert scale, on which the supervisors were asked to rate how often their subordinates were engaged in helping behaviour in their respective work groups. The use of rating frequency (e.g. 0 = Never; 5 = Always) for extra-role behaviour measurement is based on the recommendation of Spector et al. (2010).

Values were measured using the Schwartz's Value Survey developed by Schwartz (1992). There were 57 items, and each item consisted of a maximum of two words. Participants were asked to rate each item from –1 ('Opposite') to 7 ('Very important'), according to the degree of importance each word held for them. Perceived group value was adapted from organisational value, developed by Cable and Edwards (2004). For the purpose of this study, this instrument was modified to measure perceived group value by changing the phrase "my organisation" to "my work group". Participants were then asked to select one of the five Likert-scale choices from the least important to very important.

3 RESULTS

Correlations between the variables can be seen in Table 1. Benevolence is positively and significantly associated with helping behaviour ($r = 0.212$, $p < 0.05$), and perceived group power is significantly but negatively associated with helping behaviour ($r = -0.238$, $p < 0.05$). People with a higher level of benevolence are more likely to demonstrate a higher level of helping behaviour, while the more people perceive their group as holding a high value of power, the less they display helping behaviour. Higher levels of personal power are also found to correlate significantly with perceived group power ($r = 0.436$, $p < 0.05$), but not to correlate significantly with helping behaviour ($r = -0.113$, $p > 0.05$).

It was found that perceived group benevolence does not correlate with an individual's helping behaviour ($r = 0.04$, $p > 0.05$). The results also show that certain types of demographic data, that is, age and gender, do not correlate with our measurement variables. As a result, further calculations do not need to be performed with regard to these as control variables.

From Table 2, it can be concluded that only benevolence has a significant and direct influence on helping behaviour ($p < 0.05$; $SE = 0.9175$; *CI* [0.0026, 3.6454]), whereas perceived group benevolence does not significantly moderate the relationship between benevolence and helping behaviour ($p > 0.05$; $SE = 0.5058$; *CI* [−0.6524, 1.3664]).

Table 3 shows that power does not significantly affect an individual's helping behaviour ($p > 0.05$; $SE = 0.52$; *CI* [−1.1631, 0.8796]), but an individual's perception of his or her group's power affects his or her helping behaviour ($p < 0.05$; $SE = 0.16$; *CI* [−0.69, −0.0544]). In addition, we also did not find any effect of perceived group power as a moderator on the relationship between personal power and helping behaviour ($p > 0.05$; $SE = 0.1370$; *CI* [−0.472, 0.0072]).

Table 1. Statistical description and variables intercorrelation (N = 99).

	Mean	SD	1	2	3	4	5	6	7
1. Age	33.93	11.25							
2. Sex	1.33	0.47	−0.104						
3. Power	−1.15	0.92	0.190	−0.112	(0.700)				
4. Perceived group power	14.06	2.99	0.116	−0.021	0.436**	(0.728)			
5. Benevolence	0.64	0.47	−0.029	−0.050	−0.298**	−0.140	(0.750)		
6. Perceived group benevolence	17.31	2.16	−0.036	−0.014	−0.035	0.256**	0.131	(0.776)	
7. Helping behaviour	28.32	4.18	−0.055	−0.033	−0.113	−0.238*	0.212*	0.004	(0.870)

*$p < 0.05$; **$p < 0.01$; the Cronbach alpha of each variable is presented in brackets.

Table 2. The moderating effect of perceived group benevolence on the relationship between personal benevolence and helping behaviour.

	Helping behaviour (n = 99 pairs)					
Predictors	B	SE	T	P	LLCI	ULCI
Benevolence	1.82	0.92	1.98	0.0497	0.0026	3.6454
Perceived group benevolence	0.012	0.20	0.06	0.95	−0.472	0.072
Benevolence × Perceived group benevolence	0.3570	0.51	0.70	0.48	−0.6524	1.3664
$R^2 = 0.0498$; $F = 1.658$						

Table 3. The moderating effect of perceived group power on the relationship between personal power and helping behaviour.

	Helping behaviour (n = 99 pairs)					
Predictors	B	SE	T	P	LLCI	ULCI
Power	−0.1417	0.5145	−0.2755	0.7835	−1.1631	0.8796
Perceived group power	−0.3714	0.1597	−2.3260	0.0221	−0.69	−0.0544
Power × Perceived group power	−0.2001	0.1370	−1.46	0.1474	−0.472	0.0072
$R^2 = 0.0774$; $F = 2.6558$						

4 DISCUSSION

From the above results, it can be concluded that the personal value of benevolence influences helping behaviour in the workplace. These results are consistent with Schwartz's predictions (2010) about people's benevolence directly predicting their prosocial behaviour or, as highlighted in the research, their helping behaviour. People with a high level of benevolence feel the need to maintain harmony. Because helping behaviour is displayed for the purpose of establishing good relations with other people and maintaining harmony in the group (Van Dyne & LePine, 1998), this type of behaviour serves as a way for benevolent people to fulfil their personal need.

The personal value power is also shown to not have any effect on helping behaviour. In line with this, Liu and Cohen (2010) reveal that power does not affect an individual's helping behaviour in the workplace. This conclusion contrasts with that of a previous study by Arthaud-Day et al. (2012) which showed that people with a high level of power tend to display helping behaviour as a form of impression management. Judging from the similarity of our results to those generated by Liu and Cohen (2010) in China, and the difference of our results from those generated by Arthaud-Day et al. (2012) in the United States, we speculate that certain cultural factors must have played an important role in the relationship between these values and helping behaviour. It would be valuable for future research to directly compare results between two different cultures.

One interesting point is the finding that an individual's perception of the group values is able to significantly influence his or her helping behaviour. This indicates that an individual's opinion of his or her group also influences how he or she will act in the workplace. Schwartz (2006) reveals that the Indonesian people have a higher tendency towards embeddedness because his results show that they are very concerned about where they are from a group perspective and what others may think. In collectivist societies, the value of power is less favoured than the value of benevolence (Bardi & Schwartz, 2003), so expressions of this value are rarely appreciated by others. This condition forces people with a high level of power to restrain behaviours which may reveal this value. Dávila de León and Finkelstein (2011) found that high levels of collectivist values have a high correlation with helping behaviour in the workplace, while individualist values tend to correlate with helping behaviour through impression management.

One strength of this study is that it makes use of two sources of data to minimise the common method variance that may occur during data processing if it had come from only one source, such as self-reporting (Podsakoff et al., 2003). This study also demonstrates how cultural difference might influence the correlation between personal relationship values and helping behaviour by comparing the conflicting results of previous studies (Arthaud-Day et al., 2012; Podsakoff et al., 2000), as well as how group context might influence an individual's helping behaviour.

Future research should allocate more time to explaining the content of the questionnaires to each participant, especially Schwartz's values (1992). Many of our questionnaires could not be used because the participants did not follow the instructions correctly. It might be better if the researchers had read out the questionnaire items in front of the participants, so that any potential difficulties could be addressed more effectively. We also suggest that future research be carried out in companies that have different work environments. This research was conducted in a state-owned company where employees had a relatively clear and secure status, which might not be equally enjoyed by private sector employees or outsourced employees. With a broader understanding of organisational context, we might acquire a clearer and more comprehensive view of how contextual factors affect the relationships between personal values and helping behaviour.

Our findings can be applied to actual working environments in several ways. For instance, an individual's values can serve as a basis for their employment or work placement. If the latter involves tasks that are normally performed in groups, there may be value in using benevolence as one of the requirements during selection. Another practical suggestion is that, when a company establishes a work group, it should also provide thorough information about the group's goals and build cohesiveness among the members well before the group starts working. When all members have understood the purpose of the group and know each other better, any negative effects of their perceptions of the group's anticipated values can be reduced and they should be more willing to help each other.

5 CONCLUSION

Based on the above results and findings, we can conclude that, in Indonesian work culture, only the value of benevolence directly influences an individual's helping behaviour in the workplace. Moreover, perceived group values did not serve as a moderator of the relationship between personal values and helping behaviour, but perceived group power has been proven to directly and negatively influence an individual's helping behaviour in the workplace.

REFERENCES

Arthaud-Day, M.L., Rode, J.C. & Turnley, W.H. (2012). Direct and contextual effects of individual values on organizational citizenship behavior in teams. *Journal of Applied Psychology*, *97*(4), 792–807.

Bachrach, D.G., Bendoly, E. & Podsakoff, P.M. (2001). Attributions of the" causes" of group performance as an alternative explanation of the relationship between organizational citizenship behavior and organizational performance. *Journal of Applied Psychology*, *86*(6), 1285–1293.

Bardi, A. & Schwartz, S.H. (2003). Values and behavior: Strength and structure of relations. *Personality and Social Psychology Bulletin*, *29*, 1207–1220.

Bateman, T.S. & Organ, D.W. (1983). Job satisfaction and the good soldier: The relationship between affect and employee "citizenship". *Academy of Management Journal*, *26*(4), 587–595.

Beaton, D.E., Bombardier, C., Guillemin, F. & Ferraz, M.B. (2000). Guidelines for the process of cross–cultural adaptation of self-report measures. *Spine*, *25*(24), 3186–3191.

Boer, D. & Fischer, R. (2013). How and when do personal values guide our attitudes and sociality? Explaining cross-cultural variability in attitude–value linkages. *Psychological Bulletin*, *139*(5), 1113–1147. doi:http://dx.doi.org/10.1037/a0031347

Bolino, M.C. (1999). Citizenship and impression management: Good soldiers or good actors? *Academy of Management Review*, *24*(1), 82–98.

Bolino, M.C., Harvey, J. & Bachrach, D.G. (2012). A self-regulation approach to understanding citizenship behavior in organizations. *Organizational Behavior and Human Decision Processes*, *119*(1), 126–139.

Bolino, M., Valcea, S. & Harvey, J. (2010). Employee, manage thyself: The potentially negative implications of expecting employees to behave proactively. *Journal of Occupational and Organizational Psychology*, *83*, 325–345.

Bolino, M.C., Varela, J.A., Bande, B. & Turnley, W.H. (2006). The impact of impression-management tactics on supervisor ratings of organizational citizenship behavior. *Journal of Organizational Behavior*, *27*(3), 281–297.

Borman, W.C. & Motowidlo, S.M. (1993). Expanding the criterion domain to include elements of contextual performance. In N. Schmidt & W.C. Borman (Eds.), *Personnel Selection in Organizations* (pp. 71–98). San Francisco, CA: Jossey-Bass.

Bowler, W.M. & Brass, D.J. (2006). Relational correlates of interpersonal citizenship behavior: A social network perspective. *Journal of Applied Psychology*, *91*(1), 70–82.

Cable, D.M. & Edwards, J.R. (2004). Complementary and supplementary fit: A theoretical and empirical integration. *Journal of Applied Psychology*, *89*(5), 822–834.

Chiaburu, D.S. & Byrne, Z.S. (2009). Predicting OCB role definition: Exchange with the organization and psychological attachment. *Journal of Business and Psychology*, *24*(2), 201–214.

Cohen, A., Bentura, E. & Vashdi, D.R. (2012). The relationship between social exchange variables, OCB and performance: What happens when you consider group characteristic? *Personnel Review*, *41*(6), 705–731.

Dávila de León, M.C. & Finkelstein, M.A. (2011). Individualism/collectivism and organizational citizenship behavior. *Psicothema*, *23*(3).

Dovidio, J.F. (1984). Helping behavior and altruism: An empirical and conceptual overview. *Advances in Experimental Social Psychology*, *17*, 361–427.

Ehrhart, M.G. & Naumann, S.E. (2004). Organizational citizenship behavior in work groups: A group norm approach. *Journal of Applied Psychology*, *89*(6), 960–974.

Farrel, S.K. & Finkelstein, L.M. (2011). The impact of motive attributions on coworker justice perceptions of rewarded organizational citizenship behavior. *Journal of Business and Psychology*, *26*(1), 57–69.

Hayibor, S., Bradley, R.A., Greg, J.S., Jeffrey, A.S. & Andrew, W. (2011). Value congruence and charismatic leadership in CEO–top manager relationships: An empirical investigation. *Journal of Business Ethics*, *102*(2), 237–254. doi:10.1007/s10551-011-0808-y

Lavelle, J.J., Brockner, J., Konovsky, M.A., Price, K.H., Henley, A.B., Taneja, A. & Vinekar, V. (2009). Commitment, procedural fairness, and organizational citizenship behavior: A multifoci analysis. *Journal of Organizational Behavior*, *30*(3), 337–357.

LePine, J.A., Erez, A. & Johnson, D.E. (2002). The nature and dimensionality of organizational citizenship behavior: A critical review and meta-analysis. *Journal of Applied Psychology*, *87*(1), 52.

Liu, Y. & Cohen, A. (2010). Values, commitment, and OCB among Chinese employees. *International Journal of Intercultural Relations*, *34*(5), 493–506.

McAllister, D.J., Kamdar, D., Morrison, E.W. & Turban, D.B. (2007). Disentangling role perceptions: how perceived role breadth, discretion, instrumentality, and efficacy relate to helping and taking charge. *Journal of Applied Psychology*, *92*(5), 1200–1211.

Milliken, F.J., Morrison, E.W. & Hewlin, P.F. (2003). An exploratory study of employee silence: Issues that employees don't communicate upward and why. *Journal of Management Studies*, *40*(6), 1453–1476.

Ng, K.Y. & Van Dyne, L. (2005). Antecedents and performance consequences of helping behavior in work groups: A multilevel analysis. *Group and Organization Management*, *30*(5), 514–540. doi:10.1177/1059601104269107

Oh, I.S., Guay, R.P., Kim, K., Harold, C.M., Lee, J.H., Heo, C.G. & Shin, K.H. (2014). Fit happens globally: A meta-analytic comparison of the relationships of person–environment fit dimensions with work attitudes and performance across East Asia, Europe, and North America. *Personnel Psychology*, *67*(1), 99–152.

Organ, D.W. (1988). Organizational citizenship behavior: The good soldier syndrome. Lexington, MA: Lexington Books.

Organ, D.W. & Ryan, K. (1995). A meta-analytic review of attitudinal and dispositional predictors of organizational citizenship behavior. *Personnel Psychology*, *48*(4), 775–802.

Pekerti, A.A. & Sendjaya, S. (2011). Fostering organizational citizenship behavior in Asia: The mediating role of trust and satisfaction on OCB in China and Indonesia. In S. Makino, T. Kiyak (Eds.). *Proceedings of the 53rd Annual Meeting of the Academy of International Business* (p. 34). Michigan, USA: University of Michigan.

Podsakoff, P.M., MacKenzie, S.B., Lee, J.Y. & Podsakoff, N.P. (2003). Common method biases in behavioral research: A critical review of the literature and recommended remedies. *Journal of Applied Psychology*, *88*(5), 879–903.

Podsakoff, P.M., MacKenzie, S.B., Paine, J. & Bachrach, D. (2000). Organizational citizenship behaviors: A critical review of the theoretical and empirical literature and suggestions for future research. *Journal of Management*, *26*(3), 513–563. doi:10.1016/j.leaqua.2003.12.003

Podsakoff, N.P., Whiting, S.W., Podsakoff, P.M. & Blume, B.D. (2009). Individual and organizational level consequences of organizational citizenship behaviors: A meta-analysis. *Journal of Applied Psychology*, *94*(1), 122–141. doi:10.1037/a0013079
Purba, D.E., Oostrom, J.K., Van der Molen, H.T. & Born, M.P. (2015). Personality and organizational citizenship behavior in Indonesia: The mediating effect of affective commitment. *Asian Business & Management*, *14*(2), 147–170.
Rotundo, M. & Sackett, P.R. (2002). The relative importance of task, citizenship, and counterproductive performance to global ratings of job performance: A policy-capturing approach. *Journal of Applied Psychology*, *87*(1), 66–80. doi:10.1037//0021-9010.87.1.66
Sagiv, L. & Schwartz, S.H. (2000). Value priorities and subjective well-being: Direct relations and congruity effects. *European Journal of Social Psychology*, *30*(2), 177–198.
Schwartz, S.H. (1992). Universal in the content and structure of values: Theoretical advances and empirical test in 20 countries. *Advances in Experimental and Social Psychology*, *25*, 1–65.
Schwartz, S.H. (2004). Mapping and interpreting cultural differences around the world. In H. Vinken, J. Soeters & P. Ester (Eds.), *Comparing cultures, Dimensions of culture in a comparative perspective.* Leiden, The Netherlands: Brill.
Schwartz, S.H. (2006). A theory of cultural value orientations: Explication and applications. *Comparative Sociology*, *5*, 137–182.
Schwartz, S.H. (2007). Universalism values and the inclusiveness of our moral universe. *Journal of Cross-Cultural Psychology*, *38*(6), 711–728.
Schwartz, S.H. (2010). Basic values: How they motivate and inhibit prosocial behavior. In M. Mikulincer & P.R. Shaver (Eds.), *Prosocial motives, emotions, and behavior: The better angels of our nature* (pp. 221–241). Washington, DC: American Psychological Association.
Schwartz, S. (2013). Value Priorities and behavior: Applying a theory of integrated value systems. In C. Seligman, J.M. Olson & M.P. Zanna (Eds.), *The Psychology of Values: The Ontario Symposium* (Vol. 8, pp. 1–24). Hillsdale, NJ: Lawrence Erlbaum.
Schwartz, S.H. (2016). Basic individual values: Sources and consequences. In Sander, D. & Brosch, T. (Eds.), *Handbook of Value.* Oxford, UK: Oxford University Press.
Spector, P.E., Bauer, J.A. & Fox, S. (2010). Measurement artifacts in the assessment of counterproductive behavior and organizational citizenship behavior: do we know what we think we know? *Journal of Applied Psychology*, *95*(4), 781–790.
Van Dyne, L. & LePine, J.A. (1998). Helping and voice extra-role behaviors: Evidence of construct and predictive validity. *Academy of Management Journal*, *41*(1), 108–119. doi:10.2307/256902
Williams, L.J. & Anderson, S.E. (1991). Job satisfaction and organizational commitment as predictors of organizational citizenship and in-role behavior. *Journal of Management*, *17*, 601–617.

Diversity in Unity: Perspectives from Psychology and Behavioral Sciences – Ariyanto et al. (Eds)
© 2018 Taylor & Francis Group, London, ISBN 978-1-138-62665-2

Do self-monitoring and achievement orientation assist or limit leader effectiveness?

A.M. Bastaman, C.D. Riantoputra & E. Gatari
Faculty of Psychology, Universitas Indonesia, Depok, Indonesia

ABSTRACT: Many assume that leaders' traits, such as self-monitoring and achievement orientation, are related to leader effectiveness. However, previous studies have not shown consistent results on the relationships between self-monitoring, achievement orientation, and leader effectiveness. Some empirical works show that high self-monitoring and achievement-oriented leaders are perceived as more effective. By contrast, other research demonstrates high self-monitoring and high achievement orientation to be hindrances to leader effectiveness. High self-monitoring leaders are "chameleon-like" and can show excellent behavioural flexibility; they can also be perceived as manipulative and not genuine. Some studies find that achievement orientation, albeit found in many effective leaders, is negatively associated with motivation to learn and willingness to accept new ideas. Therefore, this current study aims to investigate the relationships between self-monitoring, achievement orientation, and leader effectiveness. To limit common method bias, data was gathered from two different sources: leaders and subordinates, with a counterbalancing method in place. Data was collected from 215 pairs of leaders and subordinates in the financial and hospitality industries in Indonesia, using very good scales (α between 0.75 and 0.95). Multiple regression analysis demonstrates that achievement orientation is positively associated with leader effectiveness. However, self-monitoring has no relationship with leader effectiveness.

1 INTRODUCTION

Leadership is a popular research topic that has been widely studied for more than a century (Avolio et al., 2009; Bass, 1990; Day, 2014). Among many focuses in leadership research, leader effectiveness is one of the most impactful in understanding leadership (Mumford & Barrett, 2012). The concept of leader effectiveness determines the criteria of effective leaders as a basis for studies of leadership (Yukl, 2012). Thus, various studies have been conducted to examine the role of leader effectiveness in various contexts, including military (Bartone et al., 2007; Hardy et al., 2010; Rockstuhl et al., 2011), governmental (Hooijberg, Lane, & Diversé, 2010), and corporate (Hooijberg & Choi, 2001; Kaiser et al., 2008) ones. This is not surprising given the importance of the role of leaders, especially in a corporate setting, and the increasing demands and challenges that exist in the workplace. The study of Gilley et al. (2008) demonstrates that leader effectiveness plays a significant role in company performance by driving change and innovation. A review by Jing and Avery (2008) describes similar arguments: that leader effectiveness has important relationships with organisational performance and individual performance at the organisational level. Leader effectiveness is a potential source of facilitation to the organisation in making improvements and facing challenges and changes.

However, despite the importance assigned to leader effectiveness, until now a consensus among researchers regarding the definition of leader effectiveness is still lacking. The term "leader effectiveness" is often treated as interchangeable with other similar terms, such as "leadership effectiveness" and "management effectiveness". One probable reason for this is the differences in the types of criteria used in each study to evaluate leader effectiveness (Yukl, 2012).

DeGroot et al. (2011) conclude from previous research that leader effectiveness can be reviewed in terms of "results" (leader effectiveness outcomes) or "behaviour" (leader effectiveness behaviour). Erkutlu (2008) proposes similar arguments: that leader effectiveness can be measured objectively, such as by evaluating productivity or profit gained, and subjectively, through evaluations provided by leaders, subordinates, and coworkers. In this study, we use a definition of leader effectiveness behaviour that states that effectiveness is derived from team members' evaluations of the actions or behaviours of leaders that are relevant to team performance (DeGroot et al., 2011). These behaviours consist of interpersonal aspects (communication, conflict resolution, and problem-solving) and task-management aspects (goal setting and planning) in teamwork, a concept initiated by Stevens and Campion (1994).

Many approaches have been put forward by researchers for understanding the concept of leader effectiveness. The trait-based approach dominated the early development of scientific research to discover what distinguishes effective and ineffective leaders. Despite the fact that its prominence has since dimmed in the midst of leadership theory development, this approach has regained some popularity (Zaccaro, 2007). The reappearance of this approach is supported by an argument that states that leader effectiveness cannot be separated from leaders' personal traits and qualities and thus the measurement of leader effectiveness should always involve leaders' traits (Judge et al., 2002).

The five-factor model of personality traits (Costa & McCrae, 1992) is often considered as a prominent aspect of personality (Goldberg, 1990) for its ability to integrate many traits into a frame of mind (Judge et al., 2002). However, Day and Schleicher (2006) criticised researchers' overemphasis on the Big Five personality traits, given that there are many other traits that are no less important to investigate. Day et al. (2002) found that self-monitoring is a relevant trait in understanding work-related outcomes and attitudes in an organisational context, including job performance, which is often used as criteria of leader effectiveness, and leadership emergence. Self-monitoring is defined as the extent to which an individual controls and regulates their self-presentation, expressive behaviour, and non-verbal affective display in social settings (Fuglestad & Snyder, 2013; Gangestad & Snyder, 2000; Snyder, 1974).

Another potential trait to be considered in studying leader effectiveness is achievement orientation. This is often referred to as the performance goal, which describes the desire or tendency of a person to demonstrate competence that outperforms others (Sijbom et al., 2015). Achievement orientation is one aspect of achievement goal orientation theory, which concerns an individual's perspective on the meaning of an event according to the individual's objectives: being focused either on demonstrating competence (performance) or on developing competence (mastery) (Dweck, 1986; Pekrun et al., 2009).

Both self-monitoring and achievement orientation are potential traits to be focused on in understanding leader effectiveness. Nonetheless, researchers are still arguing as to whether both traits are predictors of, or hindrances to, leader effectiveness. This leads to our research question: what is the relationship between self-monitoring, achievement orientation, and leader effectiveness?

Caligiuri and Day (2000) show that self-monitoring has a significant relationship with commitment, motivation, and interpersonal relationships in a work setting, which is related to the interpersonal aspect of leader effectiveness. Self-monitoring is also associated with subordinates' evaluation of their leaders' ability to adapt to different situations (Foti & Hauenstein, 2007; Zaccaro et al., 1991). In order to be effective, leaders need to be able to diagnose situations that are experienced by their company and identify what types of behaviour are most appropriate (Yukl & Mahsud, 2010). This self-monitoring flexibility has been shown by previous research to be related to both interpersonal and task-management aspects of leader effectiveness. In addition, in comparison to low self-monitoring individuals, high self-monitoring individuals also tend to get better performance appraisals, and are more likely to get promotions and to emerge as leaders (Day et al., 2002; Day & Schleicher, 2006).

However, research on the relationship between self-monitoring and leader effectiveness is still limited and does not show consistent results. High self-monitoring is also found to have negative sides. It might increase a person's tendency to be unauthentic and opportunist in task-based and non-interpersonal situations (Oh et al., 2013), and authenticity is important for leaders to effectively show their true selves, value, and vision to subordinates (Ilies et al., 2005). Studies conducted by Oh et al. (2013) show that high self-monitoring individuals are chameleon-like; able to behave according to the situation by suppressing expression relevant to their original personality. While beneficial and often perceived as highly effective (Gardner et al., 2009), such displays of flexibility by high self-monitoring leaders can also be perceived as inconsistent by their followers (Day et al., 2002; Day & Schleicher, 2006).

Despite this dual nature of self-monitoring, the flexibility associated with high self-monitoring individuals is fundamental to leadership, given that the change which frequently occurs in the workplace requires leaders to be flexible and able to adapt well to new situations (Yukl & Mahsud, 2010). Such leaders are able to read situations and demonstrate the appropriate behaviour for them. Thus, we propose:

Hypothesis 1: That self-monitoring is positively associated with leader effectiveness.

Achievement orientation was originally the subject of many educational studies, particularly in research about academic achievement (Ames, 1992; Dweck, 1986; Elliot & Church, 1997; Kaplan & Maehr, 2007), and it is also important that it is investigated in organisational settings. Some studies show that achievement orientation is one trait characterising effective leaders (Müller & Turner, 2010), leading to effective and superior performance (Boyatzis & Ratti, 2009). Even so, research on leaders' achievement orientation in organisational settings is still limited.

Some other studies indicate that achievement orientation has negative sides too. It is found to have positive correlations with cheating behaviour (Van Yperen et al., 2011) and negative correlations with motivation to learn (Elliot & Church, 1997). This has been attributed to interpersonal benchmarks of achievement orientation, which only emphasise the demonstration of performance and orientation towards results, instead of the process of developing skills. Both aspects can be a hindrance to leader effectiveness, given that leaders should always have the motivation to develop their skills, and that integrity is important for leaders in order that they be trusted and perceived as effective by their subordinates. Indeed, leaders' integrity is found to have relationships with the organisational commitment and work performance of their subordinates (Leroy et al., 2012). Moreover, Sijbom et al. (2015) showed that achievement-oriented leaders tend to oppose creative ideas submitted by subordinates, which could be problematic, given that communication between leaders and subordinates is reported to be the most salient interpersonal aspect of leader effectiveness (DeGroot et al., 2011).

On the other hand, Peus et al. (2015) found that achievement orientation is one factor that drives success for female leaders in holding leadership positions in the United States and some Asian countries. The majority of women leaders stress their willingness to work hard and their dedication to achieving superior levels of performance as crucial success factors in their advancement. Supporting this finding, Dragoni and Kuenzi (2012) found that leaders' achievement orientation is positively associated with unit achievement orientation. Thus, achievement orientation appears contagious, especially for those leaders who have been with their work units for a relatively long period of time. Given all of the above, we suggest:

Hypothesis 2: That achievement orientation is positively associated with leader effectiveness.

2 METHODS

2.1 *Participants and procedure*

The participants in this study were 292 pairs of leaders and subordinates from two industries (financial and hospitality), selected with the following criteria: leaders who have at least two subordinate levels, and subordinates who were two levels directly below the leaders in the company structure. Data was gathered using convenience sampling. Complete responses were received from 229 pairs of leaders and subordinates, representing a very good response rate (approximately 78%) (Cycyota & Harrison, 2006), from which 14 pairs of questionnaires were excluded due to invalid responses, leaving 215. The leaders who participated in this study were between 22 and 55 years old ($M_{age} = 41.20$; $SD_{age} = 7.82$); 62.8% were males; 88.4% were married; 82.8% had children. At least a bachelor degree had been attained by 63%, and 32.1% had worked in the company for more than 15 years. A majority of participants were from the hospitality industry (60.9%), and the most common level of interaction between leaders and subordinates was more than 15 times a week (33%).

2.2 *Measures*

To limit common method bias (Podsakoff et al., 2012), this study used two sources of data (leaders and subordinates) and a counterbalancing method (mixing measurement items, splitting them, and putting them in different sections of the questionnaire). All scales used a 6-point Likert-type scale, anchored from 1 (*Strongly disagree*) to 6 (*Strongly agree*).

Leader Effectiveness. Leader effectiveness was measured using 14 items of a leader effectiveness scale adapted from DeGroot et al. (2011) ($\alpha = 0.95$). Rather than their direct leader, we asked the participants to evaluate the leader of their leader. Sample items included "The leader of my leader greets me while passing by" and "The leader of my leader gives ideas to solve problems".

Self-Monitoring. The instrument consisted of ten items adapted from a revision of a self-monitoring scale (Lennox & Wolfe, 1984), with $\alpha = 0.75$. The items reflected the definition of self-monitoring, including: "I am sensitive to the slightest change of expression in people I talk with" and "It is difficult for me to adjust my behaviour while dealing with different people".

Achievement Orientation. Leaders' achievement orientation was measured using four items about performance goals, adapted from Sijbom et al. (2015), that were adjusted to fit the work setting ($\alpha = 0.78$). Sample items of this scale included "As a leader, I have to prove that I work better than others" and "As a leader, I need to show better performance than others", thus describing a desire to outperform others.

2.3 *Control variables*

We controlled industry type (finance vs hospitality) and some demographic variables which are theoretically linked to leader effectiveness, specifically gender (Ayman & Korabik, 2010; DeRue et al., 2011), age (Kirkman et al., 2004), marital status (Rad & Yarmohammadian, 2006), number of children (Johnson, 2005; Wallace & Young, 2008), education (Barbuto et al., 2007), and tenure (Kirkman et al., 2004). Frequency of interaction between leaders and subordinates was also controlled, given that leaders are required to create productive ties with their subordinates in order to work together to achieve company goals (Harvey et al., 2006).

3 RESULTS

The means, standard deviations, and correlations observed are presented in Table 1. This shows that leader effectiveness had a significant positive correlation with achievement orientation ($r = 0.17$; $p < 0.05$), number of children ($r = 0.15$; $p < 0.05$), and frequency of interaction between leaders and subordinates ($r = 0.19$; $p < 0.01$).

To analyse the relationship between self-monitoring, achievement orientation, and leader effectiveness, we conducted multiple regression analyses with leader effectiveness as the dependent variable, and self-monitoring and achievement orientation as two independent variables. It was revealed that achievement orientation was a significant predictor of leader effectiveness ($\beta = 0.18$; $p < 0.05$). In contrast, the effect of self-monitoring on leader effectiveness was not significant ($\beta = 0.63$; $p = 0.37$). The model predicted 4% of leader effectiveness (R^2 change = 0.04). The second model, which included additional control variables that have significant correlations with leader effectiveness (type of industry, frequency of interaction

Table 1. Means, standard deviations and correlations.

Variable	Mean	SD	1	2	3	4	5	6	7	8	9	10	11
1. Age	41.20	7.83	1										
2. Sex	–	–	−0.15*	1									
3. Material status	1.95	0.34	0.32**	0.05	1								
4. No. of children	1.69	1.03	0.35**	0.09	0.36**	1							
5. Education	4.18	1.55	−0.21**	−0.24**	−0.06	−0.16*	1						
6. Tenure	4.86	1.80	0.54**	0.10	0.21**	0.25**	−0.08	1					
7. Industry type	–	–	−0.17*	−0.32**	−0.02	−0.21**	0.55**	0.04	1				
8. Frequency of interaction between leaders and subordinates	3.56	2.02	−0.06	−0.03	0.05	−0.12	0.04	−0.11	−0.08	1			
9. Self-monitoring	4.52	0.58	0.06	0.08	−0.04	0.01	−0.01	−0.19**	−0.08	0.05	1		
10. Achievement orientation	4.77	0.82	0.22**	0.15*	0.11	0.12	−0.12	0.09	−0.18**	−0.01	0.23**	1	
11. Leader effectiveness	5.05	0.68	0.03	0.06	0.04	0.15*	0.03	0.02	−0.10	0.19**	0.11	0.17*	1

*Correlation is significant at 0.05 (2-tailed); **Correlation is significant at 0.01 (2-tailed).

Table 2. Multiple regression analysis results.

	Step 1	Step 2
Variable	Control variable	Control variable, self-monitoring and achievement orientation
Type of industry		0.05
Number of children		0.17*
Frequency of interaction between leader and subordinate		0.20**
Self-monitoring	0.063	0.06
Achievement orientation	0.175*	0.15*
R^2	0.04	0.11
F	4.16*	4.72**
df1; df2	2; 203	5; 200

*Correlation is significant at 0.05 (2-tailed); **Correlation is significant at 0.01 (2-tailed).

between leaders and subordinates, and number of children), predicted 11% of the variance of leader effectiveness (R^2 change = 1.10). It also yielded similar results: the impact of achievement orientation on leader effectiveness was significant ($\beta = 0.15$; $p < 0.10$), while the impact of self-monitoring on leader effectiveness was not significant ($\beta = 0.06$; $p = 0.42$). These results indicate that self-monitoring is not associated with leader effectiveness (Hypothesis 1 was not supported), yet provide initial support for the positive relationship between achievement orientation and leader effectiveness (Hypothesis 2 was supported). The small R^2 changes of 0.04 and 1.10 do not necessarily imply that our findings make little contribution to the literature; rather, they demonstrate that leader effectiveness is a large construct, and its relationship with other traits should also be explored.

4 DISCUSSION AND FUTURE RESEARCH DIRECTION

This study investigates the relationship between self-monitoring, achievement orientation, and leader effectiveness. We found that achievement orientation has a significant positive correlation with leader effectiveness. On the other hand, self-monitoring does not have a significant relationship with leader effectiveness.

This research contributes to leadership research by addressing the critics of Day and Schleicher (2006) and identifying other personality traits associated with effective leaders besides the so-called Big Five, although it does not explain the mechanism of how these traits actually affect leader effectiveness. The influences of general or cross-situational traits on leader effectiveness are likely to be more distant, although still significant (Zaccaro, 2007). Thus, future research should use more integrative models, which might include behavioural or situational aspects, in order to develop a more comprehensive picture of leadership.

Although this research finds that self-monitoring is not associated with leader effectiveness, it is, nonetheless, a meaningful contribution to the field of leadership studies. It contributes by showing that self-monitoring is a relevant trait to predict specific aspects of leader effectiveness, which is supported by the study results of Semadar et al. (2006). In this study we used a broad definition of leader effectiveness, which included both interpersonal and task-management aspects. Therefore, future research should focus on specific roles or tasks of a leader in order to investigate the effect of self-monitoring on the performance of such work.

This study also advances current knowledge in arguing that research on self-monitoring in the field of leadership should focus more on authenticity than effectiveness. In alignment with previous research, the chameleonic effect of high self-monitoring may not always be perceived as effective (Day & Schleicher, 2006; Oh et al., 2013). In spite of the association between self-monitoring and behavioural flexibility that benefits leaders by allowing them to adjust themselves to the situation they are in, their flexibility might also be perceived as inconsistency by their subordinates. Genuineness or authenticity is important for leaders in showing their true selves and their values and vision to subordinates in an effective manner (Ilies et al., 2005). In addition, the authenticity of a leader also increases the trust of subordinates towards them (Gardner et al., 2005; Walumbwa et al., 2008). Furthermore, the trust that subordinates have towards their leaders is correlated with subordinates' perception of leader effectiveness (Norman et al., 2010).

There are also two demographic variables which are associated with leader effectiveness: number of children and frequency of interaction between leaders and subordinates. This finding is in line with the study conducted by Wallace and Young (2008), which found that the presence of children can increase one's productivity in the workplace. This might be caused by the work value shifting from intrinsic to extrinsic in employees with children; that salary and compensation (including benefits for children) have become more valuable than the satisfaction from doing the work itself (Johnson, 2005).

The finding of a relationship between the frequency of interaction between leaders and subordinates and leader effectiveness is not surprising if one refers back to the definition of leadership itself. The higher the frequency of interaction between leaders and subordinates, the greater the opportunity for leaders to establish productive relationships with subordinates, and to direct them in achieving the objectives of the company, as well as being perceived as more effective by those subordinates (Harvey et al., 2006). Further, good superior–subordinate relationship quality can produce good outcomes for a company, among which are increased commitment to the organisation and reduced level of turnover (Joo, 2010). However, there is no significant relationship found between type of industry and leader effectiveness. This shows that the study results can be generalised across at least two types of service industry: finance and hospitality.

The research shows that achievement orientation is a trait that is associated with leader effectiveness. The findings of this research can be used by companies as a consideration in selecting and promoting employees to leadership positions. Companies are expected to recruit employees who have the potential to be leaders, and leading positions in the company can be held by employees who are competent and able to bring advancement to the company.

As with all other studies, this study also has several limitations. First, this is a cross-sectional study in which we gathered data only at one single time. This design cannot explain cause and effect of correlations. Nevertheless, cross-sectional studies remain among the most used in organisational research for their ease and efficiency. We also used two techniques to minimise the effect of common method bias in this study: different sources of data and counterbalancing.

Furthermore, our findings cannot describe the relationships between these two traits and leader effectiveness for different management levels. This was as a result of the difference of structure that each company had, which complicated our attempts to categorise managers. Management levels should form a subsequent research focus because each level has different tasks and roles; there might be different concepts of leader effectiveness at each management level. Leader effectiveness in lower-level management relies more on individual differences, while for higher-level leaders, leader effectiveness is as much a function of environmental factors as it is leaders' individual differences (Hoffman et al., 2011). As an aspect of individual differences, traits might thus lend more impact to lower-level leaders than to higher-level leaders. Hence, further studies should be conducted on different levels of management in order to see whether these traits are only associated with certain levels of management or can be generalised for all management levels.

5 CONCLUSION

The purpose of this study was to examine whether self-monitoring and achievement orientation traits are positively associated with leader effectiveness. Our findings suggest that there is a significant positive relationship between achievement orientation and leader effectiveness; leaders exhibiting a high degree of achievement orientation are perceived as more effective by their subordinates than those exhibiting low achievement orientation levels. However, no significant relationship was found between self-monitoring and leader effectiveness. These findings contribute to the limited research about the relationships between leadership effectiveness and the traits of self-monitoring and achievement orientation. These findings also have practical implications for recruitment and training programmes in companies. Further studies should be conducted in order to better characterise the concept of leader effectiveness.

REFERENCES

Ames, C. (1992). Classrooms: Goals, structures, and student motivation. *Journal of Educational Psychology*, *84*(3), 261–271.

Avolio, B.J., Walumbwa, F.O., & Weber, T.J. (2009). Leadership: Current theories, research, and future directions. *Annual review of psychology*, *60*, 421–449.

Ayman, R., & Korabik, K. (2010). Leadership: Why gender and culture matter. *American Psychologist*, *65*(3), 157.

Barbuto, J.E., Jr., Fritz, S.M., Matkin, G.S., & Marx, D.B. (2007). Effects of gender, education, and age upon leaders' use of influence tactics and full range leadership behaviors. *Sex Roles*, *56*, 71–83.

Bartone, P.T., Snook, S.A., Forsythe, G.B., Lewis, P., & Bullis, R.C. (2007). Psychosocial development and leader performance of military officer cadets. *The Leadership Quarterly*, *18*(5), 490–504.

Bass, B.M. (1990). *Bass and Stogdill's handbook of leadership: Theory, research, managerial application* (3rd ed.). New York, NY: The Free Press.

Boyatzis, R.E., & Ratti, F. (2009). Emotional, social and cognitive intelligence competencies distinguishing effective Italian managers and leaders in a private company and cooperatives. *Journal of Management Development*, *28*(9), 821–838.

Caligiuri, P.M., & Day, D.V. (2000). Effects of self-monitoring on technical, contextual, and assignment-specific performance: A study of cross-national work performance ratings. *Group & Organization Management*, *25*(2), 154–174.

Costa, P.T., & McCrae, R.R. (1992). Four ways five factors are basic. *Personality and individual differences*, *13*(6), 653–665.

Cycyota, C.S., & Harrison, D.A. (2006). What (not) to expect when surveying executives: A meta-analysis of top manager response rates and techniques over time. *Organizational Research Methods*, *9*(2), 133–160. doi:10.1177/1094428105280770

Day, D.V. (2014). *The Oxford handbook for leadership and organization.* Oxford, UK: Oxford University Press.

Day, D.V., & Schleicher, D.J. (2006). Self-Monitoring at work: A motive-based perspective. *Journal of Personality*, *74*(3), 685–714.

Day, D.V., Schleicher, D.J., Unckless, A.L., & Hiller, N.J. (2002). Self-Monitoring personality at work: A meta-analytic investigation of construct validity. *Journal of Applied Psychology*, *87*(2), 390–401. doi:10.1037//0021–9010.87.2.390

DeGroot, T., Aime, F., Johnson, S. G., & Kluemper, D. (2011). Does talking the talk help walking the walk? An examination of the effect of vocal attractiveness in leader effectiveness. *The Leadership Quarterly*, *22*(4), 680–689.

DeRue, D.S., Nahrgang, J.D., Wellman, N.E.D., & Humphrey, S.E. (2011). Trait and behavioral theories of leadership: An integration and meta - analytic test of their relative validity. *Personnel psychology*, *64*(1), 7–52.

Dragoni, L., & Kuenzi, M. (2012). Better understanding work unit goal orientation: Its emergence and impact under different types of work unit structure. *Journal of Applied Psychology*, *97*(5), 1032–1048.

Dweck, C.S. (1986). Motivational processes affecting learning. *American Psychologist*, *41*(10), 1040–1048.

Elliot, A.J., & Church, M.A. (1997). A hierarchical model of approach and avoidance achievement motivation. *Journal of Personality and Social Psychology*, *72*(1), 218–232.

Erkutlu, H. (2008). The impact of transformational leadership on organizational and leadership effectiveness: The Turkish case. *Journal of management development*, *27*(7), 708–726. *Development, 27*(7), 708–726. doi:10.1108/02621710810883616

Foti, R.J., & Hauenstein, N. (2007). Pattern and variable approaches in leadership emergence and effectiveness. *Journal of Applied Psychology*, *92*(2), 347–355. doi:10.1037/0021-9010.92.2.347

Fuglestad, P.T., & Snyder, M. (2013). Self-Monitoring. In Leary, M.R. & Hoyle, R.H. (Eds.), *Handbook of individual difference in social behavior*. New York, NY: The Guilford Press.

Gangestad, S.W., & Snyder, M. (2000). Self-monitoring: Appraisal and reappraisal. *Psychological bulletin*, *126*(4), 530.

Gardner, W.L., Avolio, B.J., Luthans, F., May, D.R., & Walumbwa, F. (2005). "Can you see the real me?" A self-based model of authentic leader and follower development. *The Leadership Quarterly*, *16*(3), 343–372.

Gardner, W. L., Fischer, D., & Hunt, J.G.J. (2009). Emotional labor and leadership: A threat to authenticity?. *The Leadership Quarterly*, *20*(3), 466–482.

Gilley, A., Dixon, P., & Gilley, J.W. (2008). Characteristics of leadership effectiveness: Implementing change and driving innovation in organizations. *Human Resource Development Quarterly*, *19*(2), 153–169.

Goldberg, L.R. (1990). An alternative description of personality: The big five factor structure. *Journal of Personality and Social Psychology*, *59*(6), 1216–1229.

Hardy, L., Arthur, C.A., Jones, G., Shariff, A., Munnoch, K., Isaacs, I., & Allsopp, A.J. (2010). The relationship between transformational leadership behaviors, psychological, and training outcomes in elite military recruits. *The Leadership Quarterly*, *21*(1), 20–32.

Harvey, P., Martinko, M.J., & Douglas, S.C. (2006). Causal reasoning in dysfunctional leader-member interactions. *Journal of Managerial Psychology*, *21*(8), 747–762.

Hoffman, B. J., Woehr, D.J., Maldagen-Youngjohn, R., & Lyons, B.D. (2011). Great man or great myth? A quantitative review of the relationship between individual differences and leader effectiveness. *Journal of Occupational and Organizational Psychology*, *84*(2), 347–381.

Hooijberg, R., & Choi, J. (2001). The impact of organizational characteristics on leadership effectiveness models an examination of leadership in a private and a public sector organization. *Administration & Society*, *33*(4), 403–431.

Hooijberg, R., Lane, N., & Diversé, A. (2010). Leader effectiveness and integrity: Wishful thinking? *International Journal of Organizational Analysis*, *18*(1), 59–75.

Ilies, R., Morgeson, F.P., & Nahrgang, J.D. (2005). Authentic leadership and eudaemonic well-being: Understanding leader–follower outcomes. *The Leadership Quarterly*, *16*, 373–394.

Jing, F.F., & Avery, G.C. (2008). Missing links in understanding the relationship between leadership and organizational performance. *International Business & Economics Research Journal (IBER)*, *7*(5), 67–78.

Johnson, M.K. (2005). Family roles and work values: Processes of selection and change. *Journal of Marriage and Family*, *67*(2), 352–369.

Joo, B.K. (2010). Organizational commitment for knowledge workers: The roles of perceived organizational learning culture, leader–member exchange quality, and turnover intention. *Human Resource Development Quarterly*, *21*(1), 69–85. doi:10.1002/hrdq.20031

Judge, T. A., Bono, J.E., Ilies, R., & Gerhardt, M.W. (2002). Personality and leadership: a qualitative and quantitative review. *Journal of applied psychology*, *87*(4), 765.

Kaiser, R.B., Hogan, R., & Craig, S.B. (2008). Leadership and the fate of organizations. *American Psychologist*, *63*(2), 96–110.

Kaplan, A., & Maehr, M. L. (2007). The contributions and prospects of goal orientation theory. *Educational psychology review*, *19*(2), 141–184.

Kirkman, B.L., Tesluk, P.E., & Rosen, B. (2004). The impact of demographic heterogeneity and team leader-team member demographic fit on team empowerment and effectiveness. *Group & Organization Management*, *29*(3), 334–368.

Kotter, J.P. (1992). *Corporate culture and performance.* New York, NY: The Free Press.

Lennox, R.D., & Wolfe, R.N. (1984). Revision of the self-monitoring scale. *Journal of Personality and Social Psychology, 46*(6), 1349–1364.

Leroy, H., Palanski, M.E., & Simons, T. (2012). Authentic leadership and behavioral integrity as drivers of follower commitment and performance. *Journal of Business Ethics*, *107*(3), 255–264.

Müller, R., & Turner, R. (2010). Leadership competency profiles of successful project managers. *International Journal of Project Management*, *28*(5), 437–448.

Mumford, M.D., & Barrett, J.D. (2013). Leader Effectiveness: Who Really is the Leader? In *The Oxford handbook of leadership*.

Norman, S.M., Avolio, B.J., & Luthans, F. (2010). The impact of positivity and transparency on trust in leaders and their perceived effectiveness. *The Leadership Quarterly*, *21*(3), 350–364.

Oh, I.S., Charlier, S.D., Mount, M.K., & Berry, C.M. (2013). The two faces of high self-monitors: Chameleonic moderating effects of self-monitoring on the relationships between personality traits and counterproductive work behaviors. *Journal of Organizational Behavior*, *35*(1), 92–111. doi:10.1002/job.1856

Pekrun, R., Elliot, A.J., & Maier, M.A. (2009). Achievement goals and achievement emotions: Testing a model of their joint relations with academic performance. *Journal of Educational Psychology*, *101*(1), 115–135.

Peus, C., Brown, S. & Knipfer, K. (2015). On becoming a leader in Asia and America: Empirical evidence from women managers. *The Leadership Quarterly*, *26*, 55–67. doi:10.1016/j.leaqua.2014.08.004

Podsakoff, P.M., MacKenzie, S.B., & Podsakoff, N.P. (2012). Sources of method bias in social science research and recommendations on how to control it. *Annual review of psychology*, *63*, 539–569.

Rad, A.M.M., & Yarmohammadian, M., H. (2006). A study of relationship between managers' leadership style and employees' job satisfaction. *Leadership in Health Services*, *19*(2), 11–28.

Rockstuhl, T., Seiler, S., Ang, S., Van Dyne, L., & Annen, H. (2011). Beyond general intelligence (IQ) and emotional intelligence (EQ): The role of cultural intelligence (CQ) on cross-border leadership effectiveness in a globalized world. *Journal of Social Issues*, *67*(4), 825–840.

Semadar, A., Robins, G., & Ferris, G.R. (2006). Comparing the validity of multiple social effectiveness constructs in the prediction of managerial job performance. *Journal of Organizational Behavior*, *27*(4), 443–461.

Sijbom, R.B., Janssen, O., & Van Yperen, N.W. (2015). How to get radical creative ideas into a leader's mind? Leader's achievement goals and subordinates' voice of creative ideas. *European Journal of Work and Organizational Psychology*, *24*(2), 279–296.

Snyder, M. (1974). Self-monitoring of expressive behavior. *Journal of Personality and Social Psychology*, *30*(4), 526–537.

Stevens, M.J., & Campion, M.A. (1994). The knowledge, skill, and ability requirements for teamwork: Implications for human resource management. *Journal of management*, *20*(2), 503–530.
Van Yperen, N.W., Hamstra, M.R.W., & van der Klauw, M. (2011). To win, or not to lose, at any cost: The impact of achievement goals on cheating. *British Journal of Management*, *22*(s1), S5–S15.
Wallace, J.E., & Young, M.C. (2008). Parenthood and productivity: A study of demands, resources and family-friendly firms. *Journal of Vocational Behavior*, *72*(1), 110–122.
Walumbwa, F.O., Avolio, B.J., Gardner, W.L., Wernsing, T.S., & Peterson, S.J. (2008). Authentic leadership: Development and validation of a theory-based measure. *Journal of management*, *34*(1), 89–126.
Yukl, G. & Mahsud, R. (2010). Why flexible and adaptive leadership is essential. *Consulting Psychology Journal: Practice and Research*, *62*(2), 81–93.
Yukl, G. (2012). *Leadership in organization* (7th edition). Upper Saddle River: Pearson.
Zaccaro, S.J., Foti, R.J. & Kenny, D.A. (1991). Self-monitoring and trait-based variance in leadership: An investigation of leader flexibility across multiple group situations. *Journal of Applied Psychology*, *76*(2), 308–315.
Zaccaro, S. J. (2007). Trait-based perspectives of leadership. *American Psychologist*, *62*(1), 6–16.

Diversity in Unity: Perspectives from Psychology and Behavioral Sciences – Ariyanto et al. (Eds)

The effect of psychological capital as a mediator variable on the relationship between work happiness and innovative work behavior

A. Etikariena
Faculty of Psychology, Universitas Indonesia, Depok, Indonesia

ABSTRACT: In this competitive business environment, organizations must have the ability to compete with other organizations. One of the effective ways to cope with this challenge is to instill innovative work behavior in the employees. Some studies found that happy workers are capable of performing productively, but some others show different results. This inconsistency of the results needs to be further investigated. Some studies found that work happiness is related to psychological capital (PsyCap). On the other hand, PsyCap also has a significant relationship with innovative work behavior. PsyCap may therefore explain the dynamics between work happiness and innovative work behavior. The participantsinvolved in this research were 135 employees of the headquarters of PT Bank Syariah X. The instruments used to measure the variables were the Innovative Work Behavior Scale (2014) having a Cronbach Alpha coefficient (α) of 0.93,the Work Happiness Scale (2013) $\alpha = 0.76$, and the Psychological Capital Questionnaire (2007) $\alpha = 0.87$. The resultsshow there was no significant correlation between work happiness and innovative work behavior ($r = 0.14$; $p > 0.12$). On the other hand, work happiness had a significant correlation with PsyCap ($a = 0.29$) and PsyCap had a significant correlation with innovative work behavior ($b = 0.33$). In other words, there was an indirect correlation between work happiness and innovative work behavior through psychological capital as a mediator variable ($ab = 0.09$). These results are expected to help organizations to have a more comprehensive understanding about those variables to reach optimum advantages.

1 INTRODUCTION

Research on innovation in organizations has been progressing rapidly because many organizations have become more aware that innovation is one of the effective ways to survive in today's competitive business environment. Innovation is one way that organizations can compete in the business environment (Schermuly, Meyer, & Dammer, 2013). Jiménez-Jiménezand Sanz-Valle (2011) found in their research that innovation is an important factor that affects the optimal organizational performance. Furthermore, the challenge is how to make the individuals in the organization implement innovation in their daily activities at work (Gailly, 2011). The level of analysis on innovation in an organization consists of three levels: organization level, group level, and individual level (West & Farr, 1989). This study focused on the individual level that isinnovativework behavior, defined as a planned effort of individuals to create, introduce, and apply new ideas in the context of work, group, or organization for the benefit of the group or organization (Scott & Bruce, 1994). This definition is also used in the research by Janssen (2000; 2003; 2004; 2005), Janssen & Van den Vegt (2006), Carmelli, Meitar & Weisberg (2006), De Jong (2007), Carmelli & Spreitzer (2009), Baunman (2011), & Xerri (2012).

Innovative work behavior is a multistage process, meaning there is a series of stages of behavior to be considered as an innovative work behavior. The stages are unidimensional, so there will be one score for innovative work behavior (Scott & Bruce, 1994). Then innovative work behavior process is divided into three important stages. The first is idea generation, which illustrates how individuals make efforts to acquire new ideas or new ways to benefit the company (De Jong & Den Hartog, 2008). The second is idea promotion, which reflects how individuals seek support from the surrounding environment to realize the idea or the way that has been found (De Jong & Den Hartog, 2008). If in the first stage new ideas emerge, then there should be an attempt to gain support from the surrounding environment so that the ideas are followed up in practice in the form of real program or application. The third is idea realization, the stage at which people realize how significant the idea is related to the development of their respective sectors. At this stage, ideas that have been obtained are realized according to the company's needs such as in the form of program or application.

Innovative work behavior is one form of productive behavior at work (Jex & Brett, 2008). It is usually used as a productivity indicator in organizations, so the antecedents that support innovative work behavior existence in organizations should be found. There are some factors that have been tested as antecedents of innovative work behavior, such as individual factors like employee diversity (Baldrige & Burnham 1975; Østergard, Timmerman & Kristinsson, 2011; Baunmann 2011), personality (Shalley, Zu & Oldham, 2004; Amo & Kolvereied, 2005; Su, Ming & Chun 2010), cognitive capability (Parzefall, Seeck & Leppänen, 2008), intrinsic motivation (Collins & Amabile, 1999) and affect (Clapham, 2001; Isen & Reeve, 2005; George & Zhou, 2007; Hennessey & Amabile, 2010). Related to the role of affect, experts in organizations have been inspired by the move toward positive psychology and have begun to pursue positive organizational behavior (Luthans, 2002; Wright, 2003). In this study, positive organizational behavior is represented by innovative work behavior. Individual affect is one of the important antecedentsof innovative work behavior (Isen & Reeve, 2005; Hennesey & Amabile, 2010). The exploration on the structure of affect, mood, and emotions consistently found that the most important dimension in describing individuals' affective experiences is the hedonic tone or pleasantness–unpleasantness (Watson *et al*, 1999 in Fischer, 2010). Describing happiness, there are some constructs in common that refer to pleasant judgments (positive attitudes) or pleasant experiences (positive feelings, moods, emotions, flow states) at work. There are also some work-related happiness constructs that focus largely on the hedonic experiences of pleasure and liking and/or positive beliefs about an object, such as job satisfaction, affective commitment, and the experience of positive emotions while working, so they are used interchangeably with work happiness (Fisher, 2010).

This study focuses on happiness that is defined as thoughts and positive feelings towards a person's life (Diener *et al*, 2008). In psychology and especially in the topic of workplace experience, happiness in the form of pleasant moods and emotions, well-being, and positive attitudes has attracted increasing attention. Happiness has important consequences for both individuals and organizations, but in the past research tended to underestimate the importance of happiness at work (Fisher, 2010). Feeling happy is fundamental to the human experience, and most people are at least happy most of the time (Diener & Diener, 1996). Happiness is the level of their own work when someone has a positive affection and satisfaction at work (Youssef & Luthans in Choi & Lee, 2015). According to Pryce-Jones (2010), happiness in the workplace is a mindset that allows a person to get maximum performance and reach his or her highest potential at work. They also claim that happy workers will be more easily promoted, earn more, achieve goals more quickly, generate better and more creative ideas, be able to interact with superiors and colleagues, be more physically and mentally fit, and be more ready to learn and accept new knowledge.

Although happiness at the workplace is important to both individuals and organizations and the effects of momentary states of happiness are largely positive, research on employee happiness in organizations remains limited (Fisher, 2010). So far, previous studies have shown that someone who works with a sense of happiness and has positive feelings will have a way to manage and influence his or her work to maximize performance and achieve job satisfaction (Pryce-Jones, 2010). Moreover, maintaining happiness at the workplace can increase employees' productivity (Quick & Quick, 2004). Previous studies (e.g. Quick & Quick, 2004; Rego & Cunha, 2008) stated that happy employees are productive employees, yet

Spicer & Cederstörm (2015) found that happiness does not automatically lead to increased productivity. They show some contradictory results about the relationship between happiness, often defined as "job satisfaction," and productivity. A survey by Silverstro (2002) found a negative correlation between job satisfaction and corporate productivity. The study has pointed in the opposite direction, saying that there is a link between feeling content with work and being productive. All these studies as a whole demonstrate a relatively weak correlation between work happiness and productivity.

Seligman (2002) suggested that (authentic) happiness is facilitated by developing and practicing character virtues, such as kindness, gratitude, optimism, curiosity, playfulness, humor, open-mindedness, and hope. Even though there is a specific definition of happiness explained by Seligman, there are common aspects about happiness explained by the theory of psychological capital (PsyCap), which are optimism and hope (Luthans *et al*, 2007). Seligman and Luthans are experts known to have built the concept in positive psychology that focuses on the positive state to develop and encourage people to be aware of their potential. PsyCap is an individual's positive psychological state of development that is characterized by having to take on and put in the necessary effort to succeed at a challenging task (self-efficacy), making a positive attribution about succeeding now and in the future (hope), persevering toward goals (optimism) and, when necessary, redirecting paths to goals in order to succeed, and when beset by problems and adversity, sustaining and bouncing back and even beyond to attain success (resiliency).

The concept of PsyCap in the individual level is aimed at promoting growth and individual performance, whereas in the level of an organization, PsyCap is aimed at encouraging organizations to reach their competitive advantage through the investment/development of employee performance (Luthans and Avolio, 2003: Luthans *et al*, 2006, in Avolio & Luthans, 2005). When an individual is constantly under pressure, he or she will think narrowly and know only one solution, experience loss of creativity, and find it hard to absorb new things. Positive emotions are necessary to ensure that individuals have an opportunity to think creatively. PsyCap is considered as a source of psychological importance because PsyCap can improve the work performance of an individual through positive cognition and processes that can motivate him or her (Luthans, Youssef & Avolio, 2007). A study by Ornek & Ayas (2015) also found that there is an indirect relationship between intellectual capital, innovative work behavior, and organizational performance. A study by Avey *et al.* (2008) also shows that innovative ideas can be performed by individuals that have a positive PsyCap. Psychological capital is also related to innovative work behavior (Jafri, 2012; Abbas & Raja 2015; Ziyae, Mobaraki & Saeediyoun 2015). To date, research that correlates work happiness with performance has been done only if the work happiness has a correlation with psychological capital (Choi & Lee 2013). Therefore, the hypothesis proposed by the researcher in this study was "PsyCap is mediating the relationship between work happiness and innovative work behavior".

2 METHOD

2.1 *Participants*

The participants were the employees of the headquarters of PT Bank Syariah X in Jakarta. It was chosen because they have an innovation program in regards to their marketing i.e. a program to gain customers. The criteria for the participants in this study were those that had worked for at least one year and with a permanent employment status.

The response rate in this study was 90%, meaning that out of 150 questionnaires distributed, 135 were filled in by the participants. Most of the participants were males (60.7%), aged approximately 25 to 45 years (79.3%), and mostly were at the level of office staff (77%). The majority of the participants (83.7%) had an undergraduate degree, worked during normal office hours (80.7%), and came from the accounting division (22.2%).

2.2 *Instruments*

2.2.1 *Innovative work behavior scale*

Innovative work behavior was measured using the scale developed by Janssen (2000) based on the definitions and scale proposed by Scott & Bruce (1994). The scale contains three items in each stage. It has a strong correlation with the three stages of innovative work behavior, and the items can be combined and be used as a uni-dimensional scale. Etikariena & Muluk (2014) have translated and adapted the innovative work behavior scale developed by Janssen (2000) into Indonesian language and have tested the adapted scale for reliability and validity among employee respondents in Indonesia. Based on these reasons, the researcher decided to use the Indonesian version of the scale of innovative work behavior by Etikariena & Muluk (2014). A reliability analysis with Cronbach Alpha for the innovative work behavior scale resulted in a coefficient of $\alpha = 0.93$.

2.2.2 *Work happiness scale*

Work happiness was measured using a scale that has been modified by Choi & Lee (2013) to make the instrument more specific in measuring work happiness. The word "work" was added into the items of the happiness at work scale by Fordyce (1988). For example, the first item "in general, how happy or unhappy do you usually feel?" was modified into "in general, how happy or unhappy do you usually feel at work?" This scale consists of two parts. In the first part, the scale consists of 11 levels from "Very very unhappy" to "Extremely unhappy". In the second part, the respondents were asked to fill in a percentage value of the time when they were happy at work. A reliability analysis using Cronbach Alpha resulted in a coefficient of $\alpha = 0.76$.

2.2.3 *PsyCap scale*

PsyCap in work environment was measured using the PsyCap questionnaire (PCQ-24), developed by Luthans *et al.* (2007) and consisting of six items containing the dimensions of hope, self-efficacy, resiliency and optimism. The responses were calculated by using a 6-point Likert scale, with 1 meaning "Disagree Strongly" and 6 meaning "Strongly Agree". A reliability analysis using Cronbach Alpha resulted in a coefficient of $\alpha = 0.87$.

3 RESULTS

The results from the Pearson Product Moment correlation analysis show that there is no significant correlation between work happiness and innovative work behavior ($r = 0.14$; $p > 0.12$). Therefore, the analysis was continued to the mediation analysis with the regression path analysis by Process from Hayes (2013). The result is presented in Table 1 below.

The results of the analysis showed no evidence that work happiness had a direct effect on innovative work behavior ($c' = -0.01$; $p = 0.81$). The results of the analysis performed

Table 1. The mediation analysis of psychological capital on the correlation between work happiness and innovative work behavior.

		Consequence						
		Psychological capital (M)				Innovative work behavior (Y)		
Variable		Coef.	SE	*p*		Coef.	SE	*p*
Work happiness	*a*	0.29	0.07	<0.00	c'	−0.01	0.05	0.81
PsyCapital		–	–	–	*b*	0.33	0.60	<0.00
Constant	i_1	90.75	4.6	<0.00	i_2	−4.35	6.66	0.52
		$R^2 = 0.13$				$R^2 = 0.19$		
		F(1; 134) = 133,00, p < 0.00				F(2; 133) = 132.00, p < 0.00		

using the ordinary least squares path analysis model 4, as seen in Table 1, show that when employees feel happy at work, then the happiness can make the employees have strong psychological capital (a = 0.29), which will lead to the emergence of innovative work behavior (b = 0.33). The results obtained from the bias-corrected bootstrap confidence intervals (lower level confidence interval/LLCI and upper level confidence interval/ULCI) for the effect of indirect relationships (ab = 0.09), performed using the bootstrap analysis (10,000 replications), are both above zero (0.05 to 0.16). It appears that psychological capital is proven to indirectly mediate the correlation between work happiness and innovative work behavior.

4 DISCUSSION

The results from this study show that there is no significant correlation between work happiness and work innovative behavior. This result supports the opinion of Spicer & Cederstörm (2015) that happiness is not automatically related to an increase in employee productivity, including innovative work behavior. Previous studies even found that job satisfaction, as one form of work happiness at work, has a negative correlation with productivity. Therefore, even when some experts believe that work happiness should have a positive influence on employee or organization performance, there are also many researchers who have spent decades neglecting the commonsense belief that "a happy worker is a productive worker." This shows that ensuring happiness among employees at work may not be wholly effective (Spicer & Coderstörm, 2015), including performing a productive behavior. Because previous studies have shown that the demand to be happy brings with it a heavy burden, a responsibility that can never be perfectly fulfilled. Essentially, when happiness becomes a duty, it can make people feel worse if they fail to accomplish it. A study by Forgan & East (2008) found that people who are in a good mood are worse at picking out acts of deception than those who are in a bad mood. Another research by van Kleef, De Dreu, & Manstead (2004) found that people who are angry during a negotiation achieved better outcomes than people who are happy. This suggests that being happy all the time may not be good for all aspects of our work or jobs that rely heavily on certain abilities. In fact, for some reason, happiness cannot always make us perform well enough, including in innovative work behavior.

Being happy at work may have different meanings for one and another (not exactly the same meaning to describe the target that someone has to achieve). There are many definitions of happiness, too, but none is neither less nor more logical than the others. Therefore, while there are still problems to develop more advanced techniques to measure emotions and predict behaviors, we have also adopted increasingly simplified notions of what it means to be human and what it means to have work happiness. Happiness is a convenient idea that looks good as an idea, but the answer as to how the happiness can bring positive advantages to perform specific behaviors at work is also hard to find.

We predict that happy workers are better workers. Positive messages about happiness are proven to be particularly popular in times of crisis and mass layoffs. Consequently, we have to redirect our expectation that work has to be happy, and we have to connect the happiness that we usually get from the really good things we experience with specific goals and focus, including work. In reality, happiness is a great thing to experience, but nothing can be willed into existence. Also, the less we seek to actively pursue happiness through our jobs, the more likely we will actually experience a sense of happiness at work. However, it is the most important that employees are better equipped to cope with work in a bad condition (Spicer & Coderstörm, 2015). The findings of this research have made it possible for further studies to continue analyzing the relation between work happiness and innovative work behavior by taking other variables into consideration and assuming the presence of a mediator variable, in this case, PsyCap.

Continued analysis found that there is a significant relationship between work happiness and work innovative behavior, with PsyCap as the mediator variable between the two variables. PsyCap as a positive psychological state helps to explain why work happiness does not have a direct correlation with innovative work behavior. As a specific goal, innovative work behavior needs employees' initiative and effort to solve problems at their work. Innovative work behavior is a way to find a new method in order to accomplish the task more effectively and efficiently. In the process, employees have to experience trial and error, so they need to remain optimistic, hopeful, confident of their capability and self-efficacy, and resilient (the ability to bounce back during failure). The positive psychology perspective helps us capture greater levels of workplace happiness. Positive psychology theorizes that we have the power to reframe our life experiences to help us become more positive and productive, including taking advantage from our happiness (Seligman, 2002; 2003).

Judge *et al.* (2001) found that job complexity is a significant moderator of the satisfaction-performance relationship, with a much stronger relationship in highly complex jobs. It can be seen that in the workplace, happiness is influenced by both short-lived events and chronic conditions in the task, job, and organization. Therefore, happiness is also influenced by stable attributes of individuals such as personality as well as the fit between what the job or organization provides and the individual's expectations, needs, and preferences. Understanding these contributors to happiness, together with recent research on self-determinant actions to improve happiness, offers some potential, and it is reasonable to think that improving happiness at work is a goal. Evidence suggests that "happy–productive worker hypothesis" may be more real than we thought. Research suggests that happiness at work is an essential ingredient for employees' psychological and physical health and work-life balance (Diener, 2000) and is related to problem-solving capability and task competence (Lyubomirsky *et al*, 2005). PsyCap is considered as an important psychological resource, which can improve employees' performance through its positive cognition and motivational processes (Luthans, Youssef & Avolio, 2007). In this study, these things are needed as the prerequisites to innovative work behavior.

Individuals may be happier than they believe, so they will perform better than usual. At the person level, meta-analytic evidence shows that happiness-related constructs such as job satisfaction, engagement, and affective commitment have important consequences for both individuals and organizations. In this study, happiness was measured related to core and contextual performance, which is innovative work behavior. When attitude measures are consistent in target and scope to behavior measures and when the attitudes in question are salient, stable, and have been formed based on personal experience, as is true of happiness at work, they can indeed predict behavior (Fischer, 2010), including innovative work behavior. This is why the relation between work happiness and innovative behavior in this study is significant when the happiness is directed as positive psychological state that makes employees have self-efficacy, optimism, positive hopes about their future, and the ability to bounce back when problems hit at work.

5 CONCLUSION

This study found that work happiness is not directly correlated with innovative work behavior. To take positive advantage of the happiness they feel, employees are expected to focus on the feeling of happiness as a psychological capital that will drive a sense of happiness, from which self-efficacy, hope, positivity, optimism and ability to bounce-back from problems emerge.

For future research, increasing the numbers of respondents is necessary for the results to be generalized to a bigger population of employees in Indonesia. On the other hand, studying this in different businesses, organizations, groups of respondents and countries or cultures might result in distinctly interesting outcomes. Taking into account other variables such as the kinds of tasks, for example creative and non-creative tasks or service and technical tasks will also be interesting.

REFERENCES

Åmo, B.W. & Lars, K. (2005) Organizational strategy, individual personality and innovation behavior. *Journal of Enterprising Culture,* 13 (1), 7–19. doi:10.1142/S0218495805000033.

Biswas-Diener, R., & Ben, D. (2007) *Positive Psychology Coaching: Putting The Science of Happiness to Work for Your Clients*. Hoboken, N.J., John Wiley and Sons.

Choi, Y., & Dongseop, L. (2014) Psychological Capital, big five traits, and employee outcomes. *Journal of Managerial Psychology,* 29 (2), 122–140. doi:10.1108/JMP-06-2012-0193.

Csikszentmihalyi, M., & Jeremy, H. (2003) Happiness in Everyday Life: The Uses of Experience Sampling. *Journal of Happiness Studies,* 4, 185–199. doi:10.1023/A:1024409732742.

Etikariena, A. & Hamdi, M. (2014) Hubungan antara Memori Organisasi dan Perilaku Inovatif di Tempat Kerja [The Relation Between Organization Memory and Innovative Behavior at Work]. *Makara Hubs-Asia,* 18 (2), 77–88.

Fisher, C.D. (2010) Happiness at Work. *International Journal of Management Reviews,* 12 (4), 384–412. doi:10.1111/j.1468-2370.2009.00270.x.

Fordyce, M.W. (1988) A Review of Research on the Happiness Measures: A Sixty Second Index of Happiness and Mental Health. *Social Indicators Research*, Available from: doi:10.1007/BF00302333.

Forgas, J.P., & Rebekah, E. (2008) On being happy and gullible: Mood effects on skepticism and the detection of deception. *Journal of Experimental Social Psychology,* 44 (5), 1362–1367. doi:10.1016/j.jesp.2008.04.010.

Janssen, O. (2000) Job demands, perceptions of effort—reward fairness and innovative work behaviour. *Joumal of Occupational and Organisational Psychology,* 73, 287–302. doi:10.1348/096317900167038.

Jex, S.M., & Thomas, W.B. (2008) *Organizational Psychology: A Scientist-practitioner Approach. 2nd* Ed. Publ. Wiley.

Jimenez-Jimenez, D., & Raquel, S.V. (2011) Innovation, organizational learning, and performance. *Journal of Business Research,* 64 (4), 408-17. doi:10.1016/j.jbusres.2010.09.010.

Luthans, F. (2002) The need for and meaming of positive organizational behavior. *Journal of Organizational Behavior,* 23, 695–706. doi:10.1002/job.165.

Luthans, F., Carolyn, M.Y., & Bruce, J.A. (2007) *Psychological Capital: Developing the Human Competitive Edge. Psychological Capital: Developing the Human Competitive Edge*. doi:10.1093/acprof:oso/9780195187526.001.0001.

Luthans, F., Kyle, W.L., & Brett, C.L. (2004) Positive Psychological capital: beyond human and social capital. *Business Horizons,* 47 (1), 45–50. doi:10.1016/j.bushor.2003.11.007.

Luthans, F., & Youssef, C.M. (2007) Emerging Positive organizational behavior. *Journal of Management,* 33 (3), 321–349. doi:10.1177/0149206307300814.

Luthans, F., Steven, M.N., Bruce, J.A., & James, B.A. (2008) The mediating role of psychological capital in the supportive organizational climate-employee performance relationship. *Journal of Organizational Behavior,* 29, 219–238. doi:10.1002/j.

Parzefall, M.R., Hannele, S., & Anneli, L. (2008) Employee innovativeness in organizations: A review of the antecedents. *Finnish Journal of Business Economics,* 2 (8), 165–182. Available from: http://lta.hse.fi/2008/2/lta_2008_02_a2.pdf.

Pryce-Jones, J. (2010) *Happiness at Work: Maximizing Your Psychological Capital For Success. Happiness at Work: Maximizing Your Psychological Capital For Success*. doi:10.1002/9780470666845.

Rego, A., Filipa, S., Carla, M., & Miguel, P.C. (2014) Hope and positive affect mediating the authentic leadership and creativity relationship. *Journal of Business Research,* 67 (2), 200–210. doi:10.1016/j.jbusres.2012.10.003.

Scott, S.G. & Reginald, A.B. (1994) Determinants of innovative behavior: a path model of individual innovation in the workplace. *Academy of Management Journal,* 37 (3), 580–607. doi:10.2307/256701.

Seligman, M.E.P. (2002) *Authentic Happiness*. New York, Free Press.

Spicer, A. & Carl, C. (2015) *The Research We've Ignored About Happiness at Work*. https://hbr.org/2015/07/the-research-weve-ignored-about-happiness-at-work [Accessed August 2016].

Van, K., Gerben, A., Astrid, C.H., Bianca, B., Daan, V.K., Christopher, O., Ilmo, V.L., Aleksandr, L.K., *et al.* (2008) The interpersonal effects of anger and happiness in negotiations. *Journal of Experimental Social Psychology,* 44 (3), 57–76. doi:10.1037/0022–3514.86.1.57.

West, M.A., & James, L.F. (1989) Innovation at work: Psychological perspectives. *Social Behaviour,* 4 (1), 15–30. doi:10.1021/ci970481e.

Diversity in Unity: Perspectives from Psychology and Behavioral Sciences – Ariyanto et al. (Eds)
 ISBN 978-1-138-62665-2

The role of work-life balance as a mediator between psychological climate and organizational commitment of lecturers in higher education institutions

V. Varias & A.N.L. Seniati
Faculty of Psychology, Universitas Indonesia, Depok, Indonesia

ABSTRACT: This study was conducted to find the role of work-life balance as a mediator between psychological climate and organizational commitment of lecturers in higher education institutions. The instruments used in the study were: (1) Organizational Commitment Scale (Meyer & Allen, 1997) developed by Seniati and Yulianto (2010); (2) Psychological Climate Scale (Kahn, 1990) developed by Brown and Leigh (1996); and (3) Work-life Balance Scale (Fisher, Bulger & Smith, 2009). The research was conducted on 328 lecturers from 11 higher education institutions in Jakarta, Tangerang, Padang, Denpasar, dan Jimbaran. The analysis method for this study was simple mediation test with Hayes' (2008) PROCESS macro. The results showed significant positive effects of psychological climate on work-life-balance ($\beta = 0.31$; $p < 0.05$), work-life balance on organizational commitment ($\beta = 0.21$; $p < 0.05$), psychological climate on organizational commitment ($\beta = 0.63$; $p < 0.05$), and work-life balance as a partial mediator in the relationship between psychological climate and organizational commitment of lecturers in higher education institutions ($\beta = 0.07$; $p < 0.05$). An implication of this study is that higher education institutions can enhance the level of work-life balance and organizational commitment by creating positive psychological climate.

1 INTRODUCTION

Organizational commitment is considered one of employees' work attitudes in the organization, more particularly referring to a characteristic of the relationship between members of the organization and the organization itself, which implicates the members' decision about whether or not to continue their membership in the organization (Allen & Meyer, 1990). Organizational commitment is an essential variable in organizations because it affects employee performance through means such as strengthening Organizational Citizenship Behavior (OCB), increasing the readiness to change the individual, determining job satisfaction, and eventually increasing the effectiveness of an organization (Madsen *et al*, 2005; Hanpachern, Morgan & Griego, 1998; Meyer & Allen, 1997; Bernerth *et al*, 2007; Meierhans, Rietmann & Jonas, 2008). Furthermore, organizational commitment can also reduce factors that could be detrimental to an organization such as the frequency of absenteeism, work stress, turnover, resistance to change, and counterproductive behavior (Wasti & Can, 2008; Luchak & Gellatly, 2007; Perryer *et al*, 2010).

Currently, research on organizational commitment is more often done in business organizations, and little is done in higher education institutions. Organizational commitment of lecturers in higher education institutions is important to study because it is relevant to the job performance of the lecturers whose scope of work is mostly based on the Three Pillars of Tertiary Education, which include education, research, and community service. The current research is intended to answer questions about organizational commitment, especially as it pertains to lecturers because the job of lecturers has unique characteristics and dynamics that distinguish it from other professions. For example, compared to other professions, lecturers have more freedom to manage how they accomplish their tasks and to make their own decisions. Moreover, a higher education institution has professional bureaucracy because as skilled educators, lecturers have a vital job (Mintzberg, 1993). Consequently, as professional workers, lecturers will have more employment alternatives to enhance their professional skills, which may cause their organization's commitment to be lower. Therefore, this current study is expected to provide some insight regarding the maintenance and enhancement of the commitment of lecturers in higher education institutions.

Allen & Meyer (1990) state that employees' commitment to the organization is based on three components: affective commitment, continuance commitment, and normative commitment. Affective commitment is an attitude shown by an employee of an organization, which is based on the positive emotional attachment towards the organization, as well as identification and involvement of the employee in the organization. Continuance commitment refers to an attitude shown by an employee of an organization, which is based on the evaluation of the impact of the economic costs and social costs that the employee would incur if he/she were to lose membership in the organization. Normative commitment is an attitude shown by an employee of an organization based on the extent to which he/she feels the responsibility and obligation to the organization to maintain his/her membership.

Allen & Meyer (1990) and Steers (1977) claim that factors that can influence organizational commitment consist of individual factors and environmental factors, such as personal characteristics, personal dispositions, organizational structure, job characteristics, and work experience. One of the factors influencing organizational commitment, according to Brown & Leigh (1996), is the environment of an organization. When employees feel that the positive environment of an organization is consistent with their values and interests, they are likely to invest and undertake greater efforts for the sake of the organization. The employees' perception of the environment of their organization is known as psychological climate. Kahn (1990) defines psychological climate as the perception and interpretation of employees regarding the environment of an organization (i.e., supportive management, clarity, self expression, perceived meaningful contribution, recognition from the organization, and job challenge). Employees see psychological climate as a factor that can psychologically influence employees to involve themselves fully in or to abstain from their job.

Positive work environment, especially one with flexible management, well-communicated norms, and provision of reward for employees for their contribution, has an important role in building the loyalty of lecturers. Parker *et al.* (2003) found that psychological climate has a significant relationship with job satisfaction, commitment, work engagement, motivation, and performance of employees. Their research suggests that the organizational commitment of employees is associated with psychological climate. Based on the findings of previous research, the researchers of the current study assume that the psychological climate in a higher education institution also contributes to the organizational commitment of lecturers, therefore formulating the hypothesis that:

Hypothesis 1: Psychological climate has a significant contribution to the organizational commitment of lectures in higher education institutions.

Seniati (2002) studied lecturers and found that the organizational commitment of lecturers is associated with the psychological climate of their workplace. It was found that psychological climate provides an indirect effect through job satisfaction in influencing the organizational commitment of lecturers. On the other hand, the research of Parker *et al.* (2003) found that psychological climate has a direct influence on organizational commitment. Therefore, we argue that there may be another variable that plays a role in the relationship between psychological climate and organizational commitment. In the present study, we argue that work-life balance may mediate the relationship between psychological climate and organizational commitment.

Fisher's (2001) statement, based on the work-life balance theory of Greenhaus & Beutell (1985), indicates that work-life balance is a person's perception of his or her ability to allocate stressors, and it consists of four components, namely time, behavior (e.g., goal accomplishment), strain, and the energy inside and outside of the workplace. In other words, the balance could be fulfilled when an individual thinks that he/she could accomplish his/her job without disturbing his/her personal life, and vice versa. Fisher (2001) and Fisher, Smith & Bulger (2009) categorize work-life balance into four dimensions that are related to the interferences and enhancements in the employees' personal lives, namely work interference with personal life, personal life interference with work, work enhancement of personal life, and personal life enhancement of work.

Additionally, a positive psychological climate could positively affect the work and personal life of employees; in fact, the more a person readily accepts the organizational climate, the higher the level of life balance that one feels (Aiswarya & Ramasundaram, 2012; Chernyak-Hai & Tziner, 2016). Psychological climate can bring a balance between work and personal lives. For example, the dimension of challenge in the psychological climate could improve the quality of a person's life and can facilitate a lecturer to improve his/her skills. Therefore, it can be said that psychological climate has an important impact on the work-life balance of lecturers. In the current study, the researchers assume that a positive psychological climate could help lecturers balance their work and personal lives, as stated in the following hypothesis:

Hypothesis 2: Psychological climate has a significant contribution to the work-life balance of lecturers in higher education institutions.

The balance between work life and personal life perceived by the lecturers also has a positive impact on the effectiveness of the organization where lectures work. Work-life balance is one of the factors that could affect the organizational commitment of employees (Kaiser *et al*, 2010; Kim, 2014). Casper, Harris, Bianco & Wayne (2011) state that work-life balance could bring a sense of attachment, loyalty, and the rise of organizational commitment. When employees feel that their job does not interfere with their personal life, and the job enhances their quality of life, they tend to be more committed to the organization. The relationship between work-life balance and organizational commitment can also be rationalized by assuming that when an employee perceives no balance between their work life and personal life and considers too much of their time is spent on the job, then the employee will contemplate finding an alternative job that allows them to balance their work and personal lives (Posig & Kickul, 2004). Therefore, the following hypothesis is proposed:

Hypothesis 3: Work-life balance has a positive effect on the organizational commitment of lecturers in higher education institutions.

Positive perception about work-life balance could bring forth the commitment of lecturers when they assume that their organization supports their work-life balance. The employees would feel a sense of responsibility and obligation to the organization that compels them to be loyal and retain their membership. If the employees feel that their expectations and needs are met and that there is balance between their work and personal lives, then there will be a sense of loyalty, hence, the rise of their organizational commitment (Casper *et al*, 2011). It can be said that positive psychological climate will result in good work-life balance which, in the end, will result in high organizational commitment. Based on that explanation, the researchers assume that work-life balance can mediate the relationship between psychological climate and organizational commitment of lecturers, as stated in the hypothesis:

Hypothesis 4: Work-life balance is a mediator between psychological climate and organizational commitment of lecturers in higher education institutions.

2 METHODS

2.1 *Participants and procedure*

Respondents for this study were comprised of lecturers from 11 higher education institutions in Jakarta, Tangerang, Padang, Denpasar, and Jimbaran. The respondents had the following characteristics: (1) employed as lecturers at their university for at least three years, (2) held positions that ranged from assistant to professor, (3) had either civil servant or non-civil servant statuses. A total of 543 questionnaires were distributed in this, study, of which 420 were returned and 328 were eligible for analysis. In this study, the necessary data were collected through questionnaires distributed to the research subjects after receiving permission from the higher education institutions as their employers. The highest age range of the respondents was 45–65 years old. The method of the data analysis used in this study was a simple mediation test with PROCESS by Hayes (2008).

2.2 *Measurement*

The instrument used for measuring organizational commitment was obtained from Meyer & Allen (1997), and it consisted 21 items to measure three components: affective commitment, normative commitment, and continuance commitment. The instrument had previously been adapted and used in a research on lecturers by Seniati & Yulianto (2010). For each item, the participants had to respond by choosing one of four response choices that ranged from 1 (strongly disagree) to 4 (strongly agree). The internal consistency of the instrument was 0.883. A sample item was the statement, "I am happy to develop my career in this higher education institution."

Psychological climate was measured using an instrument developed by Kahn (1990), which had since been modified by Brown & Leigh (1996). The instrument contained 21statements assessing six dimensions, which included supportive management, role clarity, contribution, recognition, self expression, and challenge. Each statement had to be responded to by selecting a range of options from 1 (strongly disagree) to 4 (strongly agree). the internal consistency was found at 0.890. An example of an item was, "This higher education institution recognizes the importance of the contribution that I gave."

Finally, work-life balance was measured using an instrument developed by Fisher, Smith & Bulger (2009). The instrument consisted 17 items that measured four dimensions of work-life balance, namely Work Interference with Personal Life (WIPL), Personal Life Interference with Work (PLIW), Work Enhancement of Personal Life (WEPL), and Personal Life Enhancement of Work (PLEW). The responses for the items ranged from 1 (never at all) to 4 (almost every time). The instrument's internal consistency was found to be 0.817. A sample item stated, "My personal life is energizing for me to do the job."

3 RESULTS

Before performing the mediation test based on Hayes' (2008) model, the researchers felt the need to perform correlation tests to examine the correlations among variables. Based on Allen & Meyer's (1990) and Steers' (1977) studies, which stated that gender, education, length of employment, and marital status impact organizational commitment, these demographic data in the current research were consequently controlled. Significant demographic data differences in organizational commitment were then included as the control variables in the mediation test by Hayes (2008). Table 1 shows the results of the correlation tests.

Table 1. Correlation test results.

Variable	Mean	SD	1	2	3	4	5	6	7
Gender	1.55	0.50	NA						
Length of employment	2.70	0.46	–0.08	NA					
Education	2.29	0.45	–0.07	0.20**	NA				
Marital status	1.96	0.27	–0.09	0.15**	–0.03	NA			
Psychological climate	2.94	0.34	–0.09	0.11*	0.10	0.08	NA		
Work-life balance	3.03	0.33	–0.04	0.08	0.01	0.06	0.40**	NA	
Organization commitment	2.97	0.40	–0.17**	0.23**	0.09	0.08	0.55**	0.35**	NA

* = $p < 0.05$; ** = $p < 0.01$; N = 328

The demographic data that were found to be significantly correlated with organizational commitment were gender ($r = -0.17, p < 0.01$) and length of service ($r = 0.23, p < 0.01$). As a result, the two significant demographic data were included as control variables in the mediation test.

Recalling that the current research was conducted to examine the role of work-life balance as a mediator in the relationship between psychological climate and organizational commitment of lecturers, the researchers used SPSS with add-on macro PROCESS developed by Hayes (2008). In performing the mediation test, the researchers also input data on the control variables, namely gender and length of service, both of which had previously been shown to have significant correlations with organizational commitment. The results of the mediation test are illustrated in Table 2 below.

In the test, by first inputting the control data from gender and length of service, it was shown that both psychological climate and work-life balance contribute to an approximate of 36% ($R^2 = 0.36$) of the variance in organizational commitment. Furthermore, psychological climate had a significant effect on organizational commitment, with $\beta = 0.63$, $p = 0.00$, CI [0.52, 0.73]. This result provided convincing support for the first hypothesis, which stated that psychological climate has a significant positive effect on organizational commitment for lecturers in higher education institutions.

On the other hand, psychological climate contributes to 16% ($R^2 = 0.16$) of the variance in work-life balance. In addition, psychological climate has a significant effect on work-life balance, with $\beta = 0.31$, $p = 0.00$, CI [0.23, 0.40]. This result supported for the second hypothesis, which predicted that psychological climate has a significant positive effect on work-life balance for lecturers in higher education institutions. Work-life balance was similarly found to significantly affect organizational commitment, with $\beta = 0.21$, $p = 0.00$, CI [0.07, 0.36]. As with the previous two hypotheses, the third hypothesis, which claimed that work-life balance affects a lecturer's organizational commitment in higher education institutions, was also supported.

The result of the mediation test examining the mediating role of work-life balance in the relationship between psychological condition and organizational commitment can be seen from the total effect, direct effect, and indirect effect obtained from Hayes' (2008) PROCESS as illustrated in Table 3.

Table 2. Mediation test result.

	Outcome					
	Work-life balance			Organizational commitment		
Antecedent	B	SE	P	β	SE	p
Psychological climate	0.31	0.04	0.00	0.63	0.05	0.00
Work-life balance				0.21	0.07	0.00
Gender	0.04	0.57	0.94	−1.84	0.76	0.02
Length of service	0.46	0.62	0.46	2.90	0.83	0.00
	$R^2 = 0.16$			$R^2 = 0.36$		
	$P = 0.00$			$P = 0.00$		

Table 3. The result of total effect, direct effect and indirect effect psychological climate to organizational commitment.

	Effect	LLCI	ULCI
Total effect	0.63	0.52	0.73
Direct effect	0.56	0.45	0.68
Indirect effect	0.07	0.02	0.13

The mediating effect of work-life balance on the relationship between psychological climate and organizational commitment was found to have a total effect score of 63% (Effect = 0.63, $p < 0.05$). The direct effect of psychological climate on organizational commitment, on the other hand, was found at 56% (effect = 0.56, $p < 0.05$) and a CI below 0. The value of the indirect effect of psychological climate on organizational commitment was 7% (effect = 0.07) with a CI below 0. These results showed that the work-life balance of lecturers was a significant mediator in the relationship between psychological climate and organizational commitment, therefore providing support for the fourth hypothesis of the study. In addition, as the direct effect had a larger effect value compared to the indirect effect, this implies that psychological climate had a greater direct influence (i.e., 56%) on organizational commitment. In other words, work-life balance functioned as a partial mediator because upon testing, psychological climate was still shown to have a significant direct effect on organizational commitment even when work-life balance was controlled.

4 DISCUSSION

The present study extends the previous studies on the relationship between psychological climate and organizational climate (Parker *et al*, 2003), concluding that psychological climate has a direct effect on the lecturers' organizational commitment. The findings of the current study also show that the psychological climate factor that has been found to affect organizational commitment in employees is also found among lecturers working in higher education organizations. This suggests that theories resulted from research conducted in a business organization setting may be applied to higher education settings as well. In addition, judging from the average scores of the various aspects of psychological climate, the aspect of supportive management was found to have the lowest average score, implying that higher education institutions can improve on being more attentive in providing a supportive and flexible management environment for their employees.

The results of the present research additionally show that work-life balance plays a partial yet significant role in mediating the relationship between psychological climate and organizational commitment in lecturers. This shows that the lecturers' psychological climate can affect their organizational commitment by mediating it with the balance between their work and private lives. As Casper, Harris, Bianco & Wayne (2011) state, when someone feels their work-life balance is fulfilled by a supportive environment, then attachment, loyalty, and commitment to the organization will rise. As a result, it supports that a positive psychological environment for lecturers can help form a work-life balance that can ultimately shape lecturers' organizational commitment to the higher education institution where they work.

In this research, work-life balance was found to have a small effect (7%) in mediating the relationship between psychological climate and organizational commitment. Such a result can possibly be attributed to several interrelated factors that characterize the work of lecturers. First is their flexible working time. Even with a job that is quite complex, the flexible working time of lecturers enables them to work in their own pace without any pressure from anyone. Secondly, lecturers have the luxury of managing and deciding how and when they work, reducing the pressure on how they complete their work. Thirdly, their work is colleague-oriented, whereby fellow lecturers can help and work with each other to finish their jobs, meaning that the possibility of inner conflicts is smaller when compared to the possibility of conflicts among employees in a business-oriented company. These factors are possibly the reason why the variables in the lecturers' work-life balance matter little in mediating the connection between psychological climate and organizational commitment. In the future, further research may want to investigate whether work-life balance plays a similar role as a mediator between psychological climate and organizational commitment in other jobs in the business world.

Future research can also be conducted to compare work-life balance between male and female lecturers. The role of gender in work-life balance may prove to be an interesting topic considering that women outnumber men in the profession of university teaching. A woman typically also has many additional roles and duties at home within her family (Duxbury & Higgins, 2001). Therefore, future researches can delve deeper into the female lecturers' work-life balance especially by studying those who do not only work as lecturers, but who also have responsibilities in their home and family. To obtain a more complete picture of the issue, in-depth interviews can ideally be used to assess the work-life balance of female lecturers with both their work and family responsibilities.

Pertaining to the profession of university teaching, further research can examine the lecturers' commitment to the profession and how their professional commitment relates to organizational commitment. The topic needs to be studied to better understand the process that occurs in dual-commitment, or how the components of organizational and professional commitment affect each other in creating specific outcomes, especially with lecturers in higher education institutions. Professional commitment can also be tested for its mediating effect on variables other than work-life balance, such as on the relationship between psychological climate and organizational commitment among lecturers.

5 CONCLUSION

Based on the results of the current research, the conclusions that can be drawn are as follow: (1) Psychological climate has a significant positive effect on organizational commitment; (2) Psychological climate has a significant positive effect on work-life balance; (3) Work-life balance has a significant positive effect of organizational commitment; (4) Work-life balance has a role in mediating the relationship between psychological climate and organizational commitment of lecturers in higher education institutions although the mediating effect is partial and relatively small.

REFERENCES

Aiswarya, B. & Ramasundaram, G. (2012) A study on interference of work– life conflict between organisational climate and job satisfaction of women employees in the information technology sector. *Asia-Pacific Journal of Management Research and Innovation,* 8 (3), 351–60. doi:10.1177/2319510X1200800315.

Allen, N.J. & John, P.M. (1990) The measurement and antecedents of affective, continuance and normative commitment to the organization. *Journal of Occupational Psychology,* 63 (1), 1–18. doi:10.1111/j.2044-8325.1990.tb00506.x.

Bernerth, J.B., Armenakis, A.A., Field, H.S. & Walker, H.J. (2007) Justice, Cynicism, and commitment: A study of important organizational change variables. *The Journal of Applied Behavioral Science,* 43 (3), 303–26. doi:10.1177/0021886306296602.

Brown, S.P. & Thomas, W.L. (1996) A new look at psychological climate and its relationship to job involvement, effort, and performance. *Journal of Applied Psychology,* 81 (4), 358–68. doi:10.1037/0021-9010.81.4.358.

Casper, W.J., Christopher, H., Amy, T.B. & Julie, H.W. (2011) Work-family conflict, perceived supervisor support and organizational commitment among Brazilian professionals. *Journal of Vocational Behavior,* 79 (3), 640–52. doi:10.1016/j.jvb.2011.04.011.

Chernyak-Hai, L. & Aharon, T. (2016) The 'I Believe' and the 'I Invest' of work-family balance: The indirect influences of personal values and work engagement via Perceived Organizational Climate and Workplace Burnout. *Revista de Psicologia Del Trabajo Y de Las Organizaciones,* 32 (1), 1–10. doi:10.1016/j.rpto.2015.11.004.

Duxbury, L. & Chris, H. (2001) Work-life balance in the new millennium: Where are we? Where do we need to go? In *Canadian Policy Research Networks*, 1–8.

Fisher, G.G. (2001) Work/Personal Life Balance: A Construct Development Study." Ph.D., United States. Ohio: Bowling Green State University. http://remote-lib.ui.ac.id:2073/docview/304682757/abstract/3D8E9001665C42DFPQ/1.

Fisher, G.G., Carrie, A.B. & Carlla, S.S. (2009) Beyond work and family: A measure of work/non-work interference and enhancement. *Journal of Occupational Health Psychology,* 14 (4), 441–56. doi:10.1037/a0016737.

Greenhaus, J.H. & Nicholas, J.B. (1985) Sources of conflict between work and family roles. *The Academy of Management Review,* 10 (1), 76–88. doi:10.5465/AMR.1985.4277352.

Hanpachern, C., George, A.M. & Orlando, V.G. (1998) An extension of the theory of margin: a framework for assessing readiness for change. *Human Resource Development Quarterly,* 9 (4), 339–50. Available from: doi:10.1108/eb050773.

Hayes, A.F. (2013) Introduction to mediation, moderation, and conditional process analysis. *New York, NY, Guilford,* 3–4. doi:978-1-60918-230-4.

Kahn, W.A. (1990) Psychological conditions of personal engagement and disengagement at work. *The Academy of Management Journal,* 33 (4), 692. doi:10.2307/256287.

Kaiser, S., Max, R., Cornelia, U.R. & Martin, L.S. (2010) The Impact of corporate work-life balance initiatives on employee commitment: An empirical investigation in the German consultancy sector. *Zeitschrift Fur Personalforschung,* 24, 231–65. doi:10.1688/1862-0000_ZfP_2010_03_Kaiser.

Kim, H.K. (2014) Work-life balance and employees' performance: The mediating role of affective commitment. *Global Business and Management Research: An International Journal,* 6 (1), 37–51.

Luchak, A.A. & Ian, R.G. (2007) A comparison of linear and nonlinear relations between organizational commitment and work outcomes. *Journal of Applied Psychology,* 92 (3), 786–93. doi:10.1037/0021-9010.92.3.786.

Madsen, S.R., Susan, R.M., Duane, M., Duane, M., Cameron, R.J. & Cameron, R.J. (2005) Readiness for Organizational change: Do organizational commitment and social relationships in the workplace make a difference. *Human Resource Development Quarterly,* 16 (2), 213–33. doi:10.1002/hrdq.1134.

Meierhans, D., Brigitte, R. & Klaus, J. (2008) Influence of fair and supportive leadership behavior on commitment and organizational citizenship behavior. *Swiss Journal of Psychology,* 67 (3), 131–41. doi:10.1024/1421-0185.67.3.131.

Mayer, J.P. & Natalie, J.A. (1997) *Commitment in the Workplace: Theory, Research, and Application.* 2455 Teller Road, Thousand Oaks California 91320 United States, SAGE Publications, Inc. http://sk.sagepub.com/books/commitment-in-the-workplace.

Mintzberg, H. (1993) Structure in fives-designing effective organizations. *Prentice-Hall International,* 317. doi:10.1017/CBO9781107415324.004.

Parker, C., Boris, B., Scott, Y., Joseph, H., Robert, A., Healther, L. & Julia, R. (2003) Relationship between pscyhological climate perceptions and work outcomes: A meta-analytic review. *Journal of Organizational Behavior,* 24, 389–416. doi:10.1002/job.198.

Perryer, C., Catherine, J., Ian, F. & Antonio, T. (2010) Predicting turnover intentions. *Management Research Review,* 33 (9), 911–23. doi:10.1108/01409171011070323.

Posig, M. & Jill, K. (2004) Work-role expectations and work family conflict: Gender differences in emotional exhaustion. *Women in Management Review,* 19 (7), 373–86. doi:10.1108/09649420410563430.

Seniati, A.N.L. & Yulianto, A. (2010) Pengaruh Faktor Pribadi dan Faktor Lingkungan terhadap Komitmen Profesi dan Komitmen Organisasi pada Dosen Perguruan Tinggi di Jakarta. Hasil Riset Unggulan. Universitas Indonesia, Depok, Indonesia.

Seniati, L. (2010) Pengaruh masa kerja, trait kepribadian, kepuasan kerja, dan iklim psikologis terhadap komitmen dosen pada Universitas Indonesia. *Makara Hubs-Asia,* 8 (3). http://hubsasia.ui.ac.id/index.php/hubsasia/article/view/33.

Steers, R.M. (1977) Antecedents and outcomes of organizational commitment. *Administrative Science Quarterly,* 22 (1), 46–56. doi:10.2307/2391745.

Wasti, S.A. & Özge, C. (2008) Affective and normative commitment to organization, supervisor, and coworkers: Do collectivist values matter? *Journal of Vocational Behavior,* 73 (3), 404–13. doi:10.1016/j.jvb.2008.08.003.

Diversity in Unity: Perspectives from Psychology and Behavioral Sciences – Ariyanto et al. (Eds)
© 2018 Taylor & Francis Group, London, ISBN 978-1-138-62665-2

The role of job embeddedness as a mediator in the relationship between job demand resources and turnover intentions

T.A.P. Atan & D.E. Purba
Faculty of Psychology, Universitas Indonesia, Depok, Indonesia

ABSTRACT: This study aims to investigate the mediating effect of job embeddedness on the relationship between Job Demand Resources (JD-R) and turnover intentions. Using the Conservation Of Resource theory (COR) in explaining the mediation effect, employees with high level of job resources (growth opportunities, career development, social network, and job security) are more likely to stay in the organization and have low scores in turnover intentions. The data were collected from a media organization in Jakarta (N = 210), and the mediation effect was analyzed using Hayes' PROCESS macro. Results show that JD-R negatively influences the level of employees' turnover intentions and positively influences job embeddedness. In addition, job embeddedness negatively influences turnover intentions. The mediation analysis shows job embeddedness has a partial mediation effect on the relationship between JD-R and turnover intentions. The indirect effect of JD-R on turnover intentions through job embeddedness is significant. Theoretical and practical implications are discussed. Based on the result, it is concluded that job embeddedness can mediate the relationship between JD-R and turnover intention in an online media organization.

1 INTRODUCTION

Every organization needs to have high-performing employees with low level of turnover intention (intention to leave the organization) in order to compete with other organizations (Kaur, Mohindru & Pankaj, 2013; Liu & Hu, 2010). High turnover in an organization can result in high recruitment time and costs, as well as a decrease in productivity, profitability, knowledge, skills, and organizational skills (Butali, Wesang & Mamuli, 2013; Giosan, 2003; Sinatra, 2015). Turnover intention is defined as employees' thoughtful and conscious desire to leave the organization (Tett & Meyer, 1993), and the main predictor of turnover behavior (Sousa-Poza & Henneberger, 2002). Turnover intention can be affected by many factors such as demographic factors (i.e., age, gender, job tenure, position, and level of education), work attitudes (i.e., organizational commitment, job hopping, work engagement, Organizational Citizenship Behavior or OCB), job satisfaction, and working conditions (i.e., job demand resources, pressure of work, fairness of procedures, time worked, benefits from the organization, job security, advancement, and labor-market opportunities) (Bakker & Demerouti, 2007; Bothma & Roodth, 2013; Foon, Chee-Leong & Osman, 2010; Griffeth, Hom & Gaertner, 2000).

The present study was conducted in an online media organization with a high workload due to demands for speedy news reporting and short work deadlines. According to Moss (in Chang, 1998), employees who work as reporters in media organizations are known to have a high level of turnover compared to other professions. This study focuses on the job demands and job resources (known as the Job Demand Resources (JD-R) theory (Bakker, Demerouti, Tarris, Schaufeli & Schreurs, 2003) as predictors of turnover intentions. JD-R is an one-dimensional variable wherein a low score in the JD-R indicates that the employees perceive the job to have high loads, while a high score indicates that employees perceive the job offers a lot of job resources (Bakker *et al.* 2003; Demerouti & Bakker 2011). According to Bakker *et al.* (2003), job demand consists of work overload, defined as the amount of work, mental and emotional overload, and job resources consist of growth opportunities, advancement, organizational support, and job security. Growth opportunities refer to the variety of job-tasks offered to the employees, such as the opportunity to learn new things, and to be able to complete the tasks independently. Advancement is defined as progress or improvement in employment. Job security is a level of security concerning one's future career in the organization and the sustainability of employment and occupation in an organization. Lastly, organizational support is comprised of the relationships with the supervisor, the availability of information, communication, and participation in the organization, social support given by colleagues, and contact opportunities within the organization (Jackson & Rothmann, 2005).

The current research drew on the Conservation Of Resource (COR) theory (Hobfoll, 1989) to explain the relationship between JD-R and turnover intention. The COR theory explains that people tend to search, maintain, and preserve their personal resources to protect themselves from stressful situations (Hobfoll, 1989). Based on this theory, we argue that high level of job resources (such as the opportunity to grow, career advancement, relationship with people in the organization, and organizational support) may lead to a low level of turnover intention because individuals with abundant job resources tend to protect and maintain these resources by staying in the organization. Conversely, high level job demand (such as work overloads) may lead to high level of turnover intentions in order to reduce stressful situations. Previous empirical research showed a negative relationship between JD-R and turnover variables (Bakker *et al.* 2003; Shahpouri, Namdari & Abedi, 2015; Shaufeli & Bakker, 2004) Thus, we hypothesize the following,

Hypothesis 1: JD-R will be negatively related to turnover intentions.

However, previous research found the correlations between JD-R and turnover intention were relatively small, with coefficients below 0.30 (e.g., Bakker *et al.* 2003; Shahpouri, Namdari & Abedi, 2015; Shaufeli & Bakker, 2004). Furthermore, previous research found that the indirect effects of Job Demand Resources and turnover intention were established through other variables, such as job embeddedness (Bergiel, Nguyen, Cleney & Taylor, 2009). Although Bergiel *et al.* (2009) did not specifically mention job demands and job resources in their study variables, human resource practices (such as compensation, supervisor support, growth opportunities, and training) may imply job resources needed in the employees' job. In the present study, borrowing Bergiel *et al.*'s (2009) logic, we suspect that the relationship between JD-R and turnover intention may also be mediated by job embeddedness.

Job embeddedness is defined as factors that influence the employees' decision to stay in the organization (Mitchell *et al.* 2001). Three dimensions of job embeddedness are fit, links, and sacrifice (Mitchell *et al.* 2001). Fit is an employee's perception of compatibility with an organization and also with the working environment (Mitchell *et al.* 2001). Links are formal and informal relationships that the employees have with the people in the organization. Sacrifice shows the perceived cost of psychological and material benefits that the employees would experience if they decide to leave the organization (Mitchell *et al.* 2001). There are two foci of job embeddedness, namely off-the-job embeddedness (or organizational embeddedness), and off-the-job embeddedness (or community embeddedness). For the present study, we focus on work-related factors, since JD-R is one of the on-the-job aspects that is more likely to relate to the on-the job embeddedness than to the off-the-job embeddedness (Purba, Oostrom, Born, & Van der Molen, 2016; Kiazad, Hom, Holtom & Newman, 2015).

To our knowledge, there are only a few empirical studies conducted on the relationship between JD-R and job embeddedness (e.g., Bergiel *et al.* 2009), and no study has ever been conducted in Indonesia yet. Bergiel *et al.* (2009) studied human resource practices and its influence on turnover intention, and part of human resource practices included was job resources. We argue that employees' perception of JD-R affects their level of job embeddedness. With an abundance of job resources available to employees, such as higher pay, better promotion, job security and good working conditions, and a low level of work overload, they will experience fit and feel comfortable working in the organization, develop good and healthy relationships with colleagues and supervisor, thus unwilling to sacrifice the resources if they have to leave the organization. Therefore, we hypothesize the following:

Hypothesis 2: JD-R will be positively related to job embeddedness.

We again used the COR theory to explain the mediating effect of job embeddedness in the relationship between JD-R and turnover intention. Previous research found that job embeddedness explained additional variance in turnover variables, even after controlling other variables, such as job satisfaction, organizational commitment, job search behavior, and perceived job alternatives (Crossley, Benner, Jex & Burnfield 2007; Mitchell *et al.* 2001; Ramesh & Gelfand, 2010). Hence, we conclude that job embeddedness is the proximal variable of the turnover variable compared to other predictors of the turnover variable. Employing the COR theory, individuals who perceive that their job provides more resources (e.g., the opportunity to grow, career advancement, relationships with people in the organization, and support from the organization as well as the supervisor) tend to experience fit with the organization, to develop more relationships with other coworkers, and to feel unwilling to let go of the resources if they have to leave the organization. This then leads to lower level of turnover intention. Thus, we hypothesize the following,

Hypothesis 3: Job embeddedness will be negatively related to turnover intention.
Hypothesis 4: Job embeddedness mediates the relationship between JD-R and turnover intention.

2 METHODS

2.1 *Participants and procedures*

This study was conducted in an online media organization (organization X). The participants were in early adulthood (20–40 years old) with a minimum length of service of six months. In general, an online media organization has a considerably high workload because of the demands of speedy reporting and short work deadlines. Frequently, employees have to work overtime until midnight, even during religious and national holidays due to the need to continually deliver news to the public.

The majority of the study participants were male (136 employees, 65%). The mean age of the participants was 27.61 (SD = 4.07). With regard to their length of work (job tenure), 87 employees (41.4%) had been employed for 13 to 24 months. The majority of the respondents (184 employees, 88%) have a bachelor's degree (S1).

2.2 *Measurements*

All scales were translated and back translated from English to Bahasa Indonesia by two bilingual organizational psychologists. The turnover intention was measured using 3 items of Mobley's (1978) scale. An item example is "I will leave this organization" ($\alpha = 0.82$).

The JD-R was measured using the 42-item JDRS scale developed by Rothman *et al.* (2006). The scale measures two dimensions, namely job demand and job resources. An item sample for job demand was "Do you have too much work to do? While an item sample of job resources was "Do you think that your organization pays good salaries? The alpha coefficient for job demand is ($\alpha = 0.81$), for job resources is ($\alpha = 0.89$), and the JD-R total is ($\alpha = 0.84$).

Job embeddedness was measured using the 13-item job embeddedness scale (Purba, 2015). The scale reliabilities for each dimension ranged from $\alpha = 0.62$ to $\alpha = 0.81$, with a sample item, "I feel like I am a good match for this organization".

3 RESULTS

The results from the correlation analyses showed that age and job tenure were not significantly correlated with turnover intention (r = 0.02, p > 0.05 and r = 0.06, p > 0.05, respectively). JD-R had a significant negative correlation with turnover intention (r = –0.49, p < 0.01) and a positive correlation with job embeddedness (r = 0.33, p < 0.01). Furthermore, job embeddedness was significantly negatively correlated with turnover intention (r = –0.29, p < 0.01). Because the correlations among the three variables were significant, we conducted further analyses to test the hypotheses.

The hypotheses were analyzed using Hayes' PROCESS macro on the SPSS software. Our confidence intervals are based on the bias corrected method with 5,000 bootstrap samples. Hypotheses 1, 2, and 3 were tested using the path coefficients shown on Figure 1. Hypothesis 1stated that JD-R will be negatively related to turnover intention. Our results show that JD-R was negatively related to turnover intention ($b = -0.68$, $SE = 0.10$, $p < 0.001$). Thus, Hypothesis 1 was supported. Hypothesis 2 posited that JD-R is positively related to job embeddedness. Our results in Figure 1 show that there is a positive relationship between JD-R and job embeddedness ($b = 0.26$, $SE = 0.05$, $p < 0.001$), supporting our Hypothesis 2. Hypothesis 3 posited that job embeddedness will be negatively related to turnover intention. Figure 1 shows job embeddedness to be negatively related to turnover intentions ($b = -0.26$, $SE = 0.13$ $p < 0.05$). Thus, Hypothesis 3 was supported. Lastly, Hypothesis 4 stated that job embeddedness mediates the relationship between JD-R and turnover intention. As shown in Table 2, the indirect effect of JD-R to turnover intentions through job embeddedness

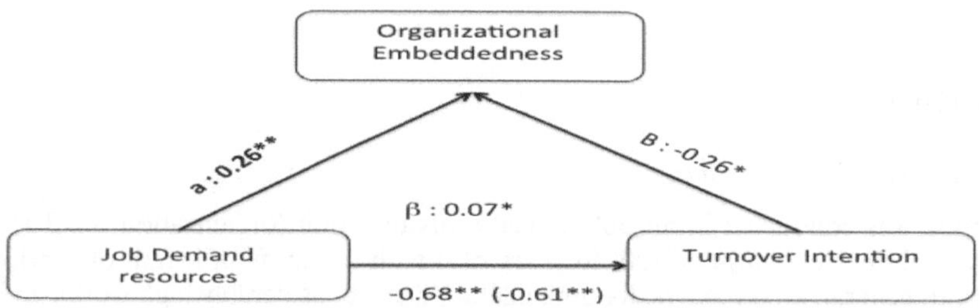

Figure 1. Mediation model.

Table 1. *Means, standard deviations,* and correlations between study variables.

Variables	Mean	SD	1	2	3	4	5
1. Age	27.61	4.07	–				
2. Job-Tenure	18.36	11.87	0.218**	–			
3. JDR	0.341	0.62	–0.11	–0.12	0.84		
4. JE	3.38	0.49	–0.12	0.03	0.330**	0.805	
5. TOI	2.80	0.85	0.02	0.06	–0.49**	–0.29**	0.875

Note: JE = Job Embeddedness. TOI = Turnover Intention. *JD-R = Job Demand-Resources. N* = 210, Age is measured in years. Job tenure is measured in month. Numbers on the diagonal are Cronbach's Alpha of each measuring instrument.

** p < 0.01, * p < 0.05.

Table 2. Direct and indirect effect JD-R with turnover intention.

	Indirect effect of X on Y				Direct effect of X on Y			
	Effect	Boot SE	BootLLCI	BootULCI	Effect	Boot SE	BootLLCI	BootULCI
JE	–0.07	0.04	–0.16	–0.01	–0.61	0.11	–0.83	–0.39

is significant (*indirect effect* = –0.07, *Boot SE* = 0.04, *95% CI* [–0.16, –0.01]), supporting Hypothesis 4.

4 DISCUSSION

The main purpose of this study is to test the mediating effect of job embeddedness in the relationship between JD-R and turnover intention. The results support all hypotheses, confirming previous studies that job resources were negatively related with turnover intention (Bakker *et al.* 2003; Shahpouri *et al.* 2015; Shaufeli & Bakker 2004), job resources were positively related with job embeddedness (Bergiel *et al.* 2009), and job embeddedness was negatively related to turnover intention (Crossley *et al.* 2007; Mitchell *et al.* 2001; Ramesh & Gelfand 2010). With regard to Hypothesis 4 that posited job embeddedness as the mediator in the relationship between JD-R and turnover intention, the analysis additionally shows a decrease in the direct effect of JD-R. This direct effect is significant and demonstrates a partial mediation effect of job embeddedness on the relationship between JD-R and turnover intention. The indirect effect of JD-R on turnover intention as mediated by job embeddedness is also statistically significant. To our knowledge, the present study is among the first to look at job embeddedness as the mediator in the JD-R and turnover intentions relationship.

Several limitations of this study need to be addressed. First, this study was conducted only in one online media organization, implying that the research results might not be generalized to the population of all online media organizations in Jakarta. Second, the cross-sectional design of the study limits our ability to confirm the causal relationship between variables. Lastly, the study employed self-report survey that may lead to common method bias. To overcome these limitations, we propose several methodological suggestions for further research. First, future research can be conducted by surveying employees in various online organizations, so that the research results can be generalized to the entire population of employees working at online media organizations. Second, further research can perhaps use a longitudinal research design in which data are collected intermittently over time from the same group of people so as to confirm the causal relationships between study variables. Third, future research may use other ratings (such as coworkers' ratings or supervisor ratings) to assess JD-R to reduce the possibility of common method bias. Finally, other research can use demographic variables (gender, education level, and income level) as control variables, especially considering that previous studies (e.g., Khatri *et al.* 2013, and Sousa-Poza & Henneberger, 2002) have found that gender, education level, and income level can affect the intention to leave an organization.

We also offer some practical implications for the organizations concerned. First, the organization should attend to its employees' level of job embeddedness. This can be accomplished by providing employees with feedback on a regular basis, offering them a variety of incentives, holding team building events (e.g., movie nights) to improve social relations within the organization, as well as offering the coaching and training of specific skills to help employees become more familiar with their roles (job tasks) in the workplace. Secondly, organizations should take into consideration labor resources such as the certainty of the employees' permanent employment status, the employees' opportunity to learn new things, and the organization's willingness to listen to the opinion of employees in decision-making situations. Furthermore, organizations could help ease the workload within the organization, such as by extending the deadlines imposed on employees and setting the frequency of administration tasks to employees, in order to foster the employees' perception of their attachment to the organization and lower their intentions to leave the organization.

5 CONCLUSION

It can be concluded that job embeddedness exerts a partial mediation effect on the relationship between JD-R and turnover intention. More particularly, the current study discovers that the relationship between JD-R and turnover intention is enhanced by the role of job embeddedness. More specifically, employees who perceive their work environment as a place that offers a lot of job resources (e.g., the opportunity to grow and career enhancement) tend to be comfortable working in an organization and experience fit with the organization, develop more connections with other coworkers, and do not want to sacrifice resources they have obtained from the organization, all of which contribute to their lower level of turnover intention.

REFERENCES

Bakker, A.B., Evangelia, D., Toon, W.T., Wilmar, B.S., & Paul, J.G.S. (2003) A multi-group analysis of the job demands-resources model in four home care organizations. *International Journal of Stress Management,* 10 (1), 16–38. Available from: http://dx.doi.org/10.1037/1072-5245.10.1.16.

Bakker, A.B., & Evangelia, D. (2007) The job demands-resources model: State of the art. *Journal of Managerial Psychology,* 22 (3), 309–328. Available from: http://dx.doi.org/10.1108/02683940710733115.

Bergiel, E.B., Vinh, Q.N., Beth, F.C., & Taylor, G.S. (2009) Human resources practices, job embeddedness and intention to quit. *Management Research News,* 32 (3), 205–219. doi: 10.1108/01409170910943084.

Bothma, C.F.C., & Gert, R. (2013) The validation of the turnover intention scale. *SA Journal of Human Resource Management/SA Tydskrif vir Menslikehul pbronbestuur,* 11 (1), 12. Available from: http://dx.doi.org/10.4102/ sajhrm.v11i1.507.

Butali, N.D., Poipoi, M.W., & Laura, C.M. (2013) Effects of staff turnover on the employee performance of work at Masinde Muliro University of science and technology. *International Journal of Human Resource Studies,* 3 (1), 2162–3058. doi: 10.5296/ijhrs.v3i1.3111.

Chang, A.L. (1998) *Job Satisfaction, Dissatisfaction of Texas Newspaper Reporter. [PhD Diss.],* Austin, The University of Texas.

Crossley, C., Rebecca, J.Benner., Steve, M.Jex., & Jennifer, L.B. (2007) Development of a global measure of job embeddedness and integration into a traditional model of voluntary turnover. *Journal of Applied Psychology,* 92 (4), 1031–1042. doi: 10.1037/0021-9010.92.4.1031.

Demerouti, E., Arnold, B.B., Friedhelm, N., & Wilmar, B.S. (2001) The job demands- resources model of burnout. *Journal of Applied Psychology,* 86 (3), 499–512. doi: 10.1037/0021-9010.86.3.499.

Demerouti, E., Arnold, B.B., Jan, D.J., Peter, P.M.J., & Wilmar, B.S. (2001b) Burnout and engagement at work as a function of demands and control. *Scandinavian Journal of Work, Environment and Health,* 27 (4), 279–286. Available from: doi: 10.5271/sjweh.615.

Demerouti, E., & Arnold, B.B. (2011) The job demands–resources model: Challenges for future research." *SA Journal of Industrial Psychology,* 37 (2), 974–979. Available from: doi: 10.4102/sajip.v37i2.974.

Giosan, C. (2003) Predictor of job embeddedness. *Psychology Society Bulletin,* 1 (1), 1–187.

Griffeth, R.W., Peter, W.H., & Stefan, G. (2000) A meta-analysis of antecedents and correlates of employee turnover: Update, moderator tests, and research implications for the next millennium. *Journal of Management,* 26 (3), 463–488. Available from: http://dx.doi.org/10.1016/S0149-2063(00)00043-X.

Hobfoll, S.E. (1989) Conservation resources: A new attempt at conceptualizing stress. *American Psychologist,* 44 (3), 513–524. doi: 10.1037/0003-066X.44.3.513.

Holtom, B.C., & Bonnie, S.O. (2004) Job embeddedness: A theoretical foundation for developing a comprehensive nurse retention plan. *Journal of Nursing Administration,* 34 (5), 216–227.

Jackson, L.T.B., & Sebastiaan, R. (2005) Work-related well-being of educators in a district of the North-West Province. *Perspective in Education,* 23 (3), 107–122. doi: 10.1002/smi.1098.

Kerlinger, F.N., & Howard, B.L. (2000) *Foundations of Behavioral Research 4th Ed.* Holt, NY, Harcourt College Publishers.

Kaur, B., Mohindru. & Pankaj. (2013) Antecedents of turnover intentions: A literature review. *Global Journal of Management and Business Studies,* 3 (10), 1219–1230.

Khatri, N., Chong, T.F., & Pawan, B. (2001) Explaining employee turnover in an Asian context. *Human Resource Management Journal,* 11 (1), 54–74. Available from: doi: 10.1111/j.1748-8583.2001.tb00032.x.

Kiazad, K., Brooks, C.H., Peter, W.H., & Alexander, N. (2015) Integrative conceptual review job embeddedness: A multifocal theoretical extension. *Journal of Applied Psychology,* 100 (3), 641–659. Available from: http://dx.doi.org/10.1037/a0038919.

Kumar, R. (2011) *Research Methodology: A Step-by-Step Guide for Beginners 3rd Ed.* London: SAGE Publications Ltd.

Labour Market framework Yukon Government. (2010) *Recruitment and Employee Retention Strategies.* Available from: http://www.labourmarketframeworkyukon.com/system/PDF/RR%20 strategies.pdf.

Liu, B., Jianxin, L., & Jin, H. (2010) Person-organization fit, job satisfaction, and turnover intention: an empirical study in the Chinese public sector. *Social Behavior and Personality,* 38 (5), 615–625. Available from: http://search.proquest.com/docview/613951267?accountid = 17242.

Mitchell, T.R., Brooks, C.H., Thomas, W.L., Chris, J.S., & Miriam, E. (2001) Why people stay: Using organizational embeddedness to predict voluntary turnover. *Academy of Management Journal,* 44 (6), 1102–1121. doi: 10.2307/3069391.

Mobley, W.H., Stanley, O.H., & Hollingswort, A.T. (1978) An evaluation of precursors of hospital employee turnover. *Journal of Applied Psychology,* 64 (4), 408–414. doi: 10.1037/0021-9010.63.4.408

Ng, T.W.H., & Daniel, C.F. (2010) Locus of control and organizational embeddedness. *Journal of Occupational and Organizational Psychology,* 84 (1), 173–190. doi: 10.1348/096317910X494197.

Ng, T.W.H., & Daniel, C.F. (2007) Organizational embeddedness and occupational embeddedness across career stage. *Journal of Vocational Behavior,* 70 (2), 336–351. doi: 10.1016/j.jvb.2006.10.002.

Purba, D.E. (2015) Employee embeddedness and turnover intentions: Exploring the moderating effects of commute time and family embeddedness. *Makara Hubs-Asia,* 19, 51–63. doi: 10.7454/mssh.v19i1.3472.

Purba, D.E., Janneke, K.O., Marise, Ph.B., & Henk, T.V.M. (2016) The relationships between trust in supervisor, turnover intentions, and voluntary turnover: Testing the mediating effect of on-the-job embeddedness. *Journal of Personnel Psychology,* 15 (4), 174–183. doi: 10.1027/1866-5888/a000165.

Ramesh, A., & Michele, J.G. (2010) Will they stay or will they go? The role of job embeddedness in predicting turnover in individualistic and collectivistic cultures. *Journal of Applied Psychology,* 95 (5), 807–823. doi: 10.1037/a0019464.

Rothmann, S., Karina, M., & Madelyn, S. (2006) A psychometric evaluation of the job demands resources scale in South Africa. *SA Journal of Industrial Psychology,* 32 (4), 76–86. doi: 10.4102/sajip.v32i4.239.

Schaufeli, W.B., Arnold, B.B. (2004) Job demands, job resources, and their relationship with burnout and engagement: A multi-sample study. *Journal of Organizational Behavior,* 25 (3), 293–315. doi: 10.1002/job.248.

Seniati, L., Aries, Y., & Bernadette N., Setiadi, B. (2011) *Psikologi Eksperimen.* Jakarta, PT. Indeks.

Shahpouri, S., Kourosh, N., & Ahmad, A. (2015) Mediating role of work engagement in the relationship between job resources and personal resources with turnover intention among female nurses. *Applied Nursing Research,* 30, 216–221. Available from: http://dx.doi.org/10.1016/j.apnr.2015.10.008.

Sinatra, M. (2015) Employee turnover: Costs and causes. *Air Conditioning, Heating and Refrigeration News* 255 (17), 30. Available from: http://search.proquest.com/docview/1718314845?accountid=17242.

Sousa-Poza, A., & Fred, H. (2002) Analyzing job mobility with job turnover intentions: An international comparative study. *Paper presented at the 7th Annual Meeting of the Society of Labor Economists (SOLE), Baltimore, May 3–4. No. 82:1–28.* Available from: http://www.genderportal.unisg.ch/~/media/internet/content/dateien/instituteundcenters/faa/publikationen/diskussionspapiere/2002/dp82.pdf

Tett, R.P., & John, P.M. (1993) Job satisfaction, organizational commitment, turnover intention, and turnover: Path analyses based on meta-analytic findings. *Personnel Psychology,* 46 (2), 259–293. doi: 10.1111/j.1744-6570.1993.tb00874.x.

Yin-Fah, B.C., Yeok, S.F, Lim, C.L., & Syuhaily, O. (2010) An exploratory study on turnover intention among private sector employees. *International Journal of Business and Management,* 5 (8), 57–64. doi: 10.5539/ijbm.v5n8p57.

Diversity in Unity: Perspectives from Psychology and Behavioral Sciences – Ariyanto et al. (Eds)
© 2018 Taylor & Francis Group, London, ISBN 978-1-138-62665-2

Positive identity as a leader in Indonesia: It is your traits that count, not your gender

Corina D. Riantoputra, Azka M. Bastaman & Hitta C. Duarsa
Faculty of Psychology, Universitas Indonesia, Depok, Indonesia

ABSTRACT: The rise of identity theory provides an opportunity for scholars to understand and explain leaders' behaviour from a new angle, emphasising the importance of positive identity as leaders. Scholars argue that leaders with positive identity are more energetic and work more wholeheartedly as their identity as leaders fits with other facets of their identities. Research has just begun to understand the factors affecting positive identity as leaders. Some argue that personality traits matter, while others are convinced that gender, male or female, is the key determinant to shaping positive identity as leaders. Based on the assumption that (Indonesian) society tends to picture the female primary role as that of mother and submissive wife, and not as leader, these scholars maintain that female leaders may be limited in building positive identity as leaders. To investigate this matter further, data was gathered from 315 people in two big cities in Indonesia: Jakarta and Denpasar. Analysis reveals that positive identity as leaders is associated with traits, especially extraversion ($\beta = 0.29$, $p < 0.05$) and conscientiousness ($\beta = 0.33$, $p < 0.05$), but not with gender ($\beta = -0.07$, $p > 0.05$), nor with neuroticism ($\beta = -0.06$, $p > 0.05$). These results shed new and promising light on the understanding of leadership behaviour in Indonesia. It demonstrates that, at least in big and multicultural cities in Indonesia, leaders are not trapped by their gender in developing their identity. Instead, their positive identity as leaders is associated with their traits, suggesting a new page of leadership in Indonesia: more egalitarian gender-roles, especially in leadership. It would be interesting to investigate whether the same pattern occurs in smaller cities in Indonesia, and in other parts of the world.

1 INTRODUCTION

In maintaining the competitiveness of companies, leaders need to be able to implement changes and drive innovation (Gilley et al., 2008), and thus the most prominent role of effective leaders is their ability to enhance team and company performance by influencing team members and facilitating the attainment of goals (DuBrin, 2016; Kaiser et al., 2008; Yukl, 2012). Therefore, it is little wonder that current discussions of leadership tend to focus on interpersonal dynamics within the leadership process, particularly on how leaders engage followers and create positive effects that extend beyond task compliance (Hannah et al., 2014; Van Knippenberg et al., 2004). Among several key factors that influence the interaction between leaders and followers, current theory focuses on identity (Johnson et al., 2012; Van Knippenberg, 2011), which refers to the knowledge that individuals have about themselves (Johnson et al., 2012) that impacts the way they feel, think, and behave in relation to the things they seek to achieve (DeRue & Ashford, 2010; Van Knippenberg et al., 2004).

Previous research has focused largely on one side of this interaction, namely, on follower identity, rather than attending to leader identity (Johnson et al., 2012; Kark & Van Dijk, 2007). Most research emphasises the identification of factors that have significant positive impacts on performance, such as the study conducted by Liu et al. (2010), which aimed to examine the relationships between employee voice behaviour, employee identification, and transformational leadership, and the study conducted by Walumbwa and Hartnell (2011), concerning relational identification, self-efficacy, and transformational leadership. However, literature which specifically focuses on leader identity is still very limited.

Identity as a leader is the result of an evaluation process of attaching positive or negative valence on an individual's role as a leader (DeRue & Ashford, 2010; Karelaia & Guillén, 2014). This evaluation depends on an individual's knowledge of how good or how fit they are as leaders (i.e. private regards) and their perception of how others think of their actions and performance as leaders (i.e. public regards) (Ashmore et al., 2004). Thus, leaders may have positive or negative evaluations of their identity as leaders. However, leaders that are energised to perform are only those with positive identity as leaders, which refers to the positive valence of the leader's own evaluations, and the perception of others' judgements in relation to their identity as leaders (Karelaia & Guillén, 2014).

Research shows that positive identity is significantly related to psychological well-being and positive work attitude (Shin & Kelly, 2013). In their empirical experiments, Karelaia and Guillén (2014) show that leaders with positive identity tend to be more energetic and to work more wholeheartedly. Leaders with positive identity are also able to motivate their followers to achieve shared goals and activate their group-based identities, which result in greater coordination among followers (Van Knippenberg et al., 2004). Because of the importance of positive identity for leaders, there is a need to examine the factors influencing it.

Gender is one potential aspect that may strongly influence the development of positive identity as a leader, because gender is known as a salient attribute influencing the way people think of leadership (Ayman & Korabik, 2010; DeRue et al., 2011), in that people evaluate male and female leaders differently. Eagly and Karau (2002) explain that people tend to perceive men's gender role as similar to a leaders' role, causing role incongruity for female leaders. Data demonstrates that, in 2015, only 14.2% of the top five leadership positions in American corporations were held by women (Egan, 2015). Globally, in 2014, women held only 24% of senior management positions (United Nations Women, 2016). Altogether, the arguments and the data indicate that gender may strongly influence the development of positive identity as leaders.

The second factor that may potentially affect the development of positive identity as leaders is traits. Traits contribute to individual differences in behaviour, shape consistency of behaviour over time, and produce stability of behaviour across situations (Feist & Feist, 2008). Previous research indicates that conscientiousness, extraversion and neuroticism are all associated with leadership effectiveness (Judge et al., 2002; Ng et al., 2008). Thus, they may also be powerful factors in assisting individuals in constructing positive identity as leaders. Therefore, this study asks, "What are the relationships between positive identity as leaders, and gender and traits?"

The current study makes use of social identity theory to explain how gender influences positive identity as a leader. Social identity proposes that the identity that individuals attach to themselves is related to the social roles to which they are assigned (Hogg & Terry, 2001; Karelaia & Guillén, 2014; Stets & Serpe, 2013). Every individual possesses multiple identities because every individual identifies him or herself with many social categories and has more than one social role. Each role has a set of meanings to which the individual attaches, and rules on how he or she should think, feel, and behave (Hogg & Terry, 2001; Stets & Serpe, 2013). Considering the extreme differences in gender-typed and leader-typed roles, this often results in role incongruity (Eagly & Karau, 2002) or identity conflict (Karelaia & Guillén, 2014). Stereotypically, people tend to expect women to be calm, warm and nurturing, which are communal traits, while they view leaders as having more agentic traits, such as being assertive, dominant and competitive (Gartzia & Baniandrés, 2015; Koenig et al., 2011).

This agency–communion paradigm, as one example of leadership stereotypes, along with the think manager–think male and masculinity–femininity paradigms (Koenig et al., 2011), tends to view leaders as having more masculine traits than feminine traits, and thus being more in accordance with males than females. This tendency is demonstrated by Cuadrado et al. (2015), who show that masculine characteristics are perceived as more important than feminine characteristics for managerial positions. Beside masculine characteristics, effective leaders are also associated with a task-orientation style of leadership, which is, again, attached to the male gender role rather than the female one (Vinkenburg et al., 2011). The existence of these stereotypes creates a double standard for women leaders. In order to be perceived as effective, women leaders have to show masculine characteristics and become more task-oriented, which does not fit their stereotypical gender role as females. This tendency is more prevalent in Eastern culture, which tends to emphasise the importance of accordance with societal norms (Jogulu & Wood, 2008). Based on the above, we posit that:

H1: Leaders' gender is significantly associated with positive identity as leaders, in that male leaders will have more positive identity as leaders than female leaders.

The trait-based approach was the first theory introduced by scientists to characterise leadership and, although it has its critics (e.g. Zaccaro, 2007), this basic approach continues to play an important role in explanations of leadership theory (Judge et al., 2002). The five-factor model and the Big Five personality traits (Costa & McCrae, 1992) are considered the most prominent personality theory for their ability to integrate many traits into an integrative frame of thinking (Judge et al., 2002). Three traits that show consistent association with leadership effectiveness are neuroticism, extraversion and conscientiousness (Ng et al., 2008). Neurotic people tend to be anxious, temperamental, self-pitying, self-conscious, emotional, and vulnerable to stress-related disorders; people who score high on extraversion tend to be affectionate, jovial, talkative, joiners, and fun-loving; conscientiousness people are described as those who are ordered, controlled, organised, ambitious, achievement-focused, and self-disciplined (Feist & Feist, 2008).

A study conducted by Judge et al. (2002) strongly supports the use of the Big Five personality traits in examining the relationship between traits and leader identity. Their study found that extraversion, openness to experience, and conscientiousness were positively associated with leader effectiveness, while neuroticism was negatively related to leader effectiveness. Agreeableness, on the other hand, demonstrated an ambivalent relationship with leader effectiveness. Another study, conducted by Ng et al. (2008), found that neuroticism, extraversion, and conscientiousness were all associated with leader effectiveness.

Arguably, if traits are related to leader effectiveness, they may also relate to the construction of positive identity as a leader, through the processes of claiming and granting identity as leaders (DeRue & Ashford, 2010). That is, leaders—those with certain traits—will be evaluated more positively by others and be granted the identity of leaders. For example, conscientious leaders who are well-organised and disciplined may show excellent performance and be assessed more positively by others. These leaders may also attach positive valence to their leadership and, thus, progress the development of a positive leader identity. On the other hand, neurotic leaders may not be liked by their followers for their emotional and temperamental personality, and thus their followers may not grant them the identity of leaders. These leaders might also have a less positive attitude toward their identity as leaders because of their incapability to handle job-related stress well. Given the relationship between traits, performance and the process of granting and claiming positive identity as leaders, we further posit that:

H2: Neuroticism is negatively associated with positive identity as leaders;
H3: Extraversion is positively associated with positive identity as leaders;
H4: Conscientiousness is positively associated with positive identity as leaders.

2 METHOD

2.1 *Participants and procedure*

We surveyed 462 leaders employed at seven public and 17 private sector organisations in two big cities in Indonesia: Jakarta and Denpasar. Usable data was obtained from 315 leaders (147 responses were excluded due to being invalid and/or incomplete), giving a response rate of 68%. This represents a very good response rate, considering that the average response rate for research in which data is gathered from executives is just 32% (Cycyota & Harrison,

Table 1. Descriptive statistics, correlations and reliabilities.

Variable	Mean	SD	1	2	3	4	5	6	7
1 Gender	–	–	1						
2 Leadership experience	4.96	1.41	0.04	1					
3 Number of levels	2.56	1.15	0.10	0.27**	1				
4 Conscientiousness	16.58	15.09	0.07	0.01	0.25**	(0.64)			
5 Extraversion	4.69	0.74	0.03	0.16**	0.14*	0.36**	(0.71)		
6 Neuroticism	3.65	1.40	0.13*	0.09	0.32**	0.42**	0.22**	(0.70)	
7 Positive identity as leader	4.59	0.58	–0.039	0.12*	0.20**	0.40**	0.44**	0.17**	(0.61)

Note: n = 315; *p < 0.05; **p < 0.01; Cronbach's alphas are in parentheses on the diagonal.

2006). Participants' mean age was 44 years old (SD = 8.46), and the majority of the sample was male (66%), married (91%), possessed Bachelor degrees (43%), had been working for more than 15 years with their current employers (46%), and had held their leadership position for 5–10 years (33%). Participants were employed in a variety of sectors (e.g. finance, tourism/hotel industries, and government offices).

Participants (the leaders) provided ratings of their own positive identity as leaders. To measure traits, participants were asked to choose the answers that they felt most appropriate to their personality. To minimise the tendency to respond in a socially desirable way, participants were urged in the instructions to answer as honestly as possible (Podsakoff et al., 2003). They were also assured that any use of data would adhere to strict requirements for confidentiality and anonymity. We also counterbalanced the order of the items to reduce eventual response bias effects related to survey design (Podsakoff et al., 2003).

2.2 *Measures*

All the scales used in this research were adapted from previous research, and underwent a back-to-back translation. All scales used a six-point Likert-type scale, ranging from 1 (Strongly disagree) to 6 (Strongly agree).

The scale for positive identity as leaders was adapted from Karelaia and Guillén (2014) and consisted of six items (Cronbach's alpha = 0.61). A sample item was, "To lead other people is something I enjoy doing". Traits were measured using an adaptation of the short version of the Big Five Inventory (Rammstedt & John, 2007), consisting of nine items (three per trait). An example from this scale is, "I rarely feel worried". The reliability of these scales is depicted in Table 1.

3 RESULTS

3.1 *Descriptive statistics and variable correlation*

Table 1 presents the means, standard deviations, reliabilities and correlations among the study variables. Consistent with previous research, length of leadership experience and the number of subordinate levels below the leaders were significantly related to positive identity as a leader ($r = 0.12$, $p < 0.05$, and $r = 0.20$, $p < 0.05$, respectively). This indicates that the leaders who have longer experience of leadership and more levels of subordinates perceive their leader identity more positively. Therefore, these two variables were controlled in the next round of statistical tests.

3.2 *Testing hypotheses*

To analyse the relationship between leaders' gender, their traits and their positive identity as leaders, we conducted multiple regression analysis (see Table 2) with leadership experience

Table 2. Results of hierarchical regression analysis.

Variable	Step 1	Step 2
Leadership experience	0.09	0.06
Number of levels	0.17**	0.09
Gender		–0.07
Conscientiousness		0.29**
Extraversion		0.33**
Neuroticism		–0.06
R^2	0.05	0.29
df	2,295	6,291

Note: *p < 0.05; **p < 0.01.

and numbers of subordinate levels below the leaders as control variables. This model explains 29% of the variance of positive identity as a leader. Table 2 also shows that conscientiousness (β = 0.29, p < 0.01) and extraversion (β = 0.33, p < 0.01) were significantly associated with positive identity as leaders (supporting Hypotheses 3 and 4). In contrast, the effect of neuroticism (β = –0.06, p > 0.05) and leaders' gender (β = –0.07, p > 0.05) were not significant (contrary to Hypotheses 1 and 2). These results indicate that leaders' gender does not predict the positivity of leader identity but some traits—specifically extraversion and conscientiousness—do. The model also shows that, in this model, length of leadership experience and the number of subordinate levels are not associated with positive identity as leaders.

4 DISCUSSION

Using surveys of 315 leaders in two big cities in Indonesia, the current research aimed to answer the question of the relationship between positive leader identity and gender and traits (conscientiousness, extraversion and neuroticism). Its results show that traits have a more prevalent role than gender in the development of positive identity as leaders. This study makes a significant contribution to the current debates concerning leadership theories in at least two areas.

First, the study provides insights to the possibility of decreasing gender stereotypes in big cities in Indonesia toward more egalitarian gender-roles (i.e. tendency to treat males and females equally). Although, theoretically, gender has a strong power over positive identity as a leader, in the current study positive identity as a leader is not associated with gender. Apparently, there is no difference between males and females in the construction of positive identity as a leader in the current data set. Although it is surprising, this result is consistent with other studies that report these stereotypes as having decreased over time (Gartzia & Baniandrés, 2015; Li Kusterer et al., 2013; Schein, 2001). The current results are also consistent with Grant Thornton's research that shows that 36% of senior managers in Indonesia are women, and the reasons many women advance themselves as leaders are their willingness to make a difference (47%) and to influence others (32%) (Priherdityo, 2016). It seems that women having aspirations to be leaders, at least in big cities in Indonesia, are becoming able to reach leadership positions and construct positive identities as leaders.

Arguably, the results may not have shown similar trends if the research had been conducted in small cities, given the higher power distance culture of smaller cities in Indonesia. A comprehensive investigation of positive identity as leaders in smaller cities in Indonesia would be of value. It would also be interesting to see whether, in big cities, people are more ready to accept male leaders who tend to be more people-oriented than task-oriented (i.e. do not inhabit such stereotypical masculine roles). O'Neill and O'Reilly (2011) observed the emerging nature of this phenomenon and termed it 'the backlash effect'. Future research may want to address this phenomenon in big cities in Indonesia.

Second, the current study extends previous research by showing that neuroticism has no relationship with positive identity as leaders. This result is not consistent with the previous research that has established a negative association between neuroticism and leadership effectiveness (Judge et al., 2002; Ng et al., 2008). One possible explanation is that neurotic people in Indonesia have effectively been trained to hide their emotions, given that the majority of Indonesian people tend to have neutral affect (Trompenaars & Woolliams, 2003). This causes the neurotic nature of their traits to not have the negative impact it usually does for people from more affective cultures (that are more expressive). Thus, the current study challenges the current understanding and questions the impact of neuroticism in different cultures: neutral vs affective ones.

Another possible explanation for the lack of significant relationship between neuroticism and positive identity as leaders is that the neuroticism scale in the current study used only three items, which may not have been sufficient to capture the nature of the neurotic behaviour of the participants. Thus, the short-version measure of the Big Five scale that was used in this research may have been one of the limitations of the current study. Only three items were used to measure each trait, and the reliability for each trait, ranging from 0.64 to 0.71, does not suggest very high reliability. Future research may want to consider the use of a larger version of the Big Five scale in order to create a comprehensive picture of each trait.

Another limitation of the current study is the possibility of common method bias (Podsakoff et al., 2003), which occurs because both the outcome variable and the predictors are measured with the same method, which is a self-reporting scale. Although this limitation is acknowledged, it should be noted that the researchers have also taken extra effort to counterbalance the scale (Podsakoff et al., 2003) to reduce the impact of common method bias. By doing so, we would argue that the measurement technique used is the best we could achieve considering the nature of the variables involved.

5 CONCLUSION

The current study, which aimed to understand factors associated with positive identity as leaders, has been able to demonstrate that positive identity as leaders is associated with traits—specifically, conscientiousness and extraversion—and is not associated with gender. By so doing, this study contributes to the current debates on leadership theories by showing that there is a possibility of decreasing gender role stereotypes in leaders' behaviour and attitudes, at least in big cities in Indonesia. The study also challenges the current wisdom as to the relationship between neuroticism and leaders' behaviour and attitudes by arguing for the possibility that neuroticism may have different impacts on leaders in neutral cultures (as compared to affective cultures). Altogether, the study advances understanding of positive identity as leaders, and stimulates more debate about leadership theories.

REFERENCES

Ashmore, R. D., Deaux, K., & McLaughlin-Volpe, T. (2004). An organizing framework for collective identity: Articulation and significance of multidimensionality. *Psychological Bulletin, 130*(1), 80–114.

Ayman, R., & Korabik, K. (2010). Leadership: Why gender and culture matter. *American Psychologist, 65*(3), 157–170.

Burke, P. J. (Ed.). (2006). *Contemporary social psychological theories*. Stanford, CA: Stanford University Press.

Costa, P. T., & McCrae, R. R. (1992). Normal personality assessment in clinical practice: The NEO personality inventory. *Psychological Assessment, 4*(1), 5–13.

Cuadrado, I., García-Ael, C., & Molero, F. (2015). Gender-Typing of leadership: Evaluations of real and ideal managers. *Scandinavian Journal of Psychology*, 56(2), 236–244.

Cycyota, C. S., & Harrison, D. A. (2006). What (not) to expect when surveying executives: A meta-analysis of top manager response rates and techniques over time. *Organizational Research Methods*, *9*(2), 133–160.

DeRue, D. S., & Ashford, S. J. (2010). Who will lead and who will follow? A social process of leadership identity construction in organizations. *Academy of Management Review*, *35*(4), 627–647.

Derue, D. S., Nahrgang, J. D., Wellman, N. E. D., & Humphrey, S. E. (2011). Trait and behavioral theories of leadership: An integration and meta-analytic test of their relative validity. *Personnel Psychology*, 64(1), 7–52.

DuBrin, A. J. (2016). *Leadership: Research findings, practice, and skills* (8th ed.). Boston, MA: Cengage Learning.

Dutton, J. E., Roberts, L. M., & Bednar, J. (2010). Pathways for positive identity construction at work: Four types of positive identity and the building of social resources. *Academy of Management Review*, *35*(2), 265–293.

Eagly, A. H., & Karau, S. J. (2002). Role congruity theory of prejudice toward female leaders. *Psychological Review*, *109*(3), 573–598.

Egan, M. (2015, March 24). Still missing: Female business leaders. *CNN Money*. Retrieved from http://money.cnn.com/2015/03/24/investing/female-ceo-pipeline-leadership/

Feist, J., & Feist, G. (2008). *Theories of personality* (7th ed.). New York, NY: McGraw-Hill.

Gartzia, L., & Baniandrés, J. (2015). Are people-oriented leaders perceived as less effective in task performance? Surprising results from two experimental studies. *Journal of Business Research*, *69*(2), 508–516.

Gilley, A., Dixon, P., & Gilley, J. W. (2008) Characteristics of leadership effectiveness: Implementing change and driving innovation in organizations. *Human Resource Development Quarterly*, *19*(2), 153–169.

Hannah, S. T., Sumanth, J. J., Lester, P., & Cavarretta, F. (2014). Debunking the false dichotomy of leadership idealism and pragmatism: Critical evaluation and support of newer genre leadership theories. *Journal of Organizational Behavior*, *35*(5), 598–621.

Hargrove, B. K., Creagh, M. G., & Burgess, B. L. (2002). Family interaction patterns as predictors of vocational identity and career decision-making self-efficacy. *Journal of Vocational Behavior*, *61*(2), 185–201.

Hogg, M. A., & Terry, D. J. (Eds.). (2001). *Social identity processes in organizational contexts*. Philadelphia, PA: Psychology Press.

Jogulu, U. D. & Wood, G. J. (2008). A cross-cultural study into peer evaluations of women's leadership effectiveness. *Leadership and Organization Development Journal*, *29*(7), 600–616.

Johnson, R. E., Venus, M., Lanaj, K., Mao, C., & Chang, C. H. (2012). Leader identity as an antecedent of the frequency and consistency of transformational, consideration, and abusive leadership behaviors. *Journal of Applied Psychology*, *9*(6), 1262–1272.

Judge, T. A., Bono, J. E., Ilies, R., & Gerhardt, M. W. (2002). Personality and leadership: A qualitative and quantitative review. *Journal of Applied Psychology*, *87*(4), 765–780. doi:10.1037//0021-9010.87.4.765

Kaiser, R. B., Hogan, R., & Craig, S. B. (2008). Leadership and the fate of organizations. *American Psychologist*, *63*(2), 96–110.

Karelaia, N., & Guillén, L. (2014). Me, a woman and a leader: Positive social identity and identity conflict. *Organizational Behavior and Human Decision Processes*, *125*(2), 204–219.

Kark, R, & Van Dijk, D. (2007). Motivation to lead, motivation to follow: The role of the self-regulatory focus in leadership processes. *Academy of Management Review*, *32*(2), 500–528.

Koenig, A. M., Eagly, A. H., Mitchell, A. A., & Ristikari, T. (2011). Are leader stereotypes masculine? A meta-analysis of three research paradigms. *Psychological Bulletin*, *137*(4), 616–642.

Li Kusterer, H., Lindholm, T., & Montgomery, H. (2013). Gender typing in stereotypes and evaluations of actual managers. *Journal of Managerial Psychology*, *28*(5), 561–579.

Liu, W., Zhu, R., & Yang, Y. (2010). I warn you because I like you: Voice behavior, employee identifications, and transformational leadership. *The Leadership Quarterly*, *21*(1), 189–202.

Ng, K. Y., An, S., & Chan, K. Y. (2008). Personality and leader effectiveness: A moderated mediation model of leadership self-efficacy, job demands, and job autonomy. *Journal of Applied Psychology*, *93*(4), 733–743. doi:10.1037/0021-9010.93.4.733

Nichols, A. L. (2016). What do people desire in their leaders? The effect of leadership experience on desired leadership traits. *Leadership & Organization Development Journal*, *37*(5), 658–671.

O'Neill, O. A., & O'Reilly III, C. A. (2011). Reducing the backlash effect: Self-monitoring and women's promotions. *Journal of Occupational and Organizational Psychology*, 84(4), 825–832.

Podsakoff, P. M., MacKenzie, S. B., Lee, J., & Podsakoff, N. P. (2003). Common method bias in behavioral research: A critical review of the literature and recommendation remedies. *Journal of Applied Psychology*, *88*(5), 879–903.
Priherdityo, E. (2016, March 8). Wanita Karir Indonesia Terbanyak Keenam di Dunia *(Female Leaders in Indonesia reach the 6th position in the world)*. *CNN Indonesia*. Retrieved from http://www.cnnindonesia.com/gaya-hidup/20160308121332-277-116053/wanita-karier-indonesia-terbanyak-keenam-di-dunia/
Rammstedt, B., & John, O. P. (2007). Measuring personality in one minute or less: A 10-item short version of the Big Five Inventory in English and German. *Journal of Research in Personality, 41*, 203–212.
Schein, V. E. (2001). A global look at psychological barriers to women's progress in management. *Journal of Social Issues*, *57*(4), 675–688.
Shin, Y. J., & Kelly, K. R. (2013). Cross-cultural comparison of the effects of optimism, intrinsic motivation, and family relations on vocational identity. *The Career Development Quarterly*, 61(2), 141–160.
Stets, J. E., & Serpe, R. T. (2013). Identity theory in J. DeLamater & A. Ward (eds.), Handbook of Social Psychology, 31–60. Springer: Netherlands.
Trompenaars, F. & Wooliams, P. (2003). *Business across cultures*. Chichester, UK: Capstone Publishing.
United Nations Women. (2016). *Women and the economy*. Retrieved from http://beijing20.unwomen.org/en/infographic/economy
Van Knippenberg, D. (2011). Embodying who we are: Leader group prototypicality and leadership effectiveness. *The Leadership Quarterly*, *2*(6), 1078–1091.
Van Knippenberg, D., Van Knippenberg, B., DeCremer, D., Hogg, M. A. (2004). Leadership, self, and identity: A review and research agenda. *The Leadership Quarterly*, *15*(6), 825–856.
Vinkenburg, C. J., Van Engen, M. L., Eagly, A. H., & Johannesen-Schmidt, M. C. (2011). An exploration of stereotypical beliefs about leadership styles: Is transformational leadership a route to women's promotion? *The Leadership Quarterly*, *22*(1), 10–21.
Walumbwa, F. O., & Hartnell, C. A. (2011). Understanding transformational leadership–employee performance links: The role of relational identification and self-efficacy. *Journal of Occupational and Organizational Psychology*, 84(1), 153–172.
Yukl, G.A. (2012). *Leadership in organizations* (7th ed.). Upper Saddle River, NJ: Pearson.
Zaccaro, S. (2007). Trait-based perspective of leadership. *American Psychologist, 62*(1), 6–16.
Zaccaro, S. J., Foti, R. J. & Kenny, D. A. (1991). Self-monitoring and trait-based variance in leadership: An investigation of leader flexibility across multiple group situations. *Journal of Applied Psychology*, *76*(2), 308–315.

Diversity in Unity: Perspectives from Psychology and Behavioral Sciences – Ariyanto et al. (Eds)
© 2018 Taylor & Francis Group, London, ISBN 978-1-138-62665-2

The role of professional commitment as a mediator in the relationship between job satisfaction and organizational commitment among lecturers in higher-education institutions

R.L. Sari & A.N.L. Seniati
Faculty of Psychology, Universitas Indonesia, Depok, Indonesia

ABSTRACT: This study was conducted in order to find the role of professional commitment as a mediator in relationship between job satisfaction and organisational commitment among lecturers in higher-education institutions. There were 328 lecturers from 11 higher-education institutions in Jakarta, Tangerang, Denpasar, Jimbaran, and Padang who answered Organisational Commitment Scale, Job Satisfaction Survey, and Professional Commitment Scale. A simple mediation analysis was conducted using PROCESS macro. The results indicate positive significant effects of job satisfaction on professional commitment, professional commitment on organisational commitment, and job satisfaction on organisational commitment. This study also found that professional commitment is a partial mediator in the relationship between job satisfaction and organisational commitment among lecturers in higher-education institutions. The implication is that higher-education institutions can enhance lecturers' professional and organisational commitment by creating a more satisfying working environment.

1 INTRODUCTION

Lecturers are essential part of higher education institutions. However, in fulfilling their duties, lecturers face various challenges and demands. They are expected to constantly expand and enhance their knowledge, to have strong willingness to do research, to further their education, to have public speaking skills, and to work with students with various characteristics and viewpoints. Lecturers also face challenges in developing creativity and innovation to formulate the right learning method and model to avoid monotony and students' lack of enthusiasm in classes (Billy, 2016). In fulfilling their obligations, oftentimes lecturers also need to bring their work home and continue their work after working hours (Seniati, 2010).

As a result of the many demands and challenges that lecturers are facing, those who find it hard to adapt may decide to quit their teaching jobs and switch to other professions. This is observable in several cases where junior lecturers decided to quit their university teaching jobs because they failed to adapt to their job (Billy, 2016). Such failures at adaptation may be experienced in several forms, such as failing to acquaint with more senior lecturers, failing to handle inattentive students, and failing to adapt to the demands of the job (Billy, 2016). In addition, the relatively low salary that lecturers receive is also one of the reasons a lecturer might decide to quit and leave higher education institution. This indicates that the lecturers' organisational commitment is still relatively low.

Meyer & Allen (1991) define organisational commitment as the psychological construction of the characteristics of the relationship between a member of an organisation and his organisation, which is implicated in the member's decision to continue his membership in the organisation. Based on this definition, organisational commitment consists of three components, namely: affective, continuance, and normative commitment. Affective commitment is defined as an individual's emotional, identification, and involvement ties with an organisation. Continuance commitment is defined as a calculation of loss and profit that one gains upon leaving an organisation. Finally, normative commitment refers to the feeling of obligation that compels one to remain in an organisation.

Lecturers' organisational commitment is believed to benefit higher education institutions, because lecturers with a higher organisational commitment are more likely to put in more effort and commitment in the teaching process (Ali & Zafar, 2006) and can improve the students' satisfaction (Xiao & Wilkins, 2015). In addition, lecturers with a higher organisational commitment tend to have better working performance (Ali & Zafar, 2006), which influences the performance and effectiveness of the faculty and university where they work (Jing & Zhang, 2014). Lecturers' organisational commitment can also reduce the negative factors that disadvantage universities, such as turnover intentions (Ali & Zafar, 2006). Therefore, a study on the factors that influence lecturers' organisational commitment is imperative.

Studies with the objective of discovering factors influencing organisational commitments have been carried out by a number of researchers. Such studies have found that factors influencing organisational commitment among lecturers include the leadership style of the universities' board of leaders (Othman, Mohammed & D'Silva, 2013), the locus of control on lecturers (Munir & Sajid, 2010), the perception of organisational support (Lew, 2009), opportunities for trainings, the perception of organisation's fairness (Ali & Zafar, 2006), job satisfaction (Malik *et al*, 2010), and the lecturers' professional commitment (Appaw-Agbola, Agbotse & Ayimah, 2013).

One of the factors that have been predicted to influence lecturers' organisational commitment is job satisfaction. Spector (1997) defines job satisfaction as an attitude that reflects one's feelings on his/her job, both comprehensively and partially on various aspects or facets of the job. Job satisfaction comprises seven aspects or facets, namely the nature of work, supervision, coworkers, work facilities, rewards, promotions, and communication in the organisation (Spector, 1997). Tett & Mayer (1993) explain that a person who is satisfied with his/her job will be more committed to his/her organisation. In contrast, an individual dissatisfied with his/her job will have a lower commitment to his/her job and will tend to be absent more often, and even think of leaving the organisation. This opinion is in alignment with the findings of Khan *et al,* (2014) that lecturers satisfied with their job, working conditions, incentives or rewards, relationships with supervisors, and promotions have a higher rate of organisational commitment and have a lower tendency to leave the university. A similar study was conducted by Malik *et al,* (2010) on lecturers in a Pakistani state university, with the results demonstrating that satisfaction on working conditions, incentives, and relationships with supervisor influence lecturers' organisational commitments.

On the other hand, research carried out by Niehoff (1997) on lecturers in a Jesuit Catholic University showed that the correlation between job satisfaction and organisational commitment is still relatively low. In fact, previous research by Curry, Wakefield, Price & Muelle (1986) found no significant relationship between job satisfaction and organisational commitment. This is in line with Luthans' (2002) view that there are many cases where employees are satisfied with their jobs but still dislike their organisations. Yet the opposite can also be true, whereby many employees who are dissatisfied with their jobs still manage to stay in the organisation where they work. As a result, the strength of the relationship between job satisfaction and organisational commitment is still in question.

Due to conflicting research findings on the relationship between job satisfaction and organisational commitment, the current study is intended to take a similar approach, yet develop it further by adding professional commitment as a mediator. Meyer, Allen & Smith (1993) define professional commitment as an individual's desire to identify oneself to a certain profession and remain as a member of that profession. Similar to organisational commitment, commitment to profession also consists of three components, namely affective, continuance, and normative commitment (Meyer *et al,* 1993). The affective professional commitment is defined as an individual's emotional attachment to the profession he/she belongs to, his/her identification with the profession, and his/her involvement in the profession. Moreover, continuance professional commitment is understood as an individual's calculation of the potential loss and profit if he/she leaves the profession; while normative professional commitment refers to the individual's sense of obligation to remain in the profession.

Professional commitment is predicted to be a potential mediator in the connection between job satisfaction and organisational commitment, in which the initial job satisfaction of an individual contributes to his/her professional commitment before finally leading to organisational commitment. As explained by Meyer & Allen (1991), when employees feel that their working experience in the organisation meets their expectations and basic needs, they will tend to have a stronger professional commitment. This is also consistent with Shukla's (2014) research findings that educators will give their best effort to be competent, thus having a high professional commitment. After their professional commitment is formed, their organisational commitment will also develop. As explained by Carson, Carson & Bedeian (1997), when an individual decides to remain in a profession (hence, a high professional commitment), other job alternatives will also decrease. With less alternatives, professional will prefer to stay in the organisation where he/she works, or in other words, will experience an increase in his/her organisational commitment. Research by Bredillet & Dwivedula (2010) supports this finding, as they also found that an employee's professional commitment can enhance his/her organisational commitment.

2 METHODS

This research involved 328 lecturers from 11 higher education institutions in Jakarta, Tangerang, Denpasar, Jimbaran, and Padang, as respondents. The lecturers were permanent lecturers who have been working at the institutions for more than two years and included those who were civil servants (PNS) and non-civil servants (Non-PNS). The higher education institutions consisted of State Universities (PTN), Private Universities (PTS), and Legal Entity State Universities (PTN-BH).

The instruments used in this study were the Organisational Commitment Scale (OCS) by Seniati & Yulianto (2010), which was based on Meyer & Allen's (1997) instrument; the Job Satisfaction Survey (JSS) by Seniati & Yulianto (2010) based on Spector's (1997) instrument; and the Professional Commitment Scale (PCS) that was based on Meyer & Allen's (1997) instrument. OCS consisted of 21 items spread across the three components as follows: affective commitment (8 items), continuance commitment (7 items), and normative commitment (6 items). An example of an item from the OCS is, "I would be very happy to spend the rest of my career in this higher-education institution". JJS also consisted of 21 items spread across seven facets that include the nature of work, supervision, coworkers, work facilities, rewards, promotions, and communication. Each facet consisted of 3 items. An example of an item from the JJS is, "I am satisfied working as a lecturer because I can interact with many people". Meanwhile, professional commitment variables are measured using PCS from Meyer & Allen (1997) by modifying the word "organisation" contained in the item into "profession". PCS consisted of 23 items distributed across three components, namely the affective commitment (8 items), continuance commitment (9 items), and normative commitment (6 items). An example of an item from the PCS is, "I do not feel emotionally attached to this profession". Each item in all three measurement tools was scaled from 1 (strongly disagree) to 4 (strongly agree).All of the instruments were reliable ($\alpha = 0.88$ for OCS, $\alpha = 0.91$ for JSS, and $\alpha = 0.79$ for PCS).

3 RESULTS

The statistical technique used to analyse data in this research is Simple Mediation Model Test using PROCESS macro introduced by Hayes (2008). Prior to carrying out a simple mediation test using PROCESS macro, information on the demographic factors that influence organisational commitment as the dependent variable had to be identified so they could be used as control variables in the mediation test. Therefore, the researchers deemed it necessary to carry out mean difference tests of the organisational commitment for each demographic factor. The following are results from the mean difference tests:

Table 1. Comparison of organisational commitment means based on demographic data.

Demographic data	Mean	Sign
Sex		
Male	3.04	F = 9.36 (p = 0.00)**
Female	2.91	
Educational level		
S2	2.95	F = 2.69 (p = 0.10)
S3	3.03	
Tenure		
Years	2.83	F = 18.00 (p = 0.00)**
>10 years	3.02	
Marriage status		
Single	2.83	F = 1.17 (p = 0.31)
Married	2.98	
Widower/Widow	3.02	

*p < 0.05, **p < 0.01

Table 1 shows a significant difference of means in organisational commitment between male and female lecturer, with male lecturers displaying a higher organisational commitment level compared to female ones. In addition, a significant difference in the means of organisational commitment based on tenure was also found, where lecturers with longer job durations demonstrated higher organisational commitment. There was no significant mean difference in organisational commitment between different educational level and marriage status. Consequently, sex and tenure will be controlled in the simple mediation test.

Prior to using a simple mediation test using PROCESS macro, a correlation test was first conducted to observe the relationships between the variables of interest.

Table 2 shows significant positive correlations between job satisfaction and professional commitment ($r = 0.47$, $p < 0.01$); between professional commitment and organisational commitment ($r = 0.66$, $p < 0.01$); and between job satisfaction and organisational commitment ($r = 0.64$, $p < 0.01$). The Simple Mediation Model Test was then carried out using PROCESS macro introduced by Hayes (2008). The model being tested was the influence of job satisfaction on organisational commitment mediated by professional commitment by controlling the demographic variables of sex and tenure.

The analysis found a significant positive influence of job satisfaction on professional commitment ($\beta = 0.36$, $p = 0.00$, LLCI = 0.28, ULCI = 0.44). In addition, professional commitment was found to have a significant positive impact on organisational commitment ($\beta = 0.58$, $p = 0.00$, LLCI = 0.48, ULCI = 0.68). Furthermore, a significant positive influence of job satisfaction on organisational commitment was discovered ($\beta = 0.41$, $p = 0.00$, LLCI = 0.33, ULCI = 0.49).

Based on the results of mediation test, a mediating effect of professional commitment was shown to play a role in the relationship between job satisfaction and organisational commitment.The results are shown below.

As shown in Table 4, the value of the indirect effect of job satisfaction on organisational commitment ($\beta = 0.21$, $p = 0.00$, LLCI = 0.15, ULCI = 0.29) suggests that professional commitment is a mediator in the relationship between work satisfaction and organisational commitment. However, the table also shows a significant direct effect of job satisfaction on organisational commitment ($\beta = 0.41$, $p = 0.00$, LLCI = 0.33, ULCI = 0.49). This implies that job satisfaction also directly influences organisational commitment without involving professional commitment. In other words, even without being mediated by professional commitment, job satisfaction can still influence organisational commitment. Furthermore, the higher value of direct effect (41%) compared to the value of indirect effect (21%) indicates that the direct influence of job satisfaction on organisational commitment is larger compared to when the effect of job satisfaction is mediated by professional commitment. In conclusion, the results suggest that the role of professional commitment as a mediator in the relationship between job satisfaction and organisational commitment is relatively low and partial.

Table 2. Correlation between research variables.

	Job satisfaction	Professional commitment
Job satisfaction		
Professional commitment	0.47**	
Organisational commitment	0.64**	0.66**

*p < 0.05, **p < 0.01

Table 3. Mediation test results.

	Outcome					
	M (Professional commitment)			Y (Organisational commitment)		
Antecedent	β	Se	p	β	Se	p
X (Job satisfaction)	0.36	0.39	0.00	0.41	0.04	0.00
M (Professional commitment)				0.58	0.05	0.00
Sex (Control)	−0.01	0.66	0.99	−1.09	0.62	0.08
Tenure (Control)	1.25	0.71	0.08	1.67	0.68	0.01
	$R^2 = 0.23$			$R^2 = 0.59$		

Table 4. Total effect, direct effect, and indirect effect of job satisfaction on organisational commitment.

	B	*p*	LLCI	ULCI
Total effect	0.62	0.00	0.53	0.70
Direct effect	0.41	0.00	0.33	0.49
Indirect effect	0.21	0.00	0.15	0.29

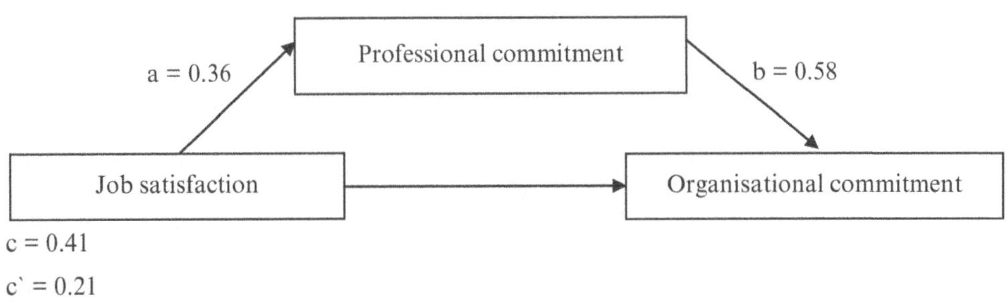

Figure 1. Result of mediation test.

4 DISCUSSION

The results indicate that job satisfaction has a positive significant influence on professional commitment. This result is consistent with the findings of previous research in Indian colleges (Shamina, 2014). Moreover, Shukla (2014) similarly reported that teachers who are satisfied with their jobs as educators (i.e., those having high level of job satisfaction) would give their best efforts to be competent and thus have higher commitment to their profession. When lecturers are able to enjoy their teaching activities in classes through which they experience happiness upon sharing their knowledge with students, satisfaction of their job would also be incurred. Subsequently, the lecturers' job satisfactions are likely to enhance the quality of learning process in class and impact their students' achievements (Qayyum, 2013). Students' satisfactory achievements could also function as a non-financial reward for the lecturers, which in turn could enhance job satisfaction among lecturers. In addition, satisfaction in interacting with their colleagues (fellow lecturers) also plays a role in lecturers' job satisfaction. When lecturers feel their job experience conforms to their expectations and fulfils their basic needs, they tend to be more committed to their profession. In other words, lecturers' job satisfaction can also influence their professional commitment rate.

This study also found that there is a significant positive influence of professional commitment on organisational commitment, which is also in alignment with other prior research findings. A study conducted by Bredillet and Dwivedula (2010) on 141 project workers, for example, demonstrated that an employee's professional commitment could enhance his organisational commitment and therefore suggested no conflict between organisational and professional commitments. In other words, the higher one's professional commitment is, the higher his/her organisational commitment. When one is satisfied with his/her job as a lecturer and is committed to the profession, subsequently his/her organisational commitment (i.e., the commitment to the university where he/she teaches) would also take form. Lecturers who decide to remain in their profession as educators (i.e., those with a high professional commitment) would need an institution wherein they are able to apply their skills and knowledge. In this case, universities are a medium for lecturers to develop their professionalism. When universities are able to provide a working environment that can facilitate the lecturers' career development, these universities would therefore acquire the lecturers' loyalty. In other words, the lecturers' rate of organisational commitment would also increase.

Furthermore, this research also discovered a significant positive influence of job satisfaction on organisational commitment. A similar research carried out on educational organisations by Anari (2012) in Iran, which involved English teachers in Senior High Schools, found a positive relationship between job satisfaction and organisational commitment. In Indonesia, research on the relationship between job satisfaction and organisational commitment in the scope of educational organisations have been conducted by Daniel & Purwanti (2015), which involved 183 administrative employees in a private university in Tangerang. The researchers reported a direct positive influence of job satisfaction on organisational commitment. In another study, Seniati (2006) also found that there was a positive influence of job satisfaction on organisational commitment among lecturers in Universitas Indonesia.

Moreover, the present study also discovered that lecturers satisfied with their jobs in aspects such as job facets, relationships with coworkers and superiors, work facilities, incentives, promotions, and communication tended to have a higher organisational commitment. Adequate work facilities and trainings provided by the university to support the lecturers' works could additionally enhance the lecturers' organisational commitment (Ali & Zafar, 2006). A possible explanation for such a trend is that lecturers who receive complete work facilities and trainings as well as adequate career development from the university feel they need to "give back" by staying loyal to the university and are thus more willing to contribute their best efforts for the university.

Finally, this research demonstrated that there is a partial role of professional commitment as a mediator in the relationship between job satisfaction and organisational commitment, as observed by the larger direct impact of job satisfaction on organisational commitment compared to when mediated by professional commitment. This is assumed to occur because professional and organisational commitments are both direct antecedents of organisational commitment. Therefore, multiple regression tests can potentially be carried out in future research to discover whether professional commitment or job satisfaction is a stronger predictor for organisational commitment.

5 CONCLUSIONS

As previously stated, job satisfaction has been found to positively influence both professional commitment and organizational commitment. In addition, professional commitment also positively influences organizational commitment. A mediation analysis showed that the relationship between job satisfaction and organizational commitment is partially mediated by professional commitment among lecturers in higher-education institutions.

REFERENCES

Ali, C.A., & Sohail, Z. (2006) Antecedents and consequences of organizational commitment among Pakistani University teachers. *Applied H.R.M. Research,* 11 (1), 39–64. doi:10.2139/ssrn.2552437.

Anari, N.N. (2012) Teachers: Emotional intelligence, job satisfaction, and organizational commitment. *Journal of Workplace Learning,* 24 (4), 256–269. doi:10.1108/13665621211223379.

Appaw, A., Esther, T., George, K.A., & John, C.A. (2013) Measuring the influence of job satisfaction on work commitment among ho polytechnic lecturers. *Research Gate,* 3 (2), 91–102.

Carson, K.D., Paula, P.C., & Arthur, G.B. (1995) Development and construct validation of a career entrenchment measure. *Journal of Occupational and Organizational Psychology,* 68 (4), 301–320. doi:10.1111/j.2044-8325.1995.tb00589.x.

Curry, J.P., Douglas, S.W., James, L.P., & Charles, W.M. (1986) On the causal ordering of job satisfaction and organizational commitment. *Academy of Management Journal,* 29 (4), 847–858. doi:10.2307/255951.

Daniel, F., & Ari, P. (2015) The impact of organizational culture and job satisfaction to organizational commitment and employees job performance. In *Research Gate*. doi:10.13140/RG.2.1.1196.4240.

Dunia Dosen Indonesia. (2016) *Ini 7 Hal Yang Harus Kamu Siapkan Kalau Mau Jadi Dosen!* April 23. [Online]. Available from: https://www.duniadosen.com/kalau-mau-jadi-dosen/.

Dwivedula, R., & Christophe, N.B. (2010) The relationship between organizational and professional commitment in the case of project workers: Implications for project management. *Project Management Journal,* 41 (4), 79–88. doi:10.1002/pmj.20196.

Hayes, A.F. (2013) Introduction to mediation, moderation, and conditional process analysis. *New York, NY: Guilford,* 3–4. doi:978-1-60918-230-4.

Jing, L., & Deshan, Z. (2014) Does organizational commitment help to promote university faculty's performance and effectiveness? *Asia-Pacific Education Researcher,* 23 (2), 201–212. doi:10.1007/s40299-013-0097-6.

Khan, M.S., Irfanullah, K., Ghulam, M.K., Shadiullah, K., Allah, N., Farhatullah, K., & Naseem, B.Y. (2014) The Impact of job satisfaction and organizational commitment on the intention to leave among the academicians. *International Journal of Academic Research in Business and Social Sciences,* 4 (2), 114–131. doi:10.6007/IJARBSS/v4-i2/562.

Lew, T.Y. (2009) The relationships between perceived organizational support, felt obligation, affective organizational commitment and turnover intention of academics working with private higher educational institutions in Malaysia. *European Journal of Social Sciences,* 9 (1), 72–87.

Luthans, F. (2002) *Organizational Behavior*. 9 edition. New York, McGraw-Hill. Available from: http://trove.nla.gov.au/work/9773091.

Mayer, J.P., & Natalie, J.A. (1997) *Commitment in the Workplace: Theory, Research, and Application*. 2455 Teller Road, Thousand Oaks California 91320 United States, SAGE Publications, Inc. http://sk.sagepub.com/books/commitment-in-the-workplace.

Meyer, J.P., & Natalie, J.A. (1991) A three-component conceptualization of organizational commitment. *Human Resource Management Review,* 1 (1), 61–89. doi:10.1016/1053-4822(91)90011-Z.

Meyer, J.P, Natalie, J.A., & Catherine, A.S. (1993) Commitment to organizations and occupations: Extension and test of a three-component conceptualization. *Journal of Applied Psychology*. doi:10.1037/0021-9010.78.4.538.

Munir, S., & Mehson, S. (2010) Examining locus of control (loc) as a determinant of organizational commitment among university professors in Pakistan. *Journal of Business Studies Quarterly,* 1 (3), 78–93.

Niehoff, R.L. (1997) Job Satisfaction, Organizational Commitment, and Individual and Organizational Mission Values Congruence: *Investigating the Relationships*. [Accessed April]. Available from: https://eric.ed.gov/?id = ED417644.

Othman, J., Kabeer, A.M., & Jeffrey, L.D.S. (2012) Does a transformational and transactional leadership style predict organizational commitment among public university lecturers in Nigeria? *Asian Social Science,* 9 (1), 165. doi:10.5539/ass.v9n1p165.

Qayyum, A.C. (2013) Job satisfaction of university teachers across the demographics: A case of Pakistani Universities. *Bulletin of Education and Research,* 35 (1), 1–15.

Seniati, L. (2010) Pengaruh masa kerja, trait kepribadian, kepuasan kerja, dan iklim psikologis terhadap komitmen dosen pada Universitas Indonesia. *Makara Hubs-Asia,* 8 (3). Available from: http://hubsasia.ui.ac.id/index.php/hubsasia/article/view/33.

Shamina, H. (2014) Impact of job satisfaction on professional commitment in higher education. *International Interdisciplinary Research Journal,* 2 (2), 1–11.

Shukla, S. (2014) Teaching competency, professional commitment and job satisfaction-a study of primary school teachers. *Research Gate,* 4 (3), 44–64. doi:10.9790/7388-04324464.

Spector, P. (1997) *Job Satisfaction: Application, Assessment, Causes, and Consequences*. 2455 Teller Road, Thousand Oaks California 91320 United States, SAGE Publications, Inc. http://sk.sagepub.com/books/job-satisfaction.

Tett, R.P., & John, P.M. (1993) Job satisfaction, organizational commitment, turnover intention, and turnover: Path analyses based on meta-analytic findings. *Personnel Psychology,* 46 (2), 259–293. doi:10.1111/j.1744-6570.1993.tb00874.x.

Xiao, J., & Stephen, W. (2015) The effects of lecturer commitment on student perceptions of teaching quality and student satisfaction in Chinese higher education. *Journal of Higher Education Policy & Management,* 37 (1), 98–110. doi:10.1080/1360080X.2014.992092.

Diversity in Unity: Perspectives from Psychology and Behavioral Sciences – Ariyanto et al. (Eds)
© 2018 Taylor & Francis Group, London, ISBN 978-1-138-62665-2

Role of job satisfaction as a mediator in the relationship between psychological climate and organisational commitment of lecturers at higher education institutions

A. Krishnamurti & A.N.L. Seniati
Faculty of Psychology, Universitas Indonesia, Depok, Indonesia

ABSTRACT: This study was conducted to investigate the role of job satisfaction as a partial mediator in the relationship between psychological climate and organisational commitment. The study involved as respondents 328 lecturers who work in various higher education institutions. The measuring instruments used in the study were: (1) organisational commitment by Seniati and Yulianto (2010), based on Meyer and Allen (1997); (2) psychological climate by Brown and Leigh, based on Kahn (1990); (3) job satisfaction by Seniati and Yulianto (2010), based on Spector (1997). The study results showed that there were significantly positive effects of psychological climate on job satisfaction ($\beta = 0.84$, $p < 0.05$), and job satisfaction on organisational commitment ($\beta = 0.47$, $p < 0.05$). Furthermore, these findings also indicate that the influence of psychological climate on organisational commitment can occur through job satisfaction in higher education institutions ($\beta = 0.23$, $p < 0.05$). On the basis of these results, we encourage higher education institutions to enhance the level of job satisfaction and organisational commitment of their lecturers by creating an improved psychological climate for them.

1 INTRODUCTION

The contemporary research on the subject of commitment within the scope of industrial and organisational psychology has been dominated by samples gathered from employees of corporate organisations. One type of organisation that has not been broadly discussed is the educational or academic organisation, such as universities, institutes and colleges. Existing studies have focused rather on the realm of education than its organisational management. Research on educational organisation is important due to its role as one of the functions that guarantees the well-being and quality of human life and social life; thus, the attention given by the educational organisation to its workers is of high importance as well. Seniati (2006) also points out that education facilitates preparation in relation to the incoming challenges that result from the development of information, globalisation and free markets, as well as problems related to national peace and unity. Every nation is obliged to support and to enhance the capabilities of teachers and lecturers at every level of education in order to prepare its future generations.

The profession of lecturer is a unique one in Indonesia, in which autonomy is conferred together with a complex role as a member of a faculty while being obligated to implement the values of *Tridharma Perguruan Tinggi* (three pillars of higher education: education, research, and community service). With globalisation and free movement of the workforce, challenges such as propositions to teach in neighbouring countries are also becoming more relevant. All of these challenges for the lecturer workforce in Indonesia have led us to study their organisational commitment.

Meyer and Allen (1997) define commitment as a psychological condition or thought process that influences an individual's tendency to maintain their membership of an organisation. The concept of organisational commitment can be characterised in three components:

emotional ties with an organisation; a perceived obligation to stay in an organisation; the perceived cost of leaving an organisation. These three components are labelled as affective commitment, normative commitment, and continuance commitment, respectively (Meyer & Allen, 1997).

We have observed that organisational commitment is one of the criteria that must be exhibited by a lecturer in order to propel their institution toward success. Research on the commitment of lecturers in the education world not only exerts a positive influence on the organisation, but also on the lecturers themselves and their students. A study within a university by Kushman (1992) revealed that a positive relationship exists between a lecturer's commitment and their students' achievement and work satisfaction, the lecturer's own satisfaction, the lecturer's perceived ability to complete tasks, and the lecturer's expectations of their students' achievement. Just as low organisational commitment on the part of an employee will have a negative effect on a company, so a low organisational commitment on the part of lecturers will also bring disadvantages to an institution or university. With low organisational commitment there is a greater possibility that lecturers will leave a university, taking with them their knowledge, research, and experience (Zafar, 2006).

Organisational commitment may develop in an individual due to several factors: work-life quality (Farid et al., 2014), empowerment (Gohar et al., 2015), autonomy (Gohar et al., 2015; Kroner, 2015), personality traits (Seniati, 2006; Hutapea, 2012) and job satisfaction (Seniati, 2006; Zafar, 2006; Bozeman & Gaughan, 2011; Daniel & Purwanti, 2015). Studies on lecturer's commitment in Indonesia also show that personal traits such as kindness may have a positive influence on that commitment (Seniati, 2006). Another antecedent that influences the organisational commitment of lecturers is the internal push from within themselves. Kroner (2015) found that instrinsic motivation brings significant positive impact to the organisational commitment of lecturers in India. This arises because lecturers enjoy the act of teaching and are less interested in its economic aspects. Kroner added that perceptions of the possession of competence and of involvement can also influence organisational commitment. In Kroner's study, perceived involvement referred to the extent to which lecturer's hard work affected the university. This aspect is similar to the concept of contribution to psychological climate that is proposed in this study, which is how far the contribution of employees to their company brings meaningful impact.

Kahn (1990) explained that psychological climate represents the engagement and disengagement that an individual has in relation to their job. This individual engagement is closely related to the affective side of a person, such as their feelings or emotions toward something in their work environment. A further example could be whether they feel their work is appreciated and what the work means to the company. The feeling of being able to express oneself safely and organise one's own job can provide comfort to an individual.

Several aspects of psychological climate were explained by Brown and Leigh (1996). A first aspect is psychological safety, which refers to the ability to reveal and engage oneself without fear and negative consequences on self-image, status, or career. A second aspect is psychological meaningfulness, which describes someone's feelings about what their physical, cognitive, and emotional investment means for the company they are working for. These are the two aspects that we assume to be indirectly influencing an individual's will to stay in an organisation.

The relationship between lecturers' psychological climate and their commitment to their educational organisation is indirect in nature. We assume that this relationship can be affected by other aspects; for example, facilities or relationships with peers or supervisors. Adequate facilities, supportive peers, or fair compensation for efforts tend to encourage lecturers to stay at the university for which they are working. Biswas and Varma (2007) observed that employees' perceived psychological climate significantly influences their organisational citizenship behaviour and job satisfaction, thus having a positive effect on their performance. We suspect that this relationship may be influenced by other aspects that need to be evaluated further. Therefore, we propose job satisfaction as the mediator variable in the relationship between lecturers' psychological climate and their organisational commitment in universities. This proposition is supported by the finding that job satisfaction may influence individual's behaviour toward a job and other job aspects (Spector, 1997). Meanwhile, as well as the external organisation, job satisfaction is also determined by internal factors, which prompt certain emotions that direct organisational commitment (Mowday et al., 1979).

According to the Law of the Republic of Indonesia No. 14 of 2005 on Teachers and Lecturers, a lecturer is a professional teacher and scientist whose main tasks include transforming, developing and disseminating science, technology and arts through education, research and community service. Every lecturer must have academic qualifications in the form of a diploma at an academic education level that suits the type, level, and unit of formal education in which they are assigned. Besides academic qualifications, lecturers and teachers must also have competency, which is defined as the set of knowledge, abilities and behaviours that must be owned, embodied and mastered by lecturers and teachers in carrying out their professional tasks. Article 45 of the Law of the Republic of Indonesia No. 14 of 2005 on Teachers and Lecturers states that a lecturer must have academic qualifications, competency, a teaching certificate, physical and spiritual well-being, and must satisfy other qualifications put forward by the higher education unit in which they are assigned, as well as the capability to achieve national education.

According to Mintzberg (1993) universities are educational organisations with a structure of professional bureaucracy. In such structures, many types of job are rather stable and predictable. However, due to the complexity of the dynamics of such jobs, there is a need for supervision by the operators of those jobs. In this context, operators refers to the people who hold the knowledge and ability to conduct the function of the job. In this case, these people are lecturers.

Organisational commitment is a psychological state that (a) characterises the employee's relationship with the organisation, and (b) has implications for the decision to continue membership in the organisation (Meyer & Allen, 1997). As we have already indicated, there are three components of organisational commitment: affective (individual emotional ties, identification, and involvement within an organisation), normative (the obligation to stay in an organisation), and continuance (the individual perception of cost of leaving an organisation).

In this study, psychological climate is the perception by lecturers of their job context that influences their psychological processes, leading either to engagement or disengagement from that job (Kahn, 1990). In this study, we refer to the development of Kahn's theory by Brown and Leigh (1996), which divides each of the psychological safety and meaningfulness aspects described above into three dimensions. An individual will feel that his/her job is meaningful if the job is challenging, yields contribution, and also results in recognition.

According to Spector (1997), job satisfaction is simply how people feel about their jobs and different aspects of their jobs. We conclude that job satisfaction is an attitude by which an individual expresses their feelings toward a job in general, or even toward specific aspects of each job. Spector (1997) divided job satisfaction into nine aspects, namely, pay, promotion, supervision, nature of work, communication, fringe benefits, rewards, operating procedures, and co-workers.

2 METHODS

This study employed a quantitative approach. Respondents were lecturers from public and private universities selected on the basis of two main criteria: those who had been employed for more than two years, and who were either civil servants (PNS) or non-civil servants (non-PNS). The selected sampling technique was a non-probability sampling, that is, sampling based on availability. We employed three main questionnaires as our measurement tools. The questionnaires concerned organisational commitment, as designed by Seniati and Yulianto (2010) with 21 items ($\alpha = 0.88$), psychological climate, as designed by Brown and Leigh (1996) with 21 items ($\alpha = 0.89$), and job satisfaction, as designed by Seniati and Yulianto (2010) with 21 items ($\alpha = 0.91$). These questionnaires use Likert-type scales of 1 (Strongly disagree) to 4 (Strongly agree). We employed a simple mediator technique to answer the research question, which was whether job satisfaction functions as a mediator in the relationship between psychological climate and organisational commitment. The data was processed and analysed statistically using IBM SPSS software with the PROCESS add-on macro by Hayes (2013).

3 RESULTS

Of the 542 questionnaires distributed, 420 (77%) were completed and returned to us. However, only 328 qualified for processing. Table 3 shows the positive correlation between psychological climate and organisational commitment ($r = 0.55$, $p < 0.05$). Similar correlation is also apparent between job satisfaction and organisational commitment ($r = 0.64$, $p < 0.05$). Finally, another positive correlation also appears between psychological climate and job satisfaction ($r = 0.74$, $p < 0.05$).

Thus, the prerequisite of the mediation test has been fulfilled with these three positive correlations amongst the studied variables. The result of our analysis concerning the mediation effect of job satisfaction on organisational commitment can be observed in Table 2.

Table 2 shows the result of our model evaluation, which indicates that job satisfaction mediates the relationship between psychological climate and organisational commitment. The result demonstrates that the value of psychological climate has a significant direct effect on job satisfaction ($p < 0.05$). The influence of job satisfaction on organisational commitment is also proven ($p < 0.05$). Interestingly, the influence of psychological climate on organisational commitment is also significant and positive ($p < 0.05$). Based on the same table, we can also explain that the relationship between psychological climate and organisational commitment mediated by job satisfaction significantly controls the demographic element of tenure ($p < 0.05$). Gender, on the other hand, appears to have insignificant effects on the tested model. Table 3 provides a better picture of the measured effects.

Based on Table 3, we may infer the mediation effect of job satisfaction from psychological climate toward organisational commitment in terms of total effect (Effect = 0.63, SE = 0.05, $p < 0.05$), direct effect (Effect = 0.23, SE = 0.07, $p < 0.05$), and indirect effect (Effect = 0.40,

Table 1. Correlation between variables (N = 328).

	Organisational commitment	Psychological climate	Job satisfaction
Organisational commitment	1		
Psychological climate	0.55**	1	
Job satisfaction	0.64**	0.74**	1

**Correlation becomes significant when $p < 0.01$ (2-tailed).

Table 2. Results of mediation.

	Outcome					
	M (Job satisfaction)			Y (Organisational commitment)		
Antecedent	Coeff.	Std. error	*p*	Coeff.	Std. error	*p*
X (Psychological climate)	0.84	0.04	< 0.05	0.23	0.07	< 0.05
M (Job satisfaction)	___	___	___	0.47	0.06	< 0.05
Gender				−1.17	0.72	> 0.05
Tenure				2.40	0.78	< 0.05
	$R^2 = 0.54$			$R^2 = 0.42$		
	$F(3, 324) = 133.11, p < 0.05$			$F(4, 323) = 64.27, p < 0.05$		

Table 3. Effects.

Effect	Coeff.	LLCI	ULCI	Remarks
Total effect: Psychological climate—Organisational commitment	0.63	0.52	0.73	–
Direct effect: Psychological climate—Organisational commitment	0.23	0.08	0.37	–
Indirect effect: Psychological climate—Organisational commitment due to Job satisfaction	0.40	0.29	0.53	Partial mediation

Notes: Coeff. is β-value; LLCI = Lower-Level Confidence Interval; ULCI = Upper-Level Confidence Interval.

$SE = 0.07$, $p < 0.06$). All three effects can also explain the role of job satisfaction as a mediator through observation of the range of the Lower-and Upper-Level Confidence Interval (LLCI and ULCI) values, both of which fall in the positive region.

The mediator role of job satisfaction can be observed from the increasing value of direct effect (0.23) with respect to indirect effect value (0.4), which suggests that the relationship between psychological climate and organisational commitment may be mediated through job satisfaction. However, this mediator role is only partial, due to an insignificant change of *p*-value either before or after the application of control to the mediator (Hayes, 2013). Therefore, job satisfaction does have a role as partial mediator in the relationship between psychological climate and organisational commitment.

4 DISCUSSION

Our study has revealed the role of job satisfaction as a partial mediator on the relationship between psychological climate and organisational commitment. This finding is in agreement with Parker et al. (2003), who found that an employee's psychological climate effect is mediated by the employee's work attitude and motivation. Kopelman et al. (1990) also revealed that the relationship between psychological climate and dominant organisational behaviour, for example, performance and organisational membership, is actually mediated by cognitive and affective conditions, such as invidual work motivation, work satisfaction, and involvement. A study by Wang and Ma (2013), based on similar concepts and focused on the role of job satisfaction on psychological climate in relation to innovation and tendency to leave the company, also showed that job satisfaction takes a role as mediator.

This study is a follow up to a previous study by Seniati (2006) about lecturers' experiences. With respect to this previous study, the additional value obtained in the present study is the finding of a direct relationship between psychological climate and organisational commitment, which was not available from the previous study. This direct effect of psychological climate on organisational commitment, which appears to be significant, is probably the reason why job satisfaction functions only as a partial mediator. Furthermore, a difference exists between the theoretical background used in this study and that in Seniati (2006), which may also be a factor in the emergence of a direct relationship between psychological climate and organisational commitment.

Compared to generic employee types, greater variation and autonomy in how lecturers carry out their work tasks may generate a greater sense of comfort in such individuals. The perception by which lecturers feel that their contribution is meaningful and significant may affect the goals of their university. More often than not, lecturers that have already developed ties with their job voluntarily apply their creativity and expertise in their workplace. As stated by Kataria et al. (2013a, 2013b), an employee's identification with their role and feelings in the workplace will facilitate a condition of psychological meaningfulness in achieving their work goals.

Through the correlation between demographic factors and organisational commitment, we put gender and working period as controlled data. Statistical analysis shows that gender, on the one hand, does not affect the relationship between psychological climate and organisational commitment that is mediated by job satisfaction. This finding is in agreement with Kaselyte and Malukaite (2013), who also found that there was no significant difference between the organisational commitment of male and female lecturers. On the other hand, working period showed a significant effect on the relationship between psychological climate and organisational commitment. This finding is consistent with the the result of English et al. (2010), who found that working period significantly affected the relationship between psychological climate and affective organisational commitment.

5 CONCLUSION

Based on our study, it can be concluded that job satisfaction functions as a partial mediator in the relationship between psychological climate and organisational commitment of lecturers in universities. The results of correlation tests also show that psychological climate exhibits a positive and significant effect on job satisfaction. Finally, we also found that psychological climate has a positive and significant influence on organisational commitment.

5.1 *Future research*

Our study leaves some room for improvement. The measurement method could be improved by taking into account the aspect of common method bias, which may be achieved by separating dependent and independent variables. More supervision could be imposed on the completion of the questionnaires. Coordination with the universities involved should be frequent, to ensure the maximum number of fully answered questionnaires. Furthermore, the process of selection of the universities to be involved in sampling must be done carefully. We also suggest that data collection should involve more than just questionnaires. A number of interviews with samples should improve the measured constructs.

It should be noted that there are several conceptual similarities between the variables of psychological climate and job satisfaction which we apply in this study. Psychological climate is a concept which is measured at an individual level. Future studies should consider other concepts that are measured at a group or organisational level: organisational climate, innovation climate and peer relationships could be considered. Doing so may help the universities involved to evaluate group or organisational aspects that need improvement.

We consider affective commitment is the most important commitment in leading an individual to stay in an organisation. Strong emotional ties to a job will incentivise people to put more effort into supporting the purpose of their organisations. Allen and Meyer (1990) showed that employees exhibiting affective commitment would apply extra effort to finishing their tasks, increasing the organisation's effectiveness and showing their best behaviour in order to preserve their relationship with the organisation for which they work. In terms of lecturers with stronger continuance commitment than affective commitment, which emphasises on aspect of cost or lack of job vacancy, we assume that lecturers decide to stay out of necessity. The productivity and work output quality of those who stay due to continuance commitment are questionable because their motivation to be involved and be part of the university is not as strong as those with strong affective commitment.

Based on this study, universities and other higher educational institutions need to pay attention to more supportive management to create a better psychological climate for their lecturers. Universities, such as those involved in this study, need to improve the work facilities provided for their lecturers. Our study shows that affective commitment is still the most important organisational commitment in explaining lecturers' commitment to universities.

Consequently, universities need to consider those aspects that promote the growth of affective commitment by creating better psychological climate conditions as well as paying attention to the job satisfaction of their lecturers.

REFERENCES

Allen, N.J. & John, P.M. (1990). The measurement and antecedents of affective, continuance and normative commitment to the organization. *Journal of Occupational Psychology, 63*(1), 1–18. doi:10.1111/j.2044-8325.1990.tb00506.x.

Biswas, S. & Arup, V. (2007). Psychological climate and individual performance in India: Test of a mediated model. *Employee Relations, 29*(6), 664–676. doi:10.1108/01425450710826131

Bozeman, B. & Monica, G. (2011). Job satisfaction among university faculty: Individual, work, and institutional determinants. *The Journal of Higher Education, 82*(2), 154–186.

Brown, S.P. & Thomas, W.L. (1996). A New look at psychological climate and its relationship to job involvement, effort, and performance. *Journal of Applied Psychology, 81*(4), 358–368. doi:10.1037/0021-9010.81.4.358

Daniel, F. & Ari, P. (2015). The impact of organizational culture and job satisfaction to organizational commitment and employees job performance: An empirical study at a university in Tangerang (Unpublished conference paper). doi:10.13140/RG.2.1.1196.4240

English, B., David, M. & Christopher, C. (2010). Moderator effects of organizational tenure on the relationship between psychological climate and affective commitment. *Journal of Management Development, 29*(4), 394–408. doi:10.1108/02621711011039187.

Farid, H., Izadi, Z., Ismail, I.A. & Alipour, F. (2014). Relationship between quality of work life and organizational commitment among lectures in a Malaysia public research university. *The Social Science Journal, 52,* 54–61. http://dx.doi.org/10.1016/j.soscij.2014.09.003

Gohar, F.R., Mohsin, B., Muhammad, A. & Faisal, A. (2015). Effect of psychological empowerment, distributive justice and job autonomy on organizational commitment. *International Journal of Information, Business and Management, 7*(1), 144–173.

Hayes, A.F. (2013). *Introduction to mediation, moderation, and conditional process analysis: A regression-based approach.* New York, NY: The Guilford Press.

Hutapea, B. (2012). Sifat-kepribadian dan dukungan organisasi sebagai prediktor komitmen organisasi guru pria di sekolah dasar (Personality trait and organisational support as a predictor of organisational commitent of man teacher in elementary school). *Makara, Sosial Humanira, 16*(2), 101–115. Retrieved from http://hubsasia.ui.ac.id/index.php/hubsasia/article/view/1496

James, L.R., James, L.A. & Ashe. (1990). The meaning of organizations: The role of cognition and values. In Schneider, B. (Ed.), *Organizational climate and culture* (pp. 40–84). San Fransisco, CA: Jossey-Bass. Retrieved from http://www.wiley.com/WileyCDA/WileyTitle/productCd-0470622032.html

Jones, A.P. & Lawrence, R.J. (1979). Psychological climate: Dimensions and relationships of individual and aggregated work environment perceptions. *Organizational Behavior and Human Performance, 23*(2), 201–250. doi:10.1016/0030-5073(79)90056-4

Kahn, W.A. (1990). Psychological conditions of personal engagement and disengagement at work. *Academy of Management Journal, 33*(4), 692–724. doi:10.2307/256287

Karatepe, O.M. (2016). The effect of psychological climate on job outcomes: Evidence from the airline industry. *Journal of Travel & Tourism Marketing, 33*(8), 1162–1180. doi:10.1080/10548408.2015.109 4002

Kaselyte, U. & Kristina, M. (2013). Antecedents of affective organizational commitment among economics and management lecturers in the higher education institutions in the Baltics. *SSE Riga Student Papers, 5*(153).

Kataria, A., Pooja, G. & Renu, R. (2013a). Does psychological climate augment OCBs? The mediating role of work engagement. *The Psychologist-Manager Journal, 16*(4), 217–242. doi:10.1037/mgr0000007

Kataria, A., Pooja, G. & Renu, R. (2013b). Psychological climate and organizational effectiveness: Role of work engagement. *IUP Journal of Organizational Behavior, 12*(3), 33–46.

Kopelman, R.E., Arthur, P.B. & Richard, A.G. (1990). The role of climate and culture in productivity. In Schneider, B. (Ed.), *Organizational Climate and Culture* (pp. 282–318). San Francisco, CA: Jossey-Bass. Retrieved from https://www.researchgate.net/publication/278965490_The_role_of_Climate_and_Culture_in_Productivity_in_Organizational_Climate_and_Culture_Ben_Schneider_edJossey-Bass_Frontiers_in_Industrial_and_Organizational_Psychology_Series

Kroner, N. (2015). What influences the organizational commitment of professors? A secondary analysis on organizational commitment of professors at German research universities and universities of applied sciences. Retrieved from https://www.researchgate.net/publication/283046426_What_influences_the_Organizational_Commitment_of_Professors_A_secondary_analysis_on_organizational_commitment_of_professors_at_German_research_universities_and_universities_of_applied_sciences

Kushman, J.W. (1992). The Organizational Dynamics of Teacher Workplace Commitment: A Study of Urban Elementary and Middle Schools. *Educational Administration Quarterly*. Retrived from http://journals.sagepub.com/doi/abs/10.1177/0013161x92028001002

Langkamer, K.L. & Kelly, S.E. (2008). Psychological climate, organizational commitment and morale: Implications for Army captains' career intent. *Military Psychology*, 20(4), 219–236. doi:10.1080/08995600802345113

Mathieu, J.E. & Dennis, M.Z. (1990). A review and meta-analysis of the antecedents, correlates, and consequences of organizational commitment. *Psychological Bulletin*, *108*(2), 171–194. doi:10.1037/0033-2909.108.2.171

Meyer, J.P. (2004). *TCM Employee Commitment Survey Academic Users Guide 2004*. London, Canada: The University of Western Ontario.

Meyer, J.P. & Allen, N.J. (1997). *Commitment in the Workplace: Theory* Research *and Application.* California: Sage Publications.

Meyer, J.P. & Natalie, J.A. (1991). A three-component conceptualization of organizational commitment. *Human Resource Management Review*, *1*(1), 61–89. doi:10.1016/1053-4822(91)90011-Z

Meyer, J.P. & Natalie, J.A. (1997). *Commitment in the workplace: Theory, research, and application*. Thousand Oaks, CA: SAGE Publications. Retrieved from http://sk.sagepub.com/books/commitment-in-the-workplace

Meyer, J.P., Natalie, J.A. & Catherine, A.S. (1993). Commitment to organizations and occupations: Extension and test of a three-component conceptualization. *Journal of Applied Psychology*, *78*(4), 538–551. doi:10.1037/0021-9010.78.4.538

Meyer, J.P., Thomas, E.B. & Christian, V. (2004). Employee commitment and motivation: A conceptual analysis and integrative model. *Journal of Applied Psychology*, *89*(6), 991–1007. doi:10.1037/0021-9010.89.6.991

Meyer, J.P., David, J.S., Lynne, H. & Laryssa, T. (2002). Affective, continuance, and normative commitment to the organization: A meta-analysis of antecedents, correlates, and consequences. *Journal of Vocational Behavior*, *61*(1), 20–52. doi:10.1006/jvbe.2001.1842

Mintzberg, H. (1993). *Structure in fives: Designing effective organizations*. New Jersey, NJ: Prentice-Hall, Inc.

Mowday, R.T. & Steers, R.M. (1979). The measurement of organizational commitment. *Journal of Vocational Behavior 14,* 224–247.

Parker, C.P., Baltes, B.B., Young, S.A. & Huff, J.W. (2003). Relationships between psychological climate perceptions and work outcomes: A meta analytic review. *Journal of Organizational Behavior*, *24,* 389–416.

Perryer, C., Catherine, J., Ian, F. & Antonio, T. (2010). Predicting turnover intentions: The interactive effects of organizational commitment and perceived organizational support. *Management Research Review*, *33*(9), 911–923. doi:10.1108/01409171011070323

Seniati, L. (2006). Pengaruh masa kerja, trait kepribadian, kepuasan kerja, dan iklim psikologis terhadap komitmen dosen pada Universitas Indonesia (The effect of tenure, personality trait, job satisfaction, and psychological climate on organisational commitment of lecturers in Universitas Indonesia). *Makara, Sosial Humaniora*, *10*(2), 88–97. Retrieved from http://hubsasia.ui.ac.id/index.php/hubsasia/article/view/33

Seniati, L. & Yulianto, A. (2010). *Pengaruh faktor pribadi dan faktor lingkungan terhadap komitmen profesi dan komitmen organisasi pada dosen perguruan tinggi di Jakarta*. (The Effect of individual and environmental factors on professional commitment and organisational commitment of higher education lecturers in Jakarta). Laporan Penelitian: Direktorat Riset dan Pengabdian Masyarakat, Universitas Indonesia.

Solinger, O.N., Woody, V.O. & Robert, A.R. (2008). Beyond the three-component model of organizational commitment. *Journal of Applied Psychology*, *93*(1), 70–83. doi:10.1037/0021-9010.93.1.70

Spector, P. (1997). *Job Satisfaction: Application, assessment, causes, and consequences*. Thousand Oaks, CA: SAGE Publications. Retrieved from http://sk.sagepub.com/books/job-satisfaction

Spector, P. (2012). *Industrial and organizational psychology research and practice* (6th ed.). New Jersey, NJ: John Wiley & Sons.

Spink, K.S., Kathleen, S.W., Lawrence, R.B. & Patrick, O. (2013). The perception of team environment: The relationship between the psychological climate and members' perceived effort in high-performance groups. *Group Dynamics: Theory, Research, and Practice*, *17*(3), 150–161. doi:10.1037/a0033552

Steers, R.M. (1977). Antecedents and outcomes of organizational commitment. *Administrative Science Quarterly*, *22*(1), 46–56. doi:10.2307/2391745

Sudiatmi, T. (2012). Pengaruh Komitmen Pada Profesi Dan Dukungan Organisasional Terhadap Motivasi Berprestasi Dosen. *Widyatama*, *21*(2), 141–153.

Wang, G. & Xiaoqin, M. (2013). The effect of psychological climate for innovation on salespeople's creativity and turnover intention. *Journal of Personal Selling & Sales Management*, *33*(4), 373–387. doi:10.2753/PSS0885-3134330402

Undang-Undang Republik Indonesia (the Law of the Republic of Indonesia) No. 14 of the year 2005 on Teachers and Lecturers.

Zafar, S. (2006). Antecedents and consequences of organisational commitment among Pakistani University teacher. *Applied H.R.M. Research* *11*(1), 39–64.

[illegible] Ministry of [illegible] and [illegible] Government of India. [illegible]

[illegible] (2006). [illegible] and consequences [illegible] *[illegible]*, [illegible]

Diversity in Unity: Perspectives from Psychology and Behavioral Sciences – Ariyanto et al. (Eds)
© 2018 Taylor & Francis Group, London, ISBN 978-1-138-62665-2

Convergent evidence: Construct validation of an Indonesian version of interpersonal and organisational deviance scales

P.T.Y.S. Suyasa
Faculty of Psychology, Tarumanagara University, Jakarta Barat, Indonesia

ABSTRACT: The phenomenon of illegitimate and potentially damaging work behaviours by employees is prominent in the Indonesian workplace and can be categorised as Counterproductive Work Behaviour (CWB). Such employees potentially harm the organisation, members of the organisation, service users, and organisations as a whole. The Interpersonal and Organisational Deviance Scales of Bennett and Robinson (2000) provide a measurement tool that is widely used in the study of CWB. Unfortunately, the adaptation of this tool in an Indonesian language version does not currently have any construct validity. In this study, a construct validation, based on previous research, is conducted of this version, using the established relationship between CWB and personality (traits). The results of this study showed that the Interpersonal and Organisational Deviance Scales in an Indonesian language version are supported by convergent evidence from conscientiousness, agreeableness, and neuroticism traits.

1 INTRODUCTION

Behaviour of an illegitimate nature that acts in opposition to the goals or interests of an organisation and that has the potential to harm the organisation or its members is defined as Counterproductive Work Behaviour (CWB). Concepts that aim to explain CWB include behavioural reactions to organisational frustration (Spector, 1975; Storms & Spector, 1987), deviant workplace behaviour (Robinson & Bennett, 1995), organisational misbehaviour (Vardi & Wiener, 1996), organisation retaliation behaviour (Skarlicki & Folger, 1997), counterproductive behaviour at work (Fox & Spector, 1999; Sackett, 2002; Sackett & DeVore, 2001), and general counterproductive behaviour (Marcus & Schuler, 2004).

Of these concepts, deviant workplace behaviour (Bennett & Robinson, 2000; Robinson & Bennett, 1995) is most frequently referred to by researchers (Belschak & Den Hartog, 2009; Brown, 2012; Chang & Smithikrai, 2010; Fox & Spector, 1999; Jensen & Patel, 2011; Langkamp Jacobson, 2009; Neff, 2009; Sackett, 2002; Sakurai & Jex, 2012; Smithikrai, 2008) in order to illustrate CWB. Deviant workplace behaviour is defined by Robinson and Bennett (1995, p. 556) as behaviour that runs contrary to the norms of, and has the potential to harm, an organisation, its members or both. In this context, norms can refer to policies, regulations or procedures, either formal or informal.

In Indonesia, CWB is an observable phenomenon in both state and private entities. Taking the example of government, of 286 reports (complaints) submitted to the Indonesian Ombudsman, protracted delays, with 165 cases, accounted for the majority (Lenny, 2011). Besides protracted delays, other complaints included abuse of authority, demands for money, goods or services, deviation from procedure, and non-compliance (RMOL, 2012a, 2012b). In the private sector, meanwhile, complaints included wasting time chatting or failing to come back from lunch within the given time, gossiping about other staff's private lives during working hours, spending time on internet sites (Facebook, news sites, YouTube, etc.), or sending private emails unrelated to professional duties (Delfina, 2013).

To further research into CWB in Indonesia, researchers recognised the need to come up with a definitive CWB scale or measurement, amalgamated from the CWB scales already in existence. Research into CWB has resulted in at least four measurement methods, namely: (a) Interpersonal and Organisational Deviance Scales (Bennett & Robinson, 2000); (b) Counterproductive Work Behaviour Checklist (CWB-C) (Spector et al., 2006); (c) Counterproductive Work Behaviour Interview (Roberts et al., 2007); (d) Past or Sustained Cases of CWBs (Oppler et al., 2008).

Of these four CWB measurement tools, Interpersonal and Organisational Deviance Scales is the most used. Several studies using Interpersonal and Organisational Deviance Scales have reported a high level of internal consistency in the measurements, as shown here: CWB-O = 0.92, CWB-I = 0.90 (Bowling & Eschleman, 2010); CWB-I = 0.84, CWB-O = 0.85 (Bowling et al., 2011); CWB-I = 0.78, CWB-O = 0.77 (Brown, 2012); CWB = 0.93 (Chang & Smithikrai, 2010); CWB-O = 0.81, CWB-I = 0.78 (Evans, 2006); CWB = 0.85 (Kwok et al., 2005); CWB-O = 0.72, CWB-I = 0.86 (Mount et al., 2006); CWB = 0.93 (Smithikrai, 2008).

Other studies—Fox and Spector (1999), Jensen and Patel (2011), Kelloway et al. (2010), Langkamp Jacobson (2009), Neff (2009), and Sakurai and Jex (2012)—have also used this measurement, but did not report on internal consistency.

As such, in this study, the researchers will adapt this frequently used tool. In adapting the Interpersonal and Organisational Deviance Scales, the researchers will retest the internal consistency of two dimensions of it, and will also test the validity of the construct, especially in relation to convergent evidence. It is hoped that through this study, measurement of the CWB phenomenon in Indonesia can better meet validity and reliability criteria.

1.1 *Deviant workplace behaviour concepts as a basic framework for the measurement of counterproductive work behaviour*

The term deviant workplace behaviour is used by Robinson and Bennett (1995) to illustrate employee behaviour with the potential to occasion organisational losses. As a concept, deviant workplace behaviour is defined by Robinson and Bennett (1995, p. 556) as spontaneous behaviour that contravenes the norms of an organisation and has the potential to harm an organisation, its members or both. In this context, norms can refer to policies, regulations, or procedures, whether formal or informal.

In their article, Robinson and Bennett employed the typology of deviant workplace behaviour with dimensions based on the gravity (minor vs serious) and the target of the behaviour (organisational vs interpersonal). Using these two dimensions, deviant workplace behaviour can be divided into four categories, namely, (a) production deviance (minor organisational deviance), (b) property deviance (serious organisational deviance), (c) political deviance (minor interpersonal deviance) and (d) personal aggression (serious interpersonal deviance).

Behaviour classified as *production deviance* (minor organisational deviance) includes: daydreaming or idling instead of working; complaining about insignificant matters at work; deliberately coming in late to work or coming back late from breaks; deliberately ignoring the instructions of superiors; not coming into work on a false pretence of illness. *Property deviance* (serious organisational deviance) includes stealing property or belongings from work, and attempting to defraud employers.

Behaviour classified as *political deviance* (minor interpersonal deviance) includes playing practical jokes at work and failing to help coworkers. Examples of behaviour in the final category, of *personal aggression* (serious interpersonal deviance), are threatening colleagues with violence, arguing at work, and verbally abusing coworkers.

Based on a study of measurement formulation carried out by Bennett and Robinson (2000), deviant workplace behaviour can also be divided into two simpler categories: interpersonal deviance (behaviour with the potential to harm individuals/members of an organisation) and organisational deviance (behaviour with the potential to harm the organisation). Behaviour classified as *interpersonal deviance* includes: (a) publicly embarrassing colleagues; (b) using aggressive or hateful language at work; (c) acting or speaking rudely to colleagues; (d) losing one's temper while at work; (e) distancing or refusing to communicate with a colleague based on their ethnicity, religion or race; (f) playing a practical joke on a colleague.

Meanwhile, behaviour considered as *organisational deviance* includes: (a) putting little effort into one's work, including arriving late, extending breaks and leaving early; (b) passing off work to colleagues; (c) working on a personal matter during working hours or using office facilities for personal matters; (d) failing to follow instructions; (e) littering the work environment; (f) gossiping or spreading unflattering rumours about the company.

1.2 *Construct validity (convergent evidence)*

Convergent evidence is one of the six types of evidence that can be used to prove the construct validity of a measurement. Convergent evidence is the condition in which a measurement: (a) bears a positive correlation with a measurement deemed to measure the same construct; (b) bears a negative correlation with a measurement deemed to measure an opposing construct; or (c) when studied, produces the same results as earlier research using the same construct (Cohen & Swerdlik, 2009).

In this study, the researchers intend to test the construct validity (convergent evidence) resulting from their adaptation of the measurement of counterproductive work behaviour. Previous research has shown a consistent relationship between CWB measurement and individual traits; the results of previous studies in this area are outlined in the remainder of this section.

1.3 *Personality and CWB*

A number of studies have identified the role of personality in counterproductive work behaviour. Based on research, the personality traits most significant to CWB are conscientiousness, agreeableness (Bolton et al., 2010; Mount et al., 2006; Salgado, 2002; Sulea et al., 2010), and neuroticism/emotional stability (Penney, 2003; Spector & Fox, 2002). Explanations of the effects of each of these personality traits on CWB are explained in the following subsections as a bar for the testing of construct validity (convergent evidence).

1.4 *Conscientiousness and CWB*

The meta-analysis carried out by Sulea et al. (2010) shows that low levels of conscientiousness are associated with high levels of CWB. This is consistent with previous research carried out by Moser et al. (1998), which showed a link between the personality trait of conscientiousness and CWB (theft and uncooperativeness). This was also shown by Salgado (2002), who noted that the lower the level of conscientiousness, the greater the chance the individual would engage in deviant CWB.

In subsequent years, CWB was categorised into two, namely organisational CWB (CWB-O) and interpersonal CWB (CWB-I). Conscientiousness has a low correlation with CWB-I but not with CWB-O. Low levels of conscientiousness result in delayed completion of tasks, a lack of discipline with regard to working hours, and absenteeism (Bolton et al., 2010; Mount et al., 2006).

The influence of conscientiousness on CWB was also explained by Yang and Diefendorff (2009), who noted that conscientiousness was a moderator of negative emotions within CWB. According to this model, which prevailed for a decade (Fox et al., 2001; Penney & Spector, 2002, 2005, 2008; Spector et al., 2006), CWB is influenced by negative emotions. According to the research of Yang and Diefendorff (2009), if an individual has high levels of conscientiousness, negative emotions will not be a significant factor in themselves.

This can be better understood by recognising the sub-trait of deliberation within the personality trait of conscientiousness. Individuals with high levels of conscientiousness tend to plan and consider their actions before they take them. When an individual with a high level of conscientiousness experiences negative emotions, they consider the effects of the CWB they are to engage in, allowing them to become aware of the harmful effects the behaviour would have on themselves, others or the organisation. Based on these considerations, individuals with high levels of conscientiousness are much less likely to escalate their negative emotions into CWB. A similar explanation results from consideration of a study showing that conscientiousness acts as a moderator between neuroticism and CWB (Bowling et al., 2011).

If the individual has a high level of conscientiousness, this trait also serves as a moderator of work stressors. Work stressors refer to the various obstacles, limits or conflicts encountered by individuals in carrying out their duties or work. Within the trait of conscientiousness there is a sub-dimension of achievement-striving (Costa & McCrae, 1992), which has other indications: high motivation in carrying out tasks, the desire to complete tasks, and the desire for future success.

Individuals with high levels of conscientiousness are strongly driven to overcome the work stressors they encounter. This desire is part of individual ambition. As such, for individuals with high levels of conscientiousness, work stressors increase drive, fuelling the desire to overcome challenges, meaning that such individuals demonstrate work behaviours opposite to CWB.

Based on the results of studies explaining the link between conscientiousness and CWB, this study (testing construct validity, convergent evidence) proposes the following hypothesis:

- H_1: There is a negative correlation between the personality trait of conscientiousness and CWB; the relationship between the personality trait of conscientiousness and CWB-O is stronger than the link between the personality trait of conscientiousness and CWB-I.

1.5 *Agreeableness and CWB*

As with the personality trait of conscientiousness, the influence of the trait of agreeableness on CWB was first tested more than a decade ago. Based on the results of a study by Moser et al. (1998), it may be stated that the lower the agreeableness of an individual, the greater the likelihood that they will engage in CWB (especially as regards uncooperative behaviour). Uncooperativeness can be understood by identifying characteristics of compliance/conformity as characteristics of an individual with the personality trait of agreeableness. Disagreeable individuals are less likely to conform to group norms.

Research by Bolton et al. (2010) and Mount et al. (2006) on the influence of agreeableness on CWB involved classifying CWB into behaviour that has a negative effect on the work system/organisation (CWB-O), and behaviour that has a negative effect on interpersonal relationships (CWB-I). Both groups of researchers explained that the personality trait of agreeableness has a direct effect on CWB-I. Meanwhile, Mount et al. also showed that the personality trait of agreeableness has an indirect effect on CWB-O.

Individuals with high levels of agreeableness tend to have high levels of trust in others (Costa & McCrae, 1992), including management figures. Besides this, they also tend to more readily feel warmth towards and satisfaction in others, and to feel that management figures are working for the good of employees. This trust/belief leads the individual employee to better accept the efforts of management figures and, thus, feelings of injustice or negative emotions are lower or better managed, and therefore do not lead to CWB-I. Agreeableness is an indication of the extent to which individuals have high levels of tender-mindedness and modesty. As such, when highly agreeable individuals experience negative emotions, these emotions are not taken out on others in the form of CWB (harsh behaviour towards other people/colleagues). Put otherwise, even when such individuals experience negative emotions, the emotions remain within the realm of their private experience, and are not realised in the form of CWB.

Based on the results of research into the link between agreeableness and CWB, this study (testing construct validity, convergent evidence) posits the hypothesis that:

- H_2: There is a negative correlation between agreeableness and CWB; the negative correlation between agreeableness and CWB-I is greater than that between agreeableness and CWB-O.

1.6 *Neuroticism and CWB*

Studies on the personality trait of neuroticism and CWB have also been ongoing for more than a decade. A study by Spector and Fox (2002) was one of the first reports into the relationship between neuroticism and CWB. Individuals with high levels of neuroticism are indicated by the characteristic sub-trait of anger, including being quick to anger and disappointment. As well as the sub-trait of anger, individuals with high levels of neuroticism also demonstrate the sub-trait of anxiety (including nervousness and worrying about future events).

Spector and Fox (2002) show that various organisational situations (organisational constraint, job stressors, organisational injustice, and violation of psychological contract) are among the factors that give rise to a range of negative emotions in individuals with high levels of neuroticism. Negative emotions or negative affectivity experienced by such individuals is further explained as a proximal cause of CWB (Dalal, 2005).

After Spector and Fox (2002), continued studies on the links between neuroticism and CWB focused on negative emotions and negative affectivity (Bowling & Eschleman, 2010; Penney, 2003). In meta-analysis studies of CWB, discussion of neuroticism as a characteristic linked to negative emotional reactions tends to use the term negative affectivity (Kaplan et al., 2009; Sulea et al., 2010). Other terms used in relation to neuroticism include emotional stability/instability (Jensen & Patel, 2011; Penney et al., 2011). The development of the use of the concept of emotional stability/instability as an alternative term to neuroticism and negative affectivity can be understood with reference to the concepts of vulnerability and impulsiveness as sub-traits of neuroticism (Costa & McCrae, 1992).

Penney (2003) uses the term negative affectivity as a predictor of CWB. Penney states that individuals with high levels of Negative Affectivity (NA) are more likely to engage in CWB thanthose with low levels. Penney also states that NA moderates the relationship between workplace incivility and CWB. Individuals with intensive levels of NA are far more likely to engage in CWB after experiencing incivility.

The effect of internal factors (negative affectivity, neuroticism) in moderating the influence on CWB of external situations and factors is supported by various studies. Thus, Bowling and Eschleman (2010) reported that the influence of work stressors on CWB was greater in employees with intensive levels of NA than in those with low levels. Similarly, Flaherty and Moss (2007) stated that in cases of low-intensity neuroticism, the influence of interactional justice on CWB was lessened.

Efforts to explain CWB in relation to susceptibility to negative emotions among individuals possessing the trait of neuroticism/negative affectivity have also been made in a number of studies. Thus, in the opinion of Bowling et al. (2011), conscientiousness and agreeableness reduce the effect of negative emotions on CWB. However, in the view of Yang and Diefendorff (2009), it is only conscientiousness (and not agreeableness) that consistently minimises the influence of negative emotions on CWB. High-intensity conscientiousness reduces the effect of regular negative emotions on CWB.

Using the term "emotional stability", Penney et al. (2011) have arrived at a somewhat different model. They state that conscientiousness has a positive correlation with CWB in cases where individuals have low levels of emotional stability. Referring to the concepts of vulnerability and impulsiveness, the terms "emotional stability/instability" can be categorised as sub-traits of neuroticism. As such, it may be understood that the influence of neuroticism on CWB is greater than the influence of conscientiousness on CWB.

Based on a summary of the results of studies into the link between the trait of neuroticism and CWB, this study (testing construct validity, convergent evidence) posits the following hypotheses:

- H_{3a}: There is a positive relationship between the trait of neuroticism and CWB;
- H_{3b}: Neuroticism functions as a moderator between conscientiousness and CWB.

2 METHOD

2.1 *Participants*

Based on the results of data collection, the number of participants in this study was 308. A majority (62%) were male. The gender of the participants is illustrated in Table 1.

In terms of education, a majority of participants (67.2%) held Bachelor degrees, with a Diploma degree being held by the smallest proportion of participants (7.5%). Participants' level of education is illustrated in Table 2.

The average age of participants was 29.77 years and the average length of working service was 3.32 years. From this, it can be seen that a majority of participants were at an early stage in their career. A majority (81%) were private-sector workers, the remainder being employed in the public sector. A majority (76%) of participants were operating at the level of regular staff (i.e. without management responsibilities). An account of participants based on position is shown in Table 3.

2.2 *Measurement*

In this study, two scales of measurement were used: a scale of counterproductive work behaviour and a scale of personality traits. The following subsection explains these two scales.

2.3 *Counterproductive work behaviour measurement*

In this study, in order to measure CWB, the researchers used the Interpersonal and Organisational Deviance Scale developed by Bennett and Robinson (2000). This measuring tool has

Table 1. Participants based on gender.

Gender	*f*	%
Female	114	37
Male	190	62
Did not respond	4	1
Total	308	100

Table 2. Participants based on education level.

Education	f	%
Junior/senior high school	30	9.7
Diploma	23	7.5
Bachelor's degree	207	67.2
Master's/doctoral degree	42	13.6
Did not respond	6	1.9
Total	308	100

Table 3. Participants based on position.

Position	f	%
Staff	234	76
Manager/supervisor	47	15
Senior manager/commissioner	27	9
Total	308	100

been used by a number of other researchers (e.g. Bowling et al., 2011; Bowling & Eschleman, 2010; Mount et al., 2006) to measure CWB. The measurement scale uses the cumulative rating scale method, which has grades of 1–6, with the figure 1 denoting extreme non-compliance; 2—non-compliant; 3—tending towards non-compliance; 4—tending towards compliance; 5—compliant; 6—extreme compliance.

This scale contains two dimensions: interpersonal deviance and organisational deviance. The dimension of interpersonal deviance was measured through 20 statements (the researchers added 12 of these to more closely classify the CWB-I in which employees engaged). The dimension of organisational deviance was measured in 28 statements (as with CWB-I, the researchers added extra statements—nine in this case—to make the CWB-O measurement more specific).

Examples of statement descriptors for interpersonal deviance (CWB-I) included: "Acting or speaking rudely towards superiors/colleagues"; "Telling colleagues about embarrassing incidents experienced by superiors/colleagues"; "Treating others with derision". The higher the score for interpersonal deviance, the more the employee does these things. The internal reliability coefficient (α) of the 20 statements of interpersonal deviance was 0.909.

In the dimension of organisational deviance (CWB-O), examples of statements included: "Pretending to be ill to superiors/colleagues"; "Attempting to use notes or receipts that are improper or invalid to obtain reimbursement from the company"; "Using working hours to browse the internet for non-work needs or ends". Again, the higher the score for organisational deviance, the more the employee does these things. The internal reliability coefficient (α) of the 28 statements of organisational deviance was 0.915.

2.4 *Personality assessment*

Personality traits were assessed through responses to 52 statements. This scale is an adaptation of the Big Five Personality Inventory (NEO-PI) developed by Costa and McCrae in 1992 (Tarumanagara University, Department of Psychology, 2012). This scale assesses five personality traits, including conscientiousness, agreeableness, and neuroticism. As with the scale used to measure CWB, this scale uses the cumulative rating scale with figures from 1 to 6. The figure 1 indicates extreme non-compliance; 2—non-compliant; 3—tending towards non-compliance; 4 tending towards compliance; 5—compliant; 6—extreme compliance.

2.4.1 *Conscientiousness*

Conscientiousness was assessed through 12 statements. Examples of the positive statements used to assess conscientiousness included: "I always store goods neatly in their proper place"; "When I make a commitment/promise to a figure of authority (superior/teacher), I always make sure I honour it". Examples of the negative statements included: "I am considered ill-disciplined and dislike being restricted by time constraints"; "I tend to be spontaneous, and dislike detailed or strict plans". The higher the score for conscientiousness, the more an individual stores thing tidily, keeps their promises, observes discipline and punctuality, and/or makes detailed, strict plans. The internal consistency reliability (α) of the 12 statements for conscientiousness was 0.809.

2.4.2 *Agreeableness*

The trait of agreeableness was also assessed through 12 statements. Examples of the positive statements used to assess agreeableness are "I am the kind of person who is always ready and willing to help" and "If my ideas/thoughts are rejected, I try to understand and accept alternative ideas from others". Examples of the negative statements used include: "I sometimes deceive people in order to obtain what I want"; "I am suspicious when someone behaves kindly or pleasantly towards me". The higher the agreeableness score, the more an individual understands and supports (the opinions of) other people, refrains from manipulating or exploiting other people for his or her own benefit, and/or believes in the goodwill of others. The internal consistency reliability (α) of the 12 statements for agreeableness was 0.724.

2.4.3 *Neuroticism*

The trait of neuroticism is assessed through responses to 11statements. Examples of the positive statements assessing neuroticism included: "I often feel stressed"; "I often experience disappointment". Examples of the negative statements included: "I rarely feel frightened/ nervous"; "I rarely feel sad". Thus, the higher the neuroticism score, the more readily an individual feels stressed, disappointed, frightened/nervous, and/or sad. The internal consistency reliability (α) for the 11 neuroticism statements was 0.781.

2.5 *Procedure*

The collection of data from the 308 participants in the study involved the assistance of 79 students from Tarumanagara University's Psychological Assessment, Psychological Assessment Tool Formulation, and History of Employee Psychology programmes. This data collection was done in tandem with the student's studies into the formulation of tools and the history of employee psychology. The topics for research carried out by the students are all linked to the professional world, including work engagement, social adjustment, and organisational citizenship behaviour.

In accordance with the research topics being conducted by the students, the researchers stressed that the basic characteristic of the participants had to be individuals with a job, while the gender, age, and length of service were not specified.

Initially, data were collected from 378 participants. However, upon examination it was found that 70 sets of data were incomplete or inconsistent, reducing the final number to 308.

To support data validity, the researchers stressed to the students that it was better for data to be collected from fewer people (for example, between two and seven), but with the data valid, than from many people, with the data subsequently proven to be invalid. As such, each student assisted by collecting data from between two and seven participants. Further, in support of data validity, the researchers reminded students to gather data from people with the time to fill in the questionnaire properly; in other words, they had to choose participants who were able to commit the time to participate by fully completing the questionnaire, having established a good rapport with them.

3 RESULTS

3.1 *Conscientiousness and CWB*

In order to prove hypothesis H_1, that there is a negative correlation between conscientiousness and CWB, the researchers used the Pearson correlation method.

Based on the results of testing using this method (Table 4), it can be seen that for conscientiousness and overall CWB, $r(306) = -0.535$, $p < 0.01$; for conscientiousness and CWB-I, $r(306) = -0.436$, $p < 0.01$; and for conscientiousness and CWB-O, $r(306) = -0.584$, $p < 0.01$.

Table 4. Testing construct validity (convergent evidence) between conscientiousness and CWB.

No.	Personality and CWB	1	2	3	4
1	Conscientiousness	*0.809*	–	–	–
2	CWB—Total	–0.535	*0.952*	–	–
3	CWB-I	–0.436	0.953	*0.909*	–
4	CWB-O	–0.584	0.953	0.817	*0.915*

Notes: CWB-O = Counterproductive Work Behaviour directed at the Organisation; CWB-I = Counterproductive Work Behaviour directed at the Interpersonal. Number in diagonal cells show the internal consistency (α) of each dimension/measurement scale. All coefficients show a significant correlation at a level of 0.01.

Thus, there is a negative correlation between the personality trait of conscientiousness and CWB, and it can also be seen that the correlation between conscientiousness and CWB-O (–0.584) is greater than that between conscientiousness and CWB-I (–0.436). This supports hypothesis H_1, that there is a negative correlation between the personality trait of conscientiousness and CWB, and that the relationship between conscientiousness and CWB-O is stronger than that between conscientiousness and CWB-I.

3.2 *Agreeableness and CWB*

In order to prove hypothesis H_2, that there is a negative correlation between agreeableness and CWB, the researchers used the Pearson correlation method.

Based on the results of testing using Pearson correlation (Table 5), it can be seen that the correlation between agreeableness and overall CWB was $r(306) = -0.596$, $p < 0.01$; for agreeableness and CWB-I it was $r(306) = -0.569$, $p < 0.01$; and for agreeableness and CWB-O, $r(306) = -0.566$, $p < 0.01$. As such, it can be said that there is a negative correlation between the personality trait of agreeableness and CWB, meaning CWB bears convergent evidence (construct validity) in relation to agreeableness. However, in Table 5, although it appears the correlation between agreeableness and CWB-I (–0.569) is higher than that with CWB-O (–0.566), the difference between these two coefficients is very small. Thus the first part of hypothesis H_2 is supported, that there is a negative correlation between agreeableness and CWB, but the correlation between agreeableness and CWB-I is quite similar to, rather than higher than, that with CWB-O.

3.3 *Neuroticism and CWB*

In order to prove hypothesis H_{3a}, that there is a positive correlation between neuroticism and CWB, the researchers again used the Pearson correlation method.

Based on the results of testing using Pearson correlation (Table 6), it can be seen that there is a correlation between neuroticism and overall CWB, $r(306) = 0.364$, $p < 0.01$; between

Table 5. Test of construct validity (convergent evidence) between agreeableness and CWB.

No.	Personality and CWB	1	2	3	4
1	Agreeableness	*0.724*	–	–	–
2	CWB—total	–0.596	*0.952*	–	–
3	CWB-I	–0.569	0.953	*0.909*	–
4	CWB-O	–0.566	0.953	0.817	*0.915*

Notes: CWB-O = Counterproductive Work Behaviour directed at the Organisation; CWB-I = Counterproductive Work Behaviour directed at the Interpersonal. Number in diagonal cells show the internal consistency (α) of each dimension/measurement scale. All coefficients show a significant correlation at a level of 0.01.

Table 6. Testing construct validity (convergent evidence) between neuroticism and CWB.

No.	Personality and CWB	1	2	3	4
1	Neuroticism	*0.781*	–	–	–
2	CWB—Total	0.364	*0.952*	–	–
3	CWB-I	0.305	0.953	*0.909*	–
4	CWB-O	0.389	0.953	0.817	*0.915*

Notes: CWB-O = Counterproductive Work Behaviour directed at the Organisation; CWB-I = Counterproductive Work Behaviour directed at the Interpersonal. Number in diagonal cells show the internal consistency (α) of each dimension/measurement scale. All coefficients show a significant correlation at a level of 0.01.

neuroticism and CWB-I, $r(306) = 0.305$, $p < 0.01$; and between neuroticism and CWB-O, $r(306) = 0.389$, $p < 0.01$. It can be seen that the correlation between neuroticism and CWB is positive, meaning that CWB bears convergent evidence (construct validity) with regard to the personality trait of neuroticism. This is in line with hypothesis H_{3a}, that there is a positive relationship between the trait of neuroticism and CWB.

In order to prove hypothesis H_{3b}, as a test of construct validity (convergent evidence) for the study conducted by Penney et al. (2011), the researchers again used the testing moderation procedure (Baron & Kenny, 1986). Here, the researchers divided participants into two groups, those with high emotional stability (Neuroticism –) and those with low emotional stability (Neuroticism +), according to the neuroticism scores. After dividing moderator variables into two levels, the researchers tested the correlation/regression of the independent variable (conscientiousness) with regard to the dependent variable (CWB).

As shown in Table 7, it appears that emotional stability/neuroticism does not function as a moderator variable of the correlation between conscientiousness and CWB. The measurement tool/scale adapted from the Interpersonal and Organisational Deviance Scales (Bennett & Robinson, 2000) in an Indonesian language version has no construct validity (convergent evidence) in relation to the research carried out by Penney et al. (2011), which reported that conscientiousness did have a positive correlation with CWB in individuals with low emotional stability (Neuroticism +).

As a validation of the effect of the moderation of neuroticism, the researchers conducted regression on CWB based on conscientiousness (predictor), neuroticism (moderator), and the interaction of predictor × moderator, as carried out by Penney et al. (2011). The results of this regression test can be seen in Table 8.

Table 7. Testing neuroticism as a moderator of the correlation between conscientiousness and CWB, by dividing neuroticism into two levels.

		Unstandardised coefficients				Correlations
Step	Predictors	B	Std. error	t	Sig.	Zero-order
All	(Constant)	0.00	0.05	0.00	1.00	
	Conscientiousness	–0.53	0.05	–11.07	0.00	–0.53
Neuroticism –	(Constant)	–0.10	0.07	–1.45	0.15	
	Conscientiousness	–0.55	0.07	–8.17	0.00	–0.54
Neuroticicm +	(Constant)	0.18	0.09	2.08	0.04	
	Conscientiousness	–0.40	0.08	–4.71	0.00	–0.37

Note: Dependent variable—Counterproductive Work Behaviour.

Table 8. Testing of neuroticism as a moderator of the correlation between conscientiousness and CWB by testing the effect of interaction between neuroticism and conscientiousness.

		Unstandardised coefficients				r		
Step	Predictors	B	Std. error	t	Sig.	Part	R	R^2
1	Conscientiousness	–0.53	0.05	–11.07	0.00	–0.53	0.53	0.2859
2	Conscientiousness	–0.47	0.06	–8.49	0.00	–0.44	0.55	0.2986
	Neuroticism	0.13	0.06	2.35	0.02	0.13		
3	Conscientiousness	–0.48	0.06	–8.70	0.00	–0.45	0.56	0.3123
	Neuroticism	0.12	0.06	2.18	0.03	0.12		
	Conscientiousness x neuroticism	0.10	0.04	2.46	0.01	0.14		

Note: Dependent variable—Counterproductive Work Behaviour.

4 DISCUSSION

In relation to the conclusions that there is a negative correlation between conscientiousness and CWB, a negative correlation between agreeableness and CWB, and a positive correlation between neuroticism and CWB, the researchers referred to the study conducted by Berry et al. (2007). Testing of the construct validity (convergent evidence) of the measurement tool/scale of Interpersonal and Organisational Deviance Scales (Bennett & Robinson, 2000) in the Indonesian language was in line with the results of the meta-analytical study carried out by Berry et al. (2007), who indicated that agreeableness and conscientiousness were the strongest predictors of the overall CWB score. Based on the results of the testing of the convergent evidence in this study, the regression coefficients of agreeableness and conscientiousness on CWB are greater than that of neuroticism.

To explain the result that showed conscientiousness to be a stronger predictor of CWB-O than of CWB-I, the researchers refer to the study conducted by Mount et al. (2006). According to Mount et al., it can be demonstrated that conscientiousness can directly predict CWB-O (CPB-O), but it cannot do the same for CWB-I (CPB-I). As agreeableness has been shown to equally strongly predict both CWB-I and CWB-O, it can be concluded that these two criteria are mediated by the variable of job satisfaction. In the results of their study, Mount et al. explain that job satisfaction has a more or less equal correlation with both CWB-I and CWB-O (job satisfaction and CWB-I = -0.040, $p < 0.05$; job satisfaction and CWB-O = -0.41, $p < 0.05$).

Conscientiousness functions as a moderator variable between negative affectivity and CWB, in line with the results of research carried out by Bowling et al. (2011). In their article, Bowling et al. (2011) suggested that the effect of negative affectivity can be reduced in individuals with high levels of conscientiousness. Although negative affectivity does influence CWB, in individuals with high levels of conscientiousness, the rate of CWB influenced by negative affectivity is much lower than in individuals with low levels of conscientiousness.

Our results showed that the function of emotional stability/neuroticism as a moderator variable between conscientiousness and CWB was not proven in this study; in others words, the results of this testing challenge the results of the study carried out by Penney et al. (2011), which suggested that conscientiousness correlates with CWB in individuals with low emotional stability (high levels of neuroticism). Our study showed that both in individuals with high levels of neuroticism and in those with low levels of neuroticism, conscientiousness always negatively correlates with CWB. The researchers suspect that this is due to the difference in the CWB measurement tools used by Penney et al. (2011) and those used in this study. Penney et al. (2011) used the CWB-Checklist (Spector et al., 2006), which is based on an emotion-centred model (Spector & Fox, 2002) in which negative emotions are a proximal cause of CWB. However, a number of figures (Berry et al., 2007; Bowling et al., 2011; Sulea et al., 2010) have shown that conscientiousness and agreeableness are stronger predictors of CWB.

4.1 *Suggestions for further research*

The authors suggest that further research might use participants with more specific characteristics; in this study, the characteristics of the participants varied in terms of education and job levels. In order to contribute additional information on the construct validity of the adapted Interpersonal and Organisational Deviance Scales (Bennett & Robinson, 2000) in an Indonesian language version, the construct validity could be tested using variables such as job satisfaction (Fox & Spector, 1999; Mount et al., 2006), perceived unfairness (Cohen-Charash & Mueller, 2007), organisational citizenship behaviour (Chang & Smithikrai, 2010; Spector, Bauer, & Fox, 2010), job stress (Brown, 2012), interpersonal justice and organisational culture (Langkamp Jacobson, 2009), normative control (Kwok et al., 2005), surveillance in the workplace (Martin et al., 2016), and workplace incivility (Bibi et al., 2013).

5 CONCLUSION

Based on analysis of the results of this study, three conclusions can be drawn regarding the construct validity (convergent evidence) of the measurement tool/scale of Interpersonal and Organisational Deviance Scales (Bennett & Robinson, 2000) in an Indonesian language version. These conclusions are:

1. There is a negative correlation between conscientiousness and CWB. The correlation between conscientiousness and CWB-O is greater than that between conscientiousness and CWB-I.
2. There is a negative correlation between agreeableness and CWB. The correlation between agreeableness and CWB-I is quite similar to the correlation between conscientiousness and CWB-O.
3. a. There is a positive correlation between neuroticism and CBW.
   b. Emotional stability/neuroticism does not function as a moderator variable between conscientiousness and CWB.

REFERENCES

Baron, R.M. & Kenny, D.A. (1986). The moderator-mediator variable distinction in social psychological research: Conceptual, strategic, and statistical considerations. *Journal of Personality and Social Psychology*, *51*(6), 1173–1182.

Belschak, F.D. & Den Hartog, D.N. (2009). Consequences of positive and negative feedback: The impact on emotions and extra-role behaviors. *Applied Psychology: An International Review*, *58*(2), 274–303. doi:10.1111/j.1464-0597.2008.00336.x

Bennett, R.J. & Robinson, S.L. (2000). Development of a measure of workplace deviance. *Journal of Applied Psychology*, *85*(3), 349–360. doi:10.1037/0021-9010.85.3.349

Berry, C.M., Ones, D.S. & Sackett, P.R. (2007). Interpersonal deviance, organizational deviance, and their common correlates: A review and meta-analysis. *Journal of Applied Psychology*, *92*, 410–424.

Bibi, Z., Karim, J. & ud Din, S. (2013). Workplace incivility and counterproductive work behavior: Moderating role of emotional intelligence. *Pakistan Journal of Psychological Research*, *28*(2), 317–334.

Bolton, L.R., Becker, L.K. & Barber, L.K. (2010). Big five trait predictors of differential counterproductive work behavior dimensions. *Personality and Individual Differences*, *49*(5), 537–541. doi:10.1016/j.paid.2010.03.047

Bowling, N.A. & Eschleman, K.J. (2010). Employee personality as a moderator of the relationships between work stressors and counterproductive work behavior. *Journal of Occupational Health Psychology*, *15*(1), 91–103. doi:10.1037/a0017326

Bowling, N.A., Burns, G.N., Stewart, S.M. & Gruys, M.L. (2011). Conscientiousness and agreeableness as moderators of the relationship between neuroticism and counterproductive work behaviors: A constructive replication. *International Journal of Selection and Assessment*, *19*(3), 320–330. doi:10.1111/j.1468-2389.2011.00561.x

Brown, T.G. (2012). *Job stress and counterproductive work behaviors: Does moral identity matter?* (Doctoral dissertation, Seattle Pacific University). Available from ProQuest Dissertations and Theses database (UMI No. 3523919).

Chang, K. & Smithikrai, C. (2010). Counterproductive behaviour at work: An investigation into reduction strategies. *The International Journal of Human Resource Management*, *21*(8), 1272–1288. doi:10.1080/09585192.2010.483852

Cohen, R.J. & Swerdlik, M.E. (2009). *Psychological testing and assessment: An introduction to tests and measurement* (7th ed.). New York, NY: McGraw-Hill.

Cohen-Charash, Y. & Mueller, J.S. (2007). Does perceived unfairness exacerbate or mitigate interpersonal counterproductive work behaviors related to envy? *Journal of Applied Psychology*, *92*(3), 666–680.

Costa, P.T. & McCrae, R.R. (1992). Normal personality assessment in clinical practice: The NEO Personality Inventory. *Psychological Assessment*, *4*(1), 5–13. doi:10.1037/1040-3590.4.1.5.

Evans, A. L. (2006). Counterproductive group behavior. *ProQuest Information & Learning, 66*(8), 4469.

Dalal, R. S. (2005). A meta-analysis of the relationship between organizational citizenship behavior and counterproductive work behavior. *Journal of Applied Psychology, 90*(6), 1241–1255. doi:10.1037/0021-9010.90.6.1241

Delfina, R.R. (2013, January 31). Sikap tidak profesional yang sering dilakukan karyawan [Unprofessional habits frequently shown by young employees]. *Wolipop, Detikcom*. Retrieved from https://wolipop.detik.com/read/2013/01/31/103449/2157294/1133/sikap-tidak-profesional-yang-sering-dilakukan-karyawan.

Flaherty, S. & Moss, S.A. (2007). The impact of personality and team context on the relationship between workplace injustice and counterproductive work behavior. *Journal of Applied Social Psychology, 37*(11), 2549–2575. doi:10.1111/j.1559-1816.2007.00270.x

Fox, S. & Spector, P.E. (1999). A model of work frustration–aggression. *Journal of Organizational Behavior, 20*(6), 915–931.

Fox, S., Spector, P.E. & Miles, D. (2001). Counterproductive work behavior (CWB) in response to job stressors and organizational justice: Some mediator and moderator tests for autonomy and emotions. *Journal of Vocational Behavior, 59*(3), 291–309. doi:10.1006/jvbe.2001.1803

Jensen, J.M. & Patel, P.C. (2011). Predicting counterproductive work behavior from the interaction of personality traits. *Personality and Individual Differences, 51*(4), 466–471. doi:10.1016/j.paid.2011.04.016

Kaplan, S., Bradley, J.C., Luchman, J.N. & Haynes, D. (2009). On the role of positive and negative affectivity in job performance: A meta-analytic investigation. *Journal of Applied Psychology, 94*, 162–176. doi:10.1037/a0013115

Kelloway, E.K., Francis, L., Prosser, M. & Cameron, J.E. (2010). Counterproductive work behavior as protest. *Human Resource Management Review, 20*, 18–25.

Kwok, C.K., Au, W.T. & Ho, J.M.C. (2005). Normative controls and self-reported counterproductive behaviors in the workplace in China. *Applied Psychology: An International Review, 54*(4), 456–475. doi:10.1111/j.1464-0597.2005.00220.x

Langkamp Jacobson, K.J. (2009). *Contextual and individual predictors of counterproductive work behaviors* (Doctoral dissertation, Arizona State University). Available from ProQuest Dissertations and Theses database (UMI No. 3357268).

Lenny. (2011). Ombudsman terima 286 keluhan warga Jakarta. *Berita Jakarta*. Retrieved from http://www.beritajakarta.com/2008/id/berita_detail.asp?nNewsId=44161

Marcus, B. & Schuler, H. (2004). Antecedents of counterproductive behavior at work: A general perspective. *Journal of Applied Psychology, 89*(4), 647–660. doi:10.1037/0021-9010.89.4.647

Martin, A.J., Wellen, J.M. & Grimmer, M.R. (2016). An eye on your work: How empowerment affects the relationship between electronic surveillance and counterproductive work behaviours. *The International Journal of Human Resource Management, 27*(21), 2635–2651. doi:10.1080/09585192.2016.1225313

Moser, K., Schwörer, F., Eisele, D. & Haefele, G. (1998). Persönlichkeitsmerkmale und kontraproduktives verhalten in organisationen. ergebnisse einer pilotstudie [Personality and counterproductive behavior in organizations: Results of a pilot study]. *Zeitschrift Für Arbeits- Und Organisationspsychologie, 42*(2), 89–94.

Mount, M., Ilies, R. & Johnson, E. (2006). Relationship of personality traits and counterproductive work behaviors: The mediating effects of job satisfaction. *Personnel Psychology, 59*(3), 591–622. doi:10.1111/j.1744-6570.2006.00048.x

Neff, N.L. (2009). *Peer reactions to counterproductive work behavior* (Doctoral dissertation, Pennsylvania State University). Available from ProQuest Dissertations and Theses database (UMI No. 3374527).

Oppler, E.S., Lyons, B.D., Ricks, D.A. & Oppler, S.H. (2008). The relationship between financial history and counterproductive work behavior. *International Journal of Selection and Assessment, 16*(4), 416–420. doi:10.1111/j.1468-2389.2008.00445.x.

Penney, L. M. (2003). Workplace incivility and counterproductive workplace behavior (cwb): What is the relationship and does personality play a role? *ProQuest Information & Learning), 64*(2–), 992.

Penney, L.M. & Spector, P.E. (2002). Narcissism and counterproductive work behavior: Do bigger egos mean bigger problems? *International Journal of Selection and Assessment, 10*(1–2), 126–134. doi:10.1111/1468-2389.00199.

Penney, L.M. & Spector, P.E. (2005). Job stress, incivility, and counterproductive work behavior (CWB): The moderating role of negative affectivity. *Journal of Organizational Behavior, 26*(7), 777–796. doi:10.1002/job.336

Penney, L.M. & Spector, P.E. (2008). Emotions and counterproductive work behavior. In N.M. Ashkanasy & C.L. Cooper (Eds.), *New horizons in management. Research companion to emotion in organizations* (pp. 183–196). Cheltenham, UK: Edward Elgar Publishing.

Penney, L.M., Hunter, E.M. & Perry, S.J. (2011). Personality and counterproductive work behaviour: Using conservation of resources theory to narrow the profile of deviant employees. *Journal of Occupational and Organizational Psychology*, *84*(1), 58–77. doi:10.1111/j.2044-8325.2010.02007.x.
RMOL. (2012a, April 2). Pelayanan pemda terbanyak dipermasalahkan masyarakat: Catatan ombudsman selama tiga tahun terakhir. *Rakyat Merdeka Online*. Retrieved from http://www.rmol.co/read/2012/04/02/59431
RMOL. (2012b, May 5). Pelayanan administrasi pemda banyak dikeluhkan masyarakat: Catatan Ombudsman Selama Januari–Maret 2012. *Rakyat Merdeka Online*. Retrieved from http://www.rmol.co/read/2012/05/05/62829
Roberts, B.W., Harms, P.D., Caspi, A. & Moffitt, T.E. (2007). Predicting the counterproductive employee in a child-to-adult prospective study. *Journal of Applied Psychology*, *92*(5), 1427–1436. doi:10.1037/0021-9010.92.5.1427.
Robinson, S.L. & Bennett, R.J. (1995). A typology of deviant workplace behaviors: A multidimensional scaling study. *Academy of Management Journal*, *38*, 555–572.
Sackett, P.R. (2002). The structure of counterproductive work behaviors: Dimensionality and relationships with facets of job performance. *International Journal of Selection and Assessment*, *10*, 5–11.
Sackett, P.R., & DeVore, C.J. (2001). Counterproductive behaviors at work. In N. Anderson, D.S. Ones, H.K. Sinangil & C. Viswesvaran (Eds.), *Handbook of industrial, work and organizational psychology: Personnel psychology* (Vol. 1, pp. 145–164). Thousand Oaks, CA: SAGE Publications.
Sakurai, K. & Jex, S.M. (2012). Coworker incivility and incivility targets' work effort and counterproductive work behaviors: The moderating role of supervisor social support. *Journal of Occupational Health Psychology*, *17*(2), 150–161. doi:10.1037/a0027350.
Salgado, J. (2002). The big five personality dimensions and counterproductive behaviors. *International Journal of Selection and Assessment*, *10*(1–2), 117–125. doi:10.1111/1468-2389.00198.
Skarlicki, D.P. & Folger, R. (1997). Retaliation in the workplace: The roles of distributive, procedural, and interactional justice. *Journal of Applied Psychology*, *82*, 434–443.
Smithikrai, C. (2008). Moderating effect of situational strength on the relationship between personality traits and counterproductive work behaviour. *Asian Journal of Social Psychology*, *11*(4), 253–263. doi:10.1111/j.1467-839X.2008.00265.x
Spector, P.E. (1975). Relationships of organizational frustration with reported behavioral reactions of employees. *Journal of Applied Psychology*, *60*(5), 635–637.
Spector, P.E. (2001). Science briefs: Counterproductive work behavior: The secret side of organizational life. *Psychological Science Agenda*, *14*(3), 8–9.
Spector, P.E. & Fox, S. (2002). An emotion-centered model of voluntary work behavior: Some parallels between counterproductive work behavior and organizational citizenship behavior. *Human Resource Management Review*, *12*(2), 269–292. doi:10.1016/S1053-4822(02)00049-9.
Spector, P.E. & Fox, S. (2005). The Stressor-Emotion Model of Counterproductive Work Behavior. In S. Fox & P.E. Spector (Eds.), *Counterproductive work behavior: Investigations of actors and targets* (pp. 151–174). doi:10.1037/10893-007.
Spector, P.E. & Fox, S. (2010). Counterproductive work behavior and organisational citizenship behavior: Are they opposite forms of active behavior? *Applied Psychology: An International Review*, *59*(1), 21–39. doi:10.1111/j.1464-0597.2009.00414.x.
Spector, P.E., Bauer, J.A. & Fox, S. (2010). Measurement artifacts in the assessment of counterproductive work behavior and organizational citizenship behavior: Do we know what we think we know? *Journal of Applied Psychology*, *95*(4), 781–790. doi:10.1037/a0019477.
Spector, P.E., Fox, S., Penney, L.M., Bruursema, K., Goh, A. & Kessler, S. (2006). The dimensionality of counterproductivity: Are all counterproductive behaviors created equal? *Journal of Vocational Behavior*, *68*(3), 446–460. doi:10.1016/j.jvb.2005.10.005.
Storms, P.L. & Spector, P.E. (1987). Relationships of organizational frustration with reported behavioural reactions: The moderating effect of locus of control. *Journal of Occupational Psychology*, *60*, 227–234.
Sulea, C., Maricuţoiu, L., Dumitru, C.Z. & Pitariu, H.D. (2010). Predicting counterproductive work behaviors: A meta-analysis of their relationship with individual and situational factors. *Psihologia Resurselor Umane Revista Asociaţiei De Psihologie Industrială Şi Organizaţională*, *8*(1), 66–81.
Vardi, Y. & Wiener, Y. (1996). Misbehavior in organizations: A motivational framework. *Organization Science*, *7*(2), 151–164.
Yang, J. & Diefendorff, J.M. (2009). The relations of daily counterproductive workplace behavior with emotions, situational antecedents, and personality moderators: A diary study in Hong Kong. *Personnel Psychology*, *62*(2), 259–295. doi:10.1111/j.1744-6570.2009.01138.x.

Diversity in Unity: Perspectives from Psychology and Behavioral Sciences – Ariyanto et al. (Eds)
 ISBN 978-1-138-62665-2

Differences in personality and individual entrepreneurial orientation between entrepreneur students and non-entrepreneur students

A. Wisudha
Department of Psychology, University of Westminster, London, UK

G.A. Kenyatta
Lumina Learning Indonesia, Jakarta, Indonesia

P.C.B. Rumondor
Faculty of Psychology, Bina Nusantara University, Jakarta, Indonesia

ABSTRACT: Previous research using a student population reported a relationship between personality and Individual Entrepreneurial Orientation (IEO). This study attempts to address this area of interest in more detail by investigating where there might be a difference between students who already have an established venture (Entrepreneur students) and those who have not as yet put their plans into action (Non-entrepreneur students). This study compares total IEO score and 4 out of 8 Lumina Spark aspects of personality between Entrepreneur students and Non-entrepreneur students. A sample of 292 students in 18–26 year age range from the Bina Nusantara University (BINUS) in Indonesia, consisting of Entrepreneur students (n = 146) and Non-entrepreneur students (n = 146) was administered IEO and Lumina Spark questionnaires. The IEO questionnaire measures Risk taking, Proactiveness, and Innovativeness. The Lumina Spark questionnaire is a psychometric instrument that uses the Big Five personality model as its cornerstone and a Jungian lens to inform the model. It measures 8 aspects: Inspiration Driven, Big Picture Thinking, Extraverted, Outcome Focused, Discipline Driven, Down to Earth, Introverted and People Focused. Independent T-tests showed statistically significant differences in total IEO, Risk-taking and Innovativeness dimensions of IEO, also in Big Picture Thinking, and Extraverted aspects of personality. In view of the practical implications that can be derived from the study, the subsequent discussion refers to the importance of awareness about IEO and personality in entrepreneurial education.

1 INTRODUCTION

Entrepreneurship has recently become the focus of attention amongst a number of disciplines. Entrepreneurship is seen as central to the enhancement of a country's economic growth (Seth, 2015). Therefore, an increase in the number of successful start-up businesses will arguably contribute to the rise in job opportunities and the generation of new wealth, leading to positive impact on measures of quality of life.

Entrepreneurship is defined as the identification and exploitation of business opportunities within the individual-opportunity nexus (Shane & Venkantraman, 2000). According to Baron (2007), entrepreneurship as a process is made up of three phases: (a) the pre-launch or opportunity identification phase in which the entrepreneur identifies viable and feasible business opportunities, (b) the launch or development and execution phase in which the entrepreneur assembles the necessary resources for starting a venture, and (c) the post-launch phase in which the entrepreneur manages the new venture in such a way that it grows and survives. Entrepreneurship brings with it a greater freedom of choice of businesses and the flexibility in which it can be executed. These are attractive factors for the young generation who seek independence and flexible choice of opportunities. This is also demonstrated by the increased percentage of new enterprises initiated by Indonesia's young generation. Data from a survey on 'Becoming an Entrepreneur' conducted by Kompas' Research and Development reported that the largest age group from which entrepreneurs emerge is between 20–40 years old (Gianie, 2015).

In Indonesia, both private and government sectors are beginning to give their support to meet the entrepreneurial demands of young people. This ranges from providing boot camps, funding opportunities, mentoring schemes, longitudinal workshops, incorporating entrepreneurial education in almost all universities across disciplines, and many other activities. The government has even dedicated 100 billion USD to help this cause. The Creative Economy Council (BEKRAF) recently launched their newest program called BEKUP that stands for "BEKRAF for Pre-startups" this year (Palupi, 2016). In Jakarta and other large cities in Indonesia, a number of educator-curator bodies have been created that provide communities for entrepreneurs to collaborate as well as give and attend workshops, such as Lingkaran.co., maubelajarapa.com, Indoestri, and many more. Moreover since 2009, the Directorate General of Higher Education (DITJEN DIKTI) has made entrepreneurship a compulsory part of courses in the curriculum regardless of the students′ major (BSI Entrepreneur Center, 2016). Many major Universities across Indonesia are committed to designing and implementing their curriculum to include the topic of entrepreneurship, as well as establishing entrepreneurial support centers for their students. Such commitment has been implemented, for example, in University of Indonesia, Institute of Technology Bandung, Pelita Harapan University, and Bina Nusantara University. All of these initiatives have been designed to support the growth in the number of successful and sustainable entrepreneurial businesses for students, regardless of their respective industry.

One model that provides a framework to study entrepreneurial success through a psychological perspective is the action-characteristic model developed by Frese & Gielnik (2014). This model suggests that personality can influence entrepreneurial success through action characteristics such as personal initiative, goals/vision, search for opportunities, information search, planning, feedback processing, social networking, seeking of niche, seeking of resources, deliberate practice and entrepreneurial orientation. In the meta-analysis conducted by the authors, entrepreneurial orientation was shown to have a relatively higher correlation with business performance compared to other action-characteristics. In accordance with their findings, the current study will focus on personality as a factor that influences entrepreneurial success, and Individual Entrepreneurial Orientation (IEO) as action characteristics that can mediate the influence of personality on entrepreneurial success.

Entrepreneurial Orientation (EO) can be described as strategic processes in organizations that focus on the actions and decisions within an entrepreneurial context (Guth & Ginsberg, 1990; Zahra & Covin, 1995; Rauch, Wiklund, Lumpkin & Frese, 2009). One strand of research into entrepreneurship has shown that EO has a remarkable influence on an organization's performance, profitability, growth and product innovation (Johan & Dean, 2003; Avlontis & Salavou, 2007; Moreno & Casillas, 2008; Tang, Kacmar & Busenitz, 2012). Studies of EO and its relation to company performance have consistently shown highly significant correlations (Fairoz, Hirobumi & Tanaka, 2010; Schillo, 2011; Mahmood & Hanafi 2013; Zulkifli & Rosli, 2013). Moreover, EO at the organizational level has been shown to correlate with entrepreneurial performance (Koenig, Steinmetz, Frese, Rauch & Wang, 2009).

EO was first measured within an organizational context and was characterized by the following factors: Autonomy, Innovativeness, Risk-taking, Proactiveness and Competitive Aggressiveness (Lumpkin & Dess, 1996). Items on the EO scale at the organizational level assess the strategic stance adopted by top managers, the CEO, or general directors (Frese & Gielnik, 2014). Organizations with a high entrepreneurial orientation outperform other firms because Autonomy, Innovativeness, Risk-taking, Proactiveness, and Competitive Aggressiveness collectively have been shown to help the company to seek and exploit new opportunities for growth (Lumpkin & Dess, 1996).

An organization, particularly a small or entrepreneurially-founded one, can be considered to result from an individual's inspiration and related behaviors. Following this, EO dimensions can also be measured at this individual level (Bolton & Lane, 2012; Frese & Gielnik, 2014). Among the five dimensions of EO, three dimensions have been identified and used consistently in the literature; those dimensions are Risk-taking, Innovativeness and Proactiveness (Miller, 1983; Bolton & Lane, 2012). This study adopts previous definitions of EO, but places an emphasis on the individual level that focuses on actions and decisions within an entrepreneurial context, measured by an IEO scale adapted from Bolton and Lane (2012). This is also in line with Frese & Gielnik's (2014) description of action characteristic as not action per se but they are rather ways of performing an action. Therefore, this study considers IEO, which consists of Risk-Taking, Innovativeness and Proactiveness, as action characteristics.

Our previous research found that there is a significant correlation between IEO and personality (Wisudha, Kenyatta, Rumeser, Rumondor, & Andangsari, 2016). The research used a 20-item IEO questionnaire in Bahasa Indonesia, adapted from Bolton & Lane (2012) and the 144-item Lumina Spark personality questionnaire (Desson, Benton, & Golding, 2014). Lumina Spark was chosen because it treats all personality traits equally. It measures all traits independently and does not infer the strength of one traits at the expense of its opposite. For instance, Introverted traits will have their own directly measured scores, which are not inferred from the scores of Extraverted traits. This therefore allowed us to explore a comprehensive set of personality traits. The results from the research showed that all of the personality aspects, as measured by the Lumina Spark model, correlated significantly with IEO dimensions, even though the degree of correlations vary from moderate to strong ($r = 0.28$ to 0.71). Four of the strongest degree of correlations were shown between IEO and the personality aspects: Big Picture Thinking; Extraverted; Outcome Focused and Discipline Driven.

Our previous research however did not look at differences between Entrepreneur and Non-entrepreneur students, unlike the study conducted by Kropp, Lindsay, & Shoham (2008) that reported significantly different scores of IEO between Entrepreneurs and Non-entrepreneurs. In this study, we aim to address whether there is a difference in the IEO scores and the personality measures between these two groups. Our findings may inform stakeholders in entrepreneurial education of IEO dimensions and personality aspects that contribute to promoting and developing entrepreneurship. It has been reported that self-awareness of one's internal state, emotion, resource and intuition is related to higher job performance in general (Joseph, Jin, Newman, & O'Boyle, 2015) as well as with entrepreneurs' business outcomes (Cross & Travaglione 1995; Ahmetoglu, Leutner, & Chamorro-Premuzic, 2011). Therefore, if action characteristics such as Risk-Taking, Innovativeness and Proactiveness, and personality aspects such as Big Picture Thinking, Extraverted, Outcome Focused, and Discipline Driven are statistically different between Entrepreneur and Non-entrepreneur students, this will enable those engaged in entrepreneurial education to help raise awareness in students and to provide guidance for educators in the preparation of the type and form of materials that are tailored to cater for both groups of students.

We hypothesized that there would be a significant difference in IEO and its dimensions (Risk Taking, Innovativeness and Proactiveness) between Entrepreneur and Non-entrepreneur students. Since personality was found to correlate with IEO (Wisudha *et al.*, 2016), we also hypothesized that there would be a significant difference between Entrepreneur and Non-entrepreneur students in the four personality aspects which indicates the strongest degree of correlation, namely: Big Picture Thinking; Extraverted; Outcome Focused and Discipline Driven in Wisudha *et al.* (2016).

2 METHODS

Our previous research investigated the correlation between IEO and Personality using the Lumina Spark personality questionnaire with 585 participants from Bina Nusantara University students across faculties and semesters (Wisudha *et al.*, 2016). There were 146 participants who reported that they had their own venture (Entrepreneurs) and 439 participants who reported no venture ownership (Non-entrepreneurs). The operational definition of 'entrepreneur' in this research was based on a self-report on whether or not they owned ventures or small businesses in various industries. Using SPSS, we took a random sample from the Non-entrepreneur students so it matched the sample amount of Entrepreneur students, which was 146 individuals.

The participants of this research were students from Bina Nusantara University, in age range of 18 to 26 years (N = 292) taken from the study. The demographic profile of the respondents was 54.5% males and 45.5% females. Meanwhile, 90% of the respondents were in the age range of 18–22 years. Other demographics showed that 58% of the respondents were in their senior years (5th semester and above). The participants were students from eight different faculties in Bina Nusantara University, namely the Faculty of Humanities (26%), Faculty of Economics and Communication (18.2%), School of Information Systems (13.4%), School of Computer Science (13%), School of Design (11.3%), Faculty of Engineering (12.7%) and School of Business Management (5.5%).

In this study, IEO was measured using a 20-item, five-point Likert scale questionnaire in Bahasa Indonesia, adapted from Bolton and Lane (2012), on three dimensions namely, Risk-Taking ($\alpha = .81$), Innovativeness ($\alpha = 0.85$), and Proactiveness ($\alpha = 0.77$). Personality was measured using the Lumina Spark personality questionnaire (Lumina Learning 2013), comprising of 144 items on a five-point Likert scale in Bahasa Indonesia. Four personality aspects were measured namely: Big Picture Thinking ($\alpha = 0.86$); Extraverted ($\alpha = 0.87$); Outcome Focused ($\alpha = 0.86$); and Discipline Driven ($\alpha = 0.80$). These four personality aspects showed the highest degree of correlation in a previous study (Wisudha *et al.*, 2016).

3 RESULTS

Independent-sample t-tests were conducted to compare the IEO between Entrepreneur and Non-entrepreneur students, and to compare the Lumina Spark measures between the two groups. As presented in Table 1, there is a statistically significant difference in the total IEO scores between Entrepreneur students and Non-entrepreneur students. This suggests that IEO in Entrepreneur students is significantly higher than Non-entrepreneur students, although the magnitude of the differences in the means varies at the dimensional level from medium for Risk-taking and Innovativeness, to small for Proactiveness. Entrepreneur students tend to take more risks and are more innovative. As for Proactiveness, there is no significant difference between the two groups.

In terms of personality, there are significant statistical differences between Entrepreneur and Non-Entrepreneur students in two out of four Lumina Spark aspects being measured in this study, i.e., Big Picture Thinking and Extraverted, as presented in Table 2. This suggests that compared to the Non-Entrepreneur students, the Entrepreneur students are more flexible and like to let the direction of behavior emerge from an evolving situation; they are more

Table 1. Differences between Entrepreneur and Non-entrepreneur students in IEO measure.

Individual entrepreneurial orientation	Non-entrepreneur (n = 146) M	SD	Entrepreneur (n = 146) M	SD	Mean difference	*t*	*p*	Cohen's d	95% confidence interval of the difference *Lower*	*Upper*
Proactiveness	3.53	0.58	3.66	0.60	0.13	−1.90	0.058	0.22	−0.004	0.269
Risk taking	3.47	0.63	3.79	0.61	0.32	−4.40	0.000	0.52	0.177	0.464
Innovativeness	3.51	0.56	3.72	0.63	0.21	−3.03	0.003	0.35	0.074	0.348
IEO	3.50	0.53	3.72	0.55	0.22	−3.52	0.000	0.41	0.098	0.345

Table 2. Differences between Entrepreneur and Non-entrepreneur Students in the personalitymeasure.

	Non-entrepreneur		Entrepreneur		Mean difference				95% confidence interval of the difference	
	(n = 146)		(n = 146)				*p*	Cohen's d		
Lumina spark	M	SD	M	SD		*t*			Lower	Upper
Big Picture Thinking	3.39	0.55	3.52	0.53	0.13	−2.01	0.05	0.24	0.03	0.25
Extraverted	3.44	0.54	3.64	0.55	0.19	−3.04	0.00	0.37	0.07	0.32
Outcome Focused	3.38	0.57	3.49	0.50	0.11	−1.71	0.09	0.20	−0.02	0.23
Discipline Driven	3.65	0.47	3.69	0.45	0.03	−0.61	0.54	0.13	−0.07	0.14

visionary, willing to make improvements and shake up the status quo; furthermore, they enjoy working with other people, and are more expressive.

4 DISCUSSION

This study shows that there are statistically significant differences in total IEO between Entrepreneur and Non-entrepreneur students. At the dimension level, Risk-taking and Innovativeness show significant differences between the two groups. However, there is no difference seen for the Proactiveness dimension. Compared to total IEO and Innovativeness, Risk Taking has the biggest effect size ($d = 0.52$). This is consistent with the findings made by Kropp *et al.* (2008) who found significant differences in Risk-taking ($d = 0.49$) between Entrepreneurs (full time entrepreneurs) and Non-entrepreneurs (employees in companies). It is also consistent with Frese & Gielnik's (2014) Model of Entrepreneurship that uses IEO as one of the predictors of entrepreneurial success. In their study, the effect size of IEO to entrepreneurship is seen to be medium ($d = 0.41$).

There are several reasons which may explain the medium size of effect in the results. Firstly, the items constructed may not be sensitive enough to measure the essence of IEO, so there is a call for a review of the items. Linked to that, in view of the small mean differences, it is suggested that the Likert measurement scale can be refined and widened from five points to seven points. Secondly, in the questionnaire, the definition of entrepreneurship was limited to owning a venture; therefore, it is noted that the study is limited to investigating the second phase of entrepreneurship, the launch phase, comprising the development and execution phase. We can argue that IEO, especially Risk Taking and Innovativeness, is needed in this second phase, launching, though by only looking at the this phase, the indicators of entrepreneurship used in this research are not sufficiently comprehensive. It is noted that the research does not cover the first phase of entrepreneurship, namely the pre-launch or opportunity identification phase as well as the third phase which is concerned with maintaining the business. It is suggested that measurements be included for these two phases, which may result in a shift in the effect size of the IEO measures and where the role of Proactiveness may also become more evident.

In terms of personality, Lumina Spark's Big Picture Thinking and Extraverted are the personality aspects that measure significantly higher in Entrepreneur students ($p < 0.05$), with Extraverted having a higher effect size. This supports findings by Zhao Seibert and Lumpkin (2010) where correlations were found between Extraversion, Conscientiousness, and Openness with entrepreneurial intent and performance. Moreover, the researchers across disciplines have suggested that Extraversion predicts Risk-taking behavior in several contexts (Nicholson, Soane, Fenton-O'Creevy, & Willman, 2005; Anic, 2007; Zafar & Meenakshi, 2011). Although our previous research shows that all personality aspects correlate with IEO

(Wisudha *et al.*, 2016), the result from our current study suggests that Entrepreneur students have different personality characteristics compared to Non-Entrepreneur students. However, because this research only considers entrepreneurial status from self-reports by the participants, we can say that the result does not imply that other personality dimensions are not important for an entrepreneur.

There are several limitations to this study. Firstly, IEO was measured using a self-report questionnaire, while most recent research reported that there are other ways of measuring IEO to get a more holistic view, namely using the methodology of Assessment Centers. The Assessment Center approach is a method that can involve a unique combination of essential elements codified in Guidelines and Ethical Considerations of Assessment Center Operations that have been used for selection, diagnosis and development in organizations (Thornton & Gibbons, 2009). It has recently been used to measure IEO (DeGennaro, Wright, & Panza, 2016). Further research can adopt the Assessment Center method to measure the aspects of IEO that cannot be measured by a self-report scale, for example by having multiple trained assessors observe overt behavior displayed by an assessee in a complex entrepreneurial context. Secondly, this study is not able to provide a meaningful breakdown of information based on business industries as the list does not account for a large percentage of responses regarding that factor. The absence of a choice led 65% of respondents to choose "Other" when describing their business industry. Future research will benefit from using a refined list of entrepreneurial business industries that is updated in view of the expansion in recent times of types of entrepreneurial ventures. This will enable a more meaningful analysis based on business industries.

This research has provided an insight that risk-taking as an action characteristic has the largest impact on business ownership. Students who have their own business take more bold actions by going into the unknown and committing significant resources to ventures in an uncertain situation, compared to students who do not have their own business. Universities with entrepreneurial courses can develop curricula with activities that help students practice risk-taking in a business context. Risk-taking can be incorporated as one of the sessions, and therefore, training such as Achievement Motivation Training can be useful as one way to teach risk-taking to students. The module to teach risk-taking can also be developed into different forms of risk-taking in a business context, such as risk-taking about time, decision making, finance, and many more. Moreover, universities can use this result as evidence for programs that aim to help increase student's awareness of their personality trait, as it is related with entrepreneurial outcome. Students who have high scores on Big Picture Thinking and Extraverted can be encouraged to perceive it as a commodity for successful entrepreneurship. As for students with lower scores on those aspects, universities can develop programs to cultivate Big Picture Thinking and Extraverted aspects of their personality and/or associated behaviors.

5 CONCLUSION

Our first hypothesis states that there is a significant difference in IEO and its dimensions (Risk Taking, Innovativeness, Proactiveness) between Entrepreneur and Non-entrepreneur students. Following the results, this hypothesis is partially supported, Entrepreneur students are reported to have higher Risk-taking and Innovativeness compared to Non-entrepreneur students. It suggests that Risk taking and Innovativeness are helpful dimensions to indicate whether or not a student is successful in starting a new business, with Risk taking having a bigger effect than Innovativeness. However, Proactiveness is shown to be statistically insignificant in determining a difference between both groups. Similarly, our second hypothesis also partially supports that the Big Picture Thinking and Extroverted aspects of Lumina Spark are higher in Entrepreneur students. The Entrepreneur students, however, do not differ from Non-entrepreneur students in terms of the Outcome Focused and Discipline Driven aspects of Lumina Spark. This result has provided supportive evidence for universities to develop curricula which include activities that help students practice risk-taking in a business context, increase students' awareness of their personality trait, and extend their range of behavior to embrace Big Picture Thinking and Extraversion aspects of their unique personality.

REFERENCES

Anic, G. (2007) The Association Between Personality and Risk Taking. *[Master's Thesis]*, University of South Florida. Available from: http://scholarcommons.usf.edu/etd/605.

Avlontis, G.J., & Helen, E.S. (2007) Entrepreneurial orientation of smes, product innovativeness, and performance. *Journal of Business Research,* 60 (5), 566–575. Available from: http://dx.doi.org/10.1016/j.jbusres.2007.01.001.

Ahmetoglu, G., Franziska, L., & Tomas, C.P. (2011) EQ-nomics: Understanding the relationship between individual differences in trait emotional intelligence and entrepreneurship. *Personality and Individual Differences* 51, 1028–1033. doi:10.1016/j.paid.2011.08.016.

Baron, R.A. (2007) Entrepreneurship: A process perspective. In Baum J.R., Frese M, Baron R.A. (Eds) *The Psychology of Entrepreneurship, 19–39*. Mahwah, NJ, Erlbaum.

Bolton, D.L., & Michele D.L. (2012) Individual entrepreneurial orientation: Development of a measurement instrument. *Education and Training* 54 (2), 219–233. doi:10.1108/00400911211210314.

BSI Entrepreneur Center. (2016) *Saatnya Kampus Dijadikan Basis Produksi Entrepreneur*. Available from: http://bec.bsi.ac.id/baca-artikel/2016/02/saatnya-kampus-dijadikan-basis-produksi-entrepreneur#.V_sNjeB942x.

Cross, B., & Anthony, T. (1995) The untold story: Is the entrepreneur of the 21st century defined by emotional intelligence? *International Journal of Organizational Analysis*11 (3), 221–228. doi:10.1108/eb028973.

DeGennaro, M.P., Chris W.W., & Nancy, R.P. (2016) "Measuring Entrepreneurial Orientation in an Assessment Center: An Individual Level-of-Analysis Study."*The Psychologist-Manager Journal,* 19 (1), 1–22. Available from: doi:10.1037/mgr0000035.

Desson, S., Stephen, B., & John, G. (2014) "Lumina Spark—Development of an Integrated Assessment of Big 5 Personality Factors, Type Theory and Overextension."*International Conference on Psychotechnology Abstract Book*. Available from: https://www.luminalearning.com/pdfs/SparkResearch.pdf.

Fairoz, F.M., Hirobumi, T., & Tanaka, Y. (2010) Entrepreneurial orientation and business performance of small and medium scale enterprises of Hambantota District Sri Lanka. *Asian Social Science,* 6 (3), 34–46. Available from: doi: 10.5539/ass.v6n3p34.

Frese, M, & Michael, M.G. (2014) The psychology of entrepreneurship. *Annual Review of Organizational Psychology and Organizational Behavior,* 1 (1), 413–438. Available from: doi:10.1146/annurev-orgpsych-031413-091326.

Gianie. (2015) *Berani Menjadi Wirausaha. Kompas Print*. Available from: http://print.kompas.com/baca/2015/04/21/Berani-Menjadi-Wirausaha.

Guth, W.D., & Ari, G. (1990) Guest editors' introduction: Corporate entrepreneurship. *Strategic Management Journal,* 11, 5–15. Available from: http://www.jstor.org/stable/2486666.

Johan, W., & Shepherd, D. (2003) Knowledge-based resources, entrepreneurial orientation, and the performance of small and medium-sized businesses. *Strategic Management Journal,* 24 (13), 1307–1314. doi: 10.1002/smj.360.

Joseph, D.L., Jing, J., Daniel, A.N., Ernest, H.O.B. (2015) Why does self-report emotional intelligence predict job performance? A meta-analytic investigation of mixed EI. *Journal of Applied Psychology,* 100 (2), 298–342. doi:/10.1037/a0037681.

Koenig, C., Holger, S., Michael, F., Andreas, R., & Zhong-Ming, W. (2007) Scenario-based scales measuring cultural orientations of business owners. *Journal Evolutionary Economics,* 17 (2), 211–239. doi: 10.1007/s00191-006-0047-z.

Kropp, F., Noel, J.L., & Aviv, S. (2008) Entrepreneurial orientation and international entrepreneurial business venture startup. *International Journal of Entrepreneurial Behavior and Research,* 14 (2), 102–117. doi:10.1108/13552550810863080.

Lumina Learning. (2013) *Lumina Spark Qualification Manual*. Lumina Learning, Camberley, UK.

Lumpkin, G.T., & Gregory, G.D. (1996) Clarifying the entrepreneurial orientation construct and linking it to performance. *Academy of Management Review,* 21 (1), 135–172. doi:10.2307/258632.

Mahmood, R.N.H. (2013) Entrepreneurial orientation and business performance of women-owned small and medium enterprises in Malaysia : Competitive advantage as a mediator. *International Journal of Business and Social Science,* 4 (1), 82–90. doi:10.1177/0266242612455034.
Miller, D. (1983) The correlates of entrepreneurship in three types of firms. *Management Science,* 29 (7), 770–791. Available from: http://dx.doi.org/10.1287/mnsc.29.7.770.
Moreno, A.M., & Jose, C.C. (2008) Entrepreneurial orientation and growth of SMEs: A causal model. *Entrepreneurship Theory and Practice,* 32 (3), 507–528. doi: 10.1111/j.1540–6520.2008.00238.x.
Nicholson, N., Emma, S., Mark, F.O.C., & Paul, W. (2005) Personality and domain-specific risk taking. *Journal of Risk Research,* 8 (2), 157–176. doi:10.1080/1366987032000123856.
Palupi, H. (2016) *BEKRAF Luncurkan Program BEKUP dalam Upaya Meningkatkan Keberhasilan Pre-Startup Indonesia.* Available from: https://www.codepolitan.com/bekraf-luncurkan-program-bekup-upaya-meningkatkan-keberhasilan-pre-startup-indonesia.
Rauch, A., Johan, W., Lumpkin, G.T., & Michael, Frese. (2009) Entrepreneurial orientation and business performance: An assessment of past research and suggestions for the future. *Entrepreneurship Theory and Practice,* 33 (3), 761–787. doi: 10.1111/j.1540-6520.2009.00308.x
Schillo, S. (2011) Entrepreneurial orientation and company performance: Can the academic literature guide managers? *Technology Innovation Management Review* 1 (2), 20–25. http://timreview.ca/article/497.
Seth, S. (2015) *Why Entrepreneurs Are Important for the Economy.* Available from: http://www.investopedia.com/articles/personal-finance/101414/why-entrepreneurs-are-important-economy.asp.
Shane, S, & Venkataraman, S. (2000) The promise of entrepreneurship as a field of research. *The Academy of Management Review,* 25 (1), 217–226. Available from: http://www.jstor.org/stable/259271.
Tempo.co.id. (2016) *Pemerintah Gelontorkan Rp 100 Triliun untuk Bantu Pengusaha Muda.* Available from: https://m.tempo.co/read/news/2016/05/24/090773725/pemerintah-gelontorkan-rp-100-triliun-untuk-bantu-pengusaha-muda.
Tang, J., Kacmar, K.M., & Lowell, W.B. (2012) Entrepreneurial alertness in the pursuit of new opportunities. *Journal of Business Venturing,* 27 (1), 77–94. doi: 10.1016/j.jbusvent.2010.07.001.
Thornton, G.C., & Alyssa, M.G. (2009) Validity of assessment centers for personnel selection. *Human Resource Management Review,* 19 (3), 169–187. doi:10.1016/j.hrmr.2009.02.002.
Wisudha, A., Gabriela, A., Johannes, R., Pingkan, C.B.R., & Esther, W.A. (2016) *The Correlation Between Personality and Individual Entrepreneurial Orientation: A Recommendation for Entrepreneurial Education.Proceeding of International Conference on Entrepreneurship (ICONENT), Tangerang, 2016.* Available from: http://iconent.global.uph.edu/news.html.
Zafar, S. & Meenakshi, K. (2012) A study on the relationship between extroversion-introversion and risk-taking in the context of second language acquisition. *International Journal of Research Studies in Language Learning,* 1 (1), 33–40. doi:10.5861/ijrsll.2012.v1i1.42.
Zahra, S.A., & Jeffrey, G.C. (1995) Contextual influences on the corporate entrepreneurship—performance relationship: A longitudinal analysis. *Journal of Business Venturing,* 10 (1), 43–58. doi:10.1016/0883-9026(94)00004-E.
Zhao, H., Scott, E.S., & Lumpkin, G.T. (2010) The relationship of personality to entrepreneurial intentions and performance: A meta-analytic review. *Journal of Management,* 36 (2), 381–404. doi:10.1177/0149206309335187.
Zulkifli, R.M., & Rosli, M.M. (2013) Entrepreneurial orientation and business success of malay entrepreneurs: Religiosity as moderator. *International Journal of Humanities and Social Sciences,* 3 (10), 264–275. Available from: http://www.ijhssnet.com/journals/Vol_3_No_10_Special_Issue_May_2013/29.pdf

Diversity in Unity: Perspectives from Psychology and Behavioral Sciences – Ariyanto et al. (Eds)
© 2018 Taylor & Francis Group, London, ISBN 978-1-138-62665-2

Intergenerational differences in shame and guilt emotions and the dissemination of cultural values among the Buginese

Z.Z. Irawan & L.R.M. Royanto
Faculty of Psychology, Universitas Indonesia, Depok, Indonesia

ABSTRACT: This research investigated the differences in shame and guilt emotions between old and young generations of the Buginese. The aim of the study was also to identify the cultural values associated with the socialisation process around shame and guilt. The research used mixed methods. Forty-five people from the older generation (mean age = 70.98) and 45 people from the younger generation (mean age = 19.31) were involved. Shame and guilt emotions were evaluated using TOSCA 3, together with interviews about the socialisation of their cultural values. The findings showed that the older generation experienced stronger shame and guilt emotions than the younger generation, and this happened because of differing cultural socialisation processes. Although both generations reported parents and school as primary agents of socialisation in developing their cultural values related to shame and guilt emotions, the older generation also felt the community to be a crucial agent of such socialisation. The community taught them about cultural values and increased their cultural knowledge. Thus, stronger shame and guilt emotions emerged as a result of deeper knowledge of cultural values promoted by a community-based socialisation process.

1 INTRODUCTION

Developments in technology, especially in telecommunications and the internet, distinguish the era of globalisation. They simplify people's lives in matters related to work and communication with others. However at a certain point, they can also have a negative impact; shifting the moral standards of society. For example, Labeodan (2009) reports that the high moral standards and religiosity of the Yoruba tribe in Africa are in decline because of globalisation. In Indonesia, similar issues are evident in the mass media as it delivers news about corruption, rape, free sex, substance abuse and even murder. Society produces sanctions to enhance moral behaviour (e.g. prison and social sanctions) and reduce immoral behaviour, but these frequently seem ineffective.

Moral emotion has a role in encouraging behaviours that are considered to uphold moral standards and repress less moral behaviours in society (Tangney, 1995). There are two moral emotions that are portrayed as negative emotions caused by a failure to perform good behaviour in society, that is, the emotions of shame and guilt (Tangney, 1995). Both emotions prompt humans to act more morally; however, shame and guilt are different. Tangney (1995) emphasises that shame is an emotion that holds the self in negative judgement, and thus the person feels small and tries to avoid others. By contrast, guilt is an emotion that is focused on behaviour, making someone feel regret, and causing them to try to apologise.

Tangney (2003) explains the difference between shame and guilt. Shame is a painful emotion that does not always involve the presence of others; nevertheless, it still can make a person feel worthless, small, helpless and exposed. Lindsay-Hartz et al. (1995) explain that an ashamed person wishes themselves not to be themselves, because they perceive themselves as a bad and horrible person. Thereafter they wish to run away and avoid others, look downwards to avoid eye contact, and lower their shoulders. Guilt is a tense and remorseful feeling caused by an action taken. It makes someone constantly think about the act and wish they had not done it, motivating the person to show reparative behaviour (i.e. admit their mistake, ask forgiveness, and try to fix the situation) (Tangney, 2003). Ferguson and Stegge (1995) explain that in situational antecedents, guilt occurs when people make mistakes intentionally or unintentionally, perceive their action as an immoral act, and feel responsible for the consequences. Tangney (1995) explains that guilt has no effect on the self because it focuses on the action, not the person. However, an individual who feels ashamed will focus on themselves (Tangney, 1995), and may carry out unfavourable actions that are not perceived as amendable, seeing them as a reflection of themselves.

Shame and guilt emotions can be varied in individuals, depending on how the emotions are internalised. Furthermore, they can also vary on a cultural basis. Wallbott and Scherer (1995) found that shame and guilt have a connection with cultural values in certain cultures. Matsumoto and Juang (2004) explain that cultures can shape emotions differently as a result of the different reality that every culture has, thus producing different psychological needs and purposes. It is clear that more research needs to be done into this cultural basis to distinguish between shame and guilt experiences. Indonesia, a nation that has a wide range of different cultures, includes the Buginese tribe in which shame is a particular focus. According to Pelras (1996), the Buginese are well known as a friendly and loyal tribe, but they can use violence to uphold their honour. Rahim (1985) notes there are five main values that guide Buginese actions in society, namely, honesty (*lempu'*), intellectuality, decency (*astinajang*), strong will (*getteng*), and striving, which are united by the core of Buginese culture, *siri'*. Matthes (1874, in Rahim, 1985) refers to *siri'* as shame, shyness, pride or disgrace. Said (2008) explains the importance of *siri'*, which moulds the Buginese as a society that has the strong belief that it is 'better to die than to be ashamed in front of other people'. Without *siri'*, someone can lose his/her social identity and honour. Moreover, Rahim (1985) explains that these values were inherited through advice (*pappangaja'*) and also by will (*paseng*). Both of them are also written about in old books of the Buginese.

Wallbott and Scherer (1995) explain that there are differences in the experience of shame and guilt in different cultures. Therefore, they believe that socialisation plays a vital role in the implementation of both emotions. The existence of socialisation as a process of constantly delivering knowledge between the generations, in particular, values, norms, and rules which are formed by the culture, is the means by which a culture is maintained by each generation. The influence of socialisation can be seen in the work of Malti and Buchmann (2010), who found that socialisation by parents is important in adolescents' moral motivation. Papalia and Feldman (2012) define socialisation as the forming process for habits, skills, values and motives. Through this process, the individual becomes more responsible and productive as a member of society. Ideally, socialisation enables a society to hold on to its values and rules; thereby living in harmony and peace.

Berns (2013) explains social culture as one of the methods by which values are socialised in society. The method involves four processes, which are group pressure, traditions, rituals and routines, and symbols. The agents of socialisation are responsible for delivering values within society through the following of methods and processes. Five agents of socialisation have an important role in the socialisation process, namely, family, school, friends, mass media, and community, but there are also contrasting realities that can influence the socialisation from time to time. The remarkable development of mass media is one example; as Henten and Tadayoni (2008) explain, the internet that began in the middle of the 1960s has had a great impact on the media, which can be seen in the present-day forms of mobile TV, digital TV and even web platforms. Acknowledging the reality of cultural differences, this study focuses on different generations to examine whether different eras have different socialisation processes that could also be manifested in their moral emotions of shame and guilt.

2 METHOD

2.1 *Participants*

This research was conducted on 90 participants, who were separated into two groups: an older and a younger generation. Both generations had equal proportions, that is, 45 participants in each group. The older generation were in their late adulthood stage (mean age = 70.98; 31 women and 14 men) and the younger generation were in their late adolescence stage (mean age = 19.31; 17 women and 28 men). All of the participants were Buginese and had been living in South Sulawesi their entire lives; thus the Buginese language had to be used. Four other participants were also used to undertake a qualitative element in this study, consisting of two participants from the late adulthood group and two from the late adolescence group. All of these were women.

2.2 *Measurement*

The instrument used to measure shame and guilt emotions was TOSCA 3 (Test of Self-Conscious Affect 3), adapted and modified by Barlian (2013) for the Indonesian language. For the qualitative approach, an interview and an observation were used. The interview method was used to examine the shame and guilt emotions discernible from the participants' knowledge, experience, and process of socialisation.

3 RESULTS

The differences between the shame and guilt emotions in the older and younger generations will be discussed first, and then the socialisation process will be explained.

As can be seen in Table 1, the shame and guilt scores for the older generation were significantly higher than those of the younger generation (shame emotions $t = 6.372$, $p = 0.000$; guilt emotions $t = 5.033$, $p = 0.000$).

The qualitative results underwent two analyses. The first one captured the emotion differences (Table 2) and the second captured the differences in the process of socialisation (Table 3).

There were differences between the triggering factors and the expressions of both shame and guilt emotions. Shame can be triggered in response to family matters, actions that have been done before, other people's actions, interpersonal relationships and apologising. On the other hand, guilt can be triggered in response to family matters and immoral actions. Both generations express shame by evasion and apology. Guilt is expressed by the older generation with an apology, and they describe feeling terrified and experiencing a pounding heart. After their apology has been accepted, they feel relief. In the younger generation, guilt is also expressed through apology and they stated that they were scared to admit their mistakes and were haunted by their own feelings and regret.

Both generations had obvious knowledge about Bugis but it was found that the older generation had clearer knowledge about Buginese culture than the younger one. The older

Table 1. Means, standard deviation, and mean differences of shame and guilt emotions between old and young generations.

Emotion	Generation	N	Mean	Mean difference	SD	t	df	p
Shame	Old	45	35.16	12.489	9.658	6.372	88	0.00**
	Young	45	22.67		8.921			
Guilt	Old	45	51.96	10.400	5.300	5.033	58.640	0.00**
	Young	45	41.56		12.807			

Table 2. Differences of shame and guilt between older and younger generations.

Difference	Old generation	Young generation
Triggers of shame	God, family, acts that have been done (false promises, inappropriate actions, making mistakes, being caught by others), ashamed of other people, meeting new persons, and not asking for an apology	Family, impolite actions, blaming other people, meeting a new person, meeting the opposite sex, people who can give evaluation or opinion about their behavior when going to ask for an apology
Triggers of guilt	Acts that have been done (making a mistake, possessing others' belongings) and family	Acts that have been done (making a mistake, telling a lie, making a false promise) and family
Shame expressions	Avoidance and apologising	Avoidance and apologising
Guilt expressions	Apologising, terrified, pounding heart, and relief after asking for apology	Apologising, terrified to admit a mistake, feeling haunted and remorseful

Table 3. Differences in the socialisation process between older and younger generations.

Difference	Old generation	Young generation
Cultural values	Have deep knowledge about *siri'* and other Buginese cultural aspects and can explain them fluently	Have slight knowledge of Buginese culture, but can only explain *siri'*
Method of socialisation	Advice (i.e. advice was given before their parents passed away, before visiting a friend or family's house, before going to school) and acquisition of ancestor books about Buginese culture	Advice (given when they were young) and observation
Content of socialisation	Shame, guilt, and rules of the society	Shame, guilt, and rules of the society
Start of socialisation	Since childhood	Since childhood
Agent of socialisation	Parents, school and community (teachers of the Quran and traditional figures, *panritta'*)	Parents and school

generation were able to give clear explanations of Buginese culture, including such terms as *siri'*, *bunga bangka'*, *mutempo*, *takabbur*, *masiri'*, and *masiri siri'*. The younger generation did not mention these and could not explain them, with the exception of *siri'*.

The method of socialisation in both generations is mostly done through advice from older people, predominantly delivered by parents. However, the younger generation also make substantial use of observation in learning the culture. Furthermore, socialisation in the older generation also happens by acknowledging their ancestors' books about the Buginese culture, which does not happen in the younger generation.

The content of socialisation were varied from shame, guilt, and rules in the society. This process occurred from an early age, and parents were the primary socialisation agents, together with others such as school. There were slight differences in the older generation because their socialisation agents were people in their community, the teachers (*guru mengaji*) who taught them to read the Quran, and traditional figures called *panritta'*.

4 DISCUSSION

The present study found that there are significant differences in shame and guilt between the older and younger generations of the Buginese. The shame and guilt emotions of the older generation were greater than those of the younger generation. This may be influenced by their age differences as Orth et al. (2010) and Ferguson and Stegge (1995) reported that self-conscious emotions increase over time. Furthermore, based on Erikson's stages of psychosocial development, the older generation is in the late development stage and will have learned more about wisdom. On the other hand, the younger generation is still in the adolescence stage, experiencing confusion about their identity (Papalia & Feldman, 2012). In this stage, they are exploring life experiences to find their identity, which might lead them to carry out thoughtless actions such as breaking the rules. Thus, adolescence tends to involve lower levels of moral emotion than in older generations.

The results also imply that the differences in shame and guilt between older and younger generations stems from a difference in their socialisation processes. As Barrett (1995) explains that socialisation happens through acknowledgement of the moral standards of society, so shame and guilt also emerge from socialisation. The older Buginese generation had different agents of socialisation compared to the younger generation. The important agents of socialisation for the older generation were not only their parents and school, but also their community. The older generation in the present study explained that the teachers (*guru mengaji*) who taught them to read the Quran, and traditional figures called *panritta'* both played important roles in their socialisation. They met their *guru mengaji* every afternoon and *panritta'* would deliver advice about values and culture on a monthly basis. These activities helped them to internalise such values through repeated exposure. In contrast, the younger generation had only their parents and school as their agents of socialisation. As a result, the younger generation has less understanding of Buginese culture and values than the older generation, including the emotions of shame and guilt. This aligns with the views of Eisenberg et al. (1998), who found that socialisation about an emotion in the family will increase that emotion in a child; thus would a child learn to understand certain emotions. Lack of awareness and comprehension regarding shame and guilt are caused by a decline in the socialisation process.

This study found that the process of socialisation in the older generation continued into adulthood, until their parents passed away. They would gather in one room to receive guidance in the moral values of life and Buginese culture, and they would be given books written in Buginese about their norms and culture. Therefore, it can be concluded that there is a different level of intensity in the socialisation of the older and younger generations.

There is also a different perception of the factors that trigger shame and guilt in the two generations, even though they both expressed shame and guilt emotions in similar ways. One of the differences is the elicitor of shame in both generations related to their perceptions of apology. The older generation explained that every time they meet the person they owe an apology to, they would feel ashamed, because they never do what they have to do (apologise). The older generation perceives that not asking for an apology is inappropriate, thus they would feel ashamed. This is explained by their values, as both generations explained that *siri'* is a value that triggers someone to feel ashamed of their inappropriate behaviour or when they make a mistake. On the other hand, the younger generation would feel ashamed when they apologise because they are not accustomed to doing so. Tangney (2003) explains that shame could be elicited without the presence of others, but it would appear in thoughts of how others perceived them. This is probably what makes the younger generation feel ashamed when they apologise, because they consider others' perceptions of them doing something they are not accustomed to. Further research is needed in this matter.

Tangney (1995) explains shame as an emotion that focuses on the self, whilst guilt focuses on behaviour, which contrasts with the Buginese value, *siri'*, which triggers someone to feel ashamed of inappropriate behaviour or making a mistake. This suggests that the Buginese implicitly teach about guilt. It also explains why Buginese people who feel ashamed are willing to apologise, which according to Tangney (1995) is the more usual behaviour when someone feels guilty.

The Buginese are expected to not be ashamed of themselves in front of other people. This pressure triggers them to feel guilt when they do feel ashamed, and this motivates them to apologise. Ausubel (1955) explains this as *moral shame*, which occurs because they are afraid of other people's negative perceptions of them. Therefore, shame that is associated with guilt is one example of *moral shame*. Although these findings assume that shame in the Buginese is linked to *moral shame*, a further study is required to verify this assumption.

There are some limitations to this study. The tool (TOSCA 3) that was adapted to Indonesian circumstances lacks common everyday situations, especially in terms of those commonly found in the society of the sample population used here. As a consequence, some of the subjects had difficulties in imagining the situations. The second limitation was the language used. The study was conducted in a district of South Sulawesi, where people rarely use Indonesian as their daily language, especially the older generation. Rather, they use Buginese as their everyday language and some of them are illiterate. Thus, extra assistance might have been needed to complete the questionnaires. The use of translators to assist the older generation could affect the results, because of differences in meaning between the text and its translation.

5 CONCLUSION

The present study found differences in the shame and guilt emotions of older and younger generations of Buginese. This is related to the different socialisation processes experienced by the two generations. The result of this study also suggests that the Buginese need to amplify the current processes of socialisation if the younger generation are to better acknowledge cultural values and rules and, as a consequence, increase their moral emotions. Parents and school have important roles in this socialisation process, but the community's role is also significant, as shown in the older generation. Thus, consideration may need to be given to community activities based on the local values of the Buginese (e.g. cultural performances or harvest feasts involving the traditional figure of the *panritta'*), which could enhance the socialisation process of the younger generation.

REFERENCES

Ausubel, D.P. (1955). Relationships between shame and guilt in the socializing process. *Psychological Review*, *62*(5), 378–390. doi:10.1037/h0042534

Barlian, I.Y. (2013). *Perbedaan Emosi Malu Dan Emosi Bersalah Pada Generasi Tua Dan Generasi Muda [Differences in shame and guilt between old and young generations]* (Dissertation, Universitas Indonesia, Depok, Indonesia). Retrieved from http://lib.ui.ac.id/naskahringkas/2016-04/S46931-Irene Yolanda Barlian

Barrett, K. (1995). A functionalist approach to shame and guilt. In Tangney, J.P. & Fischer, K.W. (Eds.), *Self-conscious emotions: The psychology of shame, guilt, embarrassment, and pride* (pp. 25–63). New York, NY: The Guilford Press.

Berns, R.M. (2013). *Child, family, school, community: Socialization and support* (9th ed.). Belmont, CA: Wadsworth, Cengage Learning.

Eisenberg, N., Cumberland, A. & Spinrad, T.L. (1998). Parental socialization of emotion. *Psychological Inquiry*, *9*(4), 241–273. doi:10.1207/s15327965pli0904_1

Ferguson, T. & Stegge, H. (1995). Emotional states and traits in children: The case of guilt and shame. In Tangney, J.P. & Fischer, K.W. (Eds.), *Self-conscious emotions: The psychology of shame, guilt, embarrassment, and pride* (pp. 174–197). New York, NY: The Guilford Press.

Henten, A. & Tadayoni, R. (2008). The impact of the internet on media technology, platforms and innovation. In Küng, L., Picard, R.G. & Towse, R. (Eds.), *The internet and the mass media* (pp. 45–64). Thousand Oaks, CA: SAGE Publishing. doi:10.4135/9781446216316.n3

Labeodan, H.A. (2009). Nigerian youths and African cultural moral values in the age of globalization: Challenges and prospects. In Adebayo, A.G. & Adesina, O.C. (Eds.), *Globalization and transnational migrations: Africa and Africans in the contemporary global system*. Newcastle upon Tyne, UK: Cambridge Scholars Publishing.

Lindsay-Hartz, J., De Rivera, J. & Mascolo, M.F. (1995). Differentiating guilt and shame and their effects on motivation. In Tangney, J.P. & Fischer, K.W. (Eds.), *Self-conscious emotions: The psychology of shame, guilt, embarrassment, and pride* (pp. 274–300). New York, NY: Guilford Press.

Malti, T. & Buchmann, M. (2010). Socialization and individual antecedents of adolescents' and young adults' moral motivation. *Journal of Youth and Adolescence*, *39*(2), 138–149. doi:10.1007/s10964-009-9400-5

Matsumoto, D. & Juang, L. (2004). *Culture and psychology.* Belmont, CA: Wadsworth.
Orth, U., Robins, R.W. & Soto, C.J. (2010). Tracking the trajectory of shame, guilt, and pride across the life span. *Journal of Personality and Social Psychology*, *99*(6), 1061–1071. doi:10.1037/a0021342
Papalia, D.E. & Feldman, R.D. (2012). *Experience human development* (12th ed.). New York, NY: McGraw-Hill.
Pelras, C. (1996). *The Bugis*. Oxford, UK: Blackwell Publishers.
Said, M. (2008). *Konsep Jati Diri Manusia Bugis: Sebuah Telaah Falsafi tentang Kearifan Bugis*. South Tangerang, Indonesia: Churia Press.
Rahim, A.R. (1985). *Nilai-Nilai Utama Kebudayaan Bugis.* Makassar, Indonesia: Lembaga Penerbitan Unhas.
Tangney, J.P. (1995). Shame and guilt in interpersonal relationships. In Tangney, J.P. & Fischer, K.W. (Eds.), *Self-conscious emotions: The psychology of shame, guilt, embarrassment, and pride* (pp. 114–139). New York, NY: The Guilford Press.
Tangney, J.P. (2003). Self-relevant emotions. In Leary, M.R. & Tangney, J.P. (Eds.), *Handbook of self and identity*. New York, NY: The Guilford Press.
Wallbott, H.G. & Scherer, K. (1995). Cultural determinants in experiencing shame and guilt. In Tangney, J.P. & Fischer, K.W. (Eds.), *Self-conscious emotions: The psychology of shame, guilt, embarrassment, and pride*. New York, NY: The Guilford Press.

Diversity in Unity: Perspectives from Psychology and Behavioral Sciences – Ariyanto et al. (Eds)
© 2018 Taylor & Francis Group, London, ISBN 978-1-138-62665-2

To be leader or not to be leader? Correlation between men's negative presumption toward women leaders and women's leadership aspirations

N.I. Muthi'ah, E.K. Poerwandari & I. Primasari
Faculty of Psychology, Universitas Indonesia, Depok, Indonesia

ABSTRACT: In Indonesia, men obtain more favourable chances to take vital roles in leadership. It is an indication that women remain vastly underrepresented as leaders, and they are also undervalued even by other women. This study is aimed at investigating how women leaders are perceived by men and how social construction could contribute to how men perceive women's leadership aspirations, as one of the determinants of women's participation as leaders. This study was conducted by distributing online questionnaires to 147 male sophomore students in state and private universities in Indonesia. An adapted version of the Gender Authority Measure (GAM) instrument was employed to measure attitudes toward female leaders. In addition, the Leadership Aspiration Subscale (LAS) was employed to measure perceptions of female leadership aspirations. The results of the study showed that there was a negative correlation between male students' attitudes toward women leaders and perceptions of female leadership aspirations ($r = -0.218$, $n = 147$, $p < 0.05$), indicating that male students with more negative attitudes toward women leaders tend to perceive that women have lower leadership aspirations.

1 INTRODUCTION

In Indonesia, women's participation in various sectors of life is still limited. The apparent level of female participation in the workforce is still very low in the country, where many women reported carrying out activities related to family responsibilities. Occupational segregation of men and women is still clearly visible. Many women perform jobs with lower wages and more limited career development prospects than men (ILO, 2015).

In terms of the status of women in employment, women's participation is still limited to a secondary position. Only a small number of women reach the top in public roles. The survey results of Grant Thornton's International Business Report for 2015show that only 20% of leadership positions were held by women, with the remainder held by men (Grant Thornton, 2015). The situation in the government sector and the civil state apparatus is very similar.

Such low percentages of female participation in leadership positions occur due to many factors. According to the International Business Report by Grant Thornton (2015), among the causes preventing women from becoming business leaders are domestic life, obligation or pressure from family, small numbers of women candidates to target for promotion, gender bias, the lack of structural support for women, few candidates who have expertise, the lack of female role models, and the conditions in companies. In the political field, the lack of participation of women in strategic positions is rooted in the patriarchal culture of Indonesia (Parawansa, 2002).

Another factor that has led to this low participation of women is, as the work of Carli and Eagly (2016) described it, the "labyrinth". The labyrinth is a term that refers to the obstacles—the series of complexities, detours, dead ends, and unusual paths—that block women's access to leadership positions with challenges that the opposite sex do not face, such as gender stereotypes that depict women as unsuited to leadership, discrimination in pay and promotion, lack of access to powerful mentors, greater responsibilities for child-rearing and other domestic responsibilities.

People who want to be leaders need to internalise a leadership identity of themselves and develop a sense of purpose (Ibarra et al., 2013). Improving the levels of women's leadership necessitates not only widening the access to women's leadership but also noticing the aspirations of women. One of the success factors for advancement to leadership positions is managerial aspiration (Metz & Tharenou, 2001). Good (1959, cited in Nanda, 2008) defines aspiration as the degree to which a person sets a realistic goal, which is associated with pre-existing mental and environmental attributes. Aspiration can also be desires, hopes and ideals related to a cause (Gray & O'Brien, 2007). For example, leadership aspirations mean the desires, hopes and ideals related to leadership in the chosen career field. Leadership aspiration is one dimension of its own in career aspiration that has been formed since childhood. But how the aspirations of women to be leaders can be developed and actualized, is in fact much influenced by the environmental situation, including in it, the acceptance and views of men.

Women in leadership positions are faced with a large number of gender-related barriers and expectations. In connection with the barriers to success, women in leadership positions frequently mention an absence of acceptance from their associates, unfairness in regard to benefits and status, problems with family life and work balance, and an absence of strong role models. This is in line with previous studies describing the existence in society of social stereotypes, one of which is based on gender stereotypes. Gender stereotypes are a collection of properties, behaviours and roles expected either of men or women (Fiske & Stevens, 1993). There is a component that specifies how women and men "should" and "should not" act. Fiske (1998, cited in Burgess & Borgida, 1999) also stated that both women and men will be sanctioned if they act contrary to traditional gender stereotypes.

The leadership domain has long been considered to be normally dominated by men or, in other words, is a male prerogative in corporate, political, military and other sectors of society (Eagly & Karau, 2002). The attitude of society that has been formed is: those who are considered appropriate to leadership positions are men. Women make an effort to fit themselves to their stereotypes. Numerous studies, including those focused on gifted women, have indicated that women have lower career aspirations than men and often select more traditional careers that underutilise their abilities (Ferriman et al., 2009; Kerr et al., 2005; O'Brien et al., 2000).

Shimanoff and Jenkins (1991) stated that there are more similarities than differences in the leadership behaviours of men and women, simply stating that they are equally effective. The only difference is derived from the attitudes of others toward them, where leadership behaviours are rated more positively when attributed to men than to women (Shimanoff & Jenkins, 1991). It is, therefore, clearly a waste if women are not given the opportunity to be leaders because of the negative presumption toward them as leaders.

The formation of women's aspirations is heavily influenced by their interactions with their mothers as role models (Surrey, 1984, cited in Jaya, 1997). This realistic aspiration is also influenced by the feminine gender stereotypes that include being pleased to serve the needs of others, being gentle and patient, and not prioritising their own needs. Meanwhile, Karmel (cited in Jersild, et al., 1978) suggested that men have higher aspirations than women in terms of education and career. This difference occurs because parents emphasise differing roles for boys and girls. As a result, boys and girls form different self-images (Jones et al., 1979), which will certainly affect their aspirations.

Theoretically, the leadership aspirations of women are the desires, hopes and ideals of women to be leaders with all the personal and social complexities that surround them, which include getting a promotion, leading and training others, and becoming known as a leader in a particular field (Gray & O'Brien, 2007). Meanwhile, perceptions regarding leadership aspirations of women is the process of acquisition, interpretation, selection and arrangement of sensory information regarding leadership aspirations held by women (Takwin, 2009). The identities shaped from childhood that are internalised by women form women's leadership aspirations.

The research of Eagly & Carli (2007) suggests that in many organisations employees tend to prefer male supervisors to female ones. The current researchers would like to further investigate how men, as the opposite sex, perceive women leaders in various sectors, as well as men's perception of women's leadership aspirations in general. Therefore, this study aims to investigate the correlation between the attitudes of men toward women leaders as a contributing factor in female leadership. The hypothesis of this research is:

- H1: That there is a significant relationship between attitudes toward women leaders and perceptions of women's leadership aspirations in male college students in Indonesia.

2 METHODS

The sample selected for this study consisted of male students enrolled as active students in state and private universities in Indonesia who had spent a minimum of three years in the universities by the academic year 2015–2016. Sophomore college students were selected as the subjects of this research because they should have started thinking about their future more seriously, taking into account the suitability of their self-identities. The sampling technique used in this research was a convenience sampling that was conducted by seeking participants willing to participate in the research. Questionnaires were distributed online to the participants by the researchers through e-mails and social media to survey college students from various universities in Indonesia.

In this research, an adapted Gender Authority Measure (GAM; Rudman & Kilianski, 2000) was employed to measure explicit attitudes toward women leaders (i.e. ten questionnaire items regarding attitudes toward women leaders). The original GAM consists of 15 items in which respondents indicate preferences for male versus female authorities in French and Raven's (1959, cited in Raven, 1959) five areas of social power (i.e. legitimate, expert, reward, coercive, and referent). Respondents express agreement with each item on a scale ranging from 1 (Strongly disagree) to 5 (Strongly agree). Further, a modification of the Leadership Aspiration Subscale (LAS; Gregor & O'Brien, 2015) was also employed, to measure perceptions of males toward female leadership aspirations. This questionnaire consisted of six items on the men's perceptions of women's leadership aspirations. Items on the measure were rated on a five-point Likert-type scale ranging from 0 (Not at all true of me) to 4 (Very true of me). The scores for attitudes toward women leaders and perceptions of women's leadership aspirations are obtained by summing all of the GAM and LAS responses. Participants were also asked to respond to four additional items that aimed to establish the circumstances of the participants in terms of their families and communities.

The measuring instruments were adapted by translating them from English into Indonesian. In this study, the adaptation process included a cultural translation too, by means of the use of two translators. One translator understood the psychological concepts in the context of the measuring instruments and the other one did not have any knowledge about the context of the instruments. The translations of the two were compared and then abstracted to select the most appropriate one. The researchers then invoked the help of a psychology professor to provide an expert judgement. After undergoing revisions in terms of language and culture, the readability of the measuring instruments was then tested with groups of people who had the same characteristics as the study sample. The measuring instruments were then finalised and disseminated via online networks and social media to male students in Indonesia.

The target population for this research was active students at the undergraduate level throughout Indonesia. In total, 147 students participated in this study. The majority of the participants were aged 21 (38.1%) or 22 years old (27.9%). The participants of this study came from different groups, ranging from the class of 2009 to that of 2013, who were still actively enrolled in undergraduate programmes. The largest group came from the class of 2012 (46.9% of the participants), closely followed by the class of 2013 (46.2%). The ages of

Table 1. Correlation between attitudes toward women leaders and female students' leadership aspirations.

Variable	N	Pearson correlation	Sig. (2-tailed)
GAM and LAS perception	147	–0.218	0.008

the study participants ranged from 18 to 27 years old, and they came from universities which were either located in Java (93.9%) or outside Java (6.1%).

2.1 *Correlation analysis between variables*

This study was conducted to determine the correlation between attitudes toward women leaders and the perception of male students throughout Indonesia toward women's leadership aspirations. Pearson's correlation was utilised to test the groups of participants, with a total number of responses of 147. The results of the calculation of the correlation between these variables are presented in Table 1.

From the calculation of Pearson correlation reported in Table 1, a significance value of 0.008, $r = -0.218$, $p < 0.05$ was obtained. This suggests a relationship between the variables of attitude toward women leaders and perception by males of female leadership aspirations. These values indicate that, in line with the hypothesis, there is a significant relationship between attitudes toward women leaders and men's attitudes toward women's leadership aspirations.

3 DISCUSSION

In this study, there was a significant negative correlation between the attitude of men toward women leaders and men's perception of the leadership aspirations of women. This negative correlation resulted from social gender roles which are learned from childhood and influence what women and men are expected to do, the talents they cultivate, and the occupational paths they pursue (Shapiro et al., 2015). Men gain advantages from these roles and perceive that leadership is a men's area. Books, media, educators, and parents provide the means by which gender roles are socialised. Another argument that can explain the finding is that men do not have enough exposure to women's leadership. Men are far less likely than women to have a female manager (Reskin & Ross, 1995, cited in Eagly & Karau, 2002).

The adaptation process of the measuring instruments could also contribute to the result of this study. At the time of adapting the measuring instruments of GAM into Indonesian, there were five items with inverted scores that were excluded from the study because the validity and reliability did not meet the study's criteria. For example, the item containing the statement, "If I am convicted by a judge, I would prefer it if the judge was a woman". Logically, when the unfavourable items were reversed, it should be seen that men will show low preference to women leaders. But the result is quite the opposite. This meant that such items were not actually measuring the intended aspect of social power, which in this case was legitimacy. In other words, participants were not indicating a preference for women leaders because women had more legitimacy but, perhaps, chose a female judge to handle the case proceedings because women are considered more loving and attentive. The differences between the culture of Indonesia and that of the place where the measuring instruments were originally created may be the reason for this difference. Men and women have negative attitudes toward women who become leaders in Indonesia.

4 CONCLUSION

From the research findings, the results show a significant negative correlation between attitudes toward women leaders and women's leadership aspirations by male participants. The more negative the participants' attitudes toward female leaders, the greater the tendency for men to perceive that women have lower leadership aspirations.

Research on leadership aspirations and attitudes toward women leaders in the future can be elaborated by taking into account several things, namely:

1. Improvement in the adaptation of the measuring instruments, which should be more concerned with the content of behavioural indicators that represent their construction and accommodate the cultural differences between the culture of the creators of the measuring instruments and the culture of Indonesia.
2. Consideration of the attitudes and aspirations of the leadership of the wider students' groups and universities, or in samples with other age criteria.
3. Introduction of an interview-based method to understand the psychological dynamics of women's leadership aspirations.

As for individuals and society, the following are suggested:

1. Create an empowering atmosphere for women to pursue their careers and advancement to leadership positions. Women need a safe space for learning and experimentation. Community plays an important role in leadership development programmes for women, enabling them to develop a stronger sense of leadership identity.
2. Educate everyone about the potential of gender bias in evaluating women as leaders because both women and men are sometimes unaware of having personally been victims or perpetrators of gender discrimination.

REFERENCES

Beaman, L., Duflo, E., Pande, R., & Topalova, P. (2012). Female leadership raises aspirations and educational attainment for girls: A policy experiment in India. *Science, 335*(6068), 582–586. doi:10.1126/science.1212382

Boatwright, K.J., & Egidio, R.K. (2003). Psychological predictors of college women's leadership aspirations. Journal of College Student Development, *44*(5), 653–669. doi:10.1353/csd.2003.0048

Brown, S.D., & Lent, R.W. (Eds.) (2004). Career development and counseling: Putting theory and research to work. NJ: John Wiley & Sons.

Burgess, D., & Borgida, E. (1999). Who women are, who women should be: Descriptive and prescriptive gender stereotyping in sex discrimination. *Psychology, public policy, and law, 5*(3), 665–692. doi:10.1037/1076-8971.5.3.665

Carli, L.L., & Eagly, A.H. (2016). Women face a labyrinth: An examination of metaphors for women leaders. *Gender in Management: An International Journal, 31*(8), 514–527. doi:10.1108/GM-02-2015-0007

Eagly, A.H., & Carli, L.L. (2007). Women and the labyrinth of leadership. *Harvard Business Review, 85*(9), 62.

Eagly, A.H., & Chaiken, S. (1993). *The psychology of attitudes.* Fort Worth, TX: Harcourt, Brace, and Jovanovich College Publishers.

Eagly, A.H., & Karau, S.J. (2002). Role congruity theory of prejudice toward female leaders. *Psychological Review, 109*(3), 573–598. doi:10.1037/0033-295X.109.3.573

Ferriman, K., Lubinski, D., & Benbow, C.P. (2009). Work preferences, life values, and personal views of top math/science graduate students and the profoundly gifted: Developmental changes and gender differences during emerging adulthood and parenthood. *Journal of Personality and Social Psychology, 97*(3), 517.

Fiske, S.T., & Stevens, L.E. (1993). *What's so special about sex? Gender stereotyping and discrimination.* (pp. 173–196). London, UK: SAGE Publications.

Grant Thornton. (2015). *Women in business: The path to leadership.* London, UK: Grant Thornton International. Retrieved from http://www.grantthornton.md/Pdf/ibr2015_wib_report_final.pdf

Gray, M.P., & O'Brien, K.M. (2007). Advancing the assessment of women's career choices: The Career Aspiration Scale. *Journal of Career Assessment, 15*(3), 317–337. doi:10.1177/1069072707301211

Gregor, M.A., & O'Brien, K.M. (2015). The changing face of psychology: Leadership aspirations of female doctoral students. *The Counseling Psychologist, 43*(8), 1090–1113. doi:10.1177/0011000015608949

Gregor, M.A., & O'Brien, K.M. (2016). Understanding career aspirations among young women: Improving instrumentation. *Journal of Career Assessment, 24*(3), 559–572. doi:10.1177/1069072715599537

Hurlock, E.B. (1973). *Adolescent development* (International Student ed.). Tokyo, Japan: McGraw-Hill Kogakusha.

Ibarra, H., Ely, R., & Kolb, D. (2013). Women rising: The unseen barriers. *Harvard Business Review*, *91*(9), 60–66.

ILO. (2015). *Tren ketenagakerjaan sosial di Indonesia 2014–2015: Memperkuat daya saing dan produktivitas melalui pekerjaan layak [Employment social trends in Indonesia, 2014–2015: Strengthening competitiveness and productivity through decent work]*. Jakarta, Indonesia: International Labour Organization Office for Indonesia. Retrieved from www.ilo.org/jakarta

Jaya, M. (1997). *Isi Aspirasi Mahasiswi Fakultas Psikologi Universitas Indonesia [The aspirations of students of the Faculty of Psychology at the University of Indonesia]* (Thesis, Universitas Indonesia, Depok, Indonesia).

Jersild, A.T., Brook, J.S., & Brook, D.W. (1978). *The psychology of adolescence* (3rd ed.). New York, USA: Macmillan Publishing Co. Inc.

Jones, R.A., Clyde, H., & Epstein, Y.N. (1979). Introduction to Social Psychology. Massachusetts, USA: Sinauer Associates.

Kerr, B., Foley-Nicpon, M., & Zapata, A.L. (2005). The development of talent in girls and young women. In Kerr, B., Kurpius, S., & Harkins, A. (Eds.), *Handbook for Counseling Girls and Women: Talent Development* (Vol. 2). Mesa, AZ: Nueva Science Press.

Lent, R.W., Brown, S.D., & Hackett, G. (1994). Toward a unifying social cognitive theory of career and academic interest, choice, and performance. *Journal of Vocational Behavior*, *45*(1), 79–122. doi:10.1006/jvbe.1994.1027

Lent, R.W, Hackett, G., & Brown, S.D. (1999). A social cognitive view of school-to-work transition. *The Career Development Quarterly*, *47*(4), 297–311. doi:10.1002/j.2161-0045.1999.tb00739.x

Metz, I., & Tharenou, P. (2001). Women's career advancement: The relative contribution of human and social capital. *Group & Organization Management*, 26(3), 312–342. doi:10.1177/1059601101263005

Nanda, J. (2008). *Education for all*. New Delhi, India: APH Publishing Corporation.

O'Brien, K.M., Friedman, S.M., Tipton, L.C., & Linn, S.G. (2000). Attachment, separation, and women's vocational development: A longitudinal analysis. *Journal of Counseling Psychology*, *47*(3), 301–315.

Parawansa, K.I. (2002). Hambatan terhadap partisipasi politik perempuan di Indonesia. In Panduan, S.B. (Ed.) *Perempuan di Parlemen: Bukan Sekedar Jumlah* [*Women in parliament: Not just a number*] (pp. 41–52). Stockholm, Sweden: International Institute for Democracy and Electoral Assistance (IDEA).

Raven, B.H. (1959). Social influence on opinions and the communication of related content. *The Journal of Abnormal and Social Psychology*, *58*(1), 119. doi:10.1037/h0048251

Rudman, L.A., & Kilianski, S.E. (2000). Implicit and explicit attitudes toward female authority. *Personality and Social Psychology bulletin*, *26*(11), 1315–1328. doi:10.1177/0146167200263001

Rudman, L.A., & Phelan, J.E. (2008). Backlash effects for disconfirming gender stereotypes in organizations. *Research in Organizational Behavior*, *28*, 61–79. doi:10.1016/j.riob.2008.04.003

Schein, V.E. (2001). A global look at psychological barriers to women's progress in management. *Journal of Social Issues, 57*(4), 675–688. doi:10.1111/0022-4537.00235

Shapiro, M., Grossman, D., Carter, S., Martin, K., Deyton, P., & Hammer, D. (2015). Middle school girls and the "Leaky Pipeline" to leadership: An examination of how socialized gendered roles influences the college and career aspirations of girls is shared as well as the role of middle level professionals in disrupting the influence of social gendered messages and stigmas. *Middle School Journal*, *46*(5), 3–13. doi:10.1080/00940771.2015.11461919

Shimanoff, S.B., & Jenkins, M.M. (1991). Leadership and gender: Challenging assumptions and recognizing resources. *Small group communication: A reader*, 101–133.

Takwin, B. (2009). *Persepsi sosial: Mengenali dan mengerti orang lain [Social perception: Recognizing and understanding others]*. In Sarwono, S.W. & Meinarno, E.A. (Eds.), Psikologi sosial (pp. 79–102). Jakarta, Indonesia: Salemba Humanika.

United Nations Development Programme. (2010). *Partisipasi perempuan dalam politik dan pemerintah: Makalah kebijakan [Women's Participation in Politics and Government: A Policy Paper]*. Jakarta, Indonesia: UNDP Indonesia. Retrieved from http://www.peacewomen.org/assets/file/Resources/UN/partpol_womensparticipationpoliticsgovernmentindonesia_undp_may2010.pdf

Diversity in Unity: Perspectives from Psychology and Behavioral Sciences – Ariyanto et al. (Eds)
© 2018 Taylor & Francis Group, London, ISBN 978-1-138-62665-2

Democratic quality as a predictor of subjective well-being

B. Takwin
Faculty of Psychology, Universitas Indonesia, Depok, Indonesia

ABSTRACT: This study examines the relationship between democratic quality and subjective well-being in order to fill the gap in the current debate about the influence of democracy on well-being. Using the proposed model of (Ringen 2010), democratic quality as a latent variable is measured by using three constructs: the quality of government, political participation, and degree of civil liberties. Subjective well-being is measured by using three constructs: life satisfaction, positive effect, and negative effect. This study was conducted using a correlational design involving 1500 participants (Male = 752, Female = 748, Mean Age = 33.85) from two of the most populous islands in Indonesia (Java and Sumatra). Respondents were selected by using multistage random sampling. Structural equation modeling supports the hypothesis that democratic quality is a predictor of subjective well-being.

1 INTRODUCTION

Life satisfaction, as a main component of Subjective Well-Being (SWB), can be achieved if people are able to choose and obtain the things they believe to be the best, which include achieving goals, feeling satisfied with their life, having what they want, and obtaining important things that they want in life (Diener *et al.*, 1985). On the other hand, democracy is a regime that has the potential to deliver the citizens to situations where they get what they believe to be the best (Dahl, 1989). Therefore, the achievement of life satisfaction is in line with the democratic system.

However, studies on the relationship between democracy and happiness or subjective well-being have conflicting results regarding the relationship between democracy and happiness. Some studies showed a high correlation between the two (e.g. Frey & Stutzer, 2000, 2002; Inglehart & Klingemann, 2000; Inglehart, 2006; Bratu, 2011), whereas others state that democracy does not necessarily contribute to happiness (Dorn *et al.*, 2007; Weitz-Shapiro & Winters, 2008; Heliwell *et al.*, 2010; Helliwell, 2012; Bjørnskov, 2010; Dolan & Metcalfe, 2012). Thus, it is still apparent that there are differences in the explanations on the relationship between democracy and happiness.

The problem lies in the notion of democracy and what is designated as democracy in political life. In previous studies, the definition of democracy is still too common and has emphasized on the presence or absence of democratic institutions without examining the extent to which these institutions function. This definition is bias as it is limited to the regime aspect of democracy. Thus, leading previous researchers to ignore the democratic purpose and observations of other units outside the regime, especially individual citizens who are the all important aspect of democracy (Ringen, 2010). To overcome this limitation, Ringen (2010) proposed a measurement model that emphasizes on measuring democratic quality, where the quality of democracy is defined as a function of democratic institutions in the community that produces a situation in which people get what they want and what they believe to be the best.

According to Ringen (2010), a strong normative explanation of democratic quality must rely on three basics. First, the measurement should begin with the observation of the regime. Secondly, it should include measurement of observations about how the potential of the regime is delivered into the lives of its citizens. Third, a description of the democratic quality should be made only through the combination of system and individual analysis.

Following Dahl (1989) and Lijphart (1977, 2008, 2011), Ringen (2010) has built a model of democracy that takes into account the goals of democracy. Regime is observed by recognizing the presence or absence of democratic apparatus. To determine the extent of functioning and the impact of democratic institution, we need to know to what extent each democratic institution delivers to its citizens. Therefore, the opinions of citizens regarding the quality of each institution are important to know. Based on these assessments by citizens, we can recognize the quality of the democratic apparatus that takes place in a democratic society.

Using the definition of democratic quality by Ringen (2010), this study will examine the relationship between democracy and the subjective well-being that are more representative to the reality of life in a democratic society. Democratic apparatus is grouped into three components: quality of government, political participation of citizens, and civil liberties.

Quality of government is necessary in a democratic system. The course of democracy depends on how the government carries out its duties. In a system of representative democracy, government plays an important role in carrying out the people's agendas and programs. Political participation is the availability of opportunities and facilities for citizens to be involved in the political process and policy making. This includes trust in the government, involvement in government programs, participation in decision-making and supervision, the right to vote, the right to compete for public office, the right to compete in the capture sound, free and fair elections, and government policy-making by public voice or option. Civil liberty is the freedom of citizens to live and choose what they regard as the best without experiencing interference. This includes the security of citizens in physical and human capital, confidence in the continuation of freedom, freedom to form and participate in organization, freedom of expression and opinion, freedom to become public officials, freedom of competition or contestation to gain support to compete for important public positions, freedom to vote, freedom of information, as well as the institutional guarantees that government policy will be based on sound support.

SWB is defined as a person's cognitive and affective evaluations of his or her life. Those evaluations include emotional reactions to events as well as cognitive judgments of satisfaction and fulfillment (Diener, Lucas & Oishi 2005). SWB includes the affective and cognitive component (Diener *et al.,* 1999). The affective component is an actual or perceived individual's hedonic balance i.e., the balance between pleasant (positive) affect and unpleasant (negative) affect (Lucas, Diener & Larsen, 2003). The cognitive component is an individual's life satisfaction, i.e., evaluations of one's life according to subjectively determined standards.

As mention earlier, life satisfaction is conceptually closely linked with democracy. It can be achieved when people choose and obtain the things believed by him/her to be the best. This is in line with the objectives of democracy that produces a situation where people get what they believe to be the best (Dahl, 1989). Thus democracy and SWB should have a positive correlation.

With a basic understanding of the relationship between democracy and SWB, this study built a model that links life satisfaction with the interaction between the democratic system

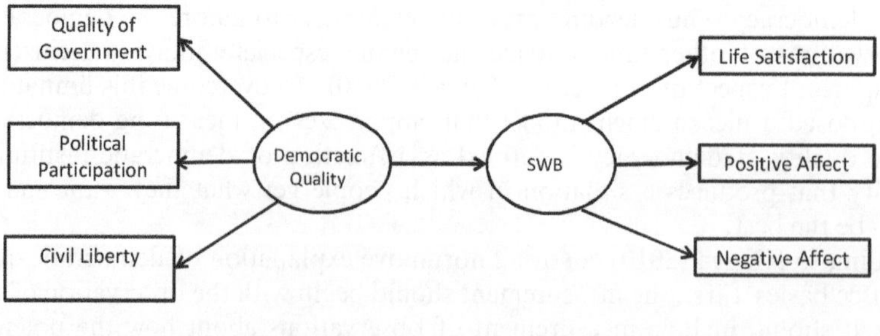

Figure 1. Proposed theoretical model of democratic quality as a predictor of subjective well-being.

(comprising the apparatus of democracy) and the democratic life (citizen's assessment on the apparatus of democracy). Empirical support for this model will demonstrate a significant positive relationship between life satisfaction and the interaction of the presence of a democratic system and its delivery to the citizens. As such, this model will show that democratic quality is a predictor for SWB.

2 METHOD

2.1 *Participants*

This study was conducted using a correlational design involving 1500 participants (Male = 752, Female = 748, Mean Age = 33.85) from two most populous islands in Indonesia, i.e. Sumatra and Java, gathered using multistage random sampling. Participants must meet the age criteria of 17–65 years old. This age is the voting age in Indonesia. The proportion of men and women who become participants is adjusted to the proportions at the population level. Participants represent districts or cities of 10 provinces in Sumatra and of 6 provinces in Java.

2.2 *Instruments*

The Satisfaction with Life Scale (SWLS). The SWLS was adapted from Pavot & Diener (1993, 2008) with a reliability coefficient of $\alpha = 0.87$. The instrument contains 6 items rated on a 7-point scale from 1 (strongly disagree) to 7 (strongly agree). The items included in the SWLS are listed as follows: (1) In most ways my life is close to my ideal; (2) The conditions of my life are excellent; (3) I am satisfied with my life now; (4) So far I have gotten the important things I want in life; (5) If I could live my life over, I would change almost nothing; and (6) Overall, I am satisfied with my life.

2.2.1 *Positive Affect and Negative Affect Scale (PANAS)*

PANAS was adapted from Watson, Clark, & Tellegen (1988) with a reliability coefficient of 0.86 for positive affect and 0.87 for negative affect. The instrument contains 20 items composed of 10 items on positive affect and 10 items on negative affect with 5-point scale that ranged from 1 (very rarely) to 5 (very often). Examples of the positive affect that are included are glad and proud, while some examples of the negative affect assessed are stressed and hopeless.

2.2.2 *Democratic Quality Scale*

Democratic Quality Scale uses the construct of democratic quality as proposed by Ringen (2010). This scale consists of 21 items that ask for the presence of the apparatus of democracy in the region surveyed, with a choice of "Yes", "Do not know/Undecided", and "No". The general question asked was "Are there any of these things (the apparatus of democracy) in your area?" Scoring is done by giving 3 to "Yes", 2 to "Do not know/Undecided, and 1 to "No ". Then, participants are asked to assess the quality of each of 21 apparatus with 4-point scale with choice options "very bad", "bad", "good" and "excellent". Scoring is done by giving 1 for "very bad", 2 for "bad", 3 for "good" and 4 for "very good". The score of each item is obtained by multiplying the scores of the presence and assessment of the apparatus of democracy. Through a confirmatory factor analysis technique it was found that this tool has three factors: the quality of government, political participation, and civil liberties.

2.3 *Procedure*

After being briefed about the purpose of the study, participants are asked to fill out all the scales used in this study, guided by an officer of a data taker. The officer answers the questions posed by participants or give an explanation to the participants if any part of the scale which is not understood by the participants.

Processing and data analysis is performed using LISREL 8.72 program. Analysis of the data of this study begins with a measurement model analysis or factor analysis to obtain measurable variables that can be indicators of a latent variable. Confirmatory factor analysis using SIMPLIS program is performed to identify the best item loading for measuring quality of government, political participation, and civil liberty. The next stage is to test the model fit of the proposed relationship using structural equation modeling.

3 RESULT

Probability value of the test of goodness of fit is 0.99 (>0:05). It shows the model is good, and the results predicted by the model fit values of other observations already qualified. Normed Fit Index (NFI) = 0.99; Non-normed Fit Index (NNFI) = 0.96; Comparative Fit Index (CFI) = 0.99; Incremental Fit Index (IFI) = 0.99; Relative Fit Index (RFI) = 0.96; and Goodness of Fit Index (GFI) = 0.99. When viewed under standardized RMR value, all variables have standardized RMR values <0.05, in accordance with good measurement model. Chi-square value of this model shows that the model proposed in this study is less in accordance with the data, i.e $\chi^2 = 32.00$ and df = 5. However from the value of RMSEA = 0.060, the proposed model fit. Generally, values less than or equal to 2.00 for the chi-square divided by degrees of freedom statistic suggest adequate fit of the model to the data (Carmines & Mclver, 1981). In contrast, values greater than 0.90 for the CFI, NFI, and NNFI indices (they generally range from 0–1) are thought to indicate relatively good fit, whereas values below 0.90 suggest the hypothesized model can be improved substantially (Hoyle & Panter, 1995).

Structural equation modeling using path analysis resulted in the following model fit.

This model provides empirical support that Democratic Quality is a predictor of Subjective Well-Being ($R^2 = 0.14$), where 14% of Subjective Well-Being is determined by Democratic Quality.

Democratic Quality is derived from three dimensions, namely Quality of Government, Political Participation, and Civil Liberty. Civil liberty has major contribution to the Democratic Quality with $R^2 = 0.70$, that means 70% of Democratic Quality is determined by the Civil Liberty. Political participation has contributed greatly to the Democratic Quality with $R^2 = 0.57$, that means that 57% of Democratic Quality is determined by the Political Participation. Quality of Government has also contributed greatly to the Democratic Quality with $R^2 = 0.38$, that means that 38% of Democratic Quality is determined by Quality of Government.

Subjective Well-Being (SWB) is obtained from three dimensions, namely Life Satisfaction, Positive Affect, and Negative Affect. Life Satisfaction contributes to Subjective Well-Being with $R^2 = 0.42$. That is, the Subjective Well-Being set at 42% by the Life Satisfaction. Positive Affect contributes to Subjective Well-Being with $R^2 = 0.21$, that means 21% of by

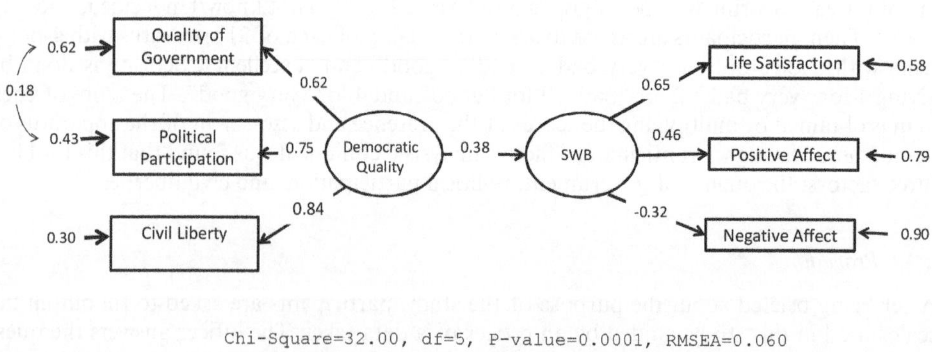

Figure 2. Confirmed theoretical model of democratic quality as a predictor of subjective well-being.

Subjective Well-Being determined by Positive Affect. Negative Affect is negatively correlated with Subjective Well-Being with $R^2 = 0.10$.

4 DISCUSSION

In this study, relationships and pathways linking quality of government, political participation, civil liberty, democratic quality and SWB were examined. Empirical support for this relationship model was obtained. This result is in line with previous studies that suggested a positive relationship between democracy and SWB (e.g. Frey & Stutzer, 2000, 2002; Inglehart & Klingemann, 2000; Inglehart, 2006; Bratu, 2011).

This study also found support for the model of democracy proposed by Ringen (2010) which emphasizes the importance of measuring the quality of democracy through the observation on the combined effect of the presence of democratic institution and its delivery to its citizens. Thus, the measurement of democracy should not be limited to identification of democratic apparatus rather it should also takes into account the delivery of those apparatus to its citizens. Therefore study of democratic quality should begin by collecting individual citizen's assessment on the democratic apparatus, which are then combined and are analyzed at collective level.

With this model of measurement, the error generated by regime bias can be reduced. It is important to address this bias in order to understand the contribution of democracy to happiness. There are at least two consequences of regime bias. First, it leads political scientists to construct a limited theory of democracy that ignores democratic purpose as important aspect of democracy. Second, it causes them to ignore the advancement in measurement theory. Further, it also causes them to ignore an individual citizen as an important and relevant empirical unit in the study of democracy. Ambiguity and gaps in previous studies on democratic quality was partly due to this bias. In line with Ringen (2010), this study has shown that in order to understand democratic quality, we need to incorporate theory of democratic potential, theory of democratic purpose, and theory of measurement.

Implication of this study is in line with the concept of democracy as proposed by Dahl (1989) that the mere existence of democratic institutions in a society does not necessarily make it a democratic society. Without good and optimally functioning institutions, we cannot generate good democratic qualities. In order to produce good democratic qualities, institutions should function as regulator and manage the community to achieve democratic purposes. These institutions should work in society as a system that moves both the government and its citizens to achieve what the citizens believe as the best destination and to accomplish well-being.

Democracy is a part of the explanation of happiness and the good life. However, it should be noted that the effect of democracy on happiness only occurs when a democratic system achieves democratic purpose, where the people are able to get what they want and achieving what they believe is best for them. Democratic purposes should be consistent at the individual and the collective levels. It is not enough when a democratic purpose is achieved only at the collective level while the individual does not feel its effect. Thus, an explanation of democracy must also involve an explanation of the attainment of individual happiness as one of the goals of democracy. A true form of democracy must strive for the attainment of happiness of everyone within the democratic system.

5 CONCLUSION

Structural equation modeling supports the hypothesis that democratic quality is a predictor of subjective well-being. Thus, SWB is determined by civil liberty, political participation, and quality of government, where civil liberty has the largest influence on SWB through democratic quality.

REFERENCES

Bjørnskov, C. (2010) How comparable are the gallup world poll life satisfaction data? *Journal of Happiness Studies,* 11 (1), 41–60, doi:10.1007/s10902-008-9121-6.

Bratu, C. (2011) Determinants of Subjective well-being: A romanian case study. *Economics,* 399, 1–23.

Dahl, R.A. (1989) *Democracy and Its Critics*. New Haven, Yale University Press.

Dolan, P. & Robert, M. (2012) Measuring subjective wellbeing: Recommendations on Measures for use by national governments. *Journal of Social Policy,* 41 (2), 409–427. doi:10.1017/S0047279411000833.

Dorn, D., Justina, A.V.F., Gebhard, K. & Alfonso, S.P. (2007) Is it culture or democracy? The impact of democracy and culture on happiness. *Social Indicators Research,* 82 (3), 505–526. doi:10.1007/s11205-006-9048-4.

Frey, B.S. & Alois, S. (2000) Happiness, economy and institutions. *Economic Journal,* 110 (466), 918–938. doi:10.1111/1468-0297.00570.

Frey, B.S. & Alois, S. (2002) *Happiness and Economics*: How the Economy and Institutions Affect Human Well-being. Princeton, Princeton University Press.

Helliwell, J.F., Christopher, P.B.L., Anthony, H. & Haifang, H. (2010) International Evidence on the Social Context of Well-Being. In: Diener, Helliwell, J.F. & Kahneman, D. *International Differences in Well-Being*. Oxford University Press.

Helliwell, J. (2012) *Understanding and Improving the Social Context of Well-Being*. NBER Working Paper 18486. National Bureau of Economic Research, Inc.

Hoyle, R.H. & Abigail, T.P. (1995) Writing about structural equation models. In: *Structural Equation Modeling: Concepts, Issues, and Applications*, pp. 158–176.

Inglehart, R. (2009) *Democracy and Happiness: What Causes What?* Chapters. Edward Elgar Publishing.

Inglehart, R. & Hans-dieter, K. (2000) Genes, Culture, Democracy, and Happiness. *Culture and Subjective Well-Being*, pp. 165–183.

Lijphart, A. (1977) *Democracy in Plural Societies: A Comparative Exploration*. New Haven, Yale University Press.

Lijphart, A. (2008) *Thinking About Democracy: Power Sharing and Majority Rule in Theory and Practice*. London, Routledge.

Lijphart, A. (2011) Democratic Quality in Stable Democracies. *Society,* 48 (1), 17–18. doi:10.1007/s12115-010-9389-0.

Pavot, W. & Ed, D. (1993) Review of the satisfaction with life scale. *Psychological Assessment,* 5 (2), 164–172. doi:10.1037/1040-3590.5.2.164.

Pavot, W. & Ed, D. (2008) The Satisfaction with life scale and the emerging construct of life satisfaction. *The Journal of Positive Psychology,* 3 (2), 137–152. doi:10.1080/17439760701756946.

Ringen, S. (2011) The Measurement of democracy: Towards a new paradigm. *Society* 48 (1), 12–16. doi:10.1007/s12115-010-9382-7.

Weitz-Shapiro, R. & Winters, M.S. (2008) *Political Participation and Quality of Life.* New York, Inter-American Development Bank [Omline] Available from: http://www.iadb.org/res/publications/pubfiles/pubWP-638.pdf.

Diversity in Unity: Perspectives from Psychology and Behavioral Sciences – Ariyanto et al. (Eds)
 ISBN 978-1-138-62665-2

Mother-child interaction in families of middle-to-low socioeconomic status: A descriptive study

R. Hildayani, S.R.R. Pudjiati & E. Handayani
Faculty of Psychology, Universitas Indonesia, Depok, Indonesia

ABSTRACT: This study aims to describe mother–child interactions in families with middle-to-low socioeconomic status (SES). Data were obtained by observing interactions between mothers and children using the Marschak Interaction Method (MIM), which assesses four domains of interaction: structure, engagement, nurture and challenge. The data were analysed qualitatively. Fifteen pairs of mothers and children participated in this study. The majority of mothers were young adults with preschool-age children. The results reveal that some of the children were willing to follow their mothers' structure. They were also willing to accept nurture from their mothers, which was expressed through physical contact. Most mothers, however, were less engaged with their children. They exhibited a lack of conversation and eye contact during the interactions. They also provided less support when their children encountered difficulties while working on given tasks. In addition, almost all mothers did not have habitual play activities that they used with their children. When they were asked to play with their children, they tended to direct the play activities into more academic activities.

1 INTRODUCTION

Fiese (1990) defines parent–child interaction as a reciprocal relationship between a parent and a child that involves certain roles being fulfilled by each of them. Jernberg (1991) has divided parent–child interaction into four dimensions: nurture, structure, engagement and challenge. *Nurture* is related to parents' ability to provide a nurturing contact, such as touching and caregiving, as well as recognising children's need to be calmed. It also measures the child's acceptance of the nurturing contact and their willingness to take care of their parents. *Structure* includes parents' ability to set limits for their children. Thus, parents are expected to be able to give instructions and make their children comply in a safe and playful way. *Engagement* describes parents' ability to be involved in their children's activity empathically. It also describes the extent to which the parents and children are in tune with each other, both physically and affectively. Finally, *challenge* is about adjusting the activities chosen by parents to their children's development. It also assesses how well children can control their frustration while working on tasks given by their parents, as well as how well parents can assist their children in overcoming their frustration and making themselves feel competent.

One factor that can influence parent–child interaction is the family's socioeconomic status (SES) (Chivanon & Wacharasin, 2012; AACAP, 2012). Pressure in the form of financial problems can lead parents to be less involved in the monitoring, caregiving, and disciplining of their children (Kalil, 2003). Past studies have demonstrated that parents of low SES were more likely to punish, be unresponsive and be insensitive, as well as showing a lack of affection, while interacting with their children (Ahmed, 2005; Conger & Donnellan, 2007; Dodici et al., 2003). SES has been found to be associated with mothers' level of education, which has an effect on their level of knowledge about their children's needs and development (Davis-Kean, 2005).

It has been shown that parent–child interaction has many positive benefits for children's development in physical, cognitive and social domains (WHO, 2004; Licata et al., 2014). For example, a study by Siu and Yuen (2010) found a positive correlation between parent–child interaction and children's social behaviours. A study by Dodici et al. (2003) shows that parent–infant/toddler interaction was related to early literacy skills. Lastly, mothers' sensitivity and responsiveness were found to be important for children's well-being in the long run (Taulbut & Walsh, 2013). These historical findings reflect the importance of parent–child interaction in children's development.

Studies of parent–child interaction in Indonesia using the observational method are still limited. Most studies used a self-reporting method for data collection. In our pilot study, we used a brief questionnaire that we distributed to mothers of middle-to-low SES. We found that about 70% of children were reported as showing non-compliant behaviours, and approximately 72% of mothers used punishment or resorted to permissiveness when dealing with their children's inappropriate behaviours. Additional data showed that all of the mothers performed parenting activities, such as feeding or bathing their children. However, only 22% of mothers played with their children. From the empirical data in our study, we concluded that the interactions between mothers and children did not occur in an optimal way.

Considering the importance of parent–child interaction on a child's development, this present study aims to further characterise mother–child interaction in middle-to-low SES families.

2 METHOD

2.1 *Design*

This study was based on a qualitative approach. Data were acquired through observation and analysed in descriptive terms.

2.2 *Participants*

The participants in this study were children and mothers who were members of a *Pos Pendidikan Anak Usia Dini* (*Pos PAUD*) (Centre of Early Childhood Education Programme) in one area of the city of Depok. This place was chosen based on a referral from *Dinas Pendidikan Kota Depok* (Depok City Education Service), which showed that, in this area, there was a lack of training for parents and teachers, the majority of families were of low SES, and the facilities to conduct early childhood education were still limited.

The majority of children were males with a mean age of 4.3 years. Most of these children had siblings and were the youngest child in their respective families. Most of the mothers were housewives who had graduated from high schools, while the fathers were blue-collar workers who had graduated from junior high schools. Both mothers and fathers were mostly still young adults.

2.3 *Setting and apparatus*

Observation was conducted in a small mosque close to the *Pos PAUD* where the children were educated. The mother–child interaction was recorded one at a time while they were interacting in a play setting. A digital camera and tripod were used to record the interaction. Eight envelopes containing task instructions were provided to the mother. In addition to the task instructions, certain objects were also placed in the envelopes. For example, in the first envelope, there were two squeaky animal toys: a fish and a bear. In the third envelope, there was a bottle of body lotion. In the seventh envelope, there were building blocks. Finally, in the eighth envelope, some snacks were placed (for more details of the task instructions, see the Procedures).

2.4 *Procedures*

Firstly, potential participants were invited for a briefing. During this briefing, information about the study was presented and consent forms were handed out.

Subsequently, we arranged a schedule of observation for each mother–child dyad. This observation was conducted to see: (1) how engaged the mothers were when they played with their children; (2) how capable the mothers were in giving structure to their children; (3) how challenging the tasks that the mothers gave to their children were; (4) how expressive the mothers were in showing affection to their children.

To assess the four dimensions of parent–child interaction, each mother–child dyad was given eight tasks describing activities that they had to do together. The tasks were given in the following order: (1) playing with a squeaky animal toy; (2) teaching the child something that they did not know; (3) applying the lotion; (4) telling a story about the child's early life; (5) leaving the child alone in the room for one minute; (6) playing something familiar to both mother and child; (7) copying the mother's block building; (8) feeding each other.

While the mother and child interacted, observers remained unseen by them. There was no time limit for completing the tasks; however, they were usually completed in approximately 30 minutes.

2.5 *Measures*

This study used the Marschak Interaction Method (MIM) that was developed by Jernberg (1991). In this method, mother–child dyads are observed while they are carrying out the tasks assigned to them. There were eight such tasks and they were assessed according to four dimensions: nurture, structure, engagement and challenge. In each task, the interaction between mother and child was assessed in terms of at least one of these four dimensions. Playing with a squeaky animal toy and playing something familiar to both mother and child are ideal tasks in which researchers can observe a mother's *engagement* when playing with her child. Teaching something that the child did not yet know is an ideal task in which researchers could observe the *challenge* that the mother gave to her child. Applying lotion, telling a story about the child's early life, leaving the child alone for one minute, and feeding each other are ideal tasks in which researchers can assess the expression of affection (or *nurture*) between mother and child. The last task, which is copying mother's block building, was assigned to assess the mother's capability to give *structure* to her child, as well as assessing whether or not the *challenge* was appropriate for the child's age.

3 RESULTS

This section provides a description of the interactions between mother and child during the performance of the tasks.

Table 1 shows that the majority of mothers displayed a lack of eye contact with their children. Mothers seemed to be reluctant to be involved in the squeaky toy play. Most

Table 1. Playing with a squeaky animal toy.

Qualities	1	2	3	4	5	6	7	8	9	10	11	12	13	14	15
Eye contact	V														V
Giving clues	V				V			V		V					
Praise	V									V					
Asking something related to academic matters	V	V	V		V	V	V	V		V	V	V		V	V
Role play		V			V										
Only squeezing	V		V	V	V	V	V	V	V	V	V	V	V	V	V
One-way conversation				V											

mothers simply asked their child to squeeze the toy animal without any clear motivation or instruction. Pretend play was only performed by two mothers, and only for a short period of time. Almost no dyadic conversation was seen between mother and child. Interestingly, most mothers asked more academically related questions of their children while playing with the squeaky toy, such as its name, its colour, or its facial features. Some mothers gave a few clues to their children so that they could answer the questions, such as "The bear's colour is wh...", meaning that the child just had to complete the sentence and say "White". Only a few mothers gave praise when their children answered correctly. Some mothers seemed to be insensitive, choosing to stop the activity even when their children still wanted to play with the squeaky toy.

Table 2 shows that more than half of the mothers failed to carry out the second task of playing something familiar to both mother and child. The main reason given by those mothers was that there was no tool to play with. Only a few mothers continued to play with their children using their bodies, such as through finger fighting or hand clapping games, and these lasted only a short time. One third of the mothers did not perform the activities in a playful way, and tended to focus instead on the academic aspects of the play, such as asking the child to count using fingers. Some children refused when their mothers asked them to perform the play activity.

Table 3 shows that the majority of mothers were involved in academic-related tasks while playing with their children, such as teaching their children to spell words, reading, asking the

Table 2. Playing something familiar to both mother and child.

Qualities	1	2	3	4	5	6	7	8	9	10	11	12	13	14	15
No play with children	V		V				V	V			V	V		V	
Child refused to play		V			V										
Playing together				V		V			V	V					V
Doing activities related to academic matters (e.g. counting the fingers)					V		V	V					V		V

Table 3. Teaching something that the child does not know.

Qualities	1	2	3	4	5	6	7	8	9	10	11	12	13	14	15
Cognitive stimulation	V			V	V	V	V	V	V	V	V	V	V		
Motoric stimulation (e.g. putting paper in an envelope)	V	V								V					
Giving a verbal prompt	V														
Giving a physical prompt	V	V								V					
Manipulating the physical environment to support the child in performing the task	V									V					
Giving examples	V									V					
Giving clues										V		V			
Praise					V	V				V					
Asking the child to remain sitting but not in a pushy way (e.g. holding the child in the mother's lap, giving clues to not squeeze the animal toy while teaching the child how to pray, asking them to wait)				V				V		V				V	
Not doing this task			V												V

number of objects, teaching parts of an animal, and teaching them how to solve additional problems. A few mothers taught motor skills or activities that were related to daily life, such as how to put a piece of paper in an envelope. When teaching something to her child, some mothers gave verbal prompts (i.e. verbally informing the child of how to do a particular task) and physical prompts (e.g. helping the child push the paper into the envelope), manipulated the physical environment to support the child in performing the task (e.g. widening the lips of the envelope so that the child could put the paper into it more easily), gave examples, and gave clues to the answers to certain questions. When the children seemed to be bored, some mothers could handle them in appropriate ways, such as putting the child in their lap or giving structure in a supportive manner. On the other hand, the mothers rarely praised their children. Moreover, two mothers did not carry out this task at all.

In general, the children did not refuse when their mothers put lotion on them (see Table 4). They were also willing to put the lotion on their mothers' arms and leg. Expressions of positive emotion, such as smiling, were displayed in this activity. Before the children started to put the lotion on their mothers, they were taught by their mothers how to apply the lotion. This instruction was given verbally or by examples. Some mothers gave physical prompts in the form of assistance. For example, they helped squeeze the lotion bottle. Some physical prompts were given by mothers when their children needed them. However, there was also a tendency for some mothers to give physical prompts without giving their child any opportunity to first try performing the task by themselves. Most mothers were involved in conversations, whose topics could be related to the lotion-application activity or not. However, these were mostly one-way conversations led by the mother. In particular, these one-way conversations occurred during the actual lotioning activity. Besides applying the lotion to each other, touching in the form of kissing the leg of the child or putting the child on the mother's lap were also exhibited during the performance of this task.

As shown in Table 5, almost half of the mothers told a story about their child's early life expressively, using body gestures. The children paid attention to their mothers' stories

Table 4. Playing something familiar to both mother and child.

Qualities	1	2	3	4	5	6	7	8	9	10	11	12	13	14	15
Lotioning each other	V	V	V		V	V			V		V	V	V	V	V
Teaching the child how to dab lotion	V		V		V	V	V	V		V		V		V	
Giving physical prompts	V			V	V	V	V	V	V	V		V	V	V	
Smiling		V		V			V		V		V		V		
Conversation during lotioning	V	V	V		V	V	V	V	V	V				V	V
Touching (except lotioning)					V					V					
Not lotioning each other				V			V	V		V					

Table 5. Telling a story about the child's early life.

Qualities	1	2	3	4	5	6	7	8	9	10	11	12	13	14	15
Interactive conversation	V							V	V	V			V	V	
Telling the story expressively	V	V										V		V	V
Child's attention to mother's story	V	V				V				V		V			V
Child smiling	V	V				V		V				V	V		V
Showing affection	V	V			V					V	V	V	V	V	V
Story not told				V	V										

and showed responses such as smiling. The mothers showed their affection by touching or hugging their children. One of the mothers sang a song while cradling her son and telling him a story. Two mothers skipped this task because their children did not want to hear their mothers' story.

In the separation task (see Table 6), half of the mothers employed reasons for leaving the child alone, such as going to the toilet, checking on the child's sibling, or looking for a pen outside. Four mothers asked for permission to leave the child without giving any reasons and only said that they would go outside for a while. Two mothers went outside without asking permission. More than half of the children followed their mothers when their mother left the room. One child did not want to be left alone and, therefore, her mother stayed in the room. Almost one third of the children completed the separation task smoothly. However, when their mother returned, only one child showed positive emotion while the rest displayed blank expressions.

Table 7 shows that the majority of children did not refuse to be fed by or to feed their mothers. In this task, one mother added a challenge for her child by suggesting a competition in eating the food. One child also fed his mother in a playful way, as if his arm was an aeroplane that would be landing at his mother's mouth. Some mothers showed sensitivity by helping the children who had difficulties in opening the food packs.

Table 8 shows that most mothers gave some instruction to their child to copy the block buildings they had built. Some mothers gave direct instructions, such as "*Please, make it in the way that I've made it*". Others gave indirect instructions, such as "*Can you build a tunnel?*" More than half of the children were willing to copy their mothers' block buildings. The others built the blocks independently of their mothers' constructions. Some children paid attention when their mothers were building something. In general, their attention persisted for longer if their mothers built something interesting or suited to the child's interests. When asking their child to build blocks, some mothers gave verbal prompts. For example, when the child placed a block incorrectly, the mother said "*The position (of the block) is wrong*", "*Look for the triangle block*", or "*Put the blocks closer to each other*". Some mothers seemed to take control of the arrangement of the blocks that their children were working on without even asking the children to do it by

Table 6. Leaving the child alone for one minute in the room.

Qualities	1	2	3	4	5	6	7	8	9	10	11	12	13	14	15
Asking for permission with various reasons	V	V				V		V			V		V	V	V
Asking for permission without any reason			V	V						V		V			
Not asking for any permission at all					V		V								
The child lets the mother leave the room	V							V			V	V		V	V
The child smiles when meeting her/his mother again after separation	V														
The child shows a blank expression when meeting her/his mother again after separation											V	V			V
The child follows the mother when she goes outside		V	V	V	V	V	V			V			V		
The child does not let their mother leave them									V						

Table 7. Feeding each other.

Qualities	1	2	3	4	5	6	7	8	9	10	11	12	13	14	15
The child is willing to be fed	V	V	V			V	V	V	V	V	V	V	V	V	V
The child is willing to feed	V	V	V			V	V	V	V	V	V	V	V	V	V
The mother gives assistance	V	V	V			V							V		V
The mother challenges the child	V														
Feeding in a playful way														V	

themselves. Another reason for this was because the children did not comply with their mothers' instructions.

Some mothers showed more sensitivity and helped their children when they saw that they were encountering difficulties, such as helping them to open the blocks' container. They also gave praise when their children copied their block successfully. However, they only said "smart boy" without explaining why this made them "smart".

Only a few mothers supported their children by saying that "*You can do it just the way I did*". One mother threatened to leave when her child did not want to copy the blocks that she had built. Some of the mother–child pairs seemed to be preoccupied with their own blocks and thus were building independently of each other. In fact, some mothers did not give any instructions to their children.

Table 8. Copying the mother's block building.

Qualities	1	2	3	4	5	6	7	8	9	10	11	12	13	14	15
Asking the child to copy the mother's block building	v		V		V	V		V	V	V	V	V	V	V	V
The child is willing to copy the mothers' block building	V		V			V		V	V	V		V		V	V
The child pays attention to what the mother does	V					V	V	V							V
The child is reluctant to copy					V			V			V				
The mother does not give any instruction to the child		V		V			V								
The mother and the child are preoccupied with their own blocks		V			V	V	V		V	V	V			V	
The mother gives verbal prompts while the child is copying her block building	V					V	V	V	V	V		V	V		
The mother gives assistance to the child when they have difficulties			V							V					
The mother gives assistance without giving the child an opportunity to perform the task by themselves					V	V			V	V			V	V	V
The resulting block building fits child's age	V						V			V		V	V	V	V
The mother asks some questions related to academic matters	V	V							V	V		V	V	V	
Praise					V		V			V		V		V	V
The mother gives support							V			V					
The mother makes threats when the child does not comply										V					

In terms of the challenge that the mothers gave, they were likely to give a challenge that was appropriate to their child's age. They asked the children to copy the block buildings in a step-by-step manner. However, when giving the task, some mothers still asked their children more academically related questions, such as about the blocks' shapes, colours and number.

4 DISCUSSION

Of the four dimensions of parent–child interaction, the dimension of nurture was the most commonly observable, demonstrated by both mothers and children. Both seemed to enjoy performing activities in this category, such as feeding and applying lotion to each other. This is understandable because these types of activities are routinely performed by both mothers and children. Furthermore, a classical study by Harlow and Zimmerman (cited in Papalia & Martorell, 2015) found that babies always look for warmth, intimacy and comfort.

In terms of activities involving unstructured play, such as playing with squeaky animal toys and playing something familiar to both mothers and children, the majority of the mothers showed some interactional problems. They displayed a lack of eye contact; they had no idea how to create play activities, especially if no tool was immediately available; they engaged in one-way conversations; and they showed little enjoyment. In other words, they showed little engagement while interacting with their children. This is in accordance with previous data that showed that mothers are rarely involved as their children's playmates.

In terms of the dimension of structure, more than half of the children were willing to receive the directions that their mothers gave them. This can be attributed to the kind of task that suits the children's stage of development. It also meant that the mothers were able to give challenges to their children that were appropriate.

An interesting finding of this study is the fact that the majority of mothers tended to associate almost all of the tasks with academic activities. This did not only apply to the tasks that were more structured, such as teaching children something that they did not know, but also to the tasks that were more unstructured and playful, such as playing with a squeaky toy or playing something familiar to both mother and child. This phenomenon could have arisen due to the parental and cultural values that are embraced by the family. Chao and Tseng (2002) explain that education is highly valued in Asian cultures, which leads parents to give much attention to learning activities by asking their children questions related to academic matters.

5 CONCLUSION

In general, mothers showed good nurture of their children, but a lack of engagement during play activity. Meanwhile, the structure and challenge dimensions of parent–child interaction were displayed to a moderate level. Most mothers associated playing activities with academic tasks.

REFERENCES

AACAP. (2012). *When children have children*. Facts for Families, No. 31. Washington, DC: American Academy of Child and Adolescent Psychiatry. Retrieved from https://www.aacap.org/App_Themes/AACAP/docs/facts_for_families/31_when_children_have_children.pdf

Ahmed, Z.S. (2005). *Poverty, family stress, and parenting.* Retrieved from http://www.humiliationstudies.org/documents/AhmedPovertyFamilyStressParenting.pdf

Chao, R. & Tseng, V. (2002). Parenting of Asians. In *Handbook of Parenting: Social Conditions and Applied Parenting*, *4*, 59–93.

Chivanon, N. & Wacharasin, C. (2012). Factors influencing Thai parent–child interaction in a rapidly changing industrial environment. *International Journal of Nursing Practice*, *18*(Suppl. 2), 8–17. doi:10.1111/j.1440-172X.2012.02024.x

Conger, R.D. & Donnellan, M.B. (2007). An interactionist perspective on the socioeconomic context of human development. *Annual Review Psychology*, *58*, 175–199. doi:10.1146/annurev.psych.58.110405.085551
Davis-Kean, P.E. (2005). The influence of parent education and family income on child achievement: The indirect role of parental expectations and the home environment. *Journal of Family Psychology*, *19*(2), 294–304. doi:10.1037/0893-3200.19.2.294
Dodici, B.J., Draper, D.C. & Peterson, C.A. (2003). Early parent–child interactions and early literacy development. *Topics in Early Childhood Special Education*, *23*, 124–136.
Fiese, B.H. (1990). Playful relationships: A contextual analysis of mother–toddler interaction and symbolic play. *Child Development*, 61(5), 1648–1656.
Jernberg, A.M. (1991). Assessing parent–child interactions with The Marschak Interaction Method (MIM). In Schaefer, C.E., Kitlin, K. & Sandgrund, A. (Eds.), *Play, diagnosis, and assessment* (pp. 493–515). New York, NY: Wiley.
Kalil, A. (2003). *Family resilience and good child outcomes: A review of the literature*. Wellington, New Zealand: Centre for Social Research and Evaluation, Ministry of Social Development.
Licata, M., Paulus, M., Thoermer, C., Kristen, S., Woodward, A.L. & Sodian, B. (2014). Mother-infant interaction quality and infants' ability to encode actions as goal-directed. *Social Development*, *23*(2), 340–356.
Martin, C.A. & Colbert, K.K. (1997). *Parenting: A life span perspective.* New York, NY: McGraw-Hill.
Papalia, D.E. & Martorell, G. (2015). *Experience human development* (13th ed.). New York, NY: McGraw-Hill.
Siu, A.F.Y. & Yuen, E.Y.H. (2010). Using the Marschak Interaction Method rating system for Chinese families: Relationship between parent–child interaction pattern and child's social behavior. *International Journal of Play Therapy*, *19*(4), 209–221.
Taulbut, M. & Walsh, D. (2013). *Poverty, parenting, and poor health: Comparing early years' experiences in Scotland, England, and three city regions*. Glasgow, UK: Glasgow Centre for Population Health. Retrieved from http://www.gcph.co.uk/assets/0000/3817/Poverty__parenting_and_poor_health.pdf
WHO. (2004). *The importance of caregiver–child interactions for the survival and healthy development of young children: A review.* Geneva, Switzerland: World Health Organization. Retrieved from http// whqlibdoc.who.int/publications/2004/924159134X.pdf

Diversity in Unity: Perspectives from Psychology and Behavioral Sciences – Ariyanto et al. (Eds)
 ISBN 978-1-138-62665-2

Time metaphors in Indonesian language: A preliminary study

D.T. Indirasari
Faculty of Psychology, Universitas Indonesia, Depok, Indonesia

ABSTRACT: Studies in the field of cognitive linguistics have shown that time is metaphorically structured in different ways among cultures. Based on the metaphor theory proposed by Lakoff and Johnson (1980c), a qualitative study was conducted to explore the use of time metaphors and concepts in Indonesian language. Data were primarily collected using the focus group discussion method. The results from 50 participants (24 men and 26 women) aged 18–33 years old show that, in the context of Indonesian language, time is metaphorically expressed in terms of moving objects, quantities, volume objects, and as a living thing. Contrary to the researcher's expectation, the 'rubber time' that is commonly used to describe Indonesian people's behaviour toward time was rarely mentioned by the participants. Compared with the metaphors of time represented in English (Lakoff & Johnson, 1980c), time metaphors in Indonesian language do not treat time as a commodity and orientational metaphors are not used. In conclusion, the study provides further evidence of the way in which the metaphorical structures placed around time vary between cultures. Further research is required to explore how each culture interprets time based on its associated metaphors and how this actually shapes the way a society thinks and acts.

1 INTRODUCTION

Time is an abstract concept. Every day we experience the flow of time, yet we cannot really see what time is. What is time? If a child asked that question of you, what would be your answer? Even though we can easily describe time in terms of hours, days, months, and so on, what the meaning of time is and what really constitutes time are difficult questions to answer. Usually, people will answer those questions using types of analogy that explain one concept through a concept from another domain. In cognitive linguistics, these types of explanation are called metaphors (Lakoff, 1993). For example, in terms of time, we could describe it according to the way we sense the flow of time while we are waiting in line ("Time is moving slowly"), or while we are running late for an appointment ("Time moves really fast"). From these experiences, we describe time as an object that has the capability to move like a train, for instance. In this way, we make an analogy between the first domain (time) and another domain (moving objects or trains).

The metaphor theory of Lakoff and Johnson (1980c) suggests that metaphors guide much of our everyday action and thinking. Metaphors are commonly used in our daily lives, and from these metaphors we can learn how people actually interpret everyday experiences. Lakoff (1993) explains that in the classic theory of metaphor, metaphorical expressions are merely seen as a linguistic form that is typically used in relation to a novel or poetic idea or experience. Lakoff (1993) also argues that the way language is used to define a concept using an analogy from another domain is actually the main area that we should focus on when we analyse a metaphor. In other words, the central point of metaphor is not in language, but how we actually make an analogy using a cross-domain mapping. Lakoff (1993, p. 2) explains the word metaphor as a "cross-domain mapping in the conceptual system". Thus, the way we think and act can actually be described in terms of the metaphors that we use in our daily lives because they are embedded in our conceptual systems. To explore the way that people think and act in a given society, a researcher should, therefore, study the metaphors that are commonly used in their language.

In Indonesian culture, there is a well-known phrase that describes the peculiarity of how Indonesian people view time, known as 'rubber time'. The term 'rubber' is used because time is perceived to have certain characteristics that mimic the concept of rubber: it is elastic and flexible. Thus, on the basis of the contemporary metaphor theory, the abstract concept of time is defined using the concept of rubber from the domain of materials. As a metaphor, the 'rubber time' phrase might explain how Indonesians actually perceive the abstract concept of time according to their daily experiences. Thus, the existence of this expression is reflected in many forms of behaviour that are encountered almost daily, especially in terms of punctuality. For example, many Indonesians are accustomed to being late when attending an event because the organiser themselves usually starts the event behind schedule (time is elastic), and sometimes the event will only begin after some distinguished guests have arrived (time is flexible) (Asyhad, 2015; Wahyuni, 2015; Wardani, 2015; Martiyanti, 2016).

Although the Indonesian language recognises the use of metaphor in daily speech, how is the concept of time actually metaphorically expressed in the Indonesian language? Are there any differences between the English and Indonesian languages in explaining the concept of time in terms of metaphor? Addressing these questions provides the principal reason for the conduct of this study. Thus, the objective of this study is, through a deeper understanding in terms of the time metaphors used in Indonesian language, to unpack some of the underlying motives and beliefs in terms of time-related behaviours seen in everyday situations.

How a culture actually defines the concept of time varies from one culture to another. Cross-cultural studies related to the concept of time show that each culture has a different philosophy about time. Moreover, this philosophy may influence how individuals within that culture interpret time and behave according to this interpretation. White et al. (2011) found that some cultures define limits on time in many different ways. There are various boundaries in defining a person as 'too early' or 'too late' when they are attending a meeting. This may also indicate that people perceive each occasion to have different priorities, and they behave accordingly. Furthermore, Sircova et al. (2014) show that every culture places different emphases in relation to perspectives of time (past, present and future). A study by Levine and Norenzayan (1999) also showed that there are different paces of life among countries in terms of how their populations respond to time. Their assessments were conducted according to three criteria: walking speed, postal speed, and accuracy of clocks. Among the 31 countries studied, Indonesia was ranked second-lowest, indicating that Indonesia has a relatively slow pace of life. In addition, the conception of time affects how we behave, and how time is used in the language within a culture also affects how people think in terms of temporal logic (Boroditsky et al., 2002; Casasanto et al., 2004).

In order to satisfy the objective of this study, it was conducted using a qualitative method. The concept of time was explored in terms of the metaphors and understandings held by individuals when discussing concepts of time. The analysis was carried out using Systematic Metaphor Analysis, as developed by Schmitt (2005). The English time metaphor analysis conducted by Lakoff and Johnson (1980b, 1980c) was used to highlight the similarities and differences between English and Indonesian language metaphors for time.

2 METHOD

The time metaphors were identified using a Focus Group Discussion (FGD) method. The total number of participants for this study was 50 (male = 24; female = 26), aged between 18 and 33 years old, and either graduate or undergraduate students from Universitas Indonesia. They all originated from Indonesia, and use the Indonesian language as their primary language. They were divided into five groups for the discussions, which were led by the researcher herself, helped by two research assistants. Each discussion lasted between one and two hours. All of the participants were given rewards upon the completion of the discussions.

The results of the discussion were further analysed to identify the time metaphors. Using Schmitt's (2005) Systematic Metaphor Analysis, the data were first analysed to determine time metaphors based on the following criteria: a) a word or phrase which can, in the context of what is being said, be understood beyond its literal meaning; b) a literal meaning which stems from an area of physical or cultural experience (the source area); and c) which is—in this context—transferred to a second, often abstract, area (the target area) (Schmitt, 2005, p. 371). Based on the results of this first stage, a categorisation was then performed as a second stage of analysis, to cluster metaphors in terms of their similarity in relation to the source and target areas.

3 RESULTS

In the discussions, the participants were asked to recall as many phrases as they could in the Indonesian language that they usually used or heard in daily conversations and that expressed time concepts. Almost half of the participants identified the time concept based on units of time: hours, days, months, dates, periods, and Islamic prayer times. Most of the responses were not described in terms of a phrase or sentence, but used only a single word to describe the time. Some of the participants expressed time in terms of metaphors, but these were infrequent. Even the concept of 'rubber time' itself was rarely articulated by the participants. However, some of the rubber-like qualities of time were mentioned in the discussions, such as *Waktunya molor* (there is a time delay) and *Jangan ngaret* (be on time), where the word *molor* means lengthening the time and *ngaret* is a slang word for rubber. This finding was quite contrary to the researcher's expectation. There are two possible explanations: 1) the participants might not use the phrase often in their daily conversations, so it does not immediately come to mind; 2) the participants might consider rubber time as a common behaviour, easily found in daily situations, and, hence, not identified as anomalous. Although the phrase 'rubber time' was not frequently expressed, thematic analysis of the responses show that phrases that describe the concepts of 'late', 'lateness' and 'punctuality' were mentioned. These concepts can be seen in the participants' responses, such as: *Terlambat* (late); *Dari mana aja baru datang jam segini?* (Where have you been? You are late); *Mana nih ga dateng-dateng* (Where is s/he? Why hasn't s/he shown up?).

The analysis of the time metaphors was done first by grouping the time responses gathered from the discussions into several categories based on their meaning. The words that were used beyond their literal meaning to explain concepts of time were analysed using Systematic Metaphor Analysis. One of the metaphors that was identified by the participants was 'time is a moving object', which can be seen in some responses as follows: *Waktu tuh cepat banget ya* (time moves so fast); *Waktu terasa lama* (it feels like long hours). Because the literal meanings of *cepat* (fast) and *lama* (long hours) refer to a moving object, the target area is time, and the source area is transport of something from one point to another. Other metaphors that were identified were grouped as follows:

1. time can be measured: *Masih banyak waktu* (we still have plenty of time), *Tidak ada waktu* (we don't have time)—the target area is time, and the source area is quantity;
2. time can be described spatially: *Waktunya sempit* (the time is narrow)—in this metaphor, time is considered as a volume object (source area);
3. time can be personified: *Makan waktu* (the time is eaten), *Waktu berjalan cepat* (time is walking fast), *Waktu terus bergerak* (time keeps moving)—in these phrases, time is treated as analogous to a human or living thing.

Even though there are some Indonesian language phrases that are commonly used to describe time as currency—such as *Terima kasih atas waktunya* (thank you for your time), *Kamu menghabiskan waktu saya* (you are wasting my time), and *Bagaimana kamu menghabiskan waktu?* (How do you spend your time?)—none of them were mentioned by the participants. Most of the metaphors that were expressed in the discussions illustrated time as a concrete object that can be measured in terms of time units, and not as a limited resource that has a value attached to it.

When the participants were asked to recall any proverb in the Indonesian language that describes time, the responses given mostly used the time phrases. There are not many proverbs in the Indonesian language describing time. One of the best-known proverbs of *Biar lambat asal selamat*, meaning 'no need to rush, being safe is the most important thing', was rarely expressed in the discussions. Another proverb, adopted from Javanese culture, *Alon-alon asal kelakon*, which indicates that one should take more time to get the best result, was also not mentioned by the participants. The fact that most of the participants were of 'Generation Y' might explain why these proverbs were not used as frequently as in previous generations.

If the results are compared with time metaphors in English (Lakoff & Johnson, 1980b, 1980c), time as a commodity was not really reflected in the phrases that were recalled. Although the Indonesian language has expressions such as *Membuang waktu* (wasting time), *Membeli waktu* (buying time) and *Membagi waktu* (time allocation), none of these phrases were uttered by the participants. Time is also regarded as a limited resource in English and this is also reflected in some Indonesian language phrases, such as *Punya banyak waktu* (have enough time) and *Habis waktu* (run out of time).

Another interesting finding is that, unlike English, the Indonesian language does not use orientational metaphors when describing time. Orientational metaphors—the use of spatial orientation reflecting body movement (up/down, in/out, on/off)—are not commonly used in the Indonesian language. Although there are some spatial metaphors that were mentioned, the term spatial is itself not identified with the way the body moves. The spatial metaphors that were uttered by the participants mostly defined the volume of a space. While in English, the phrase 'time is up' can be associated with a jar that is filled up with water, by contrast, in the Indonesian language, the translation of 'time is up' (*Waktu sudah habis*) can be described as an empty jar.

4 DISCUSSION

The results of this study indicate that, compared to English, time is perceived differently in the Indonesian language. One of the most visible findings is that time is most commonly expressed in terms of a concrete object. This conception might indicate that, as a language, Indonesian provides a different way of thinking in terms of illustrating an abstract concept. Although the use of metaphor is common in conversations, the way an abstract concept is metaphorically represented might be different from one language to another. Orientational metaphor, for instance, is rarely used in the Indonesian language. The way English language expresses a concept using spatial orientation is not common in Indonesian language. Many English phrases consist of orientational concepts that are used to emphasise the way people think spatially about an idea. The use of orientational words such as 'up', 'down', and so on, is not common in Indonesian language. According to the Sapir–Whorf perspective of linguistic relativity, this dissimilarity might provide one example in explaining how an idea or thought is articulated differently within a language or culture, and thought of in a different way too.

As already mentioned, the participants of this study were mostly drawn from Generation Y, studying in an urban area. This might explain why some of the old time-related proverbs that are usually used were not commonly articulated in the discussions. Even the term 'rubber time' was rarely mentioned by participants. These findings are quite contrary to the researcher's expectation. Future research should compare the time metaphors used by Generations X and Y. This result might derive from different patterns of socialisation by which each individual in society absorbs the concept of time. The fact that most of the participants are studying in an urban area might also influence the way they think about time. It would be interesting for future studies to compare the use of time metaphors not just between generations, but also between subcultures in Indonesia, as the country consists of many subcultures with different languages and dialects.

Furthermore, one limitation of this study is that the metaphors that were analysed were mainly extracted from the group discussions. Schmitt (2005) suggests that in order to search for the metaphors being used in a society, a researcher might need extensive materials containing references to the concept being investigated, including print media. Incorporating this approach in further research may provide a more complete understanding in terms of time metaphors because there may be other associated metaphors that were not uttered by participants in the discussions.

The results from this research can be used as an initial basis for exploring the metaphors of time more thoroughly, and they also emphasise some other topics to be considered for future research, such as the types of metaphors that are commonly used in Indonesian language, the ways time is metaphorically expressed between subcultures in Indonesia, and how these metaphors create differences between cultures in their manners of acting and thinking.

5 CONCLUSION

In conclusion, although time is a global concept, and frequently used in everyday lives, how it is understood and represented in one culture might be different to another. In the Indonesian language, the way in which time is expressed metaphorically is different from the time metaphors used in English. This finding might reflect different perceptual organisations, and how they affect the thinking process. Further studies are needed to investigate more deeply the concepts surrounding time and how these are actually reflected in everyday behaviours.

REFERENCES

Asyhad, M.H. (2015). Penumpang lion air terlantar inilah kompensasi yang harus dibayar maskapai jika pesawat delay. *Intisari-online.com*. Retrieved from http://intisari-online.com/read/penumpang-lion-air-terlantar-inilah-kompensasi-yang-harus-dibayar-maskapai-jika-pesawat-delay

Boroditsky, L. (2000). Metaphoric structuring: Understanding time through spatial metaphors. *Cognition, 75*(1), 1–28.

Boroditsky, L., Fuhrman, O. & McCormick, K. (2011). Do English and Mandarin speakers think about time differently? *Cognition, 118*(1), 123–129.

Boroditsky, L., Ham, W. & Ramscar, M. (2002). What is universal in event perception? Comparing English and Indonesian speakers. In W.D. Gray & C.D. Schunn (Eds.), *Proceedings of The 24th Annual Meeting of The Cognitive Science Society* (pp. 136–141). Mahwah, NJ: Erlbaum.

Casasanto, D. & Boroditsky, L. (2008). Time in the mind: Using space to think about time. *Cognition, 106*(2), 579–593.

Casasanto, D., Boroditsky, L., Phillips, W., Greene, J., Goswami, S., Bocanegra-Thiel, S. & Gil, D. (2004). How deep are effects of language on thought? Time estimation in speakers of English, Indonesian, Greek, and Spanish. In K. Forbus, D. Gentner, and T. Regier (Eds.), *Proceedings of the 26th Annual Meeting of the Cognitive Science Society* (pp. 186–191). Austin, TX: Cognitive Science Society.

Hansen, J. & Trope, Y. (2013). When time flies: How abstract and concrete mental construal affect the perception of time. *Journal of Experimental Psychology, 142*(2), 336–347.

Lakoff, G. (1993). The contemporary theory of metaphor. In Orthony, A. (Ed.), *Metaphor and thought* (2nd ed., pp. 203–251). New York, NY: Cambridge University Press.

Lakoff, G. & Johnson, M. (1980a). The metaphorical structure of the human conceptual system. *Cognitive Science, 4*(2), 195–208.

Lakoff, G. & Johnson, M. (1980b). Conceptual metaphor in everyday language. *The Journal of Philosophy, 77*(8), 453–486.

Lakoff, G. & Johnson, M. (1980c). *Metaphors we live by*. Chicago, IL: University of Chicago Press.

Leclerc, F., Schmitt, B.H. & Dube, L. (1995). Waiting time and decision making: Is time like money? *Journal of Consumer Research, 22*(1), 110–119.

Levine, R.V. & Norenzayan, A. (1999). The pace of life in 31 countries. *Journal of Cross-Cultural Psychology, 30*(2), 178–205.

Martiyanti, E. (2016). *Penyelesaian Pembangunan MRT Mundur dari Target*. Available from: http://www.beritajakarta.com/read/3163/Penyelesaian_Pembangunan_MRT_Mundur_dari_Target#.Vox9o_l97IU

Pierro, A., Giacomantonio, M., Pica, G., Kruglanski, A.W. & Higgins, E.T. (2011). On the psychology of time in action: Regulatory mode orientations and procrastination. *Journal of Personality and Social Psychology*, *101*(6), 13–17.

Schmitt, R. (2005). Systematic metaphor analysis as a method of qualitative research. *The Qualitative Report*, *10*(2), 358–394.

Sircova, A., van de Vijver, F.J.R., Osin, E., Milfont, T.L., Fieulaine, N., Kislali-Erginbilgia, A. & Zimbardo, P.G. (2014). A global look at time. *Sage Open*, *4*(1). doi:10.1177/2158244013515686

Sircova, A., van de Vijver, F.J.R., Osin, E., Milfont, T.L., Fieulaine, N., Kislali-Erginbilgic, A. & Zimbardo, P.G. (2015). Time perspective profiles of cultures. In M. Stolarski, N. Fieulaine & W. van Beek (Eds.), *Time perspective theory: Review, research and application* (pp. 169–187). Cham, Switzerland: Springer.

Wahyuni, T. (2015). Menghilangkan Kebiasaan Datang Terlambat karena 'Jam Karet'. *CNN Indonesia.* Retrieved from http://www.cnnindonesia.com/gaya-hidup/20150203070016-255-29176/menghilangkan-kebiasaan-datang-terlambat-karena-jam-karet/

Wardani, H. (2015). Susahnya Antri di Jalan Raya. Retrieved from http://www.kompasiana.com/benitoramio/susahnya-antri-di-jalan-raya_550aca5aa33311bb102e3a42

White, L.T., Valk, R. & Dialmy, A. (2011). What is the meaning of "On Time"? The sociocultural nature of punctuality. *Journal of Cross-Cultural Psychology*, *42*(3), 482–493.

Zimbardo, P.G. & Boyd, J.N. (1999). Putting time in perspective: A valid, reliable individual-differences metric. *Journal of Personality and Social Psychology*, *77*(6), 1271–1288.

Diversity in Unity: Perspectives from Psychology and Behavioral Sciences – Ariyanto et al. (Eds)
 ISBN 978-1-138-62665-2

Better now than later: The effect of delayed feedback from the receiver of a thank-you letter on the sender's happiness

A. Kartika, I.I.D. Oriza & B. Takwin
Faculty of Psychology, Universitas Indonesia, Depok, Indonesia

ABSTRACT: This study is an extension of findings of previous studies conducted by Seligman et al. (2005) and Toepfer et al. (2012) that indicate how writing a thank-you letter can increase happiness. This study examines the effect of delayed feedback from the receiver of the thank-you letters on the happiness of the person who wrote the letter. Every week for three consecutive weeks, 45 college participants were asked to write one thank-you letter to people who had had positive effects on their life but had not been given proper thank-you letters and delivered it personally during the same week. Participants were also randomly assigned to two experimental groups based on the time of the feedback obtained from the receivers, either directly during the meeting (n = 22) or delayed after the meeting (n = 23). The results suggest that when the feedback from the receivers was given directly, it could generate more happiness in the senders than if the feedback was delayed.

1 INTRODUCTION

> "Close your eyes. Imagine any influential person in your life or someone who had changed you for the better. Remember that person again and think why is that person very important to you. Then write it down on a piece of paper."

On that occasion, JD, a 35-year-old man, chose his mother as the most influential person in his life. After he finished writing a thank-you letter to his mother, JD was given a chance to contact her by phone and read his letter out. Even though he hesitated at first, when he actually made the call his eyes sparkled, his smile became wider, and his arm wiped his tears from his cheek. His mother responded by saying that what JD had expressed was beautiful and extraordinary. JD was one of volunteers on experimental video series by Soulpancake (201). In the end of the video, the host stated that the happiness of volunteers who read their thank-you letters or were connected to a voice box increased by 4–6% compared to previously and it went up to 19% in volunteers who were connected by phone, including JD.

Happiness is something that everyone looks for. Since 1998, happiness has been one of the topics thoroughly examined by researchers in the field of positive psychology (Compton & Hoffman, 2013). According to Seligman (2004), 'happiness', which tends to be equated with well-being, is not only a positive feeling that an individual experiences (like enjoyment and comfort), but it can also be defined as a positive activity that may generate engagement and preoccupation. Happiness is also experienced when an individual can use his or her character's strengths and virtues to serve something bigger than themselves, so that they can consider their lives as meaningful. Several studies have also proven the positive effects of happiness, such as physical health, high life satisfaction, beneficial interpersonal relationships, higher productivity and job satisfaction, achievement of life goals, and high levels of empathy (Carr, 2004; Lyubomirsky et al., 2005; Seligman, 2004; Seligman et al., 2005).

These benefits of happiness encourage people to utilise any possible way to reach happiness. However, not all methods work effectively and some specific strategies may work better (Tkach & Lyubomirsky, 2006, in Schueller & Seligman, 2010). From all possible ways to increase happiness, Lyubomirsky et al. (2005) state that only 10% of happiness variance may be explained by circumstantial factors and the remaining 50% from genetic factors. On the other hand, there is a 40% chance to increase happiness by doing simple activities intentionally that are related to the characteristics of a happy individual, also known as positive activities or intentional activities (Layous & Lyubomirsky, 2012; Lyubomirsky & Layous, 2013).

One simple positive activity like JD has done to increase his happiness was expressed in his happy and thankful feeling that occurred as a response when he received a gift and kindness from another person or, according to Peterson and Seligman (2004), can be referred to as gratitude expression. Previous studies have shown that gratitude has positive impacts on us, like increasing positive affect, helping us cope with stress, increasing health physically and psychologically, maintaining positive relationships, and inducing prosocial behaviour towards benefactors or other people (Algoe et al., 2010; Emmons & McCullough, 2003; Emmons & Shelton, 2005; McCullough et al., 2002; Peterson & Seligman, 2004). However, Watkins (2004) stated that an individual's pleasant feelings may not be completed until appreciation or gratitude has been given to the source of those pleasant feelings, including writing a letter. Seligman (2004) also explained that a thank-you letter is a form of gratitude intervention that does not only involve pleasant emotions, but also presents activities that generate gratification.

Based on Pennebaker and Seagal (1999), who stated that writing positive words will have a positive impact on an individual, Seligman et al. (2005) conducted a study that asked the participants to write out thank-you letters and deliver them personally within one week, so they could get direct feedback from the receivers. Compared to participants who were asked to write down early memories every night for a week, participants who delivered a thank-you letter had a higher level of happiness and lower level of depression as soon as the intervention period was over and one month afterwards. In accordance with the previous study, an experimental study by Toepfer and Walker (2009) of 85 college students also showed that compared to the non-writing group, the group that wrote one thank-you letter every week and sent it by e-mail every weekend for three weeks had a higher score of happiness and gratitude every week and at the end of the study. The study was replicated by Toepfer et al. (2012) using more samples. However, participants were asked to deliver a letter with an e-mail address to researchers, to be sent after the study ended to avoid feedback from receivers which could affect the findings of the study. Not only did they possess a higher score of happiness, the experimental group also had higher life satisfaction and reduced levels of depression.

Nevertheless, the studies by Seligman et al. (2005), Toepfer and Walker (2009), and Toepfer et al. (2012) were still questionable regarding the possibility that there was an effect caused by the difference in timing of the feedback given by the receivers, either immediate or delayed. According to Roberts (2004), gratitude is a social emotion. Therefore the interaction between the senders of the thank-you letter with the letter receivers that we could see as social feedback became important. When the feedback was given directly, as in a study by Seligman et al. (2005), an individual can catch and feel the non-verbal emotional response immediately from the receiver or, in other words, an emotion contagion (Hatfield et al., 1992). Parkinson (1996) added that within a reciprocal situation, the positive response from the receiver will strengthen the positive emotion that an individual feels when he writes a thank-you letter indirectly. According to Buck (2004), participants also have a higher chance of receiving returned gratitude from the receiver because of the letter or their preceding kindness. So that, participants and the receivers could have mutual relations. Gratitude returned and other positive responses that are considered to reinforce and to be given immediately by receivers, based on the operant conditioning learning principle, may also increase the frequency of the same behaviour in the future for the participants (Powell et al., 2009). On the other hand, a study by Toepfer et al. (2012) showed that writing a thank-you letter without receiving feedback from the receiver still increases happiness. However, according to Watkins et al. (2003), the advantage of this activity will be lower compared to other forms of intervention because the participants may feel anxious regarding how the receivers will respond when they receive delayed feedback.

Based on emotion contagion, reciprocity, and the operant conditioning learning principle explanation, we conclude that direct feedback from thank-you letter receivers delivers more improvement in the senders' happiness than delayed feedback. This is due to the positive verbal and non-verbal feedback that is shown directly by the receivers, which can affect the senders' positive emotions and also applies reciprocally. Participants also have a higher chance of getting reinforcers such as expressions of gratitude in return from the receivers immediately, and thus, in accordance with the operant conditioning learning principle, this can increase the emergence of the same behaviour in the future. In this study, we hypothesised that participants from the group who wrote thank-you letters and were given feedback immediately afterwards from the receivers had a significantly higher happiness gained score compared to participants from another group who were given delayed feedback.

2 METHODS

2.1 *Participants*

The participants of this study consisted of 45 college students who were randomly assigned either to a group who wrote thank-you letters followed by immediate feedback (Experimental Group 1; EG1), comprising 22 participants, or were placed in a group who wrote thank-you letters followed by delayed feedback (Experimental Group 2; EG2) and comprised 23 participants. Participants consisted of six males (13.33%) and 39 females (86.67%), and their ages ranged between 17 and 25 years old ($M = 19.09$, $SD = 1.38$).

2.2 *Research instruments*

2.2.1 *Thank-you letter writing assignment*

Every week for three consecutive weeks, all participants were given instructions by instant messenger to write one thank-you letter to people who had had positive effects on their life, but had not been given a proper thank-you letter, and to deliver the letter to the receiver on each week The content of the letter is an expression of gratitude for one or more specific actions of the receivers and how it affected the participants. However, there was a difference in the timing when receivers gave their feedback. EG1 was instructed to deliver the letter face-to-face and to ask the receivers to read and discuss the letter immediately during the meeting, so that the feedback was obtained directly from the receivers. On the other hand, participants of EG2 were instructed to ask the receivers to read the letters after the meeting and were not asked to discuss the content of the letter during the meeting; therefore the feedback was obtained in a delayed fashion.

2.2.2 *Authentic happiness inventory*

The Authentic Happiness Inventory (AHI) (Peterson, 2005) covers three components of happiness (pleasure, engagement and meaningfulness) and can be used for measuring happiness as a whole. AHI consists of 20 items that require the participants to choose one out of five statements ranging from negative (first) statement to extremely positive (fifth) statement. Every item is scored according to the selected statement, with scores ranging from one to five. The gained happiness score is identified from the difference between the average happiness score before and after the study. This study showed that AHI had a high level of reliability ($\alpha = 0.904$) (Kaplan & Saccuzzo, 2005). A validity test also showed that 19 of the AHI items were valid because they had $r_{it} \geq 0.2$ (Aiken & Groth-Marnat, 2006), while one item had $r_{it} < 0.2$ but was retained.

2.2.3 *Manipulation check*

We presented questions related to the activity based on a study conducted by Toepfer et al. (2012) and additional questions, such as the duration of the letter writing, who the receiver was, the time of delivery, how the delivery was done, and the feedback obtained. We also checked the contents and length of the letters. Every week, we also measured the positive affect which was positively correlated with the gratitude emotion (Froh et al. 2009) using the dimension of *Positive Affect Scale* with ten items by Watson et al. (1998). Participants were asked to rate the extent of their feeling or emotion after delivering a thank-you letter, ranging from 1 (Very slightly or not at all) to 6 (Extremely) and will be counted to find the total score of the positive affect. A reliability test by Watson et al. (1998) showed that this instrument was reliable ($\alpha = 0.86–0.90$). Reliability and validity test results of this instrument's adaptation by Herwibowo (2014) also showed that it was reliable ($\alpha = 0.76$) and valid ($r_{it} \geq 0.2$).

2.3 *Procedure*

We came to classes to explain the study. Participants who agreed to join the study were randomised into two experimental groups and were given research instructions using Instant Messenger software. Subsequently, participants were asked to fill in an online AHI questionnaire before the study period began (pretest). Every week, all participants were asked to deliver one thank-you letter to a different person each week for three consecutive weeks. However, the two groups were differentiated according to the timing of the feedback given, either immediately or delayed. All participants were also asked to fill in a questionnaire as a manipulation check before 24 hours had passed after the delivery of a letter every week. Participants also sent a photo of the letter to us by e-mail. One day after the study ended, participants were asked to fill in an online AHI questionnaire one more time (post-test). All participants were also given an explanation regarding the study using Instant Messenger software. We then interviewed some participants a week after the study ended to further understand the experience of the writing and delivering of the thank-you letters

3 RESULTS

3.1 *Manipulation check*

Based on the control questions, we learned that participants from both groups required 10 to 20 minutes to write a thank-you letter and to deliver the letter on the same day or one day afterwards. The letters were given to different persons every week, such as parents, friends, colleagues in organisations, and girlfriends or boyfriends. All participants also expressed gratitude in at least one paragraph in every letter.

A comparison between the positive affect score of the two groups every week immediately after delivering the thank-you letters is shown in Table 1.

As shown in Table 1, there were significant differences in the positive effect scores every week between the two experimental groups, with the group of participants who received the feedback from receivers immediately after delivering their letters (EG1) having a higher average positive effect score compared with the group of participants who received delayed feedback (EG2). This showed that the manipulation was successful.

Table 1. Comparison of positive affect as manipulation check.

	M (SD)			
Weeks	EG1	EG2	*t*/*U*	*P*
1	43.45 (8.12)	38.69 (7.39)	$t = 2.057$	0.023*
2	44.32 (5.96)	38.30 (7.81)	$U = 140$	0.005*
3	43.45 (7.30)	37.39 (7.61)	$U = 146.5$	0.007*

*significant on $p < 0.05$ (*one-tailed*).

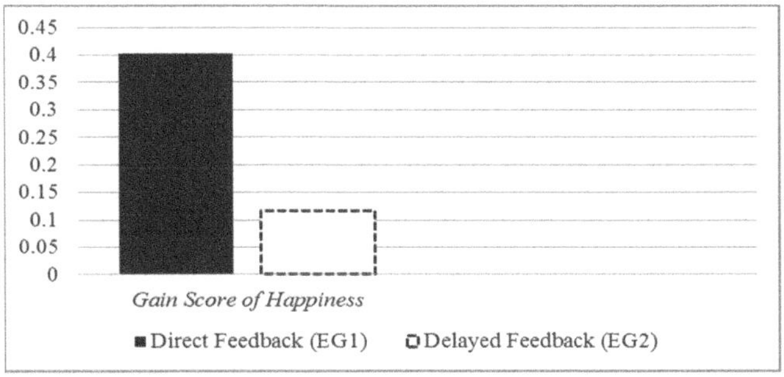

Figure 1. Comparison of gained happiness score.

Table 2. Description of the effect of delayed feedback from receivers of thank-you letters on the sender's happiness.

Groups	*n*	*M*	*SD*	*T*	*P*	r^2
Immediate feedback (EG1)	22	0.402	0.453	2.207	0.016*	0.319
Delayed feedback (EG2)	23	0.117	0.412			

*significant on $p < 0.05$ (one-tailed).

3.2 *Hypothesis test result*

Before proceeding with the hypothesis test, we checked if there was any difference between the average happiness score before the study (pre-test) of the two experimental groups. The analysis results showed that the pre-test score of the group who wrote thank-you letters with direct feedback ($M = 2.63$, $SD = 0.51$) was not significantly different from participants who wrote thank-you letters with delayed feedback ($M = 2.56$, $SD = 0.59$), $t(43) = 0.416$, $p > 0.05$. This showed that participants were successfully randomised according to their happiness baseline level. Therefore, there was no significant difference in the participants' happiness score between both groups before the study.

During the data collection period, the volume of delayed feedback received by EG2 varied. We decided to check if there was any difference between the gained happiness score for participants of EG2 based on the amount of delayed feedback from the receivers. The results of the analysis showed that the gained happiness score of participants who received delayed feedback three times ($R = 9.39$), twice ($R = 14.08$), once ($R = 10.75$), and not at all ($R = 16.00$) did not differ significantly, $H = 3.44$ (3, N = 45), $p > 0.05$ (two-tailed). This showed that the volume of delayed feedback received did not influence the gained happiness score of participants of EG2. Therefore, no further analysis was made regarding the volume of feedback.

A comparison was then made of the gained happiness score between both groups as a hypothesis test. As shown in Figure 1 and Table 2, the average increase in happiness of participants in the group who wrote thank-you letters followed by direct feedback ($M = 0.402$, $SD = 0.453$) was significantly higher compared to the participant group with delayed feedback ($M = 0.117$, $SD = 0.412$), $t(43) = 2.207$, $p < 0.05$ (one-tailed), $r^2 = 0.1017$. Based on this result, we conclude that the hypothesis of the study is supported.

4 DISCUSSION

The result of this study is consistent with previous studies conducted by Seligman et al. (2005) and Toepfer et al. (2012) that showed that the activity of writing thank-you letters may increase happiness. However, previous studies had not specifically looked into the role of the time when the feedback was given by the receivers. Therefore, the results of this study showed that direct feedback from the receivers had its own advantage in terms of increasing an individual's happiness.

According to Pennebaker and Seagal (1999), writing a thank-you letter may encourage individuals to remember their happy experience. When we interviewed 28 participants, they revealed that they felt happy because they remembered the positive experience with the receivers and realised that people around them cared and loved them. Half of the participants also reported that they learned to understand the meaning of kindness that was shown by people around them.

Based on the explanations above, both experimental groups had the opportunity to feel the same positive emotions when they wrote thank-you letters, but we argue that there are four possibilities which may explain why EG1 had a higher gained happiness score. First, an emotion contagion happens when individuals interact directly (Hatfield et al., 1992). Based on the results of interviews with 18 participants, it was found that positive feedback was observed from the response of the receivers who gave immediate feedback by smiling, hugging and crying, and it also influenced the participants' own emotion. A second possibility is related to the indirect influence from the reciprocal appraisal of one's own emotion and the emotion of other people (Parkinson, 1996). EG1 participants rated feedback given by all three receivers as positive ($M = 5.44$) and having an influence on their emotions ($M = 5.03$). Participants also reported that they were worried at first when they thought about what response the receivers would show but this worry was reduced as soon as they saw positive emotions showed by the receivers. On the other hand, EG2 participants were not able to capture and imitate the receivers' responses immediately because they received the feedback by message or phone.

Furthermore, receivers need to express their gratitude back to the letter senders to maintain equality and trust on both sides (Buck, 2004). EG1 had a higher chance of receiving a returned thank-you from the receivers during the meeting, but there were only nine participants in EG2 who received it. This may lead participants who did not get the response to feel disappointed and anxious about the receivers' response (Watkins et al., 2003), so that they received less happiness. The fourth possibility is related to the operant learning principle, especially with regards to reinforcement. Immediate positive feedback from receivers when participants of EG1 delivered their letter may be regarded as a reinforcer for the participants and encouraged them to repeat this behaviour the week after. However, it was also found that there were participants of EG2 who still felt happy and relieved after expressing their gratitude even if they did not receive any feedback. Therefore, individual differences, such as what may be regarded as reinforcer and how individual expectations become a reinforcer after writing a thank-you letter needs to be studied further.

This study also had limitations. First, we did not get enough information from previous studies regarding the specific length of time in the delayed feedback from gratitude letters receivers that may influence senders' happiness. We also recruited participants based on their willingness and commitment to join the programme, so other factors can affecting the effectiveness of writing thank-you letters on happiness and difficult to control in this study, such as motivation, conviction of the effectiveness of the activity to increase happiness, and person fit activity (Layous & Lyubomirsky, 2012), along with whether the participants were happiness-seekers (Seligman et al., 2005),. Personality factors, such as those in a study conducted by Schueller (2012), who reported extroverts receiving higher benefits when asked to write and deliver thank-you letters compared to introverts, were not measured in this study. Those variables may be confounding or moderating variables.

Individuals also have tendencies to adapt with continuous positive activity (Layous & Lyubomirsky, 2012); therefore, variety in the activity, such as variation of the receivers of letters based on closeness, is needed to avoid boredom. Then, we are aware that technology development allows gratitude expression to be given by other media, such as voice notes, video call, or social media. Therefore, the most effective media for thank-you letters needs to be investigated in further studies.

Despite these limitations, 33% of the participants reported an increase in the perception of closeness to the receivers of their thank-you letters. This conforms with a study by Algoe et al. (2008), which indicated that gratitude expression from junior to senior members in a college student organisation for a week predicted a better quality of relationship immediately after the study and one month afterwards.We see individual perceptions of the relationship's quality of the letter's writer and receiver as an interesting topic for a follow-up study.

5 CONCLUSION

Gratitude expression, such as writing a thank-you letter, can indeed increase happiness (Seligman et al., 2005; Toepfer et al., 2012). However, the timing of obtaining the feedback from the letter receivers can also be important. This study showed that thank-you letter senders who get immediate feedback from the receivers are happier than those who receive delayed feedback.

REFERENCES

Aiken, L.R. & Groth-Marnat, G. (2006). *Psychological testing and assessment*. Boston, MA: Pearson Education.

Algoe, S.B., Haidt, J. & Gable, S.L. (2008). Beyond reciprocity: Gratitude and relationships in everyday life. *Emotion, 8*(3), 425–429.

Buck, R. (2012). The gratitude of exchange and the gratitude of caring: A developmental-interactionist perspective of moral emotion. In Emmons, R.A & McCullough, M.E. (Eds).*The Psychology of Gratitude*. New York, NY: Oxford University Press Series in Affective Science.

Carr, A. (2004). *Positive psychology: The science of happiness and human strengths*. New York, NY: Brunner-Routledge.

Compton, W.C. & Hoffman, E. (2012). *Positive psychology: The science of happiness and flourishing* (2nd ed.). New York, NY: Thomsond Wardsworth.

Emmons, R.A. & McCullough, M.E. (2003). Counting blessings versus burdens: An Experimental investigation of gratitude and subjective well-being in daily life. *Journal of Personality and Social Psychology, 84*(2), 377–389.

Emmons, R.A. & Shelton, C.M. (2005). Gratitude and the science of positive psychology. In Snyder, C.R. & Lopez, S.J. (Eds.), *Handbook of positive psychology* (pp. 459–471). New York, NY: Oxford University Press.

Froh, J.J., Yurkewicz, C. & Kashdan, T.B. (2009). Gratitude and subjective well-being in early adolescence: Examining gender differences. *Journal of Adolescence, 32*(3), 633–650.

Hatfield, E., Cacioppo, J.T. & Rapson, R.L. (1992). Primitive emotional contagion. In Clark, M.S. (Eds). *Emotion and Social Behavior* (151–177). Newbury Park: Sage Publication.

Herwibowo, D. (2014). *The relationship between perceived freedom in leisure and subjective well-being in University of Indonesia students* (Unpublished Undergraduate thesis, Universitas Indonesia, Depok, Indonesia).

Kaplan, R.M. & Saccuzzo, D.P. (2001). *Psychological testing: Principles, applications, and issues* (5th ed.). 44, 1–11. Belmont, CA: Wadsworth Cegage Learning.

Layous, K. & Lyubomirsky, S. (2012). The how, why, what, when, and who of happiness: Mechanisms underlying the success of positive interventions. In Gruber, J. & Moscowitz, J. (Eds.), *The light and dark side of positive emotions*. New York, NY: Oxford University Press.

Lyubomirsky, S., King, L. & Diener, E. (2005). The benefits of frequent positive affect: Does happiness lead to success? *Psychological Bulletin, 131*(6), 803–855.

Lyubomirsky, S. & Layous, K. (2013). How do simple positive activities increase well-being? *Current Directions in Psychological Science, 22*(1), 57–62.

Lyubomirsky, S., Sheldon, K.M. & Schkade, D. (2005). Pursuing happiness: The architecture of sustainable change. *Review of General Psychology, 9*(2), 111–131.

McCullough, M.E., Emmons, R.A. & Tsang, J.-A. (2002). The grateful disposition: A conceptual and empirical topography. *Journal of Personality and Social Psychology, 82*(1), 112–127.

Parkinson, B. (1996). Emotions are social. *British Journal of Psychology, 87*, 663–683.

Pennebaker, J.W. (1997). *Opening up: The healing power of expressing emotions*. New York, NY: The Guilford Press.
Pennebaker, J.W. & Seagal, J.D. (1999). Forming a story: The health benefits of narrative. *Journal of Clinical Psychology*, *55*(10), 1243–1254.
Peterson, C. (2005). Authentic happiness questionnaire. Retrieved from http://www.authentichappiness.com
Powell, R.A., Symbaluk, D.G. & Honey, P.L. (2009). *Introduction to learning and behavior* (3rd ed.) Belmont, CA: Wadsworth Cengage Learning.
Roberts, C.R. (2004). The blessing of gratitude: A conceptual analysis. In Emmons, R.A. & McCullough, M.E. (Eds.) *The psychology of gratitude* (pp. 58–80). New York, NY, Oxford University Press.
Schueller, S.M. (2012). Personality fit and positive interventions: Extraverted and introverted individuals benefit from different happiness increasing strategies. *Psychology, 3*(12), 1166–1173.
Schueller, S.M. & Seligman, M.E.P. (2010). Pursuit of pleasure, engagement, and meaning: Relationships to subjective and objective measures of well-being. *The Journal of Positive Psychology, 5*(4), 253–263.
Seligman, M.E.P. (2004). *Authentic happiness: Using the new positive psychology to realize your potential for lasting fulfillment*. New York, NY: Free Press.
Seligman, M.E.P., Steen, T.A., Park, N. & Peterson, C. (2005). Positive psychology progress: Empirical validation of interventions. *The American Psychologist*, *60*(5), 410–421.
Soulpancake. (2013). An experiment in gratitude: The science of happiness. Retrieved from https://www.youtube.com/watch?v=oHv6vTKD6lg
Toepfer, S.M. & Walker, K. (2009). Letters of gratitude : Improving well-being through expressive writing. *Writing, 1*, 181–198.
Toepfer, S.M., Cichy, K. & Peters, P. (2012). Letters of gratitude: Further evidence for author benefits. *Journal of Happiness Studies*, *13*(1), 187–201.
Watkins, P.C. (2012). Watkins, P.C. (2004). Gratitude and subjective well-being. In R.A. Emmons & M.E. McCullough (Eds.), *The psychology of gratitude* (pp. 167–194). New York: Oxford University Press.
Watkins, P.C., Woodward, K., Stone, T. & Kolts, R.L. (2003). Gratitude and happiness: Development of a measure of gratitude, and relationships with subjective well-being. *Social Behavior and Personality: An International Journal*, *31*(5), 431–451.
Watson, D., Clark, L.A. & Tellegen, A. (1988). Development and validation of brief measures of positive and negative affect. *The PANAS Scales*, *54*, 1063–1070.

Diversity in Unity: Perspectives from Psychology and Behavioral Sciences – Ariyanto et al. (Eds)
© 2018 Taylor & Francis Group, London, ISBN 978-1-138-62665-2

The contribution of parental autonomy support and control on executive function of preschool children

I.P. Hertyas, D. Hendrawan, N. Arbiyah & R. Nurbatari
Faculty of Psychology, Universitas Indonesia, Depok, Indonesia

ABSTRACT: Research in the past decade revealed the significance of parenting in the development of Executive Function (EF) during the preschool period. Nonetheless, the influence of father parenting on children's EF development has not been well explored. Furthermore, there have been limited studies that investigate mother and father parenting concurrently in relation to children's EF development. This research aims to explore the contribution of parental autonomy support and controlling behaviour on a child's EF. Children aged 48–72 months old (N = 59) and their parents participated in a laboratory session. The child's EF was measured through three different tests, and parenting behaviour was observed during the dyadic interaction. Results indicate that higher controlling behaviour from the mother predicts poorer child EF performances. Meanwhile, after controlling for important covariates and the quality of maternal parenting, no significant contribution from the father was found. This research extends previous studies about the independent effect of the father and the detrimental effect of negative parenting on the child's executive function.

1 INTRODUCTION

Executive function, a set of higher cognitive skills that enables cognitive, emotional, and physical control to accomplish certain goals, reaches the optimal development in young adulthood (Diamond, 2016). This protracted development of EF leaves a great opportunity for environmental factors to impact the trajectory of EF development (Zelazo & Carlson, 2012; Moriguchi, 2014; Diamond, 2016). One of the prominent factors that has been considered is parenting. The parent–child relationship in the early years can provide rich and intense experiences that could shape the child's capacity. Thus, in this research, we aim to explore the influence of parenting on the performance of the child's EF, particularly the dyadic interaction between the child and its parents.

Recent findings have confirmed the essential role of parenting in the early EF development (see Fay-Stammbach et al., 2014). Parental autonomy support, characterised by the encouragement of a child's sense of autonomy, is one aspect that has been seen as influential. It has been consistently seen to be valuable to the child's EF, beyond and above the parents' education, the child's general cognitive functioning, and family economic status in children between the age of 15 to 40 months old (Bernier et al., 2010; Matte-Gagné & Bernier, 2011; Bernier et al., 2012; Roskam et al., 2014; Meuwissen & Carlson, 2015). As previous research explained, autonomy support can enhance EF development by providing a great opportunity to exercise the child's self-control and regulation capacity which are strongly related to EF.

Although autonomy support has been considered as a robust factor, little attention has been paid to exploring the father's role. Primarily, there is a striking gap in the amount of research about the mother's and father's influence on their child's outcomes in developmental studies. The father's role on the child's outcomes should not be overlooked for several reasons. First, as Cox and Payley (1997) proposed, the mother's parenting and the father's parenting complement each other in having certain consequences for their children. By only taking the mother's role into account, we would not understand the complexity of the development process that occurs. In addition, the father and mother might implement different styles of parenting that eventually have different impacts on their children, as found in previous research (Lucassen et al., 2015; Cabrera et al., 2007; Lamb, 2010). In the context of father–children interaction, toddlers commonly spend more time with their father in play and exploration activities that could serve as a potential context supporting their children's development. Therefore, it is important to study how fathers' interaction with their children could shape the developmental outcomes of the children.

There is a growing interest in the father–child EF relation. The father's sensitivity (Towe-Goodman et al., 2014), harsh parenting (Lucassen et al., 2015), emotional support (Meuwissen & Englund, 2016), and mind-mindedness (Baptista et al., 2016) are reported to be linked with their children's EF. In this connection, the father's autonomy support has started to gain considerable attention. Studies conducted by Roskam et al. (2014) fail to report the significant relations between the father's autonomy support and the children's EF. They suspect that the measurement (a three-item questionnaire about how parents ask their children to be autonomous) is not sufficient to capture the autonomy support behaviour. Meanwhile, after observing the dyadic interaction between father and child, Meuwissen and Carlson (2015) report a positive contribution from the father's autonomy support to the child's EF, which is consistent with the findings from the samples of the mothers.

However, we argue that the paternal autonomy support still needs further investigation. The only evidence from Meuwissen and Carlson (2015) was obtained without controlling for the mother's parenting, despite her role as the primary caregiver. There might be certain discrepancies in the children's outcomes caused by the mother's parenting, but they have not been measured. Children's development is not exclusively caused by one parent's behaviour. The mother's influence could intertwine with the father's in having certain impacts on the children. Furthermore, we must also remember that mother's and fathers' behaviour could possibly have different consequences for the EF of the children. Lucassen et al. (2015) report that the father's sensitivity is not correlated to the children's EF while the mother's is. Meanwhile, the father's harsh discipline is significant to their child's EF, but the mother's is not. Baptista et al. (2016) only find significant association between the mother's mind-mindedness, defined as the parents' frequency in discussing the mental state construct with children, and their children's EF and fail to find the same result for the father's. So far, no research has focused on both the mothers' and the fathers' autonomy support to reveal the different impacts that might appear. Hence, we conducted research which considered the role of both parents to obtain a more integrative insight.

Contrary to autonomy support, controlling behaviour has a negative influence on EF performance (Meuwissen & Carlson, 2015). In other studies that focused on the mother's verbalisation, researchers found that the mother's behaviour of frequently expressing direct commands to control their child's behaviour tended to hinder the child's opportunity to make choices and decisions autonomously. Meanwhile, suggestions, indirect commands, questions, and options provided by the mothers could enhance the child's self-initiative and self-regulation capacity (Bindman et al., 2013). Nonetheless, only a few studies have been conducted to examine the effects of controlling behaviour on EF. Thus, conducting more research was necessary.

As far as we know, no research about the influence of parenting on early childhood EF has been conducted in Indonesia. A survey report noted that fathers in Indonesia tend to show low involvement in caregiving (Krismantari, 2012), in contrast with Meuwissen & Carlson's (2015) report of a moderately high involvement by fathers taken from samples in the US. With less time spent interacting, we presume that the effect of the father's parenting might be diminished as well.

The current study thus aims to explore the contribution of the parents' autonomy support and controlling behaviour on EF performance of preschoolers in Indonesia. Since they have been found to be strongly correlated to EF, we also take some variables into account, including the age of the children, parents' education and the family's Social Economic Status (SES) (Lengua et al., 2007; Sarsour et al., 2011; Hendrawan et al., 2015). Parental autonomy support and control were scored by observing parent–child dyadic interaction in a guided situation; while the child's EF was measured through three different tasks that tapped into the three core components of EF: working memory, inhibitory control, and shifting component. We predict that the mother's parenting behaviour will significantly influence the child's EF, while the father's effect would be weaker than the mother's. However, because of the lack of research, no hypothesis about which parenting aspect will give a stronger effect could be built. This research does not only involve both parents, but it also measures and compares the consequences of positive and negative parenting practice concurrently. The results are expected to be beneficial for intervention programmes to improve the child's EF, especially children from low economic status families.

2 METHODS

2.1 *Participants*

For the current investigation, the total participants included 66 children with their mothers and fathers who live in Jakarta, Bogor, Depok, Tangerang and Bekasi. Seven children were excluded because of video issues, leaving 59 videos that could be analysed. The children's ages ranged from 48 to 72 months old (mean = 58.54, SD = 6.545), while the mothers' age ranged from 23 to 48 years old (mean = 32.26, SD = 4.658) and the fathers' age was between 29 and 56 years old (mean = 36.09, SD = 5.059). Most of the fathers (83.1%) and mothers (78%) had a college degree or higher level of education. Almost half of the mothers were full-time workers (46%), and the others were housewives and occasionally had freelance jobs. The families' monthly expenditures ranged from IDR 2,500,000–20,000,000 (mean = IDR 8,641,509). Besides grandparents, mothers were reported to be the main caregiver (86%).

2.2 *Measurement*

2.2.1 *Executive function*

Three different tasks were administered to measure the children's executive function performance. All the tasks used have been adapted to Indonesian culture by Hendrawan et al. (2015).

2.2.1.1 *Matahari/rumput*

This test was adapted from the Grass/Snow test of Carlson et al. (2014) to measure the EF's subcomponent, inhibitory control (Carlson et al., 2014). The property used in this test was yellow- and green-coloured cards that represented the colour of the sun (*matahari*) and the grass (*rumput*), respectively. The children were asked to point to the yellow cards when the tester said "matahari" and to the green cards for "rumput" in 16 trials. The performance was scored by the number of correct responses, ranging from 0–16.

2.2.1.2 *Backward word span*

The Backward Word Span assessed the EF's component, which is the working memory (Carlson et al., 2014; Davis & Pratt, 1995). This test was developed by Carlson et al. (2014) and has been adapted into *Bahasa* (Indonesian). In this test, the children were asked to repeat a series of words in reverse order. The number of words that must be recalled started from two to five words for each series. The scores obtained were dependent on the number of words successfully recalled (0–5).

2.2.1.3 *Dimensional change card sorting*

The Dimensional Change Card Sorting (DCCS) test was adapted from Zelazo (2006) to measure the shifting component. This test consists of three phase; pre-switch, post-switch, and border phase. The children were expected to sort six cards in pre-switch and post-switch phases and 12 cards in the border phase (total 24 cards). In each phase, different orders were applied. The scores were obtained by the number of correct cards sorted in the post-switch (0–6) and the border (0–12) phases.

2.2.2 *Dyadic interactions*

The videos of the dyadic interactions were coded by using the autonomy support coding system of Whipple et al. (2011). This scale has been used in several research studies (Matte-Gagné & Bernier, 2011; Bernier et al., 2012; Matte-Gagné et al., 2013; Meuwissen & Carlson, 2015). We adapted the scale into Bahasa with the authors' permission. The autonomy support coding system was developed to measure verbal and nonverbal behaviour of parents in a guided situation. Beside the autonomy support, the coding scheme also includes controlling scales. It consists of four scales depicting the degree of parents' intervention: 1) intervening according to child's need, and adjusting the task according to their competency to encourage autonomy; 2) giving suggestions, hints, questions, and encouragement to their child to complete the task by using a tone of voice that signifies the presence of the parents; 3) respecting the child's perspective and showing flexibility in the attempt to make their child complete the tasks (only scored if children refuse to complete the task, otherwise scored as missing); and 4) following the child's pace and ensuring that their child plays an active role in completing the tasks, and providing enough opportunity for their child to make choices. Each behaviour was coded in terms of the degree of how autonomously supportive or controlling the parents are, from 1 (not autonomously supportive/not controlling) to 5 (very autonomously supportive/very controlling). Because we observed that more than 80% of parents do not elicit behaviour representing the scale number 3, we decided to exclude it in the analysis. By adding up the score from the other three scales, the total score for each autonomy support and controlling for each parent range from 3 to 15. Supportive parents encourage their children to take an active role, respect their children's perspective, and adjust their help to their children's capability. Meanwhile, controlling parents usually intervene too much, impose their perspective on their children, and use a harsh tone of voice. The inter-class correlation (from 14 videos) for the autonomy support scale was 0.872, and 0.945 for the controlling scale.

2.2.3 *Procedure*

The participants were recruited using several kindergarten and social media announcements. The children who were detected to have severe impairments to their cognitive, emotional, physical, and social development as confirmed in the screening process were excluded from this study. Only children that lived together with both parents were included in this research. After getting their consent, the children and their parents participated in a videotaped laboratory session. The children were tested on EF tasks, comprised of *Matahari/Rumput* (the Indonesian version of the Grass/Snow task), Backward Span task, and Dimensional Card Sorting task by a trained experimenter. After completing the EF tasks, the children and their parents participated in an observation. We observed the dyadic interactions between child and father, and child and mother independently by applying a counterbalance technique to decide the order of observation (Cozby & Bates, 2012). In the observation process, the children and their father/mother worked together to complete two challenging assignments. First, they were asked to build some blocks by replicating a model that had already been printed in 5R-sized photos prepared by the researchers. In the second task, they had to complete a jigsaw puzzle. Considering that the children's age range was quite large, we prepared all of the task materials (blocks model and puzzle) in two levels of difficulty. The easier one was given to children under 5 years old, and the more difficult one was given to the older children (above 5 years old). We understood that the task difficulty could affect the amount of help offered by the parents. Therefore, it was important to ensure that every task was suitable for the child's age and ability. We used different block model and puzzle in child-father dyadic and child mother dyadic to avoid the learning effects on children. We ensured all task materials were at equal level of difficulty as we previously examined in a pilot study.

To summarise, each child worked four tasks with both parents (two blocks model and two jigsaw puzzles). Each task lasted five minutes; the total duration for observation was around 20 minutes. All materials were chosen based on our pilot study that involved 26 different children.

3 RESULTS

3.1 *Data analysis*

The EF scores were standardised and combined into composites. Two composites scores could not be obtained because of a missing score. To check the normality assumption, we ran a Shapiro–Wilk test to indicate that our data represented a normal distribution. The zero-order correlation between the mother's and father's parenting and their child's EF were examined using Pearson's correlation. To explore the association between the parents' autonomy support and control and the child's EF, we conducted a hierarchical regression where covariates were entered in the first block, mother's parenting in the second, and father's parenting in the third block.

3.2 *Main analysis*

As summarised in Table 1, the mother's parenting model explains 14% of the variance of the child's EF score after controlling child's age, family SES, and parents' education. Additionally, maternal control was the only parenting variable that independently predicted the child's EF ($r = -0.330$, $p < 0.05$) over the mother's autonomy support and covariates. Meanwhile, the father's parenting model did not significantly contribute to the regression model after controlling for the covariates and the mother's parenting.

4 DISCUSSION

This study complements previous research by exploring the contribution of autonomy support and controlling behaviour to preschoolers' executive function, not only examining the mother's parenting behaviour but also exploring the father's. Our results support the previous notion regarding the importance of mothers' behaviour in influencing their children's outcomes. Maternal control significantly predicts the EF performance above and beyond the child's age, family SES, and parents' education. This finding is consistent with previous research where Bindman et al. (2013) concluded that children that usually receive high amounts of verbal control from their mother tend to have poorer EF performance. The same result was also obtained by Meuwissen & Carlson (2015), who studied the fathers' sample. Controlling behaviour might not be recommended as an effective way to enhance children's self-regulation capacity because it limits their opportunities to propose initiatives and choice.

Many researchers have documented the positive contribution of the mothers' and fathers' autonomy support to early childhood EF (Bernier et al., 2010, 2012; Roskam et al., 2014;

Table 1. Association of parental autonomy support and control to child's EF (N = 57).

	Variable	B	T	R^2	ΔR^2
Step 1				0.257**	0.257**
	Child's age	0.450	–0.321**		
	Family SES	0.332	2.429*		
	Parents' Education	–0.074	–0.546		
Step 2				0.337**	0.140**
	Mother's autonomy support	0.091	0.557		
	Mother's control	–0.330	–0.2046*		
Step 3				0.429	0.092
	Father's autonomy support	–0.083	–0.530		
	Father's control	–0.219	–1.610		

Meuwissen & Carlson, 2015). Surprisingly, our study failed to confirm this result. As mothers' parenting (autonomy support and control altogether) causes a significant variant to the EF model, our regression analysis found that the autonomy support alone is not sufficient to predict the child's EF performances. There are some assumptions that could be addressed. First, longitudinal research conducted by Cabrera et al. (2007) showed that parenting influence on the child outcomes could differ across ages. They found that parents' supportiveness could predict early childhood emotional development (self-regulation), but the predictive power is diminished by the time children enter kindergarten. Second, most research mentioned above was done on younger children (2–3.5 year-olds). In that critical period of development, the parents' roles are still very salient. However, when their children get older and start to interact with more people, other factors could possibly intervene. Almost all of our participant children were enrolled in kindergarten or toddler classes. Moreover, about half of the mothers who participated in this research were working mothers that delegated their children's rearing to other people, such as their grandparents, nannies, or even older siblings, which is quite common in Indonesia. We suspected that stimulation or supportiveness received from other people, like teachers at school or other caregivers at home might have provided some benefits for the children's EF capacity. Nonetheless, we couldn't test this hypothesis because we did not measure the role of other caregivers. To conclude, we postulate that despite of the inadequacy of the autonomic support from mothers, the children' needs for autonomy were still possibly compensated by other people's nurturance. However, once the mothers displayed very intrusive and controlling behaviour, their children could suffer from a feeling of incompetence and also lose their chance to better develop self-control ability.

Although mother's parenting behaviour has successfully been proven to associate with the child's EF, however none of the father's is found to be significant. These findings are inconsistent with earlier studies (Meuwissen & Carlson, 2015), where the father's autonomy support and control significantly predict the preschoolers' EF. The difference between the findings might result from the statistical method used. In this study, we control for the mother's parenting to attain fathers' independent influence on children development. The results presented here might represent the independent effect of fathering on the preschoolers' EF. Nonetheless, by considering the inadequate research about the fathers' role, we suggest that more related research is conducted in the future.

The second explanation that we could offer is that the fathers' low influence is caused by the low level of their involvement in childrearing. We did not measure the amount of time the fathers spent with their children, but based on their working and commuting hours per day, ranging from 10 to 12 hours (as testified on the self-reporting questionnaire), we assume that the time available to spend with the children was limited. Though some research suggests that the quantity of time is less of a factor than the quality of interaction (Easterbrook & Goldberg, 1984; Meuwissen & Carlson, 2015), we still predict that to some extent, frequency is also essential. Therefore, another research study needs to be conducted to explore the relationship between the frequency of time the fathers spend with their children, the quality of their parenting, and the impact on their children's EF development.

This research has a number of limitations. We did not measure the degree of the parents' involvement in child rearing practice that might contribute to the relationship between their quality of parenting and the children's outcomes. We also suggest considering the roles of other caregivers in the family like grandparents, older siblings, and other relatives, as well as the school environment for the children who have already entered formal education. Finally, longitudinal and intervention studies with a larger sample size would be necessary to understand the causal relationship between parenting and the children EF function.

5 CONCLUSION

This research provides additional insights into the importance of the mothers and fathers autonomy support and controlling behaviour. These results have important implications for intervention programmes. As already noted in many research studies, mothers have a stronger impact on children (Roskam et al, 2014; Baptista, et al, 2016). Nevertheless, it is also crucial to promote better parenting for the fathers, because two involved parents must be better than one. In addition, our research also pays reasonable attention to negative parenting aspects, and concludes that it has a more significant effect on the children, so the parents should be more cautious about their negative behaviour.

ACKNOWLEDGEMENT

This research was fully funded by PITTA Research Grant, Universitas Indonesia. We would like to thank Afiyana for her help in the video coding process, and the members of the Executive Brain Function Research Group—Ditta Hestianty, Siti Nurlaila, Claudia Carolina, Fasya Fauzani and Hanifah Nurul—for their tremendous help in recruiting participants and collecting data.

REFERENCES

Baptista, J., Osório, A., Martins, E.C., Castiajo, P., Barreto, A.L. Mateus, V. & Martins, C. (2016). Maternal and paternal mental-state talk and executive function in preschool children. *Social Development, 26*(1), 129–145. doi:10.1111/sode.12183

Bernier, A., Carlson, S.M. & Whipple, N. (2010). From external regulation to self-regulation: Early parenting precursors of young children's executive functioning. *Child Development, 81*(1), 326–339. doi:10.1111/j.1467-8624.2009.01397.x

Bernier, A., Carlson, S.M., Deschênes, M. & Matte-Gagné, C. (2012). Social factors in the development of early executive functioning: A closer look at the caregiving environment. *Developmental Science, 15*(1), 12–24. doi:10.1111/j.1467-7687.2011.01093.x

Bindman, S.W., Hindman, A.H., Bowles, R.P. & Morrison, F.J. (2013). The contributions of parental management language to executive function in preschool children. *Early Childhood Research Quarterly, 28*(3), 529–539. doi:10.1016/j.ecresq.2013.03.003

Cabrera, N.J., Shannon, J.D. & Tamis-LeMonda, C. (2007). Fathers' influence on their children's cognitive and emotional development: From toddlers to Pre-K. *Applied Developmental Science, 11*(4), 208–213. doi:10.1080/10888690701762100

Carlson, S.M., White, R.E. & Davis-Unger, A.C. (2014). Evidence for a relation between executive function and pretense representation in preschool children. *Cognitive Development, 29*, 1–16. doi:10.1016/j.cogdev.2013.09.001

Cox, M.J. & Paley, B. (1997). Families as systems. *Annual Review of Psychology, 48*(1), 243–267. doi:10.1146/annurev.psych.48.1.243

Cozby, P.C. & Bates, S.C. (2012). Methods in behavioral research (11th ed, 165–166). NY: McGraww-Hill.

Davis, H.L. & Pratt, C. (1995). The development of children's theory of mind: The working memory explanation. *Australian Journal of Psychology, 47*(1), 25–31. doi:10.1080/00049539508258765

Diamond, A. (2016). Why improving and assessing executive functions early in life is critical. In P. McCardle, L. Freund & J.A. Griffin (Eds.), *Executive function in preschool-age children: Integrating measurement, neurodevelopment, and translational research* (pp. 11–43). Washington, DC: American Psychological Association. doi:10.13140/RG.2.1.2644.6483

Easterbrook, A.M. & Goldberg, W.A. (1984). Toddler development in the family: Impact of father involvement and parenting characteristics. *Child Development, 55*, 740–752.

Fay-Stammbach, T., Hawes, D.J. & Meredith, P. (2014). Parenting influences on executive function in early childhood: A review. *Child Development Perspectives, 8*(4), 258–264. doi:10.1111/cdep.12095

Grolnick, W.S. & Ryan, R.M. (1989). Parent styles associated with children's self-regulation and competence in school. *Journal of Educational Psychology, 81*(2), 143–154. doi:10.1037/0022-0663.81.2.143

Hendrawan, D., Fauzani, F., Carolina, C., Fatimah, H.N., Wijaya, F.P. & Kurniawati, F. (2015). The construction of executive function instruments for early child ages in Indonesia: A pilot study. In *Promoting children's health, development, and well-being: Integrating cultural diversity*. Jakarta, Indonesia: Faculty of Psychology.

Krismantari, I. (2012, February 22). Calling fathers back to the family. *The Jakarta Post*. Retrieved from http://www.thejakartapost.com/news/2012/02/22/calling-fathers-back-family.html

Lamb, M. (2010). How do fathers influence children's development? Let me count the ways. In Lamb, M.E. (Ed.), *The role of the father in child development* (5th ed., pp. 1–26). Hoboken, NJ: John Wiley & Sons.

Lengua, L.J., Honorado, E. & Bush, N.R. (2007). Contextual risk and parenting as predictors of effortful control and social competence in preschool children. *Journal of Applied Developmental Psychology*, *28*(1), 40–55. doi:10.1016/j.appdev.2006.10.001

Lucassen, N., Kok, R., Bakermans-Kranenburg, M.J., Van Ijzendoorn, M.H., Jaddoe, V.W.V., Hofman, A. & Tiemeier, H. (2015). Executive functions in early childhood: The role of maternal and paternal parenting practices. *British Journal of Developmental Psychology*, *33*(4), 489–505. doi:10.1111/bjdp.12112

Matte-Gagné, C. & Bernier, A. (2011). Prospective relations between maternal autonomy support and child executive functioning: Investigating the mediating role of child language ability. *Journal of Experimental Child Psychology*, *110*(4), 611–625. doi:10.1016/j.jecp.2011.06.006

Matte-Gagné, C., Bernier, A. & Gagné, C. (2013). Stability of maternal autonomy support between infancy and preschool age. *Social Development*, *22*(3), 427–443. doi:10.1111/j.1467-9507.2012.00667.x

Meuwissen, A.S. & Carlson, S.M. (2015). Fathers matter: The role of father parenting in preschoolers' executive function development. *Journal of Experimental Child Psychology*, *140*, 1–15. doi:10.1016/j.jecp.2015.06.010

Meuwissen, A.S. & Englund, M.M. (2016). Executive function in at-risk children: Importance of father-figure support and mother parenting. *Journal of Applied Developmental Psychology*, *44*, 72–80. doi:10.1016/j.appdev.2016.04.002

Moriguchi, Y., Chevalier, N. & Zelazo, P.D. (2016). Development of executive function during childhood. *Frontiers in Psychology*, *7*, 6. doi:10.3389/fpsyg.2016.00006

Roskam, I., Stievenart, M., Meunier, J.C. & Noël, M.P. (2014). The development of children's inhibition: Does parenting matter? *Journal of Experimental Child Psychology*, *122*(1), 166–182. doi:10.1016/j.jecp.2014.01.003

Sarsour, K., Sheridan, M., Jutte, D., Nuru-Jeter, A., Hinshaw, S. & Boyce, W.T. (2011). Family socioeconomic status and child executive functions: The roles of language, home environment, and single parenthood. *Journal of the International Neuropsychological Society*, *17*, 120–132. doi:10.1017/S1355617710001335

Towe-Goodman, N.R., Willoughby, M., Blair, C., Gustafsson, H.C., Mills-Koonce, W.R. & Cox, M.J. (2014). Fathers' sensitive parenting and the development of early executive functioning. *Journal of Family Psychology*, *28*(6), 867–876. doi:10.1037/a0038128

Zelazo, P.D. (2006). The dimensional change card sort (DCCS): A method of assessing executive function in children. *Nature Protocols*, *1*(1), 297–301. doi:10.1038/nprot.2006.46

Zelazo, P.D. & Müller, U. (2002). Executive function in typical and atypical development. In *Handbook of childhood cognitive development* (pp. 445–469).

Diversity in Unity: Perspectives from Psychology and Behavioral Sciences – Ariyanto et al. (Eds)
© 2018 Taylor & Francis Group, London, ISBN 978-1-138-62665-2

The effect of social distance between the benefactor and the beneficiary on the beneficiary's emotion of gratitude among female college students

L. Mardhiah & B. Takwin
Faculty of Psychology, Universitas Indonesia, Depok, Indonesia

ABSTRACT: This research examines the effect of social distance on the beneficiary's emotion of gratitude after they receive kindness. The current study uses experimental design and includes daily behaviour. Confederates were involved as a part of the stimulus to induce the treatments given (socially proximal and socially distal). A total of 51 female students from the Faculty of Psychology of Universitas Indonesia participated in this study. Participants were randomly assigned into two groups, which were socially proximal (n = 28) and socially distal (n = 23). The emotion of gratitude was measured using a Gratitude Adjective Checklist (GAC) score from each participant. The results suggest that if the benefactor and beneficiary are socially distal, it yields in stronger gratitude emotion than if they were socially proximal. Several points about this finding are discussed: (1) having higher external validity, and (2) emphasizing on assistance given by people who are known well and not known well.

1 INTRODUCTION

Over the last three decades, studies related to gratitude have been burgeoning. Emmons and McCullough (2003) found that gratitude is positively correlated with positive affect. Gratitude has also been found to be associated with individuals' intrinsic factors. McCullough et al. (2004) showed that higher mean levels of gratitude were related to positive affective traits, particularly agreeableness, and pro-social acts. The findings reveal that this personality type has a characteristic of maintaining social relationships with others. In other words, people with a higher tendency in these personality types express stronger gratitude frequently because they assume it would maintain their relationship with others.

While there are many studies of the relationship between gratitude, personality and prosocial behavior, studies that examine the relationship between gratitude and social distance have not been widely done. People who are grateful will tend to help others. On the other hand, people also tend to be grateful when helped by others. However, it is not known whether there is a difference in gratitude between when people are helped by others who are well known and when people are helped by other unknown people. To our knowledge, there is only one study which investigated the effect of relationship closeness on the feeling of gratitude. Bar-Tal et al. (1977) found that if the relationship between the two parties is close, such as between parents, siblings, and close friends, the emotion of gratitude is not as high as when individuals received kindness from others that have distant relationship with them, such as acquaintances or strangers. While the study of Bar-Tal et al. provides important empirical evidence on the influence of social distance on gratitude, the study employed hypothetical situation in which the participants were asked to imagine a certain situation during the experiment. Considering its hypothetical situation, we decided to replicate the previous study by manipulating real interaction between individuals and their friends or strangers.

The characteristics of the relationship between the giver and the recipient in their study in 1977 can be explained using the concept of social distance. Social distance can be defined as individuals' perceived closeness and similarity with others (Trope et al., 2007). Social distance at least has three forms. First, individual perceives others who are socially distal from himself as less-close, and who are socially proximal from himself as more-close (Trope et al., 2007). Second, one sees themself as subject or object indicated by how they refer to themselves, for example by saying 'I' as the first person or name as the third person (Trope & Liberman, 2010). Third, individual compares himself with their equal or superior, for instance a distance between students and their cohorts as socially proximal and the lecturer as socially distal (Stephan et al., 2010). Based on the concept of social distance, we conclude that the characteristics of the relationship between the giver and the receiver on Bar-Tal et al. (1977) are similar to the perception of the relationship to others as being near or distant.

Psychological distance is related to individual level of construal. Trope et al.'s (2007) aforementioned theory plays a role in explaining the effect of psychological distance on individual thoughts and behaviour. According to Trope and Liberman (2010), construal level theory explains that humans would form abstract mental construal through high level of construal for object that is perceived as far from themselves, such as when one is planning to spend weekend at home without any detail of activities added. Then, humans would form concrete mental construal through low level of construal for object that is perceived as near from themselves, for example when one is planning to spend weekend at home by reading few books (there is detail activity added).

Based on psychological distance and the construal level theory explanation, we infer that distant social relationship between the benefactor and the beneficiary makes the assistance more salient. Hence, it activates stronger emotion of gratitude on the beneficiary. Thus, people feel stronger emotion of gratitude when receiving help from those who are not well known to them than when they receive help from people who are. In this study, we propose an alternative hypothesis:

H_1: Individuals who receive an act of kindness from a stranger express stronger gratitude compared to those who receive an act of kindness from a close person.

2 METHOD

2.1 *Participants*

A total of 51 female students of the Faculty of Psychology of Universitas Indonesia participated in this study. We focused on female participants only because women are predicted to be more sensitive to others' kindness and show more expression about their feeling rather than men (Kashdan et al., 2009). Their ages ranged from 17 to 22 years old ($M = 20.24$; $SD = 1.35$). They were recruited via Google Forms and were told that the study was related to social perception. All participants were randomly divided into two groups of manipulation, namely socially proximal ($n = 28$) and socially distal ($n = 23$).

2.2 *Procedures*

We conducted a preliminary survey using Google Forms to determine the act of kindness that would be used in the experiment. 64 people responded to this preliminary survey, consisting of 21 male respondents and 43 female respondents. Their ages ranged from 17 to 23 years old ($M = 20.48$, $SD = 1.65$). Most respondents answered that acts of kindness during social situations were more likely to evoke the emotion of gratitude than any other situation, such as being offered a seat in the public transportation (6.25%), having someone helping them when they got lost (31.25%), and having someone giving some food (18.75%). Based on this information, we chose to focus on giving food as the act of kindness which generates the gratitude emotion in the participants. Moreover, this act of kindness was considered as the most feasible in terms of our experiment.

We then prepared the instruments used, Positive and Negative Affect Schedule (PANAS) adapted by Herwibowo (2014), Gratitude Adjective Checklist (GAC), the manipulation checks and items designed to check participants' awareness about the questions. The GAC and the manipulation checks were translated into *Bahasa* (Indonesian) by the researcher. For measuring the Perceived Awareness of the Research Hypothesis from the participants, we asked the participants about their opinion on what was being measured in this experiment.

Subsequently, a pilot study was conducted to examine the manipulation and the instruments used in this study. The pilot targeted the third or fourth year female students from the Faculty of Psychology of Universitas Indonesia. Several things were discovered from the pilot study. First, seven out of eight participants assumed receiving bread as an act of kindness from the benefactor. Second, one of eight participants responded that when completing PANAS and GAC, the affect was more influenced by the events beforehand rather than the act of kindness during the experiment. We revised the procedures based on these evaluations. Third, despite a different retort from one participant, we decided to retain PANAS and GAC. As an alternative to omitting the question about what participants thought about the hypothesis being measured or the Perceived Awareness of the Research Hypothesis query, we altered its position to be the last section of all instruments. We also added a questionnaire that required participants to make a list of objects around them and to write a brief description of each object. This questionnaire was to distract participants' attention so that it could lessen their suspicion about the role of confederates when completing the manipulation checks.

After these careful considerations, we conducted the experiment, which took three weeks and was organised in three stages: registration, treatment, and debriefing. During the registration period, they were informed that the experiment was related to social perception. At first, interested students were requested to register themselves via Google Forms and select a specific time to join the experiment.

On the experiment day, each participant was randomly paired with a confederate who was either a fellow student whom the participant knew (e.g., a classmate) or a student whom the participant had never met before. None of the confederates had family relationship with the participants in both groups. The participant and the confederate shared the same room, and was told that they were both participating in a study about social perception. Afterwards, they were given informed consent forms and the questionnaire about five objects around them and their descriptions. They were given ten minutes to complete these. At that time, confederates, who were previously briefed by the experimenter, gave a piece of bread to the participant. After finishing the questionnaire, they were asked to fill out the Gratitude Adjective Checklist (GAC), Positive And Negative Affect Schedule (PANAS), the manipulation checks, and the Perceived Awareness of the Research Hypothesis question. Finally, they were debriefed by the experimenter regarding the process of the experiment and the role of confederates.

2.3 *Measures*

In this study, all of the instruments had a six-point Likert response scale, where 1 = Strongly disagree, and 6 = Strongly agree.

2.4 *Emotion of gratitude*

We assessed the emotion of gratitude with a Gratitude Adjective Checklist adapted from McCullough et al. (2002), which asked participants to indicate to what extent they felt appreciative, thankful, or grateful. The index had a Cronbach's α of 0.65, which is considered acceptable (Walker & Almond, 2010).

2.5 *Positive and negative affect*

We measured the positive and negative affect by using the PANAS Scale (Watson et al., 1988), which was already translated into Indonesian language and validated by Herwibowo (2014). The instrument has 20 items consisting of ten items of positive affect and ten items of negative affect. The reliability testing shows Cronbach's α of 0.76 for the positive affect items and 0.83 for the negative affect items. According to Walker and Almond (2010), these reliability indexes fall into acceptable range.

2.6 *Manipulation check*

Referring to Liviatan et al. (2008), we adapted two items measuring social distance, namely closeness and similarity. The reliability index for the manipulation checks falls into an acceptable range, according to Walker and Almond (2010), as $\alpha = 0.63$.

3 RESULTS

To analyse the responses, we used IBM SPSS Version 20 and the chi-square method to examine the manipulation checks. An Analyses of the Covariates (ANCOVA) method with positive affect as the covariate was used to examine the main hypothesis because, as Emmons and McCullough (2003) stated, gratitude level has a correlation with positive affect.

3.1 *Manipulation checks*

For the preliminary analyses, from 56 participants, five participants were eliminated from the calculation because one of them did not complete the questionnaire and the other four responded correctly on the Perceived Awareness of the Research Hypothesis query. We calculated the manipulation checks by using chi-square method by converting the scale of 1–3 to 1 and the scale of 4–6 to 2.

As illustrated in Table 1, the calculation of the similarity indicator reveals that there is no difference in perceiving similarity between participants with confederates who were close person and who were strangers to them ($\gamma^2(1, n = 51) = 0.89$). It means both groups perceived confederates as similar to themselves. Next, we computed the main hypothesis by using ANCOVA and we put positive affect as the covariate.

In contrast to the calculation of the similarity, as can be seen from Table 2, there is a significant difference in perceiving closeness between participants with confederates who were close friends and confederates who were strangers to them ($\gamma^2(1, N = 51) = 0.00$). In other words, participants from a proximal social distance group perceived more closeness with the confederates than the distal group.

Hypothesis Testing. We tested the hypothesis by using ANCOVA and used positive affect as the covariate.

In support of our hypothesis, the results (see Table 3) showed that the social distance between the benefactor and the beneficiary significantly intensified the emotion of gratitude of the beneficiary $F(1, 49) = 4.76$; $p < 0.05$. This result confirms that if the benefactor and

Table 1. Calculation of the manipulation checks for similarity.

	Similarity			
Manipulation group	Less similar	Similar	Σ^2	p
Close friend ($n = 23$)	12	16	0.02	0.89
Stranger ($n = 28$)	13	10		

Table 2. Calculation of the manipulation checks for closeness.

	Closeness			
Manipulation group	Less close	Close	γ^2	p
Close person ($N = 23$)	8	20	8.65	0.00
Stranger ($N = 28$)	7	16		

Table 3. Calculation of the effect of social distance on the emotion of gratitude.

Manipulation group	*M*	*SD*	*F*	*p*
Socially proximal ($n = 28$)	14.46	1.95	4.76	0.03*
Socially distal ($n = 23$)	15.35	1.70		

*two-tailed.

the beneficiary are socially distal, it results in stronger emotion of gratitude than if both are socially proximal.

4 DISCUSSION

The results provide evidence of the effect of social distance on the emotion of gratitude. This is consistent with Bar-Tal et al. (1977) but differs in two respects. First, this study used a more realistic setting that has a higher external validity. Second, these studies used people that are well known to the participant and people that are not, as variation of the social distance between the benefactor and the beneficiary. This has different implications to the study by Bar-Tal et al. in 1977. The emphasis in this study is on the social distance between the donor and recipient to further ensure that the different strength emotion of gratitude felt by participants is influenced by social distance.

What we demonstrated in this study is in line with construal level theory and psychological distance explanation. According to Bar-Anan et al. (2006), people tend to use a high level of construal and abstract judgment if something is perceived as a distant object from them, which also applies to social distance. As noted by Liviatan et al. (2008), when someone perceives object as distant, the level of construal will be higher. As a result, individual judgment about others would be more abstract. The use of the abstract construal about others makes people use only a few mental resources and it allows greater attention to the assistance given by others. Greater attention to the assistance makes people more aware of it and in turn this activates stronger emotion of gratitude.

This study supports the hypothesis that if the benefactor and the beneficiary are socially distal, it makes stronger emotion of gratitude than if both are socially proximal. The results are consistent with the explanations that distant social relationship between the benefactor and the beneficiary makes assistance more salient and activates stronger emotion of gratitude on the beneficiary. Thus, people feel stronger emotions of gratitude when receiving help from those who are not well known to them than when they receive help from people who are well known to them. This result also implies that people consider the kindness of another person who is not well known to them to be more altruistic because they give more attention to the kindness than to the person providing kindness that is why it activates stronger emotion of gratitude or thankfulness.

In spite of the fact that we have found out how social distance affects the emotion of gratitude, several important questions have to be asked. For instance, we have asked the participants about kindness done by the confederates but we did not ask to what degree participants assumed the significance of the kindness. We also do not know how participants judge the kindness (i.e. when receiving kindness, it is because the benefactor is a kind person, as abstract judgment, or because the participants think that the benefactor knows they are hungry, as concrete judgment). All this needs to be clarified in further studies.

This study is also limited only from female college students of the Faculty of Psychology of Universitas Indonesia so that it needs to be checked in other samples involving men. We also need to study that if the focus shifts from the person who gave assistance to the condition that the person is not known well, the gratitude of the beneficiary would be lower or not. Furthermore, it should also be checked whether the divert attention to other things would reduce gratitude or not. In fact, there are many incidents of people who do not want to accept help from strangers or are suspicious when a stranger helps them. This is an important issue when considering in society people often must interact, even work together, with other people. In addition, it is necessary to help other people, not only with people who are close to them, but also to everyone. Further studies may provide insight on how to make the relationship of mutual assistance between all the people in a society.

REFERENCES

Bar-Anan, Y., Liberman, N. & Trope, Y. (2006). The association between psychological distance and construal level: Evidence from an implicit association test. *Journal of Experimental Psychology: General*, *135*(4), 609–622. doi:10.1037/0096-3445.135.4.609

Bar-Tal, D., Bar-Zohar, Y., Greenberg, M.S. & Hermon, M. (1977). Reciprocity behavior in the relationship between donor and recipient and between harm-doer and victim. *Sociometry*, *40*(3), 293–298. doi:10.2307/3033537

Emmons, R.A. & McCullough, M.E. (2003). Counting blessings versus burdens: An experimental investigation of gratitude and subjective well-being in daily life. *Journal of Personality and Social Psychology*, *84*(2), 377–389. doi:10.1037/0022-3514.84.2.377

Herwibowo, D. (2014). *The relationship between perceived freedom in leisure and subjective well-being in University of Indonesia students* (Unpublished bachelor thesis, Universitas Indonesia, Depok, Indonesia).

Kashdan, T.B., Mishra, A., Breen, W.E. & Froh, J.J. (2009). Gender differences in gratitude: Examining appraisals, narratives, the willingness to express emotions, and changes in psychological needs. *Journal of Personality*, *77*(3), 691–730. doi:10.1111/j.1467-6494.2009.00562.x

Liviatan, I., Trope, Y. & Liberman, N. (2008). Interpersonal similarity as a social distance dimension: Implications for perception of others' actions. *Journal of Experimental Social Psychology*, *44*(5), 1256–1269. doi:10.1016/j.jesp.2008.04.007

McCullough, M.E, Emmons, R.A. & Tsang, J-A. (2002). The grateful disposition: A conceptual and empirical topography. *Journal of Personality and Social Psychology*, *82*(1), 112–127. doi: 10.1037//0022-3514.82.1.112

McCullough, M.E., Tsang, J-A. & Emmons, R.A. (2004). Gratitude in intermediate affective terrain: Links of grateful moods to individual differences and daily emotional experience. *Journal of Personality and Social Psychology*, *86*(2), 295–309. doi:10.1037/0022-3514.86.2.295

Stephan, E., Liberman, N. & Trope, Y. (2010). Politeness and psychological distance: A construal level perspective. *Journal of Personality and Social Psychology*, *98*(2), 268–280. doi:10.1037/a0016960

Trope, Y. & Liberman, N. (2010). Construal-level theory of psychological distance. *Psychological Review*, *117*(2), 440–463. doi:10.1037/a0018963

Trope, Y., Liberman, N. & Wakslak, C. (2007). Construal levels and psychological distance: Effects on representation, prediction, evaluation, and behavior. *Journal of Consumer Psychology*, *17*(2), 83–95. doi:10.1016/S1057–7408(07)70013-X

Walker, J. & Almond, P. (2010). Interpreting statistical findings: A guide for health professionals and students. Berkshire, UK: McGraw-Hill.

Watson, D., Clark, L.A. & Tellegen, A. (1988). Development and validation of brief measures of positive and negative affect. *The PANAS Scales*, *54*, 1063–1070. doi:10.1037/0022-3514.54.6.1063

Diversity in Unity: Perspectives from Psychology and Behavioral Sciences – Ariyanto et al. (Eds)

The role of bystanders' psychological well-being and gender as moderators of helping behaviour in bullying incidences

R. Djuwita & F.M. Mangunsong
Faculty of Psychology, Universitas Indonesia, Depok, Indonesia

ABSTRACT: It is still debatable whether bystanders' psychological well-being plays a role on helping behaviour in bullying incidences. 4.807 students completed the Ryff psychological well-being scale and bullying vignettes. High scores in psychological well-being were found to be related with helping behaviour. In contrast, participants with low psychological well-being tended to support the perpetrator. The findings of Structural Equation Modelling (SEM) demonstrated a moderate effect of gender on psychological well-being, which was evident only in helping the victim. The moderation analysis also suggests that when the bystander was a male student, he tended to help a victim if he had a high score in his psychological well-being. It is concluded that the psychological well-being state of the bystander would determine whether a student bystander would help a victim or support the perpetrator. This study suggests the importance of maintaining the positive psychological well-being of students to reduce bullying in schools.

1 INTRODUCTION

Bullying is widely regarded as a serious problem in schools in many countries (Craig et al., 2009; Smith et al., 2002); Indonesia is no exception. Since 2011 to 2014, bullying still ranks the highest compared to other school problems (Firmansyah, 2014). Bullying is a special form of aggression which often happens in schools and has a negative impact on the victims and also the perpetrators (Olweus & Breivik, 2014). Decreasing and preventing bullying in schools are urgent goals for schools because their immense negative impacts affect victims as well perpetrators.

Students who are victims of bullying often have academic, personal and social difficulties (Hernández & Seem, 2004; Juvonen et al., 2010) and these negative impacts continue into adulthood (Adams & Lawrence, 2011; Ttofi et al., 2012). The perpetrators are at high risk of maladjustment (Lodge & Frydenberg, 2005). Research shows that bullies also have academic, personal, and social difficulties in school (Hernández & Seem, 2004; Juvonen et al., 2010); they have an increasing risk of poor health, wealth, and difficulties in building social relationship in adulthood (Wolke et al., 2013).

Prosocial and helping behaviour plays an important role in reducing aggression in schools. Prior studies showed that prosocial attitudes and behaviour have a positive impact on reducing aggression in schools (Biglan & Hinds, 2009; Caprara et al., 2000). Some scholars even believe that the key to stop bullying is to enhance bystanders' willingness to intervene and help the victims (Bennett et al., 2014; Padgett & Notar, 2013). Helping a victim of bullying is not only considered as an instrumental tool to decreasing bullying, but also as a remedy for the victims.

Bullying usually happens in front of other students and in general they can be categorised as bullying reinforces or bystanders (Cowie, 2014; Salmivalli et al., 2011). Bystanders are students who are not actively involved in the bullying incidence, but know directly or indirectly about it (Salmivalli, 2014; Stueve et al., 2006). Students as bystanders could directly help the victim by interfering in the incidence, or indirectly by being an empathetic friend or reporting the incidence to a teacher on duty (Pozzoli & Gini, 2012).

In general, scholars and practitioners agree that empowering passive bystanders is more effective than changing the behaviour of a perpetrator not to be aggressive or to change a victim to be more assertive towards the bully (Twemlow et al., 2004). Bystanders are actually potential defenders (Mishna, 2008; Padgett & Notar, 2013; Salmivalli, 2014) because most students who act as bystanders disapprove of bullying and they feel sorry for the victims (Gini et al., 2008; Rigby & Johnson, 2006). For this reason, they already have a positive attitude towards the victim and are easier to be converted from being a passive bystander to an active bystander who is willing to help and support the victims (Burton et al., 2013). Research has shown that if bystanders are willing to intervene or report a bullying incidence (Gini et al., 2008; Hoover & Anderson, 1999), they are indeed preventing bullying to happen in the future, because they are not rewarding or supporting the perpetrators (Coloroso, 2005; Salmivalli, 2014). Further, if bystanders interfere or stand up to the bullying perpetrator, the probability that the bullying incidence will stop is about 50% (Hawkins et al., 2001). These research findings has shown that bystanders' responses toward bullying have an important role in preventing bullying.

Bystanders' response to a bullying situation can be categorised either as defending the victim or supporting the bully (Thornberg, 2007). Defending a bullied victim can be done by helping the victim directly, for example by talking to the bully in a way to prevent the incidence, or by openly helping the victim in front of the perpetrator and other students who are also observing the incidence. Bullying bystanders can also help indirectly, for example by offering emotional support after the incidence (Cowie, 2014) or by reporting it to the school authorities (Pöyhönen et al., 2012). The bystanders who choose to be passive and do nothing to stop the bully or help the victim, can be categorised as supporting the perpetrator. By being passive, they give the social reward for the perpetrator to feel, that they get the attention and having the power over other victims (Coloroso, 2005; Cowie, 2014). Bystanders' support does not always have to be by actively joining the ringleader bullies or by cheering them, but also by being a passive spectator (Salmivalli et al., 2011; Twemlow et al., 2004).

Helping behaviour, as a form of prosocial behaviour, is sometimes determined by emotion and how well a person perceives his or her internal emotional state (Baron & Branscombe, 2013; DeWall et al., 2008). It is argued that the degree of well-being or psychological well-being can influence individuals' willingness to help (Baumeister et al., 2009), however, research regarding the impact of bystanders' psychological well-being towards helping behaviour, especially in bullying situation are very scarce.

In prior studies, of bullying and psychological well-being, psychological well-being were commonly used as an outcome variable caused by bullying or as an outcome if the victims receive some help either from their peers, teachers, or from formal intervention from the school like special consultation. For example, several studies showed how severe bullying affected the victims' psychological well-being state, such as depression, low self-esteem, and feeling alienated from their friends (Sapouna & Wolke, 2013; Schäfer et al., 2004). It is also argued that if victims perceived that they were supported and helped by peers, their psychological well-being improved (Kendrick et al., 2012). The feeling of being supported by peers would lessen the psychological wounds caused by victimisation and improve the victims' psychological well-being state (Rigby, 2000). This study aims to understand the role of the bystanders' psychological well-being towards their helping responses in bullying incidences.

According to Huppert (2009), 'Psychological well-being is about lives going well. It is the combination of feeling good and functioning effectively.' (p.137). Individuals with better psychological well-being would feel good about themselves and function positively coping their own demands and their social surrounding. Ryff (1995; 2014) suggests six key dimensions of psychological well-being, namely self-acceptance, positive relationships with other people, autonomy, environmental mastery, purpose in life, and personal growth.

In general, there are two different opinions about the impact of well-being towards helping behaviour. On one hand, some scholars have the opinion that individuals with a better psychological well-being—who feels that they are living a good life—would be more willing to help a person in need. Prior studies have found that individuals who were dissatisfied with their situation would decrease their helping behaviour (Balliet & Joireman, 2010; DeWall et al., 2008). In other words, individuals who are more satisfied with their life situations, would be more willing in helping others. On the other hand, other scholars are convinced that person who are not contented with their life situations, would actually be more willing to help other people. There are some studies which support their opinion. They suggest that there is a negative relationship between negative mood or lower state of well-being and willingness to help others (Cialdini & Kenrick, 1976; McGinley et al., 2009; Piff et al., 2010). Therefore, the purpose of our study is to explore whether bystanders' psychological well-being can predict their helping behaviour when the participants are put in the role of bystanders in a bullying incidence.

In this study, we will explore the role of psychological well-being as an endogenous variable (as a determinant factor) using a structural equation model (Muthén & Muthén, 2012). This study considers psychological well-being as a predictor of bystanders helping behaviour, whether they are more willing to help a victim or to support the perpetrator. The following hypothesis was formulated:

H1: Bystanders with higher psychological well-being would be more willing to help a victim. On the other hand, bystanders with lower psychological well-being would be more willing to support the perpetrator.

In former studies, female students were found to be more willing to help a victim than male students (Salmivalli & Voeten, 2004) because they perceive victims more positively, and they are more willing to support them (Gini, 2006). Accordingly, the following hypothesis was suggested:

H2: Gender would act as a moderator in the relationship between psychological well-being and the bystanders' helping behaviour towards the victim: When a bystander is a female student, the better her psychological well-being state is, the more she would tend to help a victim.

2 METHOD

This study used data collected from a larger research project which involved 4,807 high school students, from eight cities in Indonesia. Their age was between 13 and 21 (M = 16.17, SD = 0.971). 2,091students (43.5%) were male and 2,716 were female (56.5%). The samples represented every grade of high school, 33% of them were in the tenth grade, 38.5% in the eleventh grade and 28% in grade twelve. They came from public high schools (38%), private high schools (31.5%), and vocational high schools (30.5%). Each participants are given questionnaires containing an informed consent statement, a personal data form, and questions concerning their psychological well-being state and their willingness to help a victim or to support the perpetrator if they are witnesing a bullying.

The Ryff psychological well-being scale was used in this study (Ryff, 1996). The authors had permission to use the Indonesian adapted and validated Ryff psychological well-being scale version from another study (Jaya et al., 2011). The scale reliability (Cronbach Alpha) was above 0.7, and its internal validity was above 0.3. To measure the helping behaviour, participants were given vignettes with pictures illustrating some bullying and non-bullying situations in their schools. The participants were given the role of bystander and were asked what they would have done if they had observed the illustrated in the vignettes (to help the victim or to support the perpetrator). The authors also ensured the factor loadings of each psychological well-being and helping behaviour dimension to prevent overlapping loadings by calculating the Explanatory Factor Analysis (EFA) and Confirmatory Factor Analysis (CFA) (Muthén & Muthén, 2012). EFA and CFA criteria were used to select variable indicators or items, and only satisfactory items (above 0.9) were included in later analysis. The CFA results showed that items measuring the purpose in life, and the personal growth dimensions had the same factor loading, so it was decided to merge both of these dimensions into a new dimension, named 'meaning in life'.

3 RESULTS

First, we analysed the data using MPlus programme to calculate Structural Equation Modelling (SEM): a multivariate statistical analysis technique that is used to analyse relationships between measured variables and latent constructs. The calculation of SEM will show the impact of several psychological well-being dimensions on the bystanders' helping behaviour. The results of the SEM are shown as a model in Figure 1, which includes all the standardised coefficients that are significant at and below the 0.05 level. The estimated model yields a very strong fit to the data, as indicated by the following fit indices: CFI/TLI 0.973/.964, RMSEA 0.031 (90% confidence interval: 0.029–0.033). The whole model accounted for 9% of the variance for defending behaviour and 14% of the variance for supporting the bully.

As shown in Figure 1, the 'environmental mastery' and 'meaning in life' dimensions are positively correlated with helping the victim ($r = 0.15$) This means that the more students perceive they are able to manage the various demands of their environments and the more they perceive to have a meaningful life, the more they are willing to help a victim. On the other hand, the less they perceive they are able to manage their environment demands, the more they are willing to support the bully ($r = -0.07$). The other psychological dimensions are negatively correlated to supporting the bully, except the self-acceptance dimension. This means that the less the students are dissatisfied with their social relations in school ($r = -0.20$), meaning in life ($r = -0.20$), autonomy ($r = -0.09$), the more they are willing to support the bully. The self-acceptance dimension is positively associated with supporting the bully ($r = 0.12$).

Figure 2 shows the SEM analysis results if gender were included. The SEM model indicates that gender moderates the correlation between psychological well-being with helping (the victim) behaviour. This means psychological well-being of boys or girls, shows different impact in the case of helping a bullying victim.

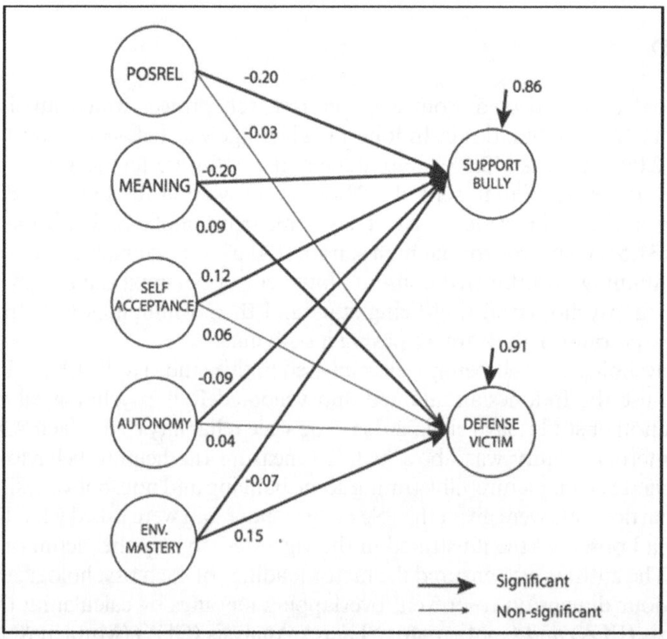

Figure 1. SEM structural model: Role of psychological well-being dimensions on bystanders' helping behaviour.

Note: The numbers refer to standardised structural coefficients; coefficients are significantly above the $p > 0.05$ level.

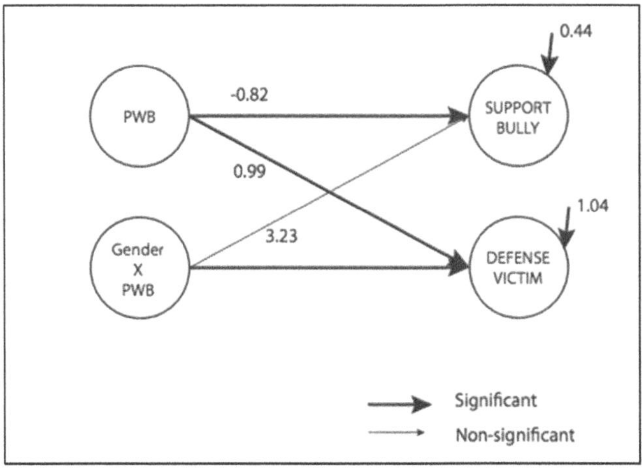

Figure 2. SEM structural model: Role of psychological well-being on bystanders' helping behaviour, moderated by gender.
Note: The numbers refer to unstandardised structural coefficients; coefficients are significantly above the $p < 0.05$ level.

The results indicate that male students with better psychological well-being conditions would have a stronger willingness to help a bullying victim. However, there is no interaction (moderation effect) between female students' psychological well-being and helping a bullying victim.

4 DISCUSSION

As hypothesised, bystanders' psychological well-being determines the students'helping behaviour in a bullying incidence, but only two dimensions of the psychological well-being have a positive impact on helping the victim, namely the 'environmental mastery' and 'meaning in life' dimensions. This means that students who feel good about themselves and perceive themselves as functioning well in their school community because they are able to manage their environment and have a meaningful life, are more likely to help a victim. The results also suggest that most dimensions (positive relation, meaning in life, autonomy and environmental mastery) show a negative association with supporting the perpetrator. Therefore, students who are not satisfied with themselves and feel that they are not able to function well in their school community prefer to support the perpetrator by reinforcing the perpetrator or just being passive onlookers. The results showed that students who perceived that they were socially supported social environment, did not know their life purposes, did not perceive to have the autonomy to make their own decisions and perceive not able to manage their environmental challenges, would tend to support the prepetrator.

The results of this study support previous research which suggests that a better psychological well-being condition would improve the willingness to help among students and a worse psychological well-being state would motivate bystanders to support the perpetrator by reinforcing him/her or just being passive and not helping the victim (Balliet & Joireman, 2010; DeWall et al., 2008). Furthermore, the results also show the importance of a bystanders' role and the degree of their psychological well-being in helping a victim in school (Boyd & Barwick, 2011; Cowie, 2014).

The second hypothesis is only partly proven. Our data indicates that the bystanders' gender does have a moderating effect on the helping behaviour of bystanders in bullying incidences, but surprisingly this only applies on male students. In other words, male students with better psychological well-being will be more willing to help bullying victims. This result is unforeseen because, in previous research done in western countries, female students have been more willing to help bullying victims. This difference of results might occur because this study used adolescents as participants, while others studies exploring the role of gender used younger participants like elementary students (e.g. Salmivalli & Voeten, 2004) or pre-adolescent students (Gini, 2006). From other studies, we know that younger children, especially girls, are more willing to help older children (Cialdini et al., 1976). Former studies also show that with the increasing of age, students-male and female-are becoming more tolerant towards peer aggression such as bullying, and this phenomenon is influenced by their peer norm (Salmivalli et al.,1998). Another possibility to explain this result is that for high school female students, hindering a bullying incidence requires assertiveness and high degree of courage. Being assertive and brave to stand up against the perpetrator might counteract with the norm of a female, especially in Indonesia. It is also long known that females feel more helpless than male students (Eagly & Crowley, 1986), so, in a high school context, it might be expected that the male students should be more responsible as a defender because they are expected to have the courage to help and intervene. If male students are brave enough to stand up to the bully, this will, moreover, enhance their self-esteem and social status among their peers.

5 CONCLUSION

It is concluded that students with a better state of psychological well-being, will be more willing to help a bullying victim. On the other hand, the less satisfied bystanders are with their psychological well-being, the more they are willing to support the perpetrator. Our study shows that only two of psychological well-being dimensions are correlated positively in helping the victim, namely the environmental dimensions and meaning in life. This study also suggests that student bystanders who are not satisfied with their social relation, meaning in life, autonomy, and environmental mastery tend to support the perpetrator. Results also suggest that gender acts as a moderator in the relations between psychological well-being and helping the victim, but only for the male students. Therefore, psychological well-being is more likely to have a stronger positive impact on the helping behaviour if the bystander is a male student.

REFERENCES

Adams, F.D. & Lawrence, G.J. (2011). Bullying victims: The effects last into college. *American Secondary Education, 40*(1), 4–14.

Balliet, D. & Joireman, J. (2010). Ego depletion reduces proselfs' concern with the well-being of others. *Group Processes & Intergroup Relations, 13*(2), 227–239.

Baron, R.A., & Branscombe, N.R. (2013). *Social Psychology: Pearson New International Edition.* Pearson Higher Ed.

Baumeister, R.F., Masicampo, E.J., DeWall, C.N. & Baumeister, R. (2009). Prosocial benefits of feeling free: Disbelief in free will increases aggression and reduces helpfulness. *PSPB, 35*(2), 260–268. doi:10.1177/0146167208327217

Bennett, S., Banyard, V.L. & Garnhart, L. (2014). To act or not to act, that is the question? Barriers and facilitators of bystander intervention. *Journal of Interpersonal Violence, 29*(3), 476–496. doi:10.1177/0886260513505210

Biglan, A. & Hinds, E. (2009). Evolving prosocial and sustainable neighborhoods and communities. *Annual Review of Clinical Psychology, 5*, 169–196. doi:10.1146/annurev.clinpsy.032408.153526

Burton, K.A., Florell, D. & Wygant, D.B. (2013). The role of peer attachment and normative beliefs about aggression on traditional bullying and cyberbullying. *Psychology in the Schools, 50*(2), 103–115. doi:10.1002/pits.21663

Caprara, G.V., Barbaranelli, C., Pastorelli, C., Bandura, A. & Zimbardo, P.G. (2000). Prosocial foundations of children's academic achievement. *Psychological Science, 11*(4), 302–306. doi:10.1111/1467-9280.00260

Cialdini, R.B. & Kenrick, D.T. (1976). Altruism as hedonism: A social development perspective on the relationship of negative mood state and helping. *Journal of Personality and Social Psychology*, *34*(5), 907–914. doi:10.1037/0022-3514.34.5.907

Coloroso, B. (2005). A bully's bystanders are never innocent. *Education Digest: Essential Readings Condensed for Quick Review*, *70*(8), 49–51.

Cowie, H. (2014). Understanding the role of Bystanders and peer support in school bullying. *International Journal of Emotional Education*, *6*(1), 26–32.

Craig, W., Harel-Fisch, Y., Fogel-Grinvald, H., Dostaler, S., Hetland, J., Simons-Morton, B., ... Pickett, W. (2009). A cross-national profile of bullying and victimization among adolescents in 40 countries. *International Journal of Public Health, 54*(Suppl. 2), 216–224. doi:10.1007/s00038-009-5413-9

DeWall, C.N., Baumeister, R.F., Gailliot, M.T. & Maner, J.K. (2008). Depletion makes the heart grow less helpful: Helping as a function of self-regulatory energy and genetic relatedness. *Personality and Social Psychology Bulletin, 34*(12), 1653–1662. doi:10.1177/0146167208323981

Eagly, A.H. & Crowley, M. (1986). Gender and helping behavior: A meta-analytic review of the social psychological literature. *Psychological Bulletin, 100*(3), 283–308. doi:10.1037/0033-2909.100.3.283

Firmansyah, T. (2014). Aduan Bullying Tertinggi (Highest Complain of Bullying). *Republika* (Republika Newspaper). Retrieved from http://www.republika.co.id/berita/koran/halaman-1/14/10/15/ndh4sp-aduan-bullying-tertinggi.

Gini, G. (2006). Bullying as a social process: The role of group membership in students' perception of inter-group aggression at school. *Journal of School Psychology, 44*(1), 51–65. doi:10.1016/j.jsp.2005.12.002

Gini, G., Albiero, P., Benelli, B. & Altoè, G. (2008). Determinants of adolescents' active defending and passive bystanding behavior in bullying. *Journal of Adolescence, 31*(1), 93–105. doi:10.1016/j.adolescence.2007.05.002

Gini, G., Pozzoli, T., Borghi, F. & Franzoni, L. (2008). The role of bystanders in students' perception of bullying and sense of safety. *Journal of School Psychology, 46*(6), 617–638. doi:10.1016/j.jsp.2008.02.001

Hawkins, D.L. Pepler, D.J. & Craig, W.M. (2001). Naturalistic observations of peer interventions in bullying. *Social Development, 10*(4), 513–527. doi:10.1111/1467-9507.00178

Hernández, T.J. & Seem, S.R. (2004). A safe school climate: A systemic approach and the school counselor. *Professional School Counseling, 7*, 256–262.

Hoover, J.H. & Anderson, J.W. (1999). Altruism as an antidote to bullying: Reclaiming children and youth. *Journal of Emotional and Behavioral Problems, 8*(2), 88–91.

Huppert, F.A. (2009). Psychological well-being: Evidence regarding its causes and consequences. *Applied Psychology: Health and Well-Being, 1*(2), 137–164. doi:10.1111/j.1758-0854.2009.01008.x

Jaya, E.S., Hanum, L. & Lubis, D.U. (2011). Indigenous psychological well-being for the elderly measurement. *Proceedings of the Second International Conference of Indigenous and Cultural Psychology*, Denpasar, 21–23 December 2011 (pp. 187–200).

Juvonen, J., Wang, Y. & Espinoza, G. (2011). Bullying experiences and compromised academic performance across middle school grades. *The Journal of Early Adolescence*, *31*, 152–173. doi:10.1177/0272431610379415

Kendrick, K., Jutengren, G. & Stattin, H. (2012). The protective role of supportive friends against bullying perpetration and victimization. *Journal of Adolescence*, *35*(4), 1069–1080. doi:10.1016/j.adolescence.2012.02.014

Lodge, J. & Frydenberg, E. (2005). The role of peer bystanders in school bullying: Positive steps toward promoting peaceful schools. *Theory into Practice, 44*(4), 329–336. doi:10.1207/s15430421tip4404_6

McGinley, M., Carlo, G., Crockett, L.J., Raffaelli, M., Torres Stone, R.A., & Iturbide, M.I. (2010). Stressed and helping: The relations among acculturative stress, gender, and prosocial tendencies in Mexican Americans. *The Journal of social psychology*, *150*(1), 34–56. doi:10.1080/00224540903365323

Mishna, F. (2008). An overview of the evidence on bullying prevention and intervention programs. *Journal of Brief Treatment and Crisis Intervention* 8 (4), 327.

Muthén, L.K. & Muthén, B.O. (2012). Mplus: The comprehensive modelling program for applied researchers: User's Guide, 5. Los Angeles, CA: Muthen & Muthen.

Olweus, D. & Breivik, K. (2014). Plight of victims of school bullying: The opposite of well-being. In Ben-Arieh, A., Casas, F., Frones, I. & Korbin, J. (Eds.), *Handbook of child well-being* (pp. 2593–2616). Norway: Springer.

Padgett, M. & Notar, C.E. (2013). Bystanders are the key to stopping bullying. *Universal Journal of Educational Research, 1*(2), 33–41. doi:10.13189/ujer.2013.010201

Piff, P.K., Kraus, M.W., Côté, S., Cheng, B.H. & Keltner, D. (2010). Having less, giving more: The influence of social class on prosocial behavior. *Journal of Personality and Social Psychology, 99*(5), 771–784. doi:10.1037/a0020092

Pöyhönen, V., Juvonen, J. & Salmivalli, C. (2012). Standing up for the victim, siding with the bully or standing by? Bystander responses in bullying situations. *Social Development, 21*(4), 722–741. doi:10.1111/j.1467-9507.2012.00662.x

Pozzoli, T. & Gini, G. (2012). Why do bystanders of bullying help or not? A multidimensional model. *The Journal of Early Adolescence*, *33*(3), 315–340. doi:10.1177/0272431612440172

Rigby, K. (2000). Effects of peer victimization in schools and perceived social support on adolescent well-being. *Journal of Adolescence*, *23*(1), 57–68. doi:10.1006/jado.1999.0289

Rigby, K. & Johnson, B. (2006). Expressed readiness of Australian schoolchildren to act as bystanders in support of children who are being bullied. *Educational Psychology*, *26*(3), 425–440. doi:10.1080/01443410500342047

Ryff, C.D. (1995). Psychological well-being in adult life. *Current Directions in Psychological Science*, *4*(4), 99–104. doi:10.1111/1467-8721.ep10772395

Ryff, C.D. (2013). Psychological well-being revisited: Advances in the science and practice of eudaimonia. *Psychotherapy and Psychosomatics*, *83*(1), 10–28. doi:10.1159/000353263

Ryff, C.D. & Keyes, C.L. (1995). The structure of psychological well-being revisited. *Journal of Personality and Social Psychology*, *69*(4), 719–727. doi:10.1037/0022–3514.69.4.719

Ryff, C.D. & Singer, B. (1996). Psychological well-being: Meaning, measurement, and implications for psychotherapy research. *Psychotherapy and Psychosomatics*, *65*(1), 14–23. doi:10.1159/000289026

Salmivalli, C. (2014). Participant roles in bullying: How can peer bystanders be utilized in interventions? *Theory into Practice*, *53*(4), 286–292. doi:10.1080/00405841.2014.947222

Salmivalli, C. & Voeten, M. (2004). Connections between attitudes, group norms, and behaviour in bullying situations. *International Journal of Behavioral Development*, *28*(3), 246–258. doi:10.1080/01650250344000488

Salmivalli, C., Lagerspetz, K., Björkqvist, K., Österman, K. & Kaukiainen, A. (1996). Bullying as a group process: Participant roles and their relations to social status within the group. *Aggressive Behavior*, *22*(1), 1–15. doi:10.1002/(SICI)1098–2337(1996)22:1 < 1::AID-AB1 > 3.0.CO;2-T

Salmivalli, C., Voeten, M. & Poskiparta, E. (2011). Bystanders matter: Associations between reinforcing, defending, and the frequency of bullying behavior in classrooms. *Journal of Clinical Child & Adolescent Psychology*, *40* (5), 668–676. doi:10.1080/15374416.2011.597090

Sapouna, M. & Wolke, D. (2013). Resilience to bullying victimization: The role of individual, family and peer characteristics. *Child Abuse and Neglect*, *37*(11), 997–1006. doi:10.1016/j.chiabu.2013.05.009

Schäfer, Mechthild, Stefan Korn, Peter K. Smith, Simon C. Hunter, Joaqún A. Mora-Merchán, Monika M. Singer, and Kevin van der Meulen. (2004). Lonely in the crowd: Recollections of bullying. *British Journal of Developmental Psychology*, *22*, 379–394. doi:10.1348/0261510041552756

Smith, P.K., Cowie, H., Olafsson, R.F.R. & Liefooghe, A.P.D. (2002). Definitions of bullying: A comparison of terms used, and age and gender differences in a fourteen–country international comparison. *Child Development*, *73*(4), 1119–1133. doi:10.1111/1467-8624.00461

Stueve, A., Dash, K., O'Donnell, L., Tehranifar, Wilson-Simmons, P.R., Slaby, R.G. & Link, B.G. (2006). Rethinking the bystander role in school violence prevention. *Health Promotion Practice*, *7*(1), 117–124. doi:10.1177/1524839905278454

Thornberg, R. (2007). A classmate in distress: Schoolchildren as bystanders and their reasons for how they act. *Social Psychology of Education*, *10*(1), 5–28. doi:10.1007/s11218-006-9009-4

Ttofi, M.M.A, Farrington, D.P.A. & Lösel, F.A.B. (2012). School bullying as a predictor of violence later in life: A systematic review and meta-analysis of prospective longitudinal studies. *Aggression and Violent Behavior*, *17*(5), 405–418. doi:10.1016/j.avb.2012.05.002

Twemlow, S.W., Fonagy, P. & Sacco, F.C. (2004). The role of the bystander in the social architecture of bullying and violence in schools and communities. *Annals of the New York Academy of Sciences*, *1036*, 215–232. doi:10.1196/annals.1330.014

Wolke, D., Copeland, W.E., Angold, A. & Costello, E.J. (2013). Impact of bullying in childhood on adult health, wealth, crime, and social outcomes. *Psychological Science*, *24*, 1958–1970. doi:10.1177/0956797613481608

Diversity in Unity: Perspectives from Psychology and Behavioral Sciences – Ariyanto et al. (Eds)
© 2018 Taylor & Francis Group, London, ISBN 978-1-138-62665-2

The effect of job satisfaction in employee's readiness for change

M.V. Azra, A. Etikariena & F.F. Haryoko
Faculty of Psychology, Universitas Indonesia, Depok, Indonesia

ABSTRACT: External and internal demands cause many organisations to change and adjust themselves. However, not all organisational change will implement effectively. One of the reasons why the implementation of change often fail, is the lack of employee's readiness for change. This research focuses on identifying the effect of job satisfaction in an employee's readiness for change. Employee's readiness for change was measured using the Readiness For Change Questionnaire which was constructed by Holt et al. (2007), while job satisfaction was measured using the Job Satisfaction Survey (JSS) by Spector (1997). Data was gathered from 36 employees who worked in dairy product company in Indonesia, working at managerial levels, given that managers are the change agents implementingchange. So, the managers are the significant part to make the organisation's change success effectively. The results show that job satisfaction had a significant influence on readiness for change ($R^2 = 0437$, $p < 0.05$). Further, not all aspects of job satisfaction had a significant influence on readiness for change. The only one which contributed to readiness for change was communication ($B = 3.94$, $t = 4.77$, $p = 0.00$).

1 INTRODUCTION

In era of globalisation, organisational change is unavoidable for various organisations. Many demands from external factors such as technological development, globalisation, changes in government policy and regulation cause the need for organisational change (Gordon et al., 2000; De Meuse et al., 2010). In addition to external factors, organisational change can also occur due to internal demands such as decline in profits, decline in employee competencies, the growth of the company itself, the opportunity to develop the company's business, and a new innovation or policy undertaken by the company (Madsen et al., 2005). Many demands from external and internal factors lead to moderate to major changes in the organisation to cope with the demands, giving organisations the ability to survive and achieve efficiency and effectiveness (Cumming & Worley, 2009).

Employees in organisations respond differently to organisation change. In general, employees' attitudes and reactions towards change can be divided into effective attitude (receiving) and ineffective attitude (rejecting). Effective attitude is shown by cooperating, supporting the circumstances, conditions, and process changes, and engaging in organisational change. On the other hand, an ineffective attitude is indicated by defending themselves in the organisation, always complaints, and openly rejecting the change (Galpin, 1996, in Rafferty et al., 2013). Based on a survey of more than 3,000 executives at some companies, Meaney & Pung (2008) revealed that two-thirds of respondents said that their company failed to show a good performance after the implementation of changes (Rafferty et al., 2013).

To be able to support the change process, people should be ready for change (Armenakis et al., 1993; Armenakis, Haris & Field, 1999). Readiness is defined as beliefs, intentions, attitudes, and behaviour that support the change and organisational capacity to achieve success. Holt et al. (2007) suggested a more comprehensive concept of readiness for change, which is defined as a comprehensive attitude that is influenced simultaneously by the content (i.e., what is being changed), the process (i.e. how the change is being implemented), the context (i.e., circumstances under which the change is occurring), and the individuals (i.e., characteristics of those being asked to change) involved. Furthermore, readiness collectively reflects the extent to which an individual or individuals are cognitively and emotionally inclined to accept, embrace, and adopt a particular plan to purposefully alter the status quo. Readiness for change is a multidimensional psychological construct that consists of five dimensions (Holt et al., 2007): 1) Organisational Valence, which describes how the company will benefit from the implemented changes; 2) Management Support, which describes employee perception about support from a variety of sources and the commitment of the parties associated with the process of change institutionalisation; 3) Change Self-Efficacy, which is an individual's belief in their ability to support organisational change 4) Discrepancy, that is when there is gap between the change situation and condition in which the change must done by organisation; and 5) Personal Valence, that is accepted by employee when organisation choose to implement the change.

Establishing employee's readiness for change, can help prevent any potential resistance from employees when change is implemented later. Therefore it is important for companies to assess individual and organisational readiness for change and also to understand the factors that influence individual and organisational readiness for change (Madsen, 2005). One factor that can influence readiness for change is job satisfaction. Locke (1976) revealed that job satisfaction is an emotional state that is pleasant or positive feeling that employee gained from their work experience, where it also includes cognitive, affective, and evaluative reactions or attitudes towards work. Approach to job satisfaction can be viewed as globally approach and as facets approach (Spector, 2000). A global approach to job satisfaction focuses on the overall feeling of employees towards work. Meanwhile, the facet approach focuses on many aspects of work, such as reward (salary or allowances and others), other individuals at work (supervisor or co-worker), working conditions, and the nature of the work itself. in this study, we used The facet approach because is useful for organisations that want to identify which aspects in organisation so that they can intervene. The facet approach gives comprehensive description to describe job satisfaction, than when we used with a global approach. This is because employees can have different feelings about different facets, which may not be identified through a global approach. Thus, in this study, we used the facet approach to job satisfaction because it could help us determine which facets of job satisfaction affect the readiness for change. Consequently, we used the concept of job satisfaction by Spector, who viewed job satisfaction as facets. Spector (1997) defined job satisfaction as perceptions, attitudes, and feelings of individuals to work, either on the whole or in terms of certain aspects of the work, which produce an unpleasant emotional state for the individual. Spector (1997) categorised job satisfaction into nine aspects, which include salary, promotion, supervision, fringe benefits, contingent rewards, operating condition, co-workers, nature of work, and communication.

Several studies have reported a positive relationship between job satisfaction and attitudes towards change (Cordery et al., 1993; Iverson, 1996). According to McNabb & Sepic (1995), employees who are comfortable with the work or have a high degree of satisfaction will have a positive attitude towards change. A positive attitude could improve employee readiness for change. Further, the research of Wanberg and Banas (2000) showed that employees who are satisfied with their work are more willing to accept change (Cordery et al., 1993; Iverson, 1996). Meanwhile, Claiborne et al. (2013). The study also showed the relationship between the readiness for change with various aspects of job satisfaction. The study revealed that not all aspects of job satisfaction contribute to the readiness for change. The results showed that only the communication aspect has an effect on the employee's readiness for change (Claiborne et al., 2013). Therefore, the communication in the company becomes an important thing to note during the implementation process of change. Further, van Dam (2005) and Oreg et al. (2011) showed a different result, namely that some employees who are not happy with their work could see organisational change as an opportunity for improvement, while other employees who are very happy with their work may resist change because they want to maintain their current status. These results are relatively different from previous studies which showed that satisfied employees are better prepared for change because they could see the positive consequences of the change (van Dam, 2005; Oreg et al., 2011).

A number of studies have been conducted to investigate whether job satisfaction influences readiness for change. However, there is still ambiguity regarding the relationship between job satisfaction and readiness for change, and many researchers still call for further research to improve our understanding and exploration of the factors in different organisational settings, with different participants and also in different cultures (Lizar et al., 2012). Accordingly, this study was conducted to answer the call to examine the effect of job satisfaction on employee readiness for change in an Indonesian dairy product company. This study was specific to the managerial level, given that managers are the change agents when companies implement change (Vakola, 2012). Further, managers are also responsible for promoting change to their subordinates and preparing their subordinates to be ready for organisational changes.

2 METHOD

2.1 *Sample*

The sample consisted of one dairy product company that had undergone some organisational changes. The numbers of participants was 36 respondents, with the following characteristics: held a position on a managerial level, were permanent employees, had worked in the company for at least one year, and were full-time employees. Purposive sampling method was used to choose the respondents in six divisions that is finance, general management, Human Resources and General Affairs (HRGA), marketing, operations and sales. The questionnaires that were administered online through Google Forms and then the email was blast into HR email address. Then, the HR staff sent it to to respondent's.

2.2 *Measures*

Data was collected through two types of questionnaires:

2.2.1 *Readiness for change*

We measured Readiness for Change with the Readiness For Change (RFC) questionnaire developed by Holt et al. (2007), translated into the Indonesian context and adapted for Indonesian's respondent. RFC consists of 22 items, and it is divided into five dimensions: 1) Organisational Valence,; 2) Management Support,; 3) Change in Self-Efficacy,; 4) Discrepancy, and 5) Personal Valence, These 22 items were measured on a six-point Likert scale: 1 = Strongly disagree, 2 = Disagree, 3 = Slightly disagree, 4 = Slightly agree, 5 = Agree, and 6 = Strongly agree. The reliability coefficient of the RFC questionnaire was 0.906.

2.2.2 *Job satisfaction*

We measured job satisfaction with a Likert scale (1 = Strongly disagree, 2 = Disagree, 3 = Slightly disagree, 4 = Slightly agree, 5 = Agree, and 6 = Strongly agree) using 36 items from the Job Satisfaction Survey (JSS). JSS was developed by Spector (1985) in Spector (1997), and it was translated and adapted to the Indonesian context. This measure consisted of nine aspects of satisfaction, which included salary, promotion, supervision, fringe benefits, contingent rewards, operating condition, co-workers, nature of work, and communication. JSS can also measure overall job satisfaction. Every aspect consisted of four statements consisting of favourable and unfavourable items, and the total number of items was 36 items. The reliability coefficient of the JSS was 0.929.

3 RESULTS

3.1 *Demographics*

The sample demographics are displayed in Table 1. The majority of the respondents (80.56%, n = 29) were male. Respondents' age ranged based on productive age for work according to Dessler (2008), with majority were establishment age, 25–44 years (94%, n = 34). The majority, 47.22% (n = 17), of respondents had worked for more than nine years. More than half (52.8%, n = 19) were employees in the operations department, and the majority of the respondents were college graduates (63.9%, n = 23).

The following table shows the results of our multiple regression analysis between job satisfaction and employee's readiness for change.

From Table 2, the results show that Job Satisfaction had a significant correlation to employee's readiness to change (r = 0.661; p < 0.000). It means that more satisfaction the employee in their work, it makes the employee more ready to face change in their organisation.

Table 1. Demographics of the respondents.

Demographic aspect	Respondent	*Frequency*	%
Sex	Female	7	19.44
	Male	29	80.56
Age	Active age for work (Dessler, 2008)		
	Exploration stage (15–24 yrs.)	0	0
	Establishment stage (25–44 yrs.)	34	94
	Maintenance stage (45–65 yrs.)	2	5.56
Length of work	Length of work stage (Cohen, 1993)		
	(1–4) Exploration/trial stage	6	16.67
	(5–8) Establishment stage	13	36.11
	(>9) Maintenance stage	17	47.22
Education level	Senior High School	1	2.8
	Diploma 1–4*	4	11.1
	Bachelor's degree	23	63.9
	Master's degree and above	8	22.2
Department	HRGA**	5	13.9
	Finance	5	13.9
	Operations	19	52.8
	Sales	5	13.9
	Marketing	1	1
	General Management	1	1

* diploma degree is the educational level that focuses on practitioner curriculum. **Human Resources and General Affairs.

Table 2. The results of our multiple regression analysis of job satisfaction and readiness for change.

Variable	R	R^2	Adjusted R^2	Sig. (2-tailed)
Readiness for change	0.661"**	0.437	0.421	0.000

Table 3. Regression model of job satisfaction aspects to readiness for change.

Aspect	Unstandardised coefficients (B)	Standardised coefficients (Beta)	T-test	Significancy
Salary	1.129	0.574	1.858	0.061
Promotion	–0.392	0.543	–0.683	0.527
Supervision	0.022	0.609	–0.267	0.792
Fringe benefits	0.281	0.594	0.296	0.769
Contingent rewards	–0.010	0.600	0.144	0.886
Operating conditions	–0.500	0.559	–1.106	0.279
Co-worker	–1.122	0.694	–1.329	0.195
Nature of work	–0.462	0.803	–0.729	0.472
Communication	3.942	0.776	4.774	0.000

From table 2, also known that Job satisfaction has influence on Readiness for Change with the impact was 42.1%. In addition, we analysed which aspects of Job Satisfaction had the greatest contribution to the Readiness for Change using multiple regression. The calculation results are as shown in Table 3.

As seen in Table 3, not all aspects of Job Satisfaction had a significant influence on Employee's Readiness for Change. Among nine aspects, only the communication aspect significantly influenced Readiness for Change (B = 0.3942, t = 4.774, p = 0.000).

4 DISCUSSIONS AND LIMITATIONS

The results showed that job satisfaction had a significant influence on employee readiness for change. This indicates that the combination of the nine aspects of job satisfaction (salary, promotion, supervision, fringe benefits, contingent rewards, operating conditions, co-workers, nature of work, and communication) contributed to the employee's readiness for change. The impact of job satisfaction on readiness for change was 42.1%. Despite cultural and organisational differences, these findings support previous research which found that job satisfaction supports individual readiness for change. For example, Wanberg & Banas (2000) found that a positive attitude towards changes correlates with an employee's satisfaction towards life in their work. In addition, this study can shed light on the contradictory results of the research carried out by van Dam (2005) and Oreg et al. (2011). The result of that study show that employees who are satisfied with the job may see change as something positive, so they would be ready for change because they could see the positive consequences of the change.

Furthermore, this study also shows that among the nine aspects of job satisfaction, only communication contributed to the readiness for change. The results support previous research by Claiborne et al. (2013), which showed that not all aspects of job satisfaction contribute to readiness for change, and only communication affects the readiness for change of employees. Organisational change often causes uncertainty for the employees about their company. This uncertainty might have caused them to not be ready for change. Therefore, it is important for a company to communicate effectively with their employees during change to avoid uncertainty. The reason why the other aspects of job satisfaction did not influence the readiness for change of the respondents was because this study was conducted on a managerial level, where employees already had a good salary and good position. This made aspects like salary, promotion, or benefit not as influential to their readiness for change. Furthermore, the majority of the respondents had worked for the company for more than nine years. They have stayed in the company for a long period, and we can assume that they felt satisfied with the nature of work, their co-workers, and the operating conditions of the company, so that those aspects had no influence on their readiness for change.

Based on these results, the satisfaction on communication aspect can be considered to prepare employee during the implementation process of organisational change. An organisation can design specific interventions which focus on communication to increase employee readiness for change. Many organisations put most of their efforts in changing the organisation strategy, structure and systems and procedures and neglect individual level issues, such as job satisfaction. To increase individual readiness for change, organisations should create intervention initiatives on an individual level. The organisation and its managers can use effective communication, which can be considered as initiatives that can increase employee readiness for change. However, these results should be interpreted with caution because the sample was from one company, represent by small sample and the questionnaire was only distributed to the managerial level. There may be different results if it were conducted in another company at a different level, with different number of sample.

There are some limitations in this research that should be taken into consideration. First, among various numbers of variables that can influence readiness for change, this research only included one variable as the predictor. Further research on the readiness for change could include several other variables, such as an individual's characteristics, leadership, managerial support, organisational culture. Second, this research was conducted with 36 respondents (N = 36) and used purposive sampling methods. For further research could use a larger sample and other sampling methods, such as random sampling, to increase the generalisability of the results. Third, it is also suggested that further research could take samples on a staff level because this research was only conducted on a managerial level. Fourth, this study was conducted in an organisation that implemented organisational changes in terms of operating procedures; however, it is not a large nor a radical type of organisational change. The findings might have different results if the study had been conducted on a large-scale organisational change, such as merger and acquisition. Further studies should therefore be conducted in organisations undertaking different types of transformation. Future work could also be conducted on different types of organisations, such as private, government, and non-government, non-profit, and other companies in many different cultures because it is also important to explore and expand our understanding of the factors that create readiness for change. Lastly, longitudinal research is also needed.

5 CONCLUSION

This research shows that job satisfaction plays an important role in organisational change. The impact of job satisfaction on readiness for change was 42.1%. These results show that employees who were satisfied with the job might see organisational change as something positive, so they would be ready for change. Further, among nine aspects of job satisfaction, only communication contributed to readiness for change. This means that employees with a high satisfaction in the communication aspect will be ready for organisational change. Based on these findings, organisation undergoing any change process needs to pay attention to this matter. An organisation can design specific interventions which focus on communication to increase employee readiness for change. This research still need constructive suggestions. So, for any advice, please contact us in arum.hidayat@gmail,com.

REFERENCES

Anastasi, A & Urbina, S. (1997). *Psychological Testing (Seventh ed.)*. Upper Saddle River (NJ): Prentice Hall.

Andersen, L. S., "Readiness for change: can readiness be primed?" (2008). Master's Theses. San Jose State University). California. USA. 3517. http://scholarworks.sjsu.edu/etd_theses/3517

Armenakis, A.A. & Fredenberger, W.B. (1997). Organizational change readiness practices of business turnaround change agents. *Knowledge and Process Management*, *4*(3), 143–152. doi:10.1002/(SICI)1099-1441(199709)4:3<143::AID-KPM93>3.0.CO;2-7

Armenakis, A.A. & Harris, S.G. (2002). Crafting a change message to create transformational readiness. *Journal of Organizational Change Management*, *15*(2), 169–183. doi:10.1108/09534810210423080

Armenakis, A.A., Harris, S.G. & Mossholder, K.W. (1993). Creating readiness for organizational change. *Human Relations*, *46*(6), 681–703. doi:10.1177/001872679304600601

Armenakis, A.A., Bernerth, J.B., Pitts, J.P. & Walker, H.J. (2007). Organizational change recipients' beliefs scale: Development of an assessment instrument. *The Journal of Applied Behavioral Science*, *43*(4), 481–505. doi:10.1177/0021886307303654

Bernerth, J. (2004). Expanding our understanding of the change message. *Human Resource Development Review*, *3*(1), 36–52. doi:10.1177/1534484303261230

Bouckenooghe, D. & Devos, G. (2008). *Ready or not…? What's the relevance of meso level approach to the study of readiness for change*. Ghent, Belgium: Vlerick Leuven Gent Management School.

Claiborne, N., Auerbach, C., Lawrence, C. & Schudrich, W.Z. (2013). Organizational change: The role of climate and job satisfaction in child welfare workers' perception of readiness for change. *Children and Youth Services Review*, *35*(12), 2013–2019. doi:10.1016/j.childyouth.2013.09.012

Cordery, J. (1993). Correlates of employee attitudes toward functional flexibility. *Human Relations*. 46 (6): 705–723.

Cummings, T.G. & Worley, C.G. (2009). Organization development and change. *Annual Review of Psychology*, *38*. doi:10.1146/annurev.ps.38.020187.002011

Cunningham, C.E., Woodward, C.A., Shannon, H.S., MacIntosh, J., Lendrum, B., Rosenbloom, D. & Brown, J. (2002). Readiness for organizational change: A longitudinal study of workplace, psychological and behavioral correlates. *Journal of Occupational and Organizational Psychology*, *75*, 377–392.
Daft, R.L. (2007). *Understanding the theory and design of organization*. South Western. Cengage Learning.
Dam, K.V. (2005). Employee attitudes toward job changes: An application and extension of Rusbult and Farrell's investment model. *Journal of Occupational and Organizational Psychology, 78*(2), 253–272. doi:10.1348/096317904X23745
De Meuse, K.P., Marks, L.M. & Dai, G. (2010). Organizational downsizing, mergers and acquisitions, and strategic alliances: Using theory and research to enhance practice. In Zedeck, S. (Eds.), *APA handbook of industrial and organizational psychology 3* (pp. 729–768). Washington, DC: American Psychological Association.
Devos, G., Buelens, M. & Bouckenooghe, D. (2007). Contribution of content, context, and process to understanding openness to organizational change: Two experimental simulation studies. *The Journal of Social Psychology*, *147*(6), 607–629. doi:10.3200/SOCP.147.6.607–630
Gordon, S.S., Stewart, W.H., Sweo, R. & Luker, W.A. (2000). Convergence versus strategic reorientation: The antecedents of fast-paced organizational change. *Journal of Management, 26*, 911–945.
Guilford, J.P. & Fruchter, B. (1978). Fundamental statistics in psychology and education. McGraw-Hill. Book-Co Singapore.
Holt, D.T., Armenakis, A.A., Feild, H.S. & Harris, S.G. (2007). Readiness for organizational change: The systematic development of a scale. *The Journal of Applied Behavioral Science*, *43*, 232–255. doi:10.1177/0021886306295295
Iverson, R.D. (1996). Employee acceptance of organizational change: The role of organizational commitment. *The International Journal of Human Resource Management*, *7*(1), 122–149. doi:10.1080/09585199600000121
Jones, R.A., Jimmieson, N.L. & Griffiths, A. (2005). The impact of organizational culture and reshaping capabilities on change implementation success: The mediating role of readiness for change. *Journal of Management Studies*, *42*, 361–386.
Judge, T.A., Thoresen, C.J., Pucik, V. & Welbourne, T.M. (1999). Managerial coping with organizational change: A dispositional perspective. *Journal of Applied Psychology, 84*(1), 107–122.
Locke, E.A. (1976). The nature and causes of job satisfaction. *Handbook of industrial and organizational psychology. IL: Rand McNally. Chichago.* (pp. 1297–1349).
Madsen, S.R., Miller, D. & John, C.R. (2005). Readiness for organizational change: Do organizational commitment and social relationships in the workplace make a difference. *Human Resource Development Quarterly*, *16*(2), 213–233. doi:10.1002/hrdq.1134
Madsen, S.R., John, C.R. & Miller, D. (2006). Influential factors in individual readiness for change. *Journal of Business Management*, *12*(2), 93–110. doi:10.1002/hrdq.1134
Matthews, J. (2012). What is a workshop? *Theatre, Dance, and Performance Training, 3*(3), 349–361. doi: 10.1080/19443927.2012.719832
McNabb, D. & Sepic, F. (1995). Culture, climate, and total quality management: Measuring readiness for change. *Public Productivity & Management Review*, *18*(4), 369–385. http://www.jstor.org/stable/10.2307/3663059
Meaney, M. & Pung, C. (2008). McKinsey global results: Creating organizational transformations. *McKinsey Quarterly*, August, 1–7. McKinsey & Company.
Melisya, C.F. (2013). Reorientasi dan Sosialisasi Perubahan sebagai Intervensi Persepsi Dukungan Organisasi dalam Meningkatkan Kesiapan untuk Berubah pada Diri Karyawan (Master Thesis, Universitas Indonesia, Depok, Indonesia). (Melisya, C.F. (2013). Reorientation and Change Socialitation as the Intervention of Perceived Organisation Support to Increase Employee' readiness to Change. Master Thesis. Universitas Indonesia. Depok Indonesia.
Miller, V.D., Johnson, J.R. & Grau, J. (1994). Antecedents to willingness to participate in a planned organizational change. *Journal of Applied Communication Research*, *22*(1), 59–80.
Mills, J.H., Dye, K. & Mills, A.J. (2009). Understanding organizational change. New York, NY: Routledge.
Oreg, S. & Berson, Y. (2011). Leadership and employees' reactions to change: The role of leaders' personal attributes and transformational leadership style. *Personnel Psychology, 64*(3), 627–659. doi:10.1111/j.1744-6570.2011.01221.x
Oreg, S., Vakola, M. & Armenakis, A. (2011). Change recipients' reactions to organizational change: A 60-year review of quantitative studies. *The Journal of Applied Behavioral Science, 47*(4), 461–524. doi:10.1177/0021886310396550
Pearson, J.C. (1983). *Interpersonal communication*. IL: Scott Foresman and Company.

Porter, L.W., Steers, R.M. Mowday, R.T. & Boulian, P.V. (1974). Organizational commitment, job satisfaction, and turnover among psychiatric technicians. *Journal of Applied Psychology, 59*(5), 603–609. doi:10.1037/h0037335

Rafferty, A.E., Jimmieson, N.L. & Armenakis, A.A. (2012). Change readiness: A multilevel review. *Journal of Management, 39*(1), 110–135. doi:10.1177/0149206312457417

Ratnasari, D. (2013). *Pengaruh Peningkatan Kepuasan Kerja Terhadap Perilaku Inovatif dengan Pemberian Pelatihan Work Resign pada Atasan (Studi pada Kantor Pusat PT. X)* (Master's Thesis, Universitas Indonesia, Depok, Indonesia).(Ratnasari, D. (2013). The Influence of the Increase Job Satisfaction on Innovative Work Behavior with Work Design Training among the Supervisor (Study at Headquarter Office of PT X. Master Thesis Universitas Indonesia, Depok, Indonesia).

Self, D.R., Armenakis, A.A. & Schraeder, M. (2007). Organizational change content, process, and context: A simultaneous analysis of employee reactions. *Journal of Change Management, 7*(2), 211–229. doi:10.1080/14697010701461129

Spector, P.E. (1985). Measurement of human service staff satisfaction: Development of the job satisfaction survey. *American Journal of Community Psychology, 13*(6), 693–713.

Spector, P. (1997). Job Satisfaction: Application, Assessment, Causes and Consequences. Thousand Oaks, CA. Sage Publications.

Spector, P.E. (2000). *Industrial and organizational psychology: Research and practice* (2nd ed.). New York, NY: John Wiley and Sons.

Wanberg, C.R. & Banas, J.T. (2000). Predictors and outcomes of openness to change in reorganizing workplace. *Journal of Applied Psychology, 85*(1), 132–142.

Weber, P.S. & Weber, J.E. (2001). Change in employee perception during organizational change. *Leadership and Organizational Development Journal, 22*(6), 291–300.

Weiss, H.M. (2002). Deconstructing job satisfaction: Separating evaluations, beliefs and affective experiences. *Human Resource Management Review, 12*(2), 173–194. doi:10.1016/S1053-4822(02)00045-1

Worley, C.G. & Lawler, E.E. (2009). Building a change capability at capital one financial. *Organizational Dynamics, 38*(4), 245–251. doi:10.1016/j.orgdyn.2009.02.004

Vakola. M. (2012). What's in there for me? Individual readiness to change and the perceived impact of organizational change. *Leadership and Organizational Development Journal, 35*(3), 195–209.

Diversity in Unity: Perspectives from Psychology and Behavioral Sciences – Ariyanto et al. (Eds)
© 2018 Taylor & Francis Group, London, ISBN 978-1-138-62665-2

The important role of leader-member exchange in the relationship between cognitive and affective trust and leader effectiveness

A. Mustika & C.D. Riantoputra
Faculty of Psychology, Universitas Indonesia, Depok, Indonesia

ABSTRACT: Studies on leader effectiveness nowadays focus on the relationship between leaders and subordinates. One of the key variables in this relationship is trust. Previous research shows that, both cognitive and affective trusts may impact leader effectiveness directly or mediated through Leader-Member Exchange (LMX). This current research aims to test the probability of these two potentials in two different organizational settings (government companies and private companies). The data were gathered from 100 employees from government companies and 131 employees from private companies. The analysis reveals that (a) cognitive trust and affective trust are positively associated with leader effectiveness, in both government companies and private companies; (b) LMX is positively associated with leader effectiveness in private companies, but it has no relationship in public companies; and (c) LMX fully mediates the relationship between cognitive trust and leader effectiveness in private companies, but only partially mediates the relationship between affective trust and leader effectiveness in private companies. The study shows that there are different quality relationships between leaders and subordinates in government companies and private companies, which will affect leader effectiveness. Theoretical and practical implications will be discussed.

1 INTRODUCTION

A leader must be able to steer his/her group in order to achieve organizational goals. Leader effectiveness can be measured from behavior and performance (DeGroot, Aime, Johnson, & Kluemper, 2011). Leader effectiveness as a behavior is the evaluation of members to the behavior of the group supervisor in accordance with the performance of the group, while the leader effectiveness as a performance or result (outcome) is the objective measurements of the leader. Previous research discusses more about internal factors as determinants of leader effectiveness, such as gender (Paustian-Underdahl, Walker, & Woerh, 2014) and traits (Hoffman, Woehr, Maldagen-Youngjohn, & Lyons, 2011). There has been research on the external factors of leader effectiveness, for example about constructive organization culture styles (the culture that improves employee satisfaction and achievement) (Kwantes & Boglorsky, 2007).

The focus of this study lies in the assessment of the effectiveness of a leader in the behavioral performance, and about the quality of the relationship between leaders and subordinates (Hannah, Sumant, Lester, & Cavarretta, 2014). The factor that influences leader effectiveness is trust. Trust is the confidence in the goodwill and competence of others and the expectation that others will reciprocate with honest efforts that are consistent with agreements, if one cooperates (Casimir, Waldman, Bartram, & Yang, 2006). McAllister (1995) distinguishes trusts into two parts: cognitive trust and affective trust. Affect-based trust refers to the emotional bond between individuals that are grounded upon the expressions of genuine care and concern for the welfare of other parties. Cognitive trust refers to the trust based on performance-relevant cognitions such as competence, responsibility, reliability, and dependability (McAllister, 1995).

Cognitive trust and affective trust can directly influence leader effectiveness by improving job performance and job satisfaction. Research shows that trust can affect leader effectiveness directly (Gillespie & Mann, 2004; Yang, 2009), but trust can cause leader effectiveness by mediating leader-member exchange (LMX) (Chen, Lam, & Zong, 2012). LMX is the exchange relationship between subordinates and supervisors (Eisenberger, Shoss, Karagonlar, Gonzales-morales, Wickham, & Buffardi, 2014). LMX theory suggests that leaders do not use the same style in dealing with all subordinates, but develop a different type of relationship or exchange with each subordinate (Liden & Maslyn, 1998). LMX occurs if there is trust between leaders and subordinates, and LMX can directly impact leader effectiveness because LMX will increase work performance.

Based on the explanations above, this study asks, "Are cognitive trust and affective trust directly related to leader effectiveness, or is the relationship mediated through LMX?"

This research is going to focus on two different organizations, government companies and private companies, which are different, both in terms of the culture and the nature of the jobs. Previous research shows that the type of organizations influences the attitude of leaders towards their subordinates (Petrick & Quinn, 2000). However, another research by Hoffman *et al.* (2011) demonstrate that the type of organizations–business and government–does not have significant differences between their leaders and employees. Our interest is to compare government and private companies in conjunction with leader effectiveness to see the quality of relationships between leaders and subordinates.

1.1 *Trust*

McAllister (1995) explains the concept of trust as interpersonal trust. Interpersonal trust is the extent to which a person believes in other people's trust, and is willing to act upon the words, actions, and decisions of other people. Gillespie & Mann (2004) wrote that trust is a psychological state comprising the intention to accept vulnerability based upon positive expectations of the intentions or behavior of another. Employees' trust in their supervisor will affect work engagement (Chunghtai *et al.* 2015), whereas cognitive trust will increase the learning organization, and affective trust will increase the knowledge sharing and social networks (Swift & Hwang, 2013).

Trust, in the end, increases the subordinates' work performances. Trust in the supervisors improves teamwork, which will then improve the performance of employees (Cho & Poister, 2014). Cognitive trust affects cognitive task performance and job satisfaction, and affective trust in the supervisor affects helping behavior (Yang, 2009). Cognitive trust makes someone think he/she believes in the leader because of the ability of the leader, and results are reflected in performance. Affective trust brings positive emotions in doing the job, so that the emergent behavior is a behavior that comes from within the individuals' inherent desires.

Hypothesis 1. There is a positive relationship between cognitive trust and leader effectiveness.

Hypothesis 2. There is a positive relationship between affective trust and leader effectiveness.

1.2 *LMX*

Leader-Member Exchange (LMX) is derived from the Social Exchange Theory, which says that exchange relationships occur between at least two people, in the form of activities, both visible and invisible (Homans, 1958). The term "LMX" was formerly known as the Vertical Dyad Linkage (VDL) theory. VDL is a reciprocal relationship between superiors and subordinates, which occurs in the dyads. Dyads are two parts that interact to constitute a unity. The LMX theory states that leaders do not treat each subordinate in the same way and the quality of the LMX can range from low to high, depending on the treatment related to work, attitude, and behavior of the subordinates (Rocksthul *et al.*, 2012). The LMX is a relationship-based leadership, where these relationships can occur at the level of "in-group" or "out-group" (Graen & Uhl-Bien, 1995). Subordinates with high quality relationships (at the level of in-group) receive a number of advantages compared to a subordinate with lower quality relationships (level out-group). The advantages gained include the increase of communication, levels of emotional support, and wider access (Dienesch & Liden, 1986; Graen & Scandura, 1987).

The LMX affects the followers' task performance (Chan & Mak's, 2012), improves job performance, lowers turnover intention (Bauer *et al.,* 2006), and improves job satisfaction, and organizational commitment (Lee, Teng, & Chen, 2015). The research by Hassan, *et al.* (2013) examines that the LMX positively influences the subordinates' perception of leader effectiveness. The interaction between leaders and subordinates makes subordinates perceive that their supervisors are able to lead effectively. The LMX also affects job satisfaction and outcomes of the effectiveness of other leaders. The research meta-analysis finds that the LMX affects outcomes, including task performance, job performance, Organizational Citizenship Behavior (OCB), distributive justice, interactional justice, job satisfaction, affective and normative commitment, and turnover intention (Deulebohn *et al.*, 2012; Rockstuhl *et al.,* 2012). If the LMX intertwines, both purposes of the group will be quickly achieved.

Hypothesis 3. There is a positive relationship between the Leader-Member Exchange (LMX) with leader effectiveness.

1.3 *LMX as a mediator*

Trust, both cognitive trust and affective trust, improves the relationship between leaders and subordinates. The LMX theory states that the leaders influence their followers through unique ways, such as having trust between two individuals (Vidyarthi, 2014). The LMX will occur after the confidence or belief of others is obtained through the process of cognition and affection. Dirks & Ferrin (2002) in their study wrote that affective and cognitive trusts are positively correlated with the formation of the LMX relationships. The process of cognition and affection that exists will make subordinates feel that their supervisor is able to perform his/her duties as a good leader. Subordinates who trust their leaders will increase their LMX relationship. Deulebohn *et al.* (2012) in their study wrote that there is a relationship between the confidence of subordinates and their leaders (trust) with the LMX. Affective trust increases emotional feelings that serve to strengthen the relationship between superiors and subordinates, and cognitive trust will make an individual improve his/her work performance in order to bring the relationship between superiors and subordinates even closer.

Some research assumes that trust influences the effectiveness of leaders through mediators. One of the potential mediators is the LMX. Chen *et al.* (2012) suggests that trust affects the LMX, and the LMX mediates trust in supervisors with work performance. Subordinates who have both cognitive and affective trust towards their leaders will help their leaders to become good leaders, as well as leaders feeling closer with subordinates. This reflects an exchange relationship between leaders and subordinates. The relationship between affective trust and cognitive trust and leader effectiveness can happen directly or indirectly through the LMX. Thus, there is an indication that the relationship between cognitive trust and affective trust with leader effectiveness is partially mediated by the LMX.

Hypothesis 4. The LMX partially mediates cognitive trust relationships with leader effectiveness.

Hypothesis 5. The LMX partially mediates affective trust relationships with leader effectiveness.

2 METHODS

2.1 *Subject*

The data were retrieved from government companies in Jakarta and private companies in Bali. We used dyadic or pairing participants in this research, which are leaders and their subordinates (two levels under the leader). The leaders in this study consisted of the director, section chief, manager or equivalent in government companies and the companies of private industry, whereas the subordinate participants included staff and supervisors. Two hundred and thirty one pairs responded to the survey, with 100 pairs from government companies (response rate 58.82%) and 131 pairs from private companies (response rate 76.16%).

2.2 *Measures*

2.2.1 *Leader effectiveness*

Fifteen items were adapted from DeGroot et al. (2011). An example item includes: "The leader of my direct leader gives ideas to solve problems". Each item was scored on a six-point Likert-type scale. The alpha for the 15 items-scale was 0.95.

2.2.2 *LMX*

The LMX was measured using a LMX7 scale compiled by Scandura and Graen (1984). An example item on this scale was "The leader of my direct leader shows how satisfied he/she is with my performances". In this study, the researchers adapted the LMX7 to transform the answer choices into a 6-point Likert-type scale. The alpha coefficient was 0.83.

2.2.3 *Trust*

The measurement of affective and cognitive trust in this study used affective-based trust scales and cognitive-based trust scales from McAllister (1995). The affective trust scale consisted of five items, and an example item was "I and the leader of my direct leader, can both freely share our ideas, feelings, and hopes", and the cognitive trust scale consisted of six items, with a sample "I can rely on the leader of my direct leader not to make my job more difficult by careless work". This scale was measured using a 6-point Likert scale. The alpha coefficient was 0.79 (affective trust) and 0.83 (cognitive trust).

2.2.4 *Control variables*

The demographic variables, such as leaders' gender, age, education, tenure, leading period, and interaction with subordinates were controlled in this study to assess the relationship between focal variables more rigorously.

3 RESULTS

3.1 *Descriptive statistics and variable correlations*

The results of the correlation test (Pearson correlation) in a government office indicated that there is a positive relationship between the dependent variable, which is leader effectiveness, and the independent variables, which are cognitive trust ($r = 0.733$, $p < 0.05$), affective trust ($r = 0.689$, $p < 0.05$), and LMX ($r = 0.718$, $p < 0.05$). The results of the correlation test in private companies indicated that there is a positive relationship between the dependent variable, which is leader effectiveness, and the independent variables, which are cognitive trust ($r = 0.690$, $p < 0.05$), affective trust ($r = 0.668$, $p < 0.05$), and LMX ($r = 0.690$, $p < 0.05$). Such correlations indicated that the higher the cognitive trust, affective trust, and LMX are, the higher the leader effectiveness is.

3.2 *Hypothesis test*

We used four stages of multiple linear regressions to analyze the data. According to Baron and Kenny (1986), to examine the relationship of each independent variable with the dependent variable and the mediator, four stages of multiple regressions should be used. The steps include: (a) the relationship between the independent variables and the mediator variable, (b) the relationship between the independent variables and the dependent variable, (c) the

Table 1. Descriptive statistics and correlations in government companies.

Variable	Mean	SD	Correlation 1	2	3	4	5	6	7	8	9	10
Gender	–	–	1									
Leader's age	6.58	1.156	–0.113	1								
Leader's education	5.81	0.647	–0.002	0.068	1							
Leader's tenure	6.45	1.192	0.032	0.175	–0.124	1						
Leader's leading period	8.23	16.081	–0.033	0.192	0.145	0.094	1					
Number of interaction	351	9.081	–0.020	–0.071	0.011	0.035	–0.001	1				
Cognitive trust	4.66	0.726	0.054	–0.069	–0.026	–0.187	–0.103	0.175	1			
Affective trust	4.68	0.690	0.104	–0.069	0.007	–0.142	–0.151	0.206*	0.750**	1		
LMX	4.55	0.664	0.114	–0.128	–0.051	–0.185	–0.319**	0.109	0.818**	0.831**	1	
Leader effectiveness	4.85	0.688	0.072	–0.056	–0.090	–0.104	–0.099	0.102	0.733**	0.689**	0.718**	1

LMX = Leader-Member Exchange * $p < 0.05$ ** $p < 0.001$.

Table 2. Descriptive statistics and correlations in private companies.

Variable	Mean	SD	Correlation									
			1	2	3	4	5	6	7	8	9	10
Gender	–	–	1									
Leader's age	4.84	1.508	0.271**	1								
Leader's education	3.50	1.628	–0.099	–0.208*	1							
Leader's tenure	4.79	2.045	0.199*	0.625**	–0.160	1						
Leader's leading period	5.19	1.314	0.290**	0.594**	–0.048	0.347**	1					
Number of interaction	8.79	21.607	0.115	–0.032	–0.023	0.024	0.077	1				
Cognitive trust	4.85	0.678	–0.069	–0.026	0.164	–0.018	–0.036	–0.036	1			
Affective trust	4.95	0.680	–0.008	–0.063	0.095	0.004	–0.018	–0.032	0.811**	1		
LMX	4.86	0.709	–0.099	–0.056	0.024	0.012	–0.051	–0.012	0.881**	0.856**	1	
Leader effectiveness	5.11	0.704	0.117	0.001	0.112	0.003	0.006	–0.156	0.664**	0.668**	0.690**	1

LMX = Leader-Member Exchange $*p < 0.05$ $**p < 0.00$.

Table 3. Regression results for testing mediation.

	Government			Private		
		Leader effectiveness			Leader effectiveness	
Factor and statistic	LMX	Step 1	Step 2	LMX	Step 1	Step 2
Cognitive trust	0.440*	0.493*	0.395*	0.564*	0.357*	0.187
Affective trust	0.498*	0.320*	0.210*	0.414*	0.379*	0.250*
LMX			0.220			0.311*
F	168.292	67.390	46.491	325.155	61.622	43.445
R^2	0.777	0.582	0.592	0.836	0.491	0.506
R adjust	0.772	0.573	0.580	0.833	0.483	0.495

LMX = Leader-Member Exchange *$p < 0.05$.

relationship between the mediator variable with the dependent variable, and (d) the correlation between the independent variables and the dependent variable. These steps test whether the relationships are significant. If the results show that the relationships are not significant, it is said to be fully mediated, but if the relationships are still significant, it is said to be partially mediated.

Table 3 shows cognitive trust has a significantly positive relationship with leader effectiveness in government companies ($\beta = 0.493$; $p < 0.05$) and private companies ($\beta = 0.357$; $p < 0.05$) (H1 is supported). Furthermore, from Table 3, affective trust has a significantly positive relationship with leader effectiveness in government companies ($\beta = 0.320$; $p < 0.05$) and private companies ($\beta = 0.379$; $p < 0.05$) (H2 is supported).

Table 3 shows different results for the relationship between the LMX and the leader effectiveness in government companies and private companies. The result shows that LMX and leader effectiveness are significantly positive in government companies ($\beta = 0.220$; $p > 0.05$), but not in private companies ($\beta = 0.311$; $p < 0.05$). These results suggest that H3 for government companies is rejected, whereas H3 for private companies is supported. It explains that in the government companies, the LMX does not mediate the relationship between leader effectiveness and cognitive trust, as well as affective trust. Thus, the H4 and H5 in government companies are not supported.

Furthermore, Table 3 shows that in private companies the relationship between cognitive trust and leader effectiveness is no longer significant ($\beta = 0.187$; $p > 0.05$), when the LMX appears as the mediator variable. This means that cognitive trust will affect leader effectiveness only through the LMX, so the LMX fully mediates the relationship between those variables (H4 is supported). The next results from Table 3 regarding private companies show that the relationship between affective trust and leader effectiveness is still significant but the score of significance changes ($\beta = 0.250$; $p < 0.05$) because of the LMX. This means that the LMX partially mediates the relationship between affective trust and leader effectiveness in private companies (H5 is supported).

4 DISCUSSION

From the results, the LMX mediates the relationship between cognitive and affective trust with leader effectiveness in private companies, but not in government companies. This research contributes to see the different qualities of relationship between leaders and subordinates in government and private companies. Government companies and private companies are different in function and nature. From the function of the organization, government companies are public oriented (Cummings & Worley, 2009). They are more administrative and have almost no competition among other government companies, so it is possible for the leaders to maintain distance from their subordinates. On the other hand, private companies, from the function of the organization, are profit oriented (Cumming & Worley, 2009). They have many competitors, so, to advance their organizations, leaders need a lot of support from their subordinates.

Government companies, according to the nature of the organization by Mintzberg (1980), can be classified as bureaucratic. A company with a bureaucratic nature is characterized by rigid organizations, workers with their respective specialties, and tendencies to have similar work tasks that follow position description. The relationship between leaders and subordinates in the bureaucratic type is not very visible, because in doing their jobs, employees in the companies have their designated specializations and do not need to do work not included in their job descriptions. On the flip side, the nature of private companies by Mintzberg (1980) is organic. In organic organizations, it is usually possible to interchange the positions, develop employees through trainings, and handle other workers' job descriptions. In private organizations, employees are given the opportunity to develop themselves and be close to their leaders. The leaders in private organizations have close relations with their subordinates because they cannot work without their subordinates. Based on this research, further research should aim to understand the characteristics of each institution, both government and private. Profiles and characteristics are needed to understand the culture and values that are applied.

This research also contributes in knowing whether the LMX fully mediates the relationship between cognitive trust and leader effectiveness in private companies. The LMX occurs when there is trust between supervisors and subordinates (Vidyarthi *et al.*, 2014). Dirks & Ferrin (2002) stated that cognitive trust is positively related to the LMX. Subordinates who believe in their leaders because of work and professionalism will improve the quality of relationships between leaders and subordinates, so that the process of the LMX becomes more pronounced. Cognitive trust improves the relations between leaders and subordinates, along with improving work performance. The relationship between leaders and subordinates improves performance and job satisfaction, which are the outcomes of leader effectiveness. As described by Chen *et al.* (2012), the LMX plays a role as a mediator between the relationship of trust and performance. Trust will cause the quality of the relationship between leaders and subordinates to increase and the LMX will improve the quality of performance and job satisfaction. Thus, cognitive trust in relation to leader effectiveness requires a mediator such as the LMX.

Nonetheless, a limitation in this research is the potential to have common method bias (Podsakoff *et al.*, 2012). This refers to the correlation coefficients between constructs, which happen when the participants make 'implicit theories' about the relation between the dependent and independent variables when answering the questions. Psychological separation method (Podsakoff *et al.*, 2003) was used to try to overcome this limitation, meaning that when gathering the data we separated the booklet of the independent variables and the dependent variable. Therefore, it is expected that the participants answered the questions without making any 'implicit theories'.

5 CONCLUSION

From this study we can conclude that the LMX mediates the relationship between cognitive and affective trust and leader effectiveness in private companies, but not in government companies. In private companies, the LMX fully mediates the relationship between cognitive trust and leader effectiveness, while the LMX only partially mediates the relationship between affective trust and leader effectiveness.

REFERENCES

Baron, R.M. & Kenny, D.A. (1986) The Moderator–Mediator Variable Distinction in Social Psychological Research: Conceptual, Strategic, and Statistical Considerations. *Journal of Personality and Social Psychology.* 51, 1173–1182.

Chan, S.C.H. & Mak, W. (2012) Benevolent Leadership and Follower Performance: The Mediating Role of Leader-Member Exchange (LMX). *Asia Pacific Journal of Management.* 29, 285–301. doi: 10.1007/s10490-011-9275-3.

Chen, Z., Lam, W. & An Zhong, J. (2012) Effects of Perceptions on LMX and Work Performance: Effects of Supervisors' Perception of Subordinates' Emotional Intelligence And Subordinates' Perception of Trust in The Supervisor On LMX and, Consequently, Performance. *Asia Pacific Journal of Management.* 29, 597–616. doi: 10.1007/s10490-010-9210-z.

Cho, Y.J. & Poister, T.H. (2014) Managerial Practices, Trust in Leadership and Performance: Case of Georgia Department of Transportation. *Public Personnel Management.* 43(2), 179–196. doi: 10.1177/009102601452316.

Cummings, T.G. & Worley, C.G. (2009) *Organization Development & Change.* 9th edition. Canada, South-western.

Cycyota, C.S. & Harison, D.A. (2006) What (Not) to Expect when Surveying Executives A Meta-Analysis of Top Manager Response Rates and Techniques Over Time. *Organizational Research Methods.* 9 (2), 133–160, doi: 10.1177/1094428105280770.

DeGroot, T., Aime, F., Jhonson, S.G. & Kluemper, D. (2011) Does Talking The Talk Help Walking The Walk? An Examination of The Effect of Vocal Attractiveness in Leader Effectiveness. *The Leadership Quarterly.* 22, 680–689. doi:10.1016/j.leaqua.2011.05.008.

Dienesch, R.M. & Liden, R.C. (1986) Leader-Member Exchange Model of Leadership: A Critique a Further Development. *The Academy of Management Review.* 11(3), 618–634.

Eisenberger, R., Shoss, M.K., Karagonlat, G., Gonzalez-Morales, M.G., Wickham, R.E. & Buffardi, L.C. (2014) The Supervisor POS-LMX-Subordinate POS Chain: Moderation by Reciprocation Wariness and Supervisor's Organizational Embodiment. *Journal of Organizational Behavior.* 35, 635–656.

Ferrin, D.L. & Dirk, K.T. (2002) Trust in Leadership: Meta-Analytic Findings and Implications for Research and Practice. *Journal of Applied Psychology.* 87 (4), 611. doi: 10.1037/0021-9010.87.4.611.

Field, A. (2009) *Discovering Statistics Using SPSS.* 3rd edition. London, Sage Publications Ltd.

Gillespie, N.A. & Mann, L. (2004) Transformational Leadership and Shared Values: The Building Block of Trust. *Journal of Managerial Psychology.* 19 (6), 588–607. doi: 10.1108/02683940410551507.

Graen, G.B. & Uhl-Bien, M. (1995) Relationship-Based Approach to Leadership: Development of Leader-Member Exchange (LMX) Theory of Leadership Over 25 Years: Applying a Multi-Level Multi-Domain Perspective. *Leadership Quarterly.*

Hannah, S.T., Sumanth, J.J., Lester, P. & Cavaretta, F. (2014) Debunking The False Dichotomy of Leadership Idealism and Pragmatism: Critical Evaluation and Support of Newer Genre Leadership Theories. *Journal of Organizational Behavior.* 35, 598–621. doi: 10.1002/job.1931.

Hassan, S., Masud, R., Yukl, G. & Prussia, G.E. (2013) Ethical and Empowering Leadership and Leader Effectiveness. *Journal of Managerial Psychology.* 28 (2), 133–146. doi: 10.1108/02683941311300252.

Hoffman, B.J., Woehr, D.J., Maldagen-Youngjohn, R. & Lyons, B.D. (2011) Great Man or Great Myth? A Quantitative Review of The Relationship Between Individual Differences and Leader Effectiveness. *Journal of Occupational and Organizational Psychology.* 84, 347–381. doi:10.1348/096317909X485207.

Lee, A.P., Teng, H.Y. & Chen, C-Y. (2015) Workplace Relationship Quality and Employee Job Outcomes in Hotel Firms. *Journal of Human Resources in Hospitality and Tourism.* 14 (4), 398–422.

McAllister, D.J. (1995) Affect and Cognition Based Trust as Foundations for Interpersonal Cooperation in Organization. *Academy of Management Journal.* 38 (1), 24–59.

Mintzberg, H. (1980) Structure in 5's: A Synthesis of The Research on Organization Design. *Management Science.* 25 (3), 322–341.

Paustian-Underdahl, S.C., Walker, L.S. & Woehr, D.J. (2014) Gender and Perception of Leadership Effectiveness: A Meta-Analysis of Contextual Moderators. *Journal of Applied Psychology.* 99 (6), 1129–1145. doi: 10.1037/a0036751.

Podsakoff, P.M., MacKenzie, S.B. & Podsakoff, N.P. (2012) Sources of Method Bias in Social Sciences Research and Recommendations on How to Control It. *Annual Review of Psychology.* 63, 539–569.

Podsakoff, P.M. & Podsakoff, N.P. (2003) Common Method Bias in Behavioral Research: A Critical Review of The Literature and Recommendation Remedies. *Journal of Applied Psychology.* 88 (5), 879–903. doi: 10.1037/0021-9010.88.5.879.

Rocksthul, T., Dulebohn, J.H., Ang, S. & Shore, L.M. (2012) Leader-Member Exchange and Culture: A Meta-Analysis of Correlates of LMX Across 23 Countries. *Journal of Applied Psychology.* 97 (6), 1097–1130. doi: 10.1037/a0029978.

Scandura, T.A. & Graen, G.B. (1984) Moderating Effects of Initial Leader-Member Exchange Status on Effects of a Leadership Intervention. *Journal of Applied Psychology.* 69 (3), 428–436.

Strang, S.E. & Kuhnert, K.W. (2009) Personality and Leadership Developmental Level as Predictors of Leader Performance. *The Leadership Quarterly.* 20, 421–433.

Swift, P.E. & Hwang, A. (2013) The Impact of Affective and Cognitive Trust on Knowledge Sharing and Organizational Learning. *The Learning Organization.* 20 (1), 20–37, doi: 10.1108/09696471311288500.

Vidyarthi, P.R., Erdogan, B., Berrin, S., Liden, R.C. & Chaudhry, A. (2014) One Member, Two Leaders: Extending Leader–Member Exchange Theory to a Dual Leadership Context. *Journal of Applied Psychology.* 99 (3), 468–483. doi: 10.1037/a0035466.

Yang, J., Mossholder, K.W. & Peng, T.K. (2009) Supervisory procedural justice: The mediating roles of cognitive and affective trust. *The Leadership Quarterly*. 20 (2), 143–154.

Diversity in Unity: Perspectives from Psychology and Behavioral Sciences – Ariyanto et al. (Eds)
 ISBN 978-1-138-62665-2

The relationship between behavioral integrity and leader effectiveness mediated by cognitive trust and affective trust

P. Maharani & C.D. Riantoputra
Faculty of Psychology, Universitas Indonesia, Depok, Indonesia

ABSTRACT: Leadership theories have shifted over the last few decades from focusing on objective measures of performance towards subordinates' evaluation of their leaders' behavior, relevant to team performance. Behavioral integrity, which refers to subordinates' perception of the patterns of word-deed alignment, is one of the most important factors that influence subordinates' evaluation of their leaders' effectiveness. We extend previous research by arguing that the importance of behavioral integrity on leader effectiveness is mediated by two forms of trust: cognitive trust, which refers to trust that is based on performance-relevant cognitions such as competence, reliability, and dependability; and affective trust, which refers to the emotional bonds between individuals that are grounded upon the expressions of genuine care and concern for the welfare of the other party. To test the hypotheses, we collected data from 215 employees in the service industry. Using parallel multiple regression by PROCESS, we find that the relationship between behavioral integrity and leader effectiveness is fully mediated by cognitive trust ($b = 0.73$, $p < 0.01$) and affective trust ($b = 0.60$, $p < 0.01$), suggesting that behavioral integrity only takes place when subordinates have cognitive and affective trust towards their leaders. This research is particularly important because it delineates the mechanism under which behavioral integrity affects leader effectiveness.

1 INTRODUCTION

Leadership theories have shifted over the last few decades from focusing on objective measures of performance towards subordinates' evaluation of their leaders' behaviors relevant to team performance (Hannah *et al.*, 2014). Leaders will have great influence if their subordinates see them as competent. An example of a leader considered effective is Ignatius Jonan, when he served as the CEO of Indonesia's train company. Before he served as the CEO of Indonesia's train company, people did not use trains. Under his leadership, trains went back to being one of the most popular forms of public transportation (Nugroho, 2012). On top of that, the company made immense profit in 2015, close to one trillion Rupiah (Khalifah, 2015). Jonan does what he says, so the Indonesian people believe and trust him (Sutianto, 2014).

Leader effectiveness refers to (1) leaders' behavior relevant to the organization's performance and (2) the outcomes of this behavior, reflected by the company's financial performance (DeGroot *et al.*, 2011). In this study, the concept of leader effectiveness is leader effectiveness behavior, which is measured subjectively through the subordinates' evaluation.

Leader effectiveness can be influenced by two factors, namely internal and external factors. The internal factors consist of personality traits (Hoffman *et al.*, 2011; Zaccaro, 2007), emotional intelligence (Riggio & Reichard, 2008; Rockstuhl *et al.*, 2011), and integrity (Kannan-Narasimhan & Lawrence, 2012; Simons *et al.*, 2015). The external factors consist of the types of company (Judge *et al.*, 2002; Hoffman *et al.*, 2011), the levels of positions in the company (Hoffman *et al.*, 2011), and the corporate culture (Kwantes & Boglarsky, 2007).

From the multitude of factors that influence the effectiveness of leaders, the relationship between integrity and leader effectiveness is very important because previous studies show that integrity is the core of leader effectiveness (Soltani & Maupetit, 2015). An effective leader is one who consistently behaves in accordance with his/her values, expectations, and priorities (Salicru & Chelliah, 2014). The alignment or consistency between someone's words and actions is the definition of behavioral integrity (Simons *et al.*, 2015).

Simons (2002) found that a pattern between the leaders' words and actions assessed by subordinates is the basis of the formation of trust between leaders and subordinates, suggesting that behavioral integrity is an antecedent of trust between leaders and subordinates. Hoffman *et al.* (2011) strengthen the argument by explaining that from 25 individual differences, integrity is one of the main factors influencing leader effectiveness. Subordinates will pay more attention to the values reflected from their leaders' actions compared to from only their words.

Although some scholars argue that behavioral integrity directly influences leader effectiveness (Hoffman *et al.*, 2011; Kannan-Narasimhan & Lawrence, 2012; Salicrue & Chelliah, 2014; Simons *et al.*, 2015), others are convinced that the relationship between behavioral integrity and leader effectiveness is fully mediated by trust (e.g., Simons, 2002). When there is trust between leaders and subordinates, subordinates are likely to believe that their leaders will behave in accordance with their spoken values and are able to fulfill their promises. This would improve their performance, which is one way to measure leader effectiveness (DeGroot *et al.*, 2011). This kind of trust refers to cognitive trust (McAllister, 1995).

Besides cognitive trust, which refers to trust that is based on performance-relevant cognitions such as competence, responsibility, reliability, and dependability, there are emotional bonds between individuals that are grounded upon the expression of genuine care and the concern for the welfare of other parties, which is called affective trust (McAllister, 1995).

Teams that have better affective relationships show better team performance than other pairs that focus more on cognitive relationship (Chua, Morris, & Ingram, 2009). Palanski, Kahai, and Yammarrino (2011) argue that affective trust partially mediates the relationship between behavioral integrity and team performance. When the interaction between leaders and subordinates happens more frequently and intensively, the relationship between them will be deep and mutual until there is an emotional bond between them. Leaders who have good interpersonal relationships with their subordinates will make their subordinates more willing to follow the leaders' direction, which will improve the team's performance. Team performance is one way to measure leader effectiveness by DeGroot *et al.* (2011).

Therefore, the study asks, "How is the relationship between behavioral integrity and leader effectiveness? Is it directly related or is it mediated by cognitive trust and affective trust?"

1.1 *Hypotheses*

In an organizational context, behavioral integrity is the extent to which employees believe a leader will walk his/her talk. Previous studies show that behavioral integrity is directly related to leader effectiveness (Simons,1999). In his study of hospitality managers, Simons (1999) found that the hotel manager with high behavioral integrity was reported to have more profit (i.e., have higher effectiveness) than the hotel manager with low behavioral integrity. Further, subordinates will pay more attention to the values that are reflected from the leaders' actions compared to from only words. The alignment between someone's words and actions is a form of behavioral integrity by Simons (2002). This hypothesis is supported by studies from Hoffman *et al.* (2011) and Parry & Protor-Thomson (2002), which argue that integrity is one of the main factors influencing leader effectiveness. Leaders with high behavioral integrity will also provide stability to subordinates by behaving in accordance to his/her values, which can influence subordinates to not only work to achieve their targets, but also to take initiative to improve overall effectiveness (Leroy, Palanski, & Simons, 2012). Thus, we hypothesize that:

Hypothesis 1: Behavioral integrity is positively associated with leader effectiveness behavior.

Although some scholars argue that behavioral integrity directly influences leader effectiveness (Hoffman *et al.*, 2011; Kannan-Narasimhan & Lawrence, 2012; Salicrue & Chelliah, 2014; Simons *et al.*, 2015), others are convinced that the relationship between behavioral integrity and leader effectiveness is fully mediated by trust. Simons (2002) states that behavioral integrity is an antecedent of trust. The patterns between the leaders' words and actions that subordinates perceive become the basis of formation of trust between leaders and subordinates (McAllister, 1995). Subordinates must perceive that their leader has high integrity, so that cognitive trust can be established between them (Kannan-Narasimhan & Lawrence, 2012).

Teams that have deep cognitive trust between leaders and subordinates show better performance than other teams that have only superficial trust (Erdem & 2003). Leaders' performance will increase if there is trust between them because subordinates acknowledge their leaders' competence. This is supported by a study from Lee, Gillespie, Mann, & Wearing (2010) that says that trust between leaders and subordinates will influence the leaders' performance. Johnson & Grayson (2005) also argue that cognitive trust between leaders and subordinates will increase subordinates' perception of leader effectiveness.

Subordinates' perceptions towards their leaders' behavioral integrity influences the establishment of trust between leaders and subordinates, which affects the positive consequences of job performance. Performance is one way of measuring leader effectiveness by DeGroot *et al.* (2011). Leaders who demonstrate alignment between their values and their actions can make subordinates understand what the leaders expect from their subordinates. Therefore, subordinates can optimize the efforts to achieve the leaders' expectations, which would increase team performance. This is supported by a study from Simons *et al.* (2015) that shows that cognitive trust fully mediates the relationship between behavioral integrity with the team's performance. Thus, we hypothesize:

Hypothesis 2: Cognitive trust fully mediates the relationship between behavioral integrity and leader effectiveness.

Some studies only use trust that focuses on the leaders' competence as a means to measure the relationship related to leaders' performance (Dirks, 2000; Lee *et al.*, 2010; Simons *et al.*, 2015), which refers to cognitive trust (McAllister, 1995). Subordinates can also believe in leaders because of an emotional bond between them, which refers to affective trust (McAllister, 1995). Cognitive and affective trust can be related to one another, but in the beginning of a relationship, it is important to establish cognitive trust first, then over time the interaction between the two will increase and become deep, so cognitive trust can become affective trust (McAllister, 1995). Asia, particularly Indonesia, has a strong culture of collectivism rather than individualism, so workers prioritize socio-emotional relationships in establishing business relationships, such as giving personal gifts, having lunch together, getting to know each other's families, etc. (Taras, Kirkman, & Steel, 2010).

Pairs of leaders and subordinates which have affective trust towards each other will make subordinates willing to follow their leaders' direction and this will eventually improve team performance (McAllister, 1995). Chua, Morris, & Ingram (2009) show that pairs of leaders and subordinates that have a better affective relationship show better team performance than other pairs that focus more on cognitive relationships. Chua, Morris, and Ingram's findings make us speculate that affective trust plays a larger role than cognitive trust in Indonesia.

The more frequent the interaction between leaders and subordinates, the bigger the chance of an emotional bond building between them, thus creating affective trust (Simons, 2002). As trust develops, subordinates will spend less time covering their backs and more time focusing on their jobs. Individuals who trust one another will develop higher quality social relationships, so they will help one another and go above and beyond the call of duty and work towards higher levels of performance (Palanski & Yammarrino, 2011). Palanski, Kahai, & Yammarino (2010) show affective trust mediates the relationship between behavioral integrity and the team leader's performance. Team performance is one way to measure the leader's effectiveness by DeGroot *et al.* (2011). Thus, we hypothesize that:

Hypothesis 3: Affective trust fully mediates the relationship between behavioral integrity and leader effectiveness.

2 METHODS

2.1 *Participants and procedure*

We tested the study hypothesis using the service industry in Jakarta and Bali. The participants (215 employees) were staff members from the service industry (finance and hotel industry) who were two levels directly below the leaders in the company structure and had been working for at least one year in their current organization. The companies in the service industry were selected because those companies have a dynamic and unpredictable environment. The greater the change in the environment, the companies will increasingly need leaders who are flexible, able to change, and ability to handle work activities and diverse problems (Chen, 2007). The original number of participants under study was 292 employees, yielding an overall response rate of 78.43 percent.

2.2 *Measures*

All scales were adapted from previous research, and back-to-back translations were applied. The data were provided by the subordinates' evaluation of their leaders. To limit common method bias, we used psychological separation (Podsakoff *et al.,* 2003), which means to separate the measurement of the predictor and the criterion variables to make it appear that the measurement of the leadership effectiveness is not connected to the measurement of the behavioral integrity, cognitive trust, and affective trust. All scales used a six point Likert scale (1 = strongly disagree to 6 = strongly agree).

2.2.1 *Leader effectiveness*

Leader effectiveness was measured using 14 items of the leader effectiveness behavior scale adapted from DeGroot *et al.* (2011). The reliability coefficient of this scale was 0.95, and a sample item of this scale was "My leader helped coordinate the team's activities".

2.2.2 *Behavioral integrity*

Behavioral integrity was measured using six items of the behavioral integrity scale adapted from Simons (2007). A sample item of this scale was "My leader delivers on promises". The reliability coefficient of this scale was 0.85.

2.2.3 *Cognitive trust and affective trust*

Trust was measured using an affect and cognition based trust scale from McAllister (1995). This scale consists of eleven items, six items assessing the level of cognitive trust ($\alpha = 0.82$), and five items assessing the level of affective trust ($\alpha = 0.82$). A sample item from this cognitive trust scale was "I can rely on the leader of my leader not to make my job more difficult by careless work", and a sample item of the affective trust scale was "We, my leader and I, can both freely share our ideas, feelings, and hopes".

2.2.4 *Control variables*

We controlled the type of industry (Hoffman *et al.*, 2011), and the leaders' age, education and tenure (Andrews, Kacmar, & Kacmar, 2015).

3 RESULTS

The mean, standard deviation, and correlation are presented in Table 1. It shows that behavioral integrity, cognitive trust, and affective trust between leaders and subordinates have significant positive correlations with leader effectiveness behavior. Behavioral integrity is positively correlated to leader effectiveness behavior ($r = 0.61$, $p < 0.01$). Cognitive trust and affective trust have significant correlations to leader effectiveness behavior ($r = 0.66$, $p < 0.01$; $r = 0.67$, $p < 0.01$).

Table 1 also shows that there is potential multicollinearity between behavioral integrity and cognitive trust ($r = 0.85$, $p < 0.01$). After undergoing a collinearity test, there is no multicollinearity between behavioral integrity and cognitive trust because the value of tolerance was above 0.1 and the value of VIF was under 10 (Field, 2009).

To analyze the relationship between behavioral integrity, cognitive trust, affective trust and leader effectiveness behavior, we conducted a parallel multiple regression analysis using PROCESS by Hayes (2013).

Table 2 shows that behavioral integrity is significantly correlated with cognitive trust ($R^2 = 0.73$, $p = < 0.01$), which means that 73% of cognitive trust could be explained by behavioral integrity. Behavioral integrity was also significantly correlated with affective trust ($R^2 = 0.60$, $p = < 0.01$), which means that 60% of affective trust could be explained by behavioral integrity.

Table 3 reveals that leader effectiveness behavior is not predicted by behavioral integrity ($\beta = -0.03$, $p > 0.05$). The relationship between behavioral integrity and leader effectiveness is fully mediated by cognitive trust ($\beta = 0.74$, $p < 0.01$) and affective trust ($\beta = 0.65$, $p < 0.01$). This indicates that the relationship between behavioral integrity and leader effectiveness

Table 1. Mean, standard deviation, and correlation.

Variable	M	SD	1	2	3	4	5	6	7
Age	4.66	1.56	1						
Education	4.18	1.56	−0.21**	1					
Tenure	4.86	1.80	0.54	−0.08	1				
Behavioral integrity	4.72	0.70	0.00	−0.01	0.01	1			
Cognitive trust	4.79	0.68	0.05	0.05	0.02	0.85**	1		
Affective trust	4.85	0.73	0.09	−0.03	0.04	0.78**	0.77**	1	
Leader effectiveness	5.05	0.68	0.03	0.03	0.02	0.61**	0.66**	0.67**	1

Table 2. Regression coefficients, standard errors, and model summary information of parallel multiple mediation.

	Consequent								
	M_1 (CT)			M_2 (AT)			Y (LE)		
Antecedent	Coeff	SE	P	Coeff	SE	p	Coeff	SE	p
X (Behavioral integrity)	0.79	0.03	<0.01	0.66	0.04	<0.01	1.36	0.13	<0.01
M_1 (Cognitive trust)	–	–	–	–	–	–	0.94	0.24	<0.01
M_2 (Affective trust)	–	–	–	–	–	–	0.99	0.21	<0.01
Constant	6.11	1	<0.01	5.21	1.12	<0.01	32.16	3.65	<0.01
	$R^2 = 0.73$			$R^2 = 0.60$			$R^2 = 0.49$		
	$F(3, 206) = 184.94$,			$F(3, 206) = 102.20$,			$F(5, 204) = 39.06$,		
	$p = < 0.01$			$p = < 0.01$			$p = < 0.01$		

Table 3. Mediation effect.

Effect	Coeff	Explanation
Total effect	1.36**	–
Direct effect	−0.03	–
Indirect effect 1 (*Behavioral integrity* and Leader effectiveness mediated by *Cognitive trust*)	0.74**	Fully mediated
Indirect effect 2 (*Behavioral integrity* and Leader effectiveness mediated by *Affective trust*)	0.65**	Fully mediated

cannot be explained simply by behavioral integrity alone, but through mediators, which in this study are cognitive and affective trust. This corresponds to what was said by Hayes (2013) that a predictor variable can be said to influence or be directly related to the criterion variable if it has a coefficient with a value greater than zero (0.00).

4 DISCUSSION

4.1 *Theoretical and practical implications*

The results of this study indicate that cognitive and affective trust fully mediate the relationship between behavioral integrity and leader effectiveness. This means that leaders who are considered to have high behavioral integrity will have cognitive and affective trust with their subordinates. The better the quality of cognitive and affective trust between leaders and subordinates, the stronger the subordinates' perception that their leaders are effective.

This study contributes to the current discussion of the relationship between behavioral integrity and leader effectiveness by showing that behavioral integrity must be mediated by trust, which is cognitive trust and affective trust. These results are in line with McAllister (1995) who states that the subordinates' evaluation of the perceived pattern of alignment between the leaders' words and actions will make subordinates willing to be in vulnerable situations because subordinates see that their leaders consistently behave in accordance with their spoken values, and are able to fulfill their promises. When subordinates feel that their leaders are not reliable or trustworthy, they become unable to work in accordance with the strategies developed by their leaders. This leads to difficulties in team performance. This is because leaders who show consistency between their values and their actions are able to make their subordinates understand what the leaders expect from them. Therefore, subordinates can optimize their efforts to achieve the leaders' expectations for them without having to guess what their leaders expect. Achieving the leaders' expectations increases team performance, which is one way to measure leader effectiveness (Simons *et al.,* 2015).

These findings are consistent with Palanski, Kahai, & Yammarino (2011), who argue that longer time working with each other results in more intense interaction between superiors and subordinates. The trust between the two will then become deeper until it reaches the stage at which the subordinates can communicate their problems with their leaders and believe that their leaders will listen. Subordinates who have emotional ties with their leaders will have higher quality of social relationships, where the leaders will help their subordinates beyond their duties as leaders. When the subordinates have reached this stage, the team will have more integrated values, so the subordinates will follow the direction of their superiors in achieving the company's goals. This will lead to an increased team performance.

The relationship between behavioral integrity and leader effectiveness fully mediated by cognitive trust suggests that to be effective in leading, leaders must consistently behave according to their values. If a leader does not consistently behave in accordance with his/her spoken values, that leader will not be perceived as an effective leader by his/her subordinates.

4.2 *Limitations*

Like all studies, this study has a limitation. The data for the predictor and criterion variables in this study were obtained from the same source. This can potentially lead to the emergence of a common method bias (Podsakoff, 2003). However, to control the possibility of a common method bias, we used psychological separation (Podsakoff, 2003) by separating the measurement into three booklet forms.

Fortunately, the strengths of this study outweigh its limitation. First, this study has a considerably large number of respondents (215 samples) with also a very high response rate (78.46%). The large number of respondents can improve generalization, and is increasingly able to describe the population. Second, the respondents of this study were leaders who work in the companies, while many leadership research use students with no work experience as respondents. Third, the research design has considered the potential of a common method bias, and has adopted strategies (e.g., psychological separation) to deal with it.

5 CONCLUSION

The purpose of this study is to examine whether cognitive trust and affective trust fully mediate the relationship between behavioral integrity and leader effectiveness. Our findings suggest that there is no significant relationship found between behavioral integrity and leader effectiveness behavior, and cognitive trust and affective trust fully mediates the relationship between behavioral integrity and leader effectiveness. This research is particularly important because it delineates the mechanism under which behavioral integrity impacts leader effectiveness, so information on the practices to improve leader effectiveness can be provided.

REFERENCES

Andrews, M.C., Kacmar, K.M. & Kacmar, C. (2015) The Interactive Effects of Behavioral Integrity and Procedural Justice on Employee Job Tension. *Journal of Business Ethics.* 126 (3), 371–79. doi:10.1007/s10551-013-1951-4.

Chua, R.Y.J., Morris, M.W. & Ingram, P. (2009) Guanxi versus Networking: Distinctive Configurations of Affect- and Cognition-Based Trust. *Networks of Chinese and American Managers.* 40 (3), 490–508.

DeGroot, T., Aime, F., Johnson, S.G. & Kluemper, D. (2011) Does Talking the Talk Help Walking the Walk? An Examination of the Effect of Vocal Attractiveness in Leader Effectiveness. *The Leadership Quarterly.* 22 (4), 680–689. doi:10.1016/j.leaqua.2011.05.008.

Detiknews. (2017) *Kisah Jonan Perbaiki KAI: Dari Rapor Merah Hingga Untung Hampir Rp 1 T.* [Online] Available from: http://news.detik.com/berita/d-2938862/kisah-jonan-perbaiki-kai-dari-rapor-merah-hingga-untung-hampir-rp-1-t. [Accessed 25th January 2017].

Dirks, K.T. (2000) Trust in Leadership and Team Performance: Evidence from NCAA Basketball. *The Journal of Applied Psychology.* 85 (6), 1004–1012.

Erdem, F. & Ozen, J. (2003) Cognitive and Affective Dimensions of Trust in Developing Team Performance. *Team Performance Management: An International Journal.* 9 (5/6), 131–135. doi:10.1108/13527590310493846.

Hannah, S.T., Sumanth, J.J., Lester, P. & Cavarretta, F. (2014) Debunking the False Dichotomy of Leadership Idealism and Pragmatism: Critical Evaluation and Support of Newer Genre Leadership Theories. *Journal of Organizational Behavior.* 35 (5), 598–621. doi:10.1002/job.1931.

Hannes, L., Michael, P. & Tony, S. (). Authentic Leadership and Behavioral Integrity as Drivers of Follower Commitment and Performance. *Journal of Business Ethics.* 107 (3), 255–264. doi:10.1007/s10551-011-10-36-1.

Hayes, A.F. (2013) *Introduction to Mediation, Moderation, and Conditional Process Analysis: A Regression-Based Approach.* 1st edition. New York, The Guilford Press.

Hoffman, B.J., Woehr, D.J. Maldagen-Youngjohn, R. & Lyons, B.D. (2011) Great Man or Great Myth? A Quantitative Review of the Relationship between Individual Differences and Leader Effectiveness. *Journal of Occupational and Organizational Psychology.* 84 (2), 347–381. doi:10.1348/096317909X485207.

Johnson, D. & Grayson, K. (2005) Cognitive and Affective Trust in Service Relationships. *Journal of Business Research,* Special Section: Attitude and Affect, 58 (4), 500–507. doi:10.1016/S0148-2963(03)00140-1.

Judge, T.A., Bono, J.E., Ilies, R. & Gerhardt, M.W. (2002) Personality and Leadership: A Qualitative and Quantitative Review. *The Journal of Applied Psychology.* 87 (4), 765–780.

Kannan-Narasimhan, R. & Lawrence, B.S. (2012) Behavioral Integrity: How Leader Referents and Trust Matter to Workplace Outcomes. *Journal of Business Ethics.* 111 (2), 165–178. doi:10.1007/s10551-011-1199-9.

Kwantes, C.T. & Boglarsky, C.A. (2007) Perceptions of Organizational Culture, Leadership Effectiveness and Personal Effectiveness across Six Countries. *Journal of International Management.* 13 (2), 204–230. doi:10.1016/j.intman.2007.03.002.

Lee, P., Gillespie, N., Mann, L. & Wearing, A. (2010) Leadership and Trust: Their Effect on Knowledge Sharing and Team Performance. *Management Learning.* 41 (4), 473–491. doi:10.1177/1350507610362036.

McAllister, D.J. (1995) Affect- and Cognition-Based Trust as Foundations for Interpersonal Cooperation in Organizations. *The Academy of Management Journal.* 38 (1), 24–59. doi:10.2307/256727.
Nugroho, S.A. (2012) Transformasi PT KAI: Mengurai Benang Kusut. *SWA.co.id.* November 29. [Online] Available from: http://swa.co.id/swa/trends/management/transformasi-pt-kai-mengurai-benang-kusut.
Palanski, M.E., Kahai, S.S. & Yammarino, F.J. (2011) Team Virtues and Performance: An Examination of Transparency, Behavioral Integrity, and Trust. *Journal of Business Ethics.* 99 (2), 201–216. doi:10.1007/s10551-010-0650-7.
Parry, K.W. & Proctor-Thomson, S.B. (2002) Perceived Integrity of Transformational Leaders in Organisational Settings. *Journal of Business Ethics.* 35 (2), 75–96. doi:10.1023/A:1013077109223.
Podsakoff, P.M., MacKenzie, S.B., Lee, J-Y. & Podsakoff, N.P. (2003) Common Method Biases in Behavioral Research: A Critical Review of the Literature and Recommended Remedies. *The Journal of Applied Psychology.* 88 (5), 879–903. doi:10.1037/0021-9010.88.5.879.
Riggio, R.E. & Reichard, R.J. (2008) The Emotional and Social Intelligences of Effective Leadership: An Emotional and Social Skill Approach. *Journal of Managerial Psychology.* 23 (2), 169–185. doi:10.1108/02683940810850808.
Rockstuhl, T., Seiler, S., Ang, S., Van Dyne, L. & Annen, H. (2011) Beyond General Intelligence (IQ) and Emotional Intelligence (EQ): The Role of Cultural Intelligence (CQ) on Cross-Border Leadership Effectiveness in a Globalized World. *Journal of Social Issues.* 67 (4), 825–840. doi:10.1111/j.1540-4560.2011.01730.x.
Salicru, S. & Chelliah, J. (2014) Messing with Corporate Heads? Psychological Contracts and Leadership Integrity. *Journal of Business Strategy.* 35 (3), 38–46. doi:10.1108/JBS-10-2013-0096.
Schaubroeck, J., Lam, S.S.K. & Peng, A.C. (2011) Cognition-Based and Affect-Based Trust as Mediators of Leader Behavior Influences on Team Performance. *The Journal of Applied Psychology.* 96 (4), 863–871. doi:10.1037/a0022625.
Simons, T. (2002) Behavioral Integrity: The Perceived Alignment Between Managers' Words and Deeds as a Research Focus. *Organization Science.* 13 (1), 18–35. doi:10.1287/orsc.13.1.18.543.
Simons, T., Friedman, R., Liu, L.A. & Parks, J.M. (2007) Racial Differences in Sensitivity to Behavioral Integrity: Attitudinal Consequences, In-Group Effects, and 'Trickle Down' Among Black and Non-Black Employees. *Journal of Applied Psychology.* 92 (3), 650–665. doi:10.1037/0021-9010.92.3.650.
Simons, T., Leroy, H., Collewaert, V. & Masschelein, S. (2015) How Leader Alignment of Words and Deeds Affects Followers: A Meta-Analysis of Behavioral Integrity Research." *Journal of Business Ethics.* 132 (4), 831–844. doi:10.1007/s10551-014-2332-3.
Simons, T.L. (1999) Behavioral Integrity as a Critical Ingredient for Transformational Leadership. *Journal of Organizational Change Management.* 12, 89–104. doi:10.1108/09534819910263640.
Soltani, B. & Maupetit, C. (2015) Importance of Core Values of Ethics, Integrity and Accountability in the European Corporate Governance Codes. *Journal of Management & Governance.* 19 (2), 259–284. doi:10.1007/s10997-013-9259-4.
Sutianto, F.D. (2014) Selama Di KAI, Jonan Pernah Jadi Kondektur Hingga Tidur Di Kereta. *Detikfinance*, October 28. [Online] Available from: https://finance.detik.com/berita-ekonomi-bisnis/d-2731867/selama-di-kai-jonan-pernah-jadi-kondektur-hingga-tidur-di-kereta.
Taras, V., Kirkman, B.L. & Steel, P. (2010) Examining the Impact of Culture's Consequences: A Three-Decade, Multilevel, Meta-Analytic Review of Hofstede's Cultural Value Dimensions. *The Journal of Applied Psychology.* 95 (3), 405–439. doi:10.1037/a0018938.
Zaccaro, S.J. (2007) Trait-Based Perspectives of Leadership. *The American Psychologist.* 62 (1), 16–47. doi:10.1037/0003-066X.62.1.6.

Diversity in Unity: Perspectives from Psychology and Behavioral Sciences – Ariyanto et al. (Eds)
 ISBN 978-1-138-62665-2

Playground breakpoint mapping of urban open spaces in DKI Jakarta province

R.K. Pratomo, M.M. Ali & Y.D. Pradipto
Faculty of Psychology, Universitas Bina Nusantara, Jakarta, Indonesia

ABSTRACT: Urban Open Spaces play a significant role in the life of city dwellers. They are also important places for children to play and explore, and to develop their cognitive, social and physical abilities, and emotional health. However, many researchers in Indonesia overlook the importance of Urban Open Spaces for children. A recent study showed preliminary research that was based on continuous research over three years to map the facilities and conditions of the Urban Open Spaces in *Daerah Khusus Ibukota* (DKI) Jakarta Province in accordance with the childrens' needs. Data from 48 Urban Open Spaces run by the local government and private sectors were analysed utilising descriptive statistics. We found that many of the 48 Urban Open Spaces were not equipped with 'A Variety of Spaces', 'Loose Parts' and 'Three-dimensional Layering' facilities. We also discovered that Central Jakarta has better Urban Open Spaces for children, based on 'desirable playground amenities'. The lack of facilities for children in many of the Urban Open Spaces in DKI Jakarta Province resulted in children having fewer opportunities to develop their abilities and potential (multiple intelligences) to the fullest.

1 INTRODUCTION

The existence of Urban Open Spaces at city parks has now entered a crucial stage, along with the surge of a growing population and the accompanying needs of the residents. Most of the Urban Open Space (RTH) was in a dire condition due to the lack of periodic maintenance. Furthermore, the city parks, located in Jakarta Province, were not considered a priority, thus they were neglected or the space reallocated to become buildings that would generate income for the government, such as public business facilities and private commercial buildings. Within an area of approximately 650,000 hectares, Jakarta only retained around 4,000 hectares of green open spaces. However, if the ideal Jakarta were to be calculated, factoring in all the other needs for the land, it should have 9,750 hectares, rather than the existing area of 3,230 hectares, of city parks.

This figure does not yet correspond with the General Regional Spatial Plan 2005, which set out that 35 per cent of spaces in Jakarta should be utilised as Urban Open Spaces, along with all of their outdoor activities. Parks functioning as an open sociopetal space can become a city resident's choice as a means to socialise, exercise, relax or have fun individually or collectively. In short, RTH located in Jakarta can become a space to get together, interact with each other and a place to hold activities for Jakarta's citizens, even for the surrounding buffer zone area residents of DKI Jakarta Province.

RTH city parks, complete with all the facilities, are one thing that people from all levels of society want to enjoy. In fact, based on the RTH facility data, the number of parks in Jakarta can be classified as somewhat lacking, both quantitatively and qualitatively. Kaplan (1987) explained that, in the context of a city that had an orientation towards public park space, it should have a recreational effect amidst the bustle of society. The city park can also reduce stress (Kaplan & Kaplan, 1989), and this is what brought us, as researchers, to the research topic.

In its development phase, the city park can also serve as a social activity media and becomes a means of social learning for its users, including children, adolescents, adults and the elderly. The activities related to children's play that occur in RTH can be categorised into: a) physical games that require the player to always be moving; b) creative games that direct the children to manipulate their fantasy and imaginary world through things in the park; c) social games, emphasising social and interpersonal relationships among its players, which can hone the children's emotional intelligence; d) sensory games, stimulated by the elements present in the RTH park, so that the children would be sensitive to the environment around them; and e) quiet games, which allows the children to rest and play by themselves. RTH is a place where children do activities, and in relation to the children's activities, RTH can become a nurturing facility for children's growth. According to Rojals del Alamo (2002), between the ages of 0–3 years old, a child will learn through their formative phase and learn to control their delicate motor skill movement, which is already present in a child of 0–3 years old.

In the next phase, which is 3–6 years old according to Rojals del Alamo (2002), a child has developed social awareness so that children will play collectively as a group. In this stage, the other intelligence types from the Gardner theory have started to form, such as: linguistic, spatial, kinaesthetic, interpersonal, intrapersonal and naturalistic intelligence (Gardner, 2011). With reference to the planning of an ideal playground, it has to be able to accommodate the child's playing and gesticulating needs. Shaw (1987) explained that there are seven dimensions of measurement of an ideal park that are seen as 'desirable playground amenities'. The first dimension is 'A Sense of Place', where the initial impression can be easily captured by the cognitive function and the five senses when we are at the location. The impression created is a perception of the whole physical milieu located inside the park, in such a way that the children feel comfortable being inside the park. The second dimension is 'Unity', in which the impression of each existing room and park area merges with one another, so that it can stimulate the child's cognitive ability and improve their ability to map the existing places. 'A Variety of Spaces' refers to the different expanse levels a playground must have between each room or area inside the park environment, so that the child can feel the mental space phenomena, which is formed from the experience of using personal capsule space. 'Key places' is the vocal point of all areas present inside the park, analyzes so the child could be able to analyzethe different phenomenon which is felt by the receiver's sense. 'A System of Pathways' is the hub or the connector between one area and another within the park environment. This facility is intended to provide the children with the option of using different spaces when they are doing activities. 'Three-dimensional Layering' is the existing park facility, which aims to avoid a monotone impression when making the landscape furniture that functions as the children's playground facility. 'Loose Parts' are supporting activity facilities in the form of playing devices or activity instruments located in the playground environment.

According to Lewin or Lewin's equation for behaviour $B = f(P, E)$, the function of the environment when interacting with the personality function has an impact on the behaviour of the individual. The intended environment is the physical environment and the social environment, and this theory shows that the environmental aspects also affect the psychological aspects of humans. Moreover, the park and open spaces are a harmonious one-whole situation that function as the activity container that can stimulate the child's gesticulation Fisher et al., 1984). An RTH park that has good facilities will produce an addicting effect to the children to redo the outside room activities from solitaire to socialising with other children, and this condition will indirectly stimulate the child to generate their maximum intelligence based on their growing phase, according to Rojals del Alamo (2002).

Therefore, the purpose of this research was to uncover what the children's playground typology image of *Ruang Terbuka Hijau* (RTH) (Open Green Space) looked like and how to create a children's playground RTH typology map, as well as to view and measure the quality and conditions of the RTH park facilities in DKI Jakarta Province. It is important to note that this current research was the initial research of a three-year plan of sustainable research, which consists of mapping, providing multiple intelligence and undertaking intervention with regards to facilities for children.

2 METHOD

This research aimed to observe the children's playground RTH's typology image and the condition of the playground or park's facilities. Based on the acquired information type, this research can be categorised as descriptive qualitative and quantitative research. The data obtained in this research was in a numeral form that would be analysed statistically, and the image acquired would be set as the blueprint reference for the ideal condition of a playground. This research sample was obtained with a non-probability sampling technique, that is, each of the population members did not have an equal opportunity to be selected as a research participant (Gravetter & Forzano, 2012). This method was more efficient in terms of both time and expenditure. The non-probability sampling technique used was convenience sampling. The questionnaire was distributed using the face-to-face method, and it was conducted at the RTH park location, which was used as a research location in the afternoons to early evenings and in the mornings. The participants in this research were 73 park users. A park user is defined as everyone that was using the park and the park's facilities. The users were the park visitors and were city residents who came from both the surrounding area and outside the area. They were doing activities that were being observed by us, and their data were collected using questionnaires that we distributed (Shaw, 1987), with an age range of between 13–60 years old. These research locations included 48 RTH parks inside the DKI Jakarta Province, which are managed by the private sector as well as the city government.

The measurement tool used in the research to measure the ideal park typology was the basic concept of 'desirable playground amenities' (Shaw, 1987). The questionnaires consisted of 17 questions in the form of the Likert scale, each with four possible answers, that is: 'Strongly disagree', 'Disagree', 'Agree' and 'Strongly agree'.

3 RESULT

Table 1 shows the number of respondents in relation to gender.

From the data above, it is explained that there were more male respondents than female respondents in this research, with a ratio of 73:27. Most respondents ranged from 12 to 20 years old, with a total of 38 people (52%). Meanwhile, the respondents ranging from 21 to 30 years old amounted to 26 people (32%) and the respondents ranging from 31 to 60 years old amounted to nine people, which was 16% of all the respondents.

From the table above, eight parks are in North Jakarta, nine parks are in East Jakarta, ten parks are in South Jakarta, 14 parks are in West Jakarta and seven parks are in Central Jakarta.

The data above was based on the calculation per dimension. The highest score inferred was 'Unity', with a mean score of 94, followed by 'Sense of Place', 'Key Places' and 'A System Pathways', with a mean score of 81. Meanwhile, the mean score of 'Three-dimensional Layering' was 72. For 'Loose Parts', the mean scored 63. The lowest mean score was on 'A Variety of Spaces', which amounted to 56.

The item 'A Variety of Spaces', which amounted to only a score of 56—the lowest mean score—showed the playground facilities distribution of RTH parks for children in Jakarta. It meant that some of the parks were lacking in a variation of styles between the predominantly green plains combined with a stretch of sand, benches, hard surfaces and floor-covered

Table 1. The subject data picture of subjects viewed from the gender of respondent.

	Frequency	Percentage
Male respondents	53	73
Female respondents	20	27
Total	73	100

Table 2. Park object research distribution table.

	North	East	South	West	Central
	Jogging 1	Jatinegara	Ayodya	Catleya	Suropati
	Jogging 2	Berkah	Lansia Langsat	Hutan Kota Sreng-seng	Situ Lembang
	Mangga	Ex pasar bypas	Martha Tiahahu	Kebon Jeruk	Cideng Barat
	Primer Mansion	Naga Raya	Honda Tebet	Kosambi	Menteng
	Green Bay	Kelapa Kopyor	Tangkuban perahu	Greenlake	Amir Hamzah
	CBD Pluit	Kelapa Sawit	Kodok	Kodam Jaya	Kodok
	Pancasila	Cijantung	Kura Kura	Surya Mandala	Petojo
	Gading Grande	Juntak Hijau Daun	Gajah	Citra Garden 5	
		Merah Delima	Barito	Empang Grogol	
			Hangtuah	Anggrek Garuda	
				Komodo	
				Aries	
				Kodok	
				Intercon	
Total	8	9	10	14	7

Table 3. RTH satisfaction based on Shaw's item list.

	Mean score
Sense of place	81
Unity	94
A variety of spaces	56
Key places	81
A system of pathways	81
Three-dimensional layering	72
Loose parts	63

Information: The above table is an explanation from mean score.

places. Viewed from the area distribution mapping, the 'A Variety of Spaces' facilities that conformed with the standard in terms of quality had more units in West Jakarta and South Jakarta, while in East Jakarta there were still many that did not conform with the 'A Variety of Spaces' standard of the dimensions of an ideal park.

After finishing the mapping from the data spread obtained, we found that the parks that had those facilities were more in number and better in quality in the Central Jakarta area. These RTH parks are managed by the DKI Jakarta Province Government. Parks that were located in this area had fulfilled the quality standard, which was based on 'desirable playground amenities', since the parks already featured playground rides for children, not only vegetation, sculptures, water elements and mere green fields. Parks in Central Jakarta apparently consisted not only of the big parks, which had the children playground provision element, but also small parks which were located on the side streets, such as Dr. Wahidin RTH park. This park, besides having vegetation, landscape furniture and footpath guide ways, already had children's playground facilities.

In other areas, such as North Jakarta, the RTH park that met the standard based on 'desirable playground amenities' was Jogging park at Kelapa Gading 1 and 2. The playground facilities for children in this RTH park were already fully-featured in both quality and quantity. Besides the vegetation and the outdoor space facilities, such as an amphitheatre and meeting room, and even the supporting facilities, such as toilets, this park had already procured a sports equipment facility specifically for children's activities.

In addition to facilities for children, RTH Menteng park was also equipped with sport facilities intended for adults, so that the caregivers could also be motivated to come to this RTH to perform their outdoor space activities. The domino effect of this condition indirectly motivated the parents to come to the park and to bring their children. Some of the RTH parks in Jakarta, such as Suropati park in Central Jakarta and Ayodya in South Jakarta, looked very spacious and had achieved the element of 'desirable playground amenities'. However, there were not any available park rides and facilities for children to play that could motivate the child to perform playing activities and to explore the space. Nevertheless, from the concept of 'A Variety of Spaces', both of these playground parks already had a good area or a between rooms landscape concept. From this condition, the child could feel the mental space phenomena created from the private capsule room experience.

'Key Places', or the main activity points in the RTH parks, already existed and had been nurtured well. From this condition, the child would have the ability to analyse the different phenomena felt by the receptor senses. The key places in question also function as the central activities in the park area. 'A System of Pathways', or a connector between one area and another area located in the same park environment, had also been constructed well.

Meanwhile, the quality of many of the RTH parks in the East Jakarta area was still not compliant with the RTH 'Three-dimensional Layering' standard of Shaw. One of the RTH parks used as a sample was RTH park in Jl.Raya Bypas Cawang-Tj.Priuk, which was located in the Jatinegara area. Even though this park was located at a strategic point for urban society activities and located in a commercial area, it was not taken care of well and was devoid of child-friendly facilities. This park also did not have the supporting aspect of a standard city park. As the Shaw's standards of RTH facility was unseen here, it cause children to feel reluctant to doactivities here. The condition of existing parks in East Jakarta nearly resembled the condition of the Jatinegara Viaduct RTH park; for example, in Barkah park and Juntak park. The absence of facilities that complied with the rules and standards of 3DL Shaw was blatantly observable in both parks.

4 DISCUSSION

This research aimed to map the children's playground facilities located in RTH Parks in DKI Jakarta. From the results, it could be inferred that, quantitatively, 'Unity' acquired the highest mean score. However, the qualitative data indicated that 'Unity' in some of the parks still appeared insufficient, based on the standard proclaimed by Shaw. There were still many playground elements that failed to have a unified impression between one facility and another. Those facilities were distantly separated from each other and not child-friendly. Meanwhile, in the qualitative review, 'A System of Pathways' in some of the parks was still absent for children. Therefore, the children would have difficulty moving around the park. The other aspect that needed to be marked was the 'Key Places' dimension. At some of the parks, the 'Key Places' were not visible due to a lack of interest from the children in going to the park. For the 'A Variety of Spaces', there were many parks with inadequate spaces for the children to do their physical and mental activities. With regards to 'Loose Parts', most of the parks did not qualify for this aspect. Referring to the points above, there was a logical consequence: if the present condition of the park is not up to the proper standard, based on Shaw, then the children did not respond as well to the external stimulus. For example, with the lack of 'Loose Parts' in the RTH, the children's fine motor skills coordination was not stimulated effectively. The present condition of the parks also did not give the children an opportunity to explore the earth's contour difference and the environmental perception of the open spaces (A Variety of Spaces). When added to the 'Three-dimensional Layering' condition, both the natural intelligence and bodily kinesthetic intelligence of the children would not develop optimally. In short, based on Gardner's multiple intelligence theory, most of the parks lacked the stimulation necessary for children's development.

5 CONCLUSION

Based on the obtained data analysis result, it could be concluded that many RTH playground parks in DKI Jakarta Province are still left unequipped. It is far from the concept of ‘desirable playground amenities The facilities that were still missing, among others, were ‘Three-dimensional Layering’, namely the facilities in RTH parks in the form of landscape furniture, such as slides, swings, children's playground rides and other furniture that could function as children's playground facilities. Other facilities that were still absent in the RTH parks were ‘Loose Parts’. which supported children activities in the playground environment. Examples of the unavailable ‘Loose Parts’ in the RTH park area were, for example, a soccer ball for children, to play with, bicycles or facilities for stimulating children's creativity.

It had become increasingly important to examine the physical conditions of the parks, based on Shaw's view, due to the fact that the park's physical conditions and facilities became the foundation for the optimal growth stimuli of the children. If the park's physical facilities are adequate, this will stimulate the multiple and kinesthetic intelligence aspects of the children optimally.

As a suggestion, we, as researchers, offered to proliferate the RTH park facilities that had a ‘Three-dimensional Layering’. They were illustrated above because the existence of those facilities would give more alternatives and motivation to the parents to bring their children along to the RTH parks. In addition, the existing 3DL should add a safety layer and more convenient equipment so that children doing activities in an RTH park would not suffer an injury. Therefore, in the second year, the research will focus on the children's multiple intelligences that are supported by the RTH facilities and the child-friendly integrated public spaces (*RPTRA*) facilities.

REFERENCES

Bell, P.A., Greene, T.C., Fisher, J.D. & Baum, A.S. (1996). *Environmental psychology*. Orlando, FL: Harcourt Brace & Company.

Fisher, J.D., Bell, P.A. & Baum, A.S. (1984). *Environmental psychology* (2nd ed.). New York, NY: Holt, Rinehart & Winston.

Gardner, H. (2011). *Frames of mind: The theory of multiple intelligence*. New York, NY: Basic Books.

Gifford, R. (1996). *Environmental psychology: Principles and practice*. Boston, MA: Allyn & Bacon.

Gravetter, F.J. & Forzano, L.B. (2012). *Research methods for the behavioral sciences*. Belmont, CA: Wadsworth.

Kaplan, S. (1987). Aesthetics, affect, and cognition: Environmental preference from an evolutionary perspective. *Environment and Behavior*, *19*(1), 3–32. doi:10.1177/0013916587191001

Kaplan, S. & Kaplan, R. (1989). The visual environment: Public participation in design and planning. *Journal of Social Issues*, *45*(1), 59–86. doi:10.1111/j.1540-4560.1989.tb01533.x

Rojals del Alamo, M. (2002). *Design for fun: Playgrounds*. Barcelona, Spain: LINKS International.

Shaw, L.G. (1987). Designing playgrounds for able and disabled children. In C.S. Weinstein & T.G. David (Eds.), *Spaces for children* (pp. 187–213). New York, NY: Springer. doi:10.1007/978-1-4684-5227-3_9

Simonds, J.O. (1998). *Landscape architecture: A manual of site planning and design*. New York, NY: McGraw-Hill.

Diversity in Unity: Perspectives from Psychology and Behavioral Sciences – Ariyanto et al. (Eds)
© 2018 Taylor & Francis Group, London, ISBN 978-1-138-62665-2

The correlation between motivational values and emotions of shame and guilt in adolescents

M. Tarisa & L.R.M. Royanto
Faculty of Psychology, Universitas Indonesia, Depok, Indonesia

ABSTRACT: This research was conducted to examine the correlations between motivational values and the emotions of shame and guilt in adolescents. The Test of Self-Conscious Affect 3 (TOSCA-3), developed by Tangney et al. (2000) and revised by Tambusai (2013) and Qonita (2013), was used to measure shame and guilt. Meanwhile, the Portrait Values Questionnaire (PVQ), developed by Schwartz (2003) and modified by Halim (2008), was used to measure motivational values. Before being used, the PVQ had been revised to meet the context of Indonesian adolescents. Approximately 500 adolescents across the Special Region of Jakarta, aged between 15 and 19, participated in this study. The results show that there were correlations between openness to change and conservation with shame, and there were also correlations between self-transcendence and conservation with guilt.

1 INTRODUCTION

In the past few years, cases of juvenile delinquency in Indonesia have continued to increase, especially in a big city like Jakarta. According to Sari (2012, cases of juvenile delinquency reported to Police Headquarter of the Greater Jakarta Region increased by 33% in 2012. There were 30 cases of juvenile delinquency in 2011, while in 2012 this rose to 41 cases (Sari, 2012). However, this number decreased by 76% in 2013, when only ten cases were found (Faris, 2013). Despite the significant decline, in reality there is an increase in some forms of delinquency acts, especially student brawls. There were 229 cases of student brawls in Jakarta throughout January to October 2013, which showed an increase of 44% compared to the year 2012 (Hermawan, 2013).

The high number of juvenile delinquency cases has led to a wide range of public opinion, particularly related to the morale of Indonesia as a nation. Society sees that in recent years, Indonesia has been experiencing moral degradation. One of the opinions that has appeared in society says that moral degradation is a social problem that occurs in the community, and which is particularly found in teenagers (Karyanto, 2013). In agreement with this notion, Reni Marlinawati, a member of the MPR RI, also argued that Indonesia is currently facing moral degradation, especially among the younger generations (Marlinawati, 2013). Furthermore, Firman Soebagyo, Vice Chairman of Commission IV of the DPR RI, said that presently, Indonesia is not experiencing an economic crisis, but a moral crisis. This moral crisis is seen as the most dangerous threat for Indonesia (Soebagyo, 2013).

This phenomenon raises the concerns of many parties in society, not only from the government and educational practitioners, but also from the general public. Most of them argued that moral education is the solution to this problem. Moral education itself is seen as guidance that can help the younger generation to act more in accordance with the norms prevailing in society. Moral education for teenagers can be undertaken in many ways, such as through school as a place of formal education, extracurricular activities, or in the family as a place of informal education. Furthermore, this form of education can be undertaken by teaching teens to have a sense of shame and guilt. For example, in school, teachers can give punishments to students who cheat, come to school late or do not do the task given by the teacher. This form of punishment is expected to foster a student's sense of shame and guilt, so they will not repeat it again in the future.

The process of growing or developing individual morals cannot be done only at the cognitive level, it is also necessary to introduce the aspects of moral virtues and identity (Nan, 2007). In the context of moral virtues, we need to consider the role of moral emotions, which includes two types of emotions: the emotions of shame and guilt. These emotions are considered as an important aspect of morals, particularly the moral motivation (Nan, 2007). Furthermore, Rest (1984), in Silfver et al. (2008), said that there are two factors that become the main source of moral motivation. The first factor is the values that are owned by individuals, which includes the moral principles that are based on those values. The existence of values is considered to be the basic impulse of one's morality. Meanwhile, the second factor is the role of empathy, shame and guilt in moral development. These emotions are seen as the core of moral motivation.

According to Silfver et al. (2008), there is a theoretical association between values and moral emotions. This is because a person's values will affect his/her behaviour, which later will be assessed by him/her or by the environment. This assessment will provoke the appearance of the emotions of shame and guilt, which are part of the moral emotions. For example, a student who upholds honesty, will not then allow their friend to cheat during exams. He/she will report this action to the teacher, because he/she judges this action as a wrong act. If he/she does not report it to their teacher, it will give rise to their feelings of shame or guilt. Nevertheless, Silfver et al. (2008) state that studies regarding the relationship between moral emotions and values are still very rare, and, therefore, more study is needed to better understand the relationship between emotions and moral values.

As mentioned before, motivational values are the foundation of individual behaviour and also act as guidance for people to decide whether or not their behaviour is good or bad. Furthermore, motivational values reflect individuals' purposes. Motivational values also act as a barrier for people's behaviour so they will not do something bad. Thus, motivational values play an important role in an individual's life.

Research regarding the relationship between motivational values and moral emotions has been conducted by Silfver et al. (2008). This research was conducted on teenagers of around 15–19 years old and military forces in Finland. Similar research was also conducted by Nan in 2007 on university students in Midwestern America. Both studies show a relationship between motivational values and the emotions of shame and guilt. Shame is related more to self-enhancement, particularly the motivational value of achievement (Silfver et al., 2008). For example, the relationship between shame and achievement could be seen from the behaviour of students who got low scores in examinations. If the student has motivational values for achievement, when they get a low score in examinations they will feel incompetent as a student, and this will enhance their feelings of shame. Another research shows that shame has a positive relationship with conservation, particularly the motivational value of tradition (Nan, 2007).

On the other hand, guilt is positively related to conservation, particularly the motivational values of conformity and tradition (Nan, 2007). The relationship between guilt and tradition can be seen through the behaviour of students who have traditional customs. If a student has a motivational value of tradition, they will act according to their traditional customs. Thus, if they act in an opposite way to their traditional customs, it will enhance a feeling of guilt. Moreover, Nan (2007) also argued that guilt is positively related to self-transcendence, particularly the motivational value of universalism.

From the above discussion, it can be concluded that it is important for this study to be conducted. This is because motivational values are the basis of human behaviour and also act as triggers for the appearance of the emotions of shame and guilt within individuals. Thus, if it can be proven that motivational values are related to moral emotions, then the result of this study can be applied in the field of moral education in adolescents through the teaching of these motivational values. In addition, this research becomes necessary because research on the relationship between moral emotions and motivational values has not previously been done in Indonesia.

2 THEORETICAL OVERVIEW

2.1 *Shame and guilt emotions*

Tangney (1999) defined shame as a painful emotion, which usually occurs together with feelings of inferiority, worthlessness and helplessness. Shame is described as a negative emotion experienced by the individual when he/she fails to comply with the rules or social standards that have been internalised, including the rules of morality or ability, as well as the aesthetic values (Tangney, 1999; Tracy & Robins, 2004, in Orth et al., 2010). Shame focuses on the personal assessment of the individual as a whole. Shame emerges in a situation that is more common or broader in nature, and not necessarily related to the moral situation. In addition, shame arises because of the exposure made by the public against such errors.

According to Tracy et al. (2007), shame is a 'painful' emotion, and has attributions that are global in nature due to its negative evaluation in targeting the core identity of the individual. Therefore, the negative evaluation is given not only to one specific behaviour, but also to the individual as a whole. Tangney & Fischer (1995) added that shame is usually followed by a desire to run away from someone, and gives the individual a strong incentive to get rid of things that lead to feelings of shame or, in other words, leads to avoidance motivation in the individual. Individuals who feel shame feel that they are not good enough, because other people cannot accept aspects of their behaviour. The situation that gives rise to shame is more difficult to control, because it is external, and makes the individual feel that others are exposed to the wrong that he/she has done.

In contrast to the emotion of shame, which focuses on the personal assessment of the individual as a whole, the emotion of guilt focuses on the assessment of the actions or behaviours of the individuals. Guilt is defined as a reaction to the behaviours that emerge that are contrary to the values that have been internalised into oneself, although no one else knew about it (Tangney & Fischer, 1995). The emergence of guilt is closely related to the moral situation and comes from the individual's conscience. In addition, guilt involves a negative evaluation of the specific individual's behaviour. According to Tangney and Dearing (2002), individuals who experience guilt are better able to empathise with others and to accept responsibility for the bad things they have done.

Lewis (1971), in Tangney and Dearing (2002), sees guilt as an emotion that does not cause as much pain or is as destructive as shame. This is because with guilt, the main concern is a specific behaviour and is apart from one's self, so that the guilt does not affect the core identity or self-concept of the individual (Lewis, 1971, in Tangney & Dearing, 2002). Furthermore, guilt involves tension and regret over the bad things that have been done. Individuals who experience guilty feelings usually think repeatedly of the bad things they have done, and hope they can behave differently or hope to erase the bad deeds. Tracy and Robins (2006) claimed that guilt focuses on specific behaviours of the individual, which are negative in nature and affect oneself or others, as people believe that guilt is caused by attribution that is internal and controllable in nature.

2.2 *Motivational value*

Schwartz (1992) defined value as a guideline professed and believed by the individual, that has a function to help select or evaluate actions and policies, whether done by themselves or by other people, and events that occur in the vicinity. The values of the individual come from three basic human needs: the need as a biological organism, the need for interaction with others, and the need to maintain the continuity and welfare of the group. According to Schwartz (2012), value has three functions, namely, (1) as the hallmark of one group's culture, society and individuals, (2) to track changes over time, and (3) to explain the basic motivation of attitudes and behaviour.

Schwartz (1992, 2012) divided values into ten types of motivational values, which are then grouped into four higher-order values, namely openness to change, self-enhancement, self-transcendence and conservation. Following are the explanations of the four higher-order values, and also their motivational values.

2.2.1 *Openness to change*

The motivational values associated with openness to change encourage people to follow their intellectual and emotional interests in an unpredictable and uncertain direction. These values are also emphasised in one's thoughts, actions and feelings, and also in one's readiness to accept changes. The motivational values belonging to the value of openness to change are as follows:

– Self-Direction

The purpose of this motivational value is the independent thinking and action that leads to selecting, creating and exploring. Self-direction comes from the organism's need for control and mastery, and is the value needed to generate autonomy and independence of the individual.

– Stimulation

The purpose of this motivational value is the excitement, novelty and challenges in life. Stimulation comes from the organism's need for diversity and stimulation in order to maintain optimal, positive and non-threatening activity. These needs may be related to the underlying needs of the value of self-control.

– Hedonism

The purpose of this motivational value is the pleasure or gratification that is sensual for oneself. Hedonism comes from the organism's need for pleasure that can give satisfaction to the individual. Hedonism is the only motivational value that can be classified into two higher-order values, namely openness to change and self-enhancement. However, on further study by Schwartz (2003), 75% of the 200 samples of the study had hedonism that was more closely associated with openness to change rather than self-enhancement. Therefore, hedonism is categorised as one of the values of openness to change.

2.2.2 *Self-enhancement*

The motivational values belonging to self-enhancement encourage individuals to advance their personal interests. These values also focus on the achievement of personal interest, either in the form of dominance over others or the success of individuals on a personal level. The motivational values belonging to self-enhancement are as follows:

– Achievement

The purpose of this motivational value is personal success, which is evident from the emergence of an individual's competence in accordance with the prevailing social standards in society. Competent performance generates the resources needed by an individual to survive, and for the achievement of the group or institution's purpose. In addition, achievement also emphasises the emergence of an individual's competence in accordance with the standards of society, resulting in recognition from the community.

– Power

This motivational value focuses on the social esteem gained when individuals are able to maintain their dominant position in the existing social system. The purposes of this motivational value are social status, prestige and control or dominance over others or existing resources. Power comes from the need for control and dominance.

2.2.3 *Self-transcendence*

The motivational values belonging to self-transcendence encourage individuals to prioritise the interest and welfare of others (near or far), and the conservation of nature. These values stress the individuals' concern for the welfare and interests of others. The motivational values belonging to self-transcendence are as follows:

– Benevolence

The purpose of this motivational value is to maintain and improve the welfare of the people who have a highly intense personal relationship with the individual concerned (in-group). Benevolence comes from the individual's need to have a positive interaction with the group and the need for affiliation.

– Universalism
The purpose of this motivational value is the understanding, appreciation, tolerance and protection of the welfare of humanity and the universe. Universalism is derived from the individual's and group's need to survive. This value appears in the interaction of individuals with their outgroup, and also with nature.

2.2.4 *Conservation*
The motivational values belonging to conservation encourage individuals to maintain the existing situation (status quo) and the certainty of their relationship to other people, institutions and traditions. These values also emphasise compliance with order, self-restraint, preservation of traditional values and resistance to change. The motivational values belonging to conservation are as follows:

– Security
The purpose of this motivational value is safety, harmony, stability within the community and relationships with others, or within oneself. The sense of security arises from basic human needs and groups.
– Conformity
The purpose of this motivational value is to refrain from actions, inclinations and impulses that may harm others and violate the prevailing norms or social expectations of society. Conformity derives from the behaviour of those individuals who tend to disturb and destroy interaction, as well as the group's function. This value also focuses on self-restraint, so that the individuals behave in accordance with the prevailing social norms.
– Tradition
The purpose of this motivational value is respect, commitment and acceptance of the values that one believes in and that thrive in one's cultural group or religion. This value requires an individual to give a response towards the expectations of the previous generation that is eternal and unchangeable in nature.

Schwartz (1992) defines structure as a form of relationship between the values that exist, not as an important indicator of the values for individuals or groups. The structure of the relationship between these values is known as the Theoretical Model of Relations Among Ten Motivational Types of Values (Schwartz, 2012). This theoretical model describes the two types of relationships between motivational values, which are conflicts and compatibility. In the model, the values are arranged in a circle, forming a continuum of motivation, as can be seen in Figure 1 below.

Furthermore, if two values have a relatively close proximity in the circle, then the underlying motivations of the values are increasingly similar, so that both motivational values tend to

Figure 1. Theoretical model of relations among ten motivational types of values (from Schwartz, 2012).

be achievable by the same efforts or acts. Meanwhile, if the two values are far away from each other in the circle, the underlying motivations of the values are increasingly dissimilar, so that both motivational values tend not to be achievable with the same efforts or acts.

3 METHODS

This study uses three variables, namely the motivational value, shame emotion and guilt emotion. Motivational value is a value that underlies the emergence of a person's behaviour (Schwartz, 1992). Meanwhile, shame is a painful emotion that usually occurs together with a feeling of inferiority, worthlessness and helplessness (Tangney, 1999), and guilt is the reaction to the emergence of behaviour that is contrary to the values that have been internalised into oneself, although no one else knows about it (Tangney & Fischer, 1995).

The number of participants in this study was 534, with the characteristics of adolescents in Jakarta aged 15–19 years. The participants came from five administrative areas of Jakarta, with a spread of 91 participants from North Jakarta, 116 participants from East Jakarta, 106 participants from Central Jakarta, 114 participants from West Jakarta and 107 participants from South Jakarta. From the 534 sets of data obtained in the data collection process, there were 28 data that could not be included in the statistical analyses due to the incompleteness of the data. The remaining 506 data were proceeded into data analyses.

The sampling technique used was convenience sampling, as the participants were included in the sample due to time availability and their willingness to fill in the given research instruments (Gravetter & Forzano, 2009). This research was quantitative, because it would generate data in the form of a numerical score that could be analysed, summarised and interpreted using statistical testing procedures (Gravetter & Forzano, 2009). In addition, this research was a correlational study, as it aimed to get a picture of the relationship between two variables, but did not try to explain the relationship between the two variables (Gravetter & Forzano, 2009).

Measurement of the emotions of shame and guilt was conducted with the Test of Self-Conscious Affect 3 (TOSCA-3) developed by Tangney et al. (2000), and modified by Tambusai (2013) and Qonita (2013). Furthermore, motivational value was measured by a Portrait Values Questionnaire (PVQ) developed by Schwartz (2003) and adapted by Halim (2008). For the purposes of this research, a PVQ adapted by Halim (2008) was further modified to suit the characteristics of this study's participants.

In this study, there were eight research hypotheses to be proved by using the partial correlation statistical technique. This technique had been selected due to the dynamics of the relationships between each motivational value, so as to be able to correlate one motivational value to another, while controlling for the other values. The eight research hypotheses were as follows:

- There is a significant relationship between openness to change and shame;
- There is a significant relationship between self-enhancement and shame;
- There is a significant relationship between self-transcendence and shame;
- There is a significant relationship between conservation and shame;
- There is a significant relationship between openness to change and guilt;
- There is a significant relationship between self-enhancement and guilt;
- There is a significant relationship between self-transcendence and guilt;
- There is a significant relationship between conservation and guilt.

4 RESULTS

4.1 *Motivational values*

Table 1 gives data regarding the estimation of the participants' motivational values score. According to the data, the conservation value had the highest average score among the other values ($\bar{x}$ = 35.8, SD = 3.83). This indicated that the conservation value plays an important role in the participants' everyday life. On the other hand, self-enhancement had the lowest average score ($\bar{x}$ = 8.46, SD = 1.58), which means that the value of self-enhancement was not a priority for the participants.

4.2 *Shame emotion*

Table 2 gives information regarding the relationship between motivational values and shame. According to the data shown in Table 2, shame was significantly related to openness to change ($r = 0.119$, $p < 0.05$) and conservation ($r = 0.106$, $p < 0.05$). This result indicated that the shame emotion score was determined by the scores of openness to change and conservation values. On the other hand, shame was not significantly related to self-enhancement ($r = -0.086$, $p > 0.05$) and self-transcendence ($r = 0.069$, $p > 0.05$), which suggested that the changes in the shame emotion score were not influenced by the scores of the self-enhancement and self-transcendence values. These results partially supported our hypothesis, which predicted that shame has a significant relationship with all of the motivational values.

4.3 *Guilt emotion*

Table 3 gives data regarding the relationship between motivational values and guilt. It shows that guilt was significantly related to self-transcendence ($r = 0.256$, $p < 0.05$) and conservation ($r = 0.304$, $p < 0.05$). This result suggested that the guilt emotion score was affected by the scores of the self-transcendence and conservation values. Meanwhile, guilt was not significantly related to openness to change ($r = -0.036$, $p > 0.05$) and self-enhancement ($r = -0.064$, $p > 0.05$). These results partly confirmed our hypothesis, which expected that guilt has a significant relationship with all of the motivational values.

Table 1. Overview of the motivational values in participants.

Higher-order values	$\bar{x}$	SD	Minimum score	Maximum score
Openness to change	28.5	4.07	18	36
Self-enhancement	20.6	2.33	7	24
Self-transcendence	8.46	1.58	4	12
Conservation	35.8	3.83	20	42

Table 2. *r* values for the partial correlation statistic of the relationship between motivational values and shame emotion.

Higher-order values	*R*	R^2
Openness to change	0.119*	0.014
Self-enhancement	–0.086	0.007
Self-transcendence	0.069	0.004
Conservation	0.106*	0.011

*Significant at $p < 0.05$ level.

Table 3. *r* values for the partial correlation statistic of the relationship between motivational values and guilt emotion.

Guilt emotion	*r*	R^2
Openness to change	–0.036	0.001
Self-enhancement	–0.064	0.004
Self-transcendence	0.256*	0.066
Conservation	0.304*	0.092

*Significant at $p < 0.05$ level.

5 DISCUSSION

The aim of the present study is to examine the relationship between motivational values and the emotions of shame and guilt. Based on the results of the questionnaire that was administered to 506 high school students in Jakarta, there was a significant relationship between shame with openness to change and conservation. This result partly confirms our hypothesis that shame is related to the values of openness to change and conservation. Furthermore, this result is also coherent with previous findings by Silfver et al. (2008), who argue that shame has a positive correlation with the value of conservation, particularly with the value of tradition. On the other hand, guilt was related to the self-transcendence and conservation values and not to the openness to change and self-enhancement values. These results are consistent with previous studies that showed a positive correlation between self-transcendence and conservation values with the emotion of guilt (Nan, 2007; Silfver et al., 2008).

The relationship between the conservation value and the emotions of shame and guilt can also be explained by looking at the priority values of the participants. According to the results of this study, the conservation value is the top priority value for the participants. This means that their behaviour was driven by the conservation value. Thus, if they do not act accordingly with this value, it will cause a negative evaluation, which later can produce emotions of shame or guilt. Furthermore, the lack of correlation between the openness to change and self-enhancement values with guilt is coherent with the previous studies conducted by Nan (2007) and Silfver et al. (2008). According to Silfver et al. (2008), guilt does not have a relationship with the values that are associated with the accomplishment of self, such as power and achievement, which are part of the self-enhancement value.

There were differences between the present study and the previous studies in motivational values and moral emotions. In the present study, there was a significant relationship between openness to change and shame. Meanwhile, the previous studies found a negative relationship between shame and openness to change (Nan, 2007; Silfver et al., 2008). These differences are due to the nature of openness to change as a value. Openness to change focuses on the thoughts, actions, feelings and the readiness to accept change (Schwartz, 1992, 2012). This is contrary to the conservation value, which is seen as a priority value for the participants. Furthermore, the conservation value itself focuses on the compliance with order, self-restraint, preservation of traditional values and resistance to change (Schwartz, 1992, 2012). Thus, if individuals have a value of openness to change, then they will generate behaviour that is incompatible with the conservation values that are upheld by society, and this will cause the appearance of shame.

Meanwhile, the low score of self-enhancement in the present study indicates that this value is not a priority for the participants. This means that the participants are not concerned with their personal achievement or success, so that when they do not have high achievements, for example in the academic field, this will not lead them to a negative evaluation. This is because they perceive such an action as a behaviour that does not deviate from the norm and culture that is applied in society. Thus, they do not feel embarrassed if they do not have high achievements in their life, which makes the self-enhancement value unrelated to the emotion of shame.

5.1 *Implications and further research*

There are some implications of the present study. First, the result of the present study has broadened the knowledge regarding the relationship between motivational values and moral emotions. Second, the result of the present study can be used as reference material for the development of a moral education syllabus for adolescents. The government might want to target the values that are associated with moral emotions and also target not only the cognitive domain but also the affective domain of students. Thus, the values that are associated with moral emotions can be internalised and become the guide to students' behaviour.

Moreover, given the importance of moral education for the moral development of individuals, such education should be given at an early stage. This moral education is not only the responsibility of the educational institutions, but also the responsibility of parents and communities. Thus, the results of this study can also be used as a guidance for parents to teach moral education to their children at home.

On the other hand, further research in this field can further explore the relationship between the ten types of motivational values with moral emotions. Thus, it can be distinguished which types of motivational values are associated with moral emotions. Moreover, further research can use the multiple regression method as the statistical procedure. This method can give a better understanding regarding which motivational values play an important role in predicting the occurrence of moral emotions.

In conclusion, it can be said that moral emotions are related to motivational values, particularly the conservation value, as it is related to both shame and guilt. Furthermore, shame is also related to the openness to change value. On the other hand, guilt is related to the conservation value and the self-transcendence value.

REFERENCES

Faris, A.F. (2013, December 28). Inilah 11 Jenis Kejahatan Di Jakarta Pada 2013. *Inilah.com*. Retrieved from http://m.inilah.com/news/detail/2059839/inilah-11-jenis-kejahatan-di-jakarta-pada-2013

Gravetter, F.J. & Forzano, L.B. (2009). *Research methods for the behavioral sciences*. Stamford, CT: Wadsworth Publishing.

Halim, R. (2008). *Hubungan Wajah Sosial Dan Nilai Motivasional Pada Etnis Tionghoa Dan Jawa* (Thesis, Universitas Indonesia, Depok, Indonesia).

Hermawan, E. (2013, November 20). Tawuran Sekolah Jakarta Naik 44 Persen. *Tempo.co*. Retrieved from https://m.tempo.co/read/news/2013/11/20/083531130/tawuran-sekolah-jakarta-naik-44-persen

Karyanto, A. (2013, May 2). Mengatasi Degradasi Moral. *Harapan Rakyat*. Retrieved from http://www.harapanrakyat.com/2013/05/mengatasi-degradasi-moral/

Kompas. (2013, March 7). Pengguna Narkoba Di Kalangan Remaja Meningkat. *Kompas.com*. Retrieved from http://regional.kompas.com/read/2013/03/07/03184385/Pengguna.Narkoba.di.Kalangan.Remaja.Meningkat

Marlinawati, R. (2013). Degradasi Moral Generasi Muda Karena Salah Sistem Pendidikan. Jakarta, Indonesia: MPR RI.

Nan, L.M. (2007). *Moral guilt and shame: An investigation of their associations with personality, values, spirituality and religiosity* (Doctoral dissertation, University of Illinois at Urbana-Champaign, IL).

Orth, U., Robins, R.W. & Soto, C.J. (2010). Tracking the trajectory of shame, guilt, and pride across the life span. *Journal of Personality and Social Psychology*, *99*(6), 1061–71. doi:10.1037/a0021342

Qonita, A. (2013). *Perbedaan Emosi Malu dan Emosi Bersalah pada Remaja yang Bersekolah di SMA Umum dan SMA Swasta berdasarkan Agama* (Undergraduate dissertation, Universitas Indonesia, Depok, Indonesia).

Santrock, J. (2014). *Adolescence*. New York, NY: McGraw-Hill.

Sari, H.R. (2012, December 27). Kasus Pembunuhan Meningkat Di Tahun 2012. *Merdeka.com*. Retrieved from https://m.merdeka.com/amp/peristiwa/kasus-pembunuhan-meningkat-di-tahun-2012.html

Schwartz, S.H. (1992). Universals in the content and structure of values: Theoretical advances and empirical tests in 20 countries. *Advances in Experimental Social Psychology*, *25*(C), 1–65. doi:10.1016/S0065-2601(08)60281-6

Schwartz, S.H. (2003). A proposal for measuring value orientations across nations. *Questionnaire Package of the European Social Survey*, 259–319. doi:10.1016/B978-0-12-411466-1.00007-0

Schwartz, S.H. (2012). An overview of the Schwartz theory of basic values. *Online Readings in Psychology and Culture*, *2*, 1–20. doi:10.9707/2307-0919.1116

Silfver, M., Helkama, K., Lönnqvist, J.E. & Verkasalo, M. (2008). The relation between value priorities and proneness to guilt, shame, and empathy. *Motivation and Emotion*, *32*(2), 69–80. doi:10.1007/s11031-008-9084-2

Soebagyo, F. (2013, September 15). Firman: Bangsa Ini Bukan Krisis Ekonomi, Tapi Krisis Moral. *Kompasiana*. Retrieved from http://www.kompasiana.com/syahandri/firman-bangsa-ini-bukan-krisis-ekonomi-tapi-krisis-moral_552bcf676ea834b4258b457b

Sofyan, E.H. (2013). Kenakalan Remaja Makin Mencemaskan. *Kompas.com*. Retrieved from https://www.google.co.id/amp/s/app.kompas.com/amp/megapolitan/read/2013/10/08/0920254/Kenakalan.Remaja.MakinMencemaskan

Tambusai, Y. (2013). *Perbedaan Emosi Malu dan Emosi Bersalah antara Remaja Jakarta dan Darah Penyangga* (Undergraduate dissertation, Universitas Indonesia, Depok, Indonesia).

Tangney, J.P. (1995). Shame and guilt in interpersonal relationships. In J.P. Tangney & K.W. Fischer (Eds.), *Self-conscious emotions: The psychology of shame, guilt, embarrassment, and pride* (pp. 114–39). New York, NY: Guilford Press.

Tangney, J.P. (1999). The self-conscious emotions: Shame, guilt, embarrasment, and pride. In T. Dalgleish & M.J. Power (Eds.), *Handbook of cognition and emotion* (pp. 542–68). Chichester, UK: John Wiley & Sons.

Tangney, J.P. & Dearing, R.L. (2002). *Shame and guilt*. Igarss 2014. doi:10.1007/s13398-014-0173-7.2

Tangney, J.P. & Fischer, K.W. (1995). Self-conscious emotions and the affect revolution: Framework and overview. In J.P. Tangney & K.W. Fischer (Eds.), *Self-conscious emotions: The psychology of shame, guilt, embarrassment, and pride*. New York, NY: Guilford Press.

Tangney, J.P., Dearing, R.L., Wagner, P.E. & Gramzow, R. (2000). *Test of self-conscious affect (TOSCA)—3 | Psychology resource centre—Hebblab*. Fairfax, VA: George Mason University.

Tangney, J.P., Stuewig, J. & Mashek, D.J. (2007). Moral emotions and moral behavior. *Annual Review of Psychology*, *58*, 345–372. doi:10.1146/annurev.psych.56.091103.070145

Tracy, J.L. & Robins, R.W. (2006). Appraisal antecedents of shame and guilt: Support for a theoretical model. *Personality & Social Psychology Bulletin*, *32*(10), 1339–1351. doi:10.1177/0146167206290212

Tracy, J.L., Robins, R.W. & Tangney, J.P. (2007). *The self-conscious emotions: Theory and research* (1st ed.). New York, NY: The Guilford Press.

Diversity in Unity: Perspectives from Psychology and Behavioral Sciences – Ariyanto et al. (Eds)
© 2018 Taylor & Francis Group, London, ISBN 978-1-138-62665-2

The use of mastering self-leadership training to improve self-leadership and innovative work behaviour

P.D. Arista & E. Parahyanti
Faculty of Psychology, Universitas Indonesia, Depok, Indonesia

ABSTRACT: This research aims to determine the relationship between Self-leadership and Innovative Work Behavior as well as how effective is the Mastering Self-leadership training to increase Self-leadership and Innovative Work Behavior. The study used action research with two research designs which are cross sectional (n = 144) and before-and-after study (n = 9). Measuring instrument used is Innovative Work Behavior Questionnaire (Jassen, 2000) which then translated into Indonesian language by Etikariana and Muluk (2014) and The Revised Self-Leadership Questionnaire (Marques-Quinteiro, Curral & Passos, 2012). The result showed that there is a significant positive correlation between Self-Leadership and Innovative Work Behavior ($r = 0.44$, $R^2 = 0.20$ ($p < 0.05$). It means that an increase in the level of Self-leadership will result in the increase of Work Innovative Behavior, and the Self-leadership explained 20% variance of Work Innovative Behavior. Moreover, among the three dimensions of Self-leadership, the constructive thought strategies is the most related dimension to the Work Innovative Behavior ($r = 0.41$). Based on the result of the difference between pre-test and post-test, it concludes that the intervention of Mastering Self-Leadership training is effectively and recommended as an intervention to increase Self-leadership and Innovative Work Behavior.

Keywords: self-leadership; Mastering self-leadership training; Innovative work behavior

1 INTRODUCTION

Organisations all around the world are facing the common challenge of rapid changes (Kalyar, 2011; Khan et al., 2012). They are not only expected to improve performance, profitability, build competitive advantages, and ensure sustainability, but also to create innovations (Abbas et al., 2012; Kalyar, 2011; Lee, 2008). Innovation is a productive behaviour that has impacts on the organisation's business performance and future successes (Jex & Britt, 2008). According to Tidd et al. (2001), innovation is more than finding or creating ideas, but it is the process of converting opportunities into new ideas, as well as applying them in practical applications. Kaboli et al. (2008) said that one of the steps that can be taken to make organisations become more innovative is to utilise their employees' ability to innovate.

Innovation in organisations is a result of the innovative work behaviour of its members (Ramamoorthy et al., 2005). Innovative work behaviour can have a direct impact on the organisation's effectiveness and productivity, as well as ensuring its long-term sustainability (Scott & Bruce, 1994; Janssen, 2000). According to Scott and Bruce (1994), innovative work behaviour is a multistage process with various activities that includes problem recognition, ideas generation (novel or adopted ideas), finding support for ideas, building coalitions, and implementing or producing ideas. Referring to Kanter (1988) and Scott and Bruce (1994), Janssen (2000) then classified innovative work behaviour into a three-staged process: idea generation, idea promotion, and idea realization. Idea generation is a process of creating new ideas, finding new work methods, new techniques or work instruments, and creating solutions for certain problems (Kanter, 1988). Idea promotion is comprised of activities in gaining support, growing enthusiasm of interested organisation members, as well as building legitimacy and supporting coalitions which provide the necessary forces to implement ideas (Kanter, 1988). The last stage is idea realization, where people produce models or prototypes that can be used in their work (Scott & Bruce, 1994). Furthermore, some of the latest studies have tried to identify factors that affect innovative work behaviour. According to Damanpour (1991), individual factors are the most significant factors in affecting innovative work behaviour.

Self-leadership is believed to be one of the individual factors that has significant effect on the innovative work behaviour (Carmeli et al., 2006; DiLiello & Houghton, 2006; Kaboli et al., 2008; Tastan, 2013; Park et al., 2014). Self-leadership is an influence-related process through which individuals (and working groups) navigate, motivate, and lead themselves towards achieving desired behaviours and outcomes (Manz, 1992). Self-leadership strategies are usually grouped into three primary categories (Manz, 1992; Anderson & Prussia, 1997; Manz & Neck, 2004). First, a behaviour-focused strategy is a strategy to apply behaviour management to improve one's self-awareness, so can be facilitated someone particularly regarding unpleasant tasks that need to be done (Anderson & Prussia, 1997; Manz & Neck, 2004). Second, a natural reward strategy is a strategy that focuses on the creation or discovery of pleasant aspects of work, thus motivating individuals to enjoy their work or activities (Anderson & Prussia, 1997; Manz & Neck, 2004). Third, a thought pattern strategy is a strategy that is focused on the development of positive thought patterns that can affect performance (Manz, 1986; Anderson & Prussia, 1997; Manz & Neck, 2004). Although itis believed to be one of the individual factors that has significant effects on the innovative work behaviour, there are very few studies that have tried to find the correlation between innovative work behaviour and self-leadership (Carmeli et al., 2006; DiLiello & Houghton, 2006; Kaboli et al., 2008; Tastan, 2013; Park et al., 2014).

This study was conducted in, PT X (pseudonym), one of the nation-wide private television companies in Indonesia. As a television company that produces innovative content, employee of PT X, especially in Production Division is demanded by management to have innovative work behaviour. Moreover, being innovative is one of the prerequisite competencies. However, PT X is currently experiencing a decline in its organisational performance which is evident from its low share price and ratings. This research then used interview, observation, and unobtrusive data on the preliminary analysis to diagnose organisation problem. The interview was conducted to Division Head, Department Head, Executive Producer, Producer, and Creative Staff. Next, the observation was conducted to creative staff and production assistant. Based on the preliminary analysis, the decline of organisational performance is caused by the low innovative work behaviour of employees in the Production Division as the division which is responsible for creating television programmes. This can be seen when the employees on the Senior Creative, Creative, and Production Assistant level do not investigate and analyse the characteristics and demands of the market, and also lack the initiative for finding necessary information for creating new programmes or developing current programmes. Furthermore, according to management, the ideas produced by employees in the Production Division are considered to be not well-thought and do not fit the market segmentation. Employees also exhibit a lack of capabilities in promoting programme ideas by using data which is lacking in validity. They also appear to be unconvincing when presenting programme ideas. Furthermore, behaviour such as paying little attention to the details of the programmes, ignoring other parties (such as the Facilities Division) who have important functions in implementing programme ideas, and lacking discipline in executing programme plans are a few factors that have caused the programme ideas to not be optimally implemented. These assessments are supported by quantitative data that show that 53% of the innovative work behaviour of the employees in the Production Division of PT X is in the lower scale.

Furthermore, based on preliminary analysis of the interview results to Division Head, Department Head, Executive Producer, and Producer, showed that the employees of the Senior Creative, Creative, and Production Assistant levels are considered to be neglecting goals and showing a lack of effort in creating programme skill. They also avoiding challenging tasks and preferring easy ones, creating programme content that lacks in detail, and leaving unfinished tasks. Furthermore, employees do not create a self-rewarding system that can improve their work motivation. Based on these findings, it can be concluded that low performance was caused by the employees' low focus on effective behaviour in achieving work goals. Moreover, most of employees in the Production Division also hold negative views towards their jobs. They felt that their current job is time-consuming, very tiring, and does not have any value and meaning. Based on that finding, it can be inferred that the employees' negative views towards their jobs were caused by their inability to create positive aspects into their work. Employees also shared negative things that they felt, had negative thoughts that they were unappreciated, and believed that they're not supported by their leaders. Therefore, it could be concluded that their low performance is caused by their negative thoughts towards their jobs and co-workers in the organisation.

Based on the previous analyses, it can be concluded that the low innovative work behaviour of the employees of the Production Division of PT X is caused by the low level of self-leadership. This conclusion is backed by the quantitative data showing that 53% of the self-leadership of the Production Division employees is in the lower scale. Thus, the correlation of innovative work behaviour and self-leadership becomes the focus of this study. We also see the need for self-leadership intervention for the Production Division employees to improve their self-leadership and innovative work behaviour. One type of intervention that can be given to improve self-leadership is training (Neck & Manz, 1996; Furtner et al., 2012; Lucke & Furtner, 2015). Training is considered as a method of human resource management intervention (Cummings & Worley, 2009). In designing the training, there are five factors that need to be considered, which are participants, goals, methods, materials, and training facilitators (Mangundjaya & Mansoer, 2010). Next, there are a few levels in assessing the effectiveness of the training: reaction, learning, behaviour, and result (Kirkpatrick & Kirkpatrick, 2006). By providing self-leadership intervention, it is expected that PT X Production Division employees' innovative work behaviour can also be improved.

The purpose of this study is to find out the correlation between self-leadership and innovative work behaviour of PT X's Production Division employees. In addition, this study also aims to determine which dimensions of self-leadership have a significant correlation with the innovative work behaviour of PT X's Production Division employees. Furthermore, the results of this study will be used as the basis for determining intervention recommendations that can be given to improve self-leadership and innovative work behaviour of the employees in the Production Division of PT X.

2 METHOD

2.1 *Research method*

This study used quantitative methods (Stangor, 2011). The results of these method were used to form a picture of the current state of self-leadership and innovative work behaviour among the Production Division employees of PT X. These method were also used to measure the correlation between self-leadership and innovative work behaviour. Furthermore, the results were then used as the basis of the action or intervention that was implemented to solve the organisation's problem.

The design used to answer the research questions consists of two: the first is a cross-sectional design, and the second is a before-and-after study (Kumar, 2005). The purpose of the first design was to find the correlation between self-leadership and innovative work behaviour. The second design aimed at assessing the effectiveness of the intervention in improving the self-leadership and innovative work behaviour.

2.2 *Population and samples*

The population of this research is the employees of Production Division of PT X that consists of the Division of Production 1, Division of Production 2, and Division of Film, Drama, and Sport that divided on the level of Senior Creative, Creative, and Production

Assistant. Based on the data from PT X's Manpower Planning from April 2016 (Man Power Planning PT X, 2016), the total number of employees in those divisions was 204 people. For the first study, this research used 144 respondents as a sample. Next, the samples used for the second study were respondents from the first study that had a low self-leadership and innovative work behaviour score. The subjects for the second study were divided into two groups: a trained group and a control group. The trained group were given self-leadership training. Meanwhile, the control group was not given any intervention and used as a comparison to measure the effectiveness of the training given to the trained group.

2.3 *Data collection*

The data collection methods used in the preliminary analysis to diagnose the organisational problem were interviews, participant observation, and unobtrusive data. Next, the data for the first study were collected using two questionnaires. The first is Janssen's (2000) Innovative Work Behaviour Questionnaire adapted into Indonesian language by Etikariena and Muluk (2014) ($\alpha = 0.94$) with a six-point likert scale. This questionnaire consisted of three dimensions and three items in each dimension. Furthermore, this is a unidimensional scale. The next questionnaire was the Self-Leadership Questionnaire from Marques-Quinteiro et al. (2011) ($\alpha = 0.91$) with a six-point likert scale. This questionnaire consisted of three dimensions with nine items in the behaviour-focused strategies and thought pattern strategy dimensions, and three items in the natural reward strategy dimension. According to Kaplan and Saccuzzo (2009), a good reliability score is in the range of 0.7 to 0.8, so both questionnaires had good internal consistency.

2.4 *Data analysis*

The descriptive analysis was done to analyse the demographic data, while the Pearson Product Moment was used to determine the correlation between self-leadership and innovative work behaviour. Next, the Wilcoxon Signed Rank Test was used to determine the mean difference of the same sample in two different conditions and to compare the mean difference between the two different sample groups (Field, 2009). In this study, the Wilcoxon Signed Rank Test was used to find out the significance of the difference between the pre-test and post-test scores of the training participants (level of learning). The level of learning was measured using an aptitude test using the training materials that consisted of ten multiple choice questions. Furthermore, this data analysis method was also used to compare the pre-test and post-test scores of self-leadership and innovative work behaviour between the trained group and control group. Before that, the significance of the difference of the pre-test score between the trained and control group was also analysed to make sure that both groups were in equivalent condition.

3 RESULTS

The table below shows the correlation coefficient between the self-leadership dimensions and innovative work behaviour.

Table 1. Correlation between self-leadership and innovative work behaviour.

Self-leadership	*Pearson correlation coefficient*	R *Square*	Sig.
Self-leadership	0.44	0.20	0.000
Behaviour-focused Strategies	0.36	0.13	0.000
Natural Reward Strategies	0.36	0.13	0.000
Thought Pattern Strategies	0.41	0.17	0.000

The table above shows that self-leadership has a significant positive correlation with innovative work behaviour ($r = 0.44$; $p < 0.05$) and 20% of the variances was predicted by self-leadership, while the remaining 80% was predicted by other factors. Moreover, each dimension of self-leadership has a significant positive correlation with innovative work behaviour. The highest correlation belongs to the thought pattern strategy dimension with $r = 0.41$. Further, both the behaviour-focused strategy and natural reward strategy dimension have a correlation coefficient of $r = 0.36$. Based on the result showed that there is a significant correlation between self-leadership and innovative work behaviour in PT X's Production Division employees.

The intervention in the form of training was divided into four sessions with the general concept of self-leadership and its three strategies given in the first session, and the detail of each strategy given in next sessions. The goal of the Mastering Self-Leadership training was to make the participants comprehend the concept of self-leadership as well as its three strategies. For this training, there were three facilitators from the Universitas Indonesia that had a psychology degree. So after several coaching sessions with the facilitator's lecturer, they were deemed to be capable and appropriate by the Faculty of Psychology, Universitas Indonesia.

The next procedure was to develop the training materials. It was designed based on the training of Furtner et al. (2012) and Lucke and Furtner (2015). In the process of designing the training materials, the significance of the correlation of each self-leadership dimension towards innovative work behaviour was considered in determining the emphasis in administering the training materials. Therefore, thought pattern strategies as the dimension that had the highest correlation coefficient were given in the first session of the training. The training intervention was conducted on Monday, 30 May 2016. It ran from 10.00 until 18.00 Indonesia Western Standard Time (IWSB). The training was done in a PT X meeting room that has a long rectangular table in the middle of the room with chairs positioned around it. This type of arrangement is called the horseshoe arrangement (Noe, 2009). This seating arrangement allowed moderate involvement of the training participants (Mangundjaya & Mansoer, 2010). The participants of the training consisted of 14 employees from the Senior Creative, Creative, and Production Assistant levels of PT X's Production Division, but only nine employees were further analysed because other participants already had a high score of self-leadership and innovative work behaviour.

The training in this study was evaluated by measuring the level of reaction, learning, and changing of perception towards self-leadership and innovative work behaviour. The reaction level was measured using 25 questions that each had a six-point response based on a Likert scale. The table below shows the mean for each aspect.

Table 2. Mean score of reaction level.

Aspect	Mean score
Implementation	4.69
Supporting instruments	4.93
Material	4.94
Facilitator	5.02

Table 3. The wilcoxon signed rank test result of the self-leadership and innovative work behaviour scores of the trained and control groups before and after the training.

	Self-leadership			Innovative work behaviour		
	Pair	Mean	Sig.	Pair	Mean	Sig.
Trained group	Pre-test score	92.78	0.013	Pre-test score	23.89	0.013
	Post-test score	103.44		Post-test score	32.67	
Control group	Pre-test score	93.00	0.673	Pre-test score	27.56	0.438
	Post-test score	93.67		Post-test score	28.44	

This result shows that from overall, training participants felt satisfied with the training. Next, the level of learning was measured using an aptitude test using the training materials which consisted of ten multiple choice questions. Based on the result, the Wilcoxon Signed Rank Test was then used to find out the significance of the difference between the pre-test and post-test scores of the training participants. The analysis showed that the significance score was $p = 0.007$ ($p < 0.05$). Thus, there was a significant difference between pre-test and post-test scores of the training participants. Next, the results of the pre-test score comparison between the trained and control group showed that there was no significant difference of the perception of self-leadership ($p = 0.889$, $p > 0.05$) and innovative work behaviour ($p = 0.172$, $p > 0.05$). Furthermore, the table below shows the results of the difference test of the perception of self-leadership and innovative work behaviour scores of the trained group and control group, before and after the training was given.

The results above show that the trained group had significant differences in the self-leadership perception ($p = 0.013$, $p < 0.05$) and innovative work behaviour ($p = 0.013$, $p < 0.05$) before and after the training. Furthermore, the control group did not show a significant difference in the self-leadership perception ($p = 0.673$, $p > 0.05$) and innovative work behaviour ($p = 0.438$, $p > 0.05$).

4 DISCUSSION

The correlation results show that self-leadership has a significant positive correlation with innovative work behaviour. This result corresponds with the results of studies by Carmeli et al. (2006), DiLiello and Houghton (2006), Kalyar (2011), Tastan (2013), and Park et al. (2014). According to DiLiello and Houghton (2006), a strong self-leadership is a predictor of better self-motivation, positive thinking, and innovation practices. Individuals with self-leadership are able to manage, direct, and motivate themselves so that they can perform effectively in order to attain established goals (Houghton & Neck, 2002).

The results show that thought pattern strategies are the dimension that has the highest correlation coefficient with innovative work behaviour compared to other dimensions of self-leadership. According to Neck and Manz (1996), individuals have control over their thought patterns; therefore, employees that are able to apply thought pattern strategies can achieve better mental performance, be more enthusiastic, be optimistic, have higher self-esteem, as well as be able to minimise negative feelings such as nervousness and anxiety (Neck & Manz, 1996).

Behaviour-focused strategies are the next dimension that has a high correlation with innovative work behaviour. In applying these strategies, employees need accurate information regarding their current behaviour and performance, so that they can establish necessary goals and effective behaviour (Neck & Houghton, 2006). However, according to an interview with the Human Capital and Organisational Development Department Head of PT X, currently, a good performance assessment system has not been developed yet. Employees are assessed by two questions regarding work quality and quantity. Moreover, assessment results were not followed up to be used as the foundation for employee development. This has caused employees to not have the necessary, well-defined information about the goals and effective behaviour that is expected from them. Therefore, this study recommends that PT X should develop a performance assessment system.

Natural reward strategies are another dimension of self-leadership that has a significant correlation with innovative work behaviour. According to Cural and Marques-Quinteiro (2009), natural reward strategies can be implemented by work and environment redesigning, introducing pleasant elements, and applying enjoyable working methods. However, based on the interview with the employees, they felt that they did not have the opportunities to explore their task. Therefore, PT X needs to provide opportunities to their employees for exploring their task and working environment, so there will be more chances for them to create positive aspects in their work.

This study designed a series of intervention programme to improve self-leadership and innovative work behaviour. The recommended intervention is the "Mastering Self-Leadership" training which included three dimensions of self-leadership. This was based on the study of Furtner et al. (2012), as well as the study of Lucke and Furtner (2015), which showed that a training that included those three dimensions had significant effects on improving self-leadership.

Based on the Wilcoxon Signed Rank Test result, there was a significant score improvement for the employees who participated in the training. In addition, the significance score shows a significant changing of perception towards self-leadership and innovative work behaviour of the training participants, while the control group did not show a significant score improvement. Therefore, Self-Leadership training is deemed to have a significant effect on the perceiving and learning process of self-leadership and innovative work behaviour.

5 CONCLUSION

The findings of this study show that there is a significant correlation between self-leadership and innovative work behaviour in PT X Production Division employees. Each dimension of the self-leadership also significantly correlated with innovative work behaviour. The pre- and post-study also show that the self-leadership and innovative work behaviour for the employees of PT X in the Production Division could be improved by having Mastering Self-Leadership training as an intervention.

REFERENCES

Abbas, G., Iqbal, J., Waheed, A. & Riaz, M.N. (2012). Relationship between transformational leadership style and innovative work behaviour in educational institutions. *Journal of Behavioral Sciences, 22*(3), 18–32.

Anderson, J.S. & Prussia, G.E. (1997). The self-leadership questionnaire: Preliminary assessment of construct validity. *Psychological Bulletin, 103*, 411–423.

Carmeli, A., Meitar, R. & Weisberg, J. (2006). Self-leadership skills and innovative behaviour at work. *International Journal of Manpower, 27*, 75–90.

Coghlan, D. & Brannick, T. (2005). *Doing action research in your own organization* (2nd ed.). London, UK: SAGE Publications.

Cummings, T.G. & Worley, C.G. (2009). *Organizational development and change* (9th ed.). OH South-Western: Cengage Learning.

Curral, L. & Marques-Quinteiro, P. (2009). Self-leadership and work role innovation: Testing a mediation model with goal orientation and work motivation. *Colegio Oficial de Psicólogos de Madrid, 25*(2), 165–176.

Damanpour, F. (1991). Organizational innovation: A meta-analysis of effects of determinants and moderators. *Academy of Management Journal, 34*, 555–590.

DiLiello, T.C. & Houghton, J.D. (2006). Maximizing organizational leadership capacity for the future: Toward a model of self-leadership, innovation and creativity. *Journal of Managerial Psychology, 21*(4), 319–337. doi:10.1108/02683940610663114

Etikariena, A. & Muluk, H. (2014). Correlation between Organizational Memory and Innovative Work Behavior. *Makara Hubs-Asia, 18*(2), 77–88.

Field, A. (2009) *Discovering statistics using SPSS* (3rd ed.). London, UK: SAGE Publications.

Furtner, M.R., Sachse, P. & Exenberger, S. (2012). Learn to influence yourself: Full range self-leadership training. *Journal of the Indian Academy of Applied Psychology, 38*(2), 294–304.

George, G. (2011). *Combining organization development and organization design: An investigation based on the perspectives of OD and change management consultants*, (Publish Doctoral Dissertation), Capella University, Minneapolis, United States.

Grant, A.M., Green, L.S. & Rynsaardt, J. (2010). Developmental coaching for high school teachers: Executive coaching goes to school. *Consulting Psychology Journal: Practice and Research, 62*(3), 151–168. doi:10.1037/a0019212

Houghton, J.D. & Neck, C.P. (2002). The revised self-leadership questionnaire: Testing a hierarchical factor structure for self-leadership. *Journal of Managerial Psychology*, *17*(8), 672–691.
Janssen, O. (2000). Job demands, perceptions of effort-reward fairness and innovative work behavior. *Journal of Occupational and Organizational Psychology*, *73*(3), 287–302.
Jex, S.M. & Britt, T.W. (2008). *Organizational psychology—A scientist practitioner approach* (2nd ed.). New Jersey: John Wiley and Sons.
Kaboli, M.R., Shaemi, A. & Teimouri, H. (2008). *The role of self-leadership in innovation and creativity employee*. Department of Management, University of Isfahan, Iran. Retrieved from http://www.ufhrd.co.uk/wordpress/wp-content/uploads/2008/06/641-the-role-of-self-leadership-in-innovation-and-creativit.pdf
Kalyar, M.N. (2011). Creativity, self-leadership and individual innovation. *The Journal of Commerce*. Hailey College of Commerce, University of the Punjab, Pakistan. Retrieved from http://joc.hcc.edu.pk/articlepdf/joc_3_3_20_28.pdf
Kanter, R.M. (1988). When a thousand flowers bloom: Structural, collective, and social conditions for innovation in organizations. *Research in Organizational Behavior*, *10*, 169–211.
Kaplan, R.M. & Sacuzzo, D.P. (2009). *Psychological testing: Principles, applications, and issue* (7th ed.). Wadsworth: Cengage Learning.
Khan, M.J., Aslam, N. & Riaz, M.N. (2012). Leadership styles as predictors of innovative work behavior. *Pakistan Journal of Social and Clinical Psychology*, *9*(2), 17–22.
Kirkpatrick, D.L. & Kirkpatrick, J.D. (2006). *Evaluating training programs: The four levels* (3rd ed.). San Francisco, CA: Berrett Koehler Publisher.
Kumar, R. (2005). *Research Methodology: A step-by-step guide for beginners*. Malaysia: Sage Publications
Lee, S.H. (2008). The effect of employee trust and commitment on innovative behavior in the public sector: An empirical study. *International Review of Public Administration, 13*(1), 27–46. doi:10.1080/12294659.2008.10805110
Lucke, G.A. & Furtner, M.R. (2015). Soldiers lead themselves to more success: A self-leadership intervention study. *Military Psychology*, *27*(5), 311–324. doi:10.1037/mil0000086
Man Power Planning PT X, (2016), Jakarta.
Mangundjaya, W.L. & Mansoer, W.D. (2010). *Human Capital Development Through Training Programme*. Jakarta, Indonesia: Swascita.
Manz, C.C. (1986). Self-leadership: Toward an expanded theory of self-influence processes in organizations. *Academy of Management Review*, *11*(3), 585–600.
Manz, C.C. (1992). Self-leadership: The heart of empowerment. *The Journal for Quality and Participation*, *15*(4), 80–89.
Manz, C.C. & Neck, C.P. (2004). *Mastering self-leadership: Empowering yourself for personal excellence* (3rd ed.). Upper Saddle River, NJ: Pearson Prentice Hall.
Marques-Quinteiro, P., Curral, L.A. & Passos, A.M. (2012). Adapting the revised self-leadership questionnaire to the Portuguese context. *Social Indicators Research*, *108*(3), 553–564. doi:10.1007/s11205-011-9893-7
Neck, C.P. & Manz, C.C. (1996). Thought self-leadership: The impact of mental strategies training on employee cognition, behavior, and affect. *Journal of Organizational Behavior*, *17*(5), 445–467.
Neck, C.P. & Houghton, J.D. (2006). Two decades of self-leadership theory and research: Past development, present trends, and future possibilities. *Journal of Managerial Psychology*, *21*(4), 270–295.
Noe, R.A. (2009). *Employee training and development* (5th ed.). New York, NY: McGraw-Hill.
Park, G.R., Moon, G.W., & Hyun, S.E. (2014). An impact of self-leadership on innovative behaviour in sports educators and understanding of advanced research. *The Standard International Journals*, *2*(3), 117–122. Retrieved from http://www.thesij.com/papers/IFBM/2014/May/IFBM-0203380402.pdf
Ramamoorthy, N., Flood, P.C., Slattery, T. & Sardessai, R. (2005). Determinants of innovative work behavior: Development and test of an integrated model. *Creativity and Innovation Management*, *14*(2), 142–150.
Scott, S.G. & Bruce, R.A. (1994). Determinants of innovative behavior: A path model of individual innovation in the workplace. *Academy of Management Journal*, *37*(3), 580–607.
Stangor, C. (2010). *Research methods for the behavioral sciences* (4th ed.). USA: Wadsworth.
Tastan, S.B. (2013). The influences of participative organizational climate and self-leadership on innovative behavior and the roles of job involvement and proactive personality: A survey in the context of SMEs in Izmir. *Social and Behavioral Sciences*, *75*, 407–419. doi:10.1016/j.sbspro.2013.04.045
Tidd, J., Besant, J. & Pavitt, K. (2001). *Managing innovation* (2nd ed.). Chichester, UK: John Wiley and Sons.

Diversity in Unity: Perspectives from Psychology and Behavioral Sciences – Ariyanto et al. (Eds)
© 2018 Taylor & Francis Group, London, ISBN 978-1-138-62665-2

The effect of negative valence on memory and perception: Negative brand names experimental study

J.E. Yulianto, C.A. Rhenardo, J. Juan & J. Pauline
Center for Consumer Psychology, Industrial-Organizational Psychology, and Social Psychology, School of Psychology, Universitas Ciputra Surabaya, Surabaya, Indonesia

ABSTRACT: While some studies suggest that a positive brand name may better predict consumers' memory of the product, other studies agree the opposite. The aims of this study are to explore how people recognise a negative brand name, and to investigate how it correlates to people's preference and willingness to buy the product. A two-stage experiment was conducted on 84 undergraduate students. The results show that when structured stimuli are exposed with attractive colourful pictures, negative brand names are more likely to have lower recognition compared to positive brand names. However, negative brand names with no colourful picture are easier to remember. The results also show that compared to positive brand names, negative brand names are more likely to have lower correlation to consumers' preference and willingness to buy the products.

1 INTRODUCTION

> 'A product can be copied by competitor, a brand is unique. A product can be quickly outdated, a successful brand is timeless' (Stephen King, as cited in Trott and Sople (2016)).

Consumers' memory is influenced by their cognitive ability, learning experiences, mental health state, and environmental situation (Pierce & Gallo, 2011). Their cognitive ability and memory capacity in remembering a brand name are limited and differ from one to another depending on how strong their short-term memory is (Pierce & Gallo, 2011). Klink (2009), for example, found that males and females differed in responding to a brand name. Kellogg (2001) stated that it was easier for people to receive visual stimuli than auditory. Furthermore, Kellogg (2001) also emphasised that because the human brain interacts continuously with the social environment, the accuracy of people in remembering a brand name is not always reliable. The interaction makes the human brain process the information to create a new meaning. These processes then explain how consumers remember and forget new brand names in marketing.

Stephen King's statement reflects the importance of selecting a suitable brand name for a business. It is important for introducing either a new product or a new business. The success of a marketing campaign often depends on how easily the consumers like and recognise a particular brand. Klink and Athaide (2012) argue that brand name is strongly related to brand personality. It can be said then that a good brand name may improve the consumers' awareness of the brand. However, failure in determining the right brand name can prevent the product's or business's development (Ghodeswar, 2008; Keller & Lehmann, 2006). Since the numbers of start-up businesses are growing rapidly, we think that research on brand names is important.

Most people agree that a brand name could influence how consumers remember and are aware of a certain product (Mccracken & Macklin, 1998; Oladepo & Abimbola, 2015). The importance of brand name challenges many entrepreneurs to determine the right brand for their business. However, which type of brand name that has the highest effect on the consumers' memory remains debatable. Some studies revealed that a positive brand name may better predict the consumers' memory of the product compared to a negative brand name (Kensinger & Corkin, 2003), but other studies showed the opposite results (Guest et al., 2016). Furthermore, those studies only focused on one type of brand, either only a positive brand name or a negative brand name.

The present studies also want to investigate consumer perception, which is measured by two aspects: the liking and intention to buy. Studies on how a brand name influences consumers' perception are extensively reviewed among scholars. Klink (2003), for example, studied how people perceive brand images through brand name, including their structure (font, size, shape, and colour) and consistency in design. Gunasti and Ross (2010) also found another structure, called alphanumeric brand name, which affects consumer preferences. In their study, they found that alphanumeric brand name activates consumer attribution and increases the liking aspect towards brands. Other scholars also found that certain brand names influence people's expectation and understanding of the product (Pavia & Costa, 1993). These findings reveal that memory and perception are two interrelated important issues in discussing a brand name.

Some studies compared negative words with other types of words. Guest et al. (2016), for example, found that negative brands and non-negative brands have no correlation with negative meaning. Another example is a study by Kensinger and Corkin (2003), which investigated the relationship between negative words and neutral words. They found that participants more likely remembered negative words rather than neutral words. It was also easier for participants to remember the font used in negative words rather than in neutral words.

Some advertisements use unusual or negative words to gain attention from the consumers. In Indonesia, for example, some food and beverage businesses are gaining attention from their consumers because they use unusual or negative words. This has become a trend in Indonesia and is well-received by Indonesian consumers. 'Nasi Goreng Mafia', 'Nasi Goreng Jancuk', and 'Mie Setan' are some brands that are reportedly succeeding in gaining consumers since their first introduction (Kompas, 2013; Thohari, 2015; Wisanggeni, 2016). Therefore, we suggest the following hypotheses:

H1: Negative brand names are more memorable than positive brand names.
H2: Consumers express a higher rate of liking and willingness to buy positive brand names.

2 METHODS

2.1 *Participants*

A two-stage experiment was given to 84 undergraduate students ($M_{age} = 19–22$; $SD_{age} = 20.34$) from several departments. Opportunity sampling was used to select participants by promoting research participation opportunities in each class. All participants signed an informed consent statement before entering the experimental room. Students who were not able to complete the two stages of the experiment could withdraw from the study without any consequences. There was no incentive given to the participants on joining the experiment. The experimental session was divided into four parallel sessions to control the number of participants in each session. Each session took place in different classrooms which had the same room setting. All instructions were in Indonesian language.

2.2 *Design and procedure*

2.2.1 *Experiment I*

In this first experiment, we investigated whether negative brand names were easier to memorise compared to positive brand names. We added particular colour, font, and logo in both types of brand names. Every slide of visual stimulus contained one positive brand and one negative logo. The duration of exposure to each slide was five seconds to explain the stimulation, and three seconds to show the experimental presentation. After that, the participants were asked to watch and remember a total of 20 slides in each session. They were not told that the stimuli were brand names, to avoid attention bias.

In the first stage, the participants were instructed to give a mark to every brand that had appeared in the previous slide. Then, the experimenter showed a total of 30 positive brand names and 30 negative brand names. Thus, the worksheet contained the logos of the brands. In the second stage, the participants were instructed to give a mark to the name of the brand that had appeared in the previous slide. Then, the experimenter showed a total of 20 negative names and 20 positive names. Thus, the worksheet only contained the names of the brands without any pictures. In the last stage, the participants were instructed to give a mark to the names of the brand that had appeared in previous slide. Then, the experimenter showed a total of ten positive brand names and ten negative brand names. Both the logos and worksheet only contained the names of the brands without any pictures.

2.2.2 *Experiment II*

In the second experiment, we explored how much the participants like particular brand names and their intention to buy the products. We used 40 products which were divided into 20 negative products and 20 positive products. A PowerPoint slide was used to show each product for ten seconds. A seven-point Likert-style scale, ranging from 1 (Not interested) to 7 (Strongly interested) was used to analyse the degree of the participants' preferences and intention to buy the product.

2.3 *Stimuli*

There were two types of stimuli used in this study. The first is the negative stimuli which are widely accepted as options to be chosen as brand names in contemporary new products and businesses. Some noticeable brands that we used in the present research include stereotypically antagonistic words. The likes of 'Demon', 'Freak', 'Nerds', and 'Mafia' were among them (see Figure 1 for examples).

The second is the positive stimuli. We used some brands that use positive words in their structure. The likes of 'Handsome', 'Smart', 'Donation', and 'Positive Energy' were chosen as stimuli (see Figure 2 for examples).

These pictures were taken from a search engine and categorised as free-to-reuse pictures. All stimuli were shown in their original colour. The researchers controlled the size of the pictures to make sure that the participants could see the stimuli without being overexposed. The resolution of the pictures was no less than 400 megapixels.

Figure 1. Examples of negative stimuli.

Figure 2. Examples of positive stimuli.

3 RESULTS

3.1 *Experiment I*

From the first experiment, we found that positive brand names like 'Chris Angel' are surprisingly easier to remember than negative brand names such as 'Blood Buster' ($F(84) = 19.25$, $\eta^2 = 0.25$, $p < 0.001$), as well as their colour ($F(84) = 12.42$, $\eta^2 = 0.87$, $p < 0.001$). However, negative brand names are still easier to memorise if their structure includes only a name with no associated logo ($F(84) = -1.93$, $\eta^2 = 0.23$, $p < 0.001$) (see Table 1).

3.2 *Experiment II*

In the second step, we want to examine whether the participants like positive and negative brands being offered, and whether they want to buy the products. We correlate the liking rating and willingness to buy rating and then compare the results between the negative brand images and positive brand images. For the negative brand names, the correlation value between liking and the willingness to buy is relatively moderate ($r = 0.657$, $p < 0.001$), whereas the positive brand names have higher correlation value of liking and willingness to buy ($r = 0.839$, $p < 0.001$) (see Figure 3).

4 DISCUSSION

The present research aims to answer two research questions: (1) Are negative brand names easier to remember than positive ones? (2) Do consumers express a higher rate of liking and the willingness to buy for positive brand names. The result shows that compared to positive brand names, negative brand names have lower recognition. This result does not support previous research that argued that negative brand names are more likely to be remembered

Table 1. Recognition data.

	Negative brand names	Positive brand names
Word recognition (% correct)	66.326%	68.784%
Partial word colour recognition (% correct)	67.321%	68.631%
Full name recognition (% correct)	48.810%	47.487%

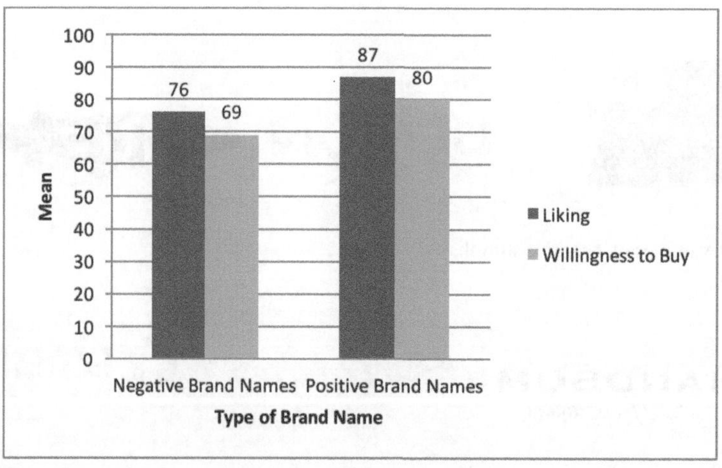

Figure 3. Mean evaluation rating: Liking and willingness to buy.

(Guest et al., 2016; Kensinger & Corkin, 2003). Negative brand names reportedly fail to activate arousal in the consumers' minds in remembering previous stimuli. The factor behind this could be the differences of arousal sources in particular words (Xu et al., 2015). Some scholars also mention that all humans have their own perception when considering a word as negative, neutral, or positive (Purkis et al., 2009; Robinson et al., 2004).

The first result also provides evidence on how negative valence tends to have higher recognition if exposed with no visual stimuli. This finding supports previous research on how the human brain tends to react better to visually colourful stimuli rather than word-based stimuli (Kuperman et al., 2014). The present finding is also in line with Adelman and Estes (2013), who found that both positive and negative words tend to activate human arousal better than neutral stimuli. However, the present finding also supports previous findings that negative brand names without pictures activate strong attention compared to both neutral and positive words (Doerksen & Shimamura, 2001; Kuperman et al., 2014).

The second experiment shows that negative brand names have lower correlation value of preference and willingness to buy compared to positive brand names. The result provides evidence that negative valence leads to negative evaluation towards the products. We interpret this result as avoidance response generated by participants through negative information from negative brand names. The result strongly supports the arguments by some scholars, that claimed that consumers tend to have automatic appraisal in perceiving negative brand names as negative information (Chen & Bargh, 1999; Krieglmeyer et al., 2010).

The implication of the findings can be applied to marketing strategy. Negative brand names should be packaged in visually colourful form rather than in merely words. In other words, business owners who want to use negative brand names as their ace brand should start to restructure, so that the logo emphasises on integrating attractive colourful pictures along with the negative brand name, rather than the negative brand name itself. Second, it is also important to consider that negative brand names probably have different effects at different levels. We are also aware of the automatic judgement among consumers in perceiving negative brand names. Thus, it is necessary to conduct implicit bias in perceiving negative brand names in future research.

5 CONCLUSION

It can be concluded that negative brand names with logos tend to be harder to remember than negative brand names. However, negative brand names with no logo are more likely to be remembered. We find that it is important for a logo to have attractive figures and colours to gain consumers' attention. The results also show that, compared to positive brand names, negative brand names tend to have lower correlation to consumers' preference and willingness to buy the product. In making a brand name, a marketer should consider the valence of negative words, since consumers tend to make automatic judgements in relation to negative brand names.

REFERENCES

Adelman, J.S. & Estes, Z. (2013). Emotion and memory: A recognition advantage for positive and negative words independent of arousal. *Cognition, 129*(3), 530–535. doi:10.1016/j.cognition.2013.08.014

Chen, M. & Bargh, J.A. (1999). Consequences of automatic evaluation: Immediate behavioral predispositions to approach or avoid the stimulus. *Personality and Social Psychology Bulletin, 25*(2), 215–224. doi:10.1177/0146167299025002007

Doerksen, S. & Shimamura, P. (2001). Source memory enhancement for emotional words. *Emotion (Washington, D.C.), 1*(1), 5–11. doi:10.1037/1528-3542.1.1.5

Ghodeswar, B.M. (2008). Building brand identity in competitive markets: A conceptual model. *Journal of Product & Brand Management, 17*(1), 4–12. doi:10.1108/10610420810856468

Guest, D., Estes, Z., Gibbert, M. & Mazursky, D. (2016). Brand suicide? Memory and liking of negative brand names. *PLOS ONE, 11*(3), 1–20. doi:10.1371/journal.pone.0151628

Gunasti, K. & Ross, W.T. (2010). How and when alphanumeric brand names affect consumer preferences. *Journal of Marketing Research*, *47*(6), 1177–1192. doi:10.1509/jmkr.47.6.1177
Huifang, X., Zhang, Q., Li, B. & Guo, C. (2015). Dissociable effects of valence and arousal on different subtypes of old/new effect: Evidence from event-related potentials. *Frontiers in Humam Neuroscience*, *9*, 1–14. doi:10.3389/fnhum.2015.00650
Keller, K.L. & Lehmann, D.R. (2006). Brands and branding: Research findings and future priorities. *Marketing Science*, *25*(6), 740–759. doi:10.1287/mksc.l050.0153
Kellogg, R.T. (2001). Presentation modality and mode of recall in verbal false memory. *Journal of Experimental Psychology: Learning, Memory, and Cognition*, *27*(4), 913–919. doi:10.1037/0278-7393.27.4.913
Kensinger, E. & Corkin, S. (2003). Memory enhancement for emotional words: Are emotional words more vividly remembered than neutral words? *Memory & Cognition*, *31*(8), 1169–1180. doi:10.3758/BF03195800
Klink, R.R. (2003). Creating meaningful brands: The relationship between brand name and brand mark. *Marketing Letters*, *14*(3), 143–157. doi:10.1023/A:1027476132607
Klink, R.R. (2009). Gender differences in new brand name response. *Marketing Letters*, *20*(3), 313–326. doi:10.1007/s11002-008-9066-x
Klink, R.R. & Athaide, G.A. (2012). Creating brand personality with brand names. *Marketing Letters*, *23*(1), 109–117. doi:10.1007/s11002-011-9140-7
Kompas. (2013, October 28). Nasi Goreng Jancuk Kembali Menuai Prestasi. *Kompas.com*. Retrieved from http://travel.kompas.com/read/2013/10/28/1629064/Nasi.Goreng.Jancuk.Kembali.Menuai.Prestasi.
Krieglmeyer, R., Deutsch, R., De Houwer, J. & De Raedt, R. (2010). Being moved: Valence activates approach-avoidance behavior independently of evaluation and approach-avoidance intentions. *Psychological Science*, *21*(4), 607–613. doi:10.1177/0956797610365131
Kuperman, V., Estes, Z., Brysbaert, M. & Warriner, A.B. (2014). Emotion and language: Valence and arousal affect word recognition. *Journal of Experimental Psychology: General*, *143*(3), 1065–1081. doi:10.1037/a0035669
Lee, Y.H. & Ang, S.H. (2003). Interference of picture and brand name in a multiple linkage ad context. *Marketing Letters*, *14*(4), 273–288. doi:10.1023/B:MARK.0000012472.11598.62
McCracken, J.C. & Macklin, M.C. (1998). The role of brand names and visual cues in enhancing memory for consumer packaged goods. *Marketing Letters*, *9*, 209–226.
Oladepo, O.I. & Abimbola, O.S. (2015). The influence of brand image and promotional mix on consumer buying decision—A study of beverage consumers in Lagos State, Nigeria. *British Journal of Marketing Studies*, *3*(4), 97–109.
Pavia, T.M. & Costa, J.A. (1993). The winning number—Consumer perceptions of alpha-numeric brand names. *Journal of Marketing*, *57*(3), 85–98. doi:10.2307/1251856
Pierce, B.H. & Gallo, D.A. (2011). Encoding modality can affect memory accuracy via retrieval orientation. *Journal of Experimental Psychology: Learning, Memory, and Cognition*. *37*(2), 516–521. doi:10.1037/a0022217
Purkis, H.M., Lipp, O.V., Edwards, M.S. & Barnes, R. (2009). An increase in stimulus arousal has differential effects on the processing speed of pleasant and unpleasant stimuli. *Motivation and Emotion*, *33*(4), 353–361. doi:10.1007/s11031-009-9144-2
Robinson, M.D., Storbeck, J., Meier, B.P. & Kirkeby, B.S. (2004). Watch out! That could be dangerous: Valence-arousal interactions in evaluative processing. *Personality and Social Psychology Bulletin*, *30*(11), 1472–1484. doi:10.1177/0146167204266647
Thohari, H. (2015). Inilah Nasi Goreng Gangster Bercitarasa Rempah, Mau? *Kompas.com*. Retrieved from http://tekno.kompas.com/read/2015/07/11/091547327/Inilah.Nasi.Goreng.Gangster.Bercita.Rasa.Rempah.Mau.?page = all.
Trott, S. & Sople, V.V. (2016). Brand equity: Indian perspective. New Delhi: PHI Learning Private Limited.
Wisanggeni, A. (2016). Pedas Setan yang Menghebohkan Lidah. *Kompas.com*. Retrieved from http://print.kompas.com/baca/gaya-hidup/santap/2016/06/12/Pedas-Setan-yang-Menghebohkan-Lidah.
Xu, H., Zhang, Q., Li, B., & Guo, C. (2015). Dissociable effects of valence and arousal on different subtypes of old. new effect: Evidence from event-related potentials. *Frontiers Human Neuroscience*, *9*, 650. doi: 10.3389/fnhum.2015.00650

Diversity in Unity: Perspectives from Psychology and Behavioral Sciences – Ariyanto et al. (Eds)
© 2018 Taylor & Francis Group, London, ISBN 978-1-138-62665-2

The effects of academic stress and optimism on subjective well-being among first-year undergraduates

M. Yovita & S.R. Asih
Faculty of Psychology, Universitas Indonesia, Depok, Indonesia

ABSTRACT: First-Year college students face various challenges as they enter a new environment. They are prone to academic stress, in which students perceive that the academic-related pressure exceeds their coping ability. Studies have found that academic stress can affect subjective well-being. One way found to reduce the influence of academic stress on subjective well-being was a high level of optimism. This study examined the effect of academic stress on subjective well-being with optimism as the moderator, among the first-year undergraduates. The subjective well-being, consisting of two components, was assessed using the Satisfaction With Life Scale for the cognitive appraisal and the Positive Affect and Negative Affect Schedule for the affective appraisal. Academic stress and optimism were assessed using the Student-Life Stress Inventory and the Life Orientation Test-Revised respectively. The overall model significantly predicted subjective well-being ($F(3, 211) = 47.653$, $p < 0.05$, adjusted $R^2 = 0.245$). Academic stress significantly decreased subjective well-being ($\beta = -0.383$, $t(213) = -6.302$, $p < 0.05$). Optimism significantly increased the subjective well-being ($\beta = 0.257$, $t(213) = 4.225$, $p < 0.05$). Optimism did not significantly moderate the effect of academic stress on subjective well-being ($t(211) = 0.491$, $p > 0.05$). In sum, academic stress and optimism affected the subjective well-being of first-year undergraduates. The findings are useful in enhancing the orientation material for new students.

1 INTRODUCTION

Starting college is an exciting experience for first-year undergraduates. It is the time when students are treated like adults. They are given the freedom to manage their own academic agenda, including the courses and the numbers of courses taken for a semester. Although students are given many choices, they are required to adapt to campus life, in which they need to be responsible in their own personal as well as academic life. These life-change experiences undergone by first-year undergraduates are likely to lead to academic stress (Fisher, 1994; Gall et al., 2000). This is because the first year of college is a transition period from a much-controlled environment, which is high school, to a less-controlled environment, namely college. Academic stress can be defined as a state when students perceive academic-related pressure as exceeding their ability to cope, causing psychological and/or biological changes (Cohen et al., 1998). Higher academic stress is also associated with less healthy lifestyle, including an increase in smoking and less consumption of healthy food (Hudd et al., 2000). Further, another impact of academic stress is a decline in Subjective Well-Being (SWB).

SWB refers to an evaluation of life satisfaction, consisting of cognitive as well as affective evaluations (Diener et al., 1999). Cognitive evaluation is an appraisal of how satisfying one's life is. Further, affective evaluation is an appraisal of a person's positive as well as negative affect. As students experience more academic stress, they would evaluate their SWB as being less satisfying (Denovan & Macaskill, 2016; Heizomi et al., 2015; Schiffrin & Nelson, 2010). A low SWB is associated with various negative impacts. As such, students with low SWB are more likely to report being depressed and demonstrate maladaptive social relationships (Park, 2004). Even more, a low level of SWB is more likely to give rise to mental disorders in students (Heizomi et al., 2015). On the contrary, a higher level of SWB is more likely to increase the students' academic achievement (Manzoor et al., 2014). Thus, it is imperative for students to maintain their SWB at an adequate level.

One way to achieve an adequate level of SWB is by being optimistic. Optimism can be defined as a general tendency to believe that someone would experience favourable outcomes (Carver & Scheier, 2001). As such, an individual would be able to be strong in times of adversity. Studies have found that an individual with a high level of optimism also has a high level of SWB (Cha, 2003; Santhosh & Appu, 2015). As such, a person with a high level of optimism believes that he or she would get favourable outcomes and would act to achieve those favourable outcomes. Related to college students, it was found that optimism is a strong predictor for achieving favourable outcomes (Cha, 2003). Further, the same study found that optimism is also a predictor of students' positive affect, in which positive affect is a part of SWB. A recent study found that students with a higher level of optimism reported less psychological problems and a higher level of SWB (Santhosh & Appu, 2015). It could be said that optimism might be a way to lessen the effect of academic stress on SWB.

Based on the explanation above, this study aimed to examine whether optimism could moderate the impact of academic stress on SWB in first-year undergraduates. For the first hypothesis (H1), it was hypothesised that a higher level of academic stress would lessen the SWB. The second hypothesis (H2) was that optimism would increase the students' SWB. Lastly, it was hypothesised (H3) that optimism would lessen the effect of academic stress on SWB.

2 METHODS

2.1 *Participants*

This study had 215 participants, recruited from a public university on the outskirts of Jakarta, the capital of Indonesia. Initially, there were 222 students recruited for the study. Seven of them were excluded for various reasons, including not giving consent to participate in the study, or that it was not their first time attending college. The final 215 participants consisted of 65 male students and 150 female students with a mean age of 18.56 ($SD = 0.683$). All participants were recruited by convenience sampling through social media and instant messaging.

2.2 *Procedures and measurements*

This study was a cross-sectional one, in which participants were assessed only once. The data collection was conducted from 28 May until 29 May 2016 through *cognitoforms.com*. Participants were asked to complete a set of online questionnaires which consisted of four instruments. Those instruments measured academic stress, optimism, affective evaluation of SWB, and cognitive evaluation of SWB.

The affective evaluation of SWB was assessed with the Positive Affect and Negative Affect Schedule (PANAS) (Watson et al., 1988). The PANAS consisted of 20 items, in which ten items were about positive affect and ten items about negative affect. The positive affect items included 'happy' and 'proud'. The negative affect items included 'disappointed', 'depressed'. This measurement used a 5-point Likert scale from strongly disagree to strongly agree. There were two total scores, one for positive affect and one for negative affect. The balance score, which was the final score, was gained by subtracting the total score of negative affect from the total score of positive affect. The possible range of affect balance score was from −40 to 40. The affect balance score shows the dominant affect of each participant. Positive score denoted positive affect and vice versa. This measurement was translated into Bahasa (Indonesian language) by Herwibowo (2014) and used for the student population in University of Indonesia. The coefficient alpha of PANAS was 0.760 for positive affect measurement and 0.826 for negative affect measurement. Based on the correlations between item scores and total scores, all coefficient correlations were above 0.2.

The cognitive evaluation of SWB was measured using the Satisfaction With Life Scale (SWLS) (Diener et al., 1985). It consisted of five items with a five-point Likert scale, ranging from 'Strongly disagree' to 'Strongly agree'. The possible range of total score was from 5 to 25, which showed how satisfied the participants were with their life. A higher score meant more satisfied, and vice versa. One item of the SWLS was 'The condition of my life is excellent'. This measurement had been translated to Bahasa (Indonesian language) by Herwibowo (2014) and used for the student population in University of Indonesia. The coefficient alpha of SWLS was 0.793. Based on the correlations between item scores and total scores, all coefficient correlations were above 0.2.

The academic stress was measured using the Student-Life Stress Inventory (SSI) (Gadzella, 1994). It consisted of 51 items grouped under two dimensions, which were stressor and reaction. This measure used a five-point Likert scale assessing the frequency of the symptoms (Never; Seldom; Occasionally; Often; Most of the time). The possible range of total score was from 51 to 255. A higher score denoted more stress with academic life. This measurement had been translated to Indonesian language by Sarina (2012) and used for the student population in University of Indonesia. The coefficient alpha of SSI was 0.906. Based on the correlations between item scores and total scores, all coefficient correlations were above 0.2.

Optimism was measured by Life Orientation Test-Revised (LOT-R) (Scheier, Carver, & Bridges, 1994). It consisted of ten items with a five-point Likert scale, from 'Strongly disagree' to 'Strongly agree'. Four of the ten items were fillers, which were excluded from the total score of optimism. Thus, the possible range of total score was from 6 to 30. A higher score meant more optimistic, and vice versa. One item of LOT-R was 'In uncertain times, I usually expect the best'. This measurement had been translated to Indonesian language by Isma (2013) and used for the student population in University of Indonesia. The coefficient alpha of LOT-R was 0.630. Based on the correlations between item scores and total scores, all coefficient correlations were between 0.29 and 0.43.

After data collection, the participants were checked for compatibility, and those who were found not compatible were eliminated. The statistical analyses were performed using SPSS version 21. Data was analysed descriptively to see the general description of participants and variables. Multiple regression analysis was performed to assess the effect of academic stress on SWB, the effect of optimism on SWB, and the effect of academic stress and optimism interaction on SWB.

3 RESULTS

The descriptive data for all variables is presented in Table 1. The mean per item was used to see the general description of SWB and academic stress subscales because each subscale did not have the same number of items. The mean per item was determined by dividing the mean total by each number of items.

Hypothesis testing for the first hypothesis showed that academic stress had a significant negative effect on SWB ($\beta = -0.383$, $t(213) = -6.302$, $p < 0.05$). It meant that the first hypothesis (H1) was accepted. Higher academic stress lessened the level of SWB.

Hypothesis testing for the second hypothesis showed that optimism had a significant positive effect on SWB ($\beta = 0.257$, $t(213) = 4.225$, $p < 0.05$). It meant that H2 was accepted. Higher level of optimism would increase the level of SWB.

The result of hypothesis testing for the third hypothesis (H3) showed that the adjusted $R^2 = 0.245$, referring to 24.5% variance of SWB, could be explained by academic stress, optimism, and their interaction. However, there was no significant interaction effect of academic stress and optimism on SWB ($\beta = 0.237$, $t(211) = 0.491$, $p > 0.05$). It meant H3 was rejected. There was no significant interaction between academic stress and optimism affecting the level of SWB. Optimism did not moderate the effect of academic stress on SWB.

Table 1. Descriptive data for all scales and subscales.

Variables	N	M	M per item	SD	Range
Subjective well-being	215	0.0265		1.616	-4-4
Positive affect	215	35.740	3.574	5.210	10–50
Negative affect	215	29.510	2.951	6.791	10–50
Satisfaction with life	215	15.400	3.080	3.164	5–25
Academic stress	215	135.41		20.36	51–255
Frustration stressor	215	19.228	2.745	3.794	7–35
Conflict stressor	215	9.177	3.059	2.059	3–15
Pressure stressor	215	13.474	3.369	2.736	4–20
Change stressor	215	8.405	2.802	2.380	3–15
Self-imposed stressor	215	22.284	3.714	3.025	6–30
Physiological reaction	215	29.130	2.081	7.323	14–70
Emotional reaction	215	12.149	3.037	3.326	4–20
Behavioural reaction	215	16.721	2.900	4.328	8–40
Cognitive reaction	215	4.847	2.423	1.600	2–10
Optimism	215	20.006		2.976	6–30

4 DISCUSSION

Based on the results, this study supported two out of the three hypotheses. It was found that there was a significant effect of academic stress on SWB. Higher academic stress decreased the level of SWB among the first-year undergraduates. Thus, the first hypothesis (H1) was supported. This result is in accordance with the theory of SWB by Diener (1984). Diener (1984) postulated that SWB is influenced by experience, both favourable and unfavourable. Academic stress could be perceived as an unfavourable experience due to its negative impact, such as being angry or getting sick. One of the most common academic stress reactions experienced by students was emotional reaction. Students reported being anxious, fearful, feeling guilty, and sad. These negative reactions decreased the first-year students' SWB. Furthermore, the results are also in accordance with previous studies in other parts of the world (Schiffrin & Nelson, 2010; Heizomi et al., 2015).

The second hypothesis (H2) was supported as it was found that optimism significantly increased SWB. This finding is in accordance with previous studies in South Korea and India (Cha, 2003; Santhosh & Appu, 2015). Students who scored highly on the optimism measure believed that they would get favourable outcomes in the future, despite any adversity. This belief positively influenced their SWB. The findings support the theory of SWB in which trait is one factor influencing social well-being (Aspinwall & Taylor, 1992; Ayyash-Abdo & Alamuddin, 2007).

It was found that there was no significant interaction between academic stress and optimism on SWB among first-year undergraduates. Therefore, the third hypothesis (H3) was not supported; optimism did not moderate the relationship between academic stress and SWB. This finding might be explained by the fact that assessment was performed during final exams. Students were very stressed during that period. Furthermore, the study was conducted in a reputable and highly competitive college. Most of the participants also reported having high levels of optimism. A more diverse sample is needed to increase the representativeness of first-year undergraduates.

REFERENCES

Aspinwall, L.G. & Taylor, S.E. (1992). Modeling cognitive adaptation: A longitudinal investigation of the impact of individual differences and coping on college adjustment and performance. *Journal of Personality and Social Psychology*, *63*(6), 989–1003. doi:10.1037/0022-3514.63.6.989

Ayyash-Abdo, H. & Alamuddin, R. (2007). Predictors of subjective well-being among college youth in Lebanon. *The Journal of Social Psychology*, *147*(3), 265–284. doi:10.3200/SOCP.147.3.265-284
Carver, C.S. & Scheier, M.F. (2001). Optimism, pessimism, and self-regulation. In E.C. Chang (Ed.), *Optimism & pessimism: Implications for theory, research, and practice* (pp. 31–51). Washington, DC: American Psychological Association.
Cha, K.H. (2003). Subjective well-being among college students. *Social Indicators Research*, *62–63*(1–3), 455–477. doi:10.1023/A:1022669906470
Clinciu, A.I. (2013). Adaptation and stress for the first year university students. *Procedia-Social and Behavioral Sciences*, *78*, 718–722. doi:10.1016/j.sbspro.2013.04.382
Cohen, S., Kessler, R.C. & Gordon, L.U. (1998). *Measuring stress: A guide for health and social scientists*. Oxford, UK: Oxford University Press.
Denovan, A. & Macaskill, A. (2016). Stress and subjective well-being among first year UK undergraduate students. *Journal of Happiness Studies*, 1–21. doi:10.1007/s10902-016-9736-y
Diener, E. (1984). Subjective well-being. *Psychological Bulletin*, *95*(3), 542–575.
Diener, E., Emmons, R. A., Larsen, R. J., & Griffin, S. (1985). The Satisfaction with Life Scale. *Journal of Personality Assessment, 49*, 71–75.
Diener, E., Suh, E.M., Lucas, R.E. & Smith, H.L. (1999). Subjective well-being: Three decades of progress. *Psychological Bulletin*, *125*(2), 276–302.
Fisher, S. (1994). *Stress in academic life: The mental assembly line*. Buckingham, UK: Open University Press.
Gadzella, B.M. (1994). Student-life stress inventory: Identification of and reactions to stressors. *Psychological Reports*, *74*(2), 395–402. doi:10.2466/pr0.1994.74.2.395
Gall, T.L., Evans, D.R. & Bellerose, S. (2000). Transition to first-year university: Patterns of change in adjustment across life domains and time. *Journal of Social and Clinical Psychology*, *19*(4), 544–567. doi:10.1521/jscp.2000.19.4.544
Heizomi, H., Allahverdipour, H., Jafarabadi, M.A. & Safaian, A. (2015). Happiness and its relation to psychological well-being of adolescents. *Asian Journal of Psychiatry, 16*, 55–60. doi:10.1016/j.ajp.2015.05.037
Herwibowo, D. (2014). The relationship between perceived freedom in leisure and subjective well being in University of Indonesia Students (Thesis, Universitas Indonesia, Depok, Indonesia).
Hudd, S., Dumlao, J., Erdmann-Sager, D., Murray, D., Phan, E., Soukas, N. & Yokozuka, N. (2000). Stress at college: Effects on health habits, health status and self-esteem. *College Student Journal*, *34*(2), 217–227.
Isma, M.N.P. (2013). The relationship between optimism and subjective well being on patients in a medical rehabilitation program (Thesis, Universitas Indonesia, Depok, Indonesia).
Manzoor, A., Siddique, A., Riaz, F. & Riaz, A. (2014). Determining the impact of subjective well-being on academic achievement of children in District Faisalabad. *Mediterranean Journal of Social Sciences*, *5*(23), 2673.
Park, N. (2004). The role of subjective well-being in positive youth development. *The ANNALS of the American Academy of Political dan Social Science, 591*(1), 25–39. doi:10.1177/0002716203260078
Santhosh, A. & Appu, A. (2015). Role of optimism and sense of humor towards subjective well being among college students. *Indian Journal of Positive Psychology*, *6*(2), 143.
Sarina, N.Y. (2012). The correlation between academic stress and psychological well-being among first-year college students in Universitas Indonesia (Thesis, Universitas Indonesia, Depok, Indonesia).
Scheier, M.F. & Carver, C.S. (1985). Optimism, coping, and health: Assessment and implications of generalized outcome expectancies. *Health Psychology*, *4*(3), 219–247. doi:10.1037//0278-6133.4.3.219
Scheier, M. F., Carver, C. S., & Bridges, M. W. (1994). Distinguishing optimism from neurotcism (and trait anxiety, self-mastery, and self-esteem): A re-evaluation of the Life Orientation Test. *Journal of Personality and Social Psychology, 67*, 1063–1078.
Schiffrin, H.H. & Nelson, S.K. (2010). Stressed and happy? Investigating the relationship between happiness and perceived stress. *Journal of Happiness Studies*, *11*(1), 33–39. doi:10.1007/s10902-008-9104-7
Watson, D., Clark, L. A., & Tellegan, A. (1988). Development and validation of brief measures of positive and negative affect: The PANAS scales. *Journal of Personality and Social Psychology, 54*(6), 1063–1070.

Diversity in Unity: Perspectives from Psychology and Behavioral Sciences – Ariyanto et al. (Eds)
© 2018 Taylor & Francis Group, London, ISBN 978-1-138-62665-2

The role of the shame (*isin*) moral value in Javanese culture and its impact on personality traits, and shame and guilt emotions of the young Javanese generation

G.S. Prayitno, H.S.S. Sukirna & C. Amelda
Faculty of Psychology, University of Indonesia, Depok, Indonesia

ABSTRACT: Shame and guilt are moral emotions that motivate ethical social behaviour and encourage normal abiding behaviours. Considering the fundamental Javanese values involved in the concept of *isin* (shame), that the Javanese tend to have a high degree of the five traits that make up a personality: openness to experience, conscientiousness, extraversion, agreeableness, and neuroticism. This study aimed to explore the relationship between the personality trait, and shame and guilt emotions. The study's participants were 165 university students whose parents were Javanese and who had been raised and lived in Yogyakarta and its surroundings, an area which is central to their culture. The NEO Five-Factor Inventory (NEO-FFI) was used to capture the personality profile, and the Guilt And Shame Proneness (GASP) Scale was used to measure the shame and guilt proneness action tendency. The result of this study showed that the conscientiousness trait merely correlated with the emotion of guilt, neuroticism only correlated with the emotion of shame, whereas agreeableness correlated with both shame and guilt emotions. Furthermore, the results also revealed that the response shown by the Javanese people when they felt guilt was mostly reparative behaviour. Only people with high agreeableness and extraversion traits showed withdrawal behaviour.

1 INTRODUCTION

Shame (or *isin*, as it is known in Javanese culture) and guilt are moral emotions that encourage people to act in accordance with accepted moral standards, and are critical for deterring unethical and antisocial behaviour (Tangney, 1995; Tangney et al., 2011). Some experts do not distinguish shame from guilt since both are feeling of distress that arise in response to personal transgressions or norm violations (Tangney & Dearing, 2002; Wolf et al., 2010). Both are considered as self-conscious emotions evoked by self-reflection and self-evaluation, aid in self-regulation (Tangney, 2003; Tracy & Robins, 2007), and emerged as feedback to non-compliance or deviation to social norms (Ausubel, 1955; Leary & Tangney, 2003; Strongman, 2003; Tangney et al., 2005). Some researchers believe that they are two different emotions.

Emotions are rooted in one's cultural experience. Su (2010) argued that culture has a profound influence on people's behaviour. Collective communities, such as Asian and Indonesian, are considered to have a shame-based culture, whereas individualistic communities have a guilt-based culture; Asian people live in collectivistic cultures in which the people's behaviour is more regulated by shame rather than guilt (Su, 2010). In most Western communities that are characterised by individualistic culture, people's behaviour is controlled more by the emotion of guilt.

People who feel ashamed after any transgression or norm violation feel more distressed compared to those who feel guilty. It was because those who feel shame will focus more on themselves as a person, whereas those who feel guilty, will focus more on specific behaviour that is incompatible to social norms (Tangney, 1991; Tangney et al., 1992; Tangney & Dearing, 2002). Shame has a detrimental, destructive effect on one's self (Tangney, 1991) as well as self-esteem (Ausubel, 1955).

In Javanese culture, shame has been the main moral emotion imposed to control behaviour. Those who violate or behave inconsistently to Javanese values or norms are considered as '*ora njawani*' (non-Javanese) or '*durung njawani*' (not Javanese yet) (Endraswara, 2010). Most of all, a mature Javanese personality is reflected in one's understanding of *isin* (Suseno, 1993). Sumantri and Suharmono (2007) and Su'udy (2009) observed that there has been a shift towards individualistic characteristics or orientation in many communities in Indonesia. Shame seems to lose its power to regulate and encourage moral behaviour. This can be reflected, for example, in the increasing prevalence of corruption or extra marital pregnancies and abortion (Takariawan, 2012). Empirical studies regarding the tendency to feel shame or guilt among Javanese youth may reveal the existence of the assumed shift in or weakened 'shame culture'.

Shame and guilt emotions are also influenced by personality (Diener & Larsen, in Strongman, 2003), and can be expressed as an individual's tendency to show a consistent pattern of thought, feeling, and action, which differentiate one individual from another (McCrae & Costa, 2006). Costa and Widiger (2002) emphasise five traits of a personality: openness to experience (O), conscientiousness (C), extraversion (E), agreeableness (A), and neuroticism (N). Based on the 'Big Five Factor' model of a personality, Hutapea (2012) found that Javanese male teachers showed high (A) and (C), moderate (E), (O), and low (N). However, no study could be found on the relationship between personality and moral emotion with regard to the Indonesian community. A study involving 332 students (Caucasians, African-Americans, Latinos and Asians) found that the traits (A), (C), and (E) had a positive correlation to shame and guilt (Nan, 2007). (O) showed a positive correlation only to guilt, whereas (N) showed no correlation to neither shame nor guilt emotion. This finding was inconsistent with the results of the studies of Abe (2004) and Wright et al. (1989) that showed (N) has a positive correlation to both shame and guilt emotions. This current study will explore the correlation between the personality traits in the five-factor model, and the moral emotions of shame and guilt emotions among the young Javanese generation.

2 METHODS

A *Participants*

This study involved 165 members of the young Javanese generation, namely those aged 18 to 24 who had Javanese parents and had lived in Central Java (Surakarta or the Special Region of Yogyakarta) since childhood. At the age of 18, it was assumed that the participants had reached the higher stage of Kohlberg's moral development (Kohlberg in Rathus, 2012). They were recruited using the incidental sampling technique.

B *Instruments*

1. The NEO Five-Factor Inventory (NEO-FFI) is a personality inventory developed by McCrae and Costa based on the 'Big Five' theory of personality. NEO-FFI measures (N), (E), (O), (C), and (A). It consists of 60 items and each trait was measured by a 12-item combination of favourable and unfavourable statements regarding each trait. The participants rated the degree to which each statement described themselves on a four-point scale: 4 = Very suitable; 3 = Suitable; 2 = Not suitable; 1 = Very unsuitable. The items validity for (O) trait = 0.38–0.526; (C) trait = 0.453–0.702; (E) trait = 0.267–0.691; (A) trait = 0.291–0.675; and (N) trait = 0.291–0.717. The usage of NEO-FFI in this study was enabled by using the licence granted to Sherly Saragih Turnip.
2. The Guilt And Shame Proneness (GASP) Scale is an instrument that was developed by Cohen et al. (2011) to measure individual differences in experiencing shame and guilt emotions following a transgression or norm violation. It consists of four sub-scales: 1) Guilt-Negative Behaviour Evaluation (Guilt-NBE) for feelings of regret or a feeling that you have committed bad behaviour; 2) Guilt-Repair for assessing an attitude or willingness to

correct mistakes; 3) Shame-Negative Self-Evaluation (Shame-NSE) to measure the feeling of shame, indicated by feeling small, helpless and feeling like a bad person; 4) Shame-Withdrawal for the response of shame, indicated by withdrawal, avoidance or escaping the situation. Each subscale consists of four items (scenarios). The 16 scenarios are to be rated on a seven-point Likert-like scale ranging from 1 for 'Impossible' to 7 for 'Very possible'. GASP has a high reliability (0.779). Each subscale also has high item validity (Guilt-NBE = 0.683–0.712, Guilt-Repair = 0.645–0.70, Shame-NSE = 0.634–0.704, and Shame-Withdrawal = 0.648–0.71).

c *Data analysis*

Descriptive statistics, frequency, mean, and median, were used to describe personality traits and the moral emotions of shame and guilt. Pearson's product moment technique was used to obtain information about the correlation between each personality trait (OCEAN) and shame or guilt proneness. Statistical analysis was conducted with SPSS for Windows 16.0.

3 RESULTS

Almost one third of the participants show high (O), (C), (E), and (N), but an almost equal percentage of the participants had high and low (A). The trait of agreeableness was not distinctively high compared to (O), (C), (E), and (N) (see Table 1).

Regarding moral emotions, the highest score was found in the Guilt-Repair scale, whereas the lowest score was found in the Shame-Withdrawal scale. This showed that violation of norms tended to elicit guilt among Javanese students followed by a behavioural tendency to repair: make a correction, extend apology and regret, and to ask for forgiveness for their wrongdoings. There was a very slight tendency that transgression-elicited shame would be followed by withdrawal (see Table 2).

Results regarding the correlation between personality traits and moral emotions (see Table 3) showed that:

a. There was a significant positive relationship between (O) and Guilt-Repair. The higher trait of (O) an individual has, the higher the likelihood that violation of the norm will elicit a Guilt-Repair response.

Table 1. Percentage of personality traits of the participants (N = 165).

	Percentages	
Personality trait	Low	High
Openness to experience	29.7	70.3
Conscientiousness	31.5	68.5
Extraversion	33.3	66.7
Agreeableness	48.5	51.5
Neuroticism	30.3	69.7

Table 2. Moral emotional-behavioural response to transgressions.

Moral response	Mean score
Guilt-NBE	5.62
Guilt-Repair	5.85
Shame-NSE	5.45
Shame-Withdrawal	3.73

Table 3. Correlation between NEO-FFI and GASP.

Personality trait	Guilt-NBE	Guilt-Repair	Shame-NSE	Shame-Withdrawal
Openness to experience	0.088	0.239**	0.012	–0.072
Conscientiousness	0.195*	0.15*	0.070	0.059
Extraversion	0.032	0.13*	–0.036	–0.274**
Agreeableness	0.27**	0.289**	0.156*	–0.183**
Neuroticism	0.019	–0.105	0.16*	0.29**

*significant at the level 0.05; **significant at the level 0.01.

b. A positive relationship between (C) and guilt existed. Those with (C) were likely to show two types of response: Guilt-NBE and/or repairing their mistakes (Guilt-Repair).
c. There was a positive relationship between (E) and guilt. The higher the trait of (E), the higher the tendency to elicit a Guilt-Repair response.
d. (E) has a strong negative correlation to shame. Individuals with a high (E) trait were very unlikely to show withdrawal behaviour as a response to the elicited shame emotion.
e. (A) has a positive relationship with both shame and guilt. Individuals with an (A) trait showed a high tendency to respond to guilt with Guilt-NBE, or made a correction to their mistakes (Guilt-Repair), and/or also responded to the elicited shame with negative self-evaluation (Shame-NSE), but showed a very slight tendency to respond with Shame-Withdrawal.
f. There were positive relationships between (N) and both shame responses of negative self-evaluation (Shame-NSE), and withdrawal (Shame-Withdrawal). Individuals with high (N) responded to the elicited emotion of shame with negative self-evaluation, such as "I am a bad person" and/or tended to avoid, escape or to show withdrawal behaviour.

Overall, more personality traits were positively related to the emotion of guilt rather than to the emotion of shame. The response of reparation (Guilt-Repair) showed a significant relationship with the four personality traits of (O), (E), (A), and (N), whereas Shame-Withdrawal had a significant relationship only with three personality traits. Shame-Withdrawal action had a negative relationship with (E), and (A), and a positive relationship with (N). This means that an individual with (E) and/or (A) tended not to make withdrawal responses to personal transgressions or norm violations, whereas an individual with (N) tended to make withdrawal responses. Shame-NSE followed by a withdrawal response was only revealed by individuals with (N).

4 DISCUSSION

Javanese students tend to be high in the four traits of personality namely (O), (C), (E), and (N). However, the percentages of participants who were high and low in (A) were almost equal. Contrary to Hutapea's (2012) findings that Javanese teachers had dominant traits of (A), there was almost an equal number of Javanese students who had high or low traits of (A); this means that (A) is less dominant among Javanese students. As for (N), in this study Javanese students were found to have a level of this trait, whereas Hutapea (2012) found this trait was low in the same population. These different findings may be explained by the different characteristics of the participants. Hutapea's study involved 26–40 years old participants, whereas this study involved younger participant with age ranges between 18 and 24 years old. Individuals from different cohorts experienced different cultural atmospheres and may have had different approaches to behaviour, relationships and parenting style. Moreover, participants in this study were university students who have been exposed to individualistic cultural norms and values (through higher education), which may well be different from the Javanese culture in which they were raised. Higher academic or education tradition demands the students to exercise assertiveness, be more open to new or different ideas and be critical to ideas or scientific concepts. This higher education culture may put lower/less importance on trait (A). In higher education, there is also a high chance of long term interaction or exposure to non-Javanese culture (non-local students) that influence the development of original (and distinctive) Javanese personality traits.

Contrary to Nan's (2007) findings in the US, Javanese students who showed high (C) tended to feel guilty after committing transgression and this emotion would be followed by repair tendencies, rather than feeling shame and thinking of people's negative evaluation of themselves. Nan's (2007) findings showed that conscientiousness was a predictor for both shame and guilt. This study yielded similar findings to Wright et al. (1989) in that the trait (N) had a positive correlation with Shame-NSE and Shame-Withdrawal. For Javanese students, the higher the (N) trait they had, the higher their tendency to feel shame and respond to the elicited shame by focusing on negative self-evaluation and/or withdrawal behaviours. The finding that the trait (O) had correlation with Guilt-Repair, corroborated the finding of Nan's (2007) study. The high trait of (O) may facilitate the process of learning to also respond to transgression with guilt, and not only with shame as learned in the Javanese family environment. The trait (E) was related to the tendency of responding with repair behaviour when one's feel guilty, which is inconsistent to Nan's (2007) findings. A Javanese individual needs to have high (E) in order to maintain harmonious interpersonal relationships between members of a community. Javanese tend to show repair behaviour, restore, or regain harmonious relationships to community members who might be in conflict or experience communication or relationship breakdowns after a transgression.

5 CONCLUSION

Personality traits of the young Javanese generation, specifically university students, were shown to have a relationship with shame and guilt emotions. The results of this study indicated a tendency among Javanese to elicit both guilt and shame self-evaluations. However, they tended to show Guilt-Repair rather than Shame-Withdrawal after commitment to any transgression. Therefore, do these results indicate that *isin*, as a Javanese moral value, is not strong enough to encourage shame emotion as expected in a collective society which is categorised as having a shame-based culture, or has there been a shift in the cultural way of responding emotionally and/or regulating behaviour among Javanese youth? This condition implies that an extensive study is needed to obtain a stronger conclusion as to whether there is a shift of moral emotion in the young Javanese generation, in which their behaviour is more controlled by guilt than by shame.

REFERENCES

Abe, J.A. (2004). Shame, guilt and personality judgement. *Journal of Research in Personality, 38*(2), 85–104.

Ausubel, D.P. (1955). Relationship between shame and guilt in the socializing processes. *Psychological Review, 62*(5), 378–390.

Cohen, T.R., Wolf, S.T., Panter, A.T. & Insko, C.A. (2011). Introducing the GASP Scale: A new measure of guilt and shame proneness. *Journal of Personality and Social Psychology, 100*(5), 947–966.

Costa, P.T. & Widiger, T.A. (2002). *Personality disorder and the five factor model of personality.* Washington, DC: American Psychological Association.

Endraswara, S. (2010). *Falsafah hidup Jawa (Javanese Philosophy of Life).* Yogyakarta, Indonesia: Cakrawala.

Hutapea, B. (2012). Sifat-kepribadian dan dukungan organisasi sebagai prediktor komitmen organisasi guru pria di sekolah dasar (*Personality and organizational support as predictor of male teacher organizational commitment at elementary school*). *Makara Seri Sosial Humaniora, 16*(2), 101–115.

Leary, M.R. & Tangney, J.P. (2003). *Handbook of self and identity.* New York, NY: The Guilford Press.

McCrae, R.M. & Costa, P.T. (2006). *Personality in adulthood: A five factor theory perspective.* New York, NY: The Guilford Press.
Mulder, N. (1996). *Pribadi dan masyarakat di Jawa (Individual and society in Java)*. Jakarta, Indonesia: Pustaka Sinar Harapan.
Nan, L.M. (2007). *Moral guilt and shame: an investigation of their associations with personality, values, spirituality and religiosity*. Urbana-Champaign, IL: University of Illinois.
Rathus, S.A. (2012). *Psychology concepts and connections* (10th ed.). Belmont, CA: Wadsworth Cengage Learning.
Strongman, K.T. (2003). *The psychology of emotion: From everyday life to theory*. Chichester, UK: John Wiley and Sons.
Su, C. (2010). *A cross-cultural study on the experience and self-regulation of shame and guilt*. Toronto, Canada: York University.
Sumantri, S. & Suharmono. (2007). *Kajian proposisi hubungan antara dimensi budaya nasional dengan motivasi dalam suatu organisasi usaha (The propotition study of relationship between national culture dimension and motivation in a business organization)* (Doctoral thesis, Fakultas Psikologi Univesitas Padjajaran and Fakultas Ekonomi Universitas Diponegoro).
Suseno, F.M. (1983). *Etika Jawa: sebuah analisa falsafi tentang kebijaksanaan hidup Jawa (Javanese ethics: A philosophy analysis of Javanese wisdom of life)*. Jakarta, Indonesia: PT. Gramedia
Su'udy, R. (2009). *Conflict management styles of Americans and Indonesians: Exploring the effects of gender and collectivism/individualism* (Doctoral thesis, University of Kansas).
Takariawan, C. (2012, Juli 18). Banyak Pacar, Banyak Aborsi (*Many Partners, Many Abortions*). Kompasiana. Retrieved from http://www.kompasiana.com/pakcah/banyak-pacar-banyakaborsi_55125cc7a33311ed56ba8498.
Tangney, J.P. (1991). Moral affect: The good, the bad, and the ugly. *American Psychological Association, 61*(4), 598–607.
Tangney, J.P. (1995). Shame and guilt in interpersonal relationship. In J.P. Tangney & K.W Fischer. (Eds). *Self-conscious emotions: the psychology of shame, guilt. embaressment and pride* (pp. 114–142). New York, NY, The Guilford Press.
Tangney, J.P. (2003). Self-relevant emotions. In M.R. Leary & J.P. Tangney (Eds). *Handbook of self and identity*, (pp. 384–400). New York, NY: The Guilford Press.
Tangney, J.P., & Dearing, R.L. (2002). *Shame and guilt.* New York, NY: The Guilford Press.
Tangney, J.P., Mashek, D., & Stuewig, J. (2005). Shame, guilt, and embarrassment: Will the real emotion please stand up? *Psychological Inquiry*, *16*(1), 44–48.
Tangney, J., P., Stuewig, J., Mashek, D., & Hastings, M. (2011). Assessing Jail Inmates' Proneness To Shame and Guilt: Feeling Bad About the Behavior or the Self? *Criminal Justice and Behavior*, Vol. 38, 710–734.
Tangney, J.P., Wagner, P., Fletcher, C. & Gramzow, R. (1992). Shamed into anger? The relation of shame and guilt to anger and self-reported aggression. *Journal of Personality and Social Psychology*, *62*(4), 669–675.
Tracy, J.L., Robins, R.W. & Tangney, J.P. (2007). *The self-conscious emotions: Theory and research.* New York, NY: The Guilford Press.
Wolf, S.T., Cohen, T.R., Panter, A.T. & Insko, C.A. (2010). Shame proneness and guilt proneness: Toward the further understanding of reactions to public and private transgressions. *Self and Identity*, *9*(4), 337–362.
Wright, F., O'Leary, J. & Balkin, J. (1989). Shame, guilt, narcissism, and depression: Correlates and sex differences. *Psychoanalytic Psychology*, *6*(2), 217–230.

Diversity in Unity: Perspectives from Psychology and Behavioral Sciences – Ariyanto et al. (Eds)
© 2018 Taylor & Francis Group, London, ISBN 978-1-138-62665-2

The association between the five-factor model of personality and the subjective well-being of Abdi Dalem of the Keraton Kasunanan Surakarta Hadiningrat

M.A. Alhad & S.S. Turnip
Faculty of Psychology, Universitas Indonesia, Depok, Indonesia

ABSTRACT: Happiness or subjective well-being is considered the most crucial motivation for individuals in their life. Personality, with regard to its stability within individuals, has been identified as an essential factor when investigating subjective well-being. The Five-Factor Model (FFM) of personality is one of the approaches taken in personality trait theory research and consists of neuroticism, extraversion, openness to experience, agreeableness, and conscientiousness. Previous studies suggest that extraversion and neuroticism are strong predictors for subjective well-being. This study aims to assess the association between the FFM of personality and the subjective well-being of the *Abdi Dalem* of the *Keraton Kasunanan Surakarta Hadiningrat*, and to identify the most influential trait in relation to subjective well-being. The results from multiple regression analysis indicate that 47.3% of subjective well-being was predicted by the FFM of personality. Agreeableness, extraversion, and openness to experience appear to be significantly influential in the subjective well-being of the *Abdi Dalem* of the *Keraton Kasunanan Surakarta Hadiningrat*.

1 INTRODUCTION

Javanese cultures are strongly related to norms and values passed on from generation to generation. The cultures are well maintained as both a legacy and a way of life. Such norms and values consist of attitudes, manners, and behaviours embedded in interpersonal relationships and in dealing with everyday life. The Javanese believe that in order to live happily, they need to be obedient to those norms and values. The *Abdi Dalem* of the *Keraton Kasunanan Surakarta Hadiningrat*, specifically speaking, are a community in Indonesia that is loyal to the Javanese norms and values. *Abdi Dalem* means the people who work for the Javanese Kingdom in the sections where they belong. Their salaries range from Rp 25,000–80,000.

This amount of salary undoubtedly does not cover living costs in Indonesia today, yet they still remain working as an *Abdi Dalem* no matter how much money they earn. They do not literally work for money; they instead seek a blessing from the King. They believe that the money they get from the King can be considered as the key for attracting other money in the future. In order to survive on their salaries, *Abdi Dalem* have side jobs, such as teaching, a Master of Ceremony for Javanese themed occasions, and other relevant jobs. However, they believe that their main job as an *Abdi Dalem* is a means to their side jobs. In interviews, they revealed that they have no intention to leave their job as an *Abdi Dalem* because their attachments to the Kingdom are priceless and the tranquillity they acquire because of it is irreplaceable. There are some spiritual matters existing in the Kingdom that make everything inside 'attached and blessed'.

This phenomena raises many questions. One of them is related to how happy they are to live as who they are. Happiness in psychological terminology is commonly called subjective well-being (Diener & Diener, 1996). Subjective well-being is defined as an individual's judgement towards events and experiences that occur in their lives, including emotional reactions in dealing with positive or negative realities (Diener, 1984). Several factors can be considered important in influencing happiness or subjective well-being, but personality traits appear to be the most crucial ones (DeNeve & Cooper, 1998). Personality traits explain the consistency of people's behavioural tendency in any situation (Feist et al., 2013). Personality traits possessed by individuals tend to be stable after reaching 30 years old (McCrae & Costa, 1991).

This research focused on the Five-Factor Model (FFM) of personality as one of the best approaches to use due to its universality (Allik & McCrae, 2004). The FFM of personality was developed by McCrae and Costa (2003) in the form of a five-trait classification of neuroticism, extraversion, openness to experience, agreeableness, and conscientiousness. These terms can be further defined: neuroticism explains emotional stability and adjustment ability; extraversion explains interpersonal relationship quantity and intensity; openness to experience explains appreciation and open-mindedness towards what happens in life; agreeableness explains interpersonal relationship quality; conscientiousness explains order, persistence, and motivation for particular life goals (McCrae & Costa, 1991).

Previous studies suggest that extraversion and neuroticism are strong predictors for subjective well-being (Diener, 1984; Elliot et al., 1997; McCrae & Costa, 1991). This study aims to assess the correlation between the FFM of personality and subjective well-being of *Abdi Dalem* of the *Keraton Kasunanan Surakarta Hadiningrat*, and also to identify the traits that significantly influence their subjective well-being.

2 METHOD

2.1 *Research participants*

Participants consisted of 200 *Abdi Dalem* aged 40 to 65. In terms of their educational backgrounds, 12 were primary school graduates, 32 were junior high school graduates, 129 were senior high school graduates, and 27 were university graduates. In terms of their gender, 58.5% were male and 41.5% were female. Participants were asked to complete two self-report questionnaires which aimed to measure their personality traits and subjective well-being. The questionnaires were distributed by the researcher simultaneously.

2.2 *Research instruments*

The scale used for measuring personality traits was the Neuroticism-Extraversion-Openness Five-Factor Inventory (NEO-FFI) (McCrae & Costa, 2004), which was adapted by Sherly Saragih Turnip and comprised of 60 items (12 items per domain). Participants were asked to respond to each of them on a four-point Likert-type scale (ranging from 1 = Strongly disagree to 4 = Strongly agree). Scores for each domain were calculated by summing all 12 items in their responses. The scale used for measuring happiness was the Oxford Happiness Questionnaire (Hills & Argyle, 2002) that had been adapted by the researchers using a procedure developed by Beaton (2000) and comprised 29 items. Participants were asked to respond to each of them on a four-point Likert-type scale (ranging from 1 = Strongly disagree to 4 = Strongly agree), with a higher score indicating greater happiness.

3 RESULTS

The Oxford Happiness Questionnaire and NEO-FFI have been used in numerous studies in the past in assessing the correlation between the FFM of personality and subjective well-being. Although previous studies suggest that extraversion and neuroticism are the two traits which significantly influence subjective well-being, this study shows a different result. Based on this study's multiple regression analysis on the population of *Abdi Dalem* of the *Keraton Kasunanan Surakarta Hadiningrat* shown in Table 1, three traits (agreeableness, extraversion, and openness to experience) were found to have a significant influence on their subjective well-being, while the other two traits (neuroticism and conscientiousness) had no significant correlation with subjective well-being. Since the population of this study is very indigenous, the analysis results expectedly differed from the previous studies which took place in western countries. Culture has an important impact on that difference because it contributes to the way people give meaning to events and experiences (Triandis & Suh, 2002), and the way personality traits are established through norms and values (Diener et al., 2003).

Based on the multiple regression analysis, it was discovered that agreeableness had significantly the strongest role in the subjective well-being of the *Abdi Dalem* population ($\beta = 0.372$, $p = 0.000 < 0.01$). The second trait of significant influence was extraversion ($\beta = 0.223$, $p = 0.001 < 0.01$). Finally, the last trait of significant impact was openness to experience ($\beta = 0.181$, $p = 0.012 < 0.05$). In general, the FFM of personality contributed to 47.3% of the subjective well-being of the population of *Abdi Dalem* of the *Keraton Kasunanan Surakarta Hadiningrat* ($R^2 = 0.473$).

Based on the t-test, it was discovered that there was no significant difference between male and female *Abdi Dalem* in their subjective well-being, nor among the three traits of agreeableness, extraversion, and openness to experience that significantly influenced subjective well-being. These findings are shown in Table 2 and confirm that gender had no significant influence on subjective well-being.

The results of Analysis Of Variance (ANOVA) shown in Table 3 indicated that having a higher level of education correlated with greater conscientiousness and less neuroticism, while having a lower level of education correlated with lower conscientiousness and higher neuroticism.

Table 1. Multiple regression of subjective well-being using the FFM of personality.

	Total samples (N = 200)		
Variables	B	95% CI	P
Neuroticism	−0.294	−0.362–0.103	0.274
Extraversion	0.223	0.202–0.726	0.001
Openness	0.181	0.088–0.687	0.012
Agreeableness	0.372	0.531–1.018	0.000
Conscientiousness	0.153	−0.195–0.613	0.309
Gender	−0.032	−2.178–1.156	0.546
Education	−0.115	−3.896–1.411	0.357

Dependent variable: Subjective well-being.

Table 2. T-test of subjective well-being using the FFM of personality based on gender.

	Male (N = 117)	Female (N = 83)		
Variable	Mean (SD)	Mean (SD)	T	Sig.
FFM				
Neuroticism	27.18 (5.23)	29.10 (6.10)	−2.385	0.018
Extraversion	34.34 (3.47)	34.13 (4.10)	0.390	0.697
Openness	32.94 (3.18)	32.73 (4.20)	0.393	0.695
Agreeableness	36.18 (3.48)	35.35 (4.04)	1.553	0.122
Conscientiousness	34.44 (5.47)	32.71 (5.90)	2.139	0.034
Subjective well-being	83.40 (6.95)	81.64 (8.76)	1.585	0.115

Level of significance = 0.05.

Table 3. ANOVA of subjective well-being using the FFM of personality based on education level.

	Primary school (N = 12)	Junior high (N = 32)	Senior high (N = 129)	University (N = 27)		
	M (SD)	M (SD)	M (SD)	M (SD)	F	P
N	38.00 (5.34)*	33.22 (4.67)*	27.05 (3.57)*	21.74 (4.45)*	67.541	0.000
E	36.67 (4.48)[ab]	33.53 (3.10)[ac]	33.50 (3.17)[bd]	37.67 (4.38)[cd]	13.491	0.000
O	30.58 (1.31)[a]	31.66 (3.36)[b]	32.40 (3.02)[c]	37.48 (3.81)[abc]	24.319	0.000
A	35.75 (6.20)	34.53 (3.53)[a]	35.54 (3.39)[b]	38.81 (2.75)[ab]	8.060	0.000
C	22.25 (1.76)*	27.34 (1.56)*	34.40 (2.62)*	43.18 (2.22)*	314.097	0.000
SWB	82.25 (6.98)[a]	80.16 (7.40)[b]	81.62 (7.21)[c]	90.85 (6.19)[abc]	14.226	0.000

Level of significance = 0.01; SWB = Subjective well-being.

4 DISCUSSION

The present study was designed to investigate the role of the FFM of personality in predicting subjective well-being of the population of *Abdi Dalem* of the *Keraton Kasunanan Surakarta Hadiningrat*. Based on multiple regression analysis, it was found that the three traits of agreeableness, extraversion and openness to experience were significantly influential in their subjective well-being. The results are relevant to previous studies: agreeableness (DeNeve & Cooper, 1998; Diener, 1984; Elliot et al., 1997), extraversion (DeNeve & Cooper, 1998; Diener, 1984; Elliot et al., 1997; McCrae & Costa, 1991), and openness to experience (DeNeve & Cooper, 1998) predict subjective well-being. Traits associated with interpersonal relationships (extraversion and agreeableness) have a significant impact on subjective well-being (DeNeve & Cooper, 1998).

The three traits mentioned above (agreeableness, extraversion and openness to experience) represent the life of *Abdi Dalem* of the *Keraton Kasunanan Surakarta Hadiningrat*. Based on observations and interviews conducted prior to this research, important explanations of how *Abdi Dalem* deal with their lives were drawn. The crucial information found and underlined here is related to their interpersonal relationships, the way they see experiences and events in life, and their spirituality. In terms of interpersonal relationships, *Abdi Dalem* prioritise other people in their social lives; they feel worthy when they can help others and share what they own; they feel their lives are more blessed when their existence in society is appreciated; personal belongings and achievements are less important than their social support system, and to obtain that support system they need to mingle with people in the society, setting aside their race, religion, culture, or other diversity, to live in harmony. The more they socialise and live with their social support system, the happier they are.

In terms of seeing events and experiences in life, *Abdi Dalem* believe that whatever happens in their lives is meant to be: they have visions and plans but believe God has a much better scenario. They reject prejudice, disappointment, anger, or any other negative emotions. What they possess are positive emotions, faith, and trust that everything happens for good reasons. They may not see the effect right away, but they have unlimited patience in waiting for good things to come to them. Because of their belief that what they do may have an impact on what they get in return, *Abdi Dalem* have very good control of attitude, manner, and behaviour. Thus, they hope that they will not hurt anyone.

Regarding spiritual matters existing in the *Keraton Kasunanan Surakarta Hadiningrat*, it is necessary to fully comprehend that the norms and values being maintained by *Abdi Dalem* have a strong connection to their faith and belief; what they do is watched by something sacred and powerful living inside the Kingdom, which is why they obey the norms and respect their values. The norms and values mentioned are not only related to their way of life, but also related to the way they speak their language, the way they dress, and the way they interact with people. All are encapsulated in the philosophy of the Javanese culture.

REFERENCES

Allik, J. & McCrae, R.R. (2004). Toward a geography of personality traits: Patterns of profiles across 36 cultures. *Journal of Cross-Cultural Psychology*, *35*(1), 13–28. doi:10.1177/0022022103260382

Beaton et al. (2000). Guidelines for the Process of Cross-Cultural Adaptation of Self Report Measures. *SPINE, 25*(24), 3186–3191.

DeNeve, K.M. & Cooper, H. (1998). The happy personality: A meta-analysis of 137 personality traits and subjective well-being. *Psychological Bulletin*, *124*(2), 197–229. doi:10.1037/0033-2909.124.2.197

Diener, E. (1984). Subjective well-being. *Psychological Bulletin*, *95*(3), 542–575.

Diener, E. & Diener, C. (1996). Most people are happy. *Psychological Science, 7*(3), 181–185. doi:10.1111/j.1467-9639.1991.tb00167.x

Diener, E., Oishi, S. & Lucas, R.E. (2003). Personality, culture, and subjective well-being: Emotional and cognitive evaluations of life. *Annual Review of Psychology*, *54*(1), 403–425. doi:10.1146/annurev.psych.54.101601.145056

Elliot, A.J., Sheldon, K.M. & Church, M.A. (1997). Avoidance personal goals and subjective well-being. *Personality and Social Psychology Bulletin*, *23*(9). doi:10.1177/0146167297239001

Feist, J., Feist, G.J. & Roberts, T.A. (2013). *Theories of Personality*. New York: McGraw Hill.

Hills, P. & Argyle, M. (2002). The Oxford Happiness Questionnaire: a compact Scale for the measurement of psychological well-being. *Personality and Individual Differences, 33*, 1073–1082.

McCrae, R.R. & Costa, P.T. (1991). Adding liebe und arbeit: The full five-factor model and well-being. *Personality and Social Psychology Bulletin, 17*(2), 227–232. doi:10.1177/014616729101700217

McCrae, R.R. & Costa, P.T. (2003). *Personality in Adulthood: A Five Factor Theory Perspective*. New York: The Guilford Press.

McCrae, R.R. & Costa, P.T. (2004). A contemplated revision of the NEO Five-Factor Inventory. *Personality and Individual Differences, 36*, 587–596.

Triandis, H.C. & Suh, E.M. (2002). Cultural influences on personality. *Annual Review of Psychology, 53*(1), 133–160. doi:10.1146/annurev.psych.53.100901.135200

Diversity in Unity: Perspectives from Psychology and Behavioral Sciences – Ariyanto et al. (Eds)
© 2018 Taylor & Francis Group, London, ISBN 978-1-138-62665-2

The effect of positive electronic word-of-mouth element variation on intention to use the TransJakarta bus

F.I. Rodhiya & B. Sjabadhyni
Faculty of Psychology, Universitas Indonesia, Depok, Indonesia

ABSTRACT: The purpose of this study is to seek how potential users of TransJakarta can be actual users by examining the effect of positive electronic Word-Of-Mouth (eWOM) element variation on intention, specifically on intention to use the TransJakarta bus. This study was an experimental study with the randomised two-group design (pretest and post-test). eWOM used in this study took the form of online reviews about TransJakarta that were given to the participants ($n = 62$) for seven days with the use of the instant messaging application, LINE. The findings of this study showed that positive eWOMs, namely text-only as well as those with visual information, increased the intention to use TransJakarta ($F (1.60) = 59.09$; $p < 0.05$). However, the increasing scores of the two experiment groups, that is the group that was given text-only online review access and the group that was given text-with-visual-information online review access, were not significantly different ($F (1.60) = 0.34$, $p > 0.05$). Therefore, it can be concluded that eWOM, with or without visual information, can become an alternative way to increase the intention to use TransJakarta.

1 INTRODUCTION

Congestion in the Special Capital Region of Jakarta (DKI Jakarta) have reached an alarming level. This can be seen by the ratio of the number of vehicles to the length of roads, which is 1,872.60 units/km for regional roads and 93,119.84 units/km for national roads (Kementrian Pekerjaan Umum, 2014). In fact, in 2015 Jakarta was designated as the city with the most congestion in the world (Castrol, 2016; Wardhani & Budiari, 2015). As a comparison, Mexico City 'won' third place and no American cities appeared in the top ten list (Toppa, 2015). In order to reduce the city's traffic congestion, DKI Jakarta's government has made serious efforts by developing various modes of public transportation; one of them is the TransJakarta bus (hereinafter referred to as 'TransJakarta)'.

TransJakarta is a Bus Rapid Transit (BRT) system which has been operating in Jakarta since 2004 (TransJakarta, 2016). However, several years after the launch of TransJakarta, traffic congestion is still a major problem in Jakarta. Passenger numbers have showed a decline: 15,200 per hour was reported in 2012, compared to 14,100 per hour in 2015 (Elyda, 2016). The decline could be attributed to passengers' past experience with TransJakarta. Lerrthaitrakul and Panjakajornsak (2014) stated that when a passenger is satisfied with their public transportation experience, they will use it again and/or share their positive experience with other people, and vice versa. This shared experience can, in turn, affect the potential use of TransJakarta.

There are some factors that influence a consumer's intention; one of them is electronic Word-Of-Mouth (eWOM). Hennig-Thurau et al. (2004) define eWOM as:

> *Any positive or negative statement made by potential, actual, or former customers about a product or company, which is made available to a multitude of people and institutions via the internet.*

A consumer may have an 'intention' after being exposed to eWOM (Ladhari & Michaud, 2015; Mauri & Minazzi, 2013; Park & Kim, 2008; Qu, 2015; Zhang et al., 2010b) and the effect is different depending on its valence (East et al., 2008; King et al., 2014; Ladhari & Michaud, 2015; Mauri & Minazzi, 2013). For example, Park and Lee (2009) found that negative eWOM has stronger effects on purchase decision. In contrast, East et al., (2008) found that positive eWOM has a stronger effect. In addition, some other studies suggest that positive eWOM will increase intention and negative eWOM will decrease it (Ladhari & Michaud, 2015; Mauri & Minazzi, 2013). Moreover, the effect of eWOM might vary according to the platform on which the messages are broadcast. For instance, Ladhari and Michaud (2015) reported that eWOM available on social networking sites may have stronger effects on intention due to trust and closeness, compared to those on an online review website. In summary, both eWOM valence and eWOM platform have an effect on a consumer's intention.

Previous studies mainly focused on the effect of the length, content usefulness or perceived quality, valence, credibility, and the layout of text-only eWOM (Archer et al., 2013; Ladhari & Michaud, 2015; Mauri & Minazzi, 2013; Pan & Zhang, 2011; Purcarea et al., 2013; Zhou et al., 2009; Zhu et al., 2014). Furthermore, previous studies only focused on the design of the platforms (website design) (Mauri & Minazzi, 2013; Qu, 2015), even though platforms now support additional visual elements, such as images, photographs, or videos. The studies of visual elements in eWOM still have mixed results. Some showed that the visual element in eWOM increased attention, product interest, and intention to buy products (Hoffman & Daugherty, 2013; Lin et al., 2012), whereas Davis and Khazanchi (2008) found that images in eWOM could not explain purchase intention.

Based on existing research, this study will test the effect of eWOM element variation on intention to use TransJakarta. eWOM will be manipulated based on its elements, namely text-only eWOM, and text-with-visual information eWOM. The intention will be measured and compared among the experimental groups.

TransJakarta is now striving to improve their service and to attract new customers. They need a way to change their potential users into actual users. Could they use eWOM as one of their methods? As stated above, eWOM is an important factor to determine consumer's behaviour. Therefore, if TransJakarta management can take advantage of eWOM, they may attract new consumers.

Thus, this study will test the following hypothesis:

H1: Positive eWOM will increase the total score of intention to use TransJakarta in potential users of TransJakarta.

Previous studies suggest that the appearance of a visual element in eWOM is important as several studies have already shown that it can increase attention, interest in product, and intention (Daugherty & Hoffman, 2014; Hoffman & Daugherty 2013; Lin et al., 2012). Visual elements (e.g. images, photographs and videos) are a pictorial representation of a product (Kim & Lennon, in Lin et al., 2012) and are well known to play an important part in advertising, but the understanding of how they play a role in eWOM is still lacking (Hoffman & Daugherty, 2013; Lin et al., 2012). Ladhari and Michaud (2015) stated that to further examine eWOM's effectiveness, all elements in eWOM need to be considered, including the visual element.

Therefore, this study will also test the following hypothesis:

H2: The mean of increasing score of intention to use TransJakarta is significantly different among experimental groups.

2 METHOD

2.1 *Research design*

This study is an experiment with the randomised two-group design (pretest and post-test) with no control group. The participants were randomly assigned to experimental groups. Both groups were given the same amount of manipulation with the same duration. The first group was given text-with-visual information eWOM, while the second group was given text-only eWOM. Intention to use TransJakarta was measured in pretest and post-test to see the differences before and after treatment. 'LINE' was used as the platform for the experiment for several reasons. Firstly, this messaging system is free and available on almost every operating system (LINE, n.d.). Secondly, the number of users in Indonesia is more than 90 million with 80% of them active users (Herman, 2016). Therefore, LINE allows a researcher to gather a greater number of participants who are already familiar with the application. Moreover, it has some features to support the research (e.g. up to 200 people per group chat room, together with an ability for photographic sharing). Most importantly, LINE does not require people to share their phone numbers, thus maintaining confidentiality.

2.2 *Participants*

Participants in this study were TransJakarta's potential users, who are defined as those who are not using TransJakarta as their daily transportation mode, but take the same route as TransJakarta during their daily activities. The participants were also identified as people who were not using TransJakarta three months prior to the start of the experiment because TransJakarta management made a lot of changes during that period, such as a significant increase in the number of buses in their fleet (Armindya, 2015), the development of a TransJakarta-related phone application (Aziza, 2015), a revamp of the service (Fenalosa, 2015; Kompas, 2015), the addition of new bus feeders (Rudi, 2015b), the replacement of old buses (Savitri, 2015), and integration with a private bus company (Yusuf, 2015). The participants also needed to be LINE messenger users because the experiments were conducted within the group chat room. The participants were recruited through convenience sampling and had given their active consent before they were invited to the group chat room.

2.3 *Dependent variable (intention to use TransJakarta)*

Intention to use TransJakarta was measured using the purchase intention scale (Jalilvand & Samiei, 2012) with a slight alteration and language adaptation. The scale employs a six-point Likert scale (1 = Strongly disagree, to 6 = Strongly agree). It is different from the original purchase intention scale, which employs a seven-point Likert scale. The scale had to be decreased to six in order to avoid participants offering a neutral response (Kaplan & Saccuzzo, 2009), for example, 'I am going to use TransJakarta's services'. The scale was chosen because it explored the purchase intention, more than just asking 'Are you going to use product x?' The scale has a Cronbach's alpha reliability coefficient of 0.78 on pretest and 0.76 on post-test, which is considered acceptable for research purposes, according to Kaplan and Saccuzo (2009).

2.4 *Independent variable (eWOM)*

This study used the online reviews made on social networking sites as eWOM. Only positive online reviews were used due to ethical reasons. There were text-only online review and text-with-visual information online reviews. Stradling et al. (2007) identified passengers' ideal characteristics of a bus ride experience, namely safety, absence of unwanted disturbance, and social interaction. There were 28 screen-captured online reviews about the perception of an ideal bus ride experience based on the characteristics defined by Stradling et al. (2007). The sample of a text-only online review is shown in Figure 1 and text-with-visual information online review in Figure 2.

2.5 *Procedure*

The nine-day experiment consisted of three parts: pretest, manipulation, and post-test, which all took place in a group chat room on LINE messenger. The first and the last day were used for pretest and post-test. On days 2 through 8, participants were exposed to two

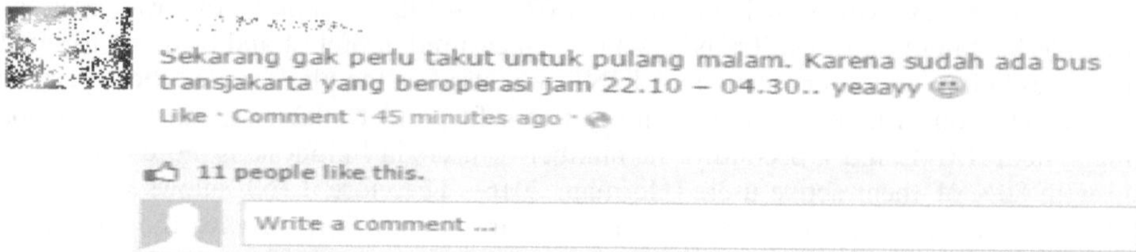

Figure 1. Sample of a text-only online review.

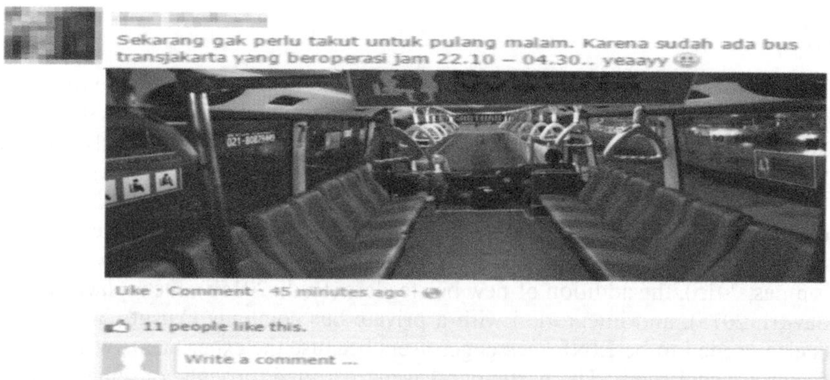

Figure 2. Sample of a text-with-visual information online review.

online reviews. The first group was given a text-with-visual information online review and the second group was given a text-only online review; the content of both texts were the same and neither of the groups was aware of the other group's existence. After the experiment, all of the participants received a reward, namely phone credit and e-commerce vouchers, and were entered into a lottery for a grocery store voucher worth IDR 200,000.

2.6 *Data analysis*

The data gathered was filtered, coded, and entered into IBM SPSS 22 statistical software. The analysis consisted of several parts. Firstly, the pretest scores between two groups were analysed to verify that there were no significant differences of intention to use TransJakarta across the groups. A normality test was then applied by means of a Shapiro–Wilk and homogeneity test using Levene's test as the assumptions for a parametric test. Descriptive statistics were used to show the dependent variable and demographic results. The latter were shown in frequency and percentage, while the general results of dependent variable were shown in mean, frequency, percentage, and standard deviation. After that, a Generalised Linear Model (GLM) Analysis of Variance (ANOVA) was conducted to test the effect of eWOM on customers' intention to use TransJakarta in the equation:

2 (Time: before and after eWOM) × 2 (eWOM: Text-with-visual information and Text-only) experiment design.

3 RESULTS

82 people agreed to participate in this study, but only 62 responded, or 31 in each group. As seen on Table 1, the participants were mostly female (75.8%), of which 67.7% had a bachelor

degree, and typically fell in the second-lowest categories of both income (51.6%) and transportation expense (40.3%).

An independent sample t-test was used to examine the mean difference before the manipulation began. It was verified that initially there were no significant differences of intention to use TransJakarta ($t(61) = 0.68$; $p > 0.05$) between the text-with-visual information group ($M = 13.06$; $SD = 3.12$) and the text-only group ($M = 12.58$; $SD = 2.43$) as shown in Table 2. The normality test using Shapiro–Wilk found that the first group had $D(31) = 0.95$ ($p > 0.05$) and the second group had $D(31) = 0.94$ ($p > 0.05$). Thus, the data distribution is normal. By using Levene's test to assess homogeneity, it was found that there were no significant variances between groups on intention to use TransJakarta ($F(1.61) = 2.32$; $p > 0.05$). Therefore, the parametric test assumptions were fulfilled and the study proceeded to GLM ANOVA.

Figure 3 displayed the results using GLM ANOVA within-subject showed that eWOM affects the intention to use TransJakarta on both treatments ($F(1.60) = 59.09$; $p < 0.05$); the first hypothesis (H1) is accepted. By using GLM ANOVA between-subject, it showed that there were no significant differences on intention to use TransJakarta between the two experimental groups ($F(1.60) = 0.34$, $p > 0.05$); the second hypothesis (H2) is rejected. However, the score of intention to use TransJakarta was slightly higher in the text-with-visual information eWOM group ($M = 15.03$, $SD = 2.77$).

Table 1. Demographics of the experimental groups.

	N	Percentage (%)
Sex		
Male	15	24.2
Female	47	75.8
Age (years)		
17	3	4.8
18	10	16.1
19	1	1.6
20	8	12.9
21	10	16.1
22	13	21.0
23	10	16.1
24	1	1.6
25	3	4.8
26	1	1.6
27	2	3.2
Income per month (IDR)		
<IDR 1,500,000	1	1.6
IDR 1,500,000–2,500,000	32	51.6
IDR 2,500,001–3,500,000	13	21.0
IDR 3,500,001–5,000,000	6	9.7
>IDR5,000,000	10	16.1
Transportation expense per month (IDR)		
<IDR 100,000	10	16.1
IDR100,000–250,000	25	40.3
IDR250,001–500,000	17	27.4
IDR500,001–750,000	8	12.9
IDR750,001–1,000,000	1	1.6
>IDR 1,000,000	1	1.6
Level of education		
High School	13	21.0
Diploma	6	9.7
Bachelor	42	67.7
Doctoral	1	1.6

Table 2. Intention to use TransJakarta results for the independent sample t-test.

	Pretest	Post-test
Group 1 (text-with-visual information eWOM)	$M = 13.06$ $SD = 3.12$	$M = 15.03$ $SD = 2.77$
Group 2 (text-only eWOM)	$M = 12.58$ $SD = 2.43$	$M = 14.87$ $SD = 2.09$

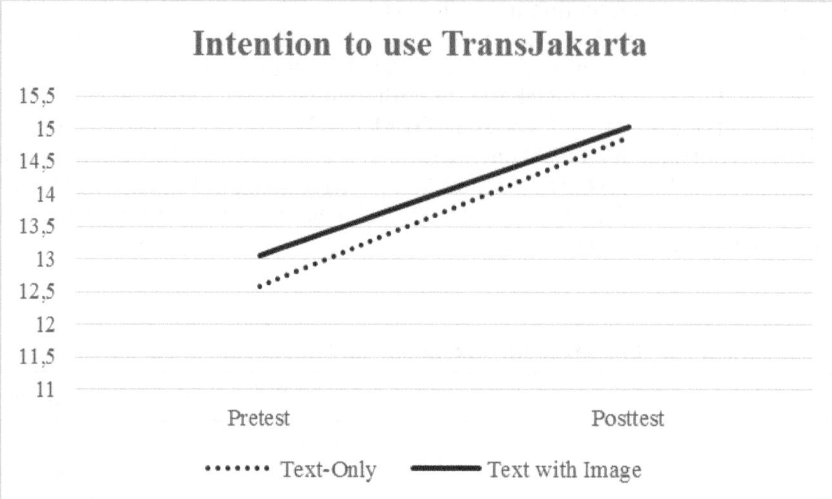

Figure 3. Intention to use TransJakarta pretest and post-test for both experimental groups.

4 DISCUSSION

The results of this study suggest that eWOM can affect consumers' intention to use TransJakarta, which supports the findings of previous studies on the effectiveness of eWOM on purchase intention (Ladhari & Michaud, 2015; Mauri & Minazzi, 2013; Park & Kim, 2008; Park & Lee, 2009; Qu, 2015; Zhang et al., 2010b). Another finding of this study was that there were no significant differences on mean of intention to use TransJakarta between the group which was given text-with-visual information online review and the group which was given text-only online review. The results were found to be similar to the findings of Davis and Khazanchi (2008) in that there were no different effects on eWOM if it was varied by its element, that is, with visual information and text-only. However, it is not consistent with other scholarly works that showed how the visual element of eWOM increases the purchase intention compared to text-only eWOM (Daugherty & Hoffman, 2014; Hoffman & Daugherty, 2013; Lin et al., 2012).

This study has several limitations that need to be acknowledged. Firstly, participants were not checked as to whether they were promotion-oriented customers or not; these types of customers are easily influenced by positive eWOM (Zhang et al., 2010a). Secondly, the study did not consider the customers' predisposition to being a verbaliser or visualiser. Verbaliser consumers tends to like and easily process text, whereas visualiser consumer tends to like and easily process visual information (Schiffman & Wisenblit, 2015). Furthermore, the participants may be low-involvement or sceptical consumers who tend to conform to the eWOM or do not care about the quality of eWOM (Lee et al., 2008; Qu, 2015).

Another thing to consider was how the study used online review screenshots, which technically, was a picture. Even though one contains only text and the other contains text and image, the participants may perceive both as an image. But there was a high chance of occurrence because the TransJakarta product is still experiential 'goods'. Peterson et al. (1997) stated that visual information can be helpful and will affect purchase intention when searching for goods (e.g. fashion items) because the visual information can increase the understanding of the products. Whereas for experiential goods (e.g. a trip), the visual information may not be sufficient to make a consumer really understand the product. Demand characteristics might have also been an influence in this study because the participants already knew that they were in an experimental group chat room.

Due to the limited number of participants and limited time to gather participants, this study was unable to test whether there was an effect of image-only eWOM on intention to use TransJakarta, and was unable to have a control group; it is important for future research to consider this matter. Another issue to address in future studies is synchronisation of the platform. The eWOM given to the participants in this research were screenshots of online reviews from a social networking site, while the experiment took place in an instant messaging group chat room. This unsynchronised platform may have had an effect on the participants' intention because a work by Lee and Youn (2009) found that there is an effect of eWOM platforms on customers' product judgement when the eWOM is positive.

5 CONCLUSION

The intention to use TransJakarta may be influenced by eWOM, even though this study found no significant differences on the presence of visual information in eWOM. Further research should consider individual differences that may be a predisposition to behaviour, for example, consumer involvement and preferences (verbalisers versus visualisers). The visual content chosen needs to match the products, for experiential goods, photographs and images may not offer sufficient information to make consumers truly understand the products.

REFERENCES

Archer, J., Chen, J. & Udo-Imeh, N. (2012). Visual cues: What combinations make for a better user review? Retrieved from: http://web.ics.purdue.edu

Armindya, Y.R. (2015). Catat, ada 500 armada baru Transjakarta hingga akhir tahun (Note it, there will be500 Transjakarta new fleets by the end of the year). *Tempo.co.* Retrieved from http://metro.tempo.co/read/news/2015/08/03/083688713/

Aziza, K.S. (2015). Ahok: Sekarang dirut PT Transjakarta sudah enggak bisa bohong lagi (Ahok: president director of PT Transjakarta cannot lie anymore). *Kompas.com*. Retrieved from http://megapolitan.kompas.com/

Castrol. (2016). *Stop-start index.* Retrieved from http://www.castrol.com/en_au/australia/car-engine-oil/engine-oil-brands/castrol-magnatec-brand/stop-start-index.html

Daugherty, T. & Hoffman, E. (2014). eWOM and the importance of capturing consumer attention within social media. *Journal of Marketing Communications, 20*(1–2), 82–102.

Davis, A. & Khazanchi, D. (2008). An empirical study of online word of mouth as a predictor for multi-product category e-commerce sales. *Electronic Markets, 18*(2), 130–141.

East, R., Hammond, K. & Lomax, W. (2008). Measuring the impact of positive and negative word of mouth on brand purchase probability. *International Journal of Research in Marketing, 25*(1), 215–224.

Elyda, C. (2016). Police, diplomats' vehicles banned from Transjakarta lanes. *The Jakarta Post.* Retrieved from: http://www.thejakartapost.com/news/2016/06/13/police-diplomats-cars-banned-transjakarta-lanes.html

Fenalosa, A. (2015). Punya keluhan soal Transjakarta, silakan mengadu ke 'call center' ini (Got a complaint about Transjakarta, contact this 'call center'). *Kompas.com.* Retrieved from: http://megapolitan.kompas.com/

Hennig-Thurau, T., Gwinner, K.P., Walsh, G. & Gremler, D.D. (2004). Electronic word-of-mouth via consumer-opinion platforms: What motivates consumers to articulate themselves on the internet? *Journal of Interactive Marketing, 18*(1), 38–52.

Herman. (2016). Pengguna line di Indonesia tembus 90 juta (Line surpasses 90 million users in Indonesia). *Beritasatu.com.* Retrieved from http://www.beritasatu.com/digital-life/383212-pengguna-line-di-indonesia-tembus-90-juta.html

Hoffman, E. & Daugherty, T. (2013). Is a picture always worth a thousand words? Attention to structural elements of eWOM for consumer brands within social media. *Advances in Consumer Research, 41*, 326–331.

Jalilvand, M.R. & Samiei, N. (2012). The effect of electronic word of mouth on brand image and purchase intention. An empirical study in the automobile. *Marketing Intelligence & Planning, 30*(4), 460–476.

Kaplan, R.M. & Saccuzzo, D.P. (2001). *Psychological testing: Principles, applications, and issues* (5th ed.). Belmont, CA: Wadsworth Cengage Learning.

Kementrian Pekerjaan Umum. (2014). Buku informasi statistik: Infrastruktur pekerjaan umum 2014 (Statistical information book: Public works infrastructure 2014). Jakarta, Indonesia: Ministry of Public Works. Retrieved from https://pu.go.id/uploads/services/infopublik20151105121638.pdf

King, R.A., Racherla, P. & Bush, V.D. (2014). What we know and don't know about online word-of-mouth: A review and synthesis of the literature. *Journal of Interactive Marketing, 28*(3), 167–183.

Kompas. (2015, May 7). Transjakarta tambah layanan Amari dan Andini (Transjakarta added services Amari and Andini). *Kompas.com.* Retrieved from http://megapolitan.kompas.com/read/2015/05/07/23000081/Transjakarta.Tambah.Layanan.Amari.dan.Andini

Ladhari, R. & Michaud, M. (2015). eWOM effects on hotel booking intentions, attitudes, trust, and website perceptions. *International Journal of Hospitality Management, 46*, 36–45.

Lee, J., Park, D.H. & Han, I. (2008). The effect of negative online consumer reviews on product attitude: An information processing view. *Electronic Commerce Research and Applications, 7*(3).

Lee, M. & Youn, S. (2009). Electronic word of mouth (eWOM): How eWOM platforms influence consumer product judgement. *International Journal of Advertising: The Review of Marketing Communications, 28*(3), 473–499.

Lerrthaitrakul, W. & Panjakajornsak, V. (2014). The impact of electronic word-of-mouth factors on consumers' buying decision-making processes in the low-cost carriers: A conceptual framework. *International Journal of Trade, Economics and Finance, 5*(2).

Lin, T.M.Y., Lu, K.Y. & Wu, J.J. (2012). The effects of visual information in eWOM communication. *Journal of Research in Interactive Marketing, 6*(1), 7–26.

LINE. (n.d.). *LINE: Panggilan dan pesan gratis (LINE: Free calls & messages)*. Retrieved from http://line.me/id/

Mauri, A.G. & Minazzi, R. (2013). Web reviews influence on expectations and purchasing intentions of hotel potential customers. *International Journal of Hospitality Management, 34*(1), 99–107.

Pan, Y. & Zhang, J.Q. (2011). Born unequal: A study of the helpfulness of user-generated product reviews. *Journal of Retailing, 87*(4), 598–612.

Park, C. & Lee, M.L. (2009). Information direction, website reputation and eWOM effect: A moderating role of product type. *Journal of Business Research, 62*(1), 61–67.

Park, D.H. & Kim, S. (2008). The effects of consumer knowledge on message processing of electronic word-of-mouth via online consumer reviews. *Electronic Commerce Research and Applications, 7*(4), 399–410.

Peterson, R.A., Balasubramanian, S. & Bronnenberg, B.J. (1997). Exploring the implications of the internet for consumer marketing. *Journal of the Academy of Marketing Science, 25*(4), 329–346.

Purcarea, V.L., Gheorghe, I.R. & Petrescu, C.M. (2013). Credibility elements of eWOM messages in the context of health care services: A Romanian perspective. *Journal of Medicine and Life, 6*(3), 254–259.

Qu, L. (2015). *Design of electronic word-of-mouth systems pays: Effects of layout of online product review webpages on consumer purchase behavior* (Doctoral thesis, The University of Hong Kong). Retrieved from http://hdl.handle.net/10722/208578

Rudi, A. (2015a). Agar warga tak gunakan mobil, Transjabodetabek akan berangkat dari perumahan (In order to encourage residents not to use cars, Transjabodetabek will depart from housing complex). *Kompas.com.* Retrieved from http://megapolitan.kompas.com/read/2015/09/07/13180291/

Rudi, A. (2015b). Bulan depan, separator 'busway' koridor 1 dan 8 ditinggikan (Next month, 'busway' separators in corridor 1 and 8 will be elevated). *Kompas.com* Retrieved from http://megapolitan.kompas.com/read/2015/09/17/13022951/

Savitri, A.W. (2015, November 19). Bus reyot di koridor 6 diganti Transjabodetabek, begini penampakannya (Here is the appearance of Transjabodetabek replacement bus in corridor 6). *Detikcom.* Retrieved from http://news.detik.com/berita/3074890/bus-reyot-di-koridor-6-diganti-transjabodetabek-begini-penampakannya

Schiffman, L.G. & Wisenblit, J.L. (2015). *Consumer behavior* (11th ed.). Harlow, UK: Pearson Education.

Stradling, S., Carreno, M., Rye, R. & Noble, A. (2007). Passenger perceptions and the ideal urban bus journey experience. *Transport Policy*, *14*(4), 283–292.

Toppa, S. (2015). These cities have the worst traffic in the world, says a new index. Retrieved from http://time.com/3695068/worst-cities-traffic-jams/

Transjakarta. (2016). *Tentang Transjakarta: Sejarah (About Transjakarta: History)*. Retrieved from http://transjakarta.co.id/tentang-transjakarta/sejarah/

Wardhani, D.A. & Budiari, I. (2015). Jakarta has 'Worst traffic in the world'. *The Jakarta Post*. Retrieved from http://www.thejakartapost.com/news/2015/02/05/jakarta-has-worst-traffic-world.html

Yusuf, M. (2015). 200 bus kopaja akan diintegrasikan ke jalur bus TransJakarta (200 Kopaja bus will be integrated in Transjakarta busway). Tribun News Retrieved from http://wartakota.tribunnews.com/2015/09/15/200-bus-kopaja-akan-diintegrasikan-ke-jalur-bus-transjakarta

Zhang, J.Q., Craciun, G. & Shin, D. (2010a). When does electronic word-of-mouth matter? A study of consumer product reviews. *Journal of Business Research*, *63*(12), 1336–1341.

Zhang, Z., Ye, Q., Law, R. & Li, Y. (2010b). The impact of word-of-mouth on the online popularity of restaurants: A comparison of consumer reviews and editor reviews. *International Journal of Hospitality Management*, *29*(4), 694–700.

Zhou, M., Dresner, M. & Windle, R. (2009). Revisiting feedback systems: Trust building in digital markets. *Information and Management*, *46*(5), 279–284.

Zhu, L., Yin, G. & He, W. (2014). Is this opinion leader's review useful? Peripheral cues for online review helpfulness. *Journal of Electronic Commerce Research*, *15*(4), 267–280.

Schultz, [illegible] (2012). [illegible] (11th ed.). [illegible] Hall [illegible]

Steadings, [illegible] M. [illegible] & [illegible] (2007). [illegible] and the [illegible] journey experience. [illegible]

[illegible] (2015). [illegible] first in the world [illegible] Retrieved from [illegible] or making [illegible]

[illegible] (2006). [illegible] [illegible] and [illegible] makers [illegible]

[illegible] D. [illegible] (2011). [illegible] The [illegible] Retrieved from [illegible]

[illegible] M. (2012). [illegible]

[illegible] (2009). [illegible]

[illegible] (2015). [illegible] research [illegible]

[illegible] (2010). [illegible] [illegible] media [illegible]

Diversity in Unity: Perspectives from Psychology and Behavioral Sciences – Ariyanto et al. (Eds)
 ISBN 978-1-138-62665-2

The relationship between system justification and belief in God: The moderating effect of cognitive style and religious system justification

M.H.T. Arifianto & B. Takwin
Faculty of Psychology, Universitas Indonesia, Depok, Indonesia

ABSTRACT: This study examined the relationship among system justification, religious system justification, cognitive style, and belief in 'God'. With a logical basis that belief in God is mainly implicit, we assumed that this belief is influenced by system justification and religious system justification in which it is accepted and preserved. This influence is bolstered by individual cognitive style. This study was a correlational study, with questionnaires, of 277 college students of the Universitas Indonesia (UI). Belief in 'God' was measured with the Belief in God Scale, while cognitive style was measured with the Cognitive Reflective Test. System justification was measured with two measurements, the General System Justification Scale and the Religious System Justification Scale, as an adapted version of the General System Justification Scale which was imbued with the context of religiosity. Using the moderated moderation regression model (model number 3) from Hayes (2013), we found that there were relationships between system justification and belief in God that are moderated by religious system justification, and that moderation effect was diluted by cognitive style.

1 INTRODUCTION

Religion is a source of cognitive aspects and culture that is maintained and has prevailed among multiple cultures (Willard & Norenzayan, 2013), where most people identify themselves as believers (Silberman, 2005). When people adhere towards a certain religion, it is predominantly because they believe in the respective god of that faith. But it is also possible that they have other reasons to be a believer, such as one that is inherited from their parents, it's compulsory, or simply because they want to be religious. Considering the possibilities, it is also possible that an individual believes in the existence of a god, but he/she does not adhere towards a religion. This is a plausible scenario, considering belief is subjective, experience-based, and often based on implicit knowledge and emotions on some matter or state of art (Pehkonen & Pietila, 2003).

According to the Global Attitudes Survey by the Pew Research Center (2014) conducted in 44 countries, there are five things that are considered to be threats to the world, one of them being religious and ethnic hatred. For example, in Indonesia 26% of the population deem that religious and ethnic hatred is the biggest threat to the world. Moreover, the same survey showed that 78% of Indonesians were worried about the problems that are caused by their religion, but still have a relatively high trust in their religion. However, based on a census conducted by the Indonesian Government, 97.1% of Indonesians still have adherence to their religion (Indonesian Statistics Agency, 2010). This situation raises a question, how do people still adhere to their religion while they consider religion as the source of problems?

One approach that could explain this phenomenon is System Justification Theory, which states that people will justify their status quo and deem it as fair, legitimate, and just to satiate their existential needs (Jost & Banaji, 1994). Based on this proposition, adherence to religion as their status quo will drive people to satiate their existential needs as it becomes mandatory, by perceiving their religion as fair, legitimate, and just. The mechanism that portrays this phenomenon is called 'system justification' (SJ), which is the psychological process whereby prevailing conditions (social, political, economic, sexual or legal) are accepted, explained and justified simply because they exist (Jost & Banaji, 1994; Jost et al., 2004; Van der Toorn & Jost, 2014). SJ operates implicitly and unconsciously (Jost & Banaji, 1994), so the unconscious nature of SJ will make the present ideology carried out and condoned by the proprietor with less cognitive interference from his/her own consciousness.

Jost & Banji (1994) postulated that the unconscious nature of SJ may allow existing ideologies to be exercised without the awareness of perceivers or targets. Because SJ operates implicitly in the mind, this study will argue that cognitive factors also contribute to the justification of religious systems. Uhlmann's work (cited in Jost et al., 2009) found that there were two dominant variables that could explain 'Belief in God' (BiG), which are cognitive aspects and existential needs. Both of these variables are implicit, intuitive and unconscious in their influence to develop, maintain and transmit. According to Kraft et al. (2015), individuals do not accept and internalise informational and contextual frames, irrespective of their predisposition, meaning that affective and cognitive reactions to external and internal events are triggered unconsciously. This tendency is followed spontaneously by the activation of associative pathways that link thoughts to feelings, intentions and behaviours, so that events, even those that occur below conscious awareness, set the direction for all subsequent processing (Custers & Aarts, 2010; Kraft et al., 2015).

Consistent with those findings, Shenhav et al., (2012) also found that religious belief is rooted in intuitive processes and conversely, religious disbelief can arise from analytic cognitive tendencies that block or override these intuitive processes. That study was based on Stanovich's 'Cognitive Style' (CS) (Stanovich, 2002, 2004, 2009; Stanovich & West, 2000), which is a configuration of domain-general abilities of directing attention, valuation, and motivation that produces a particular salient landscape within which one undertakes one's tasks. There are two variations of CS: the intuitive type (type 1) and the reflective type (type 2) (Stanovich, 2012). Intuitive processes are hypothesised to produce automatic judgements that can be overridden through the engagement of controlled or reflective processes (Inzlicht & Tullet, 2010). Just as the belief in a god may be the outcome of an intuitive belief-formation process, it may also play a supporting role in such processes; individuals who are drawn to intuitive explanations may come to believe in a god or strengthen their existing belief because believing supports intuitive explanations (Preston & Epley, 2005, 2009; Shenhav et al., 2012).

Belief in a god may enable a general class of easily accessible explanations that make sense of otherwise mysterious phenomena by appealing a god's varied and extensive causal powers (Lupfer et al., 1996; Shenhav et al., 2012), which in turn sheds light into why explanations with heuristic qualities are justified. A few studies presented that individuals with intuitive thinking style have a high tendency to rely on heuristics (Frederick, 2005; Stanovich & West, 1998), one of which was a study conducted by Shenhav et al. (2012).

Based on these explanations, it can be deduced that SJ could predict BiG in respect of an individual's particular CS with 'Religious System Justification' (RSJ) as context. The latter will activate intuitive thinking (type 1) which will, in turn, strengthen the activation of SJ that could predict the likeliness of relatively high belief in a god. This proposed relation will be tested empirically in this research.

2 METHOD

The study recruited participants online (N = 277, 67.1% female; M_{age} = 20 years; SD_{age} = 1.875; UI students only) using Google's Google Form. Participants completed a demographic survey, including questions concerning belief in a god and adherence to a certain religion. We employed continuous measures of BiG using the BiG Scale (Willard & Norenzayan, 2013) with two additional items based on correspondence with Willard and Norenzayan concerning the probability that the response will trigger the 'ceiling effect' during data gathering (initiating a relatively high response rate).

Participants also completed a three-item Cognitive Reflection Test (CRT) (Frederick, 2005), which the study used to assess CS. The three items were open-ended maths problems with intuitively attractive, but incorrect, answers. For example: 'In a lake, there is a patch of lily pads. Every day, the patch doubles in size. If it takes 48 days for the patch to cover the entire lake, how long would it take for the patch to cover half of the lake?' The correct answer is 47 days but, intuitively, many people will answer 24 days because it is the first answer that springs immediately to mind. Choosing the top of the mind response, which is more attractive intuitively, signals greater reliance on intuition and less reliance on reflection. The study analysed the number of intuitive responses given by each participant, rather than the number of correct responses, to avoid classifying non-intuitive incorrect responses (e.g. 30 days in the example above) as intuitive.

This study employed an adapted version of the General System Justification (GSJ) Scale created by Kay and Jost (2003; Jost et al., 2014) to measure the extent to which citizens feel that their society is as fair and just as it should be, thus measuring SJ. An adapted version of the four-item GSJ Scale imbued with the context of religiosity, which is called the Religious System Justification (RSJ) Scale, was also employed to measure the extent to which believers felt that their religion was as fair and just as it should be, thus measuring SJ within the context of religiosity. In these scales, the lower the scores, the higher the likelihood for that individual to have an activation of SJ. Thus, we divide the lower half of the range into 'activated system justification' and the higher half of the range into 'inactivated system justification'.

3 RESULTS

Moderated moderation regression analysis (Hayes, 2013) was used to predict belief in a god from the dynamics of SJ, RSJ and CS. As pictured in Table 1, the interaction between SJ, RSJ and CS could explain 21.1% of the variance from belief in 'God' significantly, $R^2 = 0.211$, $F(7, 269) = 10.302$, $p < 0.05$. GSJ predicts BiG significantly ($\beta = 0.548$, $p < 0.05$). RSJ could not predict BiG significantly ($\beta = 0.349$, $p > 0.05$), and CS could not predict BiG significantly either ($\beta = 0.046$, $p > 0.05$).

Interaction between GSJ and RSJ predicts BiG significantly ($\beta = -1.009$, $p < 0.05$). Both the interaction between GSJ and CS and the interaction between RSJ and CS could not predict BiG significantly ($\beta = 0.04$, $p > 0.05$ and $\beta = -0.2$, $p > 0.05$, respectively). The interaction between GSJ, RSJ and CS could not predict BiG significantly either ($\beta = 0.084$, $p > 0.05$). The details is presented in Table 2.

The moderation test of moderated moderation regression was analysed using Hayes' PROCESS (2013). This moderation test was used to calculate the moderation power from cognitive style and religious system justification in a religious context. Results from PROCESS (Hayes, 2013) were divided into three groups, which consist of ±1 Standard Deviation (SD) from the mean of CS and RSJ. These groups were divided into relatively low, relatively moderate, and relatively high for the moderators (CS and RSJ). A lower score for system justification implied activation of system justification (–1 SD, 6.465), while a lower score for cognitive style implied intuitive thinking (–1 SD, 0) and vice versa for reflective thinking (+1 SD, 2.036).

This moderation test produced significant effects in each of the groups from relatively low CS and relatively low RSJ. The relatively low CS group and the relatively low RSJ group produced a significant moderation effect of 0.274, $p < 0.05$. The relatively moderate CS group

Table 1. The effect of independent variables on BiG.

				Change statistics				
R	R^2	Adjusted R^2	Std. error of the estimate	R^2 change	F change	df1	df2	Sig. F change
0.460	0.211	0.191	5.003	0.211	10.302	7	269	0.000

and the relatively low RSJ group produced a significant moderation effect of 0.289, $p < 0.001$, while the relatively moderate CS group and the relatively moderate RSJ produced a significant moderation effect of 0.105, $p < 0.05$. Lastly, a relatively high CS and relatively low RSJ produced a significant effect of 0.305, $p < 0.01$. Detailed findings are presented in Table 3.

Table 4 is about the Johnson–Neyman significance region. The Johnson–Neyman moderation analysis used percentiles in steps of five (i.e. 0, 5, 10, 15 etc.). This analysis focused on the moderation power of the second moderator (W) on the first moderator (M), or alternatively the CS on the RSJ that moderated the influence of GSJ (X) towards BiG (Y).

The cut-off point from moderator (W) for CS that has a significant moderation effect on $X*M$ (GSJ*RSJ) towards Y (BiG) is 2.385. The scores of CS under 2.385, which are technically 2, are about 87.726% from the entire data that have significant moderation effect.

The range of cognitive style scores from CRT is 0–3. Scores that have significant moderation effects are scores under 2.385. As presented in Tables 5 and 6, for the scores of 0–0.6 (which operationally is 0), the moderation towards interaction of GSJ and RSJ is −0.026, $p > 0.05$. For the scores of 0.601–1.5 (which operationally is 1), the moderation effect between the interaction of GSJ and RSJ is −0.025, $p > 0.01$. Lastly, for the scores of 1.501–2.385 (which operationally is 2), the moderation effect towards the interaction of GSJ and RSJ is −0.025, $p > 0.05$.

Table 2. Moderated moderation regression (Model 3) results.

Model	*B*	*SE B*	*B*
(Constant)	15.925	2.732	
GSJ	0.443	0.140	0.548**
RSJ	0.266	0.189	0.349
CS	0.238	2.098	0.046
Interaction of GSJ and RSJ	−0.026	0.009	−1.009**
Interaction of GSJ and CS	0.009	0.103	0.040
Interaction of RSJ and CS	−0.057	0.134	−0.200
Interaction of GSJ, RSJ and CS	0.001	0.006	0.084

$*p < 0.05$; $**p < 0.01$; $***p < 0.001$.

Table 3. Conditional effect of X on Y at values of the moderator.

CS	RSJ	Effect	T
0.000	6.465	0.274*	2.548
0.000	13.751	0.083	1.151
0.000	21.037	−0.108	−1.060
0.957	6.465	0.289***	3.963
0.957	13.751	0.105*	2.179
0.957	21.037	−0.080	−1.124
2.036	6.465	0.305**	3.222
2.036	13.751	0.129	1.957
2.036	21.037	−0.047	−0.455

$*p < 0.05$; $**p < 0.01$; $***p < 0.001$.

Table 4. Moderator value defining the Johnson–Neyman significance region.

Value	Per cent below	Per cent above
2.385	87.726	12.274

Table 5. Conditional effect of $X*M$ on Y at values of the moderator (W).

CS	Effect	T
0.000	–0.026*	–2.510
0.150	–0.026**	–2.687
0.300	–0.026**	–2.870
0.450	–0.026**	–3.052
0.600	–0.026***	–3.219
0.750	–0.025***	–3.353
0.950	–0.025***	–3.434
1.050	–0.025***	–3.445
1.200	–0.025***	–3.383
1.350	–0.025***	–3.256
1.500	–0.025**	–3.083

*$p < 0.05$; **$p < 0.01$; ***$p < 0.001$.

Table 6. Conditional effect of X*M on Y at values of the moderator (W).

CS	Effect	T
1.650	–0.024**	–2.884
1.800	–0.024**	–2.678
1.950	–0.024*	–2.476
2.100	–0.024*	–2.287
2.250	–0.024*	–2.112
2.385	–0.024*	–1.969
2.400	–0.024	–1.954
2.550	–0.024	–1.811
2.700	–0.024	–1.683
2.850	–0.023	–1.569
3.000	–0.023	–1.465

*$p < 0.05$; **$p < 0.01$; ***$p < 0.001$.

4 DISCUSSION

Belief in God is a product of belief systems that are very common in our society, especially in Indonesian society which is quite religious. There are many conditions and factors that influence the magnitude of the belief in God, one of which is system justification. Jost et al. (2014) proved that system justification and belief in God relate in a way that the latter is justified as a religious ideology, which in turn shows that people with activated system justification will have a relatively high belief in God, and vice versa. This study also added an additional variable that assumed a role in the relationship dynamics of system justification and cognitive style because relating these variables without the context of religiosity was thought of as a liability. Therefore, religious system justification, which is a system justification, was used within the context of religiosity.

The purpose of this study was to see the relationship between system justification and cognitive style towards belief in God. From the results, it can be seen that not all variables contributed directly towards belief in God, while overall, there was a relationship among all variables. It was also found that system justification influenced belief in God significantly, while cognitive style did not. Furthermore, interaction between general system justification and religious system justification influenced belief in God significantly, which implied that the activation of system justification with religious context influences belief in God.

The statistical analysis of the moderation effect found that there were significant moderation effects among all groups of cognitive style based on ±1 SD and mean, while only the −1 SD of religious system justification has significant moderation effects. This implies that, no matter what cognitive styles people have, only those with activated system justification (in this study, the context is religiosity) will moderate the relationship between system justification and belief in God. With the Johnson–Neyman moderation analysis, we found that the moderation effects of cognitive style towards the interaction of general system justification and religious system justification were significant under a cut-off point score, which was 2.385. About 87.76% of the cognitive style scores in this study had moderation effects towards system justification.

This study could offer an alternative explanation for findings conceived by Jost et al. (2014), Uhlmann et al. (2009) and Shenhav et al. (2012). First, it affirmed the findings of Jost et al. (2014) that emphasised a relationship between system justification and belief in God and found that there was no observed relationship between cognitive style and belief in God. Second, the results contradicted the predisposition coined by Uhlmann et al. (2009) that belief in God is based on the implicit nature of intuitive thinking. Third, the results contradicted the findings of Shenhav et al. (2012) that found intuitive thinking is the major influence towards relatively high belief in God.

It has been assumed that the cultural differences between Indonesia and western countries, which produced the previous findings, have a role in these contradictions. Furthermore, there are differences in ideology between these countries. Western countries are secular, while Indonesia is much more conservative and religious. This argument is coherent with the result from the Pew Forum Survey (2008), which found that 65% of females and 44% of males in the United States thought religion to be important to one's life, while in Indonesia, 95% of the population consider that religion is important to one's life. This perspective gap gives birth to other factors that contribute to the magnitude of one's belief in God.

In relation to this study's findings, which point out that there is no specific influence from intuitive thinking towards belief in God, this is a fairly new finding. This study has discovered that no matter what cognitive style a person has, unless that individual activates the system justification, that individual will have a relatively low belief in God. It was assumed that there might be other variables in play when talking about belief in God in Indonesia, besides the recently proven system justification.

5 CONCLUSION

This study has argued that system justification predicts belief in God respective to an individual's particular cognitive style with religious system justification as a context that will activate intuitive thinking (type 1); this in turn strengthens the activation of system justification that could predict the likeliness of relatively high belief in God. The results of this study confirm the study's original argument. With a cut-off point of 2.385 in cognitive style (which, technically, are intuitive thinkers), it was found that scores under the cut-off point had significant moderation effects in relation to the proposed relationship of religious system justification as the context of system justification.

This moderated moderation model produced an effect on scores with relatively high system justification, which in turn had a significant effect on belief in God. With the moderated moderation regression model (model number 3) from Hayes (2013), it was found that there were relationships between system justification and belief in God that were moderated by religious system justification, and that the moderation effect was moderated by cognitive style.

REFERENCES

Custers, R. & Aarts, H. (2010). The unconscious will: How the pursuit of goals operates outside of conscious awareness. *Science*, *329*(5987), 47–50. doi:10.1126/science.1188595.

Finnegan, W. (2000). A slave in New York: From Africa to the Bronx, one man's long journey to freedom. *The New Yorker*, *60*, 50–61.
Frederick, S. (2005). Cognitive reflection and decision making. *Journal of Economic Perspectives*, *19*(4), 25–42. doi:10.1257/089533005775196732
Hayes, A.F. (2013). Introduction to mediation, moderation, and conditional process analysis. New York, NY: Guilford Press. doi:978-1-60918-230-4
Hayes, A.F. (2014). Comparing conditional effects in moderated multiple regression: Implementation using PROCESS for SPSS and SAS. Retrieved from http://afhayes.com/public/comparingslopes.pdf
Inzlicht, M. & Tullett, A.M. (2010). Reflecting on God: Religious primes can reduce neurophysiological response to errors. *Psychological Science*, *21*(8), 1184–1190. doi:10.1177/0956797610375451
Jost, J.T. & Banaji, M.R. (1994). The role of stereotyping in system-justification and the production of false consciousness. *British Journal of Social Psychology*, *33*(1), 1–27. doi:10.1111/j.2044-8309.1994.tb01008.x
Jost, J.T., Banaji, M.R. & Nosek, B. (2004). A decade of system justification theory: Accumulated evidence of conscious and unconscious bolstering of the status quo. *Political Psychology*, *25*(6), 881–919. doi:10.1111/j.1467-9221.2004.00402.x
Jost, J.T., Hawkins, C.B., Nosek, B., Hennes, E.P., Stern, C., Gosling, S.D. & Graham, J. (2014). Belief in a just God (and a just society): A system justification perspective on religious ideology. *Journal of Theoretical and Philosophical Psychology*, *34*(1), 56–81. doi:10.1037/a0033220
Jost, J.T., Kay, A.C., & Thorisdottir, H. (2009). *Social and psychological bases of ideology and system justification.* New York: Oxford University Press. doi:10.1093/acprof:oso/9780195320916.001.0001
Kay, A.C. & Jost, J.T. (2003). Complementary justice: Effects of 'poor but happy' and 'poor but honest' stereotype exemplars on system justification and implicit activation of the justice motive. *Journal of Personality and Social Psychology*, *85*(5), 823–837. doi:10.1037/0022-3514.85.5.823
Kraft, P.W., Lodge, M. & Taber, C.S. (2015). Why people 'don't trust the evidence': Motivated reasoning and scientific beliefs. *The Annals of the American Academy of Political and Social Science*, *658*(1), 121–133. doi:10.1177/0002716214554758
Lerner, M.J. (1980). *The belief in a just world: A fundamental delusion.* New York, NY: Plenum Press.
Lupfer, M.B., Tolliver, D. & Jackson, M. (1996). Explaining life-altering occurrences: A test of the 'God-of-the-gaps' hypothesis. *Journal for the Scientific Study of Religion*, *35*(4), 379–391.
Pehkonen, E. & Pietilä, A. (2003). On relationships between beliefs and knowledge in mathematics education. Retrieved from http://www.dm.unipi.it/~didattica/CERME3/proceedings/Groups/TG2/TG2_pehkonen_cerme3.pdf
Pew Research Center. (2008, September 17). Unfavorable views of Jews and Muslims on the increase in Europe. Retrieved from http://www.pewglobal.org/files/pdf/262.pdf
Pew Research Center. (2014, October 16). Greatest dangers in the world. *Pew Research Center.* Retrieved from http://www.pewglobal.org/2014/10/16/greatest-dangers-in-the-world
Preston, J. & Epley, N. (2005). Explanations versus applications: The explanatory power of valuable beliefs. *Psychological Science*, *16*(10), 826–832. doi:10.1111/j.1467-9280.2005.01621.x
Preston, J. & Epley, N. (2009). Science and God: An automatic opposition between ultimate explanations. *Journal of Experimental Social Psychology*, *45*(1), 238–241. doi:10.1016/j.jesp.2008.07.013
Shenhav, A., Rand, D.G. & Greene, J.D. (2012). Divine intuition: Cognitive style influences belief in God. *Journal of Experimental Psychology*, *141*(3), 423–428. doi:10.1037/a0025391
Silberman, I. (2005). Religion as meaning system: implications for the new millennium. *Journal of Social Issues*, *61*(4), 641–663. doi:10.1111/j.1540-4560.2005.00425.x
Sorrentino, R.M. & Yamaguchi, S. (2008). Motivation and cognition across cultures. In Sorrentino, R.M. & Yamaguchi, S. (Eds.), *Handbook of motivation and cognition across cultures* (pp. 1–15). San Diego, CA: Academic Press. doi:10.1016/B978-0-12-373694-9.00001-5
Stanovich, K.E. & West, R.F. (1998). Individual differences in rational thought. *Journal of Experimental Psychology: General*, *127*(2), 161–188. doi:10.1037/0096-3445.127.2.161
Stanovich, K.E. & West, R.F. (2000). Individual differences in reasoning: Implications for the rationality debate? *The Behavioral and Brain Sciences*, *23*(5), 645–665 (discussion 665–726). doi:10.1017/S0140525X00003435
Stanovich, K.E. (2002). Rationality, intelligence, and levels of analysis in cognitive science: Is dysrationalia possible? In R.J. Sternberg (Ed.), *Why smart people can be so stupid* (pp. 124–158). New Haven, CT: Yale University Press.
Stanovich, K.E. (2004). Balance in psychological research: The dual process perspective. *Behavioral and Brain Sciences*, *27*, 357–358.
Stanovich, K.E. (2009). The thinking that IQ tests miss. *Scientific American Mind*, *20*(6), 34–39.

Stanovich, K.E. (2012). On the distinction between rationality and intelligence: Implications for understanding individual differences in reasoning. In Holyoak, K.J. & Morrison, R.G. (Eds.), *The Oxford handbook of thinking and reasoning* (pp. 433–455). doi:10.1093/oxfordhb/9780199734689.013.0022

Sub Direktorat Statistik dan Keamanan. (2014). Statistik Politik 2014. *Badan Pusat Statistik*. Retrieved from http://www.bps.go.id

Uhlmann, E.L., Poehlman, T.A. & Bargh, J.A. (2009). American moral exceptionalism. In Jost, J.T. Kay, A.C. & Thorisdottir, H. (Eds.), *Social and psychological bases of ideology and system justification* (pp. 27–52). Oxford, UK: Oxford University Press. doi:10.1093/acprof:oso/9780195320916.003.002

Uhlmann, E.L., Poehlman, T.A., Tannenbaum, D. & Bargh, J.A. (2011). Implicit puritanism in American moral cognition. *Journal of Experimental Social Psychology*, *47* (2), 312–320. doi:10.1016/j.jesp.2010.10.013

Van der Toorn, J. & Jost, J.T. (2014). Twenty years of system justification theory: Introduction to the special issue on 'ideology and system justification processes'. *Group Processes & Intergroup Relations*, *17*(4), 413–419. doi:10.1177/1368430214531509

Willard, A.K. & Norenzayan, A. (2013). Cognitive biases explain religious belief, paranormal belief, and belief in life's purpose. *Cognition*, *129*(2), 379–391. doi:10.1016/j.cognition.2013.07.016

Diversity in Unity: Perspectives from Psychology and Behavioral Sciences – Ariyanto et al. (Eds)
 ISBN 978-1-138-62665-2

Comparing fear, humour, and rational advertising appeals and their effect on consumer memory and attitude centred on video-based e-commerce advertising

S.A. Kendro & E. Narhetali
Faculty of Psychology, Universitas Indonesia, Depok, Indonesia

ABSTRACT: This study examines the effect of advertising appeal of e-commerce video advertising, especially those available on YouTube. There are three appeals that will be compared, namely rational, fear, and humour. Rational appeal concentrates on the features or the benefits of a product, fear appeal expresses a threat that may happen, and humour appeal provides a boost in positive mood to people. In this study, effectiveness of an advertisement is measured by memory and attitude. Memory is measured by implicit and explicit measurements so the effect of advertising appeal to memory can be seen as a whole. This research was of a single-factor multiple-group design. Participants were 106 students from various faculties at the University of Indonesia. The results showed that fear appeal generates a score that is significantly higher for implicit memory and advertisement attitude compared to rational and humour appeal. Meanwhile, humour appeal is the only form that significantly affects explicit memory. Therefore, it can be concluded that fear appeal has the ability to persuade consumers and influence their unconscious memory, whereas humour appeal can be used to retain consumers that are already loyal to the product or brand.

1 INTRODUCTION

Rapid growth in the use of the Internet and smartphones has made Indonesia one of the fastest growing e-commerce markets in Asia (Mitra, 2014). It is also one of the most competitive industries, especially in Indonesia, where more than 140 e-commerce compete to dominate the market (Eldon, 2015). For smaller e-commerce businesses, the task becomes even more challenging with limited budgets, hence advertising online may be a cost-efficient method of gaining national or global awareness (Barnes, 2002; Brettel & Spilker-Attig, 2010; Nunan & Knox, 2011). According to Bayu Syerli, head of marketing for Bukalapak, YouTube is an ideal platform for online advertising considering its ability to reach target markets (Liputan6, 2016). This may be true as Indonesia has the most YouTube users in Asia Pacific with a growth rate of up to 600% annually (Abidin, 2015). However, these e-commerce businesses need to identify the right appeal to advertise their services online.

There is rarely any empirical research on the types of advertising appeals that are appropriate for e-commerce, especially on video advertising. Furthermore, a consensus has not been reached regarding two different types of advertising appeals, namely rational and emotional appeal. Based on previous research, service businesses relying on consumer's experience have indicated that emotional appeal in advertising is more effective for service-related products because it increases positive attitudes towards the advertisement, a desire to use the service (Zhang et al., 2014), and is interactive in nature (Li et al., 2009; Santiago & Pitta, 2011). On the other hand, Hsu and Cheng (2014) found out that YouTube, which was one of the source for video advertising, is the perfect medium to spread advertisement awareness with rational appeals based on the limitation of time so that it should show more about its qualities than emotions. Rational appeal itself is an advertising stimulus that supplies factual information about the brand, the product, or the service (Puto & Wells, 1984).

Furthermore, there has been a new debate regarding two types of emotional appeal, namely positive and negative emotional appeal. Some advertising researchers believe that positive emotional appeal, such as humour, may be used to maintain attention (Strick et al., 2009) and enhance positive attitude towards the advertised product or brand, purchase intentions, and brand awareness (Couvreur, 2015). This claim has been debated by other advertising researchers who have discovered that negative emotional appeal, especially fear, is better than positive emotional appeal for the same reasons (Ferreira et al., 2011; Pettigrew et al., 2012; Park, 2012).

The debate is further complicated by various limitations in the previous studies, which fail to shed new light on this issue. For instance, a review of the literature verified that a large portion of emotional advertising is dedicated to traditional media, such as print and television advertisements (Royo & Gutierrez, 2000; Heath & Nairn, 2005; Lin, 2011). Very few studies have focused on online advertising, as most only investigated the effective use of banner advertisements (Yoo, 2008; Porta et al., 2013; Noble et al., 2014) and pop-up advertisements (Nysveen & Breivik, 2005; Qin et al., 2011). The effectiveness of e-commerce advertisements using emotional and rational appeals, particularly in the context of video advertising, is unclear and at best contradictory.

The main objective of this study was to examine whether humour, fear, or rational appeals are more effective for e-commerce video advertisements. Effectiveness will be measured by memory of and attitude towards the advertisements. Specifically, memory of the advertisement is going to be measured by using implicit and explicit methods, in order to have a comprehensive picture of how effective the advertisements will be recalled. Implicit memory will be measured by a word-fragment completion test and explicit memory will be measured by an advertisement recognition test, whereas advertisement attitude will be measured explicitly with an attitude scale. This article strives to bridge gaps in advertising appeals and to aid practitioners in making better decisions in considering appeal in advertisements. Based on theories and previous research, the authors believe that emotional appeal will be more effective than rational appeal. Specifically, fear appeal will be the most effective in memory measurements and humour appeal will be the most effective in advertisement attitude measurements. In addition, results from implicit and explicit memory will be directly proportional.

2 METHODS

2.1 *Participants*

The participants in this study were 110 undergraduate students from the University of Indonesia, but four of them were eliminated due to failure to pass the awareness check. Thus, the total data used was 106, which consisted of a rational appeal group of 33 participants, a fear appeal group of 35, and a humour appeal group of 38. All participants were recruited through convenience sampling from all the faculties at the University of Indonesia, using social media and instant messaging. Before the experiment started, participants were requested to give their informed consent to participation in the study.

2.2 *Instruments*

Preliminary studies were conducted to identify and determine three video advertisements that most accurately represented their respective advertising appeals, namely humour, fear, and rational appeals. Real online video advertising of an e-commerce platform brand, Tokopedia, was used in order to minimise potential confusion due to brand preference, which might of affected participant's memory and attitudes. Tokopedia was selected because it has commercials that employ the three respective appeals in each of their videos. Moreover, four types of videos, namely music video, movie trailer, social project video, and fun facts video, were used as fillers and to make the experimental settings more natural (as if the participants were watching videos on YouTube). These videos were selected through surveys of 206 participants. All the videos and advertisements in this experiment were taken from YouTube.

The instrument for measuring implicit memory, namely the word-fragment completion test (see Figure 1), also underwent a pretest to decide which words were suitable for this experiment. Participants were asked to complete words that were fragmented within the specified time limits. Because all the advertisements were taken from YouTube, the target words for every appeal were different but not with the filler words. For the sake of validity, participants used control data for every appeal on word-fragment completion pretests. Five target words were selected from the pretest to represent their respective advertising appeal along with 12 filler words. Word-fragment completion tests in this study were constructed based on studies by Johnson and Saboe (2011).

Advertisement recognition for measuring explicit memory was adapted from studies by Heath and Nairn (2005). In this instrument, participants were asked to identify advertisements and specify the brand name of each advertisement that they saw when watching the videos for the first time. There were five items to be marked against regarding the five stimulus advertisements that they would watch; a checklist mark for the target advertisement and a cross mark for the filler advertisement (Figure 2).

The advertisement attitude scale used was an adaptation from Brunel et al. (2004) and was translated by Ibadurrahman (2013) in his thesis. This scale measures the tendency to like or dislike a presented stimulus advertisement. Operational definitions of this instrument were taken from the proportion of six total votes in a seven-point semantic differential scale (Figure 3).

2.3 *Procedures and measurements*

This study used a single-factor multiple-group design, comprising three experimental groups: a rational appeal manipulation group, a humour appeal manipulation group, and a fear appeal manipulation group. There were three instruments used to measure the effectiveness of each appeal. Advertisement recognition was used to measure explicit memory, a word-fragment completion test for implicit memory, and an advertisement recognition scale to determine a positive-negative attitude. The experiment was conducted from the 23 to 27 May 2016 in a classroom setting limited to ten participants.

Before the participants entered the experiment room, the experimenter drew lots as to which advertising and video that the participants would be watching; this draw was used to control whether or not the differences in video affected the advertisement effectiveness

Example:

M A _ _ _ _ R I

M A T A H A R I

Figure 1. Example of a word-fragment completion test.

Recognized	Brand
√	Aqua
X	

Figure 2. Example of an advertisement recognition test.

Interesting ___:___:___:___:___:___:___ Boring

Figure 3. Example of advertisement attitude scale.

measurements. After participants completed their informed consents, they watched the advertisement and video for five minutes. Participants then filled out a distractor sheet which asked questions about the video that they had just watched.

The experiment was started with a word-fragment test, in which the participants were asked to fill out 17 words consisting of five target words that built upon the advertisements that they had seen while watching the video, along with 12 distraction words that had no correlation with the advertisements. The scores for this test were measured by the proportion of the target words successfully completed by the participants within a certain period. The duration to fill out the word-fragment completion test sheet in this experiment was two minutes.

In the next step, the participants undertook advertisement recognition measurements. The experimenter presented five video advertisements consisting of four videos and one video distraction target; all of the videos had been censored with regard to audio and visual content for the brand name. All the distraction advertisements were placed in random order while target advertisements was placed in the middle. Participants were asked to recognise the advertisement in the interval between all the video advertising by giving a cross mark if it was not the right advertisement, and a checklist mark if it was the right advertisement along with the brand name of the advertisement. Advertisement recognition was measured by how well the participants recognised the advertisement and brand name.

The participants were asked to watch the advertisement that had been presented at the beginning of the experiment for the last time, in order to fill out the advertisement attitude scale. In this sheet, the participants gave their appraisal for the advertisement using a seven-point semantic scale. A proportion of the total six items would represent participants' attitude to the advertisement. Participants also filled out covariate data regarding familiarity and understanding of the advertisement.

For the last part, participants were asked to fill out demographic information and complete an awareness check; they were asked if they knew the purpose of the study and the stage at which they experienced the objectives of the research. This process was done to ensure that the manipulation performed in this experiment was a success, and the results obtained from the measurement of implicit and explicit memory can certainly be influenced by the manipulation given. When all the participants were finished, the experimenter announced that the experiment had been completed, explained the purpose of the study and asked that all participants maintain the confidentiality of the research.

3 RESULTS

Once the experiment was completed, the researchers checked the data obtained, and matched with the awareness check, to confirm that the data was valid. This screening was necessary to prevent the appearance of bias or other variables that may have disrupted the calculation of implicit and explicit memory, and attitudes towards the advertisements. The data that passed the awareness check was then processed by using the Statistical Package for Social Science (SPSS) software, version 22.0. In this study, there were four statistical techniques applied, namely descriptive statistics, one-way Analysis of Covariance (ANCOVA), logistic regression analysis, and Kendall's tau correlation.

Demographic information was processed to gain an overview of the background of the participants including age, gender, and faculty. Based on the data collected, the age range of the participants was 18 to 23 years (mean $M = 20.25$; standard deviation $SD = 1.358$). The majority of participants were 20 years old (28 people in number, equating to 26.4%). In this study, there were a greater number of female participants than male: 61 female participants (57.5%) compared to 45 male ones (42.5%). Based on the distribution among the faculties of the University of Indonesia, the participants consisted of 16 students from Computer Science (15.1%), 13 from Economics and Business (12.3%), seven Law students (6.6%), 11 students from Humanities (10.4%), 24 students from Social and Political Sciences (22.6%), two students from Dentistry (1.9%), three students from Public Health (2.8%), six students of Mathematics and Science (5.7%), 22 Engineering students (20.8%), and two students from

Vocational Education (1.9%). This dispersion showed that the participants were drawn from ten different faculties and the single highest proportion of participants came from the Faculty of Social and Political Sciences.

Based on the hypothesis test for implicit memory using ANCOVA techniques (see Table 1), the proportion of distraction ($F(1.101) = 0.015$, $p > 0.05$) and those that have seen the advertisements ($F(1.101) = 0.086$, $p > 0.05$) is not a significant covariate. Whereas the main effect of the appeal of advertising was found to be significant, namely $F(2.101) = 6.664$; $p < 0.05$; $\eta 2 = 0.117$. When each type of appeal of advertising was compared, the calculation results showed that fear appeal ($M = 0.4061$; $SD = 0.154$) has a proportion of implicit memory that is higher than an advertisement containing humour ($M = 0.3$, $SD = 0.153$) and an advertisement with a rational appeal ($M = 0.2629$; $SD = 0.18$). Meanwhile, the appeal of humorous advertisements has an average higher than those with rational appeal, but does not differ significantly. Therefore, the hypothesis for memory implicit and fear appeal is supported.

Analysis of the explicit memory was measured using a logistic regression, so covariates as for those who had seen the advertisements, could be eliminated statistically from the measurement of explicit memory (see Table 2). Unfortunately, the hypothesis for implicit memory and fear appeal was not found. Based on the calculation of logistic regression, it was found that watching the advertisement once does not significantly affect the explicit memory of participants ($\beta = 0.125$; $p > 0.05$). Instead, the appeal of humour in video-based e-commerce advertising proved significant to explicit memory with a value of $\beta = 2.415$; $p < 0.05$. While the appeal of fear ($\beta = -0.52$; $p > 0.05$) and rational advertising appeal ($\beta = -1.182$; $p > 0.05$) do not significantly affect explicit memory.

Furthermore, Kendall's tau correlation calculation was conducted in order to look at the relationship between implicit and explicit memory (see Table 3). Based on this calculation,

Table 1. ANCOVA of the effects of advertising appeals on implicit memory, with covariates of the proportion of filler words on the Word-Fragment Completion Test (WFCT) and having watched advertisements.

Variable	*Df*	SS	MS	*F*	*P*	η^2
Proportion of filler WFCT (covariate)	1	0.00	0.00	0.015	0.904	0.00
Have watched advertisement (covariate)	1	0.002	0.002	0.086	0.770	0.001
Advertising appeal	2	0.359	0.179	6.664	0.002	0.117
Error	101	2.718	0.027			
Total	106	14				

Table 2. Logistic regression of advertising appeal on explicit memory, with having watched advertisements as a covariate.

Variable	B	SE	OR	95% CI	Wald statistic	*p*
Have watched advertisements (covariate)	0.125	0.562	1.133	(0.376; 3.418)	0.049	0.824
Advertising appeals						
Fear	–0.52	0.583	0.950	(0.303; 2.980)	0.008	0.930
Humour	2.415	1.095	11.191	(1.309; 95.666)	4.866	0.027
Rational	–1.182	0.687	0.307	(0.080; 1.178)	2.963	0.085

Table 3. Correlation of implicit and explicit memory.

Data recognition	*T*	*P*
Advertising appeals		
Fear	0.038	0.818
Humour	0.123	0.429
Rational	–0.2	0.211

Table 4. ANCOVA of the effects of advertising appeal on attitude with having watched advertisements as a covariate.

Variable	*Df*	SS	MS	*F*	*P*	η^2
Have watched advertisements (covariate)	1	0.039	0.039	0.035	0.851	0.00
Advertising appeals	2	12.086	6.043	5.428	0.006	0.096
Error	102	113.566	1.113			
Total	106	1348.389				

it is necessary to note that implicit memory for the attractiveness of advertising based on the emotion of fear has a value of $\tau = 0.038$ ($p > 0.01$), emotional advertising appeal using humour obtains a value of $\tau = 0.123$ ($p > 0.01$), and rational advertising appeal gets a value of $\tau = -0.2$ ($p > 0.01$). The third data set shows insignificant results for each type of advertising appeal. Overall, the results show that there is no correlation between implicit and explicit memory.

In calculating the attitude towards advertising, measurements were conducted using ANCOVA (see Table 4). Levene's test results showed that the variance of the data is not homogeneous and so bootstrap sampling was used to overcome these limitations (Field, 2009) where the measurement was equal to 5000. Based on hypothesis testing using the ANCOVA technique, those that have seen the advertisements ($F(1.106) = 0.035$, $p > 0.05$) were not a significant covariate. Whereas the main effect of the appeal of advertising proved significant, namely $F(2.106) = 5.428$; $p < 0.05$; $\eta 2 = 0.096$. When each of the appeals of advertising were compared, the calculation results showed that the fear appeal ($F = 3.882$; $SD = 0.185$) has a proportion of ratings higher than an advertisement with humour ($F = 3.291$; $SD = 0.171$) and rational appeal ($F = 3.052$; $SD = 0.180$). Meanwhile, the appeal of humorous advertisements has an average higher those with rational appeal, but does not differ significantly. Through these results, the hypothesis regarding advertisement attitude and humour appeal does not support either of them.

4 DISCUSSION

It was found that e-commerce video-based advertising fear appeal is more able to influence the implicit memory but not explicit memory. In addition, there is no correlation between implicit and explicit memory. These results are consistent with previous research that found that the implicit and explicit memory do not respond in the same way even in the same experimental manipulations (Tulving et al., 1982; Law & Braun, 2000; Shapiro & Krishnan, 2001). Dissociation of memory can occur because of differences in information processing between implicit and explicit memory (Cowan et al., 1997) so the rendered output is also different. When viewed from the characteristics of the emotion of fear advertising appeal, this appeal has the effect of carrying out and hindering the effectiveness of advertising. These barriers can be a defensive response from the audience in the form of a 'dodge' (Belch & Belch, 2003). Unfortunately, as in the video advertisements on YouTube site, not all advertisements can be avoided so the only way was neglect the information provided. As a result of the neglection, it was difficult for explicit memory to recall information but implicit memory still stores the information.

These findings can be explained from the characteristics of the advertising appeal of humour too. Various studies have shown that advertisements with humour can attract and sustain attention (Belch & Belch, 2003; Eisend, 2009), but the humour can distract from the information that contains no humour which is the main subject of the advertisement (Weinberger et al., 1995). Therefore, explicit memory for an advertisement with the appeal of humour, tends to be high while implicit memory is not much different from rational appeal based advertisements. These results also show that humour is more suitable for use on a brand or product that is already known.

Rational appeal failure in influencing memory and attitude proves that emotional content, especially on video-based e-commerce advertising, is more effective. This result is in line with the findings of other studies that compare effectiveness of these both advertisement appeals (Heath & Nairn, 2005; Park, 2012; Lwin et al., 2014). This shows that consumers tend to use emotion in deciding whether to buy a product or use a service.

5 CONCLUSION

The results show that the appeal of advertising affects consumer memory for video-based e-commerce advertising. On the measurement of implicit memory, the advertisements with emotional appeal of fear was found to be remembered well than advertisements with humour or rational appeal. However, there were no significant differences in implicit memory on advertisements with humour appeal compared to the advertisements with rational appeal. During the measurement of explicit memory, participants were more significantly able to recognise advertisements with humour than the advertisements with emotional appeal and rational fear. However, there was no significance regarding the ability of the participants to recognise the advertisements with fear and rational appeal. Therefore, there was no relation between implicit and explicit memory. Furthermore, it was found that the advertisements with fear appeal tend to be more effective than advertisements with humour and rational appeal, While the advertisements with humour and rational appeal do not differ significantly.

REFERENCES

Abidin, F. (2015, October 21). Youtube users in Indonesia increased by 600 percent in 3rd quarters of 2015. *Mix Online*. Retrieved from http://www.mix.co.id/newstrend/pengguna-youtube-indonesia-meningkat-600-persen-di-q3-tahun-2015.

Bang, H.K., Raymond, M.A., Taylor, C.R. & Moon, Y.S. (2005). A comparison of service quality dimensions conveyed in advertisements for service providers in the USA and Korea: A content analysis. *International Marketing Review*, *22*(3), 309–326.

Barnes, S.J. (2002). Wireless digital advertising: Nature and implications. *International Journal of Advertising*, *21*(3), 399–420.

Belch, G.E. & Belch, M.A. (2003). Advertising and promotion: An integrated marketing communications perspective (6th ed.). New York, NY: McGraw-Hill.

Brettel, M. & Spilker-Attig, A. (2010). Online advertising effectiveness: A cross-cultural comparison. *Journal of Research in Interactive Marketing*, *4*(3), 176–196.

Brunel, F.F., Tietje, B.C. & Greenwald, A.G. (2004). Is the implicit association test a valid and valuable measure of implicit consumer social cognition? *Journal of Consumer Psychology*, *14*(4), 385–404.

Couvreur, C. (2015). *The use of humour in socially responsible advertisements: Case study of the rainforest alliance* (Master's thesis, Louvain School of Management). Retrieved from http://dial.uclouvain.be/memoire/ucl/en/object/thesis%3 A2576/datastream/PDF_01/

Cowan, N., Wood, N.L., Nugent, L.D. & Treisman, M. (1997). There are two word-length effects in verbal short-term memory: Opposing effects of duration and complexity. *Psychological Science*, *8*, 290–295.

Eisend, M. (2009). A meta-analysis of humor in advertising. *Journal of the Academy Marketing Science*, *37*, 191–203.

Eldon, A. (2015, December 8). 140 e-commerce held a huge discount sale in Harbolnas 2015. *LensaIndonesia.com*. Retrieved from http://www.lensaindonesia.com/2015/12/08/140-e-commerce-obral-diskon-besar-di-harbolnas-2015.html

Ferrerira, P., Rita, P., Morais, D., Rosa, P.J.M., Oliveira, J., Gamito, P. & Sottomayor, C. (2011). Grabbing attention while reading website pages: The influence of verbal emotional cues in advertising. *Journal of Eye-tracking, Visual Cognition, and Emotion*, *1*(1), 581–592.

Field, A. (2009). *Discovering statistics using SPSS* (3rd ed.). Thousand Oaks, CA: SAGE Publications.

Grove, S.J., Pickett, G.M. & Laband, D.N. (1995). An empirical examination of factual information content among service advertisements. *The Service Industries Journal, 15*(2), 203–215. doi:10.1080/02642069500000021

Heath, R. & Nairn, A. (2005). Measuring affective advertising: Implications of low attention processing on recall. *Journal of Advertising Research, 45*(2), 269–281. doi:10.1017/S0021849905050282

Hsu, Y. & Cheng, J.C. (2014). Maximizing advertising effectiveness. *European Journal of Business and Social Sciences, 3*(9), 147–159.

Ibadurrahman. (2013). *Implicit attitude and explicit attitude toward sensual condom ads and symbolic condom ads among adolescents* (Undergraduate thesis, Universitas Indonesia, Indonesia). Retrieved from http://lib.ui.ac.id/file?file = digital/20345281-S46913-Sikap%20implisit.pdf

Johnson, R.E. & Saboe, K.N. (2011). Measuring implicit traits in organizational research: Development of an indirect measure of employee implicit self-concept. *Organizational Research Methods, 14*(3), 530–547. doi:10.1177/1094428110363617

Laros, F.J.M. & Steenkamp, J.E.M. (2005). Emotions in consumer behavior: A hierarchical approach. *Journal of Business Research. 58*(10), 1437–1445. doi:10.1016/j.jbusres.2003.09.013

Law, S. & Braun, K.A. (2004). Product placements: How to measure their impact. *Psychology and Marketing, 17*, 1059–1075.

Li, H., Li, A. & Zhao, S. (2009). Internet advertising strategy of multinationals in China: A cross-cultural analysis. *International Journal of Advertising, 28*(1), 125–146. doi:10.2501/S0265048709090441

Life Noggin (Producer). (2014). *10 Common myths debunked!* [YouTube video]. Retrieved from https://www.youtube.com/watch?v = 3MA-K5whgZk

Lin, L.Y. (2011). The impact of advertising appeals and advertising spokespersons on advertising attitudes and purchase intentions. *African Journal of Business Management, 5*(21), 8446–8457.

Liputan6. (2016, June 8). Mas Medok The Magic Finger Warrior Achieved the Top 3 Most Popular Advertisements. *Liputan6.com.* Retrieved from http://tekno.liputan6.com/read/2526595/masmedok-pendekar-jari-sakti-masuk-top-3-iklan-terpopuler

Lohtia, R., Donthu, N. & Hershberger, E.K. (2003). The impact of content and design elements on banner advertising click-through rates. *Journal of Advertising Research, 43*(4), 410–418.

Lwin, M., Phau, I., Huang, Y. & Lim, A. (2014). Examining the moderating role of rational-versus-emotional-focused websites: The case of boutique hotels. *Journal of Vacation Marketing, 20*(2), 95–109.

Mattila, A. (1999). Do emotional appeals work for services? *International Journal of Service Industry Management, 10*(3), 292–306. doi:10.1177/109634800102500104

Mitra, W. (2014, September 16). The current statistic data regarding growth of e-commerce market share in Indonesia. *StartupBisnis.com.* Retrieved from http://startupbisnis.com/data-statistik-mengenai-pertumbuhan-pangsa-pasar-e-commerce-di-indonesia-saat-ini/

Moore, C. (Director) & Ritson, H. (Producer). (2015, August 4). *Sigala – easy love* [Video file]. New York City, New York, US: VEVO. Retrieved from https://www.youtube.com/watch?v = ozx898 ADTxM

Noble, G., Pomering, A. & Johnson, L.W. (2014). Gender and message appeal: Their influence in a pro-environmental social advertising context. *Journal of Social Marketing, 4*(1), 4–21. doi:10.1108/JSOCM-12-2012-0049

Nunan, D. & Knox, S. (2011). Can search engine advertising help access rare samples? *International Journal of Market Research, 53*(4), 523–540. doi:10.2501/IJMR-53-4-523-540

Nysveen, H. & Breivik, E. (2005). The influence of media on advertising effectiveness: A comparison of internet, posters and radio. *International Journal of Market Research, 47*(4), 383–405.

Soul Pancake. (Producer). (2012, February 8). *Soul pancake: heart attack, super soul Sunday* [Video file]. Chicago, Illinois, US: Oprah Winfrey Network. Retrieved from https://www.youtube.com/watch?v = 7VJsyEtwAUY

Park, T.N. & Purdue, S.G. (2012). Emotional factors in advertising via mobile phones. *International Journal of Human-Computer Interaction, 28*(9), 597–612. doi:10.1080/10447318.2011.641899

Pettigrew, S., Roberts, M., Pescud, M., Chapman, K., Quester, P. & Miller, C. (2012). The extent and nature of alcohol advertising on Australian television. *Drugs and Alcohol Review,* 31 (6), 797–802.

Porta, M., Ravarelli, A. & Spaghi, F. (2013). Online newspapers and ad banners: An eye tracking study on the effects of congruity. *Online Information Review, 37*(3), 405–423. doi:10.1108/OIR-01-2012-0001

Puto, C.P. & Wells, W.D. (1984), Informational and Transformational Advertising: The Differential Effects of Time. *Advances in Consumer Research, 11*(1), 638–643.

Qin, L., Zhong, N., Lu, S., Li, M. & Song, Y. (2011). Emotion and rationality in web information: An eye-tracking study. In Zhong, N., Callaghan, V., Ghorbani, A.A. & Hu, B. (Eds), *Active Media Technology. AMT 2011. Lecture Notes in Computer Science* (Vol. 6890). Berlin, Germany: Springer. doi:10.1007/978-3-642-23620-4_15

Royo-Vela, M. & Gutierrez, A.M. (2000). The effect of emotive and informative content of advertising in the evaluation of commercials. In *Marketing in the new millennium, Proceedings of the 29th European Marketing Academy Conference, Rotterdam, The Netherlands, 23–26 May* (p. 62).

Rosenfelt, K. (Producer) & Sharrock, T. (Director). (2016, April 28). *Me before you official trailer 2* [Video file]. Los Angeles, California, AS: MGM. Retrieved from https://www.youtube.com/watch?v = iJrsOPgxcig

Santiago, J.A. & Pitta, D.A. (2011). Marketing from the end of the earth: The dilemma of Puerto Nativo Lodge. *Journal of Product and Brand Management, 20*(2), 141–146.

Shapiro, S. & Krishnan, H.S. (2001). Memory-based measures for assessing advertising effects: A comparison of explicit and implicit memory effects. *Journal of Advertising, 30*(3), 1–13. doi:10.1080/00913367.2001.10673641

Stafford, M.R. & Day, E. (1995). Retail services advertising: The effects of appeal, medium, and service. *Journal of Advertising, 24*(1), 57–71. doi:10.1080/00913367.1995.10673468

Strick, M., van Baaren, R.B., Holland, R.W. & van Knippenberg, A. (2009). Humor in advertisements enhances product liking by mere association. *Journal of Experimental Psychology: Applied, 15*(1), 35–45. doi:10.1037/a0014812

Tokopedia TVC. (Producer). (2014, June 14). *More safe "shared accounts"* [Video file]. Kebon Jeruk, West Jakarta, Indonesia: Tokopedia TVC. Retrieved from https://www.youtube.com/watch?v = 8 ALqJH7be84

Tokopedia TVC. (Producer). (2014, July 3). *What is Tokopedia? "Online Open Marketplace"* [Video file]. Kebon Jeruk, West Jakarta, Indonesia: Tokopedia TVC. Retrieved from https://www.youtube.com/watch?v = SoVQdVehc6o

Tokopedia TVC. (Producer). (2015, 27 October). *Itchy?* [Video file]. Kebon Jeruk, West Jakarta, Indonesia: Tokopedia TVC. Retrieved from https://www.youtube.com/watch?v=bwOSGilNT2A

Tulving, E., Schacter, D.L. & Stark, H.A. (1982). Priming effects in word-fragment completion are independent of recognition memory. *Journal of Experimental Psychology: Learning, Memory, and Cognition, 8*(4), 336–342. doi:10.1037/0278-7393.8.4.336

Weinberger, M.G., Spotts, H., Campbell, L. & Parsons, A.L. (1995). The use and effect of humor in different advertising media. *Journal of Advertising Research, 35*, 44–56.

Yoo, C.Y. (2008). Unconscious processing of web advertising: Effects on implicit memory, attitude toward the brand, and consideration set. *Journal of Interactive Marketing, 22*(2), 2–18.

Zhang, H., Sun, J., Liu, F. & Knight, J.G. (2014). Be rational or be emotional: Advertising appeals, service types, and consumer responses. *European Journal of Marketing, 4*(4), 23–34.

Diversity in Unity: Perspectives from Psychology and Behavioral Sciences – Ariyanto et al. (Eds)

Seeking context for the theory of the enforceability of the moral licensing effect in a collectivist culture: When moral surplus leads law enforcers to get involved in corruption

N.M.M. Puteri, H. Muluk & A.A. Riyanto
Faculty of Psychology, Universitas Indonesia, Depok, Indonesia

ABSTRACT: Good deeds lead people to commit moral transgressions and crime. This research uses moral licensing theory to explain why and where law enforcers are involved in corruption, and why they commit an offence without any concern for being seen as immoral persons in the public's eyes. The study was conducted using a qualitative method through in-depth interviews with 15 practitioners working for more than ten years in the criminal justice system. Working as a law enforcer cultivates the feeling of being a moral person with a moral surplus in their moral account, without exercising kindness in real life (counter factual transgression). An orientation towards putting a value on good deeds, expecting reciprocation (*hutang budi*), allows individuals to ask for and receive bribe for their own duties. Meanwhile, law enforcers who have an orientation towards sincerity (*tanpa pamrih*) perceive the work undertaken as fulfilling the duties of a public servant, so they tend not to ask for or receive bribe. The symptoms of moral licensing and corruption are influenced by who the victims are, the impact of the transgression, as well as the support group. The proposal of this study is that the track record of the perpetrator should not be used as a consideration for leniency of punishment because it can be designed intentionally to hide criminal motives.

1 INTRODUCTION

For the past seven years, anti-corruption movements in Indonesia have successfully uncovered huge corruption cases perpetrated by law enforcers such as policemen, prosecutors, and judges (Suara Pembaruan, 2015). This study aims to explain the corruption phenomenon by law enforcers using moral licensing theory. This theory explains that people can behave inconsistently, against the morality they honour, and commit violations (Monin & Miller, 2001; Khan & Dhar, 2006; Gneezy et al., 2012). Doing good deeds will make people credit it to their moral account. When individuals perceive their moral account to be in credit, they are prone to commit offences. A moral surplus makes an individual feel, as a person with morality, that he or she then has the right and opportunity to commit violations or immoral actions (Monin & Miller, 2001; Khan & Dhar, 2006; Mazar & Zhong, 2010). Perpetrators will readily commit offences as they are not worried about being seen by others as violators, or being seen as immoral persons in the public's eyes (Effron et al., 2009; Bradley-Geistet al., 2010; Krum & Corning, 2008). The theory assumes that offences can be tolerated, if they do not make a deficit in their moral account or if they keep their moral balance (Effron & Monin, 2010; Merritt et al., 2010).

The law enforcement profession has always been connected with moral integrity so that ideally law enforcers will not be involved in crime or moral offences. Corruption among law enforcers is typically white collar crime, particularly state-authority occupational crime, a crime committed by individuals with authority as an extension of the state's roles (Green, 1997). However, this behaviour should be defined as 'comprehensive wrongfulness' (Berman, 2007, p. 301), whereas the quality of the offence can be stated as unethical behaviour (Chappell & Piquero, 2004), a moral offence (Nettler, 1976), and a crime at the same time (Gino, 2015).

Corruption is defined as a crime when it is committed in the context of a legitimate position (Clinard & Quinney, 1973), but the *esprit de corps* inside the law organisations has created individual solidarity with the group, and the group to the individuals (Waddington, 1999), so that punishments for the perpetrators tend to be weak (Sutherland, 1983; Nelken, 1995; Perri, 2011; McGee & Byington, 2009). For example, when a policeman receives bribe, it is normally not considered corruption or a serious offence (Chappell & Piquero, 2004).

Moral licensing research has been primarily based on experiments within individualistic Western societies (Effron & Monin, 2010; Merritt et al., 2010). Meanwhile, studies in moral licensing theory in collectivist societies are limited. According to Markus and Kitayama (1991), the individual self in a collectivist culture is highly related to other individuals in his or her groups (p. 227). Since collectivism highly honours harmony, people always try to show similarity with other members of their groups (Triandis, 2004). Thus, it is important to understand how good deeds cannot be directly counted in an individual's moral account, but are supposed to efface one's morality compared to other members of the group.

In Indonesian society, the act of doing good deeds can be divided into two types, doing good deeds without ulterior motives (*tanpa pamrih*) and those expecting reciprocation (*hutang budi*). When people do good deeds sincerely and without ulterior motives the good deed has been exercised without any intention of self-advantage (Sartini, 2009). On the other hand, doing a good deed in the expectation of reciprocation means one is hoping for something in return. At some point, accepting others' kindness can be a debt that they will bring to their grave (Husin, 2011; Sendra, 2008).

Good deeds with the expectation of reciprocation will cause individuals who are extending the kindness to credit their good deeds into a moral account. Hence, they will feel as if they have a moral surplus and be in a superior position which allows them to behave immorally, or licensing their moral. On the contrary, when individuals do good deeds with selfless values, they will not credit their kindness to their moral account, hence the individuals tend not to use the moral licence to commit offences.

To understand the process and the context of moral licensing in a collectivist society, this study proposes the following research questions:

1. Do moral licensing phenomena exist in a collectivist society? What kind of offences or crimes are particularly related to moral licensing? Who is the perpetrator?
2. How do these moral values (sincere VS expecting reciprocation) affect the actor's moral account?
3. What are the causes of the absence of guilt in law enforcers who engage in corruption?

1.1 *Previous studies*

A moral surplus will lead people license their moral, using *moral credit* or *moral credential* (Monin & Miller, 2001; Merritt et al., 2010). Research on moral credit proves that people who do good deeds tend to commit offences in different domains to the preceding action (Mazar & Zhong, 2010). On the contrary, moral credentials are used when people do a good deed followed by offences in the same domain (Effron & Monin, 2010; Monin & Miller, 2001).

Effron (2012) found that transgressions in the same domain can be blatantly seen as inconsistent behaviour; perpetrators will be condemned as hypocrites and get punishment. To avoid negative consequences from the public, the perpetrators find ways to get moral credentials. The first effective way is to redefine their offences as ambiguous behaviour, where the distinction between right and wrong becomes blurry (Monin & Miller, 2001; Effron & Monin, 2010), so that the offence can be rationalised (Brown et al., 2011). A study conducted by Bradley-Geist et al. (2013) proved that individuals intentionally develop a track record where they will be seen as a moral person (Merritt et al., 2010). This track record gives the perpetrator the ability not only to cover their bad motives and intentions, but also to give the impression of legitimate behaviour that may afford him impunity from the public (Merritt et al., 2012). According to Garza et al. (2011) moral credentials can be used to cover bigger crimes than moral credit offences.

When people carry out good deeds based on indebtedness values, they tend to involve in moral offences and get the punishment. But the track record from the perpetrator as an honourable person confuses the observer and triggers a moral licensing effect from the observer. Firstly, moral licensing causes the observer to tolerate offences, and as a result the observer is not able to punish the perpetrators (Effron et al., 2010; Effron & Monin, 2010). Secondly, moral licensing of the perpetrator is beneficial to free themselves from guilt (Monin & Miller, 2001; Merritt et al., 2010; Effron & Monin, 2010; Khan & Dhar, 2006; Gneezy et al., 2012). Consequently, when moral licensing in the perpetrator and the observer happens at the same moment, the perpetrators are free of guilt and do not get any punishment from the public.

Studies on morality in Indonesia have been conducted in the areas of moral hypocrisy (Rahman, 2013) and moral judgement, where corruption is categorised as a moral offence (Hakim et al., 2014). Corruption often happens because the gift receiver feel obliged to pay back the good deeds (Sendra, 2008). Furthermore, good deeds based on an expectation of reciprocation are added by the individual as a credit to their moral account that would legitimise their offences.

Oriza (2016) explains that expecting reciprocation increases one's openness to contra-normative requests, including receiving bribery. According to the sincerity value, the individual's good deeds will not be credited to their moral account. Therefore, according to moral licensing theory, individuals who have a selfless value orientation tend not to carry out moral offences.

Law enforcement is regarded as an honourable profession in Indonesian society. This status gives the law enforcer the feeling that he or she is a moral person with a moral surplus, even though he does not necessarily do good deeds in daily life. This situation can be considered as *counterfactual transgressions* or the feeling of being a good person without doing any factual good deeds (Mazar & Zhong, 2010; Tiefenbeck et al., 2013). *Counter factual transgression* could encourage people to violate moral codes (Effron et al., 2012a) with very little guilt (Merritt et al., 2010).

For Indonesians, having an inharmonious relationship with others will cause sadness in individuals (Reranita et al., 2012). Therefore, it is important for them to identify themselves as part of a group. Research on moral credentials proves that a group's good deeds can be added to an individual's moral account (Kouchaki, 2011). For the sake of protecting the group's positive image (Reese et al., 2013) it will not punish any member who commits a moral offence by the Devil Protection Effect (DPE) (Stratton, 2007).

2 METHODS

This study used qualitative methods to explore the personal interpretation of the process and situational context of moral licensing in a collectivist society.

2.1 *Participants*

Thirteen men and two women participated in this study. Most interviewees were chosen based on their knowledge, experience, and more than ten years of work in a relevant field. Seven interviewees from high and middle-management levels in the law enforcement system represented the government/law enforcement system; five subjects represented Indonesian society in social and cultural context; and three persons were offenders who have been sentenced to more than five years in prison.

2.2 *Procedures*

In-depth interviews were conducted with interview guidance, where the interviewees were given some brief information on the concept of moral licensing beforehand, with two supporting pieces of literature. Informed consent was given bye very interviewee. Three of the 15 interviewees who were interviewed refused to be recorded. The interviews lasted from 60 minutes to three hours, depending on the interviewee's availability. The in-depth interview guidelines consisted of several questions about when, where and how moral licensing effects emerge in Indonesian society. The other questions were: in what kind of offence will this phenomenon be found? Who are the perpetrators and what is the social reaction to this offence?

2.3 *Data analysis*

A written record was made of the interviews and the data was analysed based on emerging topics. To protect the subject's anonymity, excerpts from the interviews do not include the interviewee's initials. The dates of the interviews and the interviewee's profession are the only data recorded against the interview.

3 RESULTS

Moral licensing symptoms in Indonesia appear in the form of corruption among law enforcers. This behaviour is a type of moral credential because the offence (crime) and good deed (law enforcement) are happening in the same domain. The topic appears in several examples where corruption has been committed by policemen, judges, prosecutors, and chiefs of the Supreme Court. However, moral licensing theory is not applicable in crimes that involve physical harm, such as violence, robbery and murder, because the losses incurred by such crimes are obvious.

Bribery has an ambiguous quality in Indonesian society. The border between right and wrong of the offence is blurry because the definition of bribery relies heavily on the situation, context, and motive behind the event. In Indonesian Corruption Law, bribery is a form of crime that involves elements of gifts or promises to government employees or officials with the purpose of making those employees do things against their obligations. This law is contradictory to the factual condition that in several situations gift is respected as a good deed, and not an offence if people use sincerity as a moral value.

Sincerely giving gifts is an honourable behaviour as it is also related to religious values such as sincerity, where the giver believes that the good deeds will bring goodness directly or indirectly. In social relations it is not easy for an Indonesian to refuse gifts because he will be seen to be impolite, snobbish, and disrespectful. Conversely, it is hard to forbid everyone from giving something to others because it is a part of showing respect to another person. In many cases, giving gifts to each other can be done during a celebration of the life cycle (birth, marriage, and death) that is socially obligatory for Indonesians. According one of the interviewee, gift is a way to show respect, and refusing it is considered disrespectful.

> *'How could we refuse gifts that are given by someone who makes efforts to respect us........ too much has to be sacrificed in our social life if everything is defined as gratification by KPK'* (Interview with P1)

In practice, it is hard to differentiate gifts based on personal social relationships and those in the context of official status. The officials and their acquaintances may define the act of giving gifts as a form of respect, while according to the Law of Corruption Crime, the gift may be suspected as bribe and must not be accepted. It is very common for a member of society to feel well served by a public servant, and give gifts as a token of their appreciation. Hence, when a public official receives gifts for doing good deeds in the form of providing public services, the acceptance of a token gift cannot be perceived as an offence. An offence in this context is considered to be when the official is not doing his or her job (providing public services) and is asking for a certain amount of money/goods from another party, as mentioned by one of the interviewee.

> *'It is alright to accept... as long as the apparatus is doing his job... what they cannot do is to accept it but not do their work (providing public services)'* (Interview with P2)

Historically, giving to public officials or someone with a higher rank or status is encouraged; it is considered as showing decency, respect, and obedience to the leader. The behaviour of giving to an official is a practice that has been going on for a long time in the form of paying tax to the king. Putro (2010) explains that these practices were later copied by the wider society, and continue today. There is a common practice where higher officials will receive more amount or more expensive gifts from their subordinates.

> *'...if the official is holding a party, will the gifts go accordingly?'* (Interview with P3)

Almost all interviewees stated that gifts can be categorised as bribe if they are based on a reciprocation value. Therefore, the broad legal definition of bribery is potentially harming social relations within the community, since gifts are potentially criminalised. As a result, gifts that were aimed to strengthen social bonds is creating suspiciousness, questioning one's kindness, and it is even stated as a violation of the law. Hence, it can be said that bribery as a crime in Indonesia is ambiguous because it reflects a change of the moral value from giving sincerely to giving with an expectation of reciprocation. This ambiguity is then intentionally taken advantage of by offenders who accept bribery in a form of gifts.

When people do good deeds selflessly, it is defined as sincere. Therefore, the good deeds will count in their moral account. Law enforcers who have a selfless orientation perceive their work as the obligation of a servant of the state and will not expect other parties to compensate their work and will refuse gifts. On the contrary, good deeds with expectation of reciprocation will always put perpetrators in transactional social relations. Public officials may define their work as kindness in helping or giving 'quality services'. However, perpetrators will feel like they are helping others and credit it to their moral accounts. Afterwards, they will feel that they are entitled to receive something for their kindness with bribes. There is a conflict of norms, when people in everyday situations do good deeds based on sincerity but the legal perspective defines any kind of gifts made with a reciprocation value as a crime.

A professional law enforcer has an individual primary status in the society. The status is embedded as part of their self-identity, for example when member of society calls them Mr. Policemen or Ms. Judge. This honourable status gives them the sense that they are moral people and have a moral surplus. When alaw enforcer does his job as a public servant, he perceives the job as an act of kindness. As a consequence, he considers himself to be a moral person with abundant moral surplus, and that gives him the right to commit transgressions. Standard operational procedure for public service is not publicly known, so people who are accessing public services willing to pay more to public servants who are just doing their job accordingly, also known as "86".

> *'Oh yes... people may feel that they are good that way... profession, certain jobs, that conclude public position as a good thing...it is dangerous because it may be a corruption point'* (Interview with P4)

They lost their sense of guilt because, from their perspective, there is no victim. For the perpetrators, receiving bribe is a form of respecting someone who wants to do a good deed. This transaction commonly happens only once, where the perpetrator and the bribe giver do not know each other, so that the element of motive of gratification aimed at the law enforcer for committing acts that are against their duty as a state's servant will not be fulfilled. Furthermore, the lack of a victim is often interpreted as the lack of an offence. According to the interviewee, accepting gifts are corruption if done without any service provided. But, if the public servants do their work, it is acceptable to receive money or gifts from community member served.

> *'...So as long as society keep accepting services, corruption will not be seen as an unlawful behaviour...'* (Interview with P5)

Groups tend to protect corrupt perpetrators who bring them profit. There are even many violation cases by law enforcers who provide the group with legal assistance, presenting a narrative that the perpetrators are heroes or victims instead. Support from the groups also shows how they interpret rules on violations, so corruption intention cannot be seen in reality.

4 DISCUSSION

A status as law enforcers gave them the feeling that they were moral people and have moral surplus accounts. This phenomenon is called counterfactual transgression (Effron et al., 2012a). When a law enforcer provides public services, he will count the services as a credit to his moral account. This explains the opportunities for the perpetrators to double count counterfactual transgressions in their moral account (multiple counterfactual transgressions). Multiple moral surpluses create a perception by the perpetrators that they do not have to do more good deeds, and instead make them even feel like they have the moral licence to commit offences (Sachdeva et al., 2009; Khan & Dhar, 2006; Jordan et al., 2011; Effron et al., 2012a). The kindness that is followed by committing offences in the same domain is defined as moral credential (Effron et al., 2009). Corruption based on reciprocation value, where carrying out the function of a law enforcer is counted as a kindness and giving a 'special service'. Therefore, they will feel the right to get rewards from that 'special service'.

When law enforcers commit criminal offences, the offences will be seen as obvious in the public's eyes, but they will have doubts in inflicting punishment because of the moral credentials in the perpetrators' track records. This track record in the Indonesian justice system becomes something to be considered by judges to lighten the punishment. Corruption in the form of bribery is a form of ambiguous violation in Indonesian society. Giving based on the selfless values without hoping for anything in return is contrary to the Law of Corruption Crime where gift is defined as a crime, because every gift is interpreted as creating indebtedness. Consequently, the definition of gift is heavily dependent on the interpretation of the giver and receiver. However, this research shows that gift is interpreted as corruption according to the indebtedness value.

The effect of moral licensing takes the form of the removal of guilt in the law enforcers, which is strengthened by the justification that no victim has been harmed by the offence. In huge corruption cases there is support from groups that use the mechanism of Devil Protection Effect (DPE) (Stratton, 2007) that worsens the removal of guilt on the perpetrators' side. The situation only opens new opportunity for the perpetrators by the lighter punishment from the public.

5 CONCLUSION AND RECOMMENDATIONS

Law enforcement is an honourable profession that leads people in that profession to assume a moral surplus that gives them the right to commit an offence without any concern of condemnation by the public for being an immoral person. The readiness to receive bribe and the absence of guilt in receiving such among the law enforcers relates to the moral values of doing good deeds with indebtedness, a perception of the victim/loss and impact of transgression, as well as group support. This study recommends the criminal justice system not to consider a perpetrator's track record as a reason to inflict a more lenient punishment. To understand the broader context of moral licensing theory, we suggest conducting a study using different methods and in similarly honourable professions such as medicine or teaching.

REFERENCES

Berman, M.N. (2007). On the moral structure of white collar crime. *Ohio State Journal of Criminal Law*, *5*(1), 301–328.

Bradley-Geist, J.C., King, E.B., Skorinko, J., Hebl, M.R. & McKenna, C. (2010). Moral credentialing by association: The importance of choice and relationship closeness. *Personality and Social Psychology Bulletin*, *36*(11), 1564–1575.

Braithwaite, J. (1985). White collar crime. *Annual Review of Sociology*, *11*, 1–25.

Brown, R.P., Tamborski, M., Wang, X., Barnes, C.D., Mumford, M.D., Connelly, S. & Devenport, L.D. (2011). Moral credentialing and the rationalization of misconduct. *Ethics & Behavior*, *21*(1), 1–12.

Chappell, A.T. & Piquero, A.R. (2004). Applying social learning theory to police misconduct. *Deviant Behavior*, *25*(2), 89–108.
Clinard, M.B. & Quinney, R. (1973). *Criminal behavior system: A typology* (2nd ed.). New York, NY: Holt, Rinehart and Winston.
Effron, D.A. (2012). Hero or hypocrite? A psychological perspective on the risks and benefits of positive character evidence. *Bi-monthly E-journal of American Society of Trial Consultants*, *24*(4), 46–51.
Effron, D.A. & Monin, B. (2010). Letting people off the hook: When do good deeds excuse transgressions? *Personality and Social Psychology Bulletin*, *36*(12), 1618–1634.
Effron, D.A., Cameron, J.S. & Monin, B. (2009). Endorsing Obama licenses favoring whites. *Journal of Experimental Social Psychology*, *45*(3), 590–593.
Effron, D.A., Miller, D.T. & Monin, B. (2012a). Inventing racist roads not taken: The licensing effect of immoral counterfactual behaviors. *Journal of Personality and Social Psychology*, *103*(6), 916–932. doi:10.1037/a0030008
Effron, D.A., Monin, B. & Miller, D.T. (2012b). The unhealthy road not taken: Licensing indulgence by exaggerating counterfactual sins. *Journal of Experimental Social Psychology*, *49*(3), 573–578.
Gino, F. (2015). Understanding ordinary unethical behavior: Why people who value morality act immorally. *Current Opinion in Behavioral Sciences*, *3*, 107–111.
Gneezy, A., Imas, A., Brown, A., Nelson, L.D. & Norton, M.I. (2012). Paying to be nice: Consistency and costly prosocial behavior. *Management Science*, *58*(1), 179–187.
Green, G.S. (1997). *Occupational crime* (2nd ed.). Chicago, IL: Nelson-Hall.
Hakim, M.A., Karyanta, N.A. & Hardjono. (2014). Moral judgment in urban and rural context: Indigenous and socio-cultural analysis. In *Abstracts of National Scientific Meeting and Social Psychology Association Conference 2015, 23–25 January 2015, Denpasar, Bali, Indonesia.*
Husin, W.N.W. (2011). Budi Islam: Its role in the construction of Malay identity in Malaysia. *International Journal of Humanities and Social Science*, *1*(12), 132–142.
Jordan, J., Mullen, E. & Murnighan, J.K. (2011). Striving for the moral self: The effects of recalling past moral actions on future moral behavior. *Personality and Social Psychology Bulletin*, *37*(5), 701–713.
Khan, U. & Dhar, R. (2006). Licensing effect in consumer choice. *Journal of Marketing Research*, *43*(2), 259–266.
Kouchaki, M. (2011). Vicarious moral licensing: The influence of others' past moral actions on moral behavior. *Journal of Personality and Social Psychology*, *101*(4), 702–715.
Krumm, A.J., & Corning, A.F. (2008). Who Believes Us When We Try to Conceal our Prejudices? The Effectiveness of Moral Credentials with Ingroups versus Outgroups. *The Journal of Social Psychology*, *148*, 689–709.
Mazar, N. & Zhong, C.B. (2010). Do green products make us better people? *Psychol Sci*, *21*(4), 494–498.
McGee, J.A. & Byington J.R. (2009). The threat of global white-collar crime, *Journal of Corporate Accounting and Finance*, *20*(6), 25–29.
Merritt, A.C., Effron, D.A. & Monin, B. (2010). Moral self-licensing: When being good frees us to be bad. *Social and Personality Psychology Compass*, *4*(5), 344–357.
Miller, D.T. & Effron, D.A. (2010). Psychological license. When it is needed and how it functions. *Advances in Experimental Social Psychology*, *43*, 115–155.
Monin, B. & Merritt, A. (2010). Moral hypocrisy, moral inconsistency, and the struggle for moral integrity. *The Social Psychology of Morality: Exploring the Causes of Good and Evil*, *3*, 1–25.
Monin, B. & Miller, D.T. (2001). Moral credentials and the expression of prejudice. *Journal of Personality and Social Psychology*, *81*(1), 33–43.
Nelken, D. (1995). White collar-crime. In Maguire, M., Morgan, M.R. & Reiner. R. (Eds.), *The Oxford handbook of criminology* (pp. 355–392). Oxford, UK: Clarendon Press.
Nettler, G. (1976). *Lying, cheating and stealing social concern*. New York, NY: McGraw-Hill.
Oriza, I.I.D. (2016). The effect of favor, request, gratitude and indebtedness on compliance. (Unpublished doctoral thesis, Faculty of Psychology, Universit as Indonesia, Depok, Indonesia).
Perri, F.S. (2011). White-collar criminals: The 'kinder, gentler' offender? *Journal of Investigative Psychology and Offender Profiling*, *8*(3), 217–241.
Putro, Z.A.E. (2010). Javanese tolerance limit; a study of contemporary Yogyakarta. *Community Sociology Journal*, *15*(2).
Rahman, A.A. (2013). Principle of sanctity, hypocrisy, and moral integrity (Unpublished doctoral thesis, Faculty of Psychology, Universitas Indonesia, Depok, Indonesia).
Reese, G., Steffens, M.C. & Jonas, K.J. (2013). When black sheep make us think: Information processing and devaluation of in- and outgroup norm deviants. *Social Cognition*, *31*(4), 482–503.

Reranita, T., Hakim, M.A., Yuniarti, K.W. & Kim, U. (2012). Vulnerable factors of sadness among adolescents in Indonesia: Anexploratory indigenous research. *Humanitas*, *9*(1), 1–11.
Sachdeva, S., Iliev, R. & Medin, D.L. (2009). Sinning saints and saintly sinners: The paradox of moral self-regulation. *Psychological Science*, *20*(4), 523–528.
Sartini, N.W. (2009). Exploring Javanese local wisdom through idioms: Bebasan, Saloka, dan Peribahasa. *Logat, Journal of Language and Literature Science*, *5*(1), 28–37.
Sendra, I.M. (2008). Ancestorworship in Japanese and Balinese kinship system. *Analisis Pariwisata*, *8*(2).
Stratton, J. (2007). *Social identification and the treatment of in-group deviants: The black sheep effect or devil protection* (Unpublished doctoral thesis, University of Southern California, Los Angeles, CA).
Suara Pembaruan. (2015, January 14). Police generals arrested by corruption commission. *Suara Pemberuan*. Retrieved from http://sp.beritasatu.com/home/sederet-jenderal-polisi-di-tangan-kpk/75089
Sutherland, E.H. (1983). *White collar crime: The uncut version*. New York, NY: Vail-Ballou.
Tiefenbeck, V., Staake, T., Roth, K. & Sachs, O. (2013). For better or for worse? Empirical evidence of moral licensing in a behavioral energy conservation campaign. *Energy Policy*, *57*, 160–171. doi:10.1016/j.enpol.2013.01.021
Triandis, H.C. (2004). The many dimensions of culture. *Academy of Management Executive, 18*(1), 88–93. doi:10.5465/AME.2004.12689599
Waddington, P.A.J. (1999). Police (canteen) sub-culture: An appreciation. *British Journal of Criminology*, *39*(2), 287–309. doi:10.1093/bjc/39.2.287

Diversity in Unity: Perspectives from Psychology and Behavioral Sciences – Ariyanto et al. (Eds)

Exploration of moderation effect of price on the relationship between observational cues and sustainable consumption

G.C. Wajong & E. Narhetali
Faculty of Psychology, Universitas Indonesia, Depok, Indonesia

ABSTRACT: Previous research has demonstrated that the presence of observational cues increases the tendency of individuals to exhibit altruistic behavior. This research aims to explore the probability of relationship between observational cues and a form of such behavior, namely sustainable consumption, as well as the moderation effect of price on the relationship. This research employs a between-subject experimental design of 2 (observational cues: present vs. absent) × 3 (price: sustainable product > conventional product vs. sustainable product < conventional product vs. sustainable product = conventional product) and used the measurement of actual buying behavior. The analysis of 182 data collected from undergraduate students of Universitas Indonesia indicated that there was no significant effect of observational cues on increasing sustainable consumption, χ^2 (1, N = 182) = 2.348, p = 0.125. The analysis of a three-way interaction model indicated that there was no significant effect on price within the model, χ^2 (2, N = 182) = 0.11, p = 0.995. An analysis conducted on the relationship between price and product yielded significant results, χ^2 (2, N = 182) = 45.539, p = 0.001. The results called into question the generalization of effects generated by observational cues and highlighted price as an inhibiting factor in sustainable consumption.

1 INTRODUCTION

The decision made by consumer is an important one because they affect, one way or another, the decision made by companies in many regards, such as their means of production, marketing, as well as the goods and services being offered. However, the alarming number of unethical practices conducted by companies, evidence from cases of environmental degradations, child exploitation, may beg the question of whether consumers are already using the power they possess to steer companies to operate in a manner that is more ethical. If only consumers decide to only purchase goods and services that are sustainable and ethical, companies will adjust to those demands and refrain from engaging in unethical practicies. After all, consumers have the power to decide on what gets produced (Hutt, 1936). Unfortunately, a great deal of research has pointed at the existing gap between attitudes towards ethical business practices and the act of ethical consumption (Boulstridge & Carrigans, 2000; MacGillivray, 2000; Carrigan & Attalla, 2001).

Oftentimes, consumption is only viewed as an economic behavior, namely as a means of fulfilling needs (Jackson, 2005). However, there is plenty of literature which offers a moral perspective in viewing the behavior (McMurty, 1998; Sayer, 2003). The approach was sensible due to the understanding that every consumer behavior has the potential to impact on many areas, be it resources, pollution, communities, and ethics (Stern *et al.* 1997; Young *et al.* 2010); hence, there is an element of value (right or wrong) contained in the behavior. Currently, ethical issues in consumer behavior, ranging from product attributes to consumer behavior and characteristics, is being studied under the field of consumer ethics, namely a form of purchasing behavior exhibited by consumers which involves moral considerations by referring to certain moral principles (Vitell & Muncy, 1992). This is a broad concept due to the vast number of areas in which ethical concerns could identify, such as environment, community welfare, and product quality. Essentially, the concerns are similar in a sense that they are derived from the threat to sustainability, a concept which requires the present generation to ensure that their current fulfillment of needs will not compromise the ability of the future generation to fulfill theirs. Sustainable consumption falls under the field of consumer ethics because it has strong ties with issues of environmental preservation and good living conditions for humans; thus, it is a behavior which contains values of right or wrong, considering the impacts (De Pelsmacker *et al.*, 2005; Carrington, 2010). In addition to the urgent need to create a more environmentally-aware society to prevent further environmental damage, the sustainable consumption discussed in this research will be narrowed down to the one concerning environment.

The discrepancy between intention and behavior is prevalent in the realm of sustainable consumption. Surveys and research have revealed that upon entering the 20th century, there is an increasing number of consumers who are concerned with ethical issues in ways of production (Tallontire, Rentsjendorj & Blowfield, 2001; Young *et al.*, 2010). However, those concerns are not manifested into the act of purchasing sustainable goods (Folkes & Kamin, 1999; Boulstridge & Carrigans, 2000; Carrigan & Attalla, 2001).

Considering the positive impacts of engaging in sustainable consumption, namely contributing to the environmental preservation, and the fact that consumers actually have the intention to buy sustainable goods, we should be seeing a high number of sustainable goods circulating and being purchased. However, the existing attitude-behavior gap indicates that there might be something more which needs to be explored in terms of the challenges or moral dilemma that individuals are facing when faced with options of goods. The idea of purchasing behavior is a behavior which requires moral consideration is in line with the view of Burke and Milberg (1993). Individuals are exposed to the dilemma of whether they were to care about the consequences which takes place in public realm due to the consumption that they are doing in their private sphere (Webster, 1975). As a consequence, they will reconsider the essence and impacts of their purchasing behavior, for example, whether purchasing unsustainable goods implies that they are supporting unethical business practices of a particular company.

Therefore, the question that follows is in regard to how to ensure that individuals act, or, in this case, purchase morally. Research on morality has provided insights regarding the underlying mechanism which enables individuals to exhibit morally desirable behaviors. One of the most notable ones was produced through the research carried out by Duval and Wicklund (1973) which identified the existence of objective self-awareness, namely the ability of individuals to alter their behavior to be in accordance with certain moral standards when they exert inwards attention. The theory posits that by being aware of the existing moral standards, imposed on them by external or internal elements, individuals will be aware of the discrepancy between their behavior and a moral standard that they should be subjecting themselves to. It further lends the legitimacy that an internal conflict (moral dilemma) may take place before an individual decides to exhibit a behavior and that by altering certain elements, individuals will behave in a manner that is morally desirable. Research has identified several factors which can activate an individual's objective self-awareness, namely setting up mirrors or audience in front of a subject (Scheier, Fenigstein & Buss, 1974; Froming, Walker & Lopyan, 1981). Recently, observational cues have been added to the array of stimulus which can activate objective self-awareness and lead to the exhibition of morally desirable behavior (Pfattheicher & Keller, 2015). Observational cues are stimuli, oftentimes visual ones, which form a perception upon an individual that he or she is under observation.

The relationship between ethical behavior and observational cues has been probed many times before, in both laboratory and field contexts (Bateson, Nettle & Roberts, 2006; Haley & Fessler, 2005; Pfattheicher & Keller, 2015), and has yielded confirmatory results, namely the presence of observational cues increases an individual's tendency to exhibit ethical behavior. Previous research operated observational cues with images of human eyes and created the perception of being observed or watched. Such perception resulted in a moral standard that an individual perceives to be expected upon him by others, and thus he projects into himself. It can be concluded that objective self-awareness manages to be activated in a context where observational cues or eye images are present due to the strong indication emanated by such stimuli that attention is being directed to an individual (Haley & Fessler, 2005).

Another form of underlying mechanism related to the exhibition of ethical behavior in the presence of observational cues is one that is attached to humans throughout its evolution, namely reputation management. This is a mechanism which causes an individual to project a positive image when he is around other individuals by aligning himself with values or moral standards which he presumes to be owned by others (Bateson *et al.*, 2006). Such alignment is carried out in order to make sure that he will be able to interact with other members in his group which will ensure his survivability. Interestingly, the previously elaborated research on observational cues indicates that this mechanism is so inherent and automatic that humans continue to behave in a morally desirable manner even when no actual humans are present to observe their behavior; such perception is manipulated simply by using eye images. The effect which takes place due to this mechanism in conditions where observational cues are present is referred to as watching-eyes effect.

The challenges faced by individuals in resolving their moral dilemma with morally desirable behavior have been identified in a great deal of research which examines the discrepancy between attitude and behavior of sustainable consumption. Price attribute (Vantomme & Geuens, 2006; Carter, 2014) has been identified as the main factor which inhibits individuals from purchasing sustainable products. Understanding that technologies and practices used to create sustainable products are relatively new and are low in availability in contrast to that of conventional products, the price of such goods become expensive. Taking it into Indonesian context, we posit that consumers are yet to be able to bear premium cost of such goods considering their currently low purchasing capacity. Therefore, it can be concluded that high price implies more sacrifice to be made by individuals to fulfill their needs.

We postulated that there is a relationship between observational cues and sustainable consumption due to the moral consideration that individuals need to undergo prior to choosing to buy sustainable products. Observational cues are posited as effective stimuli which will enable an individual to solve his moral dilemma with morally desirable behavior, in this case, to purchase sustainable products.

Hypothesis 1: presence of observational cues will lead to sustainable consumption

Understanding that price is an attribute that is highly considered by individuals prior to a purchase, we posited that price will weaken the effect of observational cues upon sustainable consumption. The objection of consumers to purchase expensive products lead to our assumption that even in a condition where observational cues are present, most consumers will still choose cheaper goods, irrespective of the ethicality of its means of production.

Hypothesis 2: price negatively moderates the positive relationship between observational cues and sustainable consumption.

2 RESEARCH METHOD

2.1 *Participants*

182 undergraduate students of Universitas Indonesia were recruited via online prior to their participation in the research. Their ages ranged from 17 to 25 years old (M = 19.77; SD = 1.355). In parts, we viewed that this sampling could account for the current lack of literature which investigates the ethical consumption behavior exhibited by young adults (Vitell & Muncy, 2005; Maggioni, Montagnini & Sebastiani, 2013) as a group which also makes up a large portion of market.

2.2 *Procedure*

The experiment was set up as a market survey conducted by a fictional student-run shop and was carried out using a laptop, on which the pictures and prices of goods and services

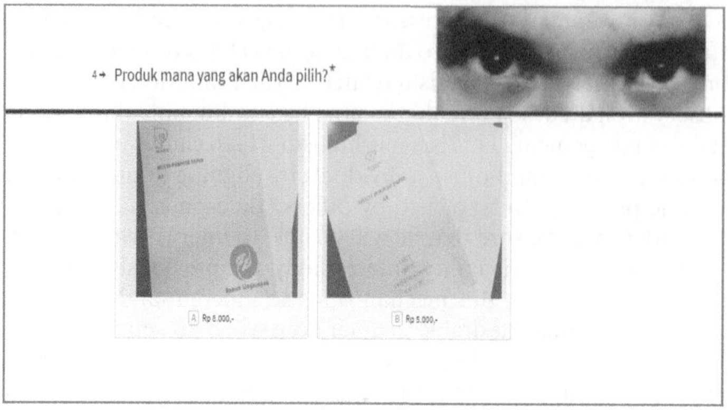

Figure 1. Condition presented in Group 1.

were displayed. The eye image employed in this research was selected by means of expert judgment in order to come up with one which could evoke anxiety, in accordance with the ones used in previous research involving observational cues (Nettle, Nott & Bateson, 2012). In controlled conditions, we used the image of the logo of Universitas Indonesia, meanwhile, image of a pair of eyes is used in experiment conditions. On each condition, the image is located on the top-righthand corner of the monitor. The products utilized in this research were pens, bottled water, ride-sharing transportation, tissue paper, and paper because of its constant use and proximity to the daily activities of undergraduate students. For each product, we offered two different brands from which the participants can choose. The manipulation was conducted on the paper products, which differed in their means of production (indicated by the presence of eco-friendly logo on one brand, and the absence of the same logo on the other) and specific sets of price. Meanwhile, the other products were still offered as a means to prevent response bias. Sustainable consumption, as the dependent variable, would be measured by the participants' choice of goods which could be viewed by looking at which picture of item that they clicked. Participants were divided into 6 different groups, and we allowed them to choose a piece of paper, on which a number was written, from a bowl as a means of random assignment. Each participant was then given a shopping coupon and instructed to spend it on a number of goods and services already prepared by researchers. Group 1–3 were put in the condition where observational cues were present, meanwhile group 4–6 where observational were absent. Another variation lay on price, in which participants of group 1 and 4 were put in the condition where the price of sustainable products was higher than conventional products, group 2 and 5 where the price of sustainable product was lower than conventional products, finally group 3 and 6 where the prices of both sustainable and conventional products were equal. Upon completion, participants were asked to undertake a manipulation check, debriefed, then given financial rewards.

3 RESULTS

Majority of participants across two conditions chose eco-friendly product (83.6%). Results of the chi-square analysis, as displayed in Table 1, provide confirmation that there is no significant difference between groups where observational cues are present and where observational cues are absent (χ^2 (1, N = 182) = 2.348, p = 0.125).

Based on loglinear analysis, we found no significant results on the model if the three-way interaction between observational cues, price, and product was eliminated (χ^2 (2, N = 182) = 3.410, p = 0.182). This finding indicates that no interaction exists between the three variables. The analysis of the main effect of each independent variable in the research using partial chi square provided additional confirmation that prices do not have an effect on the three-way interaction model of observational cues, price, and product (χ^2 (2, N = 182) = 0.11, $p = 0.995$. As a consequence, this provides statistical rejection of the second hypothesis because prices are not proven to weaken the relationship between observational cues and sustainable consumption. Results are displayed in Table 2 and Table 3.

In reference to the previous finding, we have identified the existence of a significant model of interaction between price and product. Hence we conducted an independent analysis on the relationship between price and product. The Chi square analysis indicates that there is a significant relationship between price and sustainable consumption (χ^2 (2, N = 182) = 45.539, p = 0.001). Results are displayed in Table 4.

Table 1. Chi-square analysis of two major groups.

	Sustainable product		Conventional product	
Condition	n	%	N	%
OC Present	79	43.41	11	6.04
OC Absent	73	40.11	19	10.44

*OC = Observational Cues.

Table 2. Parameter estimates, value, and goodness-of-fit index.

Effect	Λ	*z*	*p*
OC × Product × Price	−0.104	−0.351	0.726
	0.199	0.976	0.329
OC × Product	−0.315	−1.660	0.097
OC × Price	0.115	0.388	0.698
	−0.238	−1.167	0.243
Product × Price	0.493	1.668	0.095
	−0.995	−4.880	< 0.001
OC	0.271	1.425	0.154
Product	1.143	6.015	< 0.001
Price	−0.333	−1.127	0.260
	0.658	3.228	< 0.001

Table 3. Summary of hierarchical deletion steps.

Steps	Model	Df	χ^2	p	Eliminated variables	Δdf	Δχ2	Δp
1	(OC × H) (OC × P) (H × P)	2	3.410	0.182	OC × H	2	0.80	0.209
2	(OC × P) (H × P)	4	4.110	0.391	OC × P	1	2.374	0.129
3	(H × P) (OC)	5	6.484	0.262	OC	1	4.11	0.107
4	(H × P)	6	6.506	0.369	H × P	2	38.048	<0.001

Table 4. Chi square analysis.

	Sustainable Product (SP)		Conventional Product (CP)	
Condition	n	%	N	%
Price of SP > CP	35	19.23	26	14.29
Price of SP < CP	58	31.87	2	1.1
Price of SP = CP	59	32.41	2	1.1

4 DISCUSSION

The rejection of the generalization of the watching-eyes effect demonstrated in this research provides support to previous research which also denies the existence of the watching-eyes effect (Fehr & Schneider, 2010; Lamba & Mace, 2010; Antonetti, 2012). Despite the fact that a slightly higher number of participants chose conventional products in conditions where observational cues were absent, majority of the participants chose the sustainable products across all of the conditions. The finding granted support to the existence of strong reciprocity, namely a condition in which individuals behave cooperatively or altruistically despite: 1) the existing opportunity to cheat, 2) no apparent benefits of demonstrating such behavior (Fehr 2003). In addition, strong reciprocity postulates that individuals are willing to sacrifice in order to give credits to fair behaviors and, conversely, punish unfair behaviors. This is oppositional to the underlying assumption of the watching-eyes effect, namely that humans focus on themselves and are prone to unethical behaviors to meet their needs. The finding reveals that individuals do not need additional input of information to alter their behavior into one that is more ethical because they are inherently ethical. This finding also supports previous research regarding the existence of strong reciprocity in Indonesian samples (Cameron *et al.*, 1999, 2009).

An alternative explanation to this finding might derive from the fact that students are used as a sample in this research. Similar to Cameron's research (1999), which also used student participants, and eventually found proof of strong reciprocity, the idealistic nature and more knowledge on ethical issues that they posses, might account for the results. Knowledge is predicted as a factor based on the responses elicited during manipulation check concerning the environmental knowledge that students possessed. In addition to that, the characteristic of Indonesian people might also contribute to the results; that is Indonesians tend to behave collectively, something which is in stark contrast to Western people (Triandis *et al.*, 1986). As a result, Indonesians tend to abide by a collective norm, namely alturism, and internalize such norm as personal value (Hofstede, 2015).

Price did not significantly contribute to the three-way interaction model of the variables probed in this research. However, an analysis of relationship between price and sustainable consumption yielded significant a result; the higher the price of the product, the lesser participants will choose that particular product. This is in line with previous research concerning price, namely that higher price will deter consumers from purchasing sustainable goods.

This research managed to contain the threat of social desirability by employing the measurement of actual buying behavior. This measurement is considered able to bridge the gap of intention behavior which is prevalent in surveys or self-report instruments; thus, is more reflective of the actual behavior of population. Another important note for future research on observational cues is in regard to the salience of the investigated stimulus. In this research, nearly half of the participants in the experiment group claimed that they were not aware of the existence of the eye image. This research is in line with previous research which used all of the data collected despite the awareness of observational cues. We have not made any analysis in the light of such discovery; therefore, we suggest future research probes on the different awareness concerning the cues to further illuminate the implicit mechanism which might take place in participants who are unaware of the observational cues, yet choose to purchase sustainable products. Future research can also benefit from exploring many other dependent variables in relation to observational cues and different samples (based on occupations, levels of education, geographical location) from the population of developing countries, especially considering the scarce amount of literature on observational cues being tested in developing countries.

Practical considerations which may be derived from the findings of this research could be beneficial to improve the strategy of marketting sustainable products. Understanding that individuals already have the inclination to exhibit sustainable consumption, those stakeholders could do more by eliminating price barriers which might inhibit the behavior. For example, government could provide subsidy for sustainable production and limit the power of conventional companies who might already have the benefit of economies of scale. Dissemination of information concerning sustainable products to increase knowledge could also be improved in order to create and strengthen the norm that buying sustainable goods is desirable. Considering that currently certification of sustainable goods is issued on a voluntary basis, government would also benefit from increasing the exposure of such products by creating a mandatory certification of products, so that people will be more aware of the existence of sustainable products and able to distinguish sustainable and conventional products.

REFERENCES

Anggreyani, S.P. (2015) Pengaruh Informasi Keramahan Lingkungan Pada Produk, Tugas Kognitif, dan Religiusitas pada Preferensi Konsumen Dewasa Muda. Undergraduated thesis, Universitas Indonesia.

Antonetti, P. (2012) The Role of Guilt and Pride in Consumers' Self-Regulation: An Exploration on Sustainability and Ethical Consumption. Master's thesis, Cranfield University.

Bateson, M., Nettle, D. & Roberts, G. (2006) Cues of Being Watched Enhance Cooperation in a Real-World Setting. *Biology Letters,* 2 (3), 412–414.

Bedford, T. (1999) Ethical Consumerism: Everyday Negotiations in The Construction of an Ethical Self. [Online] Available from: http://discovery.ucl.ac.uk/1318018/ [Accessed 3rd July 2016

Boulstridge, E. & Carrigan, M. (2000) Do Consumers Really Care about Corporate Responsibility? Highlighting the Attitude-Behaviour Gap. *Journal of Communication Management,* 4 (4), 355–363.

Bourrat, P., Baumard, N. & McKay, R. (2011) Surveillance Cues Enhance Moral Condemnation. *Evolutionary Psychology,* 9 (2), 193–199.

Burke, S.J., Millberg, S.J. & Smith, N.C. (1993) The Role of Ethical Concerns in Consumer Purchase Behaviour: Understanding Alternative Processes. *Advances in Consumer Research,* 20, 119–122.

Cameron, L. (1999) Raising the Stakes in the Ultimatum Game: Experimental Evidence from Indonesia. *Economic Inquiry,* 37 (1), 47–59.

Carrigan, M. & Attala, A. (2001) The Myth of the Ethical Consumer—Do Ethics Matter in Purchase Behaviour? *Journal of Consumer Marketing*, 18 (7), 560–578.

Carrington, M.J., Neville, B.A. & Whitwell, G.J. (2010) Why Ethical Consumers Don't Walk Their Talk: Towards a Framework for Understanding the Gap between the Ethical Purchase Intentions and Actual Buying Behaviour of Ethically Minded Consumers. *Journal of Business Ethics,* 97 (1), 139–158.

Carter, K. (2014) Product and Consumer Characteristics as Moderators of Consumer Response to Sustainable Products. PhD diss., University of South Carolina.

Duval, S. & Wicklund, R.A. (1973) Effects of Objective Self-Awareness on Attribution of Causality, *Journal of Experimental Social Psychology,* 9 (1), 17–31.

Fehr, E. & Henrich, J. (2003) Is Strong Reciprocity a Maladaptation? On the Evolutionary Foundations of Human Altruism. *Genetic and Cultural Evolution of Cooperation*, 55–82.

Flurry, L. & Swimberghe, K. (2016) Consumer Ethics of Adolescents. *Journal of Marketing Theory and Practice,* 24 (1), 91–108.

Folkes, V. & Kamins, M. (1999) Effects of Information About Firms' Ethical and Unethical Actions on Consumers' Attitudes. *Journal of Consumer Psychology,* 8 (3), 243–259.

Haley, K.J. & Fessler, D.M.T. (2005) Nobody's Watching? Subtle Cues Affect Generosity an Anonymous Economic Game. *Evolution and Human Behavior,* 26 (3), 245–256.

Hofstede, D.G. (2015) *Cultural Insights-Geert Hofstede*. Helsinki, Finsko.

Hutt, W.H. (1990) Economists and The Public: A Study of Competition and Opinion. New Jersey, Transaction Publishers.

Jackson, T. & Michaelis, L. (2003) *Policies for Sustainable Consumption*. London, Sustainable Development Commission.

Jungbluth, N., Tietje, O. & Scholz, R.W. (2000) Food Purchases: Impacts from the Consumers' Point of View Investigated with a Modular LCA. *The International Journal of Life Cycle Assessment,* 5 (3), 134–142.

Kotler, P. & Keller, K.L. (2009) *Marketing Management. Organization*, 22.

Kuester, S. (2012) MKT 301: Strategic Marketing & Marketing in Specific Industry Contexts. University of Mannheim, 110.

Lamba, S. & Mace, R. (2010) People Recognise When They Are Really Anonymous in an Economic Game. *Evolution and Human Behavior*, 31 (4), 271–278.

Nettle, D., Nott, K. & Bateson, M. (2012) Cycle Thieves, We Are Watching You': Impact of a Simple Signage Intervention against Bicycle Theft. *PLoS ONE,* 7 (12).
Nielsen. (2012) *Annual Report March 2012*. [Online] Available from: http://www.nielsen.com/content/dam/corporate/us/en/reports-downloads/2012-Reports/Nielsen-Global-Social-Responsibility-Report-March-2012.pdf. [Accessed 3rd July 2016]
Pelsmacker, P.D.E. (2005) Implicit Attitudes Toward Green Consumer Behaviour. *Psychologica Belgica,* 45 (4), 217–239.
Pfattheicher, S. & Keller, J. (2015) The Watching Eyes Phenomenon: The Role of a Sense of Being Seen and Public Self-Awareness. *European Journal of Social Psychology,* 45, 560–66.
Sayer, A. (2003) (De)commodification, Consumer Culture, and Moral Economy. *Environment and Planning D: Society and Space,* 21 (3), 341–357.
Scheier, M.F., Fenigstein, A. & Buss, A.H. (1974) Self-Awareness and Physical Aggression. *Journal of Experimental Social Psychology*, 10 (3), 264–73.
Stern, P.C., Dietz, T., Ruttan, V.W., Socolow, R.H. & Sweeney, J.J. (1997) *Environmentally Significant Consumption: Research Directions*. (eds.) *National Academies Press*.
Tallontire, A., Rentsendorj, E. & Blowfield, M. (2001) Ethical Consumers and Ethical Trade: A Review of Current Literature. *Policy Series*, 12, 34.
Vitell, S.J. & Muncy, J. (1992) Consumer Ethics: An Empirical Investigation of Factors Influencing Ethical Judgements of the Final Consumer. *Journal of Business Ethics,* 24 (4), 297–311.
Webster, Jr., Frederick, E. (1975) Determining the Characteristics of the Socially Conscious Consumer. *Journal of Consumer Research,* 2 (3), 188.
Young, W., Hwang, K., McDonald, S. & Oates, C.J. (2010) Sustainable Consumption: Green Consumer Behaviour When Purchasing Products. *Sustainable Development*, 18 (1), 20–31.

Diversity in Unity: Perspectives from Psychology and Behavioral Sciences – Ariyanto et al. (Eds)
© 2018 Taylor & Francis Group, London, ISBN 978-1-138-62665-2

The end justifies the terrorist means: Consequentialist moral processing, involvement in religious organisations, and support for terrorism

J. Hudiyana & H. Muluk
Faculty of Psychology, University of Indonesia, Depok, Indonesia

M.N. Milla
Faculty of Psychology, State Islamic University Sultan Syarif Kasim Riau, Riau, Indonesia

M.A. Shadiqi
Faculty of Psychology, University of Indonesia, Depok, Indonesia

ABSTRACT: Terrorism is an act aimed at achieving a desired end. Terrorist supporters may justify terrorism as a moral act with certain goals, such as defending their religion. Here, we propose that a preference for consequentialist morality (a moral tendency to prioritise consequences) predicts support for terrorism. A total of 453 Indonesian Muslims participated in the survey. It was found that a higher adherence to consequentialist moral processing is positively associated with support for Islamic terrorists. This relationship is stronger in those affiliated with religious organisations. The discussion focuses on how consequentialist moral thinking is associated with terrorism support and how religious organisations may shape support for terrorism.

1 INTRODUCTION

'*The end justifies the means*'—Niccolo Machiavelli

Often, radical groups find terrorism to be a reasonable choice to draw greater public support. By engaging in terrorist acts, radical groups attain much greater impact, such as gaining publicity, weakening targeted governments, inducing sympathy, and inspiring followers, as well as spreading public fear. In this case, terrorism is an instrument or tool used to reach certain goals. Not only is it a useful means by which to achieve, but it is also quite efficient. The acts of terror can often be designed or planned with minimal time and monetary resources. Terrorism is comparatively pragmatic in this sense (Crenshaw, 1981). Thus, terrorist supporters may justify terrorism simply because it is a powerful and efficient tool to achieve a desired end, such as defending their religion or gaining political influence (Kruglanski & Fishman, 2006). While a considerable amount of research has been done to explain the psychological factors of why people support or conduct terrorist acts (Silke, 2008; Horgan, 2008; Kruglanski & Fishman, 2009), only a handful of research explains terrorism in the sense of a pragmatic means to achieve a desired end.

If it is true that terrorism is an act seen as a path towards achieving greater good, one might assume that certain moral justification processes may play a role in this matter. According to Albert Bandura's moral disengagement theory, people do not ordinarily support or engage in violent or immoral behaviour until they justify it as worthy or purposeful (Bandura, 1999). These justifications lead individuals to view themselves as moral agents and they will be disengaged with any moral responsibilities held. Consequently, victims of violent conduct become dehumanised. It should also be noted that certain individuals may be more likely to be predisposed to the tendency towards the moral disengagement process compared to others (Bandura, 1999).

In this study, we argue that individuals tend to support terrorism not only when they can justify it as a moral act, but also when this justification of terrorism is supported with a moral belief that endorses 'the realisation of the desired end'. Here, the support for terrorism might be linked more with a certain moral processing, namely consequentialist moral processing, in which individuals focus on the consequences of their decisions (Greene, 2013). As Kruglanski and Fishman (2006) have noted, terrorism is more of a 'tool' to achieve political gains than an ideologically derived act. With this in mind, terrorism might be seen as a way to achieve certain goals such as defending an individual's own religion or gaining political power. Rather than observing religious rules or sacred texts, these individuals view terrorism positively simply because it may help to achieve their own or their group's interests. Those who prefer consequentialist moral processing tend to make decisions in this way. Our present study aims to answer the question: Is support for terrorism more associated with consequentialist moral processing than deontological moral processing?

1.1 *Support for terrorism*

Due to the fact that terrorism is defined as symbolic violence used to attain political and social influence (Kruglanski & Fishman, 2009), we define support for terrorism as the positive attitudes towards symbolic violence used to attain the aforementioned goals. Attempts to explain terrorism within individuals and their demographic backgrounds have been fairly inconsistent. Those who have positive attitudes towards terrorism do not come from one specific gender, educational or social-economic status (Sageman, 2005). Some religious ideologies, such as the Salafi, have been accused of promoting violence when only small minorities support terrorism in the name of religion (Silke, 2008). Any attempt to explain support for terrorism based solely on ideology will not lead to a satisfactory outcome.

In general, social and political influence may play quite an important role. Individuals who are socially marginalised, isolated, and discriminated against often exhibit greater support for extremism (Sageman, 2005; O'Leary, 2007). Additionally, individuals who perceive themselves to be treated unfairly by the out-groups tend to have stronger support for terrorism (Silke, 2008) and a higher degree of intergroup violence (Schaafsma & Williams, 2012). In other words, perceived disadvantage derived from intergroup relations often manifests in the social and political conditions. This is most evident in the Middle East where suicide bombings are regarded as important tools to achieve political gains (Kruglanski et al., 2013). However, it is doubtful that these socio-political dynamics will happen in a context where there are fewer political or religious conflicts, such as Indonesia.

It has also been presumed that terrorism is linked with certain cognitive factors. Among these are fundamental attribution error, cognitive closure (close-mindedness), need for cognition, and cognitive complexity (Kruglanski & Fishman, 2006). It has also been considered that certain moral styles may play a part in the act of terrorism. According to Kruglanski and Fishman (2006), terrorists and their supporters may feel morally justified because the achieved consequences justify the act of terrorism itself. To our knowledge, no studies have empirically examined the relationship between different moral styles and support for terrorism.

1.2 *Moral processing style and support for terrorism*

In his theory on moral decision making, Joshua Greene differentiates between consequentialist morality and deontological morality (Greene, 2013). These two moral paradigms were inspired by a long standing philosophical discourse. Consequentialist morality places an emphasis on moral decisions with regard to their consequences for the many rather than the few. For example, in the infamous trolley dilemma game, killing one person to save five people is considered the best decision. In contrast, deontological morality places an emphasis on doing what is right according to rules and imperatives. In the trolley dilemma, deontologists strongly disagree with killing one person to save many (Young et al., 2013). The specific rule here is never to kill people. Thus, deontological morality is more resistant to relativism as it tend to apply universal rules in every situation. Although these two moral approaches might be used interchangeably, there is a stable preference in each individual (Lombrozo, 2009). The stable preference that favours consequences over rules is called consequentialist moral processing, whereas the opposite preference is called deontological moral processing.

Previous works have attempted to find an association between the moral processing style and variables such as religiosity, religious fundamentalism, and political conservatism. Despite the absence of correlations among religiosity and the two moral processing styles (Hauser et al., 2007), Young et al. (2013) have found that individuals with a higher degree of religious fundamentalism and political conservatism tend to exhibit a greater preference towards deontological moral processing. Adherence to rules is indicative of a higher degree of reliance on the divine and sacred texts or God's commandments. Thus, one can expect that religious-related attitudes such as towards support for terrorism may be explained by the same moral processing.

Terrorism itself may not be ideologically seen as a derived activity but rather as a 'tool' to achieve certain goals, such as attaining political power or defending a religion from its enemies (Kruglanski & Fishman, 2006). Moreover, disadvantageous social and political conditions are of paramount importance in shaping positive attitudes towards terrorism (Silke, 2008). Keeping this in mind, one can expect people who are supportive of terrorism as a means to achieve their desired political consequences would have a certain degree of consequentialist moral processing. In other words, support for terrorism may be positively associated with a stronger preference towards consequentialist moral processing.

1.3 *Involvement in religious organisations and support for terrorism*

Group dynamics are essential in shaping more positive attitudes towards terrorism. For example, individuals gradually internalise the ideologies of their fellow group members (Silke, 2008). For an individual, the groups to which they belong are the providers that grant personal significance (Kruglanski et al., 2013) and self-esteem (Pyszczynski et al., 2002, 2006). In other words, groups give individuals the sense that their life has a certain meaning. Individuals who join, and identify themselves as members of, certain groups tend to be more supportive of the group's ideologies. A study by Pyszczynski et al. (2006) confirms this notion. They found that Muslim students who are reminded of death (the condition of insignificance) tend to show more positive attitudes towards their fellow Muslim students in the same college who support terrorism. Therefore, we argue that there will be a high degree of support for terrorism among individuals who join or are involved in religious organisations. While support for terrorism itself can be predicted using consequentialist moral processing as an indicator, we can find more profound effects on those who are involved in religious organisations.

1.4 *Current study*

We have argued that terrorism may be seen more as a tool than as a meaningful religious activity for its supporters. For them, terrorism is perceived to be an activity that aims to protect religion from its enemies or to achieve political power. Those who support terrorism may be influenced more by consequentialist moral processing than by deontological moral processing. Thus, we hypothesised (Hypothesis 1) that the increase in support for terrorism by means of defending religion from its enemies (dependent variable) can be predicted from by a preference towards consequentialist moral processing (independent variable). The effect should be stronger for those who are involved in religious organisations (Hypothesis 2). Previous studies have linked morality with political ideologies, religiosity, and religious fundamentalism. But to date we have not found empirical data linking moral processing with attitudes towards terrorism.

2 METHOD

To test our hypotheses, we conducted an online survey along with an in-person survey. We provided two power banks and eight prepaid mobile phones with a credit of IDR 50,000. Before the survey, all participants were told that at the end of the study, ten participants will win either one power bank or one prepaid mobile credit in a lucky draw. Only Indonesian people are allowed to participate. After informed consents were given, the participants who were willing to take part in the survey completed a series of questionnaires. All participants could ask about the purpose of the study via the email contact provided by researchers.

2.1 *Participants*

The survey was completed by 552 Indonesian participants: 332 females (60.1%), 220 males (39.9%), mean age = 28.28 (Standard Deviation (SD) = 9.74). More than 80% of participants were Muslims (453), followed by Protestants (44), Catholics (21), Buddhists (14), and Hindus (7). Thirteen participants were not affiliated with any of these religions, while one participant did not state their religious affiliation. Due to the multicultural nature of Indonesia, we have also taken into account ethnic diversity. Of these participants, a majority of 48.9% (270) were of Javanese descent. Other ethnicities consisted of 56 (10.1%) Sundanese, 33 (6.0%) Betawinese, 29 (5.3%) Chinese Indonesian, 29 (5.3%) mixed ethnicity, 24 (4.3%) Bataknese, 23 (4.2%) Minangnese, 11 (2.0%) Malay, and 10 (1.8%) Buginese. The rest of the participants (51 or 9.2%) were descendants of other ethnic groups, while 16 participants (2.9%) did not state their ethnicity. Overall, the proportion of participants was in line with the wider Indonesian population, in terms of those of Javanese descent being the ethnic majority (Na'im & Syaputra, 2010) and Islam being the religion followed by most people (see 2010 census data by *Badan Pusat Statistik Indonesia*). For the purpose of this study, we analysed only the Muslim participants: total = 453; female = 274; mean age = 28.51 (SD = 9.78). Among these Muslim participants, 77 stated that they had religious involvements while one participant did not state whether they were involved in any religious organisations.

2.2 *Materials*

All the questionnaires were administered in Indonesian language. We conducted back-translations of all the questionnaires in order to fulfil the requirements for cross-cultural adaptation (Beaton et al., 2000). The actual questionnaires were administered as a series of many scales measuring different variables. For the purpose of this study, we focused only on three variables: moral processing style, support for terrorism, and religious involvement.

Moral processing style was measured with the four items adapted from Young et al. (2013). Every item contained one scenario, and participants rated their agreements on a Likert scale (1 for Strongly inappropriate, to 7 for Strongly appropriate). A higher score indicates a stronger preference towards consequentialist moral processing. The reliability of the scale is quite satisfactory ($\alpha = 0.83$). One of the item samples was:

'A runaway trolley is heading down the tracks towards five workmen who will be killed if the trolley proceeds on its present course. You are on a footbridge over the tracks, in between the approaching trolley and the five workmen. Next to you on this footbridge is a stranger who happens to be very large. The only way to save the lives of the five workmen is to push this stranger off the bridge and onto the tracks below where his large body will stop the trolley. The stranger will die if you do this, but the five workmen will be saved. Is it appropriate for you to push the stranger on to the tracks in order to save the five workmen?'

Support for terrorism was measured by a single item adapted from Cherney and Povey (2013). The item is a question about how likely it is that someone will support a violent act in order to protect Islam. More specifically, the item is:

'Some people think that suicide bombing and other forms of violence against civilian targets are justified in order to defend Islam from its enemies. Other people believe that, no matter what the reason, this kind of violence is never justified. Do you personally feel that this

Table 1. Summary of the regression analysis of support for terrorism as a dependent variable and consequentialist moral processing as a predictor.

Conditions	B	Std. Error	Beta	T	R^2	F
All participants	0.02*	0.01	0.09	1.97	0.01	3.86*
Participants involved in religious organisations	0.04*	0.02	0.23	2.01	0.05	4.03*
Participants not involved in religious organisations	0.01	0.01	0.07	1.34	0.01	1.79

Note: *p 0.05 at one-tailed hypothesis.

kind of violence is often justified to defend Islam, sometimes justified, rarely justified or never justified?'

Participants responded to this question using the Likert scale (1 for 'It can never be justified', to 4 for 'It can often be justified'). A higher score indicates stronger support for terrorism.

Religious involvement was measured by a demographic question asking whether they were affiliated with any religious organisations at the time. The specific question is: 'Are you currently involved in any religious organisation?' The participants responded with either a 'yes' or 'no' dichotomous response. After that, they were asked the names of any organisations they were a part of. This question served as a control of whether they were actually in the organisation.

3 RESULTS

Of all the participants, the majority (62%) stated that terrorism can never be justified to defend their religion. The mean score for the four-item moral processing style was 13.22 (SD = 6.22), while the mean score for the terrorism-support item was 1.75 (SD = 1.16). Zero-order correlation analysis resulted in a positive correlation between scores for moral processing and terrorism support ($r = 0.09$, $p < 0.05$) and in an even higher positive correlation between scores for moral processing and terrorism support in those who are involved in religious organisations ($N = 77$, $r = 0.23$, $p < 0.05$). Consistently, we also found a weaker correlation between the two variables in those who are not involved in any religious organisations ($N = 375$, $r = 0.07$, $p > 0.05$). The more the individuals support terrorism, the stronger the degree of preference they have towards consequentialist moral processing. The result is stronger for those who are currently in religious organisations.

Results from the regression analysis confirm our one-tailed hypothesis. With terrorism support as a dependent variable, we found that a preference for consequentialist moral processing significantly predicts support for terrorism ($B = 0.02$, $p < 0.05$) and explains a significant proportion of the variance in support for terrorism scores (see Table 1). Consistent with our second hypothesis, a preference for consequentialist moral processing significantly predicts support for terrorism among those who are involved in religious organisations ($B = 0.04$, $p < 0.05$). A significant proportion of the variance in support for terrorism can also be explained with consequentialist moral processing (see Table 1) of those joining religious organisations.

4 DISCUSSION

The results suggest that how individuals process the moral dilemma predicts the degree of support for terrorism. Although religion is often associated with rule adherence in processing moral problems (Young et al., 2013; Hauser et al., 2007), the data shows an opposite trend where terrorism supporters tend to prefer a consequence-based approach to processing moral problems. This supports our first hypothesis in which terrorism is perceived more as tool than as meaningful ideological rule by its supporters. The data also supports the second hypothesis. We found a greater effect on those who are involved in religious organisations. This is consistent with the finding that group dynamics play an important process in shaping individuals' attitudes and support.

The findings are consistent with the assertions of Kruglanski and Fishman (2006) about terrorists and their supporters exerting 'the end justifies the means' morality. On the contrary, this data casts doubts on the claim that rules from ideologies adhered to by fundamentalists become the sole factor that shapes their supports for terrorism (Allen, 2009; Saroglou et al., 2009). Even though this data offers new stepping stones in understanding the relationship between morality and terrorism, it is far from comprehensive. Future research should also consider the role of religious fundamentalism in the dynamics of moral processing and support for terrorism. The research by Young et al. (2013) employed US Christian fundamentalists as its participants. It may shed new information if future research investigates the role of religious fundamentalism in those with different religious identities.

An act of terror is derived from a moral justification (Bandura, 1999) aimed at achieving political ends. We found that a preference for consequentialist moral thinking is helpful to predict the degree to which people think terrorism should be justified. With this in mind, those who tend to process the moral dilemma in the consequentialist way may be prone to supporting violence when it can be justified. Although our self-reported data positively confirms the hypotheses, we should note that the findings are predictive at best and contain no cause-and-effect relationship. Thus, one should be careful not to interpret consequentialist moral processing as the factor that causally affects support for terrorism. Future research should determine the exact relationship between the two variables.

Consequentialist morality itself has been viewed as beneficial for human cooperation and presumed to be superior to deontological morality (Greene, 2013). The findings reject this assertion that those who approach moral problems with consequentialist processing may justify violence in order to achieve the desired end. This indicates the dark side of consequentialist moral processing acting as a factor that may make individuals vulnerable to violence justification. The data from this study further supports the notion that Islam may not always go hand-in-hand with terrorism. Rather, the key point is how individuals process the moral dilemma inside their head. Note that the majority of Muslim participants do not support terrorism (62% stated that they would not support terrorism even if it is used to defend their religion). Thus, certain justifications are needed for Muslims to support terrorism. These justifications are enhanced by consequentialist moral processing.

The present study also highlights the role of groups in the relationship between moral processing and terrorism support. For those who belong to religious organisations, we found an even stronger correlation ($r = 0.23$) compared to those who are not affiliated to any religious organisations ($r = 0.07$) and all participants combined ($r = 0.09$). Following this result, group dynamics seem to be the key factor that enhances the strength of relationship between consequentialist moral processing and support for terrorism. Consistent with the previous assertion, people in religious groups tend to gradually adopt the existing beliefs or ideologies of the other group members. It is in prospect for future researchers to examine further what factors in groups may strengthen this relationship. Research on group identification (Pyszczynski et al., 2006) and quest for significance (Kruglanski et al., 2013) may lead to the desired answer.

Finally, future studies should also address the technical limitations in our study. One possible limitation could be the online survey that we administered. Even though it was efficient, we could not control how the participants filled in the questionnaires. Additionally, when we administered the in-person surveys, several participants reported difficulty in reading the statements, especially those with low educational backgrounds. Further research should consider the language barrier faced by those with lower levels of education.

5 CONCLUSION

Present study hypothesise that support for terrorism can be predicted by the degree of consequentialist moral processing. This relationship becomes stronger in those who join religious organisations. This study may shed light upon individuals who are more likely to support terrorism. However, future research is certainly needed in order to establish this claim.

REFERENCES

Allen, C. (2009). *God's terrorists: The Wahhabi cult and the hidden roots of modern jihad.* Philadelphia, PA: Da Capo Press.

Badan Pusat Statistik Indonesia. (2010). *The religion of Indonesia citizens.* Retrieved from http://sp2010.bps.go.id/index.php/site/tabel?tid=321andwid=0

Bandura, A. (1999). Moral disengagement in the perpetration of inhumanities. *Personality and Social Psychology Review, 3*(3), 193–209.

Beaton, D.E, Bombardier, C., Guillemin, F. & Ferraz, M.B. (2000). Guidelines for the process of cross-cultural adaptation of self-report measures. *Spine, 25*(24), 3186–3191.

Brandt, M.J. & Van Tongeren, D.R. (2015). People both high and low on religious fundamentalism are prejudiced toward dissimilar groups. *Journal of Personality and Social Psychology, 112*(1), 76–97.

Cherney, A. & Povey, J. (2013). Exploring support for terrorism among muslims. *Perspectives on Terrorism, 7*(3).

Crenshaw, M. (1981). The causes of terrorism. *Comparative Politics, 13*(4), 379–399.

deMause, L. (2002). The childhood origins of terrorism. *The Journal of Psychohistory, 29*(4), 340–348.

Greene, J. (2013). *Moral tribes: Emotion, reason and the gap between us and them.* New York, NY: The Penguin Press.

Hauser, M., Cushman, F., Young, L., Jin, R.K.X. & Mikhail, J. (2007). A dissociation between moral judgments and justifications. *Mind & Language, 22*(1), 1–21.

Horgan, J. (2008). From profiles to pathways and roots to routes: Perspectives from psychology on radicalization into terrorism. *The ANNALS of the American Academy of Political and Social Science, 618*(1), 80–94.

Johnson, K.A., Hook, J.N., Davis, D.E., Van Tongeren, D.R., Sandage, S.J. & Crabtree, S.A. (2016). Moral foundation priorities reflect U.S. Christians' individual differences in religiosity. *Personality and Individual Differences, 100*, 56–61.

Jonathan, E. (2008). The influence of religious fundamentalism, Right-Wing authoritarianism, and Christian orthodoxy on explicit and implicit measures of attitudes toward homosexuals. *The International Journal for the Psychology of Religion, 18*(4), 316–329.

Koleva, S.P., Graham, J., Iyer, R., Ditto, P.H. & Haidt, J. (2012). Tracing the threads: How five moral concerns (especially purity) help explain culture war attitudes. *Journal of Research in Personality, 46*(2), 184–194.

Kruglanski, A.W., Bélanger, J.J., Gelfand, M., Gunaratna, R., Hettiarachchi, M., Reinares, F. & Sharvit, K. (2013). Terrorism—a (self) love story: Redirecting the significance quest can end violence. *American Psychologist, 68*(7), 559–575.

Kruglanski, A.W. & Fishman, S. (2006). The psychology of terrorism: 'Syndrome' versus 'tool' perspectives. *Terrorism and Political Violence, 18*(2), 193–215.

Kruglanski, A.W. & Fishman, S. (2009). Psychological factors in terrorism and counterterrorism: Individual, group, and organizational levels of analysis. *Social Issues and Policy Review, 3*(1), 1–44.

Lombrozo, T. (2009). The role of moral commitments in moral judgment. *Cognitive Science, 33*(2), 273–286.

Martensson, U., Bailey, J., Ringrose, P. & Dyrendal, A. (2011). *Fundamentalism in the modern world volume 1, fundamentalism, politics and history: The State, globalization, and political ideologies.* New York, NY: I.B. Tauris.

Na'im, A. & Syaputra, H. (2010). *Citizenship, Ethnicity, Religion, and Language among Indonesian Population: Results of the 2010 Census.* Indonesia: Badan Pusat Statistik.

O'Leary, Brendan (2007). IRA: Irish Republican Army (Oglaigh na hEireann). In Hieberg, M., O'Leary, B. & Tirman, J. (Eds.), *Terror, insurgency and the State* (pp. 189–228). Philadelphia, PA: University of Pennsylvania Press.

Pyszczynski, T., Abdollahi, A., Solomon, S., Greenberg, J., Cohen, F. & Weise, D. (2006). Mortality salience, martyrdom, and military might: The great Satan versus the axis of evil. *Personality and Social Psychology Bulletin, 32*(4), 525–537.
Pyszczynski, T.A., Solomon, S. & Greenberg, J. (2002). *In the wake of 9/11: The psychology of terror*. Washington, DC: American Psychological Association.
Rogers, M.B., Loewenthal, K.M., Lewis, C.A., Amlôt, R., Cinnirella, M. & Ansari, H. (2007). The role of religious fundamentalism in terrorist violence: A social psychological analysis. *International Review of Psychiatry, 19*(3), 253–262.
Sageman, M. (2005). Understanding terror networks. *International Journal of Emergency Mental Health, 7*(1), 5–8.
Saroglou, V., Corneille, O. & Cappellen, P.V. (2009). Speak, Lord, your servant is listening: Religious priming activates submissive thoughts and behaviors. *International Journal for the Psychology of Religion, 19*(3), 143–154.
Schaafsma, J. & Williams, K.D. (2012). Exclusion, intergroup hostility, and religious fundamentalism. *Journal of Experimental Social Psychology, 48*(4), 829–837.
Silke, A. (2008). Holy warriors: Exploring the psychological processes of jihadi radicalization. *European Journal of Criminology, 5*(1), 99–123.
Wylie, L. & Forest, J. (1992). Religious fundamentalism, right-wing authoritarianism and prejudice. *Psychological Reports, 71*, 1291–1298.
Young, O.A., Willer, R. & Keltner, D. (2013). Thou shalt not kill: Religious fundamentalism, conservatism, and rule-based moral processing. *Psychology of Religion and Spirituality, 5*(2), 110–115.

Diversity in Unity: Perspectives from Psychology and Behavioral Sciences – Ariyanto et al. (Eds)
© 2018 Taylor & Francis Group, London, ISBN 978-1-138-62665-2

The need for cognitive closure and belief in conspiracy theories: An exploration of the role of religious fundamentalism in cognition

A.N. Umam, H. Muluk & M.N. Milla
Faculty of Psychology, Universitas Indonesia, Depok, Indonesia

ABSTRACT: Latest research shows that belief in conspiracy theories as an ideological trait is manifested in both the Need For Closure (NFC) and religious fundamentalism. There are some indications that the need for closure and religious fundamentalism interact in predicting certain ideological traits. This research aims to explore the interaction between NFC and fundamentalism in predicting one's belief in conspiracy theories. By gathering data from 211 participants, the results show that NFC can predict the extent of people's belief in conspiracy theory but only in control of information conspiracy theme, a conspiracy notion that says there is a scheme to conceal important information from public. There is no interaction between NFC and fundamentalism in predicting belief in conspiracy theories. The findings of this research theoretically contribute to the role of NFC in the manifestation of cognitive traits.

1 INTRODUCTION

While there are various conceptualisations of simple thinking for making sense of the world, there is a specific concept for the belief that there are unknown entities that control world events, known as conspiracy theories (Januszewski & Koetting, 1998). Moreover, conspiracy beliefs seem to serve political purposes, such as protecting one's own social identity. In Western countries, individuals on the far right or far left of the political spectrum are prone to believe in conspiracy theories (Van Prooijen, 2015), as well as having the conviction that a simple solution is capable of solving complex political problems. Whereas in the Arab region, ideological inchoateness and the aimlessness of the society have made conspiracy theories a common popular explanation of many phenomena (Gray, 2011) where people are predisposed to blame Western countries for many misfortunes occurring in their home countries (Gray, 2010).

Meanwhile in Indonesia, conspiracy theories exist mainly in notorious issues that are related to Islam, such as the accusation that presidential candidate Joko Widodo was backed by Western countries, allied himself with the Vatican, was of Chinese descent, and thus was believed to have some sort of plan for exploiting Indonesia's natural resources and people to serve those particular parties suspected as behind him (Halim, 2014). Another conspiracy narrative became widespread when it was suggested that a famous mobile game, Pokemon GO, had a hidden agenda to glorify Jews and secretly derogate Islam (Romdlon, 2016); it was suggested that some characters in the game were named with an ancient language that meant 'I am a Jew'. The particular narratives of conspiracy theories in Indonesian Muslim society resembled global phenomena, it can be seen on the suspicion towards outgroup which states a political standing in conspiracy theory (Van Prooijen, 2015), and the theories are also used as a simplified attempt to comprehend ambiguous situation (Gray, 2011).

Research on conspiracy beliefs has not yet reached a unified conclusion, since there are various conspiracies and numerous societies that uniquely believe in certain kinds of conspiracy. Nevertheless, most researchers agree that a belief in conspiracy theories be considered as individual differences in which some people are less prone to it and some people are more likely to subscribe to it (Goertzel, 1994; Swami et al., 2011; Brotherton et al., 2013). It has been suggested that the belief derives from a limitation in thinking processes (Drinkwater et al., 2012; Landau et al., 2015). Furthermore, most findings on the topic show that conspiracy theories predominantly have polarised political standpoints, which are expressed by suspicion or disbelief of other groups (Bessi et al., 2015; Van Prooijen, 2015).

Regarding the cognitive antecedents of belief in conspiracy theories, an individual's need for closure to make sense of the world (Webster & Kruglanski, 1994) has been found to be linked to a belief in conspiracy theories (Landau et al., 2015), and there is a link between religious fundamentalism and the need for cognitive closure in predicting prejudice towards an outgroup (Brandt & Reyna, 2010). However, despite the fact that religion and fundamentalism are frequently associated with conspiracy thinking in political studies (Al-Azm, 2011; Gray, 2011), there has not yet been any psychological investigation that addresses fundamentalism as one of the contributing factors in conspiracy beliefs. With reference to Brandt and Reyna's (2010) findings, this research aims to investigate the interaction between the need for closure and religious fundamentalism in predicting conspiracy beliefs.

1.1 *Need for closure and conspiracy beliefs*

Belief in conspiracy theories was previously discussed as an alarming issue in political studies that started at the beginning of the information era when people around the world could access information and discuss the content openly in public (Januszewski & Poetting, 1998). With reference to a specific region, conspiracy theories became a popular explanation for political occurrences in Arab society when the region experienced series of foreign interventions. Moreover, conspiracy theories in this region are also used as instruments to mobilise the masses in order to fight certain political opponents (Gray, 2010; Gray, 2011).

An early discovery in psychological discussion by Goertzel (1994) suggests and confirms that belief in conspiracy theories is a generalised ideological trait, which all people have to a certain extent. In the same article, Goertzel also found that people who tended to believe in conspiracy theories were youths and minorities who were considered to have a low capacity to understand their surroundings and felt threatened by current conditions. Regarding the political nuance in conspiracy theories, Van Prooijen (2012) suggests that the belief is a desire mechanism to make sense of a phenomenon or a situation that has threatened one's social foundation. Therefore, people who believe in one conspiracy theory are prone to believe in other conspiracy theories. Brotherton et al. (2013) confirm this notion and have formulated a set of theories in a *Generic Conspiracist Belief* (GCB) scale, which reliably divides those who tend to believe in it and those who do not.

In the *lay epistemic theory,* Kruglanski (2011) states that people inherently desire to understand themselves and their surroundings. Among several aspects in making sense of the world, such as content, ability and motivation, Kruglanski (1989) highlights the motivation aspect as a basis for the construct of *Need For Closure* (NFC), a construct that measures the extent of need for answers about a particular topic in the face of confusion and ambiguity, and this desire is an individual unique feature that is mostly stable in each person (Webster & Kruglanski, 1994). This particular construct embodies the spectrum of people's desire in relation to definitive conclusions on given problems.

Research findings on the need for cognitive closure foreshadow its relationship with belief in conspiracy theories. NFC has been found to correlate positively with conservative values including security, conformity, and tradition, while it also correlates negatively with an openness to change values including stimulation and self-direction (Calogero et al., 2009). It is also positively related to ideological traits that have predisposition to conformity towards collective norms such as political conservatism (Jost et al., 2003; De Zavala & Van Bergh, 2007), religiosity (Saroglou, 2002), and religious fundamentalism (Brandt & Reyna, 2010). This frequent connection between NFC and conservatism, especially in religious matters, reflects how conspiracy theories are used in some Muslim societies to comprehend things that might threaten them.

As presented earlier in this article, common conspiracy theories mostly demonstrate how a group is being threatened by outside entities, and also denote how certain conspiracy narratives are believed by a certain group of people with similar beliefs. Research on the need for cognitive closure indeed shows that an individual with a high NFC might view a homogenous group as a closure provider where they can find uniformity and stability (Kruglanski et al., 2002, 2006). Moreover, people who are likely to believe in conspiracy theories do not bother to actually understand what has happened; they are motivated to know what has happened, but then an information deficit leads them to bolster their self-agency and seek a clear explanation of action-outcome even if the explanation is perfunctory (Landau et al., 2015).

Group salience in general, and political standing in particular, seems to be the strongest issues in conspiracy theory compared to other themes. A corpus analysis of Facebook comments in Italian shows that people are most persistent in commenting on geopolitical conspiracies compared to environment, diet and health issues (Bessi et al., 2015). In Malaysia, belief in a common Jewish conspiracy to exploit Malaysian resources and people was indicated as an attempt to serve ideological needs; it portrayed society's perceived ability to understand threat and to shape socio-political dynamics within (Swami, 2012). Thus, it can be inferred that conspiracy theories are the expression of cognitive closure, and they are manifested notoriously on religious and political topics as has happened in some Muslim societies (Al-Azm, 2011; Gray, 2010; Swami, 2012; Muslimin, 2011; Romdlon, 2016). It can be concluded that there is a stable connection between the need for closure and belief in conspiracy theories, but frequent religious influence in the connection is yet to be established.

1.2 *Need for closure and fundamentalism in conspiracy beliefs*

For specific cases in the Arab region, or in Muslim society in general, discussions in political studies have identified fundamentalism as the primary contributor to belief in conspiracy theories. Religious fundamentalism was believed to be the fuel of hatred towards Western culture and society, which later fostered conspiracy theories in Arab countries (Al-Azm, 2011; Gray, 2010) as well as in the Muslim society in Indonesia (Muslimin, 2011; Pringle, 2010). Although the role is not definitive, the rise of Muslim conservatism in Indonesia (Ichwan, 2014) might point to a cultural specificity between Indonesia and Arab countries. An investigation in the neighbouring country of Malaysia indicated that conspiracy beliefs in relation to Jews were a form of ideological need (Swami, 2012) and it hinted that religious fundamentalism was also an ideological need that led to conspiracy beliefs.

The term 'religious fundamentalism' itself has roots in Protestant Christianity's movement in the beginning of the 19th century that promoted the belief in the divinity of a sacred book and was heavily against any opposing ideas (Altemeyer & Hunsberger, 2005). It later becomes a generalised term that describes religious groups' resistance towards anything that disturbs the tradition of their holy scriptures (Hood et al., 2005). Nevertheless, not all religious organisations agree with this definition; Muslim scholars object to the label of fundamentalism because if fundamentalism is defined as the belief that the holy book is the word of god then all Muslims who base their faith on the word of Allah in the Quran would be called fundamentalists (Ruthven, 2004). Hood et al. (2005) formulated the term 'intratextual fundamentalism', which describes the tendency of an individual to make sense of the world and form knowledge through their religion's absolute truth that is written in the religion's holy scripture. More specifically, intratextual fundamentalism is a reinstatement of the original concept as an attitude towards the sacred texts, and it has been found to be psychometrically valid and reliable in various religions in various countries (Williamson et al., 2010).

Similar to the need for cognitive closure, the construct of religious fundamentalism was frequently related to conservatism and ethnocentrism. A meta-analysis of 28 studies found that fundamentalism consistently has a close relationship with authoritarianism (McCleary et al., 2011), which means that traditionalism in religion goes hand in hand with submission to perceived legitimate authorities. A study in an Islamic society shows that fundamentalism acted as a basic identity that dictates one's decision making, any cost-benefit analysis in decision making only allowed within the boundary of one's religious worldview (Monroe &

Kreidie, 1997). Furthermore, a study in a Christian society shows that fundamentalism, along with political conservatism, leads individuals to rely on their religion to formulate their moral views and refuse to accept other points of view (Young, Willer & Keltner, 2013).

Religious fundamentalism is frequently found to be related to the desire for closure. When faced with death, fundamentalists trust in their prayers and faith more than in medical treatment because they believe that the religious approach is more effective (Vess et al., 2009). Fundamentalists are known to rely for their personal opinion on authorities (Altemeyer & Hunsberger, 2005), and they also have been known to restrain their personal thoughts by relying on the internalisation of explicit information provided by preachers in religious sermons (Schjoedt et al., 2013). Religious fundamentalism has indeed a direct positive relationship with need for closure, moreover, there was also an interaction between fundamentalism and the need for cognitive closure in predicting prejudice towards outgroups (Brandt & Reyna, 2010). It suggested that there is also an interaction between need for closure and fundamentalism in one's belief in conspiracy theories.

1.3 *The current research*

Regarding the analysis of previous research, the key to predicting the degree of belief in conspiracy theories may lay in the cognitive motivation to reach closure. Moreover, findings in political studies indicate that religious fundamentalism may serve as a facilitator, or as a variable that may elevate the relationship between the need for cognitive closure and belief in conspiracy theories, as can be seen in psychological research where people with a high need for closure are prone to believe in conspiracy theories when they are provided with a source of closure such as sacred texts. We believe that religious fundamentalism per se is not enough to predict a belief in conspiracy theories, as the latest understanding states that fundamentalism is only manifested in the preservation of sacred text-based thinking and not cognitive dynamics or even closed-mindedness (Hood et al., 2005).

The current research focuses on Muslim society in Indonesia where a widespread belief in conspiracy theories with religious narratives occurs (Mashuri & Zaduqisti, 2013). On this basis, we hypothesised that *Need For Closure* (NFC) is a significant predictor of *Generic Conspiracist Belief* (GCB). We also hypothesised that *Intratextual Fundamentalism* (IF) plays a role as a moderator of the relationship between NFC and GCB.

2 METHOD

2.1 *Participants and procedure*

The *Jajak Pendapat* (*Jakpat*) survey application was used to recruit 211 people from the province of Jakarta and its surrounding provinces, West Java and Banten. The survey was open to the public, and the participants received credits that could be used for online purchases. There were 111 males (52.4%) and 100 females (47.6%) with an age range of 18 to 47, and all participants were Muslim. Of the participants, 117 (55.45%) were high school graduates and 93 (44.08%) were college graduates, and one participant did not mention their level of education. The participants each filled out a set of questionnaires with a total of 25 questions that assessed their *Need For Closure* (NFC), *Intratextual Fundamentalism* (IF), and *Generic Conspiracist Belief* (GCB).

The set of questionnaires consisting of the three variables described above was registered in the *Jakpat* survey application with an expectation of 200 participants. Anyone with a *Jakpat* user account could participate in the survey as long as the maximum number of participants had not been exceeded and the participant fitted the demographic target, which was Muslims in DKI Jakarta, Banten and West Java provinces. Participants were then instructed to read the instructions to rate all the statements in the questionnaire from 1 (*Strongly disagree*) to 6 (*Strongly agree*) for all 25 statements. The questionnaire started with the assessment of GCB, followed by the assessment of NFC and IF respectively. After finishing all of the items, participants were given digital points in the *Jakpat* application, which could be redeemed later for digital transactions. The application administrator closed the survey when the number of participants reached 211.

2.2 *Materials*

Generic Conspiracist Belief (GCB). The scale was developed to summarise the main themes in existing conspiracy theories. This originally consisted of five main themes: government malfeasance, extraterrestrial cover-up, malevolent global conspiracy, personal well-being, and control of information (Brotherton et al., 2013). The scale was adapted to the Indonesian language with the exception of the extraterrestrial cover-up theme because it was considered irrelevant to the existing Indonesian Muslim conspiracy narratives. The scale measured the degree of participants' agreement to statements such as 'Groups of scientists manipulate, fabricate, or suppress evidence in order to deceive the public' in the control of information theme and 'A small, secret group of people is responsible for making all major world decisions, such as going to war' in the global malevolence theme. The Indonesian language-adapted scale used in this research has a very good reliability of Cronbach $\alpha = 0.91$.

Need For Closure (NFC). The original scale of *need for cognitive closure* developed by Webster and Kruglanski (1994) consisted of five facets in the construct, which were preference for order, dislike of ambiguity, decisiveness, preference for credibility, and close-mindedness. After years of objections to the trait of decisiveness, because it was considered as an ability and not relevant to describe motivation, Roets and Van Hiel (2007) redeveloped the scale with a revised decisiveness element. This research used an Indonesian adaptation of Roets and Van Hiel's (2007) eight-item scale that was developed by Bagaskara (2009); two items were later removed to maintain the scale's overall reliability. The items in this scale measured participants' agreement with statements such as 'I think that having clear rules and order at work is essential for success' in the preference for order element. The scale has a fair reliability with Cronbach $\alpha = 0.59$.

Intratextual Fundamentalism (IF). This scale consisted of five elements and consequently five items, which represent each element, such as divine, inerrant, privileged, authoritative, and unchanging (Hood et al., 2005; Williamson et al., 2010). In this study, we used an Indonesian language adaptation of the scale that was also developed by Bagaskara (2009); there were some changes in the wording including the specification of the Quran as a sacred text in the scale. The scale assessed participants' agreement with statements such as 'Everything written in the Quran is an absolute truth without question' in the inerrant element and 'The truth in the Quran will never be outdated, but will always apply equally well to all generations' in the unchanging element. The item that represented the divine element was removed to drastically improve the scale's reliability. The final IF scale used in this research has a good reliability with Cronbach $\alpha = 0.78$.

3 RESULTS

3.1 *Main analysis*

To investigate the conceptual framework that NFC accounted for both IF and GCB, we conducted Pearson correlation analysis on all variables. The result showed that neither NFC, $r(199) = 0.13$, $p = 0.06$, nor IF, $r(199) = 0.04$, $p = 0.53$, was significantly correlated with GCB. Nevertheless, in a more specific evaluation, the 'control of information' theme was positively correlated with NFC, $r(199) = 0.18$, $p = 0.01$. IF was also found to be positively correlated with NFC, $r(199) = 0.27$, $p = 0.00$, and there was no significant correlation between IF and the control of information conspiracy $r(199) = 0.06$, $p = 0.43$. The correlation analysis confirmed the current conceptual framework: that the need for closure is positively linked with both religious fundamentalism and belief in conspiracy theories, and that religious fundamentalism per se was inadequate to explain the tendency towards conspiracy thinking.

Table 1. Correlation between measures.

Measure	GCB	NFC	IF	GM	MG	PW	CI
Generic conspiracy belief		0.127	0.044	0.876**	0.860**	0.892**	0.894**
Need for cognitive closure	0.127		0.269**	0.031	0.134	0.114	0.180**
Intratextual fundamentalism	0.044	0.269**		0.037	0.032	0.032	0.055

Note: *$p < 0.01$; **$p < 0.001$. GCB = Generic conspiracy belief; NFC = Need for cognitive closure; IF = Intratextual fundamentalism; GM = Government malfeasance; MG = Global malevolence; PW = Personal well-being; CI = Control of information.

Regarding the preliminary correlation analysis, we focused the conspiracy evaluation on the *control of information* conspiracy theme as a criterion variable. In order to test the first hypothesis, a linear regression was conducted with NFC as a predictor and control of information conspiracy theme as a criterion. The result showed that NFC was a significant predictor of conspiracy belief in the control of information theme, $b = 0.14$, $t(199) = 2.65$, $p = 0.00$, need for closure also explained a significant proportion of variance in information control conspiracy belief, $R^2 = 0.03$, $F(1, 199) = 7.03$, $p = 0.01$.

3.2 *Moderator analysis*

A moderated multiple regression analysis was conducted to test the attribution of NFC and the moderating effect of IF on the variability of control of information conspiracy beliefs. Referring to the interaction terminology of Aiken and West (1991), the mean of NFC and IF score was centred, and both predictors were put simultaneously in the first block of hierarchical multiple regression analysis and interaction between the two in the second block. Results indicated that only NFC significantly predicted control of information conspiracy belief, $b = 0.13$, $t(199) = 2.52$, $p = 0.00$, while IF did not, $b = -0.05$, $t(199) = 0.92$, $p = 0.93$. Subsequently, there was no interaction between NFC and IF in predicting belief in a control of information conspiracy, $b = -0.02$, $t(198) = -1.75$, $p = 0.08$.

4 DISCUSSION

Research in psychology has shown the magnitude of the need for cognitive closure in many manifestations of thinking, where abstract and careful thinking reflected a low need for cognitive closure, while concrete and instant thinking reflected a high need for cognitive closure. In this article, we investigated how the need for closure manifested itself in a belief in conspiracy theories and how it interacted with fundamentalism in predicting conspiracy thinking. The result showed that a need for closure was able to significantly predict a belief in conspiracy theories, but only in the control of information theme. Religious fundamentalism was unable to predict a belief in conspiracy theories, and there was no interaction between need for closure and religious fundamentalism in predicting belief in conspiracy theories, either. However, a closer look at the result gives us some insights regarding previous findings on conspiracy theories.

Our findings highlighted the existence of conspiracy theory beliefs in Indonesia. As seen from the correlation among themes, people who believe in one conspiracy theory tend to believe in other conspiracy theories. This is consistent with early psychology conceptualisations of belief in conspiracy theories as an ideological trait (Goertzel, 1994). The belief seems to exist as a global phenomenon as it has been found in various kinds of society (Bessi et al., 2015; Swami 2012; Van Prooijen, 2015) and does not exclusively occur in conflict areas such as the Middle East region as explained by Al-Azm (2011) and Gray (2010). However, the result showed that religious fundamentalism did not predict a belief in conspiracies, nor did it interact with NFC. Although religious fundamentalism was frequently found to be related to discrimination against outgroups (Brandt & Reyna, 2010), the suspicion towards an outgroup was non-existent when there is no definitive target group as suspicion object.

Furthermore, the 'control of information' conspiracy theme was the only theme that was significantly predicted by need for cognitive closure. It strengthens the notion that belief in conspiracies served as a defence mechanism for someone who tries to make sense of a phenomenon or situation (Van Prooijen, 2012). When the person feels that the information available to understand a particular topic is limited, they can fall into simple explanations and then confidently hold them (Landau et al., 2015).

Although religious fundamentalism and belief in conspiracy theories were both positively related to the need for cognitive closure, people with a high level of fundamentalism did not necessarily believe in conspiracy theories. Despite the fact that belief in one conspiracy will lead to belief in another conspiracy, the prevalence of conspiracy in a society was apparently thematic. It can be seen in Swami's (2012) findings on conspiracy belief in Malaysia; the research found that general conspiracy ideation did not predict a belief in the Jewish conspiracy. The research also suggested that the Jewish conspiracy was a manifestation of racial tension within the Malay ethnic group (the participants in the research) towards the Chinese ethnic group because the conspiracy was associated with modern racism towards Chinese, right-wing authoritarianism, and social dominance orientation. Therefore, it can be assumed that Jewish conspiracies, which frequently occurs in the Muslim society in Indonesia, has similar patterns to Malaysia: that it was a manifestation of other variables and not merely a generic conspiracy belief. Nevertheless, the main limitation in this research was the absence of a Jewish conspiracy, the recurring conspiracy theme in Indonesia. Further research needs to address this specific conspiracy theme to properly explain the phenomena and achieve a better comprehension of belief in conspiracy theory conceptualisation. There seems to be different cognitive patterns of believing unreliable information in general conspiracy beliefs and certain thematic conspiracies, especially when it is related to religion narratives. Some studies indicated that religious narratives activated certain kinds of reasoning that are different from one's normal reasoning (Halverson et al., 2011; Schjoedt et al., 2013). Subsequently, religion-based conspiracies might well have different cognitive antecedents compared to general conspiracy theories.

Finally, it is also important for further studies in psychology and other disciplines to address religious fundamentalism attentively. Despite the fact that religious fundamentalism is frequently related to other conservative traits, this particular sacred text-based thinking might have different behaviour manifestations and attitude expressions compared to other conservative traits.

REFERENCES

Aiken, L.S. & West, S.G. (1991). *Multiple regression: Testing and interpreting interactions* (p. 224). Thousand Oaks, CA: SAGE Publications.

Al-Azm, S.J. (2011). Orientalism and conspiracy. In Graf, A., Fathi, S. & Paul, L. (Eds.), *Orientalism and conspiracy: Politics and conspiracy theory in the Islamic world*. London, UK: I.B. Tauris.

Altemeyer, B. & Hunsberger, B. (2005). Fundamentalism and authoritarianism. In *Handbook of the psychology of religion and spirituality* (pp. 378–393). New York, NY: the Guilford Press.

Antonenko, Y., Olga, R.W. & Keltner, D. (2013). Thou shalt not kill: Religious fundamentalism, conservatism, and rule-based moral processing. *Psychology of Religion and Spirituality*, *5*(2), 110–115.

Bagaskara, S. (2009). *Fundamentalisme and closed-mindedness: Religiosity, intolerance of uncertainty, and need for closure in religious fundamentalism* (Unpublished master's thesis, Universitas Indonesia, Depok, Indonesia).

Bessi, A., Zollo, F., Vicario, M.D., Scala, A., Caldarelli, G. & Quattrociocchi, W. (2015). Trend of narratives in the age of misinformation. *PLoS ONE*, *10*(8).

Brandt, M.J. & Reyna, C. (2010). The role of prejudice and the need for closure in religious fundamentalism. *Personality & Social Psychology Bulletin*, *36*(5), 715–725.

Brotherton, R., French, C.C. & Pickering, A.D. (2013). Measuring belief in conspiracy theories: The generic conspiracist beliefs scale. *Frontiers in Psychology*, *4*, 279.

Calogero, R.M., Bardi, A. & Sutton, R.M. (2009). A need basis for values: Associations between the need for cognitive closure and value priorities. *Personality and Individual Differences, 46*(2), 154–159.

De Zavala, A.G. & Van Bergh, A. (2007). Need for cognitive closure and conservative political beliefs: Differential mediation by personal worldviews. *Political Psychology, 28*(5), 587–608.

Drinkwater, K., Dagnall, N. & Parker, A. (2012). Reality testing, conspiracy theories and paranormal beliefs. *The Journal of Parapsychology, 76*(1), 57–77.

Goertzel, T. (1994). Belief in conspiracy theories. *Political Psychology, 15*(4), 731–742.

Gray, M. (2010). *Conspiracy theories in the Arab world: Sources and politics*. New York, NY: Routledge.

Gray, M.M. (2011). Political culture, political dynamics, and conspiracy in the Arab Middle East. In Graf, A., Fathi, S. & Paul L. (Eds.), *Orientalism and conspiracy* (pp. 105–128). London, UK: I.B. Tauris.

Halim, A. (2014). Panggil aku Wie Jo Koh alias Jokowi, antek asing dan aseng [Call me Wie Jo Koh a.k.a. Jokowi, the agent of foreign countries and China]. *VOA Islam*. Retreived from http://www.voa-islam.com/read/liberalism/2014/05/20/30232/panggil-aku-wie-jo-koh-alias-jokowi-antek-asing-dan-aseng/#sthash.8rcjFzjY.dpbs

Halverson, J.R, Goodall, H.L. & Corman, S.R. (2011). *Master narratives of Islamist extremism*. New York, NY: Palgrave Macmillan.

Hood, R.W., Jr., Hill, P.C. & Williamson, W.P. (2005). *The psychology of religious fundamentalism*. New York, NY: Guilford Press.

Ichwan, M.N. (2014). Menuju Islam moderat puritan: Majelis Ulama Indonesia dan politik ortodoksi keagamaan [Into puritan moderate Islam: Indonesian Ulema Council and the politics of religious orthodoxy]. In Van Bruinessen, M. (Ed.), *Conservative turn: Islam Indonesia dalam ancaman fundamentalisme* (pp. 101–156). Bandung, Indonesia: Mizan.

Januszewski, A. & Koetting, J.R. (1998). Debunking the conspiracy theory: Understanding ideology in explanations in historical studies of educational technology. *Tech Trends, 43*(1), 33–36.

Jost, J.T., Glaser, J., Kruglanski, A.W. & Sulloway, F.J. (2003). Political conservatism as motivated social cognition. *Psychological Bulletin, 129*(3), 339–375.

Kruglanski, A.W. (1989). *Lay epistemics and human knowledge: Cognitive and motivational bases*. New York, NY: Springer Science+Business Media.

Kruglanski, A.W. (2004). *The psychology of closed mindedness*. New York, NY: Psychology Press.

Kruglanski, A.W. (2011). Lay epistemic theory. In Van Lange, P.A.M., Kruglanski, A.W. & Higgins E.T. (Eds.), *Theories of social psychology* (Vol. 1, pp. 201–223). London, UK: SAGE Publications.

Kruglanski, A.W., Shah, J.Y., Pierro, A. & Mannetti, L. (2002). When similarity breeds content: Need for closure and the allure of homogeneous and self-resembling groups. *Journal of Personality and Social Psychology, 83*(3), 648–662.

Kruglanski, A.W., Pierro, A., Mannetti, L. & De Grada, E. (2006). Groups as epistemic providers: Need for closure and the unfolding of group-centrism. *Psychological Review, 113*(1), 84–100.

Landau, M.J., Kay, A.C. & Whitson, J.A. (2015). Compensatory control and the appeal of a structured world. *Psychological Bulletin, 141*(3), 694–722.

Mashuri, A. & Zaduqisti, E. (2013). The role of social identification, intergroup threat, and out-group derogation in explaining belief in a conspiracy theory about terrorism in Indonesia. *International Journal of Research Studies in Psychology, 3*, 35–50.

McCleary, D.F., Quillivan, C.C., Foster, L.N. & Williams, R.L. (2011). Meta-analysis of correlational relationships between perspectives of truth in religion and major psychological constructs. *Psychology of Religion and Spirituality, 3*(3), 163–180.

Monroe, K.R. & Kreidie, L.H. (1997). The perspective of Islamic fundamentalists and the limits of rational choice theory. *Political Psychology, 18*(1), 19–43.

Muslimin, J.M. (2011). Polemics on "orientalism" and "conspiracy" in Indonesia: A survey of public discourse on the case of JIL vs. DDII (2001–2005). In Graf, A., Fathi, S. & Paul L. (Eds.), *Orientalism and conspiracy* (pp. 129–140). London, UK: I.B. Tauris.

Pringle, R. (2010). *Understanding Islam in Indonesia*. Singapore: Editions Didier Millet.

Roets, A. & Van Hiel, A. (2007). Separating ability from need: Clarifying the dimensional structure of the need for closure scale. *Personality and Social Psychology Bulletin, 33*(2), 266–280.

Romdlon, N. (2016). Apakah benar Pokemon berarti aku Yahudi? Ini penjelasannya. [Is it true that "Pokemon" means I am Jew? The explanation]. *Brilio.net*. Retrieved from https://www.brilio.net/gadget/apakah-benar-pokemon-berarti-aku-yahudi-ini-penjelasannya-1607183.html

Ruthven, M. (2004). *Fundamentalism: The search for meaning*. New York, NY: Oxford University Press.

Saroglou, V. (2002). Beyond dogmatism: The need for closure as related to religion. *Mental Health, Religion & Culture, 5*(2), 183–194.

Swami, V. (2012). Social psychological origins of conspiracy theories: The case of the Jewish conspiracy theory in Malaysia. *Frontiers in Psychology*, *3*, 280.
Swami, V., Coles, R.V., Stieger, S., Pietschnig, J., Furnham, A., Rehim, S. & Voracek, M. (2011). Conspiracist ideation in Britain and Austria: Evidence of a monological belief system and associations between individual psychological differences and real-world and fictitious conspiracy theories. *British Journal of Psychology*, *102*(3), 443–463.
Schjoedt, U., Sørensen, J., Nielbo, K.L., Xygalatas, D., Mitkidis, P. & Bulbulia, J. (2013). Cognitive resource depletion in religious interactions. *Religion, Brain & Behavior*, *3*(1), 39–86.
Swami, V. & Coles, R. (2010). The Truth is out there. *The Psychologist*, *23*(7), 560–563.
Van Prooijen, J.W. (2012). Suspicions of injustice: The sense-making function of belief in conspiracy theories. In Kals, E. & Maes, J. (Eds.), *Justice and conflicts: Theoretical and empirical contributions* (pp. 121–132). Berlin, Germany: Springer.
Van Prooijen, J.W. (2015, February 24). Voters on the extreme left and right are far more likely to believe in conspiracy theories. *LSE EUROPP Blog.* Retrieved from http://blogs.lse.ac.uk/europpblog/2015/02/24/voters-on-the-extreme-left-and-right-of-the-political-spectrum-are-far-more-likely-to-believe-in-conspiracy-theories/
Vess, M., Arndt, J., Cox, C.R., Routledge, C. & Goldenberg, J.L. (2009). Exploring the existential function of religion: The effect of religious fundamentalism and mortality salience on faith-based medical refusals. *Journal of Personality and Social Psychology*, *97*(2), 334–350.
Webster, D.M. & Kruglanski, A.W. (1994). Individual differences in need for cognitive closure. *Journal of Personality and Social Psychology*, *67*(6), 1049–1062.
Williamson, W.P., Hood, R.W., Ahmad, A., Sadiq, M. & Hill, P.C. (2010). The intratextual fundamentalism scale: Cross-cultural application, validity evidence, and relationship with religious orientation and the big 5 factor markers. *Mental Health, Religion & Culture*, *13*(7–8), 721–747.
Young, O.A., Willer, R., & Keltner, D. (2013). "Thou shalt not kill": Religious fundamentalism, conservatism, and rule-based moral processing. *Psychology of Religion and Spirituality, 5*(2), 110–115, doi: 10.1037/a0032262

[illegible] (2005). [illegible] adjacent [illegible] conditions [illegible] In [illegible] Journal of [illegible]
[illegible] ... *Biological Psychology* [illegible]
[illegible]
[illegible] (2011). [illegible]
[illegible]

Diversity in Unity: Perspectives from Psychology and Behavioral Sciences – Ariyanto et al. (Eds)
© 2018 Taylor & Francis Group, London, ISBN 978-1-138-62665-2

Non-normative collective action in Muslims: The effect of self-versus group-based emotion

M.A. Shadiqi & H. Muluk
Faculty of Psychology, Universitas Indonesia, Depok, Indonesia

M.N. Milla
Faculty of Psychology, Islamic State University Sultan Malik Kasim, Riau, Indonesia

J. Hudiyana & A.N. Umam
Faculty of Psychology, Universitas Indonesia, Depok, Indonesia

ABSTRACT: This research examines the relationship between self- and group-based emotions and non-normative collective action. We hypothesise that self-based emotions (anger, contempt, pride and bravery) do not affect non-normative collective action, but group-based emotions (anger, contempt, pride and bravery) do. We also hypothesise that contempt and bravery are predictors of non-normative collective action. Participating in this study were 462 Muslim students, 292 of whom affiliate to a particular Islamic group. The research results support the hypotheses and show that, among Muslim students, non-normative collective action is significantly predicted by group-based emotion rather than by self-based emotion. Specifically, group-based contempt significantly predicts non-normative collective action.

1 INTRODUCTION

Protest activities can take place in various ways. Some protests are conducted peacefully and other protests involve violence. In 2012, the general leader of the Islamic Students Association (*Himpunan Mahasiswa Islam-HMI*) and 11 other members were arrested for vandalising the offices of the Indonesian Ministry of Foreign Affairs and the US Embassy in Jakarta. They were engaged in a violent confrontation with police officers. Their protest was related to the Israeli–Palestinian conflict (Sari, 2012). This activist action can be classified as a collective action: an action by group members in which they act as the representatives of the group and whose action is directed at improving the condition of the entire group (Wright et al., 1990). More specifically, the type of their collective action was non-normative collective action involving violence (Becker & Tausch, 2015; Tausch et al., 2011; Thomas & Louis, 2014; Wright, 2009). In this study, we will focus on non-normative collective action.

Many factors can predict collective action, and one of them is the emotional factor. In collective action studies, the emotion factor can be divided into two types: self-based emotion and group-based emotion (Becker et al., 2011). Group-based anger, as one of the important factors, indicates a reaction to disadvantaged conditions in a group (van Zomeren, 2013). Another type of emotion that specifically predicts non-normative collective action is contempt (Becker et al., 2011; Becker & Tausch, 2015; Tausch et al., 2011; van Stekelenburg & Klandermans, 2013). In addition, there is evidence that positive emotions, such as pride in the success of collective action in the past, can predict involvement in collective action in the future, with group efficacy the mediation (Tausch & Becker, 2013). In contrast, group-based positive emotions (such as pride, relief, and feeling emboldened) reduce the intention of collective action with low self-investment as a moderator (Shepherd et al., 2013). Other empirical findings show that self-based positive emotions do not predict collective action in the future (Becker et al., 2011). From the evidence, it can be seen that there has not been any specific explanation about which positive emotions affect non-normative collective action. This is because many studies strictly focus on the general collective action.

This study aims to prove that emotional factors such as group-based negative (anger and contempt) and positive emotions (pride and bravery) can predict non-normative collective action, while self-based emotions cannot predict the same action. Our study also attempts to prove that *contempt* is the group-based negative emotion which affects non-normative action.

1.1 *Collective action*

This is a group related action, and it depends on the individual's decision whether or not to get involved (van Stekelenburg & Klandermans, 2009). The action of an individual can be categorised as a collective action if it is a part of a group action which involved individuals' categorisation into certain collective identity (Wright, 2001, 2009). However, Wright (2001, 2009) also argues that a number of individuals participating in the same action may not indicate collective action due to different self-interests.

Collective action and social protest can be defined as specific intergroup strategies to improve the position of the group. They differ from individual actions which are meant to enhance or solidify one's personal position (Wright, 2009). Collective action can be conducted in two ways (Becker & Tausch, 2015; Tausch et al., 2011; Thomas & Louis, 2014; Wright, 2009): first, moderate and without violence (normative), such as peaceful demonstrations, petitions, and participation in actions of non-compliance; second, radical (non-normative) such as sabotage, violence, and terrorism. Collective actions can become non-normative when the more socially acceptable normative actions are not successful (Hoog & Abrams, 2007). Many social psychology researchers focus more on moderate or normative action (Becker & Tausch, 2015).

Collective action has been defined differently by many researchers. Some of the definitions are related to the attitudes, intentions, and action tendencies associated with participation in collective action, reports about past participations, and the actual behaviours (van Zomeren & Iyer, 2009). Based on those various explanations, a collective action is an action conducted collectively by individuals, who categorise themselves as part of a group, to achieve the group's goals. In this study, we focus on the intention to engage in non-normative collective action.

1.2 *Emotion as a predictor of collective action*

Emotional factors can be categorised into two types when explaining collective action, namely self-based emotion and group-based emotion. Self-based emotion is linked with the emotional experiences of the individual (Becker et al., 2011). Group-based emotion is in line with the Intergroup Emotion Theory (IET) (Mackie et al., 2008). Group-based emotion occurs when an individual considers their emotional experience in response to the events associated with the group (Goldenberg et al., 2014).

Many studies have proved that group-based anger is a predictor of collective action (Shepherd et al., 2013; Shi et al., 2014; Stewart et al., 2015; van Zomeren et al., 2011). This evidence is relevant for collective action in general, not only for normative action. Becker & Tausch (2015) criticised the results of this social psychology research due to the narrow reference to normative collective action.

Several other studies attempted to prove group-based anger on a more specific level: different types of collective action. Group-based anger is more likely to cause normative collective action (Becker et al., 2011; Becker & Tausch, 2015; Tausch et al., 2011; van Zomeren, 2013; van Stekelenburg & Klandermans, 2013). On the other hand, the type of negative emotion that was proven to cause non-normative collective action was contempt (Becker et al., 2011; Becker & Tausch, 2015; Tausch et al., 2011; van Stekelenburg & Klandermans, 2013).

There have been numerous studies that focused on negative group-based emotions, but only a few of them studied the effects of positive emotions on collective actions. The evidence showed that positive self-based emotions did not predict collective action (Becker et al., 2011). Pride in past collective action would predict a desire to get involved in future action on the basis of group efficacy (Tausch & Becker, 2013). Other findings stated that the anticipation of positive group-based emotions (pride, relief, and feeling emboldened) reduces the intention of collective action, moderated by low self-investment (Shepherd et al., 2013). The inconsistency suggests a lot of future studies to examine the role of positive emotions in collective action.

The empirical findings of the impact of negative group-based emotions on collective action is quite clear. This is in contrast with positive group-based emotions, which are not widely or specifically discussed in relation to non-normative collective action. On the other hand, self-based emotions are not likely to lead to collective action. In our study, emotional factors will be categorised into group-based emotions and self-based emotions in the form of positive emotions (pride and bravery) and negative emotions (anger and contempt).

1.3 *Hypotheses*

H1: Both positive (pride and bravery) and negative (anger and contempt) self-based emotions, whether together or independently, can predict non-normative collective action.

H2: Both positive (pride and bravery) and negative (anger and contempt) group-based emotions, whether together or independently, can predict non-normative collective action.

2 METHODS

2.1 *Participants and procedure*

A total of 462 Muslim students from Banjarbaru, Banjarmasin, Makassar, Depok, Jakarta and some other cities in Indonesia participated in this study. We collected data from self-reported (245 participants) and online surveys (217 participants). There were 194 male participants (42%), 253 female participants (54%), and 15 participants who did not state their gender. There were 292 (63%) participants who were a part of Islamic student organisations and 91 of them claimed to have been involved in demonstrations.

The research procedure began with an adaptation of the measuring instruments. After the measuring instruments were ready, we collected data from Muslim students who were a part of particular organisations, and collected online data from Muslim students who might or might not be a part of certain online Muslim organisations. The data was analysed by multiple regression to prove that certain group-based emotions (anger, contempt, pride, and bravery) predict non-normative collective action; and to prove that self-based emotions, with the four specific emotions, predict non-normative collective action.

2.2 *Measurement*

The measuring instruments used in this study were developed from previous studies. The instruments used the Indonesian language with adaptation processes (translation, back-translation, adjustment to the research context, expert judgement, piloting/testing, and statistical calculation).

2.3 *Self-based and group-based emotions*

The measuring instrument for negative emotions was developed by Tausch et al. (2011) and Becker et al. (2011) in the context of the Palestinian conflict, while the measuring instrument for positive emotions was developed by ourselves. The response scale used a Likert-type scale (1 = Totally inappropriate, to 7 = Very strongly appropriate). The measuring instrument consisted of 15 items. In self-based emotion with negative and positive emotions: anger ($r = 0.67$, $\alpha = 0.80$), contempt ($r = 0.58$, $\alpha = 0.73$), pride ($r = 0.59$, $\alpha = 0.73$), and bravery ($r = 0.69$, $\alpha = 0.81$). In group-based emotion: anger ($r = 0.76$, $\alpha = 0.86$), contempt ($r = 0.66$, $\alpha = 0.79$), pride ($r = 0.69$, $\alpha = 0.81$), and bravery (1 item). The emotion-based measurement began with a short explanation about the Palestinian conflict, and for self-based emotion the introductory sentence was: 'Please concentrate on yourself as an individual and think about the things that describe your uniqueness as a person who is different from others. Once you read and understand the situation of the Muslim Palestinian conflict, you feel...'. On the other hand, the group-based emotion measuring instrument began with: 'Now, focus your attention and mind to be aware that you are part of the Muslim community. Once you read and understand the situation of the Muslim Palestinian conflict, the feelings of Muslims are...'. Some examples of the items: (1) 'I feel angry with the Palestinian conflict situation' (self-based emotion of anger) and (2) 'The Palestinian conflict situation makes Muslims angry' (group-based emotion of anger).

2.4 *Non-normative collective action*

This measuring instrument was developed by Tausch et al. (2011), and asked for responses to eight action options ($r = 0.73$–0.92, $\alpha = 0.96$). The response scale used a Likert Scale (1 = Totally disagree, to 7 = Very strongly agree). The measurement began with the instruction: 'How likely are you to participate in the Palestinian conflict in the future with some of these actions. All actions have occurred in student protests in the past: ...' Some examples of items are: (1) vandalising and burning symbols of other parties (such as flags or portraits); (2) blocking the highway; (3) burning used tyres; (4) throwing stones or bottles as part of a protest; (5) attacking the police; (6) attacking the responsible parties; (7) destroying and burning government buildings when there was little or no support from the government in relation to the Palestinian conflict; (8) destroying and burning properties owned by the parties responsible for the military action in Palestine (e.g. the US Embassy, the related parties of the US and Israel).

3 RESULTS

3.1 *The effect of self-based emotions on non-normative collective actions*

We tested H1 with multiple regression testing for anger, contempt, pride, and bravery in self-based and non-normative collective action (Table 1). The result showed that self-based emotions did not significantly predict non-normative collective actions, $R^2 = 0.01$, $F(4, 457) = 1.26$, ns. In addition, independently, anger ($\beta = -0.05$, $t(461) = -0.92$, ns), contempt ($\beta = 0.10$, $t(461) = 1.89$, ns), pride ($\beta = -0.06$, $t(461) = -1.15$, ns), and bravery ($\beta = -0.002$, $t(461) = -0.04$, ns) did not significantly predict non-normative collective actions. Thus, H1 was rejected, which means that both positive (pride and bravery) and negative (anger and contempt) self-emotions, whether together or independently, cannot predict non-normative collective actions.

3.2 *The effect of group-based emotions on non-normative collective action*

We also used multiple regression testing to test H2 (Table 2). The result showed that anger, contempt, pride, and bravery in group-based emotions significantly predicted non-normative collective action, $R^2 = 0.03$, $F(4, 457) = 3.21$, $p < 0.05$. Independently, contempt significantly predicted non-normative collective actions ($\beta = 0.15$, $t(461) = 2.75$, $p < 0.01$). Pride also negatively and significantly predicted non-normative collective actions ($\beta = -0.15$, $t(461) = -2.51$ $p < 0.05$). However, anger ($\beta = -0.01$, $t(461) = -0.15$, ns) and bravery ($\beta = 0.08$, $t(461) = 1.36$, ns) did not significantly predict non-normative collective actions. Thus, a part of H2 is accepted: together, positive (pride and bravery) and negative (anger and contempt) group-

Table 1. Self-based emotion as predictor of non-normative collective action.

	B	*SE B*	*B*
Constant	18.78	3.55	
Anger	–0.28	0.31	–0.05
Contempt	0.41**	0.22	0.10
Pride	–0.32*	0.28	–0.06
Bravery	–0.01	0.23	–0.002
R^2	0.01		
F	1.26		

Note: dependent variable is non-normative collective action.
*p < 0.05; **p < 0.01

Table 2. Group-based emotion as predictor of non-normative collective action.

	B	*SE B*	*B*
Constant	16.93	3.29	
Anger	–0.07	0.30	–0.01
Contempt	0.56	0.21	0.15**
Pride	–0.81	0.32	–0.15*
Bravery	0.61	0.45	0.08
R^2	0.03		
F	3.21*		

Note: dependent variable is non-normative collective action.
*p < 0.05; **p < 0.01

based emotions can significantly predict non-normative collective actions. Another finding is that contempt significantly predicted non-normative collective actions. Also, pride was the negative predictor to non-normative collective action; when pride increases, non-normative collective action decreases.

4 DISCUSSION

This study showed that self-based emotions did not predict non-normative collective action. However, non-normative collective action was significantly predicted by group-based emotions. More specifically, in group-based emotions, contempt significantly influenced an individual's willingness to act in a non-normative way, while pride decreased willingness to be involved in action.

4.1 *Participation in collective action and self-based vs. group-based emotions*

Group-based emotions were first discussed in Intergroup Emotion Theory. This theory explains that the intergroup relationships could explain the power of motivation that emerges from group members' emotions towards ingroups and outgroups (Mackie et al., 2008; Mackie & Smith, 2014). The specific consequences of intergroup emotions lead to specific tendentious action. Emotion is considered as the primary response to social structures that affect collective actions (Becker, 2012). Our study discovered that when an individual is involved in collective action, they are more likely to feel the group's emotion, but not their own emotions.

Several empirical studies that are in line with our findings explain that when performing collective action, the emotion felt by an individual is the group's emotion as a response to the situation experienced by their group, especially anger (Shepherd et al., 2013; Shi et al., 2014;

Stewart et al., 2015; van Zomeren et al., 2011). Other studies found that outgroup-directed emotion plays a more important role in future protesting behaviour (Becker et al., 2011).

Muslims who are sympathetic towards Palestinian Muslims internalise emotions felt by the group and subsequently move to participate in a non-normative collective action. We believe that the way in which Palestinian Muslims are being threatened and devastated has compelled other Muslim groups to take action. Tausch et al. (2011) found that the perception of injustice is strongly correlated with emotion, especially anger and hatred. Based on the collective action definition, our research shows that the action is contextually more related to the group. It means that the active emotions are also group-based.

4.2 *Contempt and pride in non-normative collective action*

Our research focused on anger, contempt, pride, and bravery. Our study shows that when someone wants to be involved in non-normative collective action, the emotion that influences him is contempt. This is in line with the research findings of Becker and Tausch (2015), Tausch et al. (2011), and van Stekelenburg and Klandermans (2013). On the other hand, contempt is more closely related to the lack of desire to make peace (Koomen & Van Der Pligt, 2016) and the tendency to be involved in radical action (Becker et al., 2011). It explains how hatred towards the out group leads people to be involved in violent action.

Our study shows that in Muslim groups, when someone is no longer proud of his group, he is more likely to be involved in non-normative action. If Becker et al. (2011) found that self-based positive emotions did not predict future action, our findings strengthen the notion that group-based emotions have more influence on collective action. The result of our study is consistent with Tausch & Becker's (2013) findings that state that pride predicts the intention to be involved in actions through the reinforcement of group efficacy. In other words, the absence of pride within the group decreases group efficacy. Becker & Tausch (2015) show that the lower group efficacy is, the more likely an individual is to be involved in non-normative collective action. To further elaborate on our findings, high pride will lead to the decrease in the intention to take part in non-normative collective action.

4.3 *Limitation and future research*

This research only focuses on emotions. It can be seen that the influence of emotions is relatively small despite the fact that it is significant. Other factors such as social identity and group efficacy (van Zomeren, Postmes & Spears, 2008), and morality (van Zomeren, 2013) need to be taken into account to explain the dynamics of non-normative collective action. The finding that shows that group-based emotions predict non-normative collective action reinforces the idea of developing further investigation by considering other moderating or mediating factors on Muslim groups' actions. Our findings that pride lowers the intention to be involved in action can be more comprehensively elaborated if group efficacy is also measured as one of the contributing factors. This research is also limited to explaining emotional factors assessed by survey in non-normative actions. Further research needs to investigate causality by using experimental methods that include other potential contributing factors.

The weakness of the self-report measurement of the intention to carry out collective action in this research is that it shows only the tendency instead of real action. Further research could consider observing the actual action in the field by using survey methods. Some research on collective action has tried to explain the dynamics of real action right after the action occurred. Longitudinal methodology could be considered in future research, as has been done by Becker et al. (2011). Tausch & Becker (2013) conducted longitudinal research at two different timelines in order to explain the protest action. Experimental studies based on web-based surveys could also be considered (Becker et al., 2011). In addition, technological advancement also influences collective action. Moreover, researchers could use a correlational study to further explore the explanation on collective actions (Alberici & Milesi, 2013).

5 CONCLUSION

This study found that collective actions are more influenced by collective emotions than individual ones. When people want to act collectively using violence, the act itself was based on group emotions rather than their own emotions. Our findings also show that contempt strongly influences non-normative collective actions. On the other hand, pride was found to reduce the willingness to be involved in non-normative collective actions. Reducing pride means reducing group efficacy. When this happens, violent collective actions are likely to occur.

REFERENCES

Alberici, A.I. & Milesi, P. (2013). The influence of the internet on the psychosocial predictors of collective action. *Journal of Community and Applied Social Psychology, 23*, 378–388. doi:10.1002/casp.2131.

Becker, J.C. (2012). Virtual special issue on theory and research on collective action in the European Journal of Social Psychology. *European Journal of Social Psychology, 42*, 19–23. doi:10.1002/ejsp.1839

Becker, J.C. & Tausch, N. (2015). A dynamic model of engagement in normative and non-normative collective action: Psychological antecedents, consequences, and barriers. *European Review of Social Psychology, 26*(1), 43–92. doi:10.1080/10463283.2015.1094265

Becker, J.C., Tausch, N., Spears, R. & Christ, O. (2011). Committed dis(s)idents: Participation in radical collective action fosters disidentification with the broader in-group but enhances political identification. *Personality and Social Psychology Bulletin, 37*(8), 1104–1116. doi:10.1177/0146167211407076

Goldenberg, A., Saguy, T. & Halperin, E. (2014). How group-based emotions are shaped by collective emotions: Evidence for emotional transfer and emotional burden. *Journal of Personality and Social Psychology, 107*(4), 581–596.

Hoog, M.A. & Abrams, D. (2007). Intergroup behavior and social identity. In *The Hogg, M. A., & Cooper, J. SAGE handbook of social psychology* (pp. 335–360). London, UK: SAGE Publications.

Koomen, W. & Van Der Pligt, J. (2016). *The psychology of radicalization and terrorism.* New York, NY: Routledge.

Mackie, D.M. & Smith, E.R. (2014). Intergroup emotions. In Dovidio, J. & Simpson, J. (Eds.), *APA handbook of personality and social psychology: Vol. 2. Interpersonal relationships and group processes* (pp. 263–294). Washington, DC: APA Press.

Mackie, D.M., Smith, E.R. & Ray, D.G. (2008). Intergroup emotions and intergroup relations. *Personality and Social Psychology Compass, 2*, 1866–1880. doi:10.1111/j.1751-9004.2008.00130.x

Sari, H.R. (2012). *The list of 12 HMI Activists Arrested for Alleged Troublemaking*. Retrieved from https://www.merdeka.com/peristiwa/daftar-12-aktivis-hmi-yang-ditangkap-karena-buat-ricuh.html

Shepherd, L., Spears, R. & Manstead, A.S.R. (2013). This will bring shame on our nation: The role of anticipated group-based emotions on collective action. *Journal of Experimental Social Psychology, 49*(1), 42–57. doi:10.1016/j.jesp.2012.07.011

Shi, J., Hao, Z., Saeri, A. & Cui, L. (2014). The dual-pathway model of collective action: Impacts of types of collective action and social identity. *Group Processes and Intergroup Relation, 18*(1), 1–21. doi:10.1177/1368430214524288

Stewart, A.L., Pratto, F., Zeineddine, F.B., Sweetman, J., Eiche, V., Licata, L.,…van Stekelenburg, J. (2015). International support for the Arab uprisings: Understanding sympathetic collective action using theories of social identity and social dominance. *Group Processes and Intergroup Relations, 19*(1), 6–26. doi:10.1177/1368430214558310

Tausch, N. & Becker, J.C. (2013). Emotional reactions to success and failure of collective action as predictors of future action intentions: A longitudinal investigation in the context of student protests in Germany. *British Journal of Social Psychology, 52*, 525–542. doi:10.1111/j.2044-8309.2012.02109.x

Tausch, N., Becker, J.C., Spears, R., Christ, O., Saab, R., Singh, P. & Siddiqui, R.N. (2011). Explaining radical group behavior: Developing emotion and efficacy routes to normative and non-normative collective action. *Journal of Personality and Social Psychology,101*(1), 129–148. doi:10.1037/a0022728

Thomas, E.F. & Louis, W.R. (2014). When will collective action be effective? Violent and non-violent protests differentially influence perceptions of legitimacy and efficacy among sympathizers. *Personality and Social Psychology Bulletin, 40*(2), 1–14. doi:10.1177/0146167213510525 van Stekelenburg, J. & Klandermans, B. (2009). Social movement theory: Past, present and prospects. In Ellis, S. & van Kessel, I. (Eds.), *Movers and shakers: Social movements in Africa.* Leiden, The Netherlands: Brill.

van Stekelenburg, J. & Klandermans, B. (2013). The social psychology of protest. *Current Sociology*, *61*(5–6), 886–905. doi:10.1177/0011392113479314

van Zomeren, M. (2013). Four core social-psychological motivations to undertake collective action. *Social and Personality Psychology Compass*, *7*, 378–388. doi:10.1111/spc3.12031

van Zomeren, M. & Iyer, A. (2009). Introduction to the social and psychological dynamics of collective action. *Journal of Social Issues*, *65*, 645–660. doi:10.1111/j.1540-4560.2009.01618.x

van Zomeren, M., Postmes, T. & Spears, R. (2008). Toward an integrative social identity model of collective action: Qualitative research synthesis of three socio-psychological perspectives. *Psychological Bulletin*, *134*(4), 504–535. doi:10.1037/0033-2909.134.4.504

van Zomeren, M., Postmes, T., Spears, R. & Bettache, K. (2011). Can moral convictions motivate the advantaged to challenge social inequality? Extending the social identity model of collective action. *Group Processes and Intergroup Relation*, *14*(5), 735–753. doi:10.1177/1368430210395637

Wright, S.C. (2001). Restricted intergroup boundaries: Tokenism, ambiguity, and the tolerance of injustice. In Jost, J.T. & Major, B. (Eds.), *The psychology of legitimacy: Emerging perspectives on ideology, justice, and intergroup relations.* New York, NY: Cambridge University Press.

Wright, S.C. (2009). The next generation of collective action research. *Journal of Social Issues*, *65*, 859–879. doi:10.1111/j.1540-4560.2009.01628.x

Wright, S.C., Taylor, D.M. & Moghaddam, F.M. (1990). Responding to membership in a disadvantaged group: From acceptance to collective protest. *Journal of Personality and Social Psychology*, *58*, 994–1003. doi:10.1037/0022-3514.58.6.994

 ISBN 978-1-138-62665-2

Author index